T0142255

Advances in Intelligent Systems and Computing

Volume 1045

Series Editor

Janusz Kacprzyk, Systems Research Institute, Polish Academy of Sciences, Warsaw, Poland

Advisory Editors

Nikhil R. Pal, Indian Statistical Institute, Kolkata, India

Rafael Bello Perez, Faculty of Mathematics, Physics and Computing, Universidad Central de Las Villas, Santa Clara, Cuba

Emilio S. Corchado, University of Salamanca, Salamanca, Spain

Hani Hagras, School of Computer Science and Electronic Engineering, University of Essex, Colchester, UK

László T. Kóczy, Department of Automation, Széchenyi István University, Gyor, Hungary

Vladik Kreinovich, Department of Computer Science, University of Texas at El Paso, El Paso, TX, USA

Chin-Teng Lin, Department of Electrical Engineering, National Chiao Tung University, Hsinchu, Taiwan

Jie Lu, Faculty of Engineering and Information Technology, University of Technology Sydney, Sydney, NSW, Australia

Patricia Melin, Graduate Program of Computer Science, Tijuana Institute of Technology, Tijuana, Mexico

Nadia Nedjah, Department of Electronics Engineering, University of Rio de Janeiro, Rio de Janeiro, Brazil

Ngoc Thanh Nguyen, Faculty of Computer Science and Management, Wrocław University of Technology, Wrocław, Poland

Jun Wang, Department of Mechanical and Automation Engineering, The Chinese University of Hong Kong, Shatin, Hong Kong

The series "Advances in Intelligent Systems and Computing" contains publications on theory, applications, and design methods of Intelligent Systems and Intelligent Computing. Virtually all disciplines such as engineering, natural sciences, computer and information science, ICT, economics, business, e-commerce, environment, healthcare, life science are covered. The list of topics spans all the areas of modern intelligent systems and computing such as: computational intelligence, soft computing including neural networks, fuzzy systems, evolutionary computing and the fusion of these paradigms, social intelligence, ambient intelligence, computational neuroscience, artificial life, virtual worlds and society, cognitive science and systems, Perception and Vision, DNA and immune based systems, self-organizing and adaptive systems, e-Learning and teaching, human-centered and human-centric computing, recommender systems, intelligent control, robotics and mechatronics including human-machine teaming, knowledge-based paradigms, learning paradigms, machine ethics, intelligent data analysis, knowledge management, intelligent agents, intelligent decision making and support, intelligent network security, trust management, interactive entertainment, Web intelligence and multimedia.

The publications within "Advances in Intelligent Systems and Computing" are primarily proceedings of important conferences, symposia and congresses. They cover significant recent developments in the field, both of a foundational and applicable character. An important characteristic feature of the series is the short publication time and world-wide distribution. This permits a rapid and broad dissemination of research results.

** **Indexing: The books of this series are submitted to ISI Proceedings, EI-Compendex, DBLP, SCOPUS, Google Scholar and Springerlink** **

More information about this series at http://www.springer.com/series/11156

Ashish Kumar Luhach · Janos Arpad Kosa ·
Ramesh Chandra Poonia · Xiao-Zhi Gao ·
Dharm Singh
Editors

First International Conference on Sustainable Technologies for Computational Intelligence

Proceedings of ICTSCI 2019

 Springer

Editors
Ashish Kumar Luhach
Department of Electrical
and Communication Engineering
The Papua New Guinea University
of Technology
Lae, Papua New Guinea

Ramesh Chandra Poonia
Amity University
Jaipur, Rajasthan, India

Dharm Singh
Department of Computer Science
Namibia University of Science
and Technology
Windhoek, Namibia

Janos Arpad Kosa
Neumann János University
Kecskemét, Bács-Kiskun, Hungary

Xiao-Zhi Gao
School of Computing
University of Eastern Finland
Kuopio, Finland

ISSN 2194-5357   ISSN 2194-5365 (electronic)
Advances in Intelligent Systems and Computing
ISBN 978-981-15-0028-2   ISBN 978-981-15-0029-9 (eBook)
https://doi.org/10.1007/978-981-15-0029-9

© Springer Nature Singapore Pte Ltd. 2020, corrected publication 2020
This work is subject to copyright. All rights are reserved by the Publisher, whether the whole or part of the material is concerned, specifically the rights of translation, reprinting, reuse of illustrations, recitation, broadcasting, reproduction on microfilms or in any other physical way, and transmission or information storage and retrieval, electronic adaptation, computer software, or by similar or dissimilar methodology now known or hereafter developed.
The use of general descriptive names, registered names, trademarks, service marks, etc. in this publication does not imply, even in the absence of a specific statement, that such names are exempt from the relevant protective laws and regulations and therefore free for general use.
The publisher, the authors and the editors are safe to assume that the advice and information in this book are believed to be true and accurate at the date of publication. Neither the publisher nor the authors or the editors give a warranty, expressed or implied, with respect to the material contained herein or for any errors or omissions that may have been made. The publisher remains neutral with regard to jurisdictional claims in published maps and institutional affiliations.

This Springer imprint is published by the registered company Springer Nature Singapore Pte Ltd.
The registered company address is: 152 Beach Road, #21-01/04 Gateway East, Singapore 189721, Singapore

Preface

The First International Conference on Sustainable Technologies for Computational Intelligence (ICTSCI 2019) targeted state-of-the-art as well as emerging topics pertaining to Sustainable Technologies for Computational Intelligence and their implementation in engineering applications. The objective of this international conference is to provide opportunities for the researchers, academicians, industry persons, and students to interact and exchange ideas, experience, and expertise in the current trend and strategies for information and communication technologies. Besides this, participants will also be enlightened about vast avenues, and current and emerging technological developments in the field of advanced informatics and their applications will be thoroughly explored and discussed.

The First International Conference on Sustainable Technologies for Computational Intelligence (ICTSCI 2019) was held during March 29–30, 2019, at Sri Balaji College of Engineering and Technology, Jaipur, India, in association with Namibia University of Science and Technology, Namibia, and technically sponsored by CSI Jaipur Chapter, MRK Institute of Engineering and Technology, Haryana, India, and Leafra Research Pvt. Ltd., Haryana, India.

We are highly thankful to our valuable authors for their contribution and our technical program committee for their immense support and motivation for making the first edition of ICTSCI 2019 a success. We are also grateful to our keynote speakers for sharing their precious work and enlightening the delegates of the conference. We express our sincere gratitude to our publication partner, Springer AISC Series, for believing in us.

Lae, Papua New Guinea
Kecskemét, Hungary
Jaipur, India
Kuopio, Finland
Windhoek, Namibia
April 2019

Ashish Kumar Luhach
Janos Arpad Kosa
Ramesh Chandra Poonia
Xiao-Zhi Gao
Dharm Singh

About This Book

The book reports on new theories and applications in the field of Sustainable Technologies for Computational Intelligence. Written by active researchers, the different chapters are based on contributions presented at the First International Conference on Sustainable Technologies for Computational Intelligence, ICTSCI 2019, held in Jaipur, India, during March 29–30, 2019, and other accepted papers on R&D and original research work related to the practice and theory of technologies to enable and support Sustainable Technologies for Computational Intelligence.

All in all, the book provides academics and professionals with extensive information and recent developments in the field of computing and computational intelligence, and it is expected to foster new discussions and collaborations among different groups.

Contents

About the Editors

Dr. Ashish Kumar Luhach received his Ph.D. in Computer Science from the Banasthali University, India, and was a postgraduate at the Latrobe University, Australia. Dr. Luhach serves as a senior lecturer at the PNG University of Technology, Papua New Guinea. With more than a decade of teaching and research experience, he has published over 40 research articles in reputed journals and conference proceedings. He has also been editor/conference co-chair for various conferences such as ICAICR and ICTSCI and currently serves on the editorial boards of various journals. He is a member of the CSI, ACM, and IACSIT.

Dr. Janos Arpad Kosa is a member of the GAMF Faculty of Engineering and Computer Science at Neumann Janos University, Hungary. He is a seasoned researcher and has published extensively in reputed journals. He was the first researcher worldwide to develop DC flux transfer and AC flux transfer between independent iron cores in power systems. He has chaired various conferences and been an Invited Speaker to universities around the globe, e.g. Michigan State University, USA, and Ankara University, Turkey.

Dr. Ramesh Chandra Poonia is an Associate Professor at Amity Institute of Information Technology, Amity University Rajasthan, Jaipur, India; and a Postdoctoral Fellow at the Cyber-Physical Systems Laboratory (CPS Lab), Department of ICT and Natural Sciences, Norwegian University of Science and Technology (NTNU), Alesund, Norway. With substantial research experience, he has organized various conferences and published more than 100 research papers in journals and conference proceedings. He is an active member of the IEEE, CSI, and ACM.

Prof. Xiao-Zhi Gao is currently serving at the University of Eastern Finland. He has published more than 350 articles in reputed journals and conference proceedings. He also serves on the editorial boards of many journals published by Springer and Elsevier.

Prof. Dharm Singh is a Professor of Computer Science at Namibia University of Science and Technology (NUST). He is the author of more than 140 peer-reviewed articles and the author or editor of 16 books. His research interests include multimedia communications, wireless technologies, mobile communication systems, edge, roof computing, software-defined networks, network security, and the Internet of Things. The recipient of more than 19 prestigious awards, he is also a Fellow of The Institution of Engineers (I), Computer Society of India and Chartered Engineer (I), a Senior Member of the IEEE, Distinguished ACM Speaker, and IEEE CS DVP Speaker.

Emoticon and Text Sarcasm Detection in Sentiment Analysis

Shaina Gupta, Ravinder Singh and Varun Singla

Abstract A lot of work has been attempted in the area of sentiment analysis (SA)/opinion mining of natural language texts (NLT) and social media. One of the major objectives of such tasks is to allocate polarity either positive (+ve) or negative (−ve) to a part of the text. But, at a similar time, the problem of assigning the degree of positivity and negativity of particular text occurs. The problem becomes more difficult in the case of text gathered from social sites, as these sites contain a number of emoticons and sarcasm words that have hidden meaning along with the expressions. In this paper, we have presented an emoticons and text sarcasm detection system. The value of the uploaded text document is generated by removing the stop words and data filtration process. Here, two types of polarities are identified, namely positive and negative for both sarcasm and emoticons with 100% accuracy. For classifying the polarities, artificial neural network (ANN) is used as a classifier. At last, the comparison between proposed work and existing work is discussed.

Keywords Sentimental analysis · Polarity · Emoticons · Sarcasm · ANN

1 Introduction

The increasing use of the Internet and online activities such as chat, conference, monitoring, booking, online transactions, e-commerce, social media communications, blogs, and microblogging, etc leads us to extract, convert, load, and analyze a large amount of data very quickly [1]. The sentiment analysis (SA) is a computational study of sentiments, emotions, and attitudes expressed in terms of text. SA is also

S. Gupta · R. Singh (✉) · V. Singla
Department of Computer Science and Engineering,
Lovely Professional University, Phagwara, India
e-mail: ravinder.17750@lpu.co.in

S. Gupta
e-mail: shainagpt@gmail.com

V. Singla
e-mail: Varun.17705@lpu.co.in

© Springer Nature Singapore Pte Ltd. 2020
A. K. Luhach et al. (eds.), *First International Conference on Sustainable Technologies for Computational Intelligence*, Advances in Intelligent Systems and Computing 1045, https://doi.org/10.1007/978-981-15-0029-9_1

1

known as opinion mining, comment mining or assessment extraction, attitude analysis which is used to detecting task, extracting and classifying opinions, emotions, and attitudes on different topics, such as text input [2].

An opinion analysis mainly comprises of three components, namely opinion, opinion holder, and object.

i.   Opinion holder: An individual or institute, which holds a particular opinion on a specific object.
ii. Object: It is defined as the thing on which opinion has to be expressed.
iii. Opinion: It is a review taken from the opinion holder. It may be positive or negative [3].

In this research work, a hybrid classification scheme is developed in order to enhance the accuracy of the polarity detection system of the text. Also, along with positive and negative emotions, the sarcasm in the text is also detected [4]. After detecting the sarcasm present in the sentence, the polarity (positive and negative) has to be modified accordingly. Sarcasm is a sharp, bitter, or commentary; bitter taunting. Sarcasm might be used as ambivalence, although sarcasm is not essentially ironic. The most striking thing in the spoken language is that sarcasm is mainly used to differentiate through inflection with which it is spoken, and it depends to a large extent on context [5]. Emoticon polarity detection is used in SA and other natural language program (NLP) tasks which are provided as features to an artificial neural network (ANN). Emoticons denoted by ☺ and ☹ are regularly used online in social media such as Facebook, Twitter, Instagram, etc. [6]. The defined emoticons are most commonly used in online communications that are the direct representation of sentiment or emoticons represented in the form of text. Emoji sentiment ranking lexicon detection system is used to enhance the emoticon polarity detection [7]. An emoji is a next step, which is designed with modern communication scheme and delivers more expensive messages. An emoji is a symbolic representation of ideogram, which symbolizes the facial expressions along with the conception and ideas behind the expression [8].

Emoji became popular worldwide with the regular use of the Internet. Emoji can be used in smartphone, emails, and social applications. As per the data obtained in 2015, the reported emojis are about half of the text used on Instagram [9].

In this research work, the polarity of sentimental analysis such as positive and negative of individual user's interest is determined. The emotion does not belong to any one of the categories. Therefore, in this research work, we mainly focus on the two categories of emoticons along with sarcasm [10]. There are two groups of emoticons that were labeled by the majority of the participants as positive or negative.

2   Related Work

Wang et al. presented an algorithm to establish the relationship between emotions and sentimental polarity [11]. The frequency of emoticons has been measured for large 'twitter data set'. After this to find out the relationship between the polarity of sentiments and emotions, the emoticons are grouped by using clustering technique. The parameters such as accuracy, precision, and recall are measured separately for positive and negative sentiments (with emoticons and without emoticons). Ghiassi et al., introduced a scheme for decreasing the features of text using the n-gram approach and have designed a twitter-specific lexicon for analyzing sentiments [12]. It has been shown that lexicon technique using SVM provides better accuracy than traditional lexicon method. The data have been collected from Twitter API v1.0. Prabowo et al., proposed a hybrid algorithm, namely a rule-based approach, supervised and machine learning for testing reviews of movies, products, and the comments of MySpace. The performance in terms of measure, micro-, and the macro-average has been measured [13]. Rodrigues et al. presented some norms for emoji provided by several users. The data have been gathered from LEED (Lisbon emoji and emoticon) that includes 238 stimuli, 85 number of emoticons, and 153 numbers of emoji [14]. The result in terms of mean standard deviation (MSD) and confidence interval (CI) has been measured. Patra et al. proposed a new algorithm for the recognition of sentiment degree for figurative language text [15]. A number of sentimental features are measured like intensifiers, sentiment abruptness for finding the changing from +ve (positive) to −ve (negative) in sentiments. The effectiveness of the system has been measured in terms of cosine similarity and MSE (mean square error). The values observed for MSE and cosine similarity are 2.170 and 0.823, respectively. Nadal et al. presented a novel 'Sarcastic tweets detection' (STD) approach for recognizing sarcastic tweets with higher accuracy at the hashtag level [16]. SHC (Sarcasm hashtag classifier) has been designed for categorizing tweets namely sarcasm and non-sarcasm on the basis of hashtag sentiment analysis. The function of SHC depends upon the SHI 9Sarcasm hashtag indicator) and the relation aiming the hashtag and tweets. Mhatre et al. presented a pre-processing scheme applied before the testing of the text [17]. This scheme helps to clear and makes the data by decreasing the length; by removing HTML tags, hands punctuations, slangs; and by removing stop words. The main goal of using pre-processing is to provide more accuracy. And it has been observed that the accuracy of the system obtained with the pre-processing algorithm is higher than the system without a pre-processing algorithm.

3 Artificial Neural Network

ANN is a classification approach used to obtain higher accuracy by finding the sentiments of sarcastic sentences. ANN has shown outstanding capabilities for modeling difficult word composition in one sentence. A sarcastic text is a sequence of words or combination of words. ANN is used to store these text signals into its temporary storage (Fig. 1).

ANN is a computational algorithm that is used to solve the difficult problem in a similar fashion as a brain does. The inputs are provided to the input layer, and weight is added individually to the inputs in the hidden layer. The weight is added

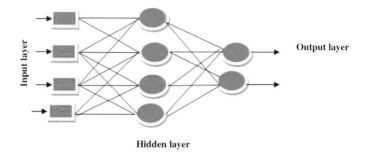

Fig. 1 ANN architecture

or removed as per the bias function. The weighted values along with the input are provided to the activation function. There are different types of activation functions. In the proposed work, we have used the step function. This function has compared the threshold value with the input data. If the input data are higher than the defined threshold value, then the output is 1 otherwise the output is zero. The algorithm of a neural network is listed below.

Algorithm 1: Artificial Neural Network

Input: Train n number of test data
Output: Classified sarcasm
Categorize the training class into training and validation. Determine the accuracy of polarity detection
Set parameters
Training data = optimized
data Class = positive and negative sarcasm
Epoch = 100
Training algorithm = Levenberg Marquardt
Performance parameter = Mean square error (MSE)
Neurons = 50
Total neuron (Trained data, class, neurons)
Classification = {Emoticons and text sarcasm as per requirement; if properties are matched}
Return {classified positive and negative sarcasm and emoticons}

4   Methodology

The steps used to find the polarity of the emoticons and sarcasm used in the social network are defined below.

The uploaded text undergoes through various processes such as filtering, removal of stop words, value generation system, and classification. The flow of work is shown in Fig. 2.

Step 1 Upload the raw test data which are also known as unfiltered data.

Step 2 Apply to stop word removal algorithm to remove the stop words from the uploaded test data.

Step 3 Apply the filtration technique and generate values for the filtered text document.

Step 4 Train neural network by using a positive and negative generator and classify the polarity.

Step 5 If the polarity of the generated text is positive, then load positive emoticons otherwise load negative emotions.

Step 6 For both positive and negative polarity values, testify emoticons in the text document.

Step 7 Trained neural network again for both positive and negative sarcasm obtained after testifying the emoticon in the text data.

Step 8 After training, again classify whether text data are sarcastic or not.

Step 9 Calculate performance in terms of true positive, false positive, and confusion metrics.

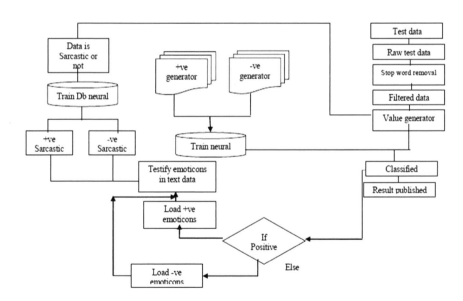

Fig. 2 Flow chart of proposed design

5   Result Analysis

In this section, the results obtained after the simulation of code in MATLAB simulator tool are discussed in detail. For measuring the performance of the system, the parameters such as true positive rate, false positive rate, and confusion matrix are determined.

The ROC (receiver operating characteristic curve) of the proposed work is shown in Fig. 3. It is the graph plotted between the true positive rate (T_p) and false positive rate (F_p) at different threshold values.

From the above figure, it is clear that initially false positive rate is 0 and the T_p rate is 0.2. After that, T_p and F_p rates increase linearly and become maximum at $t_p = f_p = 1$ (Fig. 4).

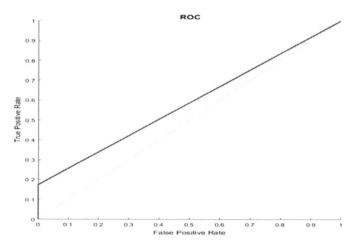

Fig. 3 ROC curve of proposed work

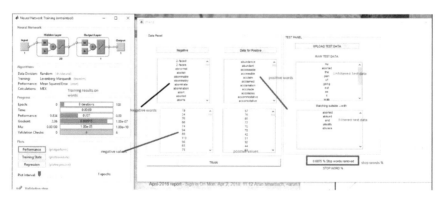

Fig. 4 GUI and training of the neural network

The GUI along with the training of the neural network is shown in the figure above. The left-hand side image represents the training of the proposed neural network and the right-hand side image shows the GUI of the proposed research. The GUI comprises of two panels, namely data panel and test panel. Data panel is again divided into two sub-panels such as negative and positive data. The negative values of the uploaded text are shown under the negative panel and positive values of text are loaded under positive data panel along with their weighted values, namely positive values and negative values shown in the figure above. Uploaded test data are again categorized into two sub-parts, namely raw test data and matching outside with. Raw test data contain unfiltered data, whereas matching outside with panel consists of filtered test data. The removal of stop words in the uploaded text is indicated in terms of percentage (%). In the proposed work, the percentage of stop words removal determined is 68.75% (Fig. 5).

The performance of the neural network used in the proposed work is represented by the confusion matrix table. X-axis defines the target class and y-axis depicts the output class. Here, the center cell consists of 6786 number of samples, and 100% indicates that all the loaded samples are classified and hence accuracy obtained is 100% (Fig. 6; Table 1).

The above figure defines the performance parameters of the proposed sarcasm detection framework measured in terms of precision, recall, and F-measure. Here, blue bar, red bar, and green bar lines represent the precision, recall, and F-measure values observed for 7 number of test data uploaded in the testing panel. The average

Fig. 5 Confusion matrix

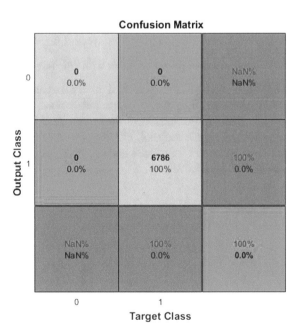

Fig. 6 Performance
parameters

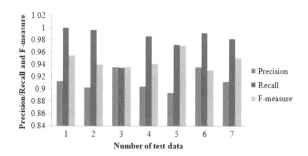

Table 1 Precision, recall, and F-measure

Number of test data	Precision	Recall	F-measure
1	0.9132	1.000	0.9546
2	0.9024	0.996	0.94025
3	0.9356	0.935	0.9358
4	0.9047	0.9862	0.9412
5	0.8936	0.972	0.9702
6	0.9356	0.991	0.9302
7	0.9125	0.982	0.95012

values of precision, recall, and F-measures observed for 7 number of test data are 0.9139, 0.9803, and 0.9460, respectively (Fig. 7; Table 2).

The above figure represents the accuracy of the proposed work. The average accuracy observed for the proposed work is 96.34.

A. Comparison of proposed work with existing work

In existing work [18], authors have used Lexicon enhanced rule-based approach to classify between the positive and negative emoticons posted as reviews in different product sites. The authors have worked on three types of a dataset taken from drug, car, and hotel sites having 350,273 and 412 total number of reviews, respectively. The accuracy measured from three datasets drug, car, and hotel are 90.216%, 88.167%, and 83.295, respectively.

Fig. 7 Accuracy

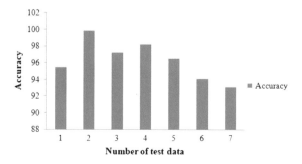

Table 2 Accuracy of proposed work

Number of test data	Accuracy
1	95.46
2	99.85
3	97.25
4	98.23
5	96.47
6	94.08
7	93.09

In the proposed work, we have applied an artificial neural network to enhance the accuracy of the prediction system. ANN is used to classify the emoticons along with sarcasm in the loaded text. The accuracy of the system is measured by using a matrix, which is about 100% and identified 6786 total number of words. Hence, it is clear that the proposed work has outperformed as compared to the existing work.

6 Conclusion

In this paper, a detection model has been introduced for detecting the sarcasm and emoticons in a social media network. As the use of the Internet is increasing day by day, therefore, managing and organizing the loaded sarcasm and emoticons in social media has become essential. The polarity of these texts has been measured by using sentimental and opinion mining process. In this research, a hybrid approach in which the polarity of emoticons along with the polarity of sarcasm has been determined. This helps to increase the accuracy of the proposed work. The detection accuracy of uploaded emoticons and sarcasm measured is 100%. There is an enhancement of 9.8% in the accuracy of the proposed work when compared with the existing work.

In the future, to enhance the accuracy of the designed framework, different processes such as noise reduction, breaking sentence, and spell correction can be used. Also, neutral sarcasm can also be detected along with positive and negative polarities.

References

1. Bouazizi, M., Ohtsuki, T.: Opinion mining in Twitter: how to make use of sarcasm to enhance sentiment analysis. In: Advances in Social Networks Analysis and Mining (ASONAM), 2015 IEEE/ACM International Conference on, pp. 1594–1597. IEEE (2015)
2. Liu, B.: Sentiment analysis and opinion mining. Synth. Lect. Hum. Lang. Technol. **5**(1), 1–167 (2012)
3. Kobayashi, N., Inui, K., Matsumoto, Y.: Opinion mining from web documents: extraction and structurization. Inf. Media Technol. **2**(1), 326–337 (2007)

4. Zheng, L., Wang, H., Gao, S.: Sentimental feature selection for sentiment analysis of Chinese online reviews. Int. J. Mach. Learn. Cybernet. **9**(1), 75–84 (2018)
5. Etter, M., Colleoni, E., Illia, L., Meggiorin, K., D'Eugenio, A.: Measuring organizational legitimacy in social media: assessing citizens' judgments with sentiment analysis. Bus. Soc. **57**(1), 60–97 (2018)
6. Asghar, M.Z., Kundi, F.M., Ahmad, S., Khan, A., Khan, F.: T-SAF: Twitter sentiment analysis framework using a hybrid classification scheme. Expert Syst. **35**(1) (2018)
7. Asghar, M.Z., Khan, A., Khan, F., Kundi, F.M.: RIFT: a rule induction framework for twitter sentiment analysis. Arab. J. Sci. Eng. **43**(2), 857–877 (2018)
8. Zhang, S., Wei, Z., Wang, Y., Liao, T.: Sentiment analysis of Chinese micro-blog text based on extended sentiment dictionary. Future Gener. Comput. Syst. **81**, 395–403 (2018)
9. Fouad, M.M., Gharib, T.F., Mashat, A.S.: Efficient twitter sentiment analysis system with feature selection and classifier ensemble. In: International Conference on Advanced Machine Learning Technologies and Applications, pp. 516–527. Springer, Cham (2018)
10. Chen, Z., Ma, N., Liu, B.: Lifelong learning for sentiment classification (2018). arXiv preprint arXiv:1801.02808
11. Wang, H., Castanon, J.A.: Sentiment expression via emoticons on social media. In: Big Data (Big Data), 2015 IEEE International Conference on, pp. 2404–2408. IEEE (2015)
12. Ghiassi, M., Skinner, J., Zimbra, D.: Twitter brand sentiment analysis: a hybrid system using n-gram analysis and dynamic artificial neural network. Expert Syst. Appl. **40**(16), 6266–6282 (2013)
13. Prabowo, R., Thelwall, M.: Sentiment analysis: a combined approach. J. Informetrics **3**(2), 143–157 (2009)
14. Rodrigues, D., Prada, M., Gaspar, R., Garrido, M.V., Lopes, D.: Lisbon emoji and emoticon database (LEED): norms for emoji and emoticons in seven evaluative dimensions. Behav. Res. Methods **50**(1), 392–405 (2018)
15. Patra, B.G., Mazumdar, S., Das, D., Rosso, P., Bandyopadhyay, S.: A multilevel approach to sentiment analysis of figurative language on twitter. In: International Conference on Intelligent Text Processing and Computational Linguistics, pp. 281–291. Springer, Cham (2016)
16. Nadali, S., Murad, M.A.A., Sharef, N.M.: Sarcastic tweets detection based on sentiment hash-tags analysis. Adv. Sci. Lett. **24**(2), 1362–1365 (2018)
17. Mhatre, M., Phondekar, D., Kadam, P., Chawathe, A., Ghag, K.: Dimensionality reduction for sentiment analysis using pre-processing techniques. In: Computing Methodologies and Communication (ICCMC), 2017 International Conference on, pp. 16–21. IEEE (2017)
18. Asghar, M.Z., Khan, A., Ahmad, S., Qasim, M., Khan, I.A.: Lexicon-enhanced sentiment analysis framework using rule-based classification scheme. PLoS ONE **12**(2), e0171649 (2017)

Toward Development Framework for Eliciting and Documenting Knowledge in Requirements Elicitation

Halah A. Al-Alsheikh, Hessah A. Alsalamah and Abdulrahman A. Mirza

Abstract Requirements engineering (RE) is the most critical factor in the success or failure of project. A project's success completely depends on the knowledge and experience of the stakeholders. RE is a team-based process and collaborative task, which involves a huge volume of deliberation and discussion between stakeholders, and comes from their tacit knowledge. During the RE process, the presence of unclear or hidden tacit knowledge causes ambiguity, incomplete, and incorrect requirements. To address these issues, we intended to propose a collaborative knowledge management system framework for identifying, categorizing, eliciting, and managing the requirements tacit knowledge that are required for software development. This work proposes a framework that is based on a combination of rationale-based model and domain ontology. In this paper, a comparison between some of methods that have been used for requirements elicitation and tacit knowledge elicitation with the proposed framework as an evaluation method was presented.

Keywords Requirements engineering · Requirements elicitation process · Tacit knowledge · Rationale-based model · Domain ontology

1 Introduction

With the rapid growth of applications, products, and services from all sectors, software has become a vital part of this growth. The quality of the delivered software

H. A. Al-Alsheikh (✉)
Information Systems Department, Al Imam Mohammad Ibn Saud Islamic
University (IMSIU), Riyadh, Saudi Arabia
e-mail: hamshaikh@imamu.edu.sa

H. A. Alsalamah
Information Systems Department, Al Yamamah University, Riyadh, Saudi Arabia
e-mail: h_alsalamah@yu.edu.sa

A. A. Mirza
Information Systems Department, King Saud University, Riyadh, Saudi Arabia
e-mail: amirza@ksu.edu.sa

© Springer Nature Singapore Pte Ltd. 2020   11
A. K. Luhach et al. (eds.), *First International Conference on Sustainable Technologies for Computational Intelligence*, Advances in Intelligent Systems and Computing 1045,
https://doi.org/10.1007/978-981-15-0029-9_2

depends on the requirements from which the system has been developed [1]. In the software development community, the requirements engineering stage has the highest influence on the quality of the final product. Requirements engineering (RE) is the process of eliciting software requirements, analyzing requirements, documenting in a specification, validating that specification to ensure it fulfill the user needs and managing the requirements [2]. In the requirement elicitation process, the required information and knowledge is extracted and collected from the stakeholder [2]. However, the process of eliciting correct, complete, unambiguous, and consistent requirements has a great number of serious and inherent difficulties that can have an effect on a software project's success [3, 4]. The intricate problems and difficulties in the elicitation and discovery of requirements result from the process being a knowledge-intensive and collaborative activity [5]. Any software development process implies multiple stakeholders, including clients, users, developers, and analysts, who collaborate with a common goal. The backgrounds, and expectations or assumptions of the stakeholders, known as their tacit knowledge, may different depending on their different experiences. This difference often results in missing needs, misinterpretation, occurrence of ambiguity or incorrect requirements [3, 4, 6]. The presence of unhandled tacit knowledge is one of the major factors in the failure of software systems [7, 8].

Unclear or hidden tacit knowledge creates difficulties in fulfilling the desires and needs of the stakeholders [6]. Moreover, if stakeholders "know more than they can tell" this will lead to poorly defined requirements and poor systems [9]. However, 90% of an organization's corporate knowledge and the valuable history of the organizations are stored in employees' minds [10, 11]. Therefore, tacit knowledge needs to be identified, elicited, and made explicit through representations [12].

Requirements elicitation methods, that are currently used to elicit requirements, document only the final agreed requirements [13]. Meanwhile, huge volumes of useful knowledge within the requirements and extensive stakeholders' deliberations often remain undocumented as part of the software requirements specification.

The process of tacit knowledge elicitation is normally referred to as knowledge management systems (KMS). Capturing tacit knowledge concerns two aspects: (i) identifying and collecting the knowledge categories (that represent the knowledge artifacts) and (ii) eliciting knowledge elements (to obtain more detailed information about specific artifacts) [14]. The appropriate elicitation methods should be concerned with these two aspects, how to identify and collect the knowledge artifacts and how to present the detailed information. Expressing that knowledge directly in words is very difficult. However, ontology can be used to identify a significant part of tacit knowledge [15].

In RE, ontology is used to improve the quality of the elicited requirements by decreasing the ambiguity through providing a common understanding among stakeholders and filling gaps in the knowledge [16]. Therefore, the tacit knowledge artifacts that should be elicited during the REP can be identified by using the concepts and relationships in domain ontology [17].

The characterization of tacit knowledge depends on an individual's analytical abilities, suggestions, assumptions logical analysis, and creativity. It is reflected in ideas, solutions, innovation, and design [18]. Tacit knowledge in RE may take many forms of personal expression. Accordingly, considerable amounts of this knowledge in this field arise in the form of rationale behind decisions or domain assumptions [19, 20]. Tacit knowledge can be documented by using a rationale-based model [21]. The rationale can be structured based on alternatives, reasons, and justifications. The rationale is the only way to discover why the system was analyzed, designed, and implemented in the way it was. Moreover, it consists of encountered problems, investigated options, the criteria selected to evaluate options, and the debate that led to making decisions [22].

However, requirements rationale knowledge management is beneficial but also costly, and considerable resources, time, and effort are required before the benefits can be felt [23]. On the other hand, capturing tacit knowledge in the form of a requirements rationale is expected to be useful in many cases, such as providing greater help in understanding the complexity of the software application, justifying the necessity of the requirements, explaining the requirements by supplying additional information, for maintenance or for future evolution of the software, to assess and to predict possible changes [13, 21, 24]. Furthermore, a rationale can help in times of reuse, when similar issues occur and thus it aids in the success of the next project [24]. Additionally, capturing rationale prevents valuable tacit knowledge that remains hidden from being lost when people leave the organization [25]. This documentation can also be an aid in building a cumulative base of tacit knowledge, which would be a useful learning tool to both students of design and practicing analysts and developers.

The motivation of this paper is to fill a research gap in this area by proposing a collaborative KMS framework for identifying, eliciting, and managing the tacit knowledge required for software development. The framework combines aspects of a rationale-based model with domain ontology in a collaborative platform. The framework primarily depends on direct interaction between the stakeholders to elicit the knowledge. The objective of the framework is to identify, categorize, elicit, and manage tacit knowledge through the requirements elicitation process to minimize the ambiguous, incomplete, and incorrect requirements. In this paper, the framework was compared with some of the major methods used directly to elicit tacit knowledge. The remainder of this paper is structured as follows. Section 2 contains a description of the proposed framework. Section 3 presents the comparison results between the proposed framework and some of the tacit knowledge elicitation methods that are currently used in organizations during RE. Later, a discussion of the framework presented in Sect. 4. Finally, Sect. 5 concludes this paper.

2 Framework

From the literature, we have chosen to adopt and modify the rationale-based models to organize and represent the tacit knowledge elements. The proposed framework will also integrate domain ontology to identify and categorize the main artifact of the tacit knowledge. These two technologies will be integrated to formulate our proposed framework. The framework is divided into three processes: (1) artifacts identification, (2) knowledge elements elicitation, and (3) query processor. The framework parts are shown in Fig. 1.

Firstly, the valuable source of tacit knowledge is found in the minds of stakeholders. It includes cultural beliefs, values, attitudes, as well as analysis skills, capabilities, and expertise. Externalizing the tacit knowledge is conducted through a communication platform such as face-to-face and knowledge virtual community. Each stakeholder will be allowed to enter an initial statement of need. Each statement of need is processed to identify the artifacts.

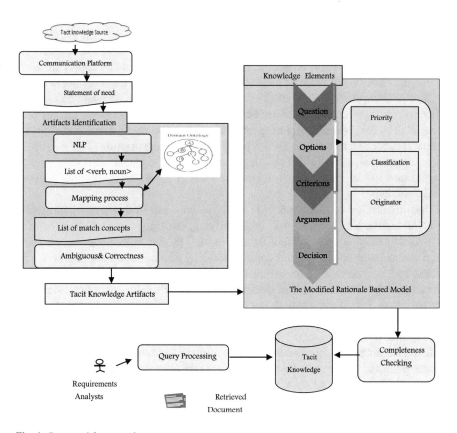

Fig. 1 Proposed framework

2.1 Artifacts Identification

The NLP component facilitates the processing of natural language of the statement of need to identify the tacit knowledge artifacts. Each statement of need will be parsed to obtain a list of verbs and several nouns that represent the artifacts of the domain. This list of verb and nouns will be used to identify and categorize the related tacit knowledge artifacts.

The domain ontology (DO) contains the basic terms and their relationships comprising the vocabulary of an application domain with set of precise definitions [26]. In the framework, each term in the DO will present an artifact of tacit knowledge where the knowledge will be elicited around that artifact in the context of the statement of need.

Mapping process will match the verbs and nouns that have been extracted from the statement of need to the concepts in DO. The framework will use a simple parsing framework for the mapping. This mapping is aimed to define the scope of requirements to minimize incorrect requirements. Also, any ambiguity in the requirements will be minimized through the mapping process by specifying one interpretation for each concept in the requirement statement. Moreover, the statement of need is provided by stakeholders, and it may include ambiguous concepts. If the concept in the statement of need is not mapped to any concept in the DO, this means the statement of need has some concept not related to the domain or has a different meaning. In this case, the statement of need will be rephrased by the stakeholder. In addition, if the concept in the statement of need is mapped into several concepts that have no semantic relationships in the domain ontology, this means the statement is incorrect. In this case, the statement of need will be split by the stakeholder. At the end of the artifacts identification process, an exhaustive list of artifacts that identify the main categories of tacit knowledge of a specific domain are identified.

2.2 Knowledge Elements Elicitation

The stakeholders will start the deliberation and discussion for each statement of need to obtain more detailed information about specific artifacts. This deliberation and discussion represent the tacit knowledge around the statement of need. The deliberation details will be represented as the rationale of that statement of need. The rationale nodes (questions/issues, options, criterions, decision options, originator, classification, and the priority) will represent the tacit knowledge element that should be elicited from the stakeholders (see Table 1). A question/issue about the artifact can be raised by the stakeholder. After that each question/issue will have a list of options that represents its solutions. The option for each question will be shown as a requirement statement. Each option represents the tacit knowledge of the person who raised it. The option can be classified under the knowledge perspective such as how to perform a task, way of documentation, organization policy and role. The

Table 1 Key tacit knowledge elements

Type of requirement rationale	Description
Questions/issues	Clarification of questions to related statement of need
Options	All possible alternative solutions for the question/issue
Criterions	Represent the qualities that are used to evaluate options in a certain context. Criteria are non-functional requirements, such as reliability, cheapness, and performance
Arguments	Represent the knowledge of the participants. Arguments can support or oppose
Decision options	The fully agreed option
Originator	The person who raised the knowledge
Classification	Classification of the knowledge perspective such as task, way of documentation, organization policy, and role
Priority	The level of strength for an option in range of -1 to 1. An option with negative level signifies that the oppose arguments more than the support arguments and options with positive level signify that the support arguments more than the oppose arguments. An option with 0 level signifies indecisiveness where the oppose arguments equal to the support arguments

stakeholders assigned priority for each option based on their expertise and knowledge of topic. The criteria are defined by the stakeholders and used to evaluate the options. The criteria represent the non-functional requirement of the system. Furthermore, the stakeholders can express their support for, or oppose, arguments about each option. The stakeholders will select one of the options for each question/issue to be a decision requirement statement. The selection based on the arguments and the priority of the option. At the end of knowledge elements elicitation process, each tacit knowledge artifact of a specific domain is linked with its rationale.

In the framework, the method for detecting the completeness and correctness of the requirements list will be performed by using the semantic relationships (e.g., detecting contradiction between concepts in single requirement statement or detecting the missing requirements). Finally, the tacit knowledge about each artifact (concept) in the domain ontology will be elicited and documented in the form of a rationale-based element and can be stored in the form of a network structure or table.

2.3 Query Processing

The proposed framework is capable of processing analyst's queries over the tacit knowledge repository. The query processor helps users to formulate queries using tacit knowledge artifacts to obtain the discussion that has been engaged on this artifact.

3 The Framework Evaluation

Knowledge elicitation methods can be divided into two categories: direct and indirect [27]. Direct methods involve direct interaction between the stakeholders to elicit the knowledge, such as interview, concept mapping and card sorting, while indirect methods elicit knowledge with no interaction between the stakeholders, such as data mining and other computer-aided techniques [27]. From the comparative analysis done between the proposed framework and some of the major existing methods used directly to elicit tacit knowledge (see Table 2), it was observed that the most of the knowledge elicitation methods do not have a high level of interaction between

Table 2 Comparison of the proposed framework with the major tacit knowledge elicitation methods

Selected methods		Interview	Sorting methods	Mapping methods	Storytelling	Proposal model
Description		It consists of interviewing the stakeholders with a set of prepared questions	It allows us to know how the stakeholder mentally organizes the information, categories, and priorities this in his mind	This method can help to understand the relationships between concepts. It is simple, and uses hierarchies, and visual representation	It consists of the description of a sequence of actions and events for a specific case in the domain	It depends on a high level of interaction between stakeholders and consists of categorizing the tacit knowledge into concepts (artifacts) and eliciting more detailed elements of knowledge
Parameters	Categories	X	✓	✓	X	✓
	Detailed elements	✓	X	X	✓	✓
	Level of interaction	High	Low	Low	Low	High

stakeholders. Moreover, the knowledge elicited from interview and storytelling cannot organize the elicited knowledge under any categories for easy reuse. On the other hand, sorting and mapping method that are organized the elicited knowledge have a lack of detailed knowledge elements. However, the proposed framework will overcome these problems when eliciting the knowledge during the requirements elicitation process.

4   Discussion

The proposed framework in Fig. 1 focuses on identifying, categorizing, eliciting, and managing the tacit knowledge in requirements elicitation process. The framework consists of three main processes that aim to minimize the ambiguous, incomplete, and incorrect requirements by handling the tacit knowledge to enhance the requirements elicitation process. It will start with providing a set of artifacts that identify the tacit knowledge categories of a specific domain. After that the arguments for and against the initial requirements are elicited, alternative solutions for debate are presented and final requirements decisions are recorded. Finally, after documenting the deliberation and discussion about the initial requirements the requirements analysts can formulate queries using tacit knowledge artifacts for reuse.

The framework will allow stakeholders to see the trade-offs between different requirements options and will solve the problems that are related to the presence of unhandled tacit knowledge in requirements elicitation process. Moreover, it will minimize the ambiguity of requirements by having a unique interpretation for the main concepts in the initial requirements by all stakeholders by using the DO concepts. Furthermore, through the DO, the framework will increase the percentage of necessary functions specified and detect the incorrect requirements by searching for requirements that are not mapped to nodes in ontology.

The rationale-based model that has been used in this framework is the question, option, and criteria (QOC) rationale-based model. QOC represents rationale as a space of alternatives and evaluation criteria after decisions are made the rationale is reconstruct [28]. Moreover, the framework proposed some other elements to represent more tacit knowledge elements.

5   Conclusion and Future Work

In conclusion, discovering unknown and un-elicited tacit knowledge within requirements is necessary to mitigate many risks that can negatively affect system design and project cost. Important experiences, in the form of stakeholders' tacit knowledge, could be lost and this paper presents a collaborative framework to identify, elicit, and manage tacit knowledge in requirements elicitation process. The framework can be integrated into the organization's requirements elicitation procedures. The research aims to improve the efficiency of the RE processes and, ultimately, the entire software development task. This research requires more evaluation methods as a next step to examine the effectiveness of the proposed framework by performing experiments on students' projects.

References

1. Wong, L.R., Mauricio, D.S., Rodriguez, G.D.: A systematic literature review about software requirements elicitation. J. Eng. Sci. Technol. **12**(2), 296–317 (2017)
2. Femmer, H.: Requirements engineering artifact quality: definition and control, Ph.D. Thesis, Technische Universität München (2017)
3. Sukumaran, S. Chandran, K.: The unspoken Requirements-eliciting tacit knowledge as building blocks for knowledge management systems. In: Uden, L., et al. (eds.) Knowledge Management in Organizations 2015, LNBIP, vol. 224, pp. 26–40, Springer, Cham (2015)
4. Ferrari, A., Spoletini, P., Gnesi, S.: Ambiguity and tacit knowledge in requirements elicitation interviews. Requirements Eng. **21**(3), 333–355 (2016)
5. Vasanthapriyan, S., Tian, J., Xiang, J.: A survey on knowledge management in software engineering. In: 2015 IEEE International Conference on Software Quality, Reliability and Security-Companion (QRS-C). IEEE (2015)
6. Sánchez, K.O., Osollo, J.R., Martínez, L.F., Rocha, V.M.: Requirements engineering based on knowledge: a comparative case study of the KMoS-RE strategy and the DMS process. Rev. Fac. de Ingeniería **77**, 88–94 (2015)
7. Emebo, O., Varde, A.S., Daramola, O.: Common sense knowledge, ontology and text mining for implicit requirements. In: International Conference on Data Mining (DMIN). The Steering Committee of The World Congress in Computer Science, Computer Engineering and Applied Computing (WorldComp) (2016)
8. Sandhu, R.K., Weistroffer, H.R.: A review of fundamental tasks in requirements elicitation. In: Wrycza, S., Maślankowski, J. (eds.) SIGSAND/PLAIS 2018, LNBIP 333, pp. 31–44. Springer, Cham (2018)
9. Polanyi, M.: The Tacit Dimension. Routlege and Kegan, London (1966)
10. Coppedge, B.B.: Transferring tacit knowledge with the movement of people: a Delphi study. In: Business Administration, Ph.D. Thesis, University of Phoenix (2010)
11. Kotz, T., Smuts, H.: Model for knowledge capturing during system requirements elicitation in a high reliability organization: a case study. In: Proceedings of the Annual Conference of the South African Institute of Computer Scientists and Information Technologists, p. 305–312. ACM, Port Elizabeth, South Africa (2018)
12. Emebo, O., Olawande, D., Charles, A.: An automated tool support for managing implicit requirements using analogy-based reasoning. In: IEEE Tenth International Conference, Research Challenges in Information Science (RCIS), pp. 1–7. IEEE (2016)
13. Mohamed, A.H.: Facilitating tacit-knowledge acquisition within requirements engineering. In: Proceedings of the 10th WSEAS International Conference on Applied Computer Science (ACS'10), World Scientific and Engineering Academy and Society (WSEAS), Stevens Point, Wisconsin, USA (2010)
14. Pantförder, D., Schaupp, J., Vogel-Heuser, B.: Making implicit knowledge explicit—Acquisition of plant staff's mental models as a basis for developing a decision support system. In: Stephanidis, C. (ed.) HCII Posters 2017, Part I, CCIS, vol. 13, pp. 358–365. Springer, Cham (2017)
15. Hao, J., Zhao, Q., Yan, Y., Wang, G.: A review of tacit knowledge: current situation and the direction to go. Int. J. Softw. Eng. Knowl. Eng. **27**(05), 727–748 (2017)
16. Castaneda, V., Ballejos, L., Caliusco, M., Galli, M.: The use of ontologies in requirements engineering. Glob. J. Res. In Eng. **10**(6), 2–8 (2010)

17. Mohamed, K.A., Farhan, M.S., Elatif, M.M.A.A.: Ontology-based concept maps for software engineering. In: Computer Engineering Conference (ICENCO), 2013 9th International. IEEE (2013)
18. Schneider, L., Hajji, K., Schirbaum, A., Basten, D.: Knowledge creation in requirements engineering-a systematic literature review. In: International Proceedings on Proceedings Wirtschaftsinformatik, pp. 1829–1843 (2013)
19. Turban, B.: Rationale management and traceability in detailed discussion. In: Tool-Based Requirement Traceability between Requirement and Design Artifacts, pp 159–258. Springer Vieweg, Wiesbaden (2013)
20. Kurtanović, Z., Maalej, W.: On user rationale in software engineering. Requirements Eng. **23**(3), 357–379 (2018)
21. Maalej, W, Thurimella, A.K.: DUFICE: guidelines for a lightweight management of requirements knowledge. In: Maalej, W., Thurimella, A.K. (eds.) Managing Requirements Knowledge, pp. 75–91. Springer, Heidelberg (2013)
22. Dutoit, A.H., Paech, B.: Rationale-based use case specification. Requirements Eng. **7**(1), 3–19 (2002)
23. Liang, P., Avgeriou, P., He, K.: Rationale management challenges in requirements engineering. In: 2010 Third International Workshop on Managing Requirements Knowledge (MARK), pp. 16–21. IEEE, Sydney, Australia (2010)
24. Thurimella, A.K., et al.: Guidelines for managing requirements rationales. IEEE Softw. **34**(1), 82–90 (2017)
25. Maalej, W., Thurimella, A.K.: An Introduction to Requirements Knowledge. In: Maalej, W., Thurimella, A.K. (eds.) Managing Requirements Knowledge, pp. 1–20. Springer, Heidelberg (2013)
26. Bugaite, D., Vasilecas, O.: Framework on application domain ontology transformation into set of business rules. In: International Conference on Computer Systems and Technologies—CompSysTech (2005)
27. Haron, A., et al.: Understanding the requirement engineering for organization: the challenges. In: 8th International Conference on Computing Technology and Information Management (NCM and ICNIT), pp. 561–567. IEEE, Seoul, South Korea (2012)
28. MacLean, A., et al.: Questions, options, and criteria: elements of design space analysis. Hum. Comput. Interact. **6**(3–4), 201–250 (1991)

An Effective MAC Protocol for Multi-radio Multi-channel Environment of Cognitive Radio Wireless Mesh Network (CRWMN)

M. Anusha and Srikanth Vemuru

Abstract CRWMN is an emerging method which makes efficient utilization of spectrum by allocating opportunistically, within the Industrial-Scientific-Medical (ISM) bands or licensed bands. Enthusiasm for the improvement of cognitive radio wireless mesh network is a result of utilizing under-usage of spectrum by the licensed users, by permitting unlicensed user to utilize. The main aim of CR in WMN is to encourage the unlicensed users for utilizing the licensed spectrum without affecting the licensed user's communication. In this paper, a MAC protocol was developed for allocating spectrum opportunistically. Additionally, an in-depth experimental analysis was performed on the proposed MAC to validate the efficiency of the proposed protocol. As a result, we allocate spectrum opportunistically by focusing on reducing access delay, interference, and hidden terminal problems and also it enhanced the throughput.

Keywords Cognitive radio · WMN · MRMC

1 Introduction

In recent years, the increase of new approaches demands new challenges for upcoming wireless combination technologies. The main challenge is to meet the demands of spectrum utilization by allocating opportunistically. In the previous decades, some techniques have been proposed for efficient utilization spectrum. CRWMN's is a developing innovation which is proposed for upgrading radio spectrum allocation. Co-operative communication and MC-MR frameworks enhance the data-rate without out bounds of upcoming wireless communications technologies. The combination of MC-MR environment with co-operative communication enhances the performance of upcoming wireless networks. Be that as it may, the combination of these

M. Anusha (✉) · S. Vemuru
Department of Computer Science & Engineering, KLEF, Vaddeswaram, Guntur dt,
Andhra Pradesh, India
e-mail: anushaaa9@gmail.com

© Springer Nature Singapore Pte Ltd. 2020
A. K. Luhach et al. (eds.), *First International Conference on Sustainable Technologies for Computational Intelligence*, Advances in Intelligent Systems and Computing 1045,
https://doi.org/10.1007/978-981-15-0029-9_3

approaches brings new challenges in wireless networks that should be tended to. This study tends to design new spectrum allocation techniques for MC-MR environment.

TDMA (Time Division MAC Access) technique is used for enabling MRMC for transmitting the data on the same channel with different radios. This method splits the radio signal into different slots. Again every time-slot is partitioned into two parts, i.e., to transmit control and data information [1]. The aim of the spectrum allocation scheme is to allocate the spectrum without interference and access delay by allocating resources opportunistically.

In this focus, we proposed a new TDMA based cluster MAC protocol (TC-MRMC-MAC) that can be used for MRMC environment in CRWMN's. This protocol integrates the approaches of cluster management and TDMA for allocating the spectrum opportunistically. The main objective of this approach is to allocate spectrum opportunistically and to allocate resources effectively by considering the issues interference access delay and hidden terminal problem.

2   Related Work

The TDMA MAC protocols use the wireless resources efficiently which avoids transmission collapse. According to time schedule, the channel resources can effectively allocate to the entire user. It has high efficiency and feasibility. The hybrid MAC protocols utilize game-theoretic dynamic spectrum access. As the players increase the negotiation problem may increase and delay the network performance. Additionally, it was difficult to synchronize among the users which lead to packet collisions among players. As a result, we understand that the TDMA MAC protocols enhance the network performance by allocating spectrum opportunistically by using scheduling. Some other existing TDMA based MAC protocols are studied as follows.

The author developed a P-MAC protocol [2] which has no common-control-channel. The selection of channel functionality is done on the basis of a historical prediction model. The protocol optimizes the consumption of energy, but due to channel errors, the updated patterns may not receive for the nodes. Some collisions may occur due to idle listen, which causes high traffic intensity. The author developed a CA-MAC protocol [3] which has a dual transceiver with the single-hop. It allows various communication pairs to transmit data concurrently.

The author developed cluster-based MAC protocol [1] which has a single transceiver with no CCC. It was a single-hop multi-channel protocol. The main functionality of the MAC protocol is to choose communication channels by forming static clusters using an experienced database. The author developed a single transceiver MAC for CRAHN protocol [4] with a dedicated CCC. It was a single-hop multi-channel protocol which works on the functionality of providing QoS for the nodes by optimizing the assignment of resources.

The author developed a multi transceiver C-MAC protocol [5] with a dedicated CCC. It was a multi-hop multi-channel protocol which balances the load of the network by reserving the traffic information. The author developed a CREAM-MAC-protocol [6] which has single/multiple transceivers with a common-control-channel. It combines sensing information and schedules the packets at physical and MAC layer respectively.

The author developed a DSA-MAC protocol [7] which consists of multiple transceivers and channel with no CCC. Whenever the user requested for spectrum, this spectrum allocation approach served as a means of guarantying the quality of service spectrum. The author developed an MMAC-CR protocol [8] with multiple transceivers, single-hop and multi-channel. The network was enabled by CR's with a distributed control mechanism. In order to utilize the best spectrum opportunities, CR users have to jump from one to another channel.

The author developed a Cog Mesh protocol which [1] is a multi-channel single-hop protocol that has multi-channel. In this approach, clusters are formed by using neighbors sensing information and communicate among clusters, allowing for the construction of the network. In the network, for achieving high-performance nodes are restricted for a limited transmission, so there is a need to cluster the network. We can attain stability, scalability, primary-user detection, by using the clustering mechanism in the network.

3 Methodology

In this division, we explain our CRWMN model and discussed our assumptions. Then we proceed to the approach for TDMA based cluster Multi-Channel Multi-Radio (TC-MRMC-MAC) to allocate Opportunistic Spectrum. Here, our intention is to design effective MAC protocol by considering the issues access delay, interference, hidden terminal problem, and resource allocation. Here, we are considering maintaining the network by static topology with dynamic channel allocation.

A CRWMN's is designed as a $G = (V, E)$, where V symbolizes as set.of.nodes and E for symbolizes edges. The set of nodes is denoted as $N \epsilon \{N_1, N_2....N_n\}$, as well as every node, is equipped with multiple-radios. And also the set of radios are denoted as $R \epsilon \{r_1, r_2 ... r_i\}$ for various channel allocation. The available channel allocations are denoted as $C \epsilon \{c_1, c_2 ... cm\}$, where $(C \geq R_i)$ which have common communication capabilities. The network has bi-directional communication capability, so that, transceivers will facilitate to identify the same channel for communication. In general, nodes involved in the network have the limited transmission range, so before channel allocation, the network will divide into clusters. Formation of cluster makes easier to maintain the network and also it improves the scalability by stabilizing end-to-end communication.

The TSA-MAC [1] approach enhances the throughput of the communication channel by which decides and allocates free channels for SU's. The basic idea behind this method is to allocate spectrum opportunistically according to the availability at different times-slots. In the network, nodes have restricted communication range; in the case of larger networks, there is a need to partition the network into clusters for achieving better network throughput. In the proposed work, each node calculates the distance to the BS for identifying the CH. The node which has a minimum distance to the BS will elect as Cluster Head (CH). At first base station (BS) assigns a random number between "0 and 1" for each of "n" nodes; each node calculates the distance to the base station. Additionally, each node calculates the threshold value which is given by

$$\text{Threshold Value} = \frac{p}{(1 - p(r \bmod 1/p)}, \quad \text{if } n \in G \qquad (1)$$

Here "p" is the expected % of CH, "r" is the random number; "G" is the set.of.nodes which has not CHs for the earlier period $1/p$ iterations. The process of the distance-based algorithm is as follows.

Algorithm D-Select

1. In the network, let O be the set.of.nodes
   $O = \{O1, O2, O3 \dots Os\}$
2. Now, calculate the distances from one node to all the nodes

   For m = 1 to s do
   For n = 1 to s do
   dmn = distance from Sm to Sn
   End for
   End for

3. Now calculate the entirety distance from one to every node

   For m = 1 to s do
   For n = 1 to s do
   Dm = Dm + dmn
   End for
   End for

4. Compute the distance for all nodes from the BS to the node

   For m = 1 to s do
   D_BSi = Distance from BS to Sm
   End for

5. Calculate the total distances from base station to every node

   For m = 1 to s do
   TD_BSm = D_BSm + Dm
   End for
   Step 6: List out the nodes with minimum distance
   TD_BSm = Min (TD_BS1, TD_BS2, TD_BS3, TD_BS4 … TD_BSk)

End.

The node with the shortest distance to the BS and the random number less than the threshold value will be elected as the cluster head. Now in the network, all the picked CH's broadcast their existence as CH's. Every node will calculate the distance from CH and connects to the nearest CH by sending a join-msg acknowledgment. Suppose if the distance between CH to the node is greater than the distance between BS to the node, then the node directly communicate with the BS. Hence the clusters are formed. If the size of the cluster is smaller the predefined threshold then the clusters can be re-organized. Once receiving all CH-join msgs from every node of the cluster, CH assigns a time-slot to every MN. Every MN has the capacity to communicate with their CH's as well as with neighboring nodes. CH also communicates with the base station and neighboring CH's. Then only each CH and MN's can make local decisions and transfers to BS through CHs. The MN senses the spectrum and broadcast to CH and its neighboring nodes. The nodes in the network co-operate to each other to make effective channel allocation. Figure 1 shows co-operative mechanism.

After formation of clusters, without loss of generality, for efficient opportunistic spectral allocation, we assume that all the channels "C" is in the same frame structure. The time-slot duration of every frame is fixed as "D-slots". Prior to transmitting a data packet, every node has to confirm the condition of that channel and transmit data when the channels are not busy. So, every node in the cluster has to continuously sense the primary network.

At the beginning of the first slot of every channel, SU's sense for active transmissions on their channels of interest, this was always busy with primary transmission. The process of sensing over a period of time "T_s" is known as "Broad-Sensing

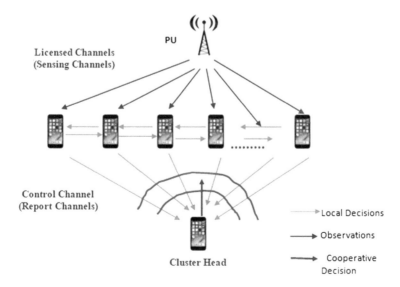

Fig. 1 Co-operative mechanism

(T_{Br-S})". As a result, the time required for broad-sensing should be equal to the time-slot duration, i.e., $T_{Br-S} = T_s$. In the remaining slots, CR users including BS and CH sense for a fraction of time in the slot "T_{ns}" to ensure that the channel is busy or idle for transmitting. The process of sensing for a fraction of time is known as "Narrow-Sensing (T_{ns})".

For recognizing whether the channel was idle or busy, the methodology of uniting received signal with a predefined-bandwidth "B" for a time-period T_s is utilized. CR user detects energy D_E and compares with the predefined- threshold value. It can be represented as idle

$$K_s = \begin{cases} 0, & \text{if } D_E > Th_E \text{ , } H_o \rightarrow \text{inactive} \\ 1, & \text{else } H_1 \rightarrow \text{Active} \end{cases} \tag{2}$$

Here K_s is the result of the sensing-period.
Therefore, the probability of false-alarm and miss-detection are

$$P_{fa} = P(0 > \lambda | H_o) \tag{3}$$

$$P_m = P(0 < \lambda | H_1) \quad \text{Respectively} \tag{4}$$

The least time to sense the channel can be calculated as

$$T_{s,\min} = \frac{1}{\lambda^2 S_f} (Q^{-1}(P_{fa}) - (Q^{-1}(P_{td}) \sqrt{2\lambda + 1})^2 \tag{5}$$

Here λ is average SNR of the PU, S_f is frequency-sampling, Q, P_{fa} and P_{td} is the probability-of-Gaussian-tail, false-alarm, and channel detection.

Suppose if the particular time-slot is busy at the end of the narrow time-slot, MN chooses next channel and continuous for the rest of the slots. The whole process will be continued for all the slots from slot-3 to slot-D until a slot is founded for transmitting.

While allocating channel there is a need to reduce access delay, interference, and hidden terminal problem. Let M_R, M_{ch}, M_N represents n-radios, Channel count, mesh nodes. Assume the communication-link b/w the nodes m and n, if and if the two nodes are within the range of each other and they have a radio-frequency operating on the same channel k [9]. The multi-objective function for the MC-CMR WMN can be formulated as:

$$\text{Minimum}\big(\text{Dem}_B / \text{Cap}^k_{(x,y)}\big) | \text{Interference Nodes } (n_{i-1}, n_i) \tag{6}$$

$$k \in \text{Availchs}(n_{i-1}, n_i)$$

Let L^k_{ij} represents the link-channel

$$L_{ij}^k = 1,$$
$$= 0, \text{ otherwise Max } L_{ij}^k \tag{7}$$

S.t. $\sum L_{ij}^k \leq \eta$, where η is the user-specified a max no.of.links.

Let $M(l)$ represent estimation of load of all radio channels. It can be represented as:

$$\text{Optimize } M(l) = N_l.C_{\text{ch}}/N_{\text{IR}} \tag{8}$$

where N_l: no.of avail channels, C_{ch}: Channel capacity, N_{IR}: links in the range.

4 Parameters Considered

The following are the parameters considered for evaluating the proposed methodology.

4.1 Link Load Estimation

Estimation of the link load is the sum of the data transmission on the link per unit time

$$\text{IC} * (m, n, k) = \lambda * \text{IC}(m, n, k) + \omega * \text{IC}(m, n, k)$$
$$0 \leq \lambda, \omega \leq 1 \tag{9}$$

where IC*: Interference within the channel, ω: Weight of the link, λ: Weight of the channel

$$\text{Max} \sum \text{IC}_{(m,n,k)}^* / C^k < 1 \tag{10}$$

4.2 Bandwidth-Satisfaction-Index (BSI)

BSI was used to measure the level of the index of bandwidth for network flow. It was assumed as "1" if there is free network flow or else "0".

$$\text{BSI} := 1,$$
$$\text{If } T_{\text{ach}}.T_{\text{rec}}^{-1} > 1 = T_{\text{ach}}T_{\text{rec}}^{-1} > 1, \text{ Otherwise}$$

Here T_{ach} represents achieved throughput; T_{rec}^{-1} represents demanded throughput.

Also, $0 \leq BSI\_value \leq 1$

Algorithm-CMR-MCWMN

R_{reqp}: Packet request; Srcnc: The adjacent neighbor node; N_{id}: Unique ID of R_{reqp}
Procedure
Suppose if a node n_i receives a request from the node n_j then get the initial nearest covered set as S_{rcnc} to each R_{reqp}:
Inflist = getInferencesets (Src_{nc})
For each pair of Inflist id does: Let A = First_ id; B = Second_id;
A sends broadcasting information along with message and node B decides to rebroadcast the message to other nodes.
$S_{A \cap B}$ (A) = Area of the Intersection region
Boolean flg = Check (Ch ($S_{A \cap B}$ (A)) and its neighbours
If flg = false within the area then Communicate the request, else
Solve the Multi-Objective Linear programming model using constraints defined in (6), (7) & (8).
If a feasible solution found-Accept the flow allocation
Else, compute the new radio signal to overcome the hidden problem:
Compute = Max (0, r-(rs-transrate)/$S_{A \cap B}$ (A)), where rs-radio-signal
Prob = (e-e $_{A \cap B}^{(1-S \ (A)/e-1))/}$prob (A)
If the Prob > Θ then Update radio signal distance r = r + d;
Where Θ: WMNs parameters for threshold
Compute new probability to the new channel allocation:
Prob = (e $- e_{A \cap B}^{(1-S \ (A)/e-1))/}$ prob(A)
End if,
Else,
Block the request,
End if,
End if,
Exit.

While allocating the channels there is a need to maintain the available resources efficiently. Here, we developed a procedure ICD-PBR. In this approach, CH first identifies the idle channels list. Then the Secondary Users (SU) which involved in the communication identifies the idle channel to them. Now the information is passed to the CH and then the information is passed to the Base Station (BS) through CH. At that point, BS will analyze the state of channels and assigns idle channels to SU. While allocating the channels BS will consider the priorities and assigns channels to SU's. An algorithm ICD-PBRA (Idle Channel Detection and Priority Based Resource Allocation) was developed for dynamic resource allocation based on the requirement. Here, the process of resource allocation was done according to their priority of accessing channel. The algorithm is as follows.

Algorithm-ICD-PBRA

Inputs: Priorities P, constraints C, Channels M
Step 1: Initialize channels Vector (V).
Step 2: SU detects idle channels from M.
Step 3: SU populates V and sends to CH.
Step 4: CH sends V to BS.
Step 5: BS analyzes the state of channels from V.
Step 6: BS assigns idle channels to SU.
Step 7: SU is notified of suitable idle channel.
Step 8: Thus multiple SUs get notifications simultaneously.
Step 9: Each SU considers P and C and make transmission strategy.
Step 10: Based on the C and P analysis BS allocates resources required to SU.
Step 11: The SU which is allocated with resources optimally participates in co-operative communication.
Step 12: Repeat from 2–11 for further transmissions.

5 Estimation of P-I

In wireless networks, different types of data can be transferred through various nodes. The data may be normal text, real-time or it may be secret data, etc. Each data type has its own requirement which leads to priority. So they should check the priority before allocating. Priority can be computed by considering the data type, number of packets and waiting-time. The data which have highest Priority Index, those packets have to transfer immediately. For calculating the P-I, the three main elements are Data Type (TD), Queue Length (LQ), and Packet Delay (DP) and index are given between 0 and 3.

The P-I can be as computed as:

$$P\text{-}I = 3*(TD + DP) + LQ \tag{11}$$

The proposed solution ICD-PBRA effectively utilizes the available resources according to their requirement which enhances the utilization of the channels and transmission throughput. By using these approaches, a TC-MRMC-MAC protocol (TDMA based co-operative Multi-Radio Multi-Channel-MAC) allocates an opportunistic spectrum for the Multi-Radio Multi-Channel environment of CRWMN. In addition, this protocol implementation reduces access delay, interference, and hidden terminal problems. And also it effectively utilizes the resources according to their requirement, which additionally enhances the throughput, packet–delivery ratio,

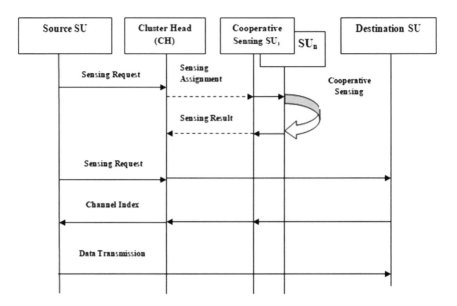

Fig. 2 Sequence diagram

compared to the methodologies in existence. The steps involved in the proposed MAC protocol are shown in Fig. 2.

6 Experimental Results

Through a comprehensive simulation, TC-MRMC-MAC protocol was evaluated. We used LabVIEW Communications Design Suite, NI USRP 2920, NI USRP 2952R and desktop computer to design test-bed. Every desktop computer was utilized for controlling the USRP devices for transmitting and receiving parameters. Following are results obtained by using the MAC protocol. Figure 3 shows the allocation of opportunistic spectrum by using the proposed MAC protocol TC-MRMC-MAC. In the graph, Amplitude was indicated on the x-axis and Time on the y-axis.

Figure 4 shows that our proposed TC-MRMC-MAC protocol has more throughputs when compared to the existing MAC protocols. The throughput of the proposed system with a variable number of nodes and compared with the existing protocol, and it is shown in Fig. 4. The throughput of the proposed MAC protocol outperforms in comparison with existing protocols.

Further, the packet–delivery ratio of proposed TC-MRMC-MAC protocol is analyzed by varying number of nodes and compared it with the existing MAC protocols and it is shown in Fig. 5. It is clear that the proposed TC-MRMC-MAC protocol reduces the packet loss in comparison with existing protocols, that's why its packet–delivery ratio is more in comparison with existing protocols.

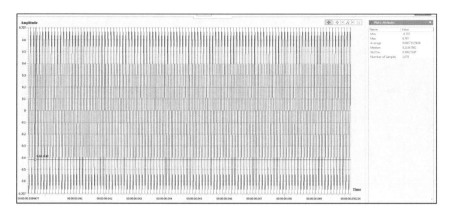

Fig. 3 Opportunistic spectrum allocation

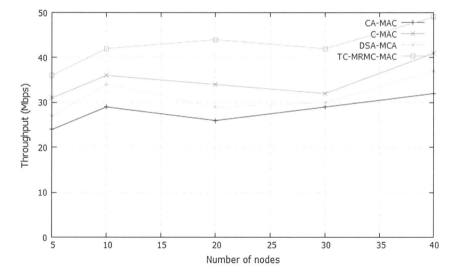

Fig. 4 Throughput

The access delay of proposed work was analyzed with respect to a number of channels, and it is shown in Fig. 6. Proposed TC-MRMC-MAC protocol access delay is less in comparison with existing MAC protocols because proposed MAC protocol utilizes the spectrum effectively.

The proposed TC-MRMC-MAC protocol utilizes the unused channel effectively than the existing MAC protocol which is shown in Fig. 7.

The throughput of the proposed protocol is computed under the varying number of unused channel, and it is shown in Fig. 8. The throughput of proposed work is more, as its utilization of unused channel is more in comparison with existing protocols.

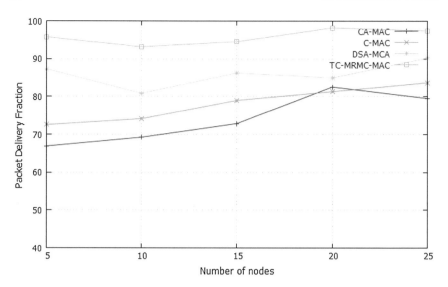

Fig. 5 Packet–delivery ratio

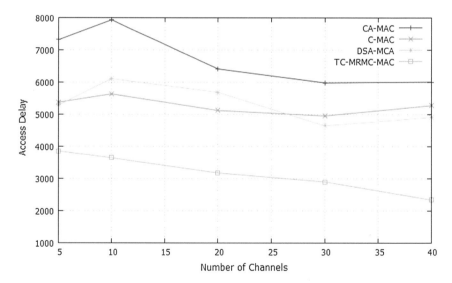

Fig. 6 Access delay

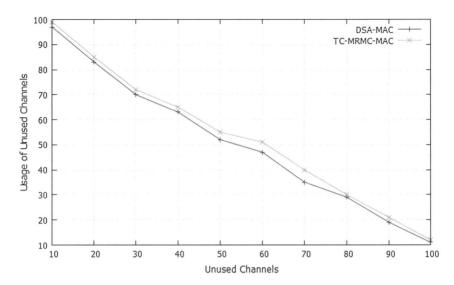

Fig. 7 Utilization of unused channels

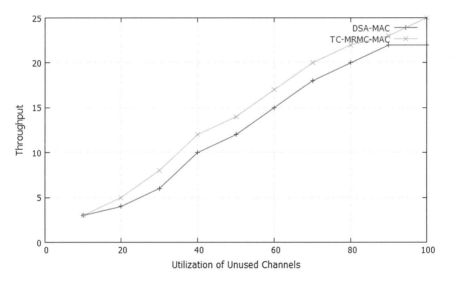

Fig. 8 Increasing of throughput while utilizing unused channels

7 Conclusion

This paper represents a TC-MRMC-MAC as a TDMA based Clustered MAC protocol for Multi-Channel Multi-Radio environment for CRWMN's, in which the protocol is used to allocate the spectrum opportunistically by considering the access delay, interference, and hidden terminal problem. We also explain a TDMA mechanism for data transmission simultaneously by utilizing distinctive channels. The proposed approach enhances the usage of the spectrum. The communication test-bed is set in the host desktop with LabVIEW using the hardware USRP. The simulation results explain that TC-MRMC is able to allocate spectrum opportunistically by scheduling resources, as well as meeting the challenges.

Acknowledgements Acknowledgements One of the authors Anusha M is very much thankful to Department of Science and Technology (DST), Government of India, New Delhi, for awarding her with a Women Scientist's scheme under DST–WOS (A) File No. SR/WOS-A/ET- 1071/2015 Programme.

References

1. Chen, T., Zhang, H., Maggio, G.M. Chlamtac, I.: CogMesh: a cluster-based cognitive radio network. In: 2007 2nd IEEE International Symposium on New Frontiers in Dynamic Spectrum Access Networks, pp. 168–178. Dublin (2007)
2. Joshi, G.P., Nam, S.Y., Kim, S.W.: Decentralized predictive MAC protocol for ad hoc cognitive radio networks. Wirel. Pers. Commun. **74**(2), 803–821 (2014)
3. Timalsina, S.K., Moh, S., Chung, I., Kang, M.: A concurrent access MAC protocol for cognitive radio ad hoc networks without common control channel, EURASIP J. Adv. Sign. Proc. **69** (2013)
4. Passiatore, C., Camarda, P.: A MAC protocol for cognitive radio wireless ad hoc networks. In: International Workshop on Multiple Access Communications MACOM 2011: Multiple Access Communications, pp 1–12 (2011)
5. Cordeiro, C., Challapali, K.: C-MAC: a cognitive MAC protocol for multi-channel wireless networks. In: 2007 2nd IEEE International Symposium on New Frontiers in Dynamic Spectrum Access Networks, pp. 147–157. Dublin (2007)
6. Su, H., Zhang, X.: CREAM-MAC: an efficient cognitive radio-enabled multi-channel MAC protocol for wireless networks. In: 2008 International Symposium on a World of Wireless, Mobile and Multimedia Networks, pp. 1–8. Newport Beach, CA, (2008)
7. Joe, I., Son, S.: Dynamic spectrum allocation MAC protocol based on cognitive radio for QoS support. In: 2008 Japan-China Joint Workshop on Frontier of Computer Science and Technology, pp. 24–29. Nagasahi (2008)
8. Timmers, M., Pollin, S., Dejonghe, A., Van der Perre, L., Catthoor, F.: A distributed multichannel MAC protocol for multihop cognitive radio networks. IEEE Trans. Veh. Technol. **59**(1), 446–459 (2010)
9. Anusha, M., Vemuru, S.: An efficient MAC protocol for reducing channel interference and access delay in cognitive radio wireless mesh networks. Inter. J. Commun. Antenna Propag. (IRECAP) **6**(1) (2016)
10. Li, Xiaoyan, Fei, Hu, Zhang, Hailin, Zhang, Xiaolong: A cluster-based MAC protocol for cognitive radio ad hoc networks. Wirel. Pers. Commun. **69**(2), 937–955 (2013)

11. Marouthu, A., Vemuru, S.: Cluster-based opportunistic spectrum access in cognitive radio wireless mesh network's using co-operative mechanism. Wirel. Pers. Commun. (Springer-SCIE) **99** (2), 779–797 (2018)
12. Zekavat, S.A., Li, X.: User-central wireless system: ultimate dynamic channel allocation. In: First IEEE International Symposium on New Frontiers in Dynamic Spectrum Access Networks, 2005. DySPAN 2005, pp. 82–87. Baltimore, MD, USA (2005)

IoT-Based RFID Framework for Tracking, Locating and Monitoring the Health's Rituals of Pilgrims During Hajj

Eman Abdulrahman Binhotan and Abdullah Altameem

Abstract Internet of Thing (IoT) is a procedure and process of interlinked computing devices, software technologies, sensors, and network connectivity, mechanical and digital machines, objects, animals or human beings which can be supplied with unique identifiers. Hajj is a completely unique gathering with Mecca and Kaaba being spiritually essential to many faiths across the globe, specifically Muslims. The form of communication that we see now could be either human–human or human–device; however, the Internet of Things (IoT) guarantees a top-notch destiny for the Internet. Radio-frequency identification system (RFID) is an automated technology and aids machines or computers to become aware of items, record metadata or manage individual goal through radio waves. This paper proposes an IoT based on RFID in order to engage the pilgrims within the slice to monitor the vital health rituals during Hajj framework according to the research gaps found in literature review. The results of the proposed framework will facilitate a high detection and monitoring the pilgrim personal RFID data. The future research tends our effort to test the proposed framework using a pre-survey as well as analyze, design, and implement the proposed framework.

Keywords Internet of things · Radio-frequency identification system · Health management system · Pilgrim · Health rituals

1 Introduction

The term Internet of Things (IoT) is firstly proposed in 1982 by Kevin Ashton. IoT is an aggregate of hardware and software program technologies alongside embedded

E. A. Binhotan (✉) · A. Altameem
Information Systems Department, College of Computer and Information Sciences,
Al Imam Mohammad Ibn Saud Islamic University (IMSIU), Riyadh, Saudi Arabia
e-mail: eman1hh@gmail.com

A. Altameem
e-mail: altameem@imamu.edu.sa

© Springer Nature Singapore Pte Ltd. 2020   37
A. K. Luhach et al. (eds.), *First International Conference on Sustainable Technologies for Computational Intelligence*, Advances in Intelligent Systems and Computing 1045,
https://doi.org/10.1007/978-981-15-0029-9_4

systems that permits to offer and facilities to any individual, every time, everywhere required using any network [1].

IoT is the network of physical systems or "things" embedded with digital systems, software program technologies, sensors, and network connectivity, which lets in those items to acquire and trade data for availing numerous services. It is a concept demonstrating an associated set of something, any individual, any time, any location, any provider, and any community connection [1]. Surely we will say that IoT is an idea of connecting any device or tool with an on and off switch to the Internet (and/or to each other). This consists of the whole lot from smartphones, smart coffee and tea makers, lamps, washing machines, headphones, wearable systems, and almost something else you could recollect (nano to macro devices) [2].

Radio frequency identification (RFID) is a wireless primarily based technology for identification of items and objects. The RFID technology is now producing notable hobby inside the market due to its robust application skills [3]. RFID utilizes electro-magnetic waves for transmitting and getting records saved in a tag to or from a reader [4]. RFID generation has an component over different identification systems which include bar code systems, smart cards, magnetic stripe cards, and biometrics as it requires no observable pathway for correspondence, sustains harsh bodily environ-ments, takes into consideration synchronous character, has first-rate records storage, huge read range and is power and cost efficient.

The modern traits in the clinical management systems may be labeled in more than one ways primarily based at the perspective of the technology, functionality, and the benefits. There may be a trend taking place with the convergence of purchaser devices and clinical devices [5].

Certainly, this requires a system to follow the vital signs of pilgrims during Hajj. So that doctors can obtain positive results and rely on them in dispensing medicines and suppositories to diseases. Constant following and checking of group is the need of any religions use. Investigating of tremendous gatherings may be practiced through taking care of the image got of pioneers in holy zones. Following and checking of pioneers status may be practiced by techniques for using different sensors [6]. The constant region of someone can be trailed by strategies for the utilization of the Global Positioning System (GPS) and Global System for Mobile correspondence (GSM) period.

Monitoring individuals and tracking, locating in crowded area and dense surround-ings is already a difficult issue. Regardless of the ever enhancing technology within the modern-day world, it has no longer been able to solve the most basic difficulty in any crowded occasions, such as the Hajj pilgrimage in Mecca. Hajj is a com-pletely unique gathering with Mecca and Kaaba being spiritually essential to many faiths across the globe, specifically Muslims. As we recognize that Hajj requires a excessive attempt to carry out the rituals and maximum pilgrims of a excessive age, which require a big remedy monitoring, correct because the aged pilgrims are always uncovered to numerous forms of sicknesses and infections with the diseases that exist for a number of the pilgrims.

Pilgrimage has a brilliant importance in Mecca, Saudi Arabia. Each pilgrimage session attracts a big crowd. Usually the pilgrims circulate simultaneously in a massive organization. Getting out of place in crowd could be very not unusual. Finding the misplaced individuals among large number of pilgrims reasons plenty of problem for his family and the authorities. One of these setup poses a actual assignment to the government in handling the gang, and tracking/identifying human beings. What makes it even extra hard is that each one pilgrims glide at the equal instances and to the same places. At the same time as such occasions are a completely unique religious enjoy for all pilgrims, it poses fundamental demanding conditions of each kind to the authorities accountable for facilitating this annual occasion. The following are a number of the not unusual troubles confronted by the usage of the pilgrims and the government like identification of pilgrims (lost, lifeless, or injured), medical emergencies, guiding out of place pilgrims to their respective camps, congestion manipulate, and control.

2 Literature Review

Monitoring and finding pilgrims and their followers in a cluster is difficult and important. Pilgrims and the government authorities confronted the issue like finding pilgrims' locations, health and clinical emergencies, guiding misplaced pilgrims, congestion control. Various technologies and devices used for monitoring and tracking pilgrims.

Various frameworks have been planned with the asset of the analysts utilizing GPS of cell phone, RFID, independent GPS with remote discussion for following the pilgrims. A few structures produce other correspondence conventions for identifying the nearness of individuals in a spot like Wi-Fi and Bluetooth now not especially following pioneers; anyway, they might be modified to follow pilgrims.

Regarding the monitoring of pilgrims [6] proposed a device for the monitoring and identity of pilgrims in the holy regions, in Makkah Saudi Arabia, within the path during Hajj. The proposed device consisted of 3.5G network which included the location via using numerous service organizations. 3.5G is a gathering of different cell communication and records innovation intended to offer higher execution than 3G structures, as an interim advance toward arrangement of full 4G capacity. With the asset of solicitation or occasionally the Unified Identification Number (UID) is ship to the server. A server maps the range and longitude data on a Google map or any topographical records framework. In the event that the Web association is lost the cell phone stores the territory data in its memory until the Web association is reestablished, at that point it sends all put-away area data and clears this information from memory. The developed framework can be utilized to follow a specific explorer. As a substitute, any pilgrim can demand crisis help the utilization of the indistinguishable framework. The propelled framework works in a joint effort with a RFID recognizable proof framework [6].

To identify the pilgrims locator during hajj [7] implemented an application known as, hajji locator, which had been structured to monitor, locate, and tracking the pilgrims for the duration of hajji. The tool makes use of the GPS constructed in cell phone. The information transfer is carried out ideally the usage of 3G or Wi-Fi, in any case, the instrument additionally might be utilized to convey the records in crisis circumstances. A pilgrim is needed to the application on his telephone, which gives numerous functions and sends his vicinity to the server [7]. The recurrence of spot supplant is resolved carefully as it impacts the power utilization and the system data transmission. The territory is refreshed after a specific term or after the customer had moved a positive separation. Picking time-based way to deal with supplant the region guaranteed that we had the refreshed area after a chose period. Be that as it may, in the event that a customer is stationary or he is moving gradually, at that point the successive region updates may be excess and could deplete the battery and development the network traffic. The spot basically based strategy, constantly put away the track of the contemporary capacity and the last dispatched job. In the event that the hole among contemporary capacity and the remaining dispatched place is in excess of a limit expense sent to the cutting-edge job and spared it as shutting obvious region. In the event that a client moved extremely quick, at that point remove-based strategy could supplant more prominent often. The server utilized PHP and MySQL; government experts could be signed in into the Web server utilizing a program to get the situation of pilgrims that is shown in either a forbidden shape or Google maps. The proposed framework inferred that the separations based absolutely update is recommended for observing pilgrim [7]. Additionally an option is outfitted to the person to convey his area physically. The exactness in open or semi-opened to places is phenomenal, however, at this point and again the server did never again get save of any insights as a result of inaccessibility of a zone reestablish. In a few case, the vicinity received from GPS may be very a protracted manner from consumer actual place. The use of previous location of a client, those defective locations have been adjusted and corrected.

RFID is used for a tracking pilgrims [8] proposed a device designed for monitoring the pilgrims at some stage in Hajj. On this device, an RFID tag is given to each pilgrim. Pilgrims having cellular smartphone with GPS used location-based totally offerings through installing in an application. Those services protected area family' people or pals, soliciting for pressing assist, a map of vital locations. A good way to transmit the cutting-edge role, the app should be running with the cell telephone. RFID readers had been moreover hooked up in awesome regions to experiment the tags. The control center gives abilities like imagining his area of the majority of the pilgrims on a guide hunting down explorers essentially dependent on various gauges like area, age, etc. Sending warning to the versatile instrument, protecting the database of spots like medical clinics, region data, and individual realities around travelers. Mobile phones utilize the Web administrations and the RFID readers use middleware programming to interface with the control focus [8]. Those structures confronted issues with RFID labels and readers. The investigation assortment of the RFID readers is low; it is likewise tormented by the natural components. The flag between the RFID tag and the readers is getting obstructed with the asset of the label holders individual body, besides the examination assortment hit by viewpoint with readers. So they chose to now not have a wristband RFID labels.

RFID-based on GPS is used for a monitoring pilgrims [9] proposed a gadget every explorer is given a sensor unit which contained a GPS for getting the present-day region of the pioneer, a microcontroller which executes this framework to send the spot to consistent sensor frameworks, a battery to vitality this matchbox estimated sensor unit. It likewise incorporates Zigbee radio it truly is utilized to change the actualities to a network of enduring handle structures conveyed inside the area of activity. The followed cell frameworks might be a whole parcel more than steady detecting hubs. The planners of those structures had propelled a RFID-based absolutely structures in the past which provided least difficult the distinguishing proof of pilgrims. This gadget is intended to follow pioneers notwithstanding distinguish them [9]. The Wi-Fi sensor hubs transmit their specific distinguishing proof range, its cutting edge include as procured from the GPS subterranean insect the time intermittently or on solicitation from observing station. The information switch takes region opportunistically utilizing impromptu system. The sensor gadgets transmit the data the use of flooding convention, so the equivalent data is sent to the majority of the near to relentless sensors. This actualities is put away for some time and after that multi-bounce directing is utilized to exchange this information. The framework structured can endure disappointment of a couple of consistent sensor hubs. The places of explorers had been mapped onto Google map simply like other framework.

For questioning the pilgrim's area, the server sends inquiries through utilizing the most critical heading using the remaining perceive locale. The framework additionally helps directing two or three inquiries in parallel. Battery controlled remote sensors gadgets need power green equipment and programming program. The records change need to likewise be insignificant. The ventured forward recurrence of refreshing the spot influences quality utilization and transmission capacity, however, it lessens the time taken to discover a pioneer [10].

To track the pilgrims during hajj [11] proposed a integrate work of Assisted Global Positioning Systems (AGPS)/Wi-Fi and cell identification to improve the execution of following framework. AGPS is to be had on most cellular phones and can be used to precisely find out a person, yet, has a few disadvantages. Even though, AGPS can substantially lessen the electricity consumption. Most cell phones will come up short on battery in depend of hours if AGPS place fixes have been performed frequently [11]. Portable recognizable proof situating has less power ingesting and is accessible both inside and out of entryways, yet estimation of locale is significantly less exact. Hence, AGPS/Wi-Fi gadget is proposed along thought of portability mapping. Wi-Fi specifically used to go over the nearness of individual. The drawback of utilizing Wi-Fi is that it specifically works appropriately inside.

RFID-included smartwatch is used for a tracking pilgrims [12] proposed the RFID approach to control and control swarm. The issue identified with personality of pilgrims is understood. Each time a stampede or a hearth passed off, distinguishing severely disfigured our bodies. Some other stop result of overcrowding is that hundreds of pilgrims were disintegrated from their corporations or from spouse and children for days or maybe weeks (and a number of them may additionally moreover in no way observed). Reuniting the pilgrims with their agencies may additionally moreover take large time due to terrible identity mechanisms in region. In instances, while pilgrims were misplaced and do now not understand their location of live, there may be no manner offending their info, without looking for the assist. If pilgrims do now not cross returned to their bases, there may be definitely no generation in area to tune them. In the end, the RFID and "smartwatch" structures sup-ported via the backend database might be capable of track the missing individuals, and discover all of the extraordinary data is proposed in [12]. Because the biometric scans for every pilgrim could also be saved inside the backend database, It'd assist identity in case the loss of the RFID tags and "smart watch." As soon as the humans with fitness dangers were recognized, they may be monitored at some stage in the pilgrimage with the aid of way of the installed RFID and "clever watch" tool. [12].

Healthcare applications in IoT structures have been receiving growing hobby due to the fact they help facilitate some distance flung tracking of sufferers, reference [13] exploited a strategic feature of such gateways at the threshold of the network to offer higher-degree offerings including real-time local data processing, embedded data mining, local storage, etc., supplying as a result a clever e-fitness gateway. Reference [14] exploited applicability of IoT in healthcare and treatment by way of presenting a holistic structure of IoT e-Health ecosystem, also reference [15] proposed a reliable one M2M-based IoT device for non-public healthcare systems.

Table 1 summarizes the comparison of the work done in some of related research and papers with our proposed work in order to fill the research gaps.

3   Research Methodology

In order to develop a framework of IoT based on RFID in order to engage the pilgrims with the slice to monitor the vital signs and health signs during Hajj, the methodology shall be achieved by: firstly, highlighting the lack of current research identified in the previous research by reviewing and summarizing the existing frameworks, algorithms, and techniques used in tracking and locating people in crowded area and secondly, the proposed framework that consists of two main modules and each module has different components have to be formulated.

4   Suggested Framework

The suggested framework is shown in Fig. 1. It consists of two main modules and each module has different components.

The descriptions of these modules and their components as follows:

Table 1 Literature review summary

Reference	Work done	Our work
[6]	Proposed a pilgrim identification system based on WSN. The author practically tested their proposed system on pilgrims	Pilgrim tracking and locating using IoT-based RFID and WSN
[7]	Proposed a framework based on mobile phone to identify the pilgrim location	Intelligent hand-watch integrated with the RFID and GPS to identify the pilgrim location
[8]	Evolved smartphones to provide highly accuracy, monitoring, and tracking of the pilgrims and offer them with location-based services for Hajj	Intelligent hand-watch integrated with the RFID and GPS to identify the pilgrim location
[9]	Proposed a system using WSNs by deploying fixed stations at various parts of the holy mosque	RFID technology can be used effectively to help pilgrims while they are busy performing their ritual at Hajj
[10]	Developed and stamped detection-based on image processing via Matrix Laboratory	Integration of GPS/GSM modems which offer definite location of pilgrims and high-speed communication instead of using separate GSM and GPS modems
[11]	Introduces an energy-efficient cellular and AGPS/Wi-Fi tracking system based on the construction of a mobility map from a person's historical location data	Since AGPS consumes a lot of power and Wi-Fi coverage is often spotty, so we used GPS/GSM instead
[12]	Proposed RFID and "smartwatch" in managing the people in crowed area	Proposed IoT/RFID and "smartwatch" for tracking and locating the people in crowed area
[13]	Concept smart e-Health gateways and fog computing in the context of IoTs-based totally healthcare systems changed were offered	Cloud computing technology will be used if needed
[14]	Presented a holistic architecture of IoT e-Health ecosystem	Since the ecosystem needed a multi-layer architecture to be work efficiently, so the idea will be complicated
[15]	Proposed a reliable one M2M-based IoT system for per-sonal healthcare devices	Proposed IoT-based RFID for healthcare devices

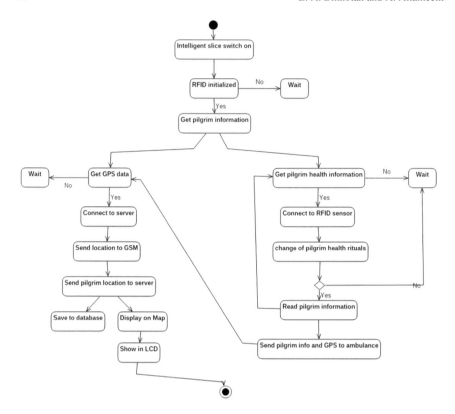

Fig. 1 Proposed framework

4.1 Pilgrim Module

This module has two components:

(1) Intelligent slice (hand-watch): this watch has to be designed and engineered using IoT devices and RFID sensors, readers and tags. As well as this watch be able to detect the change of pilgrims' health rituals such as arthritis, heart attack disease, chronic lower respiratory diseases, osteoporosis, fall, shingles, and all the dangerous diseases that needs high health caring.
(2) GPS: which should be attached to the hand-watch to detect the place of the pilgrim that has an health rituals changes detection for better and fast healthcare services.

4.2 Health Detection and Monitoring Unit Module

(1) GSM: global system for high-speed communication is used to send the data related to pilgrims' location that integrated with the GPS and reading health rituals for all pilgrims.
(2) LCD: to shows the location's signals for all pilgrims.
(3) Unified Database: that consists of previous heath profile for pilgrims that have chronic diseases.

5 Conclusion

The proposed framework of IoT devices primarily based on RFID sensors and tags with the intention to interact the pilgrims with the slice to display the crucial signs and symptoms and health symptoms during Hajj uses the integration of GPS/GSM modems, which provide genuine area of pilgrims and excessive speed communication. The proposed framework can be detected ahead to take essential movement in time to prevent it and accordingly offer security to pilgrims. A part from monitoring, tracking and detecting, the proposed framework will have an excellent gain. It gives an choice for a pilgrim in case of clinical emergency. The proposed framework automatically monitors the body conditions of the pilgrim, which helps in getting scientific useful resource right away. Thus, the proposed framework will fulfill the want of pilgrim with none objection.

In the future, we will continue our effort to test the proposed framework using a pre-survey as well as provide comprehensive analysis and design to support the integration of such technologies for the good of authorities and pilgrims. In the future also, the research will further be refined by including all the holy places for Hajj and performing simulation and later creating a prototype to conduct real-time tests on the pilgrims.

References

1. Islam, S.M.R., Kwak, D., Kabir, M.D.H., Hossain, M., Kwak, K.-S.: The internet of things for health care: a comprehensive survey. IEEE Access **3**, 678–708 (2015)
2. Feki, M.A., Kawsar, F., Boussard, M., Trappeniers, L.: The internet of things: the next technological revolution. Computer **46**, 24–25 (2013)
3. Cerlinca, T.I., Turcu, C., Turcu, C., Cerlinca, M.: RFID-based information system for patients and medical staff identification and tracking. In: Sustainable Radio Frequency Identification Solutions: InTech (2010)
4. Rida, A., Yang, L., Tentzeris, M.M.: RFID-enabled sensor design and applications. British Library, Artech House, London (2010)
5. Pande, P., Padwalkar, A.R.: Internet of things—A future of internet: a survey. Inter. J. Adv. Res. Comput. Sci. Manage. Stud. **2**(2) (2014)

6. Mohandes, M.: Pilgrim tracking and identification using the mobile phone. In: 2011 IEEE 15th International Symposium on Consumer Electronics (ISCE), pp. 196–199 (2011)
7. Mantoro, T., Jaafar, A.D., Aris, M.F.M.: Hajjlocator: A hajj pilgrimage tracking framework in crowded ubiquitous environment. In: 2011 International Conference on Multimedia Computing and Systems (ICMCS), pp. 1–6 (2011)
8. Mitchell, R.O., Rashid, H., Dawood, F., AlKhalidi, A.: Hajj crowd management and navigation system: people tracking and location based services via integrated mobile and RFID Systems. In: 2013 International Conference on Computer Applications Technology (ICCAT), pp. 1–7 (2013)
9. Mohandes, M., Haleem, M., Deriche, M., Balakrishnan, K.: Wireless sensor networks for pilgrims tracking. IEEE Embed. Syst. Lett. **4**, 106–109 (2013)
10. Barahim, M.Z., Doomun, M.R., Joomun, N.: Low-cost bluetooth mobile positioning for location-based application. In: 3rd IEEE/IFIP International Conference in Central Asia on Internet (ICI 2007), pp. 1–4 (2007)
11. Wang, X., Wong, A.K.-S., Kong, Y.: Mobility tracking using GPS, Wi-Fi and Cell ID. In: International Conference on Information Networking (ICOIN) (2012)
12. Yamin, M., Mohammadian, M., Huang, X., Sharma, D.: RFID technology and crowded event management. In: 2008 International Conference on Computational Intelligence for Modelling Control & Automation, pp. 1293–1297 (2008)
13. Rahmani, A.M., Gia, T.N., Negash, B., Anzanpour, A., Azimi, I., Jiang, M., Liljeberg, P.: Exploiting smart e-Health gateways at the edge of healthcare internet-of-things: a fog computing approach. Future Gener. Comput. Syst. **78**, 641–658 (2018)
14. Farahani, B., Firouzi, F., Chang, V., Badaroglu, M., Constant, N., Mankodiya, K.: Towards fog-driven IoT eHealth: promises and challenges of IoT in medicine and healthcare. Future Gener. Comput. Syst. **78**, 659–676 (2018)
15. Woo, M.W., Lee, J., Park, K.: A reliable IoT system for personal healthcare devices. Future Gener. Comput. Syst. **78**, 626–640 (2018)

IoT in Education: Its Impacts and Its Future in Saudi Universities and Educational Environments

Shahid Abed, Norah Alyahya and Abdullah Altameem

Abstract Currently, there has been a growing concern in the Internet of Things especially in education that allows physical objects (things), sensors and actuators to communicate via the Internet, which has changed universities and educational institutions. By using Internet of Things, universities can create a sophisticated environment for its learners and staff. So, this paper proposes a literature review to illustrate the impact of Internet of Things on the educational environment. It also aims to examine user acceptance toward this technology within Saudi universities and educational institutions by using the Technology Acceptance Model which encourage them to use this technology. Overall, the results reveal that Internet of Things has been a positive impact on the educational process and university campuses. Also, there is a strong agreement among the study individuals on the future of the Internet of Things in Saudi Arabia regarding the perceived usefulness and the ease of use of this technology.

Keywords Internet of things · IoT · Education · KSA · Saudi university and educational institutions

1 Introduction

During the last decade, technologies have been developing rapidly. The different tools, strategies, and technologies of social media, learning, internet, etc. drives the innovation in education. The internet technology, which is a category that drives innovation in education, supports it in different ways, considering that Internet of Things (IoT) is a subcategory of this technology [1, 2]. IoT is one of innovation

S. Abed · N. Alyahya (✉) · A. Altameem
Imam Mohammad Ibn Saud Islamic University (IMSIU), Riyadh, KSA, Saudi Arabia
e-mail: n-m-y-1@hotmail.com

S. Abed
e-mail: skda1952@hotmail.com

A. Altameem
e-mail: altameem@imamu.edu.sa

© Springer Nature Singapore Pte Ltd. 2020
A. K. Luhach et al. (eds.), *First International Conference on Sustainable Technologies for Computational Intelligence*, Advances in Intelligent Systems and Computing 1045,
https://doi.org/10.1007/978-981-15-0029-9_5

47

that modifies the world from simple objects to interconnected objects making a complicated infrastructure [1].

The IoT is a revolutionary technology which enables connection between devices (things), people and environments to collect data by embedding actuators and sensors and transmitting it to specialized applications to create useful and actionable information [1, 3]. Several devices, for examples: camera, smartwatches, digital displays, and audio recorder can communicate with each other via private or public internet networks [1, 4]. It is useful in various areas, such as education, agriculture, medicine, cities, businesses, and others [1, 5].

College students are moving toward technologies such as tablets, smartphones, and laptops, which makes the process of education more efficient for teachers [6]. Nowadays, most of the devices such as laptops, PCs, TVs, smartphones, and tablets connected to the internet [5]. Gartner, Inc. [7] predicts that 8.4 billion connected devices will be in use across the world in 2017 and 20.4 billion of connected things will be used in 2020. Furthermore, Gartner said that in 2020, the IoT will generate incremental revenue surpassing $300 billion [8]. Also, a report from CITC [9] states that the number of Saudis who used the Internet reached about 21.6 million in 2015.

The campuses need to embed IoT technologies to create a sophisticated environment within universities. In fact, the rise of efficiency in energy, the unique communication between learners and lecturers, and the decrease in operating cost are some of the most benefits of IoT in education [1]. For example, New Richmond schools in Ohio, US are saving almost $128,000 every year through utilizing a web-based system which manages all mechanical tools within the schools [6]. Several universities have used IoT devices within their classrooms, laboratories, and campuses, which are useful to enhance teaching and learning, automate education process, improve student outcomes, manage students' healthcare, and control security of campuses [1, 5]. Some of these campuses use the Radio Frequency Identification (RFID) readers, which are parts of entrance systems to measure humidity, air pollution, temperature, etc. [5]. Furthermore, the "Educators' and Learners' Internet of Things (ELIoT)" [10] which known for supporting education with the IoT by giving workers and students access to various sources of learning and displaying information in a way that improves education [1] perceives the significance of this technology in educational environments. IoT also can increase the security of the campus. Universities can use VT Alerts which is an emergency notification system that notifies staff and learners about an emergency situation on campus [11]. All of these are examples of the importance of IoT in education are applied in non-Saudi schools and universities.

Through what we can benefit from IoT in educational environments, it explains to us the positive impact of applying this technology on the educational process and university campuses [1, 5]. However, this technology hasn't yet applied in all Saudi university and educational environment, and there is no paper conducting the acceptance towards this technology in this country. Therefore, this paper aims to illustrate the impact of IoT on educational environment and examines user acceptance toward this technology within Saudi universities and educational institutions by using the Technology Acceptance Model (TAM), which encourage them to use this technology. This paper focuses on the following questions:

1. What is the impact of IoT technology in education?
2. Is there any potential to move towards IoT technology in Saudi educational environments?

2 Literature Review

Technology affects how people live, play, work and most importantly learn. It has played an imperative role in education. However, for many universities and educational institutions, implementing IoT technology has become a growing requirement. It improves education at all levels comprising school, college, and university. From teacher to learner, campus to a classroom, everything can obtain benefit from this technology [12, 13].

IoT is different technologies and devices that work with each other in tandem. These devices must be able to gather and process data, then transmit information to humans commonly via smart devices or other devices, receive information, operate according to specific conditions and support communication [14]. It enables educational institutions and universities to gather a huge amount of data from sensors, devices, and actuators to create useful information and perform meaningful actions accordingly [1]. A lot of universities have used IoT devices, such as security cameras, light power, and temperature-controlled devices within their classrooms and campuses. The education could be enhanced in the classrooms by using smart devices. These technologies are useful to save energy and time, reduce cost, optimize classroom and campus environments, monitor safety and student's health, improve collaboration and allow students presence remotely [1, 2]. Universities also can utilize connected objects to observe their learners, teachers, clerks, and resources in a cost-effective way [6]. Implementation IoT to smart campus within university helps lecturers and learners to automate education process [5]. Students can access the material at any place and any time and teachers can use the devices to improve learning and increase student satisfaction [1]. As of that, QR codes and embedded sensors technologies enable learners to explore their university more efficiently [1].

IoT modifies approach in education. Teachers save time by using Kahoot as an example of online IoT processes to grade students and Google documents and Telegram to share marks. Students get useful internet resource (such as information) from teachers. They also can evaluate teacher competence and level of knowledge, and management will check and observe the situation and make a required action based on provided feedback. Moreover, Without IoT technologies in university, there are some problems, such as students' and staff management, security, energy, heating, and water excess usage [5].

A study [2] has categorized IoT applications in education into 4 categories: (1) monitoring student's healthcare which enables university students to access a healthcare service, (2) improving teaching and learning which provides a lot of experiences for students, (3) classroom access control which creates a secure campus within universities, and (4) real-time eco-system monitoring and energy management which provides sustainable energy.

As more of realizing the significance of IoT, Curtin University developed Building 215 by using IoT technology, which gives learners an environment to discover engineering theory through testing a static building and improve learner experiences. This building gives students the ability to monitor temperature, thermal efficiency, moisture, air, and cooling/heating water. It also collects data on student attendance and their movement. This university also created a sophisticated educational environment for learners with particular educational needs by using smart devices to manage and control equipment and parking remotely by touch, sound or gestures, RFID, text-to-speech modules, sensor indicators for orientation and others [14]. Also, Carnegie Mellon University researchers have created two IoT apps: (1) Impromptu, which is a system that accesses applications when students are needed. (2) Snap2It, which is a system that lets students to take a smartphone photo of a projector or printer to link it easily [15].

Using IoT to enhance education, the research group at Hague Security Delta (HSD) Campus has opened an Internet of Things forensic laboratory (IoT Forensic Lab), which has specialized hardware, software, and knowledge. Chip off is a good example of particular hardware, which is used to extract data from memory device even when it is protected by a password or damaged by fire or water. In the IoT Forensic Lab, teachers accompany the bachelor students who work with advanced technologies of digital forensic during their internship [16]. Furthermore, UW-Madison IoT Lab is a lab on the University of Wisconsin–Madison campus, which allows students and staff from several academic disciplines to research, enhance collaboration between universities and the industry and create projects to learn, understand, discover and innovate by using IoT technology in several domains [17]. Also, implementation of IoT Flipped Classroom to the university as a part of intelligent campus which gives chance for students to access material and learn subject anywhere at any time. A study [5] has illustrated that students who study with traditional approach have worse results than those who study with IoT Flipped Classroom. Samsung produced a gateway using IoT technology [18] that can be used in universities, schools or office to automate light control, automate attendance recording or track location to improve accuracy. There are many advantages of the smart educational institution campus: (1) promote smart energy management to reduce energy consumption. (2) Improve communication between campus communities. (3) Produce useful application to enhance automation of business process. (4) Provide real-time warning services and disaster response to create safe learning environment. (5) Provide map information services of campus [19].

IoT can be used to improve and manage building efficiently and economically [20]. Smart building is one of IoT application used in smart city [21] and smart university and school [20]. It enables them to save cost and time, automate maintenance, protect the environment, achieve efficient parking and improve energy management [20]. University of Brescia developed their campus with smart educational buildings to improve thermal prosperities of the building and achieve reduction in energy consumption [20, 22] (Table 1).

Table 1 Comparison between educational institutions with and without IoT

	Educational institutions with IoT	Saudi Arabia educational institutions without IoT
Attendance	Automatically by utilizing different attendance systems [1]	Manually and takes time and effort
Students tracking	Easy to track students on campus [1, 5]	Hard to track students on campus
Campus	Intelligent, safe [1, 4, 5] and easy to control	Unsafe and difficult to control
Energy	Easy to manage energy consumption [20]	Difficult to manage energy consumption
Parking	Easy to find a parking lot for university communities' cars [20]	Difficult to find a parking lot for university communities' cars
Building	Smart and easy to manage [20, 22]	Difficult to manage
Learning	Personalize learning and students can learn remotely [23]	Lecturers are recourse of information

3 Theoretical Framework and Hypothesis

This paper studies user acceptance toward IoT technology within Saudi universities and educational institutions by examines the intention to use this technology as the dependent variable and it is affected by perceived ease of use and perceived usefulness which are independent variables. These two independent variables are Technology Acceptance Model (TAM) variables created by Davis in 1989 and used by researchers to study the behavior of the human [24]. This model is a validated instrument and has been widely applied to research of the user acceptance of different types of technology [24]. It has been utilized in several empirical studies. It is proven to be of statistically reliable. Furthermore, many empirical studies have discovered that this model illustrates the variance in usage behaviors and intentions with a variety of technologies. Also, it used the perceived ease of use and perceived usefulness to subrogate the subjective norm [20]. Therefore, this model has been used in this study to examine user acceptance toward IoT technology in Saudi universities and educational institutions. Figure 1 will present the conceptual model of the framework.

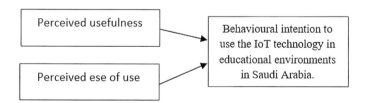

Fig. 1 Conceptual model

Based on the conceptual model of the framework, evolving the following proposed hypothesizes:

H1- There is a positive relationship between perceived usefulness and behavioral intention to use the IoT technology in educational environments in Saudi Arabia.
H2- There is a positive relationship between ease of use and behavioral intention to use the IoT technology in educational environments in Saudi Arabia.

4   Research Method

The researchers used a questionnaire to collect data and TAM model variables to evaluate the acceptance and identify the potential of moving towards IoT technology in Saudi educational environments through statistical analysis. The questionnaire targeted students and teachers, and employees at all educational environments in Saudi Arabia to gather information using an electronic form that requires its completeness to be counted. 479 of respondents have been collected and measured using 5-point rating scale with endpoints anchored by strongly disagrees and strongly agrees. Statistical packages for Social Sciences (SPSS) used to calculate the statistical measures.

5   Results and Dissection

As for internal validity of study tool, to make sure of the sincerity of the internal consistency of the instrument study, the researchers calculate the Pearson correlation coefficient between each paragraph and the total score of the dimensions as follows.

Tables (2 and 3) show that all the statements are significant at the level of (0,01), this refers to the high internal consistency as well as high and adequate validity indicators that are trusted when applying the current study.

To check the reliability of the study tool, the researchers used Alpha Cronbach's stability coefficient, as follows.

Table 2 Pearson correlation coefficients of pivot paragraphs: (perceived usefulness) with the pivot total score

Items	Person correlation	Items	Person correlation
1	0.689^{**}	7	0.662^{**}
2	0.740^{**}	8	0.610^{**}
3	0.659^{**}	9	0.653^{**}
4	0.625^{**}	10	0.694^{**}
5	0.656^{**}	11	0.681^{**}
6	0.639^{**}	–	–

** Correlation is significant at the 0.01 level

Table 3 Pearson correlation coefficients of pivot paragraphs: (ease of use) with the pivot total score

Items	Person correlation	Items	Person correlation
12	0.681**	17	0.629**
13	0.665**	18	0.545**
14	0.606**	19	0.604**
15	0.722**	20	0.596**
16	0.678**	21	0.607**

** Correlation is significant at the 0.01 level

Table 4 Alpha Cronbach's for measuring the study tool stability

No.	Pivot	Reliability coefficient
1	Perceived usefulness	0.867
2	Ease of use	0.823
Overall reliability		0.913

Table 4 shows that the study scale enjoys statistically acceptable stability. The total stability coefficient value (alpha) amounted to (0.913) which is a high degree of stability. The stability coefficients of the study tool ranged between (0.823 and 0.867) which are high and trustful when applying the present study.

5.1 The Demographic Features of the Study Participants

Table 5 shows the distribution of study individuals according to their demographic characteristics. The results show that (430) of respondents represent (89.8%) of females, while (49) of respondents are (10.2%) of males.

5.2 Results of Questionnaires

What is the future of IoT educational environment in Saudi Arabia? To identify the future of learning environment for the Internet of things in Saudi Arabia, repetitions, percentages, arithmetic averages, and standard deviation are used in the responses of the study members, as follows in Table 6 for perceived usefulness and 7 for ease of use.

Table 6 shows that the future theme of the educational environment for the IoT in Saudi Arabia in connection with the perceived usefulness includes (11) phrases in which arithmetic averages are ranged between (3.87 and 4.57). These averages come within the fourth and fifth categories of the Pentagram gradient scale categories which indicate the degree of response (agree–strongly agree).

Table 5 The distribution of the study sample according to their demographic characteristics

Gender	Frequency	Percent
Male	49	10.2
Female	430	89.8
Total	479	100.0

Age	Frequency	Percent
18 years or less	43	9.0
19–24 years	234	48.9
25–35 years	79	16.5
36 years or more	123	25.7
Total	479	100.0

Category	Frequency	Percent
Student	310	64.7
Teacher	73	15.2
Educational environment employee	96	20.0
Total	479	100.0

Degree	Frequency	Percent
Student	66	13.8
Bachelor	368	76.8
Master	21	4.4
PHD	24	5.0
Total	479	100.0

The general arithmetic average is (4.27) with a standard deviation of (0.53). This indicates that there is strong agreement among the study individuals on the future of the learning environment for the IoT in Saudi Arabia regarding the perceived usefulness, which is represented in the study individuals' agreement on the following: (Internet of things technology will save them time to acquire information relevant to their jobs, such as students' attendance and the condition of the equipment in the classroom, and using internet of things technology during teaching is helpful, in addition to using internet of things technology through smart devices is easy for them, and using internet of things technology will increase the education efficiency).

Figure 2 represents the number of respondents who responded strongly agree, agree and the statements mean of perceived usefulness elements. The statement number 6 received the highest mean which is (The IoT technology will save my time to acquire information related to my jobs, such as students' attendance and the condition of the equipment in the classroom.). There were (313) respondents who responded "strongly agree" with this statement, and (132) who responded "agree". The mean is 4.57. Table 7 refers to the term ease of use repetitions, percentages, arithmetic averages, and standard deviations.

Table 6 Frequencies, percentage means and standard deviation of the responses to the perceived usefulness

N	Items	Approval degree										Mean	SD	Ranking
		Strongly agree		Agree		Neutral		Disagree		Strongly disagree				
		F	%	F	%	F	%	F	%	F	%			
1	Using IoT technology will increase education efficiency	274	57.2	162	33.8	35	7.3	5	1.0	3	0.6	4.46	0.73	4
2	Using IoT technology will improve education performance	243	50.7	182	38.0	45	9.4	8	1.7	1	0.2	4.37	0.74	5
3	I find it is helpful to use the IoT technology during teaching	295	61.6	157	32.8	22	4.6	5	1.0	0	0.0	4.55	0.63	2
4	I believe that using IoT technology through smart devices is easy for me	273	57.0	168	35.1	31	6.5	5	1.0	2	0.4	4.47	0.70	3
5	I will be able to control all elements through an easy and clear way through the phone screen	226	47.2	173	36.1	69	14.4	10	2.1	1	0.2	4.28	0.80	6
6	The IoT technology will save my time to acquire information relevant to my jobs, such as students' attendance and the condition of the equipment in the classroom	313	65.3	132	27.6	30	6.3	3	0.6	1	0.2	4.57	0.66	1
7	Using Internet-connected technologies will increase the security inside the campus	219	45.7	147	30.7	88	18.4	21	4.4	4	0.8	4.16	0.93	8

(continued)

Table 6 (continued)

N	Items	Approval degree										Mean	SD	Ranking
		Strongly agree		Agree		Neutral		Disagree		Strongly disagree				
		F	%	F	%	F	%	F	%	F	%			
8	Using Internet- connected technologies to reduce the overuse of water and air conditioning will preserve the environment	178	37.2	128	26.7	133	27.8	33	6.9	7	1.5	3.91	1.03	10
9	I believe that the credibility of the information presented by IoT will be high	139	29.0	180	37.6	122	25.5	33	6.9	5	1.0	3.87	0.95	11
10	The IoT technology introduces the best solution for my demands	177	37.0	214	44.7	75	15.7	11	2.3	2	0.4	4.15	0.80	9
11	IoT technology may provide me with individual attention and address my own needs	197	41.1	195	40.7	76	15.9	10	2.1	1	0.2	4.20	0.79	7
	Mean for dimension											4.27	0.53	–

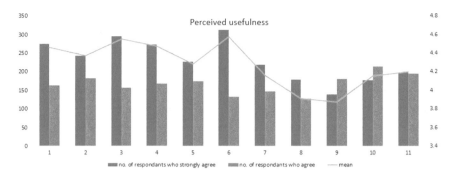

Fig. 2 Information about perceived usefulness elements

Table 7 shows that the future of IoT in the educational environment in Saudi Arabia in connection to the ease of use includes (10) phrases in which arithmetic averages are ranged between (3.88 and 4.45). These averages fall within the fourth and fifth categories of the Pentagram gradient scale categories which indicate the degree of response (agree–strongly agree).

The general arithmetic average is (4.19) with a standard deviation of (0.55). This indicates that there is agreement among the study individuals on the future of the learning environment for the internet of things in Saudi Arabia regarding the easy accessibility, which is represented by the study individuals' agreement on the following: (it is interesting to have technologies connected to the internet using the smartphone to find a car parking lot., and use devices connected to internet of things technologies if I have received the necessary technical support, in addition to using the IoT often, and it is interesting to use the IoT to record the attendance instead of the traditional way).

Figure 3 represents the number of respondents who responded strongly agree, agree, and the statements mean of ease of use elements. The statement number 20 received the highest mean which is (I find it is interesting to have technologies connected to the internet using the smartphone to find a car parking lot). There were (258) respondents who responded "strongly agree" with this statement, and (184) who responded "agree". The mean is 4.45.

Consequently, (1) There is strong agreement among the study individuals on the future of the learning environment for the IoT in Saudi Arabia regarding the perceived usefulness, which is represented in the study individuals' agreement on the following: (IoT technology will save them time to acquire information relevant to their jobs, such as the attendance of students and the condition of the equipment in the classroom, and using IoT technology during teaching is helpful, in addition to using IoT technology through smart devices is easy for them, and using IoT technology will increase the education efficiency). (2) This indicates that there is strong agreement among the study individuals on the future of the learning environment for the IoT in Saudi Arabia regarding the ease of use, which is represented in study individuals' agreement on the following: (how interesting to have technologies connected to the

Table 7 Frequencies, percentage means and standard deviation of the responses to the ease of use

N	Items	Approval degree										Mean	SD	Ranking
		Strongly agree		Agree		Neutral		Disagree		Strongly disagree				
		F	%	F	%	F	%	F	%	F	%			
12	I will often use IoT technology	179	37.4	189	39.5	92	19.2	15	3.1	4	0.8	4.40	0.68	3
13	I feel that using IoT technology is important to improve the student's academic attainment	195	40.7	128	26.7	82	17.1	53	11.1	21	4.4	4.09	0.87	7
14	I think that someday universities and educational environment will use the electronic education instead of the traditional education	214	44.7	182	38.0	68	14.2	12	2.5	3	0.6	3.88	1.19	9
15	Using mobile screens to operate the equipment of laboratories will be much easier for me	193	40.3	197	41.1	79	16.5	8	1.7	2	0.4	4.24	0.83	5
16	In my opinion, using the IoT will have a positive impact.	177	37.0	177	37.0	95	19.8	19	4.0	11	2.3	4.19	0.80	6
17	I find it is interesting to use IoT to observe the student's health in the classroom	158	33.0	172	35.9	95	19.8	40	8.4	14	2.9	4.02	0.97	8
18	I have already used Internet—connected devices previously.	248	51.8	149	31.1	62	12.9	14	2.9	6	1.3	3.88	1.05	10
19	I find it is interesting to use the IoT to record attendance instead of the traditional way	279	58.2	151	31.5	38	7.9	8	1.7	3	0.6	4.29	0.89	4

(continued)

Table 7 (continued)

N	Items	Approval degree										Mean	SD	Ranking
		Strongly agree		Agree		Neutral		Disagree		Strongly disagree				
		F	%	F	%	F	%	F	%	F	%			
20	I find it is interesting to have technologies connected to the internet using the smart phone to find a car parking lot	258	53.9	184	38.4	30	6.3	4	0.8	3	0.6	4.45	0.76	1
21	I can use IoT technologies if I have received the necessary technical support	179	37.4	189	39.5	92	19.2	15	3.1	4	0.8	4.44	0.71	2
Mean for dimension												4.19	0.55	–

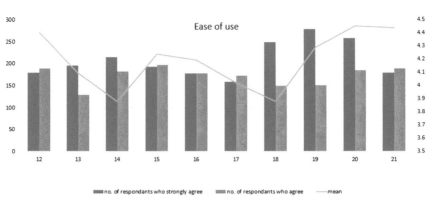

Fig. 3 Information about ese of use elements

internet using the smartphone to find a car parking lot, They can use IoT technologies if they have received the necessary technical support, in addition to using the IoT often, and it is interesting to use IoT to record the attendance instead of the traditional way).

6 Conclusion

Use of IoT technology in the education field has increased innovative ideas, so both students and teachers can get ease and betterment in their lives. This technology can be leveraged to collect and process data to provide actionable information to enhance efficiencies on campus, reduce the use of energy and improve public safety. This paper presents a literature review useful for displaying the impact of IoT technology in education and some of the current practical applications of this technology and using the TAM model variables to evaluate the acceptance and identifying the potential of moving towards IoT technology in Saudi educational environments. This kind of research aims to support and encourage Saudi educational environments to use this technology. Usage of IoT within university helps teachers and learners to enhance the communication, enhance education process, save money and enhance student experiences, utilizing the new IoT technology in campus life, optimize classroom and campus environments, monitor safety and student's health, and enable students' presence remotely. After analyzing the responses of the questionnaire, the researchers conclude that there is strong agreement among the study individuals on the future of the IoT in Saudi Arabia regarding the perceived usefulness and the ease of use of this technology.

References

1. Majeed, A., Ali, M.: How Internet-of-Things (IoT) making the university campuses smart? QA higher education (QAHE) perspective. In: 2018 IEEE 8th Annual Computing and Communication Workshop and Conference (CCWC), pp. 646–648, Las Vegas, NV (2018)
2. Bagheri, M., Movahed, S.: The effect of the internet of things (IoT) on education business model. In: 2016 12th International Conference on Signal-Image Technology & Internet-Based Systems (SITIS), pp. 435–441, Naples (2016)
3. Dutton, W.H.: Putting things to work: social and policy challenges for the internet of things. Info **16**(3), 1–21 (2014)
4. Hudson, F.: The internet of things is here, EDUCAUSE REVIEW, (2016), https://er.educause.edu/articles/2016/6/the-internet-of-things-is-here. Last Accessed 18 Jan 2019
5. Ban, Y., Okamura, K., Kaneko, K.: Effectiveness of experiential learning for keeping knowledge retention in IoT security education. In: 2017 6th IIAI International Congress on Advanced Applied Informatics (IIAI-AAI), pp. 699–704, IEEE, Hamamatsu (2017)
6. Meola, A.: How IoT in education is changing the way we learn, Business Insider, 2016, https://www.businessinsider.com/internet-of-things-education-2016-9. Last Accessed 18 Jan 2019
7. Meulen, R.: Gartner says 8.4 billion connected "Things" will be in use in 2017, Up 31 percent from 2016, Gartner (2017). https://www.gartner.com/en/newsroom/press-releases/2017-02-07-gartner-says-8-billion-connected-things-will-be-in-use-in-2017-up-31-percent-from-2016. Last Accessed 18 Jan 2019
8. Gartner: Internet of things installed base will grow to 26 billion units By (2020). https://www.mcit.gov.sa/en/media-center/news/91424, Last Accessed 18 Jan 2019
9. Al Rwais, A.: Annual Report 1436H/ 1437H 2015. http://www.citc.gov.sa/en/MediaCenter/Annualreport/Pages/default.aspx. Last Accessed 18 Jan 2019
10. Lounkaew, K.: Explaining urban–rural differences in educational achievement in Thailand: Evidence from PISA literacy data. Econ. Educ. Rev. **37**, 213–225 (2013)
11. VT Alerts.: Virginia Tech. https://www.alerts.vt.edu/, Last Accessed 18 Jan 2019
12. Mareco, D.: 10 reasons today's students NEED technology in the classroom, securedge (2017). https://www.securedgenetworks.com/blog/10-reasons-today-s-students-need-technology-in-the-classroom. Last Accessed 18 Jan 2019
13. Elsaadany, A., Soliman, M.: Experimental evaluation of internet of things in the educational environment. Int. J. Eng. Pedagogy (iJEP). **7**(3), 50–60 (2017)
14. McRae, L., Ellis, K., Kent, M.: Internet of things (IoT): education and technology, NCSEHE (2018)
15. Spice, B.: CMU leads Google expedition to create technology for "internet of things, Carnegie Mellon University (2015). https://www.cmu.edu/news/stories/archives/2015/july/google-internet-of-things.html. Last Accessed 18 Jan 2019
16. Henseler, H.: IoT Forensic Lab: exploring the connected world at HSD Campus, the hague security delta. https://www.thehaguesecuritydelta.com/cyber-security/innovation-projects/project/95-iot-forensic-lab-exploring-the-connected-world-hsd. Last Accessed 18 Jan 2019
17. UW-Madison's Internet of Things Lab Reaches New Heights of Creativity and Innovation: Inwisconsin (2015). https://inwisconsin.com/blog/uw-madisons-internet-of-things-lab-reaches-new-heights-of-creativity-and-innovation/. Last Accessed 18 Jan 2019
18. Samsung IoT Solutions Changing the way the world works. https://www.samsung.com/global/business/networks/solutions/iot-solutions/. Last Accessed 18 Jan 2019
19. Muhamad, W., Kurniawan, N., Suhardi, Yazid, S.: Smart campus features, technologies, and applications: a systematic literature review. In: 2017 International Conference on Information Technology Systems and Innovation (ICITSI), pp. 384–39. Bandung (2017)
20. Abuarqoub, A., Abusaimeh, H., Hammoudeh, M., Uliyan, D., Abu-Hashem, M., Murad, S., Al-Jarrah, M., Al-Fayez, F.: A survey on internet of things enabled smart campus applications. In: Proceedings of the International Conference on Future Networks and Distributed Systems - ICFNDS '17, pp. 1–7. Cambridge, United Kingdom (2017)

21. Asghari, P., Rahmani, A., Javadi, H.: Internet of things applications: a systematic review. Comput. Netw. **148**, 241–261 (2019)
22. De Angelis, E., Ciribini, A., Tagliabue, L., Paneroni, M.: The brescia smart campus demonstrator. Renovation toward a zero energy classroom building. Procedia Eng. **118**, 735–743 (2015)
23. Mj, M., Ga, H.: Do indoor pollutants and thermal conditions in schools influence student performance? A critical review of the literature. Indoor Air **15**(1), 27–52 (2005)
24. Davis, F.: Perceived usefulness, perceived ease of use, and user acceptance of information technology. MIS Q. **13**(3), 319–340 (1989)

A Nobel Approach to Detect Edge in Digital Image Using Fuzzy Logic

Utkarsh Shrivastav, Sanjay Kumar Singh and Aditya Khamparia

Abstract Edge detection is a popular technique to find out the boundaries of different objects in a digital image. These edges are searched on the basis of gradients that are available in the image. The gradients depend upon the intensity and value of pixels. In this paper, a new technique is proposed for edge detection which is based on the fuzzy rule-based system. Fuzzy-based system requires mainly three steps, conversion of inputs to fuzzy linguistic variables, then a set of rules that can be applied on inputs and final step is again converting back from fuzzy logic output to crisp output. Horizontal and vertical gradient vectors have been considered as inputs for the fuzzy system. Gaussian membership functions and triangular membership functions are used for converting crisp inputs into fuzzy input. For defuzzification purpose centroid method is applied. The Mamdani model has been used to develop the system. The result of proposed system is better than most of the popular conventional edge detection techniques. The proposed system also minimizes the noisy details and can also be used variety of images.

Keywords Edge detection · Fuzzy logic · Mamdani model · Prewitt · Sobel · Canny · LoG · Roberts

1 Introduction

The edge detection is the most important and very fundamental thing to be used when talking about image processing and analysis. The edge detection techniques are very helpful in performing segmentation, medical image analysis, surveillance, computer vision, human vision, and pattern recognition. Edge detection is mostly used to find out the important information like shape of the object, and reflectance in an image. An image has different intensity values at each pixel, edge detection techniques utilize this property of image to locate the boundaries. The techniques that

U. Shrivastav · S. K. Singh (✉) · A. Khamparia
Department of Computer Science and Engineering, Lovely Professional University,
Jalandhar, Punjab, India
e-mail: Sanjayksingh.012@gmail.com

© Springer Nature Singapore Pte Ltd. 2020
A. K. Luhach et al. (eds.), *First International Conference on Sustainable Technologies for Computational Intelligence*, Advances in Intelligent Systems and Computing 1045,
https://doi.org/10.1007/978-981-15-0029-9_6

we use under this kind of segmentation use a threshold value with which the intensity value of the pixel is compared. If the first derivative of intensity is greater than the threshold then an edge is detected. In the edge-based approach, first, the threshold is applied over the whole image and one by one the above technique is processed continuously. Once all the edges have been detected, they all are joined together to convert those edges into the boundaries of the various objects in an image, with the help of which the different regions or sections can be visible, there are different kind of edge detection techniques which uses different approach and different threshold values.

Researchers studied about the palmistry and decided to create a fuzzy-based expert system [1]. It was a very unique work in which the personality of any individual was decided by just taking his lines of palm as an input. In this particular work the heart line, lifeline and headline were taken as an input to extract feature, based on which the rules were created and with the help of which the personality was predicted.

Researchers also proposed their work to check the defaulter risk in bank sector, for this they developed a fuzzy rule-based expert system in which four inputs are given which are different banking related terms like CIBIL Score(CS), Loan to Value Ratio(LVR), Already Availed Loan(AAL), and Income Ratio Factor(IRF) [2], and they will be passed to the fuzzy inference system, where the rules will be triggered according to the given values, this will distribute the result into defaulter and non-defaulter.

A research paper was presented in which medical expert system for gynecology was created, this system was helpful for the detection of inflammatory diseases [3]. In this expert system, a user interface was created where the suitable data taken from a dataset which was collected from a hospital, was feed and then, the results were generated according to the rules created earlier for the fuzzification process. Intrusion detection system using fuzzy rule interpolation system developed [4]. An expert system proposed [5] for studying and detection for the symptoms of migraine headache in a person, they used a fuzzy rule-based expert system to detect the symptoms and which was trained on LFE algorithm. The accuracy varied for each kind of dataset. The Mamdani model was used and for which the centroid method for defuzzification was used.

A neuro fuzzy-based expert system was developed [6] for a very noble work of detection of thyroid. In this expert system, three phases were introduced, first step was to feature extraction, and second step was to pass the output from step one into a neuro-fuzzy network and the last step was that of evaluation of the result. The results were very effective and accurate as the results were purely based upon the good extraction of data which was feed to network and then was applied on the fuzzy system.

Heart disease detection was made with the help of an expert system which was based on the fuzzy rule-based system [7] In this work, 13 input parameters and 1 output parameter was identified and based on which three major phases were operated, which were fuzzification, rule-based, and then the final one which is defuzzification. Centroid method is used for the process of defuzzification. This also gave a good output accuracy.

With the growth in technology, one of the major changes that society has witnessed is a type of shopping. Now B2C E-commerce websites are being used for the shopping purpose rather than old fashioned traditional way of shopping.

Few researchers designed an expert system for evaluating the trust of customers on online shopping portals. The input was collected from the users in form of survey and then according to the survey inputs the rules were designed, which eventually led to the foundation of fuzzy-based expert system [8]. An expert system was designed for the detection of breast cancer [9], it is very well known that breast cancer tissues are very rare to detect and with help of creating rules for such a dangerous disease.

The prevention and protection from diseases like cancer will be easy. Tuberculosis or Tb, a very dangerous disease which is detected using x-ray image was taken as input for the proposed work which was presented in a paper [10]. This paper discussed the design and implementation of a system for the detection of TB in a person with the help of edge detection technique, few classifiers were also used for the purpose of classification. For the segmentation of underwater objects, a new edge detection method which was based on the wavelet transformation was introduced [11], and the accuracy was satisfactory for this method.

Canny edge detection algorithm used [12] for the detection of edge of spine in a CT scan image. This technique was based on the length of edge and the magnitude which provided a satisfactory result. A new technique was introduced [13] based on canny edge detection technique and mathematical morphology for weld-image edge detection. Mathematical morphology technique used [14] to find out the edges on a remote sensing image, in this technique generally the noise of the image was reduced and the edges find out based on the discontinuity find out on the image. The edge detection on SAR images was performed [15] with the help of localization technique.

A sequential approach based hybrid method was introduced [16] for the color-based image segmentation. An object segmentation technique for a moving picture was introduced by few researchers which was based on the wavelet transformation and wavelet transformation technique [17]. Modified version of Sobel edge detector technique used [18] first, morphological operations were applied to smooth the image then the vertical and horizontal gradient in the filters were expanded so as to provide better results. An improved canny operator was introduced [19] by using a scale multiplication. Bee colony optimization was used [20] for edge detector technique. In this paper, it was supposed that the onlooker and scout bee carry edge details with them, which indeed turned out to be true in the end.

2　Proposed Methodology

The proposed methodology contains various techniques to perform edge detection, the techniques have been carried out in various phases. Figure 1 shows the block diagram for the proposed methodology. The proposed work was performed using

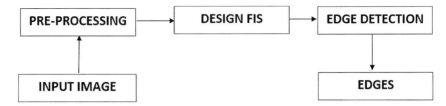

Fig. 1 Proposed methodology

MATLAB [21] tool on a 64-bit, 4 GB RAM, 1 TB HDD, Windows 8.1, and Nvidia 2 GB graphic card system.

2.1 Input Image

This is the first step in which an image is taken as the input, for performing this experiment an RGB image of rose which was captured using a smartphone camera is being used as the input.

2.2 Pre-processing

The next step is of pre-processing, before performing the edge detection, it is very important to convert the RGB image into a grayscale image. Once done than we enhance the quality of image by performing histogram equalization on the image. The following Fig. 2 shows a subplot of original image, converted image and then the enhanced image.

2.3 Designing FIS

This phase is the most important part of this technique, where the fuzzy rule-based system is implemented for the edge detection of an image.

Fuzzification. In this phase, all the crisp values are converted into fuzzy values, here two variables have been taken as the input parameters and one variable is set as the output variable. The two input variables are $\times 1$ and $\times 2$, which shows the horizontal and vertical gradients persisting in the input image. For each input and output variable, there are two membership functions defined, the overlapping section of the membership functions shows the fuzziness in both the input and output variables. The membership functions of input and output are in Table 1.

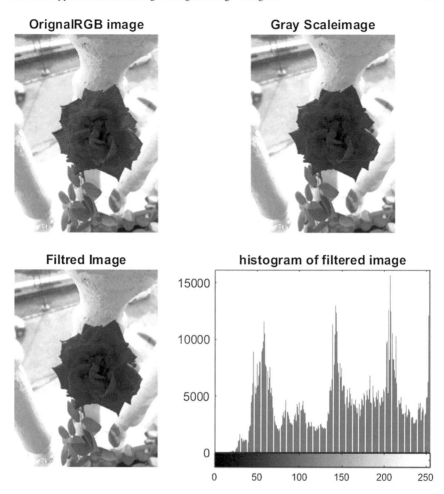

Fig. 2 Input image

Figure 3 shows the membership functions and their overlapping for the input variable × 1, and Fig. 4 shows the membership function and overlapping for output variables.

Fuzzy Rule Construction. After the fuzzification process, the next step is to construct the rules for the fuzzy system that will be helpful in decision-making capabilities of the expert system. The rules are written in form of "IF….THEN" which are separated by or linked with the help of AND in between. For two input parameters and one output parameters, there are four rules which have been created, they are as follows:

If (Ix is zero) and (Iy is zero) then (Iout is black)
If (Ix is not zero) and (Iy is not zero) then (Iout is white)

Table 1 Membership function in input and output

Linguistic variable		Zero			One		
		Membership function	Params	Range	Membership function	Params	Range
Input	Ix	Gauss MF	[0.5, 0]	[−2.5 to 3]	Gauss MF	[0.5, 1]	[−2.5 to 3]
	Iy	Gauss MF	[0.5, 0]	[−3 to 3]	Gauss MF	[0.5, 0]	[−3 to 3]
Linguistic variable		Black			White		
		Membership function	Params	Range	Membership function	Params	Range
Output	Iout	Triangular mf	[0, 0, 0.3]	[0 to 1]	Triangular mf	[0.11, 1, 1]	[0 to 1]

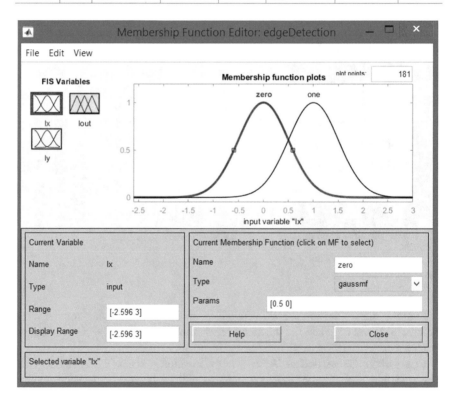

Fig. 3 Membership function in input variable Ix

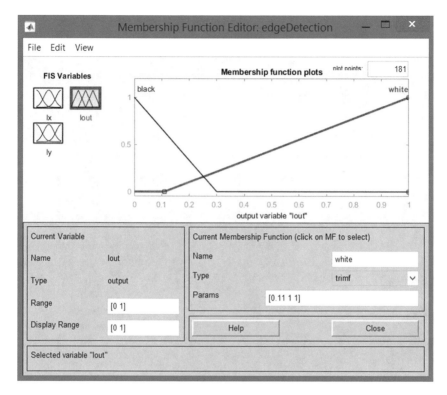

Fig. 4 Membership function in output variable Iout

If (Ix is not zero) and (Iy is zero) then (Iout is black)
If (Ix is zero) and (Iy is not zero) then (Iout is black)

Fuzzy Inference Process. For developing the proposed expert system, the Mamdani system has been used. The Mamdani inference system for the proposed work is shown in Fig. 5, which consist of two input variables (Ix and Iy) and one output variable (Iout), the inference system generated was consist of four rules which are shown in Fig. 6. And the surface view of the fuzzy rules used in the above system is shown in Fig. 7.

Defuzzification. It is a process which converts all the fuzzy values into the crisp value in a fuzzy system. In the proposed work, the centroid method has been used for converting the fuzzy values into crisp values.

3 Result and Discussion

Figure 8 shows a comparison of results of various edge detection techniques with proposed fuzzy-based system.

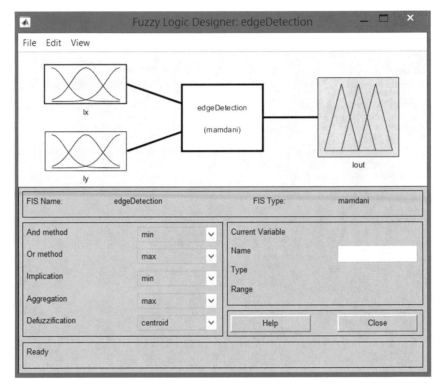

Fig. 5 Mamdani model for the proposed system

From Fig. 8 it is visible that the canny and log edge detectors have been able to detect the edges but the image is very noisy, and noisy image is not good for processing. The Prewitt, Roberts and Sobel edge detectors have also performed decently but still there is a lot more information present which is needed, the proposed method has done pretty well in finding the edge and also it has managed to minimize the noise in the result.

4 Conclusion

The proposed method describes an edge detection technique which is developed using fuzzy rule-based system. Two input variables horizontal and vertical gradients are taken as the parameter for fuzzy system. Four rules are created for the purpose of detecting an edge and with the help of which the new pixel values were decided and that results into the detected edges.

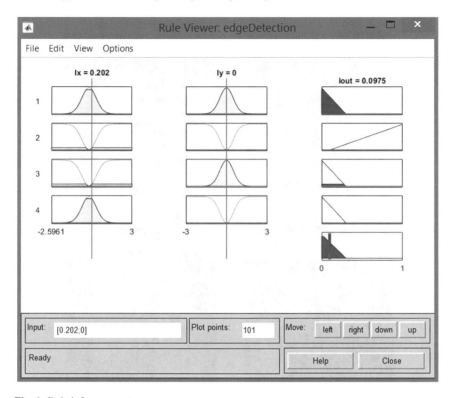

Fig. 6 Rule inference system

Fig. 7 Surface view

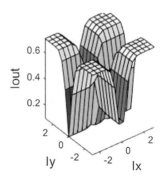

Some other parameters and more rules may give a better understanding of images. The fuzzy-based expert system can be used to detect an edge based on horizontal and vertical gradients.

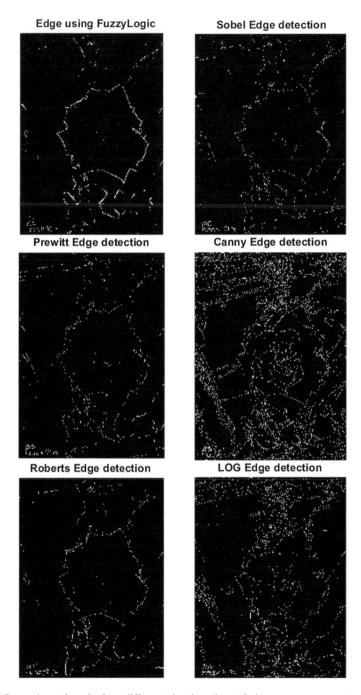

Fig. 8 Comparison of results from different edge detection techniques

References

1. Singh, S.K., Sharma, M., Agrawal, P., Madaan, V., Dhiman, A.: Expar: a fuzzy rule based expert system for palmistry. Int. J. Control Theory Appl. **9**(11), 5207–5214 (2016)
2. Kaur, A., Madaan, V., Agrawal, P., Kaur, R., Singh, S.K.: Fuzzy rule based expert system for evaluating defaulter risk in banking sector. Indian J. Sci. Technol. **9**(28), 1–6 (2016)
3. Sethi, D., Agrawal, P., Madaan, V., Singh, S.K.: X-Gyno: fuzzy method based medical expert system for gynaecology. Indian J. Sci. Technol. **9**(28) (2016)
4. Naik, N., Diao, R., Shen, Q.: Dynamic fuzzy rule interpolation and its application to intrusion detection. IEEE Trans. Fuzzy Syst. **26**(4), 1878–1892 (2018)
5. Khayamnia, M., Yazdchi, M., Vahidiankamyad A., Foroughipour, M.: The recognition of migraine headache by designation of fuzzy expert system and usage of LFE learning algorithm. In: 2017 5th Iranian Joint Congress on Fuzzy and Intelligent Systems (CFIS), pp. 50–53. Qazvin (2017)
6. Biyouki, S.A., Turksen, I.B., Zarandi, M.H.F.: Fuzzy rule-based expert system for diagnosis of thyroid disease. In: IEEE Conference on Computational Intelligence in Bioinformatics and Computational Biology (CIBCB), pp. 1–7. Niagara Falls (2015)
7. Kasbe, T., Pippal, R.S.: Design of heart disease diagnosis system using fuzzy logic. In: International Conference on Energy, Communication, Data Analytics and Soft Computing (ICECDS), pp. 3183–3187. Chennai (2017)
8. Kaur, B., Madan, S.: A fuzzy expert system to evaluate customer's trust in B2C E-Commerce websites. In: International Conference on Computing for Sustainable Global Development (INDIACom), pp. 394–399, New Delhi (2014)
9. Ohri, K., Singh, H., Sharma, A.: Fuzzy expert system for diagnosis of Breast Cancer. In: 2016 International Conference on Wireless Communications, Signal Processing and Networking (WiSPNET), pp. 2487–2492. Chennai (2016)
10. Ramya, R., Babu, P.S.: Automatic tuberculosis screening using canny Edge detection method. In: 2nd International Conference on Electronics and Communication Systems (ICECS), pp. 282–285. Coimbatore (2015)
11. Xiaoheng, D., Minghang, L., Jiashu, M., Zhengyu, W.: Edge detection operator for underwater target image. In: IEEE 3rd International Conference on Image, Vision and Computing (ICIVC), pp. 91–95. Chongqing (2018)
12. Punarselvam, E., Suresh, P.: Edge detection of CT scan spine disc image using canny edge detection algorithm based on magnitude and edge length. In: 3rd International Conference on Trendz in Information Sciences & Computing (TISC2011), pp. 136–140. Chennai (2011)
13. Lu, J., Pan, H., Xia, Y.: The weld image edge-detection algorithm combined with Canny operator and mathematical morphology. In: Proceedings of the 32nd Chinese Control Conference, pp. 4467–4470. Xi'an (2013)
14. Kaur, B., Garg, A.: Mathematical morphological edge detection for remote sensing images. In: 3rd International Conference on Electronics Computer Technology, pp. 324–327, Kanyakumari (2011)
15. Wang, W., Xu, H., Liu, X.: Edge detection of SAR images based on edge localization with optical images. In: 3rd International Asia-Pacific Conference on Synthetic Aperture Radar (APSAR), pp. 1–4. Seoul (2011)
16. Chinu, Chhabra, A.: A hybrid approach for color based image edge detection. In: International Conference on Advances in Computing, Communications and Informatics (ICACCI), pp. 2443–2448, New Delhi (2014)
17. Zhang, X., Zhao, R.: Automatic video object segmentation using wavelet transform and moving edge detection. In: International Conference on Machine Learning and Cybernetics, pp. 3929–3933. Dalian, China (2006)
18. Zhang, Y., Han, X., Zhang, H., Zhao, L.: Edge detection algorithm of image fusion based on improved Sobel operator. In: IEEE 3rd Information Technology and Mechatronics Engineering Conference (ITOEC), pp. 457–461, Chongqing (2017)

19. Bao, P., Zhang, L., Xiaolin, W.: Canny edge detection enhancement by scale multiplication. IEEE Trans. Pattern Anal. Mach. Intell. **27**(9), 1485–1490 (2005)
20. Liu, Y., Tang, S.: An application of artificial bee colony optimization to image edge detection. In: 13th International Conference on Natural Computation, Fuzzy Systems and Knowledge Discovery (ICNC-FSKD), pp. 923–929. Guilin (2017)
21. MATLAB. http://www.mathworks.com

Sentiment Analysis Techniques for Social Media Data: A Review

Dipti Sharma, Munish Sabharwal, Vinay Goyal and Mohit Vij

Abstract The world is going to digitize day by day. *A* lot of data generated by the social website users that play an essential role in decision-making . It is impossible to read the whole text, so sentiment analysis make it easy by providing the polarity to the text and classify text into positive and negative classes. Classification task can be performed by using different algorithms results in a different level of accuracy. The purpose of the survey is to provide an overview of various methods that deal with sentiment analysis. The review also presented a comparative analysis of various sentimental analysis techniques with their performance measurement.

Keywords Sentiment analysis · Opinion mining · Decision-making

1 Introduction

Sentiment analysis (SA) is a process of studying public opinion about an entity. Opinion mining can be used in place of SA. Both terms are interchangeable. Opinion is a judgment of a person regarding an entity that varies from one another and tells about the choice of opinion holder [1]. In this era, Social Media is an important platform for communication and interaction. A lot of people also found innovative information on social media and due to that social media become treasures of information.

Sentiment analysis plays an important role in decision-making and the recommender system [2]. Decision-making includes purchasing a product, making an

D. Sharma (✉) · M. Sabharwal · V. Goyal
Chandigarh University, Mohali, Punjab, India
e-mail: dipi.sharma150@gmail.com

M. Sabharwal
e-mail: smunish.cse@cumail.in

V. Goyal
e-mail: hod.cse@cumail.in

M. Vij
Skyline University, Dubai, UAE
e-mail: dr.mohit.vij@gmail.com

© Springer Nature Singapore Pte Ltd. 2020
A. K. Luhach et al. (eds.), *First International Conference on Sustainable Technologies for Computational Intelligence*, Advances in Intelligent Systems and Computing 1045, https://doi.org/10.1007/978-981-15-0029-9_7

investment. Users are always interested in seeking the experience of their colleague while making an investment or purchasing a product. Nowadays, there are a lot of reviews on social media, which are impossible to read by an investor or a buyer. Sentiment analysis makes this task easy because it describes the polarity of review so that a buyer can directly know whether a given review is positive or negative without reading the whole sentence which helps in decision-making. Three levels of SA are aspect level, sentence level, and document level [3].

To classify sentiment, various steps need to follow that are data collection, data pre-processing, feature extraction, sentiment classification, and evaluation. Data is collected from various source that is in raw form. To find the sentiment it needs to maintain in a structured form. This can be done using the pre-processing of data [4]. After pre-processing, feature extraction is performed. Once the feature of data has been extracted, now the task of sentiment classification has to be performed. To perform classification different approaches or methods of sentiment classification can be used like: Lexicon-based, Machine learning, and hybrid method. Section 2 describes the basic step for SA. Sections 3 and 4 discusses the feature extraction and sentiment classification methods, respectively. Comparison table, discussion and conclusion, and future work are presented in Sects. 5, 6 and 7, respectively.

2   Sentiment Analysis

The Sentiment analysis (SA) which is commonly known as opinion mining or con-textual mining, is used in the Natural Language Processing (NLP), computational linguistics, text analysis which helps in identify, systematically extract and quantify, the subjective information [5]. The sentiment analysis actually works widely in the form of a customer's voice like reviews or responses on any material or item. Example: Suppose a customer wants to buy any item online, so before buy that item the customer generally reads reviews about that item or product and this will help to take the right decision about that item [6, 7].

Sentiment analysis uses three terms to define sentiment. These are, object about which opinion is given, features of that object, opinion holder who give his opinion about the object. Sentiment analysis handles various challenges such as identification of the object, feature extraction and finds the orientation of opinion. Sentiment Analysis performs the classification task in 3 steps:

- Document level
- Sentence level
- Feature level or Aspect level

The document level of classification is used where the task is to find the overall polarity of a topic irrespective of opinion holder. Document-level sentiment analysis assumes opinion about the single entity is expressed by the document. This is true in case of product review, movie review, etc., where a document expresses the opinion about a single movie or a single product. The sentence is a shorter form of document as

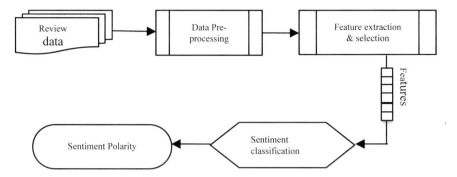

Fig. 1 Sentiment analysis process

collection of sentence makes a document [3]. Sentence level classification assumes each sentence holds a single opinion. Here classification includes two subtasks: subjectivity detection and opinion detection. At the feature level or aspect level, the analysis of various features of an object is performed. Example: Suppose a customer buy a Samsung Mobile Phone, then he observed that the camera quality of the cell phone is fair but the sound quality of the cell phone is not fair. So to analyze, the various aspect of an entity aspect level analysis is performed.

Sentiment Analysis includes Data Pre-processing, Feature Selection, and classification then find the polarity of data as shown in Fig. 1. Data pre-processing includes tokenization, stop word removal, stemming, lemmatization, etc. Tokenization is a task of breaking a sequence of words into individual words called tokens. Stop words are the words (is, am, are, in, to, etc.) which do not hold any opinion, so it is beneficial to remove them. Stemming is a task of converting word's variant forms to its base form like helping to help.

3 Features Extraction in Sentiment Analysis

The extraction of the feature from the text is a very basic task in the Sentiment analysis. In this technique, the text has to be converted into the feature vector with the help of the data-driven approach. Here below, we have seen some features which are commonly used in a Sentiment analysis (SA). Term Presence versus Term Frequency, *N*-gram Features, Parts of Speech, Term Position [8].

3.1 Term Presence Versus Term Frequency

The "Term Frequency" used to find the term count occur in the corpus. The "Term Presence" is actually a binary-valued feature vector, which indicates that the term occurs in the sentence or not. 1 represents the presence of term and 0 represents the absence of the term. Pang and Lee [9] show that the "term presence" is more important than the "term frequency" in the Sentiment analysis. Here we also saw that the occurrence of rare words contains more reliable information as compared to the occurrence of frequent words. The phenomenon which is used in this process is known as Hapax Legomena.

3.2 N-*Gram Features*

The *N*-gram Features are widely used in NLP. Number of terms occur together in a text known as *n*-gram. When only one term is taken as a feature known as unigram, for two-term it is bigram. Here the Pang and Lee [9] experimented that the unigrams outperform the bigrams with the sentiment polarity whereas Dave et al. [10] found that bigrams and trigrams performed better.

3.3 *Parts of Speech (POS) Tagging*

Verbs, adjectives, and adverbs mainly contain the opinion of a person in the English language. POS tagging helps to find these tagged words in a corpus. Adjective, adverb, and verbs can be considered as features and irrelevant words can be removed from the corpus so that vocab size can be reduced.

3.4 *Negation*

Negative words when words come with the positive opinion, invert the polarity, positive to negative like "not good movie" has " good" with positive polarity but " not" change the polarity to negative.

4 Sentiment Analysis Methods

Sentiment analysis methods are machine learning-based, lexicon-based and hybrid method. In machine learning method labeled dataset is used where the polarity of a sentence is already mentioned. From that dataset, we extract the feature and that features help to classify the polarity of the unknown input sentence. Machine learning methods divided into supervised learning and unsupervised learning (Fig. 2).

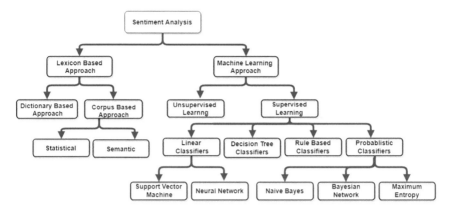

Fig. 2 Sentiment analysis methods

4.1 Supervised Learning

This approach is used when there is labeled data available for training the model. Two steps are used in supervised learning: first is to train the model and another is prediction [11]. During training, data set with its labels are fed to the classification algorithm which gives a model as an output. After that test data is fed into the model to predict the category. There are various supervised classification algorithm are as follows:

Naïve Bayes. It is a probabilistic classification algorithm. It considers each word independent as it does not consider the location of a term in the sentence. Naïve Bayes based on Bayes theorem to calculate the probability of each term which corresponding to a label.

$$p(\text{label}|\text{features}) = \frac{p(\text{label}) * p(\text{features}|\text{label})}{p(\text{features})} \tag{1}$$

$p(\text{label})$ is the prior probability of the label in the dataset. $p(\text{feature}|\text{label})$ is the prior probability of a feature related to a label. $p(\text{feature})$ is the prior probability of a feature that is occurred. Geol et al. [12] used SentiWordNet Lexicon with Naïve Bayes that improve the classification of twitter dataset as it provides the score of positive and negative tweets.

Bayesian Network. As the Naïve Bayes classifier treat each word as independent so it is not able to find a semantic relationship between the words whereas Bayesian Network can. Bayesian network strongly considers the words dependency on each other. The Bayesian network represents dependency in term of a directed graph which is acyclic where each node represent the word as a variable and edges represent the dependency between the variables. As a sentiment classifier, Al-Smadi et al. [13] used Bayesian networks, finding competitive output and sometimes high, as compared to other classifiers.

Support Vector Machine (SVM). SVM is initialized first time to solve the problems of binary classification. Its focuses on determining best hyperplanes which act as a separator to describe the decision boundaries among the data points which are from different classes. A hyperplane should be selected which can maintain the maximum distance between two support vectors of different classes as shown in the figure. The SVM has the capability to manage the linear, and non-linear classification tasks.

Zainuddin and Selamat [14] used SVM for classification with various weighting schemes like TF-IDF, term occurrence, Binary Occurrence. He uses chi-square as a feature selection which is used for dimensionality reduction and noise removal. With the help of the experiment, he showed that the use of chi-square feature selection with SVM improve the accuracy.

Artificial Neural Network. Artificial Neural Network (ANN) mimic the neuron structure of the human brain. The basic unit for the neural network is neuron. ANN comprises an input layer, hidden layer, and an output layer. A vector "a (i)" is given as input to neuron, vector denotes the frequency of a word in a document. There is a weight " A ", corresponding to each neuron which is used to calculate the function. Neural network use linear function is x (i) = A. (a (i)). The sign of x (i) is used to classify the class.

In artificial neural networks training of model consist of two steps: forward propagation and backward propagation. In forward propagation, the input is given at the input layer of neurons which is multiplied by the weights which are random numbers. Functions are used to normalize the output value between 0 and 1. Then the output is compared with the target value, if there is a difference (error) between two values then backward propagation is performed. During backpropagation input is multiplied by error value so that weight can be adjusted. Hence learning depends upon error. The author [15] used a neural network for face classification which has given a high accuracy rate.

Vega and Mendez-Vazquez [16] proposed a Dynamic Neural Network (DNN) Model where he used competitive and Hebbian learning for the learning process. He compared the baseline approach with DNN and showed that DNN performs in a better way than baseline methods. Patil et al. [17] proposed a technique where he used latent semantic analysis (LSA) with a convolution neural network (CNN). LSA is a technique for converting word to vector. Weighting in LSA performed with TF-IDF algorithm. His model provides 87% accuracy.

Decision tree. It is a tree-like structure where the non- terminal nodes represent a feature and terminal node represents the label. The path is taken on the basis of a condition. This is a recursive process and ultimately reach a terminal node which gives a label to an input.

The main challenge in the decision tree is to find which attribute is to be chosen as a root node. This can be solved by using some statistical approach such as information gain and Gini index. A decision tree is a good method for sentiment analysis because it also provides a good result on a large amount of data. Commonly used decision tree algorithms are CART, CHAID, and C5.0.

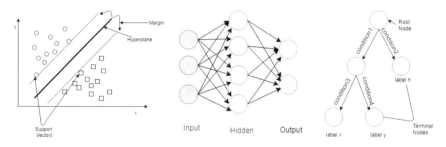

Fig. 3 Structures of SVM, ANN and decision tree respectively

Decision tree divides the training data hierarchically. For the division of data, a condition is used that is on the attribute value. Condition on the basis of whether a word is present or absent. The division process is continued until terminal nodes represent the small numbers of features which are used for the classification task. Kotenko et al. [18] used a decision tree to block the false content on the web site. He used TF-IDF for weighting the word which tells about the importance of the word and a binomial classifier which tell about whether a word belongs to a specific category or not (Fig. 3).

Rule-Based Classifier. Model produced by the rule-based classifier is in form of set of rules. On the basis of these rules prediction for new information is driven. Rules are always in the form of antecedent and consequent. IF (antecedent) is in the left-hand side represents the conditions while right-hand side (consequent) represent the prediction of class. A rule form can be seen below [19].

$$\{w_1/\backslash w_2/\backslash w_3\} \rightarrow \{+|-\}$$

Word in a rule expresses the sentiment shown below.

$$\{Good\} \rightarrow \{+\} \ \{Bad\} \rightarrow \{-\}$$

In text classification IF part represents the features set that may be term presence and THEN part represent the label. The rule-based classifier uses two terms to define the rule: Confidence and Support. Number of the instance in a training data set related to rule is defined by the support. The conditional probability of a label if a feature set occurs represented by confidence in a rule.

Buddeewong and Kreesuradej [20] proposed Association Rule-based Text classifier algorithm (ARTC). In his work, he made two itemsets: One for those words which did not overlap with other class and other for those which overlap with other classes. Then with the help of frequent itemset, he generated the rules. He used the Apriori algorithm for rule generation. He has experimented 95.08% accuracy rate.

4.2 Unsupervised Learning

This method is used when the reliability of labeled data is difficult. It is easy to collect the unlabeled data than labeled data. The sentence is categorized on the basis of keyword lists of each category. In order to analyze the domain-dependent data, it is easier using the unsupervised approach. Unnisa et al. [21] performed sentiment analysis using the unsupervised approach where tweets were clustered into the positive and negative cluster using spectral clustering approach. Spectral clustering outperforms Naïve Bayes, SVM, and Maximum Entropy.

4.3 Lexicon-Based Method

The words which express the opinion are most important for sentiment analysis. The positive opinion is the desired label where negative opinion is an undesired label for an entity. Lexicon is a collection of the predefined word where a polarity score is associated with each word. It is the easiest approach for sentiment classification. This classifier makes use of a lexicon and performs word matching which used to categorize a sentence. The performance of this classification approach depends upon lexicon size. There are two approaches used under the lexicon-based method explained in the subsection.

Dictionary-Based Approach. In dictionary-based approach, some words are selected as a seed word and these words are used to find the synonym to enlarge the size of word set. Online dictionaries are used to expand size. Seed words are the opinion words that are unique and important in a corpus. These seed words and new expanded words are used as a feature for performing sentiment analysis. There are various dictionaries like WordNet, SentiWordNet, SentiFul, SenticNet.

Park and Kim [22] proposed a method for building thesaurus lexicon by using dictionary-based approach. He used three online dictionaries for building a thesaurus and only store the words that are co-occurred in the lexicon that enhanced the reliability of lexicon. The expansion of lexicon is done by synonym and antonyms of seed words. Expended thesaurus enhances the accuracy of the classification task. He selected the seed word using TF-IDF methods. He also mentioned that this approach is a time-consuming approach.

Corpus-Based approach. In corpus-based approach, we do not only find the label of a word but also the context orientation. In this approach firstly a list of seed words is prepared and then the syntactic pattern of these listed words is used to generate new subjective words from the corpus. Syntactic pattern means a word which occurs with each other or together. This approach further works in two way:

- Statistical based approach.
- Semantic-based approach.

The author [23] demonstrate both lexicon-based approaches. He observed that corpus-based approach with SVM provides high accuracy for the light-stemmed data. He also declared that with the increase of lexicon, the accuracy of lexicon-based approach is also increased.

5   Comparison of Different Sentiment Analysis Technique

The work that has been performed by the number of researchers in the field of sentiment analysis and analyzing the performance using various techniques is described in Table 1.

6   Discussion

This review paper covers the basic understanding about sentiment analysis and methods used for classification. The systematic review has been explored the various sentiment analysis methods with their performance parameters. It has been observed that high accuracy of classification depends upon the quality of selected features and classification algorithm used. In recent years a lot of work done in order to find the semantic relationship using word embedding methods and classification using artificial neural networks [32, 36]. The semantic relationship is required to check as related words mostly express the same polarity. SVM and Naïve Bayes are used by the researchers as a reference model for comparing their proposed work. These two algorithms provide high accuracy with feature selection techniques.

Lexicon-based approach is used by the researchers to solve sentiment analysis problems as it is scalable and also computationally efficient. This approach can solve high complex tasks and also performed very well as experimented in [27, 29, 33]. It is also observed that researchers mostly used SentiWordNet [33, 34] lexicon in order to find the polarity score of the words. The datasets used in the analysis are mostly movie reviews, tweets. Mostly researchers performed SA using English language but here we can see some researchers [27, 29, 36] used non-English languages for solving SA problem which also provides compatible results. Hence with respect to above discussion, we can say that more good is our domain-dependent inputs (dataset, lexicon, feature extraction/selection, and classification algorithms), much better output (results) can be achieved.

7   Conclusion and Future Work

This paper discussed the methods for sentiment classification and comparison of algorithms experimented by different researchers on different datasets along with performance measures. It is concluded that Naïve Bayes and SVM are the most frequently used algorithm for classification. These two algorithms are used by researches

Table 1 Comparison of different analysis techniques

Author	Description	Technique used	Dataset	Performance measurement
Singh et al. [24]	The author has implemented SentiWordNet approach with different variations of linguistic features, scoring schemes and aggregation thresholds for sentiment analysis	Lexicon-based and machine learning	Movie review data	• SentiWordNet(SWN) = 65% • Naïve Bayes = 82% • SVM = 77%
Luo et al. [25]	The author demonstrates how can exploit social and structural textual information of Tweets and improve Twitter-based opinion retrieval	Support vector machine	Twitter	Accuracy = 82.52%
Socher et al. [26]	The author introduced a Sentiment Treebank for sentiment detection and evaluation resources. They used Recursive Neural Tensor Network for Treebank	Deep learning	Stanford sentiment treebank	Recognition Rate = 80.70%

(continued)

Table 1 (continued)

Author	Description	Technique used	Dataset	Performance measurement
Wanxiang et al. [27]	The author proposed a framework of adding a sentiment sentence compression (Sent Comp) step before performing the aspect-based sentiment analysis. They applied a discriminative conditional random field model, with certain special features, to automatically compress sentiment sentences	Lexicon-based approach	Chines blog dataset	• 88.78% accuracy for No_comp_ssc • 88.04% accuracy for anual_comp_ssc • 87.95% accuracy for auto_comp_ssc
Yan et al. [28]	The author developed the Tibetan sentence sentiment tendency judgment system based on maximum entropy and test it on the corpus which contains 10000 Tibetan sentiment sentences	Maximum entropy classifier	Blogs	F-value = 82.8%
Sharma et al. [29]	The author purposed the sentiment analysis of Hindi tweets.	Lexicon-based approach	Tweet on JAIHIND and #worldcup2015	• 73.53% accuracy for "JAIHIND" • 81.97% accuracy for "#wprldcup2015"

(continued)

Table 1 (continued)

Author	Description	Technique used	Dataset	Performance measurement
Zimbra et al. [30]	The author proposed a method to brand-related sentiment analysis using feature engineering and artificial neural network	Artificial neural network	Twitter dataset	• 86% accuracy for three class problem • 85% accuracy for five class problem
Kale et al. [31]	The author considered Semantic analysis of his work and comparison between algorithms also performed	Naïve Bayes and maximum Entropy	Tweets	• 63.9% accuracy for Naïve Bayes • 27.8% accuracy for Maximum Entropy
Jianqiang et al. [32]	The author performed Twitter Sentiment analysis by introducing word embedding obtained by unsupervised learning and then integrated it into deep convolution neural network	Deep convolution neural network	Stanford twitter sentiment dataset	Accuracy = 87.36%
Alshariet al. [33]	The author used SentiWordNet (SWN) to find the polarity of the non-opinion word and propose new method Senti2Vec	Lexicon-based approach	Movie review dataset	• 85.4% accuracy for positive data • 83.9% accuracy for negative data

(continued)

Table 1 (continued)

Author	Description	Technique used	Dataset	Performance measurement
Bandana [34]	The author described the Heterogeneous feature such as machine learning–based and lexicon-based and supervised learning algorithms like Linear Support Vector Machine and Naïve Bayes for purposed model	Hybrid approach (SentiWordNet + Naïve Bayes + support vector machine)	Movie review dataset	• Using 250 training and 100 testing dataset (89% for Naïve Bayes, 76% for SVM) • Using 300 training and 150 testing dataset (84% for Naïve Bayes, 79% for SVM)
Ghosh and Sanyal [35]	The author used three feature selection techniques which are Information gain, Chi-Square and Gini index in combination in order to increase the performance of four classifiers	Sequential Minimal Optimization + Multinomial Naïve Bayes + Random forest + Logistic regression	Movie (IMDb) Electronics product Kitchenware	• 90.18(F-measure) for SMO • 88.18 accuracy for MNB • 87.73 accuracy for RF • 87.32 accuracy for LR
Sumit et al. [36]	The author experimented sentiment analysis in Bangla language using a continuous bag of word and Word2Vec Skip gram word embedding methods with word to index model and also compared them	Artificial neural network	Bangladeshi Facebook pages	• 83.79% accuracy for Skipgram • 82.57% accuracy for CBOW • 54.40% accuracy for Word to Index

for comparing their proposed work. After studying these researches, it is very clear that the expansions in sentiment classification and feature selection algorithms are still required and hence an open area of research.

For sentiment analysis data is taken from blogs, social media website like Facebook, Twitter, Amazon, flip kart, etc. People freely express their view on these media about certain topic, product, and politics. By analyzing these reviews one can extract the information about their area and can do improvement. Since so much research has been done in the field of sentiment analysis, still it faces many challenges. Sometimes people express their views in a sarcastic way that is hard to detect. Due to these challenges, sentiment analysis still remains an area of research. In order to improve the classification result deep data analysis is required based on context.

References

1. Liu, B.: Web data mining: exploring hyperlinks, contents, and usage data. Springer (2006)
2. Tatemura, J.: Virtual reviewers for collaborative exploration of movie reviews. In: Proceedings of the 5th International Conference on Intelligent user interface. ACM, pp. 272–275 (2000)
3. Liu, B.: Sentiment analysis and opinion mining. In: Synthesis Lectures on Human Language Technologies, pp. 1–167 (2012)
4. Maks, I., Vossen, P.: A lexicon model for deep sentiment analysis and opinion mining applications. In: Decision Support Systems, vol. 53, pp. 680–688. Springer (2012)
5. Contratres, F.G., Alves-Souza, S.N., Filgueiras, L.V.L., DeSouza, L.S.: Sentiment analysis of social network data for cold-start relief in recommender systems. In: World Conference on Information Systems and Technologies, pp. 122–132. Springer, Cham (2018)
6. Neri, F., Aliprandi, C., Capeci, F., Cuadros, M., By, T.: Sentiment analysis on social media. In: 2012 IEEE/ACM International Conference on Advances in Social Networks Analysis and Mining, pp. 919–926 (2012)
7. Etter, M., Colleoni, E., Illia, L., Meggiorin, K., D'Eugenio, A.: Measuring organizational legitimacy in social media: assessing citizens' judgments with sentiment analysis. Bus. Soc. **57**(1), 60–97 (2018)
8. Mejova, Y., Srinivasan, P.: Exploring feature definition and selection for sentiment classifiers. In: Proceedings of the Fifth International AAAI Conference on Weblogs and Social Media (2011)
9. Pang, B., Lee, L.: Opinion mining and sentiment analysis. Found. Trends Inf. Retrieval **2**, 1–135 (2008)
10. Dave, K., Lawrence, S., Pennock, D.: Mining the peanut gallery: opinion extraction and semantic classification of product reviews (2003)
11. Kaur, J., Sabharwal, M.: Spam detection in online social networks using feed forward neural network. In: RSRI Conference on Recent Trends in Science and Engineering, vol. 2, pp. 69–78 (2018)
12. Goel, A., Gautam, J., Kumar, S.: Real time sentiment analysis of tweets using Naive Bayes. In: 2nd International Conference on Next Generation Computing Technologies (NGCT), pp. 257–216. IEEE (2016)
13. Al-Smadi, M., Al-Ayyoub, M., Jararweh, Y., Qawasmeh, O.: Enhancing aspect-based sentiment analysis of Arabic hotels' reviews using morphological, syntactic and semantic features. Inf. Process. Manag. (2018)
14. Zainuddin, N., Selamat, A.: Sentiment analysis using support vector machine. In: International Conference on Computer, Communications, and Control Technology (I4CT), pp. 333–337. IEEE (2014)

15. Sachdeva, K., Kaur, A.M. Sabharwal.: Face recognition using neural network with SURF technique. In: International Conference on Futuristic Trends in Computing and Networks, vol. 2(1), pp. 256–261 (2018)
16. Vega, L., Mendez-Vazquez, A.: Dynamic neural networks for text classification. In: International Conference on Computational Intelligence and Applications (ICCIA), pp. 6–11. IEEE (2016)
17. Patil, S., Gune, A., Nene, M.: Convolutional neural networks for text categorization with latent semantic analysis. In: International Conference on Energy, Communication, Data Analytics and Soft Computing (ICECDS), pp. 499–503. IEEE (2017)
18. Kotenko, I., Chechulin, A., Komashinsky, D.: Evaluation of text classification techniques for inappropriate web content blocking. In: 8th International Conference on Intelligent Data Acquisition and Advanced Computing Systems: Technology and Applications (IDAACS), pp. 412–417, IEEE (2015)
19. Xia, R., Xu, F., Yu, J., Qi, Y., Cambria, E.: Polarity shift detection, elimination and ensemble: a three-stage model for document-level sentiment analysis. Inf. Process. Manage. **52**, 36–45 (2016)
20. Buddeewong, S., Kreesuradej, W.: A new association rule-based text classifier algorithm. In: 17th IEEE International Conference on Tools with Artificial Intelligence (ICTAI'05) (2005)
21. Unnisa, M., Ameen A., Raziuddin, S.: Opinion mining on twitter data using unsupervised learning technique. Inter. J. Comput. Appl. **148**(0975–8887) (2016)
22. Park, S., Kim, Y.: Building thesaurus lexicon using dictionary-based approach for sentiment classification. In: IEEE 14th International Conference on Software Engineering Research, Management and Applications (SERA) (2016)
23. Abdulla, N.A., Ahmed, N.A., Shehab, M.A., Al-Ayyoub, M.: Arabic sentiment analysis: lexicon-based and corpus-based. In: IEEE Jordan Conference on Applied Electrical Engineering and Computing Technologies (AEECT) (2013)
24. Singh, V.K., Piryani, R., Uddin, A., Waila, P.: Sentiment analysis of movie reviews and blog posts. In: 3rd IEEE International Advance Computing Conference (IACC), pp. 893–898 (2013)
25. Luo, Z., Osborne, M., Wang, T.: An effective approach to tweets opinion retrieval. Springer J. World Wide Web, pp. 545–566 (2013)
26. Socher, R., et al.: Recursive deep models for semantic compositionality over a sentiment Treebank. In: Proceedings of the Conference on Empirical Methods in Natural Language Processing (EMNLP), pp. 1631–1642 (2013)
27. Che, Wanxiang, Zhao, Yanyan, Guo, Honglei, Zhong, Su, Liu, Ting: Sentence compression for aspect-based sentiment analysis. IEEE/ACM Trans. Audio, Speech, Lang. Process. **23**, 2111–2124 (2015)
28. Yan, X., Huang, T.: Tibetan sentence sentiment analysis based on the maximum entropy model. In: 10th International Conference on Broadband and Wireless Computing, Communication and Applications (BWCCA), pp. 594–597. IEEE (2015)
29. Sharma, Y., Mangat, V., Kaur, M.: A practical approach to sentiment analysis of Hindi tweets. In: 1st International Conference on Next Generation Computing Technologies (NGCT), pp. 677–680. IEEE (2015)
30. Zimbra, D., Ghiassi, M., Lee, S.: Brand-related twitter sentiment analysis using feature engineering and the dynamic architecture for artificial neural networks. In: 49th Hawaii International Conference on System Sciences (HICSS), pp. 1930–1938. IEEE (2016)
31. Kale, S., Padmadas, V.: Sentiment analysis of tweets using semantic analysis. In: International Conference on Computing, Communication, Control, and Automation (ICECUBE). IEEE (2017)
32. Jianqiang, Z., Xiaolin, G., Xuejun, Z.: Deep convolution neural networks for twitter sentiment analysis, pp. 23253–23260. IEEE Access (2018)
33. Alshari, E.M., Azman, A., Doraisamy, S., Mustapha, N., Alkeshr, M.: Effective method for sentiment lexical dictionary enrichment based on Word2Vec for sentiment analysis. In: Fourth International Conference on Information Retrieval and Knowledge Management (CAMP) (2018)

34. Bandana, R.: Sentiment analysis of movie reviews using heterogeneous features. In: 2nd International Conference on Electronics, Materials Engineering & Nano-Technology. IEEE (2018)
35. Ghosh, M., Sanyal, G.: An ensemble approach to stabilize the features for multi-domain sentiment analysis using supervised machine learning. J. Big Data. 5(1) (2018) (Springer)
36. Sumit, S.H., Hossan M. Z., Muntasir, T.A., Sourov T.: Exploring word embedding for Bangla sentiment analysis. In: International Conference on Bangla Speech and Language Processing (ICBSLP). IEEE (2018)

Towards a Model-Driven Framework for Data and Application Portability in PaaS Clouds

Kiranbir Kaur, Sandeep Sharma and Karanjeet Singh Kahlon

Abstract Cloud computing paradigm has gained momentum during last decade owing to the numerous benefits (Cost, Scalability, etc.) it offers. Several developers and organizations have steadily got interested in PaaS (Platform as a Service) clouds since it allows deploying applications without any requirement for a consecrate infrastructure, installations of dependencies or server configurations. However, there still exist some issues that need to be resolved to make Cloud computing and PaaS widely adopted. Portability of applications along with the portability of databases (DBs) among heterogeneous cloud platforms is one of the issues. To address this, a middleware platform has been proposed in this paper in which the application layer and the DB layer of that application shave been decoupled from each other. Both the layers can be ported from one platform (PaaS) to another PaaS independent of each other. Model-Driven Engineering (MDE) approach has been leveraged to make the middleware flexible, further making application layer and DB layer portable, productive and reusable. MDE has been proved to be a better alternative to standardization in the literature.

Keywords Cloud computing · Application portability · PaaS clouds · Model-driven engineering

1 Introduction

The chance of standardization initiative becoming a reality is a distant dream as these are against the market cloud vendors' interest. The obvious reason for this is, they tie their customers to their proprietary technology and tools (which is termed as vendor lock-in). Although a plethora of standards have been proposed for the cloud computing portability and interoperability majority of them target IaaS level rather than PaaS or SaaS (and even Database as a Service DBaaS). Several solutions for the interoperability and portability have been suggested by the researchers, namely

K. Kaur (✉) · S. Sharma · K. S. Kahlon
Guru Nanak Dev University, Amritsar, Punjab 143001, India
e-mail: kiran.dcse@gndu.ac.in

© Springer Nature Singapore Pte Ltd. 2020
A. K. Luhach et al. (eds.), *First International Conference on Sustainable Technologies for Computational Intelligence*, Advances in Intelligent Systems and Computing 1045,
https://doi.org/10.1007/978-981-15-0029-9_8

Standards, Abstraction layers, middlewares, MDE, etc., [1]. MDE approach is a new hope towards the solution of vendor lock-in as it has its solid and established roots in Software engineering. The concept of portability applies to all the 3 layers of cloud computing reference model viz IaaS, PaaS, and SaaS. Stravoskoufos et al. [2] defined the portability for:

IaaS—migrating and running VMs (Virtual machines) among disparate infrastructure providers. Data and configurations also get migrated.
PaaS—deploying applications across heterogeneous platform providers. This also includes different VMs.
SaaS—Here, clients are able to shift their data from SaaS application. This basically is the concept of Data Portability.
Baudoin et al. [3] have a different viewpoint which says that application portability may be required at all the 3 layers and is facilitated by:
IaaS—OVF (Open Virtualization Format) and the availability of LINUX operating system [4] has done some work on it.
PaaS—2 trends that ease the situation are
Adopting standard open-source PaaS like Cloud Foundry, OpenShift, Morpheus.
Espousing containerization technologies which provide subdivision and independent deploying of different parts of the application, for example, Dockers.
SaaS—Shifting from 1 SaaS application to another similar SaaS application results in changes in the user interfaces as well as any application belonging to the cloud consumer that is using any API (Application Programming Interface) provided by that SaaS provider.

This paper emphasizes particularly the facet of portability in the context of Platform as a Service which involves the feasibility of an application and its database to be ported among heterogeneous platforms. Even though most of the major platform providers (Google App Engine, and Microsoft Azure) support similar languages (JAVA, .NET, Python, and Ruby) but when it comes to port an application developed and deployed with one provider to another, it incurs change in the code and requires efforts. So, the application ramped up for PaaS platforms poses the intense challenges for application portability.

The aim of this paper is to introduce the basic concepts of vendor lock-in, interoperability, and portability issue in Cloud computing and to introduce the proposed architecture for tackling the trouble in portability an application and its DB from one cloud provider to another with the help of MDE approach. Analysis of the research work done in the field of application portability and DB portability is also done.

1.1   Vendor Lock-in and "Interoperability and Portability" Issues in Cloud Computing

Whereas both [5, 6] has done a survey and categorized the solutions for vendor lock-in, [5] has presented a critical view of this prominent deterrent to the Cloud computing paradigm's growth. A befitting definition of the "Vendor Lock-in" is

> "Vendor lock-in is the restricted or proprietary use of a technology, solution or service developed by a vendor or vendor partner. This technique can be disabling and demoralizing because customers are effectively prevented from switching to alternate vendors."[1]

Cloud Providers offer various technologies providing tremendous advantages (cost saving, scalability, and agility) to the users but they do so at a price of sturdy restrictions of specified hypervisors, infrastructure and (Application Programming Interfaces) APIs. This apprentice on the providers' specific technology locks the customer to that vendor. There may arise some situations when a consumer wants to shift to some other cloud, for example[2]

Cloud Provider going out of business.
Cloud provider does some substantial changes in the product that do not suit customer's business.
Cloud provider's oblations do not fulfill customer's needs.

Vendor lock-in impedes the portability and interoperability, therefore is considered dissenting by the customers. But organizations embrace it volitionally as it gives them an edge over their competitors. In case of PaaS vendor lock-in, suppose a customer has developed an application on Microsoft Azure, utilizing some basic cloud services such as compute, storage, DBs networking and also some specialized services like Azure's m/c Learning, analytics and bot services. Now as every provider has their branded specifications and proprietary APIs, it becomes very difficult to shift to any other provider.

SQLs and NoSQLs Databases
The NoSQL DBs have evolved gradually to tackle the explosive increase of data of modern applications. However, they are not there to sack the tralatitious relational stores which provided structured data storage as well as declarative querying [7]. Rather, both SQL and NoSQL systems are coordinated in many enterprises to accomplish complementary data management requirements. So, these arisen needs of having some mechanisms for porting data from SQL to NoSQL DBs or vice versa or from one type of NoSQL to another type as well as sometimes integrating both types to fulfill the enterprises' DB needs. Even though NoSQL DBs share the following characteristics [8], it's not a trivial task to port data from one type to another due to the deviation in reference to their respective models:

---

[1] https://www.techopedia.com/definition/26802/vendor-lock-in.
[2] https://www.thorntech.com/2017/09/avoidingcloudvendorlockin/.

Schemaless Data structures—Presuming that the structures may evolve and new attributes may be appended, most of the NoSQL store data without any defined schema.

Replication—It refers to the copying of stored data in at least two disparate nodes of a network to enhance the reliability of the system and minimize the loss of data in case of any node's or cluster's failure.

Sharding—this method gives horizontal scalability in which several nodes hold some of the original data called shard. Nodes can perform the requisite operation in their respective shards.

CAP theorem—Seth Gilbert and Nancy Lynch gave the CAP (Consistency, Availability, and Partition tolerance) theorem in contrast to ACID (Atomicity, Consistency, Isolation, and Durability) model.

Consistency—This property assures that a client should always get the latest and updated version of data.

Availability—This refers to the availability of the system and response even if a few nodes have failed.

Partition Tolerance—The nodes of the system should not be communicating with each other. The partitions should keep working either in the case of the temporal loss of connectivity or simply loss of a packet.

Some notable research work towards data portability is discussed in this section. Shirazi et al. [9] presented an effective technique of design patterns to shift data from a columnar DB (HBase) to graph DB (Neo4J) and vice versa. However, this work seems to be just a proposal as no implementation work is given either in this paper or any other papers published by the author (to the best of our knowledge). Hill and Humphrey [10] proposed an abstraction layer for providing common storage abstraction to cater heterogeneous cloud providers. The layer also renders a namespace to be used by programmers for supporting blobs, tables, and queues. Alomari et al. [11] focused the challenge of data portability and proposed a framework called "CD-Port" which is equipped with tools for conversion, transformation and data exchange among disparate data storage models. Beslic et al. [12] used an ontology called KDM (Knowledge Discovery meta model) to extract system knowledge and the used several predefined patterns to guide the users throughout the application migration from one cloud platform to another. Scavuzzo et al. [13] proposed an architecture called "Hegira4Cloud" which provides an intermediate metamodel for Columnar DBs especially. The authors also focused fault tolerance feature of the NoSQL portability of Big Data applications. Bansel et al. [14] proposed a framework to facilitate data portability among heterogeneous NoSQL DBs using metamodel techniques. An effective mapping has been implemented between the source and target NoSQL DBs (particularly Azure Table, MongoDB, and Neo4J) focused both the data representation as well as data migration.

2 Application Portability

Baudoin et al. [3] defines "Application portability" as "Application portability is the ability to easily transfer an application or application components from one cloud service to a comparable cloud service and run the application in the target cloud service". Cloud services being offered by the providers have proprietary and incompatible interfaces which make it difficult to port application to some other platform. The solutions which facilitate application portability decimate the vendor lock-in problem by obviating the program's dependence on branded interfaces as well as protocols.

Various scenarios in which application portability is required are discussed as follows:

Cloud to Cloud. Due to numerous reasons like better quality, cost or unexpected termination of 1 cloud makes the user find options of other clouds [15]. In this situation, it becomes necessary to port the complete application and services to some other cloud.

Multi-Cloud. This scenario involves migrating some parts of an application to one cloud and other parts to different cloud (which needs the reconfiguration of the deployment plan). This scenario involves more interoperability than portability issues.

Hybrid Clouds. These are similar to the 'Multi-cloud' concept but involve a private cloud and a public cloud. Reduced capital expenditure (CAPEX) and operational expenditures (OPEX) lures enterprises to deploy applications in the clouds. However, there still remain issues relating to privacy, technicality, and legality which forbid the deployment of the whole software in the cloud [16].

Legacy to cloud. According to [17] "Legacy systems are software systems that have been built for more than decades with old technologies and methodologies." Owing to the advantages like increased scalability, high fault tolerance, cost and energy efficiency, etc., the IT departments deliberate to migrating such applications to cloud environment.

Cross cloud. It involves integrating ready third party applications through Application Programming Interfaces (APIs) [18]. Here, applications are not directly developed for branded cloud platforms. Instead, specific APIs for certain services are mashed up together to create an application. For example, an application which requires to read data from a DB to produce some analytics can consume APIs for these specific services from different providers [19].

2.1 Approaches for Handling the Challenge of PaaS Application Portability

After exploring the literature which discussed the application portability, these two major solutions come up as the solution for the challenge of application portability [20].

Standardization. If all the cloud service providers embrace the standards, then the developers would be able to create the applications agnostic of the particular cloud environment or even the application developed for one environment would be effortlessly ported over another environment. Several organizations have taken an endeavor to institute users' trust for various cloud computing services by proposing standards (OVF, OCCI, UCI, CIMI, CDMI, TOSCA, CAMP, etc.) associated with the operation of cloud services. A plethora of research efforts has been done to describe and discuss standards [all standard folders]. However, standardization of services among different cloud providers is least likely to happen; users must consider other approaches to facilitate interoperability and portability of applications among disparate clouds. Moreover, the focus of most of the active cloud standards is on the IaaS instead of the PaaS level [1].

Intermediation. Besides standardization, another alternative to facilitate portability issue is intermediation which detaches the application's development from any particular platform's APIs and supported formats [20]. Intermediation solutions further encompass three types: Library based (JClouds, LibClouds, Pkgclouds, etc.), middleware solutions and Model-Driven Engineering-Based solutions.

Some of the research work done towards finding the solutions for application portability in clouds is discussed further. Kolb and Cedric [21] propose unified interfaces for assisting a user to take decision about the most appropriate cloud platform and also the management and deployment of cloud applications among disparate cloud platforms. Hossny et al. [22] proposes the development of a generic API which is semantically annotated for the automatic generation of a specific provider's adapter. Leymann et al. [23] described how to move an application to the cloud by splitting the application (manually or with the help of optimization algorithms). Cunha et al. [24] defined a distributed architecture for creating and exposing services through standardized APIs. Jimenez-domingo et al. [25] presented a complex application interoperability language established on the semantic technologies [15]. To tackle the application portability issue, this paper proposes a common model for heterogeneous Platform as a Service platforms by defining three-layered structure viz. Infrastructure, Platform, and Management. Jonnalagedda et al. [16] shed some light on the design issues while developing an application whose some components would be deployed on a cloud platform and some remain in-house (On premises). Rafique et al. [26] presented an approach for supporting hybrid clouds with the help of a middleware architecture. Cunha et al. [27] proposed PaaSManager to abstract the deviations in application deployment and life cycle management of disparate cloud providers. For this purpose, the authors defined a set of fundamental operations and aggregated them to define a common API. Ranabahu and Sheth [28], Ranabahu et al.

[29] presented a user-driven perspective leveraging semantics and Domain Specific Language (DSL) for application portability. Kamateri et al. [30], D'Andria et al. [31], Loutas et al. [32], Loutas et al. [33] presented the Cloud4SOA project which is based on broker architecture to deal with the semantic interoperability issues emerging at the PaaS layer. Martino et al. [34], Cretella and Martino [35], Cretella and Martino [36], Martino et al. [37], Cretella and Martino [38], Martino et al. [39], Markoska et al. [40] leveraged the concepts of Design patterns and Semantic technologies to handle cloud application portability in the context of mOSAIC project. Petcu et al. [41], Petcu [42], Petcu et al. [43] describe European Union research project mOSAIC which is a middleware platform for developing provider agnostic cloud applications along with monitoring and scalability capabilities. Darko thesis [44] focused on semantic annotation of APIs and Web Services to enable portability of applications across providers. Various ontologies were developed and Artificial Intelligence Planning was used to detect various interoperability problems. Hamdaqa et al. [45] paper presented a cloud application meta-model capable of describing the important components of a cloud application. Vijaya and Neelanarayanan [46, 47] introduced an eclipse plugin named DSkyL leveraging MDE for the development of CRM SaaS applications. Gonidis [18], Gonidis et al. [20], Gonidis et al. [20], Gonidis et al. [19] leveraged MDE based technique to enable the development of cloud applications which can take services from various provider platforms. Giove et al. [48] developed a Java library called CPIM (Cloud Provider Independent Model) to encapsulate various PaaS-level services to provide a mediation layer hiding the specificities of various PaaS providers.

3   Our Approach

In this section, the details of our proposed approach are discussed to prevent vendor lock-in and migrating applications between different cloud platforms. We recommend using middleware approach where we will be developing a library which acts as an adapter between the application and cloud platforms. This application will handle all the database-related operations and other application feature, e.g., Mailer, Task Queues, and Message Queues.

Our main objective is to provide the flexibility to the user to migrate the application between clouds while using the data stores of some different cloud. For example, a user wants to host the application on Azure but wants to use the AWS RDS and AWS Redshift database. Our library will provide the flexibility to do so. We will develop a prototype library for Java platform, which will enable Java programmers to flexibly migrate applications among clouds. This library can be ported into other programming languages also such as C#, PHP, Python, etc. But right now, it is being developed for Java only. This library is based on Model-Driven Engineering concepts. Basically user's application will handle all the data in the form of objects of Entities. So, all the CRUD operations that user's application performs will send/receive the object of respective entities. To tweak the library for different data stores, user will create configuration files for each data store. These configuration files will contain connection strings, credentials and a list of entities that are being stored in respective data store. For example, if user's application wants to store an object of Student entity in an SQL based database, it will send the object to the library. The library will read the configurations to check that which data store is to be used for the said entity and then generate the corresponding code/query to insert the received object. In this case, Student object will be stored in a table named Student.

3.1 Architecture

The library will accept the objects and convert the object into the corresponding query/document for the storage. It will also receive the emails, messages to send and maintain them in queues. Event logging will also be handled by the library: both internal and external logging. Internal means the logs of the exceptions and messages of the library. External means that the user can create its custom logs. The library will act as middleware between the user's application and cloud services as shown in Fig. 1. And our library's architecture is shown in Fig. 2.

Fig. 1 Position of proposed library with respect to User application and Cloud Services

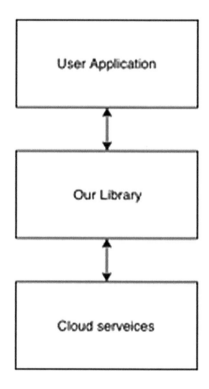

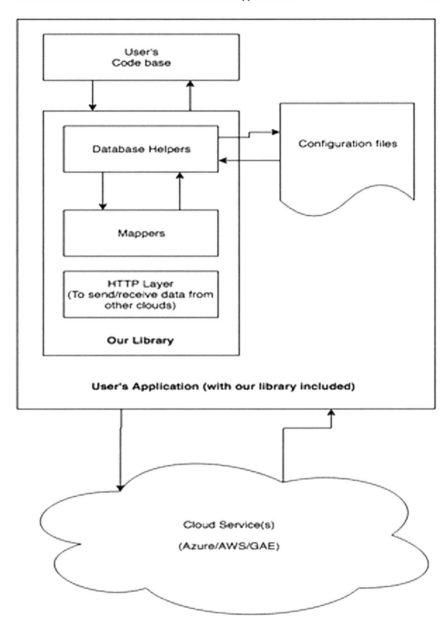

Fig. 2 Architecture of proposed library

The library consists the code to interact with cloud services (Azure, AWS and GAE for now) and the data stores in them. We are focusing on SQL, DocumentDb, and GraphDb right now.

3.2 Configuration Files for Settings

The user has to only modify the configuration files in order to modify the behavior of the library according to the requirement of its application. The configuration file will enable the user to let the library know which cloud service to use, which data store to use and also contain the email/message settings.

If there are more than one cloud services being used, user will be able to define the configuration settings for each cloud service.

If there are more than one data stores being used which are spanned over various cloud services, user will have to provide the data store configuration(s) in respective cloud service(s) configuration.

How the library will work?

When an object is received by the library, first its type will be determined.

If it is a message/email object, it will be sent to the email/message queue. The email/message queue will access the configuration file for the settings required to send the email/message. The email/message will be sent in a queue along with the settings provided by user in the configuration files.

If the object is a database related object, then the library will determine the data-store type in which the object is to be converted. The configuration file will contain the list of every entity and its data store type. For example, Imagine a cab service application. Customer, Cab and Driver data is stored in SQL. The live location data is logged in a DocumentDb (noSQL). In above case, the configuration file will have two data stores: SQL and DocumentDb. SQL will have its own settings to contain the information about the cloud service on which the datastore is located, the con-nection string and the entities list that will contain Customer, Cab, and Driver. The DocumentDb portion of the configuration file will contain the cloud service name, the connection string and the entity list (LocationLog in this case). When the library will determine that the passed object is database related object, it will access the configuration file to determine that in which form should it be converted: SQL or noSQL Document. After the conversion is done, the library will send the data to the datastore associated with the entity. When the data is to be retrieved, user will mention the name of entity from which the data is to be fetched along with the search parameters. The library will determine the datastore type and cloud service on which it is hosted; then prepare the query and send the query to the associated cloud service. Then it will retrieve the data to be sent to the cloud service and convert the data into the objects list of the mentioned entity.

Why The Middleware Approach?

Middleware is one of the approaches used by the programmers to inject a feature or functionality in the application. All the modern frameworks use the application packages to make their framework flexible. User can add the packages in the pipeline of the project as required. Making a library that can be added as a middleware makes it easy for the user to manage the pipeline of its application. Being a middleware, this library can be used in a console application, a GUI Desktop application, Web Forms-based web application and an MVC application. Basically, it can be used in every application that uses messaging, email, logging and database operations. Using a middleware is also convenient for the user because the user can replace one library with another library if it finds the other library more effective. It is then the user's responsibility to develop its application in such a way that it is not closely coupled with any middleware. We have developed our library in such a way that it is not dependant on any other third party libraries (except for database drivers). All the code will use Java only (and in future it will use only that language in which it is written). It will make the library to run on any cloud service that provides the support for Java.

As much as the middleware is easy to add in an application, it is as easier to upgrade also. The user just has to copy the new build of the library to its application and it is done. There is no need to change any code. However, if there are new features added in the library (that are not being used in the application), then the user might have to write new code if he/she wishes to incorporate those new features.

Isn't the middleware a lock-in in itself?

One would say that the middleware is a lock-in in itself. While it might seem true, but it is not entirely true. That's because our library is not involved in the business layer or the presentation layer of the application. It interacts with the data abstraction layer only. Our library requires an object as input and returns an object or list of objects as output. It can be replaced with any other ORM and messaging/email client library.

It is the user's responsibility to make its application loosely coupled so that it is ready to replace our middleware with other middleware of similar kind.

In our case, the user can perform all the business logics in its business layer. When the objects are ready to be stored, user will send those objects to our library and our library will handle the data manipulation operations (Figs. 3 and 4).

So, if everything is divided among layers, we can see that our library doesn't cause any lock-in. User can write his/her own piece of code that can interact with the cloud service(s) or databases.

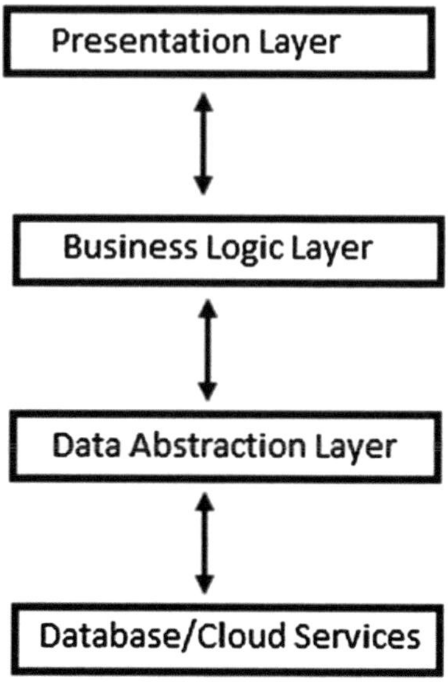

Fig. 3 An ideal application structure

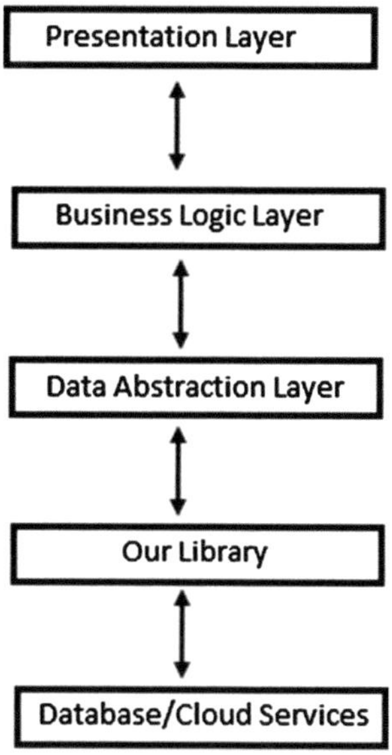

Fig. 4 An ideal application structure with our library

4   Conclusion

In this paper, a novel approach is introduced to facilitate the portability of an application from one PaaS provider to another. Not only the application can be shifted among various platforms, but the database layer is also detached so that the customer is not bound to use the database of the platform provider on which he/she has deployed the application. The customer has the liberty to put the database on any data store of any platform of his/her choice.

References

1. Kaur, K., Sharma, S., Kahlon, K.S.: Interoperability and portability approaches in inter-connected clouds: a review. ACM Comput. Surv. **50**(4), 40 (2017)
2. Stravoskoufos, K., Preventis, A., Sotiriadis, S., Petrakis, E.G.: A survey on approaches for interoperability and portability of cloud computing services. In: CLOSER, pp. 112–117 (2014)
3. Baudoin, C., Dekel, E., Edwards, M.: Interoperability and portability for cloud computing: a guide. Cloud Stand. Cust. Counc. **2014**, 1–8 (2014)
4. Polo Sony, I.: Inter-Cloud application migration and portability using Linux containers for better resource provisioning and interoperability Ivin polo sony (2015a)
5. Tian, F., Opara-Martins, J., Sahandi, R.: Critical review of vendor lock-in and its impact on adoption of cloud computing. In: International Conference on Information Society (i-Society 2014) Critical, pp. 92–97 (2014)
6. Silva, G.C., Rose, L.M., Calinescu, R.: A systematic review of cloud lock-in solutions a systematic review of cloud lock-in solutions. In: CloudCom.2013. https://doi.org/10.1109/CloudCom.2013.130
7. Gottlob, G., Grasso, G., Olteanu, D., Schallhart, C.: LNCS 7968—Big Data Gerhard Goos, Juris Hartmanis, & Jan van Leeuwen, eds., (2013)
8. Pulgatti, L.D.: Data migration between different data models of NOSQL databases (2017)
9. Shirazi, M.N., Kuan, H.C., Dolatabadi, H.: Design patterns to enable data portability between clouds' databases. In: 12th International Conference on Computational Science and Its Applications Design, pp. 5–8 (2012) https://doi.org/10.1109/ICCSA.2012.29
10. Hill, Z., Humphrey, M.: CSAL: a cloud storage abstraction layer to enable portable cloud applications. In: Proceedings of 2nd IEEE International Conference on Cloud Computing Technology and Science (2010)
11. Alomari, E., Barnawi, A., Sakr, S.: CDPort: a portability framework for NoSQL datastores CDPort: a data portability framework for. Arab. J. Sci. Eng. Dec (2015). https://doi.org/10.1007/s13369-015-1703-0
12. Beslic, A., Bendraou, R., Sopena, J., Rigolet, J.-Y.: Towards a solution avoiding Vendor Lock-into enable Migration between cloud platforms. In: 2nd International Workshop on Model-Driven Engineering for High Performance and Cloud computing (MDHPCL 2013), pp. 5–14 (2013)
13. Scavuzzo, M., Tamburri, D.A., Di Nitto, E.: Providing big data applications with fault-tolerant data migration across heterogeneous NoSQL databases. In: 2016 2nd International Workshop on BIG Data Software Engineering Providing, pp. 26–32 (2016)
14. Bansel, A., Gonzalez Velez, H., Chis, A.E.: Cloud-based NoSQL data migration. In: 24th Euromicro International Conference on Parallel, Distributed, and Network-Based Processing Cloud-based. pp. 224–231 (2016). https://doi.org/10.1109/PDP.2016.111
15. Kolb, S., Wirtz, G.: Towards application portability in platform as a service. In: IEEE 8th International Symposium on Service Oriented System Engineering Towards, pp. 218–229 (2014). https://doi.org/10.1109/SOSE.2014.26

16. Jonnalagedda, M., Jaeger, M.C., Hohenstein, U., Kaefer, G.: Application portability for public and private clouds. In: 1st International Conference on Cloud Computing and Services Science(CLOSER-2011), pp. 484–493 (2011). https://doi.org/10.5220/0003394104840493
17. Zhang, W., Berre, A.J., Roman, D., Huru, H.A.: Migrating legacy applications to the service cloud. In: Research Gate, pp. 59–67 (2015)
18. Gonidis, F.: A Framework Enabling the Cross-Platform Development of Service-based Cloud Applications. The University of Sheffield (2015)
19. Gonidis, F., Gkasis, P., Lazouras, L., Stamatopoulou, I.: Infusing research and knowledge in South-East Europe. In: 8th Annual South-East European Doctoral Student Conference, pp. 1–525 (2013)
20. Gonidis, F., Paraskakis, I., Kourtesis, D.: Addressing the challenge of application portability in cloud platforms. In: In 7th South-East European Doctoral Student Conference, pp. 565–576 (2012)
21. Kolb, S., Rock, C.: Unified cloud application management. In: IEEE World Congress on Services Computing, pp. 1–8 (2016). https://doi.org/10.1109/SERVICES.2016.7
22. Hossny, E., Khattab, S., Omara, F.A., Hassan, H., Randy, H., David, A.: Semantic-based generation of generic-API adapters for portable cloud applications. In: Proceedings of the 3rd Workshop on CrossCloud Infrastructures & Platforms, ACM (2016). https://doi.org/10.1145/2904111.2904117
23. Leymann, F., Fehling, C., Mietzner, R., Nowak, A., Dustdar, S.: Moving applications to the cloud: an approach based on application model enrichment. Int. J. Coop. Inf. Syst. **20**(3), 307–356 (2011). https://doi.org/10.1142/S0218843011002250
24. Cunha, D., Neves, P., Sousa, P.: Interoperability and portability of cloud service enablers in a PaaS environment. In: Proceedings of 2nd International Conference on Cloud Computing and Services Science, CLOSER. pp. 432–437 (2012)
25. Jimenez-domingo, E., Gomez-berbis, J.M., Colomo Palacios, R., García-Crespo, Á.: CARL: a complex applications interoperability language based on semantic technologies for platform-as-a-service integration and cloud computing. J. Res. Pract. Inf. Technol. **43**(3), 227 (2011)
26. Rafique, A., Walraven, S., Lagaisse, B., Desair, T., Joosen, W.: Towards portability and interoperability support in middleware for hybrid clouds. In: Computer Communications Workshops (INFOCOM WKSHPS), IEEE Conference, pp. 7–12 (2014)
27. Cunha, D., Neves, P., Sousa, P.: A platform-as-a-service API aggregator. In: Advances in Information Systems and Technologies, pp. 807–818, Springer Berlin Heidelberg (2013). https://doi.org/10.1007/978-3-642-36981-0
28. Ranabahu, A., Sheth, A.: Semantics centric solutions for application and data portability in cloud computing. In: 2nd IEEE International Conference on Cloud Computing Technology and Science, pp. 234–241 (2016). https://doi.org/10.1109/CloudCom.2010.48
29. Ranabahu, A., Maximilien, E.M., Sheth, A., Thirunarayan, K.: Application portability in cloud computing: an abstraction-driven perspective. IEEE Trans. Serv. Comput. **8**(6), 945–957 (2015)
30. Kamateri, E., et al.: Cloud4SOA: a semantic-interoperability PaaS solution for multi-cloud platform management and portability. In: European Conference on Service-Oriented and Cloud Computing, Sept, pp. 64–78 (2013). https://doi.org/10.1007/978-3-642-40651-5
31. Di Martino, B., Esposito, A., Cretella, G.: Mapping design patterns to cloud patterns to support application portability: a preliminary study. In: Proceedings of the 12th ACM International Conference on Computing Frontiers, p. 50. ACM (2015). https://doi.org/10.1145/2742854.2747280
32. Loutas, N., Peristeras, V., Bouras, T., Kamateri, E., Zeginis, D., Tarabanis, K.: Towards a reference architecture for semantically interoperable clouds. In: 2nd IEEE International Conference on Cloud Computing Technology and Science, pp. 143–150 (2010). https://doi.org/10.1109/CloudCom.2010.38
33. Loutas, N., Kamateri, E., Tarabanis, K.: A semantic interoperability framework for cloud platform as a service. In: Cloud Computing Technology and Science (CloudCom), IEEE Third International Conference on, pp. 280–287. IEEE (2011)

34. Di Martino, B., Cretella, G., Esposito, A., Sperandeo, R.G.: Semantic representation of cloud services: a case study for microsoft windows azure. In: International Conference on Intelligent Networking and Collaborative systems. pp. 647–652 (2014). https://doi.org/10.1109/INCoS.2014.76

35. Cretella, G., Di Martino, B.: Towards a semantic engine for cloud applications development. In: Sixth International Conference on Complex, Intelligent, and Software Intensive Systems. pp. 198–203 (2012a) https://doi.org/10.1109/CISIS.2012.159

36. Cretella, G., Di Martino, B.: Towards automatic analysis of cloud vendors APIs for supporting cloud application portability. In: Sixth International Conference on Complex, Intelligent, and Software Intensive Systems. pp. 61–67 (2012b). https://doi.org/10.1109/CISIS.2012.162

37. Di Martino, B., Cretella, G., Esposito, A.: Advances in applications portability and services interoperability among multiple clouds. IEEE CLOUD Comput. **2**(2), 22–28 (2015)

38. Cretella, G., Di Martino, B.: A semantic engine for porting applications to the cloud and among clouds. Softw. Pract. Exp. **45**(12), 1619–1637 (2015). https://doi.org/10.1002/spe.2304

39. Di Martino, B., Esposito, A., Cretella, G.: Semantic representation of cloud patterns and services with automated reasoning to support cloud application portability. IEEE Trans. Cloud Comput. **2015**, 1–15 (2015). https://doi.org/10.1109/TCC.2015.2433259

40. Markoska, E., Ackovska, N., Ristov, S., Gusev, M., Kostoska, M.: Software design patterns to develop an interoperable cloud environment. In: 23rd IEEE Telecommunications Forum Telfor (TELFOR), pp. 986–989 (2015)

41. Petcu, D., Macariu, G., Panica, S., Crăciun, C.: Portable cloud applications—From theory to practice. Futur. Gener. Comput. Syst. **29**(6), 1417–1430 (2013). https://doi.org/10.1016/j.future.2012.01.009

42. Petcu, D.: Portability and interoperability between clouds: challenges and case study. European Conference on a Service-Based Internet, pp. 62–74. Springer, Berlin Heidelberg (2011)

43. Petcu, D., Di Martino, B., et al.: Experiences in building a mOSAIC of clouds. J. Cloud Comput. Adv. Syst. Appl. **2**(12), 1–22 (2013)

44. Darko, A.: Application programming interfaces (APIs) based interoperability of cloud computing. University of Zagreb (2015)

45. Hamdaqa, M., Livogiannis, T., Tahvildari, L.: A reference model for developing cloud applications. In: 1st International Conference on Cloud Computing and Services Science, pp. 98–103 (2011). https://doi.org/10.5220/0003393800980103

46. Vijaya, A., Neelanarayanan, V.: A model driven framework for portable cloud services: proof of concept implementation. Int. J. Educ. Manag. Eng. (4), 27–35 (2015). https://doi.org/10.5815/ijeme.2015.04.04

47. Vijaya, A., Neelanarayanan, V.: Framework for platform agnostic enterprise application development supporting multiple clouds. In: 2nd International Symposium on Big Data and Cloud Computing (ISBCC'15), pp. 73–80. Elsevier Masson SAS (2015b). https://doi.org/10.1016/j.procs.2015.04.063

48. Filippo, G., Davide, L., Yancheshmeh, M.S., Ardagna, D., Di Nitto, E.: An approach for the development of portable applications on PaaS clouds. In: 3rd International Conference on Cloud Computing and Services Science (CI-2013), pp. 591–601 (2013). https://doi.org/10.5220/0004511605910601

Tweet Sentiment Classification by Semantic and Frequency Base Features Using Hybrid Classifier

Hemant Kumar Menaria, Pritesh Nagar and Mayank Patel

Abstract The technique of sentiment analysis is considered as the most powerful tools in the field of natural language processing as it comes up with large number of possibilities to perceive the sentiments of people's on several distinct topics. The concept of aspect-based sentiment analysis aims to figure out it further and determines what an individual is talking about, and explains whether she/he likes it or not. A practical example of an ideal realm in context to this topic discussed represents millions of possible schemes of Indian welfare planning. The government has launched such type at all the possible levels in schools, center and state level. These schemes work with an association of both the state and the central government. Such welfare-based schemes are generally introduced for various distinct levels on the basis of peoples (individuals) and their behavior or lifestyle. The schemes are launched for the purpose of developing the minority and the weaker section of the society. Whereas some of the welfare schemes are mainly introduced for girls and women only and it helps in empowering the status of the women by providing financial help as well as the basic need and requirements. There are several distinct ways to handle this major issue by the mechanism of machine learning process. In this paper, a labeled form of data is mainly used based on the polarity, preprocessing of the Tweets, which further extracts unigram features after the process of Tweet-based preprocessing methodology. In case of preprocessing, the data with huge noise is removed with the help of tokenization process; stop word removal process and stemming (deriving) these processes to clear the redundant data such as repeat emoji, words, and hashtags. These label and features are usually learned by SVM, KNN and a hybrid of KNN. In proposed experiment Hybrid approach shows improvement in precision and accuracy than other.

Keywords Machine learning · Support vector machine · Twitter sentiment analysis

H. K. Menaria (✉) · P. Nagar · M. Patel
Computer Science and Engineering, Geetanajli Institute of Technical Studies, Udaipur, India
e-mail: hemantmenaria88@gmail.com

© Springer Nature Singapore Pte Ltd. 2020   107
A. K. Luhach et al. (eds.), *First International Conference on Sustainable Technologies for Computational Intelligence*, Advances in Intelligent Systems and Computing 1045,
https://doi.org/10.1007/978-981-15-0029-9_9

1 Introduction

The great impact of social media worldwide has led to the discovery of sentiment analysis. The recent developments of smart technologies using mobile-based communication have entailed massive amount of data creation. Social media provides an ability to share thoughts, opinions, and emotions. The term sentiment analysis (SA) is popularly known as opinion mining which is a process of emotion classification usually conveyed by a text that may be positive, negative or neutral. The available data on social media has contributed to vast research using sentiment analysis. The twitter-based social media represents a gold-mine approach for analyzing the performance of the brand. Large opinions of the people are found over Twitter that are honest, informative, and casual as compared to the formal type of data-survey analysis using magazines or reports. Millions of people share and express their sentiments over the media discussing the brands whom they interact with. When such type of sentiments is identified over the media, then the information gained from such sentiments represents fruitful results benefiting large companies or organizations. This data is very helpful to monitor performance of different brands and to locate time periods and aspects receiving polar sentiments [1, 3]. The brands can be celebrities, political parties or events, products, etc. Approximately more than 500 million Tweets are generated over daily basis which represents a huge/vast collection of data for the process of analyzing the brand performance used by the members or teams of companies on manual basis. The tweets diversity cannot be captured probably by using constant or fixed rules. It is very difficult to measure tweet sentiment analysis due to its complex behavior as compared to a well-formatted documentary. The tweets do not rely over any formal type of language or over any formal language word. The symbols and punctuations are basically used to express opinions such as emoticons, smileys, etc. So, the thesis work presents the supervised learning approaches and natural language processing techniques for understanding the concept of tweets based on its characteristics and patterns including sentiment-based queries.

1.1 Sentiment Analysis

The concept of sentiment analysis is understood by combining the terms "Sentiment" and "Analysis". The word sentiment represents feeling that can be joyful, confusing, irritating, distracting. The sentiments are the feelings based on certain attitudes and opinions rather than facts due to which sentiments are of subjective nature. The sentiment implies an emotion usually motivated by opinion or perception of a person [5]. The psychologists attempt to present multitude of emotions classified into six distinct classes: joy, love, fear, sadness, surprise, and anger. The emotions based on sadness and joy are experienced on daily basis at different levels. We are mainly concerned about sentiment analysis detecting a positive or negative response or opinion. The major significance of sentiment analysis is that every emotion is linked to human perception forming an ingrained part of all humans which means that every human has the potential to generate different opinions acting as a tool for sentiment analysis. Sentiment analysis refers to the analysis automation of a known text determining the distinct types of feelings conveyed. The term sentiment analysis and opinion mining can be used interchangeably. Sentiment analysis as defined as information extraction and natural language processing task with an aim to gain the feelings of writer expressed positively or negatively based on requests, comments or questions analyzing large datasets or documents. It basically intends to define writer's feeling regarding a specific topic based on writer's own opinion. It models a branch that can help in providing a judgement over distinct fields [9]. The measurement of sentiments is a biased technique with it is really complex to achieve high accuracy of automated systems.

A. **Types of Sentiment Analysis (SA)**: Various types of SAs have been discussed in the section below:

(a) *SA based on Document Level:* This kind of SA is usually applied over the entity which helps in identifying the negative or positive views on an individual entity with the help of using documents [10].

(b) *SA based on Comparative Level:* In most of the cases, the users of the system generally expresses their sentiments or views on comparing it with similar entity or product. The main objective of this level is the identification of the opinions with the help of comparative sentence. For example, "I drove the Verna, it does not handle better that Honda city superb."

(c) *Aspect-based SA*: Document-level and sentence-level analysis gives good results when they are used on single entity but when we want to analyze the multiple entities then we need aspect-based sentiment analysis. For example, "I am a Samsung phone lover. I like the look of the phone. The screen is big and clear. The camera is fantastic [14]. But, in any case, there are a couple of drawbacks as *well;* the life of battery isn't up to the check and access to WhatsApp is troublesome". The classification of the negative and positive sentiments of this study masks the product-based valuable amount of information. In case of SA in terms of aspect-based analysis, it usually helps to analyze the negative and positive item-based aspect. This kind of analysis is mostly domain-specific. First, the concept-based aspects are searched and then aspect-based location are searched. In the end, the polarity of such a view defines such an aspect.

B. **Online Social Media:** In this present scenario, the concept of social media is developed each and everywhere as well as for everyone, it forms a heavy part of generation-based lifestyle. It presents an internet (web)-based communicational tools empowering the people to discuss and share as well as to consume (employ) their day-to-day activities on distinct levels of social media which encounters either in a formal or in an informal way. Large number of online platforms of networking is available, for instance, facebook, twitter, and

Flicker, YouTube, etc. It consists of blogging and gatherings, writing articles of the news that allows an individual or a group of people get involved in healthy talks or discussions over the web-based social media. For the users, it usually provides a platform of great importance for significance to share and express their certain feelings, opinions, joy, issues, views, emotions, and struggle [12]. At the personal level, the process of online-based networking helps to provide various facilities to link with friends for performing inventive things, picks up the information in the field of education, to generate interests in newly developed things, and as the entertainment source. Whereas at the professional level, the user of the system mainly helps in utilizing the concept of online networking in order to model expertized network worldwide to increase the knowledge in a particular area. At the organization-based level, the concept of online networking usually allows them to conduct a live communicational network with the audience which may take place from remote areas, get reaction in form of feedback of consumer and it can further boost their brand. The concept of social media [5] carries lots of specific features as discussed below:

- *Scalable*—The process of online-based networking is considered to be custom-made in order to meet the requirements of any type of association and it can be further integrated into an organization or an office-based correlation and struggling procedure.
- *Interaction*—The network based on an online basis helps in encouraging both the association and association between the organization and the group, thereby upgrading the correlation and connections.
- *Immediacy*—This type of feature helps in controlling the part of discussions and is capable to react fast in response to the data of wrong type.
- *Audience*—With the help of PCs and cell phones, different group of persons can model or create several locales of online networking.

B.1 *Social Media Characteristics:* These are known to be conventional unbinding services.

- It authorizes the users from all corners of the world and helps to establish communicational process among each other.
- It helps in providing a place for web-based/online, mass or interpersonal communication.
- These mainly specializes in a singular form of hard-back topic or mask a large number of several themes.
- It authorizes the users to advance or forward on their own while accomplishing the process of online-based fingerprint in terms of who they are, and what they may crave to be.
- The full extension of such locales may offer simple features that get updated on a constant basis.
- It usually provides a new give and take ideas and a freeway for communicational process.

B.1 *Social Media Types:* The concept of online-based networking is flourished everywhere by the technology of Internet. The users of the system have several encouraging methods in order to take advantage of such outlets of online networking. Different types of social media are discussed below [7]:

(a) *Blogs:* The concept of blog represents a streamline for personal journal or diary. It helps to provide a good platform for expressing passions and thoughts to the practical world. Blog is of dynamic nature which makes it significantly efficient. It can further be remodeled and it authorizes the guest users to transmit through the remarking segment joined to each and every single post.

(b) *Wikis:* It forms an aggregation-based site which helps in allowing each user to modify or create a page by using his/her Web-based browser. Wikis mainly offer a powerful yet resilient communitarian-particularized apparatus for creating content-particular Web sites.

(c) *Social bookmarking:* This represents the process of storing the bookmarks along with keywords in response to an open internet-based web site. It involves the participation of each and every one. The stored amount of information and data can be obtained from the resources of tagged nature. It eases the diffusion of source listing, journals, records, and distinct type of assets under study.

(d) *Social networking sites:* In short, the social networking sites are usually abbreviated as SNS which forms an expression for describing any type of website which allows the people to grab the benefits of services related to social networking services and it presents various challenges in terms of its definition. It may carry several features such as: (a) communication from person to person (b) user-generated content (UGC) to be considered as the life of social networking sites, (c) client's service (CS)-particular profiles which is usually kept by the association-based SNS, and (d) social network administration of interpersonal communication helps in supporting the advancement of online groups/communities by coordinating the profile of a customer with various packs and/or individuals [1].

(e) *Services based on Status-update:* These are also known as micro-blogging administrations, for instance, the technology based on Twitter helps in allowing the end-users to share or post small type of tweets regarding with reference to events, people, and to view the updates framed by the others. The technique of micro-blogging represents a broadcasting medium which mainly exists in blogging form. A microblog usually variates from a traditional as the contents of site are small in size. This technique allows the users to use small type of substance-based texts, for example, "singular images, video joins, or small tenses".

(f) *Media-sharing sites*: It grants users to post photographs or recordings on Pinterest, YouTube, etc. Further, the users can share the media with others all over the world or it may just select a class or group of friends. Several types of destination based on the process of media-sharing grants a person to place the media over distinct sites [4]. In order to post in an easy way, Web 2.0 impressive technologies have been started which offer media over numerous informal type of groups, communities, and other modern stages. The concept of Media-based sharing has provided a great impact both for the organizations and the people to further establish their working methodologies.

C. *Role of twitter*

In this paper, our main focus is over Twitter for the examination of informational data, where Twitter presents internet-based regulating administration which helps in empowering the clients to read and send short messages of 140-character known as tweets. Yet, its expanding exposure, the technology of Twitter is always open for the clients that are not registered to screen most of the tweets, not like the concept of Facebook where the users can easily protect and control their own profiles. It usually presents a hefty wide-range interpersonal type of communication presenting a microblog webpage. The obscene data provided by Twitter, for instance, tweeted messages, data of client's profile and the number of system devotees assuming a large part in information-based examination, which therefore makes explores and examines at distinct procedures of investigation in order to handle the growing uti-lizing innovations [8]. The concept of retweet in methodology of twitter presents the activity of assertion with respect to a specific tweet. The user of the process forwards data to his people-based gatherings in order to clearly express the reaction on a particular tweet. The retweeting instrument comes up with an assumption of having an unambiguous part or section in the dispersion of data. The rate of retweet of the primary tweets and the noticing number of the tweets helps in researching whether the retweet quantity and noticing number are determined with a similar type of system or not. Extra work by the ability of retweet was considered by transmitting two different underlining concepts. One is the Content (hashtags and URL), and the component of Contextual form (adherents number after and the recording period from 74 million kinds of tweets [9].

D. Twitter Sentiment Analysis

Sentiment analysis is a fairly growing field. Approximately 81% of the web-users usually 60% of the Americans have performed research on a product online analyzing that each year articles with different text-domain forms are targeted over years. One such example experimented was comparison of consumer confidence-based Gallup polls and Twitter sentiment. The obtained results were positive and the value of correlation was 0.804, suggesting impeding that one can use Twitter for measuring different public opinions. So, the study user analysis of Tweets to extract distinct opinions, determining the polarity of the tweets on real-time basis. A popular micro-blogging site named Twitter allows its users to write entries or texts up to 140 characters popularly known as Tweets. Twitter has approximately 302 million users on monthly basis. Out of which 88% of the users have freely readable tweets and around 80% of the users have placed their location over the profiles. The Twitter created data is usually available through Twitter's API representing information on

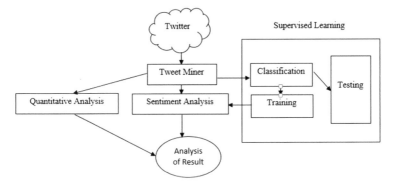

Fig. 1 Twitter sentiment analysis [3]

real-time basis as a stream opinionated data form [10]. The tweets can be easily filtered using both the publishing time and the location. This has designed a new sentimental analysis sub-field named Twitter sentiment analysis (TSA). The process of implementing natural language processing over textual form of data from the twitter media represents certain new challenges due to data informal nature (Fig. 1).

Tweets generally contain spelling mistakes and the problem of character limitation resulting in abbreviation mistakes. Many unconventional methods on linguistic basis like words elongation or capitalization are used. In addition, tweets consist of unique features like hashtags and emoticons having an analytical value. The hashtags are used for categorization and mechanism of searching represented as "#" whereas the emoticons usually express different emotions expressed as ":-)". A replying form if tweet as directed to another user, a symbol "@" is mentioned in front of the person's that is to be replied.

2   Related Work

For the purpose of analyzing the sentiments, a vast number of lexicon-based and machine learning approaches are estimated in modern analysis. Recently, several types of lexicons are made to strengthen the classification based on polarity; we favor to offer an impressive review of a vast number of modernized lexicon-based and machine learning approaches.

Krishna et al. [1] proposed a methodology by proposing a model over the fuzzy logic for gaining the sentiment analysis and feature-based sentiment mining. The opinions or sentiments are mainly used for decision making in order to select any type of alluring topic. The proposing methodology is mainly used for extracting several features from the tweets. It is completed by the use of fuzzy-based and machine learning approaches. The sentiment analysis classification as well as review is done energetically by such kind of approaches. Shidaganti et al. [2] investigated over a method that presented a combined process of machine learning and data mining. The work proposed was done over tweeter data for the purpose of analyzing the tweet to collect the opinions of the user in regard to a specific type of topic or issue. The platform of tweeter is mainly used by individuals for expressing their sentiments (views) in a short type of message in response to several types of products, brands, celebrities, along with political criticisms. In this type of work, clustering and TF-IDF algorithm are mainly discussed with their effective efficiency. Rout et al. [3] investigated the social-media-based unstructured data such as tweeter for sentiment, emotion, and blogs analysis. Such type of work takes place over both the supervised as well as the unsupervised type of approach on distinct type of databases. The approach based on unsupervised form is mainly used for the process of automatically identifying the sentiments for several tweets. Distinct type of algorithms based on the mechanism of machine learning such as maximum entropy and SVM are used for identifying the sentiments. The POS, unigram, and bigram features are effectively in formation of a tweet. Mumtaz and Ahuja [4] viewed and expected a methodology presenting a combined form of lexical-based and machine learning approach. The hybrid approach, i.e., proposed provides high amount of accuracy than the classical-type of lexical method and it helps in providing the enhanced form of redundancy than the approach of machine learning. The approach, i.e., proposed is mainly used for opinion/sentiment mining through natural language processing (NLP) which helps in extracting the opinions/sentiments from the text associated with an entity. Al-Smadi et al. [5] inspected the term sentiment analysis on the basis of hotel-based reviews in Arabic language using the methodology of memory-based neural networks on long-term basis. It is further carried out in two of the levels like and aspect-based and character-level with polarity classification and random field classifier. The approach proposed provides better results by 39% of enhancement. Chen et al. [6] viewed and further proposed a visualization-based approach known as TagNet mainly used for the purpose of sentiment analysis. Such kind of approach usually links various tag clouds with an upgraded form of nodal link diagrams presenting the heterogeneous time-varying informational data. The algorithm proposed enhances large dataset scalability. Fouad et al. [7] scheduled a miniature for analysis of tweeter-based sentiments describing weather the is or negative or positive form through the machine learning concept. The model proposed uses distinct type of methods for labeling the training phase input by using distinct forms of datasets. The method of classification was performed by the use of different types of classifiers in order to correlate and compare certain performances. The methodology of information gain and feature-based selection is mainly used in such type of work. Bifet et al. [8] focused on faced challenges of Twitter informational technology, focus over the ordering issues, and further it considers the streams for sentiment analysis and supposition mining. For the management of emitting unequal type of classes, the author has proposed a sliding or rolling window-based Kappa measurement for assessing informational streams that are of time-changing nature. The utilization of such type of measurements, the experts have investigated the information of Twitter that further utilizes the calculations for informational streams. Zahrotun [9] has examined several techniques of clustering used in the process of mining the data. The technique of clustering presents the categorization of essential data which usually belongs to similar class or it may come under similar type of group. Cosine and Jaccard similarity, along with combinational analysis of both the processes is used for getting the flawless value-based similarity. The Cosine-based similarity helps in measuring the similarity provided within two of the non-zero type of vectors on the basis of inner space of product which evaluates the cosine angle between the product spaces. From the combinational analysis, the values by two of the titles are expected to increase. Such type of methodology was performed on practical basis in Dahlan under the Department of Informatics Engineering. This results gained from such type of study are of cosine-based similarity providing the appropriate approximations than the combinational analysis of both as well as from Jaccard itself. Agarwal.et al. [10] researched several documents in the form of semi-structured, structured and unstructured data. In order to gather the data in bulk amount and the methodology to divide them is of challenging nature in accordance with respective type of domain. In order to overcome such kind of issue, domain-based clustering of two algorithms is mainly, i.e., cosine and Jaccard techniques of similarity-based algorithm in order to determine the type of similarity within two types of documents. The similarity of Cosine form within two type of documents provides fast results due to the clustering form of generation which is of steady form in comparison with the coefficient of Jaccard. Jaccard based coefficient mainly used more complex form of mathematical design in order to compute the similarity between two of the documents used in the process. So, the Cosine-based similarity provides more reliable and accurate results. Medhat et al. [11] conferred about various applications of SA, recent modernized advancements in algorithm which were presented and investigated briefly on paper. Recently, the articles were reviewed gathering the reader's interest in technology offered by sentiment analyses (detection of emotions, resource building up, and transfer learning). Various surveys took place in context to several algorithms of SA providing a sophisticated distribution. The algorithm based on Emotion detection was used for analyzing and enhancing emotions, it could either be implicit or explicit. Several types of algorithm were used for presenting the emotions and sentiments. Some of them are Point-wise Mutual information, Latent Semantic Indexing, Chi-square. The opinion-based techniques of classification were disjointed into hybrid, lexicon-based, and machine learning approaches. The area of research in this paper is based on FS and SC algorithms. Most commonly used ML type of algorithms is used for determining the problems of SC. Virmani.et al. [12] has explained the sentiment analysis collaboration with summarization, sentiment extraction, and further maintains the document of each student. In order to have the collaborated and enhanced opinion (sentiment) about the student helps in modifying the current algorithm. For analyzing the sentiments, a database of a sentiment word has been used in the process. First, it sets the score for any kind of sentiment or opinion word. In this process, when an opinion word is encountered in sentence, it performs the operation of matching with the set of database and further it sets the score as per the requirement. Then from such type of scores the value based on cumulative opinion was evaluated. The algorithm provides an opinion-based numerical value. If in case the value of numerical score is large then it provides a positive remark conclusion and if in case the value is low then it shows a negative remark. For instance, if the remarks provided by two teachers are extremely high and correspondingly the remarks provided by one teacher is very low then the process of collaboration occurs which will help in providing an averaging score. The performance on overall basis usually depends on the remarking states of both the teachers, the opinion word being used by faculty does not match with database word, which further affects the score on overall basis. Balachandran and Kirupananda [13] explained the present status about selecting the right type of institute which is considered as the most demanding task. In order to get an approximated idea about a specific institute each and every student of the class performs net surfing over the social media networking sites for knowing the ratings and the reviews about specific institutions. But this task is of challenging nature in the process of analyzing the statistical form of aspect from the viewed reviews. In case of aspect-based, the process of sentiment analysis is implemented directly over the reviews providing positive and negative reviews of the specific institution. Numerous techniques have been used for the purpose of aspect-based identification approaches like Machine Learning, Dictionary-based, NLP-based technique, unsupervised technique, Corpus-based approaches. The suitable and the best possible result analytics is provided by the ML and the NLP type of classifiers in order to classify each and every aspect into various respective forms of category.

3   The Proposed Method

A newly built model has been proposed with the help of using three distinct types of algorithms based on the concept of data mining such as Support Vector Machine (SVM), k-nearest-neighbor (KNN), hybrid method. The main objective of such type of model proposed is to classify the opinions of public. The model proposed helps in predicting the tweet-based sentiment polarity by using three distinct type of classifiers based on data mining techniques. The figure shown below represents the proposed model design which is further implemented with the help of using Python-based PyCharm tool.

A. *Proposed steps*

The informational amount of data is viewed and collected for performing several experiments being taken from Tweeter and it is further stored in the database for preprocessing methodology. The section below helps in describing the model-based methodology proposed and it also describes various techniques with detailed/deep analysis.

Step1: Data Collection: The data provided to the proposed model in the form of input is gathered from Tweeter, in regard to government welfare-based schemes. Due to demanding information, the data is fetched from social media in unstructured form. Some of the preprocessing steps are used for extracting favorable informational data from the twitter-based dataset.

Step 2: Fetching and storing the data: The tweets that are retrieved are mainly stored in the form of .csv format files, and further these files are retrieved in python-based PyCharm tool. Approximately 3000 tweets are stored for testing and training purpose. The algorithm of data mining such as KNN, Hybrid, and SVM are used for the testing and training of the retrieved tweets.

Step 3:Preprocessing of data: In this type of step, the process of twitter data cleaning is done. Preprocessing of Twitter helps in removing the redundant, noisy data from the data of raw type and then it makes the trained form of dataset for more work. Various steps to clean the data are discussed as follows:

- The uppercase is usually converted into lowercase.
- Discard all the slangs of internet from data.
- Removal of all the list-based stopping words.
- Elimination of all the white spaces additionally.
- Restrict the duplicity of words.
- All of the hashtags are eliminated yet the texts based on hash tag are reserved.

Step 4: Applying various mining techniques: To classify the data in various categories data mining techniques are used which are based on the following aspects:

- 1st aspect: decrease/increase fund
- 2nd aspect: growth in improvement/no growth/growth
- 3rd aspect: works/works really fast/hard fix
- 4th aspect: good work/incredible work/no work

These parameters are used for training and testing of data.

The trained sets of data are mainly used for training the model-based on machine learning process. Further over the trained form of data, the algorithm is mainly implemented for the purpose of classification. The data-based testing is designed or prepared.

Step 5: Result optimization: To check the model that rules are learned by training dataset or not error rate is computed. According to the dataset cross-validation is used to get the accurate result. To build the model in Python, Support Vector Machine (SVM) and K-Nearest-Neighbor (KNN) is used and hybrid of both is used to train and test the data. The approach of K-nearest-neighbor is depended upon the neighbors. The tweets are then classified, predict the nature of the tweets and give us the optimal value.

B. *Proposed methodology: Flowchart*

This section includes the proposed methodology based on the steps proposed in the earlier section (Fig. 2).

C. *Algorithm Used*

Naïve Bayes classifier is based on the Bayesian theorem of probability. It is a Supervised Learning algorithm which is used for classification. It solves the problem in continuous as well as categorical value attributes [7].

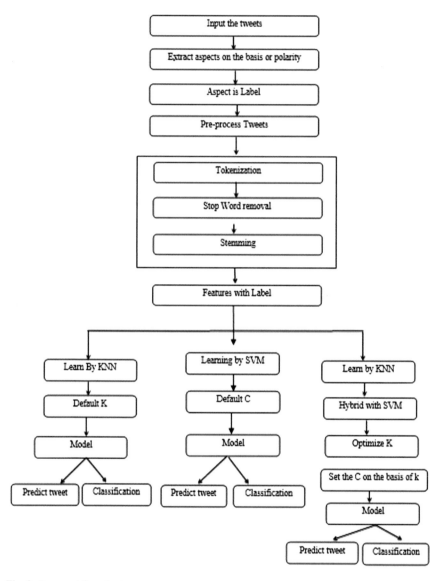

Fig. 2 Proposed flowchart

- It is mainly used in text classification and spam filtering.
- Recommendation based systems also used this classifier.

1. *Bayes's theorem*: It is the base of the classifier on which it is developed. In this theorem, conditional probability that an event y belongs to class n can be calculated from the conditional probabilities of finding particular event in each class and the unconditional probability of the event in each class. In given data $y \, \varepsilon \, Y$ and C classes where y is a random variable. Conditional probability that an event y belongs to class n is computed by using following formula (1)

$$P(C_n|y) = P(C_n)\frac{P(y|c_n)}{P(y)} \tag{1}$$

This equation is used for pattern classification and finds the probability of the given data y belongs to class n and gives the optimum class with high probability among all classes.

For Statistically Independent Features perform the given equation:

$$P(y|C_n) = \prod_{i=0}^{p} P(y|C_n) \tag{2}$$

Here, y is a p-dimensional vector data $y = (y_1, y_2, \ldots y_n)$
Summarized the Naïve Bayes result:

$$n = \operatorname{argmax}_n P(c_n) \prod_{i=0}^{n} P(y_i|C_n) \tag{3}$$

2. *Random Forest*: For classification and regression decision tree models are used by it. If results are variable then it provides more information and accuracy. In this classifier class distribution is maintained and training dataset is used to randomly set the data features. Gain Index method is used in it to split the data. This method is very helpful in predicting the missing data and it performs very efficiently dataset which is large in size. Classification of these features is done by Random forest method and is also used for support vector machine feature selection. Accuracy in result is 85% in identification.

Following steps are performed on image when classification is done by using random forest method. The feature space of M-dimensional $\{A_1, A_2 \ldots A_m\}$ is considered in this method. The weights $\{W_1, W_2 \ldots W_m\}$ are calculated in every feature in the space in this method. Each decision tree is grown by improved algorithm in random forest method using weights.

3. *Feature Weight Computation*: To register the feature weight, it is utilized to gauge the usefulness of each information feature A as its connection to the class

Y. The high estimation of the weight shows the question in preparing information corresponds with the estimations of features. Accordingly, *A* is useful to the class names of new objects. Amaratunga utilized a two-example t-test as the feature weight so this strategy must be utilized as a part of two class information.

4. *Support Vector Machine*: Vapnik introduced for support vector machine, and is popular tool for supervised machines learning methods which are based on the minimization of the structural risk. The SVM basic characteristics is the original non-linear data into data class and the separation margin among itself is maximized and typing points nearer from the support vectors.

The training sample is

$$n = \{(u_i, v_i) | i = 1, 2, \ldots, m\}$$

where

m	Sample no.
$\{u_i\} \in r_k$	Input vector set
$v \in \{-1, 1\}$	Desired corresponding input vector

Then, optimal classification of existing hyper-plane has the following condition to meet:

$$\begin{cases} \omega^t u_i + B \le 1, v_i = 1 \\ \omega^t u_i + B \le -1, v_i = -1 \end{cases}$$

where

ω^τ Super plane omega vector,
B Offset quality

Then, the decision function is classified as:

$$F(u_i) = \text{sgn}(\omega^t u_i + B)$$

SVM classification model is described with optimization model $\min_{\omega, \xi, B} P(\omega, \xi)$

$$\min_{\omega, \xi, B} P(\omega, \xi_i) = \frac{1}{2} \omega^t \omega + \frac{1}{2} \gamma \sum_{i=1}^{m} \xi_i^2$$

$$v_i [\omega^t \phi(u_i) + B = 1 - \xi_i, i = 1, 2, \ldots, m$$

$$\xi = (\xi_1, \xi_2, \ldots \xi_m)$$

where

ξ_i Slack variable

B   Offset
ω   Support vector
γ   Classification parameter for balancing the model complexity and fitness error

Transforming the optimization problem into dual space and for solving it, Lagrange function is introduced:

$$l(B, \omega, \alpha, \xi) = \frac{1}{2}\omega^t\omega + \frac{1}{2}\gamma\sum_{i=1}^{m}\xi_i^2 - \sum_{A=1}^{m}\alpha_i\{v_i[\omega^t\phi(u_A) + B] - 1 + \xi_i\}$$

where

α_i   Lagrange multiplier

Then, describing the classification decision function:

$$F(x_i) = \text{sgn}\left(\sum_{i=1}^{m}\alpha_i v_i A(u, u_i) + B\right)$$

4   Result Analysis

The performance of application to core mapping derived from proposed heuristic was evaluated in terms of dynamic communication latency with applied traffic load on the NoC simulation framework. The application to core mapping derived from the proposed Branch and Bound heuristic is anticipated to reduce the system's dynamic communication latency for the reason that the proposed heuristic moves through the search tree that corresponds to the solution space to find an optimal mapping where the applications are mapped keeping their communication characteristics into consideration.

A.   *Result Analysis:* In this chapter the results of the method used are presented and also there is a comparison of the result obtained by the method used and the classical method. 5-fold and 10-fold methods are used to verify the result obtained by the experiment. SVM, KNN, and Hybrid algorithm are used to analyze the result. Accuracy, precision, and recall are the measures used for result analysis. Precision of 5-cross validation model trained by SVM classifier and dataset with the accuracy is shown in Fig. 3. 0.47 is the average accuracy obtained over the test dataset. Precision of 5-cross validation model trained by KNN classifier and dataset with the accuracy is shown in Fig. 3. 0.44 is the average accuracy obtained over the dataset in this test. In Fig. 4 the model is trained by the hybrid classifier and it shows precision of 5-cross validation model and the dataset with the accuracy. In this test the average accuracy obtained over the test dataset is 0.526. KNN and Hybrid classifiers accuracy is depicted by

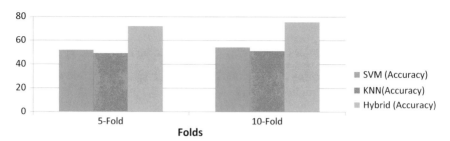

Fig. 3 Accuracy on different classifiers

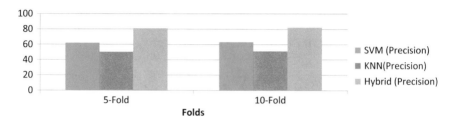

Fig. 4 Precision on different classifiers

Fig. 3. The *X*-axis and *Y*-axis on graph represent the validation fold and values of accuracy respectively. 5-fold and 10-fold validation testing process shows the maximum and minimum accuracy when hybrid algorithm is used. 49.23 is the minimum accuracy obtained when KNN in 5-fold validation process is used.

SVM, KNN, and Hybrid classifiers precision is shown in Fig. 4. Validation fold and values of Precision are represented on *X*-axis and *Y*-axis, respectively. Precision in 5-fold and 10-fold validation testing process is maximum in hybrid algorithm and KNN in 5-fold validation process shows the minimum precision of 50.23. True positive and true negative are improved by the accuracy which in turn minimizes the error and further improves the accuracy. True negative directly shows impact on errors and Precision and Recall show the impact on true positive. If it increases then its significance reduces the error. If it reduces then it increases the error. The average of precision and recall is F-measure.

In Fig. 5 it depicts the comparison of the different classifiers that are SVM, KNN and Hybrid. The *X*-axis on the graph represents the classifiers and *Y*-axis represents the values of Accuracy, Precision, Recall, and F-measure. The hybrid algorithm shows the maximum results on all parameters and KNN shows the minimum result among all.

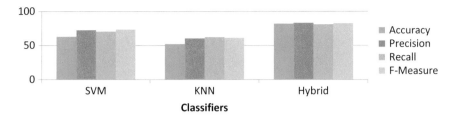

Fig. 5 Comparison of different classifiers

References

1. Krishna, R.: Fuzzy aspects in sentiment analysis and, pp. 7750–7755 (2016)
2. Shidaganti, G., Hulkund, R.G., Prakash, S.: Analysis and exploitation of twitter data using machine learning techniques. In: International Proceedings on Advances in Soft Computing, Intelligent Systems and Applications, pp. 135–146. Springer, Singapore (2018)
3. Rout, J.K., Choo, K.K.R., Dash, A.K., Bakshi, S., Jena, S.K., Williams, K.L.: A model for sentiment and emotion analysis of unstructured social media text. Electron. Commer. Res. **18**(1), 181–199 (2018)
4. Mumtaz, D., Ahuja, B.: A Lexical and Machine Learning-Based Hybrid System for Sentiment Analysis, vol. 713. Springer, Singapore (2018)
5. Al-Smadi, M., Al-Ayyoub, M., Al-Sarhan, H., Jararwell, Y.: An aspect-based sentiment analysis approach to evaluating Arabic news affect on readers. J. Univers. Comput. Sci. **22**(5), 630–649 (2016)
6. Chen, P.T., Sessions, T.: PacificVAST Program PacificVis Program (2018)
7. Fouad, M.M., Gharib, T.F., Mashat, A.S.: The International Conference on Advanced Machine Learning Technologies and Applications (AMLTA2018), vol. 723, no. January (2018)
8. Bifet, D.S., Khan, F.H., View, S.A.: TOM : twitter opinion mining framework using hybrid classification scheme, no. January (2014)
9. Zahrotun, L.: Comparison jaccard similarity, cosine similarity and combined both of the data clustering with shared nearest neighbor method. Comput. Eng. Appl. **5**(11), 2252–4274 (2016)
10. Agarwal, N., Rawat, M., Vijay, M.: Comparative analysis of jaccard coefficient and cosine similarity for web document similarity measure. Int. J. Adv. Res. Eng. Technol. **2**(5), 18–21 (2014)
11. Medhat, W., Hassan, A., Korashy, H.: Sentiment analysis algorithms and applications : a survey. Ain Shams Eng. J. **5**(4), 1093–1113 (2014)
12. Virmani, D., Malhotra, V., Tyagi, R.: Sentiment analysis using collaborated opinion mining. arXiv Prepr. arXiv1401.2618, no. January (2014)
13. Balachandran, L., Kirupananda, A.: Online reviews evaluation system for higher education institution: an aspect based sentiment analysis tool. In: 2017 11th International Conference on Software, Knowledge, Information Management and Applications, pp. 1–7 (2017)

Using RFID Technology in Vaccination Cards for Saudi Children: Research Study

Thahab Albuhairi and Abdullah Altameem

Abstract The Kingdom of Saudi Arabia (KSA) seeks to improve processes in all fields and sectors using the latest technologies and take care of its citizens. Healthcare sector plays an essential role in the country's growth. This research paper is focused on the children healthcare in KSA, specifically on the vaccination process of children in Saudi Arabia, how it can be improved, and how can avoid the delays and mistakes committed during vaccination. One of the main problems in this sector is using a traditional paper card for tracking the vaccination. This paper suggests using the Radio Frequency Identification (RFID) technology in the smart card for replacing the traditional vaccination cards. The RFID technology is made up of two components: a passive RFID tag for storage of children's health information regarding vaccination. The second part is the RFID reader to read and interpret the data from the RFID tag. The research presented questionnaires to 167 people, 87% of them favor to use this technology. According to some advantageous as it made the vaccination process easier, faster, and it reduced errors, and the major reasons are an avoidance of loss of information and increase in the information availability at any time.

Keywords RFID · Vaccination · Healthcare · KSA

1 Introduction

Today, governments in all the world are trying to improve their processes in various fields and sectors to increase effectiveness and efficiency regarding increasing usability, accuracy, and reducing the required effort and time.

The health sector considered as one of the major sectors that governments should work to improve and use by adopting the most recent and sophisticated technologies. The health sector in the Middle East has a lot of scopes to improve when compared

T. Albuhairi (✉) · A. Altameem
Imam Mohammad Ibn Saud Islamic University (IMSIU), Riyadh, KSA, Saudi Arabia
e-mail: amalbuhairi@imamu.edu.sa

A. Altameem
e-mail: altameem@imamu.edu.sa

© Springer Nature Singapore Pte Ltd. 2020
A. K. Luhach et al. (eds.), *First International Conference on Sustainable Technologies for Computational Intelligence*, Advances in Intelligent Systems and Computing 1045,
https://doi.org/10.1007/978-981-15-0029-9_10

to the developed countries which have adopted some emerging technologies for the healthcare of their citizen. The key healthcare challenges in the Kingdom of Saudi Arabia (KSA) are long wait times and data entry errors in vaccination. Furthermore, according to statistics published by the General Authority for Statistics in the Kingdom of Saudi Arabia by the Legatum Institute in 2016, Saudi Arabia is ranked fifth in the Arab world and 45th in the world [1]. In 2017, the number of hospitals in KSA increased to 487 hospitals [2]. In the same year, *Makkah* newspaper published an article about six problems facing health centers, such as the acute shortage of medical personnel in the villages. Unavailability of some medicines has forced health centers to turn away patients from hospitals. Moreover, there is poor coordination between centers and hospitals. Consequently, the aim of restructuring solves these problems by linking the centers and hospitals to a central electronic database [3].

Children are an integral part of Saudi's society. The population of KSA increased to more than 20 million people in 2017, according to statistics published by the General Authority for Statistics in the Kingdom of Saudi Arabia [4]. This followed the birth of 300,000 more children in 2017 than in 2016. The number of children under four years is more than 2.1 million, which equates to about 11% of the total population of KSA [4]. Consequently, this age group is important in Saudi society as it constitutes and accounts as a large percentage of the total population. Also, because of the vulnerability of children, such a large number calls for concerted efforts to ensure that the worrying trend is reversed before it gets out of control or the children contract diseases that may negatively affect the quality of life.

Vaccination is one of the main ways to protect children from diseases. A doctor named Omar AlShaikh—a member of the Medicine Faculty at King Saud University—argued that many reasons warrant the need to use vaccines [5], such as:

- About 10 million children under the age of five die each year around the world. A quarter of those deaths occur from vaccine-preventable disease.
- Today, vaccination can save 2 to 3 million people every year. It is one of the most successful public health interventions.
- Measles vaccination in the Middle East led to a 90% reduction of measles deaths in 2007.
- Polio incidence rates worldwide declined by 99% in 2008.

In KSA, children under two years visit the hospital eight times to take the vaccines. It is either through an injection under the skin or by mouth as the Ministry of Health directed [6]. Vaccination in KSA is facilitated by a papered card and extracted by the hospital where the mother gave birth to her baby. The papered card contains information about the baby, her/his parents, and the date of birth. The card also has a national vaccination schedule and visit periods [6]. However, the disadvantage of this card is that it could be lost or damaged over time, and requesting another card is time-consuming.

This paper focuses on children and babies' health, especially the aspect of vaccination through the traditional card that is prone to damage or inaccurate filling. The paper proposes the use of new technology to improve the vaccination process and enhance safety and flexibility. Radio frequency identification (RFID) technology offers a practical solution to the above challenges. RFID exists in many applications of our daily lives, such as preventing the theft of vehicles or goods [7]. This term formulated for radio technology in short range mainly used to communicate the digital information between "stationary location and a movable object" [7]. RFID is considered one of the automatic identification techniques, such as barcodes and smart cards. Some researchers describe the automatic identification technology as the "new way of controlling information and material flow" [8]. Also, these researchers define the RFID technology as "a way of collecting data about a particular item without the need of touching or seeing a data carrier, by using the electromagnetic waves" [8]. Furthermore, RFID is a fast-improving technology that uses radio waves to collect and transform data. It can capture data efficiently and automatically way without human intervention [9].

RFID technology has many uses in healthcare and in hospitals, such as managing the patients' medical processes and monitoring the treatment plan for outpatients. This technology may increase the level of patients' safety and accelerate critical therapy. Moreover, it is useful for reducing costs whether direct or indirect [10].

This research seeks to address many questions. Does the use of papered card affect parents in any way? What are the benefits of using RFID in vaccination cards for the children? Do parents agree to use this technology and support converting from paper cards to electronic cards? Do they support browsing vaccination information online for their child? What are their reasons?

This research proposes the use of electronic cards supported by RFID technology to overcome the limitations of using the papered card in vaccination. It also provides a general overview of how this technology works, its benefits and clarifies the ease access to the medical file, reduces the waiting time of the parents and ensures that vaccination takes place in time.

2   Literature Review

This section clarifies a brief historical view of RFID, studies of RFID in multi-sectors, and some studies conducted about RFID technology in hospitals and the healthcare sector.

2.1   Rfid

The first person who researched about RFID is Harry Stockman in 1948 in the research titled "Communication by Means of Reflected Power". The era of the 1950s focused on the exploration of RFID techniques such as "the long-range transponder systems of identification". The first commercial use of RFID was in electronic article surveillance (EAS) equipment in the late 1960s to counter the theft of goods. In 1991, the RFID is introduced in the first electronic tolling system of a highway [7].

RFID consists of two main parts, namely tags and readers, both connected to a computer or server. It used radio frequencies to communicate at a range of between 100 kHz and 10 GHz [7].

RFID tags have many classifications as mentioned below [8]:

- A passive tag or "purely passive" takes power from the reader. The reader sends electronic waves while the tag reflects it and adds information.
- A semi-passive tag gets power from a battery to maintain the memory.
- An active tag uses a periodically replaceable internal battery.

The memory can be read-only—where the memory cannot be modified, and the data are static—or it can be read/write, and the data can be modified [8].

2.2 RFID in Multiable Sectors

This subsection shows the application of RFID technology in fields other than hospitals and the healthcare sector—in the business field, especially in the shipping of perishable goods.

One of the benefits of this application is the ability to evaluate the status of shipment or content without opening it. RFID technology in conjunction with built-in sensors offers more benefits [11].

This should be done by preparing each container with an integrated circuit (IC) including an RFID tag and a transmitter beside sensors such a temperature sensor to check if the container was maintained. The RFID tag contains encrypted data about the original goods with an electronic signature. When the container is delivered, the receiver or the customs can evaluate the state of the container without opening it [11].

Australia is planting RFID devices on military personnel to monitor them. The tag contains personal information and the country identifier. A study looks into the effect or danger of implanting RFID tags for vaccination management and finds that RFID tag implants does not poses any risk associated with vaccination [12].

The KSA has applied the RFID technology in various fields, one being the use of RFID in Hajj, where each pilgrim wears a bracelet containing a chip of RFID to count the total number of pilgrims and ensure that every pilgrim has the Hajj permit [13]. Also, KSA and most countries around the world use FRID in the ATM cards. It provides more flexibility and increases speed when used in the purchasing process, for example. Furthermore, it adds an additional safety feature [14]. However, some researchers suggest that using RFID technology in ATM cards may be exposing it to violation and piracy [15].

A review study published in 2017 focuses on applications that use sensors and passive antenna which supported with RFID technology. This study concluded that there many features such the physical sense of objects and inkjet printing. The RFID reader can measure the sensing data whether direct or indirect, and other of applications [16].

2.3 RFID in Health Sector

Improvements in the healthcare sector are paramount. Use of modern technologies and the Internet of things (IoT) have drastically improved healthcare [17]. Alain Yee-Loong Chong, Martin J. Liu, and their comrades conducted a study about the influence of RFID technology on the healthcare sector, especially on individual differences for physicians and nurses [17]. They interviewed 252 Malaysian physicians and nurses and indicated that the worker personality plays a key role in accepting the use of RFID technology. The other determinant of acceptability is the age of workers, whereby younger workers are more flexible and adaptive to change compared to older workers. Gender is also a critical factor in the sense that men readily accept to use RFID compared to women. RFID increases work performance and efficiencies and reduce human errors [17].

An empirical study published in 2012, conducted by Yee-Loong Chong, Liu, and others, indicates that the application of RFID technology on organizations in the healthcare sector can help to improve operations and gain a competitive advantage over the rest of other competitors [18]. This study uses data collected from 182 health care organizations. The diffusion of RFID consists of three stages: the initial stage, the adoption stage, and evaluation stage. This study concluded that there are three main categories of factors that determine the application of RFID in the healthcare industry [18]:

– Technological factors: The use of RFID and barcodes may lead to some complexity, thus mandating the need for training of hospital staff [19]. These factors have a significant influence on adopting and evaluating stages [18].
– Organizational factors: This consists of support by the top management, organization size, technological knowledge, and financial resources. Support by the top management is crucial to avoid resistance to change because the implementation of RFID technology mandates some changes in the healthcare organizations [20].
– Environmental factors are competitive pressure and market trends, where both are correlated positively with the evaluation and adoption stages of RFID implementation [21].

A study published in 2004 indicated that finding hospital equipment consumes 30 percent of the staff, whereas 10% of the equipment is lost every year [22]. Therefore, the application of RFID technology could solve these problems [23]. Also, RFID is used for some medical states of "eyeball pressure" by using a sensor and transmitter. In addition, a study published at 2017 revolves around using RFID tags with the medical equipment in the operations room to avoid forgetting this equipment inside the patient by measuring the magnetic distance for each equipment [24].

Some Arabic countries use electronic cards for vaccination management, such as Bahrain and the United Arabic Emirates (UAE). The vaccination management process is better in the UAE where hospitals use a virtual reality (VR) to reduce the children's fear and pain during vaccination [25].

Yao, Chu, and Li add that RFID can lead to the doctors' and other hospital staff members efficiently accessing the medical data of the patient in addition to monitoring and managing people and equipment in the real time, thus reducing medical errors and increasing patients' safety, also, time-saving. It is also time-saving because it takes less time. Furthermore, this technology can store a large amount of data, and the memory can be able to read/write not as the barcodes [8]. Also, in 2017, a survey focused on application of RFID tags in the healthcare sector to measure its benefit on the patients, which is linked with Wi-fi sensor. It proved that the RFID be more helpful in emergency conditions, supervising the patient's health status and weight [26].

The major barriers to its adoption are high cost and the violation of privacy [9]. Another barrier that limits the adoption of RFID in the healthcare sector is the fact that RFID reader may be unsuitable in the power, especially "the lack of standardization of the protocols for RFID". Another study established that collaborating partners need to agree on the standards to use during communication [8]. The infrastructure cost of RFID may be around 200$ to 600$ to run a system of medium-sized hospitals. The other barriers include the cost of hardware and software, security and trust, and the overlap between RFID signals and medical equipment signals [9]. Moreover, the signals may be crashed during the exchange of information simultaneously [8].

To solve these barriers, hospitals can use programs to improve data quality, as well as educate staff and patients about RFID technology and increase their understanding [9].

2.4 Ways to Improve the Vaccine Process

Some studies provide ways to improve the vaccine process. A study published in 2016 suggests the use of a mobile application to notifies parents regarding the details of vaccination. Parents in Canada found the use of mobile application due to its time saving benefit [27]. A survey published in 2018 focused on using CAN Immunize application designed to enable newcomers in Canada to manage their vaccination records, where the largest category comes from Arabian countries. Most of these newcomers did not use the health electronic applications, perhaps they are having smartphones, but it is still useful [28].

A study was published in 2015 discussing the use of the mobile phone to improve the children vaccination in Bangladesh, especially those who lived in rural areas and are difficult to reach. The researchers notified mothers living in rural areas regarding vaccination for their 0–11-month-old kids using the application called 'mTika' [29].

In February 2017, a Systematic Review published under the title of "Using Mobile Phones to Improve Vaccination Uptake in 21 Low and Middle-Income Countries" [30]. It stated the benefits of using mobile health (mHealth) to improve vaccination in 21 countries with low- and middle-income household by sending text messages to the parents on their mobile phone [30].

An application called VAccApp developed in Vienna-enabled parents to learn more about the vaccination status of their children. A program of hospital-based quality management in co-operation with "the Robert Koch Institute" validated this application during 2012–2014 in Germany [31].

All of the above studies have used mobile phone applications and indicated a lot of benefits, but no one focused on using RFID technology on the vaccination cards for the children.

3 Research Method

This research uses a survey research methodology. The questionnaire is used for collecting the data from various parents in different areas in KSA and they are the target groups. The questionnaire consists of pre-defined questions within defined alternatives and multiple-choice answers. It is a primary resource used in this research to clarify the major reasons or motivations for using RFID technology in the vaccination process for the children in KSA. An eight closed questions with basic personal information like: gender and age are included in the questionnaire in Arabic language and English to ensure quality responses, cover largest range and avoid parents' ignoring. This methodology aims to discover problems in the use of the traditional paper card in vaccination and access the parents' opinion on the RFID technology, and the rationale behind it. The respondents are 167 in total.

4 Discussion and Results

This section contains the discussion and the results of the research. The ratio of the number of men and women respondent was 1:1. Among the total survey respondents, 69 respondents (~41%) aged more than 46 years old, whereas the percentage of the respondents with age between 45 and 36 years was 26%. The rest were less than 36 years old with a percentage of 33% as shown in Fig. 1.

The distribution of respondents as per their geography was as follows:

About 89% of the total respondents were from Riyadh which equals 148 respondents, while the rest came from various cities such as Qassim, Dammam, Abha, and others as shown in Fig. 2.

As for the number of children, the highest percentage of respondents is 47% and equal to 79 respondents of them had one child to three children, while 38 percent had four to six children. About 20 respondents have more than six children with a percentage of 12% as shown in Fig. 3.

As shown in Fig. 4, about ninety-three percentage of the respondents denied the filling of incorrect information in the vaccination card. However, 7% of them faced this problem. Considering the importance of healthcare especially of children, 7% of respondent are still relevant.

Fig. 1 Count of age

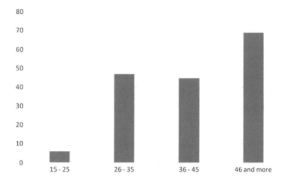

Fig. 2 Count of city

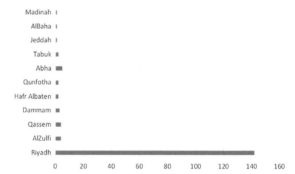

Fig. 3 Number of children

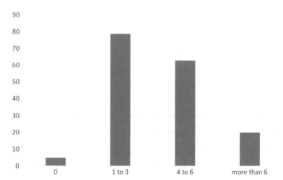

One of the questions asked during the survey was whether their child's vaccination date was postponed or delayed in the past, 93 respondents (55%) said that they had their children's vaccination appointment delayed due to the vaccination being unavailable as shown in Fig. 5. Another question asked to the parents was whether they ever lost or damaged their child's vaccination card. As shown in Fig. 6, eighteen percent of the parents responded agreed that their children's vaccination cards were lost and 9% responded reported the damage of the cards.

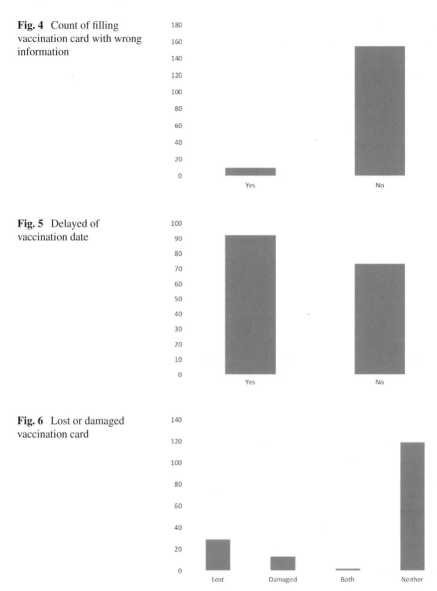

Fig. 4 Count of filling vaccination card with wrong information

Fig. 5 Delayed of vaccination date

Fig. 6 Lost or damaged vaccination card

Another question in the questionnaire how long it took to get a replacement card where three respondents (2%) said it took them more than one month to extract get a replacement, whereas 11 (7%) of them replaced the card within one week to one month. As shown in Fig. 7, 22% reported that it took less than a week to replace the vaccination card. Following the previous question, 37 respondents (22%) had to visit more than one hospital to fill the replacement card with the previous vaccines

Fig. 7 Period to extract
replacement card

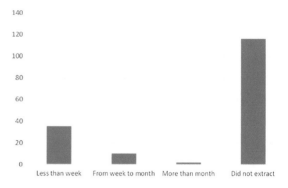

Fig. 8 Visit more than one
hospital to extract the card

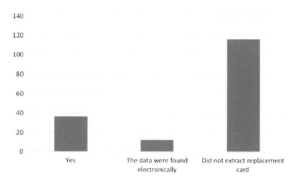

information as shown in Fig. 8. It reveals the lack of coordination among hospitals in KSA.

The total of 127 respondents (74%) confirmed that they would like to see their child's vaccination information and the latest updates in electronic form, as shown in Fig. 9. This point supports the applied technologies in the literature review part to improve this process. The core question of this questionnaire was if parents favor changing the card from paper to a smart card with the RFID technology almost similar to ATM cards. In response to this question, 145 respondents (87%) were in favor

Fig. 9 Availability of
electronic form

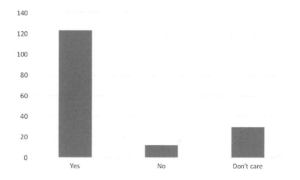

Fig. 10 Using RFID technology in vaccination cards

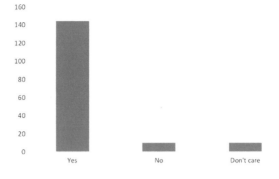

Fig. 11 Reasons to favor this technology

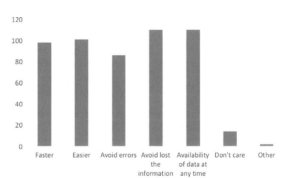

of the use of this technology as shown in Fig. 10. The respondents wanted to avoid loss of information and desired to access it any time considering the advantages of the RFID technology. Specific advantages of this technology were the speed, easy to use, and error free as agreed by the questionnaire respondents. One of the significant reasons why parents agreed to use the RFID was to avoid the loss of the vaccination information of their children and access it anytime they want (see Fig. 11). The questionnaire included a negative question with a statement saying, "This technology is not useful for them", but more than 60% of the respondents were unsupportive of this statement.

From these findings, the research concludes that the vaccination card supported by RFID technology can facilitate the vaccination process more rapidly and efficiently. It reduces the wait time and the loss of information.

5 Conclusion

This preliminary survey research using a questionnaire for 167 parents represents a very small sample of the total parents in KSA. In general, a large proportion agreed to apply the RFID technology on vaccination cards and create a central electronic database that all hospitals and parents can access at any time to avoid the loss of information and delays in vaccination. The vaccination process can be faster, easier, and efficient than before with paper cards. Briefly, this technology consists of two components a passive RFID tag and an RFID reader. The vaccination card would contain the passive RFID tag when parents pass it on the RFID reader; the information will be updated, and they can view it later from the central database.

Future research should have more respondents to cover more parents and include the children and doctors. Also, future work will clarify how this technology should be applied in detail along with the estimated cost.

References

1. Legatum-Institute: The Legatum Prosperity Index™ Rankings (2016). https://www.li.com/docs/default-source/publications/2016-legatum-prosperity-index-pdf.pdf. Last Accessed 4 Mar 2019
2. Ministry-of-Health: Health Indicators for the year (1438). https://www.moh.gov.sa/Ministry/Statistics/. Last Accessed 4 Mar 2019
3. Albogomy, T.: 6 problems face the healthcare sector. May 2017. https://makkahnewspaper.com. Last Accessed 4 Mar 2019
4. General-Authority-for-Statistics: Population Characteristics surveys. https://www.stats.gov.sa/sites/default/files/population_characteristics_surveysar.pdf. Last Accessed 4 Mar 2019
5. AlShaikh, O.: Ten facts of vaccination. http://ssdds.org/main/index.php?option=com_content&view=article&id=24:2012-04-20-01-04-05&catid=12:2012-04-19-16-02-21&Itemid=160. Last Accessed 4 Mar 2019
6. Ministry-of-Health: Ministry of Health. https://www.moh.gov.sa/en/HealthAwareness/EducationalContent/HealthTips/Documents/Immunization-Schedule.pdf. Last Accessed 4 Mar 2019
7. Landt, J.: The history of RFID. IEEE POTENTIALS, pp. 8–11 (2005)
8. Ilie-Zudor, E., Kemény, Z., Egri, P., Monostori, L.: The RFID technology and its current applications. In: Proceedings of The Modern Information Technology in the Innovation Processes of the Industrial Enterprises, pp. 29–36 (2006)
9. Yao, W., Chu, C.-H., Li, Z.: The use of RFID in healthcare: benefits and barriers. In: Program for the IEEE International Conference on RFID-Technology and Applications, pp. 128–134 (2010)
10. Visich, J.K, Li, S., Wicks, A.M.: Radio frequency identification applications. In: Hospital Environments. Hospital Topics, pp. 2–8 (2010/9)
11. Todorovic, V., Neag, M., Lazarevica, M.: On the usage of RFID tags for tracking and monitoring of shipped perishable goods. Procedia Eng. 1345–1349 (2014)
12. Nicholls, R.: Implanting military RFID. IEEE Technol. Soc. Mag. 48–51 (2017/3)
13. Alasmari, A.: Radio frequency identification technology. (2008). http://www.aleqt.com/2008/12/04/article_170995.html, last accessed 2019/3/4
14. Alrajhi: New Al Rajhi ATM Card with an additional safety feature for greater protection. https://www.alrajhibank.com.sa/ar/personal/accounts/pages/smart-chiaspx. Last Accessed 4 Mar 2019
15. Paget, K.: ShmooCon2018. In: CYBERSECURITY CONFERENCES, Washington Hilton (2018)
16. Zhang, J., Tian, G.Y., Marindra, A.M.J., Sunny, A.I., Zhao, A.B.: A review of passive RFID tag antenna-based. Sensor 1–33, 29 (2017/1)
17. Chong, A.Y.-L., Liu, M., Luo, J., Keng-Boon, O.: Predicting RFID adoption in a healthcare supply chain. Int. J. Prod. Econ. 66–75 (2014)

18. Chong, A.Y.-L.Y.-L., Chan, F.T.: Structural equation modeling for multi-stage analysis on Radio Frequency Identification (RFID) diffusion in the health care industry. Expert Syst. Appl. 8645–8654 (2012)
19. Brown, I., Russell, J.: Frequency identification technology: an exploratory study on adoption in the South African retail sector. Int. J. Inf. Manage. 250–265 (2007)
20. Jeyaraj, A., Rottman, O., Lacity, M.: A review of the predictors, linkages and biases in IT innovation adoption research. J. Inf. Technol. 1–23, (2006/2)
21. Wang, Y.-M., Wang, Y.-S., Yang, Y.-F.: Understanding the determinants of RFID adoption in the manufacturing industry. Technol. Forecast. Soc. Change 803–815 (2010)
22. Glabman, M.: Room for tracking: RFID technology finds the way. Mater. Manage. Health Care 26–38 (2004)
23. Wicks, A.M., Visich, J.K., Li, S.: Radio frequency identification applications. In: Hospital Environments. Hospital Topics, pp. 3–8 (2010/8)
24. Kusuda, K., Masamune, K., Muragaki, Y., Ameya, M., Kurokawa, S., Ohta, Y.: Measurement of magnetic field from radio-frequency Identification antenna for use in operation room. IEEE CAMA, pp. 414–415 (2017)
25. Bokhtamain, M.: New technology to reduce the children fear and pain during the vaccination. https://www.alittihad.ae/article/18650/2018/. Last Accessed 4 Mar 2019
26. Khan, S.F.: Health care monitoring system in Internet of Things (IoT) by using RFID. In: International Conference on Industrial Technology and Management, pp. 198–204 (2017)
27. Atkinson, K.M., Westeinde, J., Ducharme, R., Sarah, E.: Can mobile technologies improve on-time vaccination?. Hum. Vaccines Immunotherapeutics, 2654–2661 (2016)
28. Paradis, M., Atkinson, K.M., Hui, C., Ponka, D.: Immunization and technology among new-comers: a needs assessment survey for a vaccine-tracking app. Hum. Vaccines Immunotherapeutics 1660–1664 (2018)
29. Uddin, M.J., Shamsuzzaman, M., Horng, L., Labrique, A., Vasudevan, L., Zeller, K.: Use of mobile phones for improving vaccination coverage among children living in rural hard-to-reach areas and urban streets of Bangladesh. Vaccine 276–283 (2015)
30. Oliver-Williams, C., Brown, E., Devereux, S., Fairhead, C.: Using mobile phones to improve vaccination uptake in 21 low and middle-income countries. JMIR mHealth uHealth 1–15 (2017/2)
31. Seeber, L., Conrad, T., Hoppe, C., Obermeier, P., Chen, X.: Educating parents about the vaccination status of their children: a user-centered mobile application. Prev. Med. Rep. 241–250 (2017/1)

Towards a Real IoT-Based Smart Meter System

Anas Bin Rokan and Yasser Kotb

Abstract Smart grid grants a more reliable and cost-effective electricity system for transporting electricity from power stations to homes, business and industry. It has effectively and efficiently evolved into a smarter system of grid connecting peers and devices in an intelligent manner. In this paper, we give an overview about the smart grid and smart meter technologies, what are the functionalities of those systems, and then propose and implement a new prototype of such smart grid system in order to show their benefits in different domains. The proposed prototype system is a device with relative similarity to the smart meter contains two different sensor modules that are constructed to measure electricity voltage, current, and power. The data obtained from these sensors transmit to a Microcontroller Unit to process it. Two different platforms were integrated to send and receive data from and to the server. The data displayed on an online platform that is also acted as a control panel page for connecting devices. The effects on the current, voltage and power meter were then compared remotely. Additionally, an SMS would be sent to the consumer's mobile device to notify them as to when the high load has been reached. The results gained from the proposed prototype show that the real smart meters will have great benefits in different domains such as environmental, economic, and benefits for the service providers (utility) and the customers.

Keywords Smart meter · Smart grid · ESP8266 · Arduino

A. Bin Rokan · Y. Kotb (✉)
College of Computer and Information Sciences, Information Systems Department, Imam Mohammad Ibn Saud Islamic University, Riyadh, Saudi Arabia
e-mail: ykkotb@imamu.edu.sa

A. Bin Rokan
e-mail: aabinrokan@sm.imamu.edu.sa

Y. Kotb
Computer Science Division, Department of Mathematics, Faculty of Science, Ain Shams University, Cairo, Egypt

© Springer Nature Singapore Pte Ltd. 2020
A. K. Luhach et al. (eds.), *First International Conference on Sustainable Technologies for Computational Intelligence*, Advances in Intelligent Systems and Computing 1045,
https://doi.org/10.1007/978-981-15-0029-9_11

139

1 Introduction

In the traditional power grid, performance focuses on the generation, transmission, distribution the power with general control of the transportation process of electricity. The communication is only one way, which means the control and recovery processes are done with a minimal extent of control [1]. Furthermore, the traditional power grid suffers from different issues such as a huge amount of power losses, the lack of the visibility of the medium electricity network, and emission of harmful gases, with these issues and the huge increase of power demand, a solution is needed. The traditional power grid can be upgraded to the smart grid using ongoing technological advancements. This upgrade is beneficial for both the customer and service providers and leads to other benefits such as national, economic, and environmental benefits.

Electricity is the backbone of any civilization. As the demand for electricity rises in every aspect of our lives. The current energy demand and a huge amount of power losses due to the traditional power grid have led people in the electrical industry to look for ways to improve the energy efficiencies with the help of different technologies and with the consumers' involvement. The traditional electric meters installed to measure the amount of electric energy that is consumed so that billing can be done accordingly. The meter reader comes once a month so that the billing units can be measured, and the consumer can be billed for energy that has been consumed. This process needs to be improved due to the errors come with it and the long-time its take to perform the required work. This has eventually led to the need for smart meters that could measure the amount of energy consumed while updating the energy provider so that they can produce the desired amount of energy at the right time and can have a great amount of data that help in load forecasting, load management, etc. and reduce the need of send an employee physically to each electric meter which lead to saving the transportation cost also.

This work aims to introduce a new smart grid prototype to demonstrate how the smart meter will help in reducing and/or eliminating the flaws existed in the traditional power grid. The impact of implementing the real smart meter on the environment, services providers and customers will be exposed. Moreover, the efficient methods to install the smart meters with low trigger energy will explain and how IoT services can be used alongside smart meters. The proposed prototype developed through a dynamic webpage giving real-time electricity data readings. It is represented these recorded data in visualization and graphical forms so that it can be easy to assimilate. The data like peak usage time, an hourly breakdown of the energy consumed, and comparison of the monthly-recorded data may help in reducing the overall units billed by the consumer. The deployment of this webpage has an automation platform that giving the users control over the connected device to simulate Smart Appliance and Smart Meter concepts.

The work of this paper consists of an introduction and four sections. In the second section, we introduce briefly the literature review for smart grid and smart meter technologies. Section 3 discusses the methodology and the main steps of implementing the proposed smart grid prototype. Section 4 contains the discussion and the experimental results of the designed prototype. Finally, we conclude our work in Sect. 5.

2   Literature Review

2.1   Background

Industrial Internet of Things (IIoT) is the part of IoT which produce benefits for the industrial sector; this branch leads to the efficiency of the industry-based procedure and equipment and makes the processes and procedure easier. Automation is one of the major benefits resulted in IoT, which is an integral factor in many industrial organizations. The Industrial Internet Consortium has defined IIoT as "machines, computers, and people enabling intelligent industrial operations using advanced data analytics for transformational business outcomes" [2]. IIoT depends on the use of different machines and advanced analytics. IIoT network contains several devices connected by communications technologies lead to a system that can monitor, collect, exchange, analyze, and deliver valuable new insights which give the decision-makers in industrial companies the capabilities to make smarter and faster business decisions.

2.2   Electricity Grid

The electricity grid process starts from generation plant which generates the electric power in medium voltage from 11 to 33 kV. After generation, the electric power is passed across a step-up transformer to increase the voltage to be higher than 100 kV so that it can be transferred over long distance through high voltage transmission lines. After that, the electricity is passed through a step-down distribution transformer to reduce the level of electricity to a level that can be safely distributed to end customers [3]. Traditional grids have several drawbacks such as:

- The electric power is being wasted with an estimated range of 7–10%. During peak demand hours, the energy wasted reaches up to 20 or 30% [4].
- Upgrading or changing old equipment will be costly process.
- The visibility of the medium electricity network is unclear.
- It is difficult to predict power outages due to limited visibility.

The companies must deal with the increased demand by improving the whole process, starting from generating the power and until it has reached the end customers as well as integrating it with the production of renewable energy and other sources of energy in an efficient manner, which could be done through the moving to Smart Grid.

2.3 Smart Grid

The smart grid was proposed to change the pattern of the power grid in favor of obtaining economic efficiencies, robust control, and environmental benefits. Smart grids use an efficient network control system such as *distribution automation systems* and *smart meters* that improves reliability, sustainability, security, flexibility, and control. The Information and Communication Technology (ICT) has helped pave the way to the smarter world and the integration of the grid for better and easier monitoring and control [5]. The success of smart grids lies largely on different aspects such as using the information provided by the smart meter and control systems to manage and control electricity network.

This grid could be seen as one of the largest instantiations of the IoT network. The entire power grid chain, from the power generation plant to the final electricity consumers such as houses, building, factories, including transmission and distribution electricity networks, will be prepared with intelligence and via two-way communication to provide monitoring and controlling capabilities to the electricity grid, which also means having good visibility of the electricity network and equipment in real-time and with high accuracy.

Smart Grid Functionalities. The smart grid gained its major popularity by proposing the following functionalities:

Self-Healing. Those systems are smart enough to judge the problem arising in an intelligent manner. This allows the smart grid to analyze, react and identify any major fault in it. Any trouble that might be caused because of a failure in any component in the grid can be both identified and acted upon in a short timeframe and in a more advanced system it has the capability to respond or solve the problem without human intervention [6].

Participation of customers. Traditionally, consumers were only provided with services on a fixed rate. With the introduction of tariffs, they can actively participate in the selection of appropriate tariffs and control the usage of their appliances in a smart and cost-friendly fashion. With the growing demand in solar power panels and their easy installment, it is predicted by some studies that the customers can actively participate in selling electric power to the utilities in the peak/surge time. It can be done by using a two-way communication where the service provider signals the need for electric supply which will be met by customers with alternate energy sources such as solar energy.

Smart appliances. Smart appliances are devices that programmed to respond to the signals sent by the customer or the service provider. These appliances enable customers to control their appliances using the Internet to participate in voluntary demand response during peak hours or when the grid overloaded or by enforcing them to reduce their consumption using different tariffs during peak time [7]. Another thing in some countries they have agreements with the big customers with huge power load like universities to control some of their load like the air conditioners machines during the peak time by control the smart appliances for a period of time to allow the grid to stabilize and to avoid the loss of electricity services for other customers.

Flexibility in network topology. Due to the two-way communication the energy flow could be in two directions, which allows the integration with other energy sources such as wind turbines, solar panels, and other sources.

2.4   Smart Meter

The electric meter has been evolved over the years until reach to the smart meter, at the beginning in most traditional electricity meters where the overall consumption information was collected on-site manually after a specific period of time (mostly once a month), then the meter evolved to *Automatic Meter Reading* (AMR) which differ from the old one by the automatic collection of electrical energy consumption, diagnoses, and sends data to the utility (a one-way communication) [8]. AMR system helps reduce the expense of sending an employee to physical locations to read the electricity meters.

Then the *Advanced Metering Infrastructure* (AMI) systems appeared which is a system with a combination smart meters, communication networks, and data management systems which has the capacity of operating in two-way communication between utilities and customers [9]. This system refers to the measurement, collection, and analysis of the usage of electric power or energy and the effective communication between metering equipment for various energy. A smart meter in AMI Systems has the ability to:

- Automatically and remotely measure electricity usage.
- Connect and disconnect services remotely.
- Theft Detection.
- Identify and isolate outages.
- Support *Multiple Tariff.*
- Update the smart meter firmware remotely.
- Support Post or Pre-Payment
- Interface with In-home Display (IHD)
- Business Intelligence and Data Analytics.

Smart Meter Benefits. With all features mentioned before is obvious that implementing smart meters will generate benefits in different domains as described below:

Environmental Benefits. The smart meter will reduce utilities' vehicle needs in meter reading process thus lead to reduce the emission of harmful gases used for the combustion of fuel in cars. In electricity generation unit, smart meters can help in producing only the desired amount of energy by predicting an estimated amount consumed per day and per hour which allows the power generation units to alter their production patterns, which also leads to energy saving both in the production and transmission phases. This could be seen as a reduction of losses as well as the reduction of the emission of harmful gases that result from the overproduction [10].

Benefits for Service Provider. There are different benefits that smart meter systems will generate for the electricity service providers one of those benefits it will help in saving overall transportation cost (in terms of fuel and other costs) and service restoration time. Other benefit the ability to Tamper detection while the smart meter sent its data periodically to the utility, thus any unusual activity can be easily detected, last but not least smart meter will help in improving the outage management due to capability to send last gasp messages and data to outage management systems about which lead to resolving issues in a shorter time interval and cost-effective manner [11].

Benefits for Customer.

- Improve customer satisfaction: Provide the customer with the ability to view the consumption at any time and can see if they are being billed based on their consumption. Customer trust and satisfaction will greatly increase.
- Active consumption management: By allowing the customers to monitor the consumption in real-time, they can easily manage their usage due to the integration between the smart appliances and the smart meter.

National Benefits.

- Optimizing energy use: From a broader perspective, smart meters not only benefit consumers or utilities, but their advantages are also on a global level or countries level. By optimizing energy usage in every household or building which leads to different benefit for countries.
- Commercial or Economic benefits: Smart meter installation programs may create new economic opportunities for energy suppliers or third-party businesses that are likely to be based on selling smart meter data or consumers' behavioral data, and delivering targeted behavioral advertising to consumers.

3   Methodology and Implementation

In this section, we will show how our smart grid prototype system will build. This section is broken down into the following parts: *Hardware, Platforms, Control management website SMS services* and *Project Workflow*. This system simulates the work of an IoT-based smart meter. The system is designed by combining different sensors, using different platforms that support the transfer of data from one side to another, developing a webpage for having control over the designed device (prototype), presenting the values in a user-friendly and understandable manner along with having a database that keeps a record of the timestamps. In addition, the proposed system will be able to construct visualized tables to display different information about consumption during peak time, consumption over the last twelve months, and an average of monthly consumption. The prototype will display real-time, voltage, and current values of the connected circuitry.

3.1 Hardware

The hardware selection part is done by looking at various components then studying and comparing them so that the best of the component fitting for the project's needs, we have selected the following component:

- 240 V AC power supply
- Power converters (220 AC to 5 V DC)
- Zmpt101B—Voltage sensor
- ACS712—Current sensor
- Four bulbs (which simulate smart devices/appliance in a customer's home)
- A Four-channel relay module to control the four bulbs
- Arduino Nano
- Esp8266
- LCD screen 2 * 16

Figure 1 shows the Circuit Diagram of the designed prototype and how these components are connected to each other. With regard to Internet connectivity, ESP8266 connected to Arduino Nano. Other than that, the current sensor, the voltage sensor, the relay channel, and the LED screen have their connections to the Arduino Nano Board.

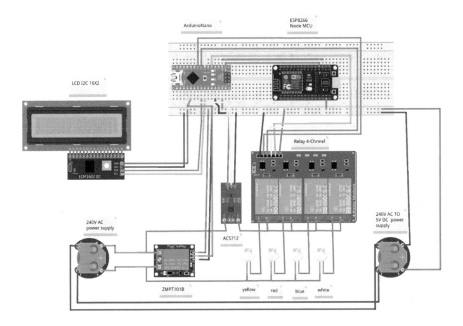

Fig. 1 Circuit diagram of the designed system prototype

3.2   Platforms

The proposed system uses different IoT platforms:

Thing Speak. Thing Speak is an open IoT analytics platform that is available online could be used to display sensor data in meaningful representation. In our project Thing Speak took the data from our server because the esp8266 sent the data to the server then transferred it to the Thing Speak then display the data on charts.

IFTTT. "If This Then That" is a free web-based service that helps make automation services easier to embed in a project. In this study, the IFTTT works as a translator to trigger the HTTP request and translate it into MQTT messages. We have created several Applets using the webhook service in IFTTT. These applets are triggered when we press buttons on a webpage they send MQTT messages to Adafruit we have designed these applets to use the MQTT broker's which basically cuts or supplies the power to the loads, which are the bulbs in our case.

Adafruit IO. Adafruit.io is commonly referred to as a solution for the construction of applications of IoT by using the MQTT protocol and MOTT broker. In our project, we have used the MQTT broker provided by Adafruit to publish/subscribe messaging transport protocol for sending bulbs ON/OFF commands, but there is no direct way to send MQTT protocol messages from the webpage. Thus, some intermediate service was required, and we chose the IFTTT website that we have mentioned before.

3.3   Control Management Website

A webpage has been developed to display the data collected from sensors and to control the device remotely to simulate the smart meter concept with the smart appliance that enables the user which represent the house owner to control his devices remotely which, in our case, are the four bulbs. The following section will elaborate on our website that is constructed to control remotely our smart grid prototype. This website contains four pages: *Home, Customer, Company,* and *Contact*. The main pages among them are the *Customer* and *Company pages*. This website is accessible through this URL; http://www.arokan.sa.

   Customer page. From this page, the customer could control their devices (bulbs) through different buttons and could monitor the following:

- Current power: which is the real-time reading
- Total power consumption.
- Month power which represents the consumption during the current month compared with the average consumption.
- A form enables the user to show the power consumed during a specific time.
- The power consumed over the past 12 months.

Company pages. From this page, the company or service provider employee could control customer smart meter by switching the power on/off through different buttons and to monitor the following:

- Current power, voltage, and current intensity of the selected customer.
- Total power consumption, the existing Month power consumed and Peak time.
- The power consumed over the past 12 months.

3.4 SMS Service

When the maximum consumption has been reached (which we identified as 53 Watts), the system automatically sends a message using an API. The message sent to notify the user of their high consumption in real-time.

3.5 Project Workflow

We used a combination of two different MCU. (1) Arduino Nano selected to connect with the ACS712 Current Sensor Module and the Zmpt101B Voltage sensor module. The sensor data was collected initially at the Arduino Nano, (2) ESP8266 which connected with Arduino Nano to provide Internet connectivity. Using the Internet, the data is transferring wirelessly to the online server then transfer to the database and Thing Speak. The ESP8266 was also connected to the Adafruit.io (subscribe) which would transmit MQTT messages from the Adafruit Cloud that would be received at our device.

Apart from this, the Arduino Nano was connected to the 4-load relay circuit which is connected to 4 bulbs (Red, Yellow, White, Blue). The MQTT broker hosted by Adafruit.io was selected to transmit the commands from the online server to the Arduino Nano, but an intermediate service was required to transmit messages from the webpage (HTTP to MQTT). For this purpose, the IFTTT open platform was selected. It provided support like a translator to comprehend the trigger of the HTTP request. The HTTP request was set as the trigger which looked for the applet having HTTP request and did the action (in our case, control the button) defined in the applet. The IFTTT had several applets using webhook services which sent MQTT messages to Adafruit. From there, the Adafruit sent a command to the esp8266 then it was transmitted to Arduino which transferred it to the relay module and the control was achieved. HTML WebPages were developed to display charts from the Thing Speak platform and from the database. The webpage also has ON/OFF (control) buttons that trigger events on IFTTT using webhook. Triggered events then send data to Adafruit MQTT broker. The messages transferred to the hardware are done using the ESP8266 MQTT library to subscribe to the MQTT broker. Whenever there is a new MQTT message (publish), ESP8266 receives (subscribes) that message and forwards

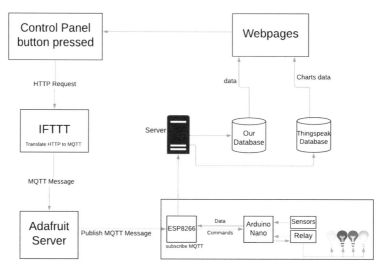

Fig. 2 Workflow of proposed system

it to Arduino Nano. From here, Arduino Nano checks the received command and acts accordingly. Figure 2 illustrates the whole workflow described in the previous paragraphs:

4 Discussions and Results

This section will visualize what exactly the work is done

4.1 The Customer Page

Then customer page, as seen in Fig. 3, shows the control buttons and two gauges for measurement of current power (Watts because we use only bulbs), total consumption (Watts-hour), month power (Average), the past 12 months consumption and a form to display the consumption of specific period of time. The buttons control the four bulbs: red, yellow, white and blue. For each bulb, there are two buttons that are meant to switch the bulbs on and off. The remaining two buttons control all connected devices (bulbs), which would either switch on or off all the bulbs eventually. The monitoring and controlling are the crux of the project as they provide smart services to the user with a friendly interface, so the customer can know their consumption and control their devices remotely.

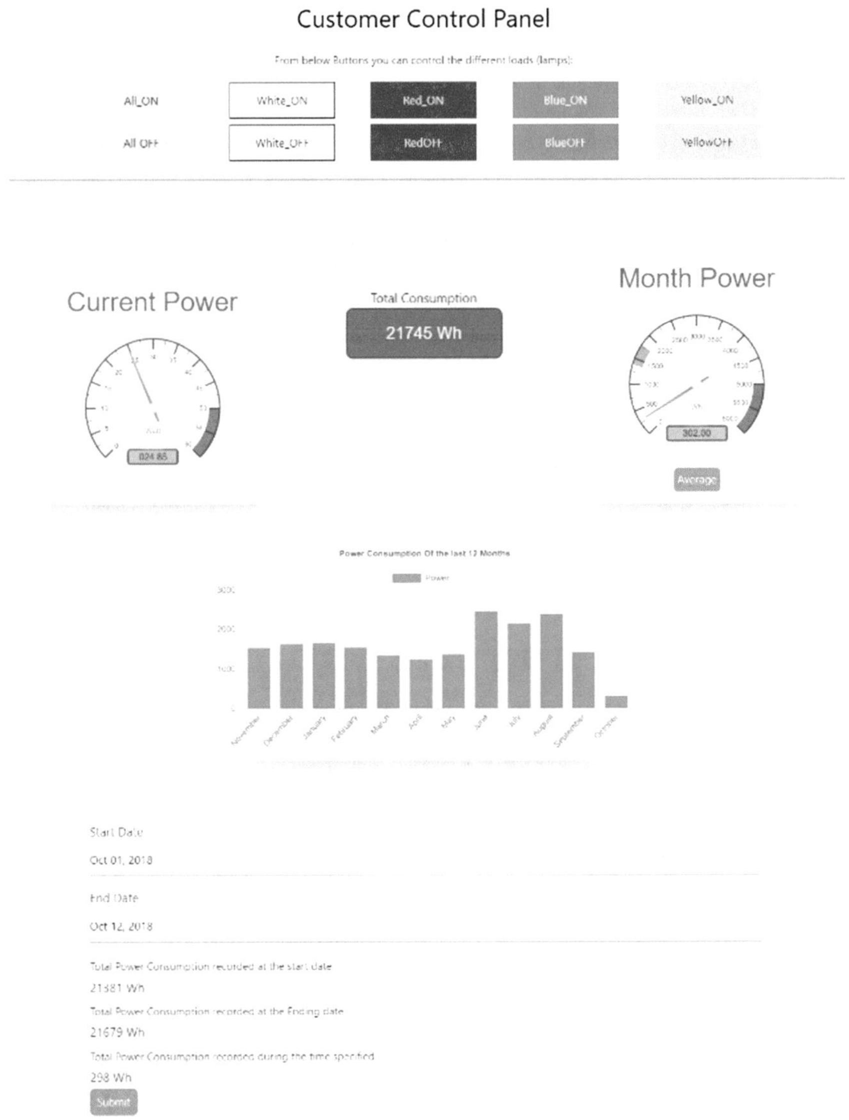

Fig. 3 Customer page

Current Power. For the experiment, the prototype was tested with varying loads connected so that the effects on the current (watts) can be analyzed. Figure 4 represents the result when a single bulb (device) is connected and when four devices are connected.

SMS Message. As previously mentioned in the last chapter, the customer will be notified after energy has reached its threshold level so that they can act effectively (Power > 53 watts). See Fig. 5.

Monthly Power. Figure 6 shows the month power gauge displays the total amount of power consumed by the customer at a particular time of month as well as an indication of the average usage per month (highlighted in green).

Total Consumption. Figure 7 shows how the total consumption, representing the total amount of energy consumed from that time when the smart meter has been placed.

Fig. 4 Current power

Fig. 5 SMS notification after reaching a high consumption

Text Message
Today 8:42 PM

Dear customer: Your home consumption has reached a high load, please reduce consumption.

Fig. 6 Monthly power indicator

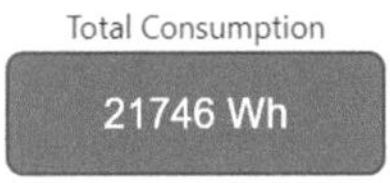

Fig. 7 Total consumption as
in smart meter

Total Consumption

21746 Wh

Monthly Consumption. Figure 8 displays the form used to display the amount of power consumed during a given period. The user chooses the start and end date then the power consumption for the selected period displayed with the estimation cost. (the estimated cost calculated by multiply the power consumed by 0.18 Saudi Riyal "halalas")

The Power Consumption of the Past 12 Months. Figure 9 gives a comparison analysis of the power consumption used over the past 12 months. This graph helps the readers understand the consumption behavior during different months.

Fig. 8 Detailed
consumption history with
dates

Start Date

Sep 12, 2018

End Date

Oct 12, 2018

Total Power Consumption recorded at the start date

20574 Wh

Total Power Consumption recorded at the Ending date

21679 Wh

Total Power Consumption recorded during the time specified

1105 Wh

Estimated cost for the selected timeline

198.90 SR

Submit

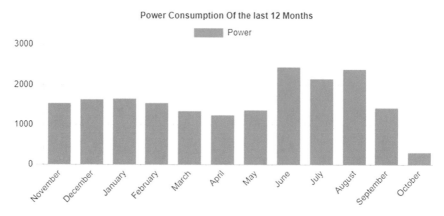

Fig. 9 Bar-graph showing consumption per year

4.2 The Company Page

In the company page, we have created a page that simulates the Smart Meter Dashboard which represents the geographical area that the company dispatcher has access to and choose which meter to display. The selected customer meter page will then appear; it contains the control buttons that represent the company's ability to cut off or return the power supply remotely, and four gauges for measurement of current power (Watts), total consumption (Watts-hour), month power (Average), current intensity, the past 12 months consumption, and the peak time chart which helps the company to predict the user's behaviors and plan accordingly. Figure 10 is a screenshot from the company page. It shows how the page will look like for the company dispatcher (employee).

Peak Time. Figure 11 displays the recorded usage of power over the day mapped over Watt Hour. The peak time graph helps in analyzing the peak usage over of the electricity in the day so that if the consumers are using most of the energy in the peak time or higher tariff times if applied, they can reduce their consumption. This will not only help in the reduction of their bills but also help companies (services providers) to estimate the peak time and respond accordingly.

5 Conclusion

The work was designed to be automation for the implementation of a smart meter as well as the working and learning of the smart grid. We worked in collaboration with different sensors, different platforms, and different processing units. We achieved a successful working prototype that was able to display real-time data and control the devices connected wirelessly via relay switch, providing information that can

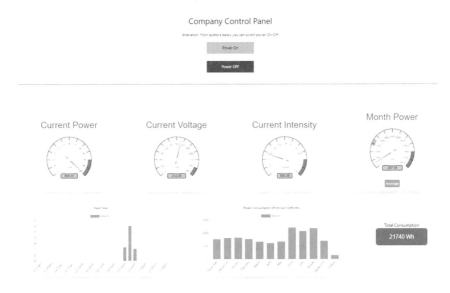

Fig. 10 Customer page as displayed via company control panel

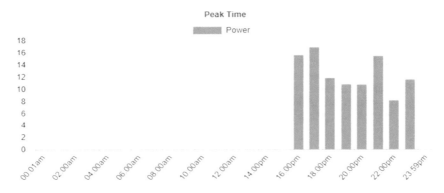

Fig. 11 Power consumed distribution of the day (peak time)

help estimate the highest load during the day and the power consumed during a specific period, etc. As future work, we plan to integrate further sensors to make the system convenient on a wider scale and introduce security measures between the messages and data transmitted remotely. A refinement of the system can be done by the development of iOS- and Android-based applications downloadable over the Internet providing an easier and friendlier interface.

References

1. Gupta, B.B., Akhtar, T.: A survey on smart power grid: frameworks, tools, security issues and solutions. Ann. Telecommun. **72**(9–10), 517–549 (2017)
2. LI, H.: Introduction about Industrial Internet (IIoT) and relevant standardization activities. Technology and Standards Research Institute, CAICT (2017). [Online]. Available: https://www.itu.int/en/ITU-D/Regional-Presence/AsiaPacific/SiteAssets/Pages/Events/2017/Oct2017CIIOT/CIIOT/12.Session5-2IndustrialInternet(IIoT)-李海花.pdf. Accessed: 08 Oct 2018
3. Desai, R.: ELECTRICITY. http://drrajivdesaimd.com (2012). [Online]. Available: http://drrajivdesaimd.com/2012/09/04/electricity/. Accessed 10 Nov 2018
4. Kevin, B.: The big smart grid challenges, MIT Technology Review (2009). [Online]. Available: https://www.technologyreview.com/s/414386/the-big-smart-grid-challenges/. Accessed 20 Oct 2018
5. Ma, Z., Asmussen, A., Jørgensen, B.: Industrial consumers' smart grid adoption: influential factors and participation phases. Energies **11**(1), 182 (2018)
6. Reka, S.S., Dragicevic, T.: Future effectual role of energy delivery: a comprehensive review of Internet of Things and smart grid. Renew. Sustain. Energy Rev. **91**(June 2017), 90–108 (2018)
7. Kobus, C.B., Klaassen, E.A., Mugge, R., Schoormans, J.P.: A real-life assessment on the effect of smart appliances for shifting households' electricity demand. Appl. Energy **147**, 335–343 (2015)
8. Saleem, Y., Crespi, N., Rehmani, M.H., Copeland, R.: Internet of things-aided smart grid: technologies, architectures, applications, prototypes, and future research directions. arXiv Prepr. arXiv1704.08977, pp. 1–30 (2017)
9. U.S. Department of Energy.: Advanced metering infrastructure and customer systems: results from the smart grid investment grant program. Off. Electr. Deliv. Energy Reliab. 1–98 (2016)
10. Keeping, S.: Carbon emission reductions by the implementation of a smart grid. NOJA Power (2013). [Online]. Available: http://www.nojapower.com/dl/library/carbon-emission-reductions-by-the-implementation-of-a-smart-grid.pdf. Accessed 16 Nov 2018
11. The Department of Energy -Office of Electricity Delivery and Energy Reliability.: Last gasp message—Outage notification (2010). [Online]. Available: https://www.smartgrid.gov/files/Last_Gasp_Message_Outage_Notification_V30.pdf. Accessed 16 Nov 2018

Proposing Model for Recognizing User Position

Biky Bikas Tirkey and Baljit Singh Saini

Abstract In the progressive world of Internet, online banking or any other transaction is not secure because of methods such as brute force technique, dictionary attacks, and spoofing. It is because traditional methods of authentication require either password or pin which is not secure because it can be cracked easily using the methods mentioned above. Therefore, biometric authentication can be a viable option which can be used for identification of a genuine user. In this paper, we propose a model for identifying the user's position that plays a vital role in identifying the genuinity of any user. This paper revolves around the effects on verification of genuine user if the data collected from user is done in both controlled and uncontrolled environment.

Keywords Physical biometrics · Behavioral biometrics · Keystroke dynamics · False acceptance rate (FAR) · False rejection rate (FRR) · Equal error rate (EER) · Random forest algorithm · J48 · Reduced error pruning

1 Introduction

PCs have turned into valuable equipment of the advanced society. In mid-2011, the web attacks on an organization have brought about their system to shutdown and traded off their passwords and individual data of a large number of their clients. Therefore, authentication Burger-Paul [1] mechanism is one such way to authorize the genuinity of a user's Identity or to authorize a user to use a system. There are different mechanisms that are introduced by various authors to be used at different levels of authentication:

B. B. Tirkey (✉) · B. S. Saini
Lovely Professional University, Delhi G.T. Road, Jalandhar 144411, Punjab, India
e-mail: vick.tirkey@gmail.com

B. S. Saini
e-mail: baljit.22078@lpu.co.in

© Springer Nature Singapore Pte Ltd. 2020
A. K. Luhach et al. (eds.), *First International Conference on Sustainable Technologies for Computational Intelligence*, Advances in Intelligent Systems and Computing 1045, https://doi.org/10.1007/978-981-15-0029-9_12

- Traditional methods used for verification mechanism were the use of password and pin by Mizrah-Len [2] for the authentication of a user. Traditional methods were futile when an attacker uses brute force algorithm according to Karawash [3], and therefore, it is necessary for the user to use of strong passwords that contain a mixture of alphabets, special characters and numbers which makes the password unpredictable and unbreakable no matter the algorithm or methods used by the attackers to crack the password.
- Token and smart card was later introduced by Deo-Seidensticker [4] in order to strengthen authentication mechanism that is used apart from using the password or pin, but there were various drawbacks of using this mechanism because there is a chance of token or smart card being stolen or either getting locked after number of failed attempts which makes it a very useful but still very naive to be used for a full-scale authentication purpose.
- The introduction of biometrics led to an era of more precise authentication mechanism without the fear of loss in the privacy of the genuine user. The mechanism was more accurate and precise compared to the previous authentication mechanisms and had baffled most authors by its accuracy.

1.1 Biometrics

As we can see from Fig. 1 how biometrics is the most applauded invention in the era of security. There are two types of biometrics: (a) physical biometrics and (b) behavioral biometrics, Ravi [6]. Physical biometrics include such mechanism that helps authenticate a user using his/her physical traits such as: (a) fingerprint, Maltoni-Maio [7], (b) iris, Daugman [8], (c) palm, Saeed-Werdoni [9], (d) face, Zhao [10]. Behavioral biometrics is another method which is more concrete technique which

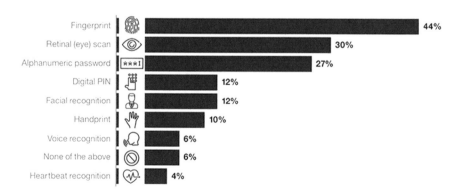

Fig. 1 Biometric preference [5]

uses behavioral traits of the users such as (a) keystroke dynamics, (b) voice, (c) handwriting, and (d) signature, these techniques are certainly unbreakable because of each and every user has their own behavioral characteristics.

1.2 Keystroke Dynamics

Keystroke dynamics Monrose [11] estimate and differentiate among clients on various properties for instance: span of a keystroke, dwell time (between keystroke times), power of keystrokes (pressure on the keystroke), flight time (keystroke between two keys). There are basically two methods involved in the verification of clients. Static keystroke examines the templates created for user identification. Dynamic analysis implicates a nonstop or periodical pursuit of stored templates created for user identification.

1.3 Data Acquisition

It was Mickey Spillane from United Nations agencies which first instructed the deployment of keyboards for keystroke dynamics to authenticate the user. Throughout the enrollment process, a temporal arrangement pattern and key pressure of the distinctive user are to be kept along with their corresponding password. The stored information is compared with the secret key and keystroke dynamics that can be used to authenticate the user. The keystroke dynamic information can be acquired from different types of sources:

– Laptop keyboard
– Desktop keyboard
– Customized pressure sensitive keyboard
– Virtual keyboard
– Customized pressure-sensitive keyboard virtual keyboard special-purpose numerical input device
– Cellular phone
– Smart phone (qwerty)

1.4 Classification of Methods

The users keystroke dynamics are extracted for the purpose of authentication during log-in period and are distinguished as well as categorized with the template in the database, if the user's keystroke is among the endurance bound the person is verified,

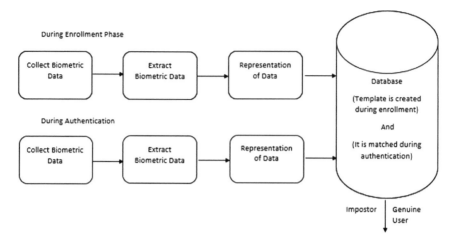

Fig. 2 Biometric system [12]

if the user is not genuine, then the machine will determine in case to lock up or not to lock up the system. The next sectors consists of different categories

- **Statistical method**

  – Mean and variance
  – Mean, standard deviation and digraph

- **Neural networks**

  – Back propagation
  – Sum-of-products (SOP)

- **Pattern recognition techniques**

  – K-means algorithm
  – Bayesian

Figure 2 shows the working of the biometric system, and the process includes the collection of data, extraction of data, representation of data and finally storing data in the database as a template. During authentication, the stored template is compared with the user credentials.

1.5 Data Classifier

In this research, three classifiers are being used for analyzing the data and helping to identify the user position, and they are:

- **Random forest algorithm**: The random forest calculation was advanced by Leo Breiman and Adele Cutler. Random decision forests are outfit kind of taking in the technique for the arrangement, the classification and regression, which will work by the building a huge number of the decision trees at the season of preparing and the class yielding that is method of classes (classification) or the mean expectation (regression) of the client trees.
- **Decision tree algorithm J48**: J48 classifier is basic C4.5 kind of the choice tree for characterization. It makes a paired tree. The methodology of tree most valuable in order issue. With this methodology, a tree is worked to the model procedure of characterization. When a tree is constructed, it is expected that each tuple in the database and the outcomes are inside order for those tuples.
- **REP tree algorithm**: Reduced-error pruning (REP) tree is a machine learning technique that trims the certain parts of the tree that are difficult to classify, therefore reducing the complexity and improve the accuracy.

2   Related Work

Significant element extraction techniques and highlights/characteristics are reported in this area. Gaines [13] utilized measurable importance tests between eighty-seven letter inter-key latencies to ascertain whether the ways for keystroke interim times area unit are equivalent. Garcia [14] made a model utilizing the mean and covariance of keystroke interim occasions. Joyce-Gupta [15] devised two additional login sequences: the users first name and the last name feature set. This progressed the execution considerably. Young et al. [16] included time span of the keystroke, length of time to type a string and pressure to be measured while typing. Obaidat-Sadoun [17] recommended the use of flight time and dwell time instances to be stored using terminate and stay resident (TSR) in MS-DOS domain. Revett [18] suggested the use of motif signature to be grouped. Burger [1] introduced a mechanism to authorize the genuinity of a user's Identity or to authorize a user to use a system.

3   Proposed Work

Identifying user position plays a major role in the authentication of the user. Identifying user's posture in which he/she is registering their data helps us to identify whether the user is genuine or an impostor because of their typing behavior and differences in there in different features that are used for the collection of the data very minute differences can help us to identify the impostor, it is therefore very important to collection the data in both environment (Controlled and uncontrolled) which makes it much easier to identify the user (Genuine or Impostor). Therefore, proposing a model for identifying user's posture is as shown in Fig. 3. The dataset used in the research work is collected from dif-

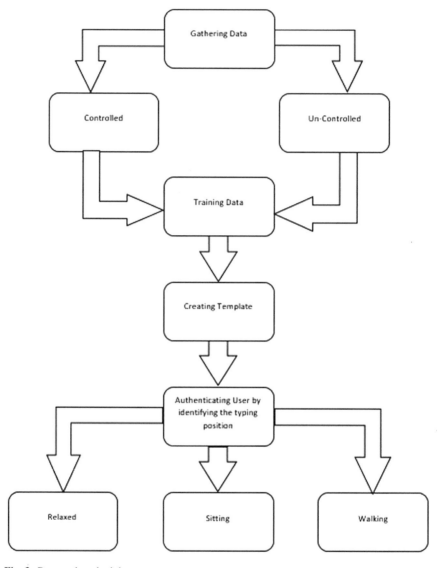

Fig. 3 Proposed methodology

ferent users using an android application in both the environment-controlled and uncontrolled, the purpose of which is to understand the differences caused by the dataset during verification by identifying user's position. Collected data may contain some outliers that could be removed by using mean, median, standard deviation, K-mean nearest neighbor and regression, then the classified data is applied to the above-mentioned anomaly detection algorithm, K-means nearest neighbor and Naive Bayes then both the algorithms are compared to identify which algorithm obtain the best performance on basis of EER. The collected data then would be used to identify user position using the three classifiers: random forest algorithm, decision tree algorithm J48 and REP tree algorithm.

4   Conclusion

It may come when a certain user dataset may match with the impostor during verification but because of the difference in position of an impostor and a genuine user which is caused because of difference in values of the already created dataset, it becomes the need to identify every user's position during verification, since using only keystroke dynamics is sufficient because there may be some cases where an impostor may get incorrect validation because of the endurance limit during authentication. Therefore, the proposed model helps to identify user's posture that increases the probability for verification of genuine user. The implementation of this system would be done by the data collected from both controlled and uncontrolled environment to ensure no chance of either FAR or FRR or ERR error metrics. The model not only simplifies to understand the working of the system but also simplifies the necessity to identify the user's posture during authentication.

References

1. Burger, M.P.: Biometric authentication system. Google Patents, vol. 17 (2001)
2. Mizrah, L.L.: Method of one time authentication response to a session-specific challenge indicating a random subset of password or pin character positions, Mar. 16, 2010, uS Patent 7,681,228
3. Karawash, A.: Brute force attack
4. Deo, V., Seidensticker, R.B., Simon, D.R.: Authentication system and method for smart card transactions, Feb. 24, 1998, uS Patent 5,721,781
5. Berislav Kucan, Z.Z., Mirko Zorz: Biometric preference (1998)
6. Das, R.: An introduction to biometrics. Mil. Technol. **29**(7), 20–27 (2005)
7. Maltoni, D., Maio, D., Jain, A.K., Prabhakar, S.: Handbook of Fingerprint Recognition. Springer Science & Business Media (2009)
8. Daugman, J.: How iris recognition works. In The Essential Guide to Image Processing, pp. 715–739. Elsevier, Amsterdam (2009)
9. Saeed, K., Werdoni, M.: A new approach for hand-palm recognition. Enhanced Methods in Computer Security. Biometric and Artificial Intelligence Systems, pp. 185–194. Springer, Berlin (2005)
10. Zhao, W., Krishnaswamy, A., Chellappa, R., Swets, D. L., Weng, J.: Discriminant analysis of principal components for face recognition. In Face Recognition. Springer, pp. 73–85 (1998)
11. Monrose, F., Rubin, D.A.: Keystroke dynamics as a biometric for authentication. Future Gener. Comput. Syst. **16**(4), 351–359 (2000)
12. Karnan, M., Akila, M., Krishnaraj, N.: Biometric personal authentication using keystroke dynamics: a review. Appl. Soft Comput. **11**(2), 1565–1573 (2011)
13. Gaines, R.S., Lisowski, W., Press, S.J., Shapiro, N.: Authentication by keystroke timing: some preliminary results. Tech. Rep, Rand Corp Santa Monica CA (1980)

14. Garcia, J.D.: Personal identification apparatus, Nov. 4, 1986, uS Patent 4,621,334
15. Joyce, R., Gupta, G.: Identity authentication based on keystroke latencies. Commun. ACM **33**(2), 168–176 (1990)
16. Young, R.J., Hammon, W.R.: Method and apparatus for verifying an individual's identity. Google Patents vol. 14 (1989)
17. Obaidat, M.S., Sadoun, B.: Verification of computer users using keystroke dynamics. IEEE Trans. Syst. Man. Cybern. Part B (Cybernetics) **20**(2), 261–269 (1997)
18. Revett, K.: A bioinformatics based approach to behavioural biometrics. In 2007 Frontiers in the Convergence of Bioscience and Information Technologies, pp. 665–670 (2007)

Popularity-Based Detection of Malicious Content in Facebook Using Machine Learning Approach

Somya Ranjan Sahoo and B. B. Gupta

Abstract In this world, people are encircled with various online social networks (OSNs) or media platform, various websites and applications. This brings media contents like texts, audio, and videos in daily basis. People share their current status and moments with their belongings to keep in touch by using these tools and software like Twitter, Facebook, and Instagram. The flow of information available in these social networks attract the cybercriminals who misuse this information to exploit vulnerabilities for their illicit benefits such as stealing personal information, advertising some product, attract victims, and infecting user personal system. In this paper, we proposed a popularity-based method which uses PSO-based feature selections and machine learning classifiers to analyze the characteristics of different features for spammer detection in Facebook. Our detection framework result shows higher rate of detection as compared to other techniques.

Keywords Online social networks · PSO · Facebook · Machine learning

1 Introduction

In day-to-day life of human being, online social network or media become an integral part for sharing of knowledge, thoughts, and personal communication among belongings and friends. Social networks like Facebook, Twitter, Instagram, and other media-related networks are used by teenagers frequently in their work schedule. The popularity of networks leads to get benefited by posting certain advertising, blogs, and posts. Leading industries realize the usefulness of OSN sites for brand management and directly communicating with users for their benefits. Therefore, online social networks are the origin of user's personal information and commercial content

S. R. Sahoo · B. B. Gupta (✉)
Department of Computer Engineering, National Institute of Technology, Kurukshetra, Kurukshetra, Haryana, India
e-mail: gupta.brij@gmail.com

S. R. Sahoo
e-mail: somyaranjan.sahoo@gmail.com

© Springer Nature Singapore Pte Ltd. 2020   163
A. K. Luhach et al. (eds.), *First International Conference on Sustainable Technologies for Computational Intelligence*, Advances in Intelligent Systems and Computing 1045,
https://doi.org/10.1007/978-981-15-0029-9_13

that can be used in many ways. Due to the flooding on content like user information, photos, blogs, and other relevant information, there is the chance for cyberattack and it degrades the performance and user experience and provides negative impact at server-side activities like mining of the data and behavioral analysis of the user. To provide the security and managing the quality of social interaction is an increasing challenge.

At present, several malicious contents spread by malicious users and different threats have been reported [1–4]. One of the most vernacular problems in online social network is social spam bots and spammers. Spammers spread malicious content in the form of link, hashtag, and fraudulent content among legitimate users. Spam are propagated by social spammers in the form of advertising, viruses in the form of link, phishing contents, and form of fake accounts in many pages to attract people. The spam contents are spread in autonomous fashion and act like machine to steal personal information of the user. Most common category of pages that attract the user to visit and spread the content by like and share are Government portal, athletic sites, public figures, TV and shows, artist contents, and various company-related matters. Due to the above concern, it becomes extremely essential to figure a method/framework that identifying the different activities of the user in social platform to detect the spammers and the characteristics. This paper presents an approach which can detect spammers in social network sites like Facebook and the behavior of account based on the popularity-based machine learning model.

The rest of the paper contains the following main contributions. In Sect. 2, we describe the various state-of-the-art techniques for prevention and detection of spammer contents in Facebook. In Sect. 3, we describe the proposed model that detects the spammer in Facebook by analyzing various posts by the user based on collected dataset by our crawler. We studied various features' analysis based on machine learning classification techniques and comparative analysis describes in Sect. 4. In Sect. 5, we conclude our work with future research direction.

2   State-of-the-Art Techniques for Detection and Prevention

Spammer detection in online social network becomes a trending topic in various fields like industries and academicians. Many methods/frameworks are developed to identify and trace spammers on Facebook accounts in current trends, including machine learning-based feature analysis, optimization-based method, and graph-based analysis. Sometimes various researchers and antivirus development companies built various algorithms and software to detect spammers in OSN. The different service providers also inbuilt certain detection technique to control spamming activity accounts in their end. Facebook reports several behavior of the content that defines spamming activity in various forms.

- Sending harmful messages in the form of links including malware and phishing sites.

- For attention toward specific accounts, use follower or following.
- Sending unwanted messages with @ and # to the users.
- Creating fake account to gain the credentials of the user accounts.
- To grab the attention of the user sending repeated messages.

Sami et al. [5] designed a framework to detect the fake accounts in Facebook by applying machine learning technique. The framework trained Bayesian classifier to identify fake accounts by identifying key traits in the user account. Sohrabi et al. [6] proposed one framework that selects the different features of the profile and analyzed the content through supervised and unsupervised learning method to detect spam content in the Facebook platform. The selection of the features are based on the PSO algorithm to analyze the content in the Facebook and generate the specific content. Campos et al. [7] proposed an algorithm to classify the account as a human-made, legitimate robot or a malicious account generated by bot in social network. The algorithm based on the concept of discrete wavelength transforms to identify the writing content in the post. They use two different datasets for their analysis to detect malicious content accounts. Gurumurty et al. [8] proposed a FrAppE tool using SVM machine learning classifier to detect the malicious content present in the account. The tool detects the malicious content that is generated from a source but the source is not the legitimate one. Zhou et al. [9] analyze the spam content in the social network platform based on the different perspectives like viability of the account, different sequence of the transaction, and the correlation between the accounts. The detection rate of spammer in this technique is quite impressive as comparison to the other techniques. Talukder et al. [10] introduced one the technique base on the user questionnaires called 'AbuSniff' to detect and defense Facebook accounts against spammers and abusive friends. They impose certain questions based on the profile activity to detect the account as abuse or not by imposing supervised learning algorithms. Wang et al. [11] introduced Katz similarity-based unsupervised method, semi-supervised method and graph embedding method to detect multiple accounts in OSN. The detection of Facebook accounts content is analyzed in the form of groups. The malicious activity contents that are present in the account can be detected with the cluster form approach. Dewan et al. [12] proposed a technique called Facebook Inspector (FBI) to detect malicious account online. This technique works on empirical finding upon pre-trained model that captures the difference between malicious and benign post. Such techniques are efficient to detect the malicious content and the account which they have seen in the past.

However, none of the techniques classify the different category of post shared by the user in their time line that content malicious link or post. For efficient detection of malicious content in Facebook, our method utilizes various features that are associated with the content. This approach eliminates the dependencies in the post-similarity. The accuracy of detection of malicious content is more as comparison to other detection system.

3   Proposed Method/Model

The overall procedure of our spam detection model based on the popularity of content and PSO-based selection of features described in Fig. 1. We first collected different datasets based on the profile content that shared by the users in Facebook. Then PSO-based extracted features are processed with the data contents. After that, based on the selected features, we process the content with the help of supervised learning to detect the spammer activity. In the following, we introduced our dataset and the extracted features concept and the targeted result what we are looking for.

3.1   Data Collection

For our experiment, we collected data shown in Table 1 from various profiles based on the popularity of the contents like job alert, athletes post, public figure, shows related to TV or stage performances, politician blogs, artistic related contents, various product advertising, and fresh news-related contents based on the popularity. We

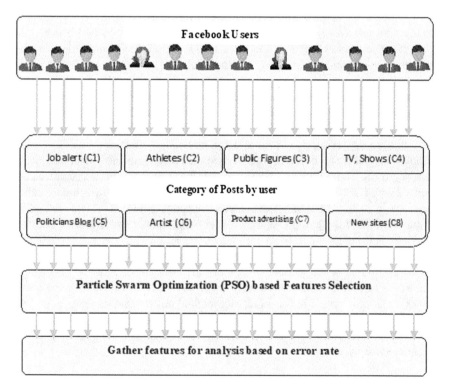

Fig. 1 Proposed model for data collection and feature extraction

Table 1 Collected dataset based on the category

Category of pages	Total number of profile pages	Mutual likes between page contents	Average mutual likes per profile
Politicians blogs	5908	41,729	7
Job alert	14,113	52,310	4
Artist	50,515	819,306	16
New sites	27,917	206,259	7
Product advertising	7057	89,445	12
TV, Shows	3892	17,256	4
Public figures	11,565	67,114	5
Athletes	1386	86,858	62

calculated the popularity index based on Eq. (1).

$$\text{Popularity Score (PS)} = \frac{\text{Number of Likes } (N^{\text{Likes}})}{\text{Number of Followers } (N^{\text{Followers}})} \tag{1}$$

3.2 Preprocessing and Collected Features

To get the resultant more exact, preprocessing is the steps to modeling the data is an inevitable part of data analysis. This process helps to find missing place values and replace those places with some values used in different features and normalized the content by removing useless attributes in the dataset. After preprocess the generated content, we selected certain profile based and content based features for observation and these features are the input parameter of the PSO algorithm. The details of collected features and their uses describes in Table 2.

We gather other profile related information's like profile name, profile image, cover photo, date of join and other relevant information by our crawler. The various content generated based on the features are provided as input parameter to the PSO algorithm for better decision making toward selecting best suitable feature for our operation to detect spammer accounts based on the profile content shown in Table 2. We also calculated fraction of message with URLs, replies, message with spam words etc.

3.3 Feature Selection Based on PSO Algorithm

The feature selection of the proposed system based on the particle swarm optimization (PSO) algorithm is described in algorithm 1. PSO is a self-adaptive stochastic

Table 2 Collected dataset based on the category

Features	Description	% as compared to features	
		Spammer	Legitimate
Friends	The peoples those who are connected with each other directly and their presence is in the list of each other consider a friend in the online social network	Less friends	More friends
Following	It is a feature by which the user can see the contents of a profile without being a friend	More	Less
Followers	If the user allows other users to view their contents that means the user is your follower. The user can see the blogs and post shared by you in their timeline	Less	More
Like	People put their positive opinion regarding a post such as image, video, or certain comments. By this phase, the owner of the account notices the number of friend visible the content in their wall post	Less	More
Comment	It is similar to the feature likes with opinion as an explanation regarding the content. It is visible only to the friends of their account not to others in the network	More	Medium
Reply to comment	Spammer uses this feature to reply a lot to attract the user toward their content further	More	Less
#tag	Facilitate toward identifying a specific subject #tag is used. Spammer uses many number of hashtag in a single post or comments $$F_{\text{fbhashtag}} : \left\{ \sum_{i=1}^{N} \text{msg}_{\text{hashtag}_i} \right\} \ (1) \ (\text{No. of message with one hashtag})$$ $$F_{\text{hashtagfb}} : \left\{ \sum_{i=1}^{N} \text{hashtag}_{\text{msg}_i} \right\} \ (2) \ (\text{No. of hashtag in each message})$$	More	Less
URLs	These are the hyperlinks toward a specific page in the network by redirecting the users. Spammer uses the URL feature to spread malicious content in the network. A message with URL is the high probability of spreading malicious content $$F_{\text{fbURL}} : \left\{ \sum_{i=1}^{N} \text{msg}_{\text{URLs}_i} \right\} \ (3) \ (\text{No. of message}$$ with one URLs) $F_{\text{URLfb}} : \left\{ \sum_{i=1}^{N} \text{URL}_{\text{msg}_i} \right\} \ (4)$ (No. of URL in each message.)	More	Less

(continued)

Table 2 (continued)

Features	Description	% as compared to features	
		Spammer	Legitimate
Share	For sharing the content like photograph, video, and other advertising content in the network user uses the share feature in the Facebook to attract the people toward their accounts	More	Less
Messages with spam words	The content share in the social network for communication and content sharing in the form of text, image, or video is called message. The spammers always share certain content to attract the user toward their account. The spammer always used certain hashtag and URL associated with that content to redirect the user to other pages $F_{\text{fbspamword}} : \left\{ \sum_{i=1}^{N} \text{msg}_{\text{spamword}_i} \right\}$ (5) (No. of spam word in a single message)	More	Less
Stories	It is the visual information in the form of news feed	Less	More
No. of Group joined	For content gathering and sharing, people join different groups on their interest. Spammers joined more group as compared with legitimate one for attracting users toward their accounts	More	Less

optimization technique based on the idea of simulation modeling for searching of foods of a swarm of birds. In this process, each particle in the swarm "i" moved toward the optional point by using addition with a velocity factor with its own value. The velocity is influenced with various factors like different post, URLs, media content, number of hashtag, feedback to the comments, etc. The component called inertial simulates the activity of the features to move toward the same direction used previously. The features are moved around multidimensional search until find the solutions. By using the above discussion, the PSO for feature selection is as follows.

$$V_{ij}^k = w * C^1 * r^1 * \left((P_{\text{best}})_{ij}^{k-1} - X_{ij}^{k-1} \right) + C^2 * r^2 * \left((G_{\text{best}})_i^{k-1} - X_{ij}^{k-1} \right), \quad (2)$$

where

$i = 1, 2, 3, \ldots N_{\text{Attribute}}$
$j = 1, 2, 3, \ldots N_{\text{Particle}}$

The equation for the position update is

$$X_{ij}^k = X_{ij}^{k-1} + V_{ij}^k \tag{3}$$

where

$i = 1, 2, 3, \ldots N_{\text{Attribute}}$
$j = 1, 2, 3, \ldots N_{\text{Particle}}$

And,

K	Number of iteration count
V_{ij}^k	(kth) iteration of attribute (i) of the velocity of particle (j)
X_{ij}^k	(kth) iteration of attribute (i) of the position of particle (j)
W	Weight of inertia
C^1, C^2	Acceleration co efficient
$(P_{\text{best}})_{ij}^k$	Until iteration (k), attribute (i) of the own best position of particle (j)
$(G_{\text{best}})_i^k$	Until iteration (k), attribute (i) of the best particle
$N_{\text{Attribute}}$	Attribute (i) of the best particle
N_{Particle}	Number of particle present in the swarm
r^1, r^2	Uniformly generated distributed random number from the range 0 to 1

Based on the error rate, we selected certain features out of all features for our experimental approach shown in Fig. 2. The selection of the more features leads to a better result in terms of accuracy and error rate but it leads to decrease the time and memory in terms of efficiency. Permutation approach is used after selecting the best suitable features for analysis. The implementation phase uses a random key and a meta-heuristic approach for selecting features based on the permutation. Selecting the features having lower error rate leads to better result with less number of features. We selected twelve numbers of features for our observation to detect malicious accounts based on the pages collected.

Algorithm 1 The multi-objective PSO-based proposed algorithm for feature selection [13].

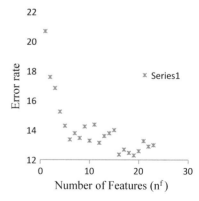

Fig. 2 Feature selection based on error rate

```
data=LoadData();
nx=data.nx;

BestSol=cell(nx,1);
S=cell(nx, 1);
BestCost=zeros(nx,1);
for nf=1:nx
    disp(['Selecting ' num2str(nf) '
    feature(s) ...']);
      results=RunPSO(data,nf);
       disp(' ');
    Best Sol{nf}=results.BestSol;
      S{nf}=BestSol{nf}.Out.S;
BestCost(nf)=BestSol{nf}.Cost;
End
```

4   Experimental Analysis

The proposed framework detects spammer accounts in Facebook by using our dataset shown in Table 1. We selected randomly 200 posts from each categories contains malicious and benign accounts as a total of 1600 profile posts. Out of all the posts, more than 700 contents are spammer contents as per our analysis by using different features selected based on the PSO. We did the experiment using Python for selecting suitable features from all selected features and pass the contents through Weka tool [14], a machine learning platform for detecting spammer content by using some classification techniques. For all types of classification, we use tenfold cross-validation technique to get more suitable result. The result in terms of accuracy is the number of instances correctly predicted from all instances selected for experiment. The selected values for predicted result based the error rate and the different parameter that supported for successful execution.

4.1   Performance of Experiments

The performance reports attained by our framework for detecting spammer contents on our test set are based on the averages on tenfold cross-validation. It shows the content generated in the form of advertising and job alerts having higher percentage of spammer content as comparison to other form of posts. The detection rate gives the higher accuracy as 99.5% by using JRip classifier. The error detection rate and the time to build the model for our dataset also described in Tables 3 and 4. Our feature

Table 3 Different measures of our analysis based on dataset

Measures		TP rate	FP rate	Precision	Recall	F-measure	MCC	ROC area	PRC area	Correctly classified instances (Accuracy)
Classifiers	Random forest	0.9936	0.0032	0.9972	0.9935	0.9956	0.989	1.0000	1.0000	99.4771
	Random tree	0.9961	0.0074	0.9943	0.9966	0.9955	0.989	0.9955	0.9924	99.4771
	Bagging	0.9900	0.0066	0.9955	0.9907	0.9936	0.984	0.9984	0.9973	99.1955
	JRip	0.9951	0.0041	0.9961	0.9954	0.9967	0.990	0.9951	0.9952	99.5173
	J48	0.9933	0.0062	0.9955	0.9935	0.994	0.986	0.9955	0.9945	99.3162
	AdaBoost	0.9863	0.0269	0.9786	0.9865	0.982	0.960	0.9982	0.9961	98.029

Table 4 Error rate and model timing of different classification

Statistics	Classifier	Random forest	Radom tree	Bagging	JRip	J48	AdaBoost
Kappa statistics		0.9895	0.9895	0.9838	0.9903	0.9862	0.9606
Mean absolute error		0.0095	0.0052	0.0148	0.0072	0.0089	0.0248
Root mean square error		0.0652	0.0723	0.0791	0.0699	0.0082	0.1269
Relative absolute error		1.9055	1.0519	2.9704	1.4512	1.7842	4.9952
Root relative square error		13.08	14.50	15.84	14.02	16.515	25.4524
Model building time in second		1.08	0.8	0.37	0.33	0.19	0.28

selection model based on the PSO selected only 9 features for the experimental analysis based on the error rate and popularity of content spread by the user. We observed that, malicious contents are spread in the network by advertising content mostly with 20.33%. The percentages of malicious content based on the different categories are shown in Fig. 3.

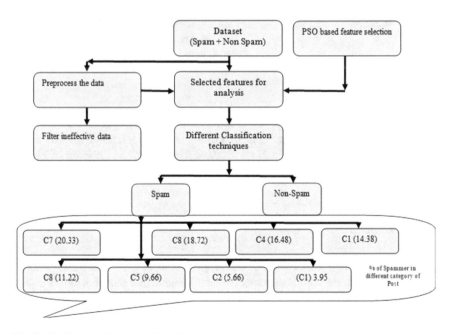

Fig. 3 Machine learning-based detection system

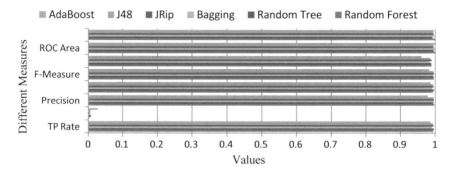

Fig. 4 Different measures based on classification

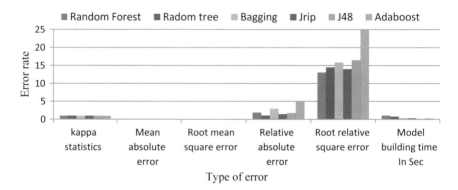

Fig. 5 Error rate based on classifications

As per experimental analysis, our proposed model gives better result with more accuracy for classifying the content as spam and non-spam compared to other methods. The graphical analysis of various classifications' result is depicted in Figs. 4 and 5.

4.2   Comparative Analysis

To detect spammers in various OSNs, our proposed model gathers and analyzes different posts associated with Facebook content. Our detection model achieves higher accuracy as compared to other states-of-the-art techniques described in Table 5.

Table 5 Comparative analysis with existing approaches

S. No	Research article	Dataset used	Accuracy
1.	Automatic detection of cybersecurity-related accounts on online social networks [6]	424 users information	97.17
2.	A feature selection approach to detect spam in the Facebook social network [15]	2 Lakh wall post	91.20
3.	Our proposed model	Different categories of wall posts (More than 5 Lakh)	99.6

5 Conclusion and Future Work

Nowadays, spammer's attacks on OSNs are the serious concerns for the users to protect their content. The specific features of social networking sites and the behavior of cybercriminals makes attract to classify malicious users. This paper introduced a spammer detection mechanism in Facebook by using machine learning-based model. Our research based on feature selection by using PSO algorithm from set of features and some supervised learning methods. The selected features process through different classifiers that give better accuracy in terms of ROC and TP rate. Our detection system also classifies the percentage of malicious content in each category of post. The best advantage of our overall research comprises in the step-down of various features and selected some specific features based on the error rate for detecting spammer activity. Our experimental approach for various model designs takes less time as compared to other researches. The result shows by using tenfold cross-validation, the popularity class prediction is more than 99% and the detection rate also. To achieve the detection rate more precision, PSO algorithm-based feature selection attain a very good selection rate.

Acknowledgements This research work is funded by Ministry of Electronics & Information Technology, Government of India under YFRF scheme and under the project Visvesvaraya Ph.D. Scheme which is implemented by Digital India Corporation.

References

1. Adewole, K.S., Anuar, N.B., Kamsin, A., Varathan, K.D., Razak, S.A.: Malicious accounts: dark of the social networks. J. Netw. Comput. Appl. **79**, 41–67 (2017)
2. Gupta, B.B. (ed.): Computer and Cyber Security: Principles, Algorithm, Applications, and Perspectives, pp. 666. CRC Press, Taylor & Francis, (2018)
3. Zhang, Z., Gupta, B.B.: Social media security and trustworthiness: overview and new direction. Future Gener. Comput. Syst. **86**, 914–925 (2018)
4. Gupta, B., Agrawal, D.P., Yamaguchi, S. (eds.): Handbook of Research on Modern Cryptographic Solutions for Computer and Cyber Security, p. 589. IGI Global, USA (2016)

5. Sami, M., Memon, S., Baloch, J., Bhatti, S.: An automated framework for finding fake accounts on Facebook. Int. J. Adv. Stud. Comput. Sci. Eng. **7**(2), 8–16 (2018)
6. Sohrabi, M.K., Karimi, F.: A feature selection approach to detect spam in the Facebook social network. Arab. J. Sci. Eng. **43**(2), 949–958 (2018)
7. Campos, G.F., Tavares, G.M., Igawa, R.A., Guido, R.C.: Detection of human, legitimate bot, and malicious bot in online social networks based on wavelets. ACM Trans. Multimed. Comput. Commun. Appl. (TOMM) **14**(1), 26 (2018)
8. Gurumurthy, S., Sushama, C., Ramu, M., Nikhitha, K.S.: Design and implementation of intelligent system to detect malicious Facebook posts using support vector machine (SVM). In: Soft Computing and Medical Bioinformatics (pp. 17–24). Springer, Singapore (2019)
9. Zhou, Y., Wang, X., Zhang, J., Zhang, P., Liu, L., Jin, H., Jin, H.: Analyzing and detecting money-laundering accounts in online social networks. IEEE Netw. **32**(3), 115–121 (2018)
10. Talukder, S., Carbunar, B.: AbuSniff: automatic detection and defenses against abusive Facebook friends (2018). arXiv preprint arXiv:1804.10159
11. Wang, X., Lai, C.M., Hong, Y., Hsieh, C. J., Wu, S.F.: Multiple accounts detection on Facebook using semi-supervised learning on graphs (2018). arXiv preprint arXiv:1801.09838
12. Dewan, P., Kumaraguru, P.: Facebook Inspector (FbI): Towards automatic real-time detection of malicious content on Facebook. Soc. Netw. Analy. Min. **7**(1), 15 (2017)
13. Sohrabi, M.K., Karimi, F.: A clustering based feature selection approach to detect spam in social networks. Int. J. Inf. Commun. Technol. Res. **7**(4), 27–33 (2015)
14. WEKA tool, http://www.cs.waikato.ac.nz/ml/weka
15. Aslan, Ç.B., Sağlam, R.B., Li, S. (2018). Automatic detection of cyber security related accounts on online social networks: Twitter as an example

Computational Management of Alignment of Multiple Protein Sequences Using ClustalW

Riddhi Sharma and Sanjay Kumar Dubey

Abstract The astounding and multidisciplinary field of bioinformatics has had its origin in the year 1900, a decade before the concept of DNA sequencing became feasible and applicable to researchers. The era of the "new biology" that has a subtle link with the modern-day computers emerged and was accompanied by the birth as well as the development of other streams of sciences such as bioinformatics (it conceptualizes biology in reference to molecules and then applies "informatics techniques" (derived from computer science in this paper) to organize and understand the information that is related to these molecules. The several categories in bioinformatics focus on genomics (gene study), proteomics (protein study: which will be discussed in this paper), metabolomics (study of metabolic pathway within a cell), etc. and computational biology. Bioinformatics focuses on the application of IT and statistics. It is very significant to note here that bioinformatics does not have to do anything in the real world. Rather, it creates a virtual image of our living system having computerized biological knowledge, sequence, and structural information fed in the CPU and by using virtual-based techniques like mathematical modeling and through stimulation that helps in basic principles leading to its practical application. In bioinformatics, a sequence is subjected (or susceptible towards) to certain range of analytical methods in order to understand its function, or features, or structure, or evolution. Analysis of sequences can be very efficiently used to assign roles to certain genes and proteins as a result of the study of the various similarities between the compared sequences. In this research paper, comparison and alignment of multiple protein sequences are provided. It will help to generate a phylogenetic tree and also draw the relationships between the aligned sequences.

Keywords Multiple sequence alignment · Dynamic programming · Pairwise alignment · ClustalW · Phylogenetic tree

R. Sharma (✉) · S. K. Dubey
Amity University, Noida, Uttar Pradesh, India
e-mail: riddhisharma2602@gmail.com

S. K. Dubey
e-mail: skdubey1@amity.edu

© Springer Nature Singapore Pte Ltd. 2020
A. K. Luhach et al. (eds.), *First International Conference on Sustainable Technologies for Computational Intelligence*, Advances in Intelligent Systems and Computing 1045, https://doi.org/10.1007/978-981-15-0029-9_14

1 Introduction

If focused on one of the cornerstones of modern bioinformatics, then in their finest vision, it is the comparison or going by the terms of bioinformatics, the alignment of multiple protein sequences which seems significant, as it allows in analyzing and determining evolutionary relationships through the concept of homology between sequences. These, in turn, help in building the phylogenetic trees (the details of which will be mentioned later). Constructing a precise multiple sequence alignment (MSA) for proteins remains a difficult task since the required computational complexities emerge and extend rapidly with the growing sequence length and the number. The multiple sequence alignment generates a brief but comprehensive, information-rich summary of the sequence data (information carried by a protein/gene in computer language). The sequence data can be a primary data, a secondary or a tertiary data. In this paper, we will be working on the primary/basic/raw sequence data. MSA is used to illustrate the similarity and dissimilarity among a group of sequences, and sometimes the alignments are treated as models that are used to test an earlier hypothesis. With the support of protein multiple sequence alignments or MSAs, biologists are easily comprehending and working on the various patterns of sequences of biomolecules which were conserved through evolution for so many years and also the ancestral relationships between different organisms. Now, a program that is widely used for the global MSAs of sequences is derived from the Clustal series (ClustalW (that will be discussed in the paper later), ClustalV, ClustalOmega) of programs. A scientist "Des Higgins" in the year 1988 wrote his first program which was designed in a way that it specifically worked efficiently only on PCs. That time, it had a fragile computing power in comparison with today's standards. The one designed had an advantage that it could combine a "memory-efficient" algorithm of dynamic programming along with the pairwise alignment strategy. ClustalW's Alternative 1: If ever, arises a situation where a 3rd protein sequence is aligned with the first two, then whenever a gap is to be introduced for the improvement of the alignment, each one of these two entities have to be treated as two sequences as individuals. ClustalW's Alternative 2: And if otherwise, two separate (individual) sequences have to be shown that they align together, then it should be noted that the first PA is placed to one side and only then the PA of the other two should be carried out.

The most closely related sequences or the most distantly related sequences are categorized in order to align them. This is done directly from the guide tree and is called sequence weighing. In terms of group of secretly related sequences that contain all similar type of information and are therefore down weighed. Therefore, divergent sequences have high weights. The most divergent sequences are the most complex, and hence they are aligned after the alignment of the easiest closely related sequences. Condition: Weights < 1, ClustalW thereby avoids any sort of overweight of inessential data from existing close duplicate sequences.

2   Literature Review

Nowadays, the use of multiple sequence alignment in the multifaceted field of bioinformatics is widely being increased in various organizational sectors. This section encompasses the work done related to bioinformatics, MSA and the keywords related to it in bioinformatics, and the newer issues and challenges faced by researchers during their work in the 4 years from 2004 to 2018.

Bioinformatics is a field that is pledged to the analysis and interpretation of the biotic data (i.e., the protein sequences in this paper) using computational techniques. Bioinformatics and computational biology are the topics that require skills from a variety of other significantly involved fields to enable the amassing and stowing, handling and analyzing, interpreting and unfurling and unrolling of biological information. It requires the usage of top notch as well as neoteric computers and inventive software tools to superintend and organize sizeable quantities of proteins and nucleic acids. Requirement of an intelligent system to be able to classify the incoming protein into an isolated and discrete super family which happens through using various algorithms, classification programs under the sub-topic multiple sequence alignment [1]. A multiple sequence alignment (MSA) gives a cognizance to the three "ESF" relationships among the closely related proteomic sequences. Using the operators of a genetic algorithm, it is easy to locate the optimized alignment after a string of iterations of this algorithm. Thus, a protein alignment using stochastic algorithm (PASA) has been unfolded. The popular ClustalW uses a progressive alignment strategy in order to align protein sequences. This approach focuses on a multiple alignment that builds up "progressively" (as it goes by the name) from a series of pairwise alignments by utilizing a phylogenetic tree as the reference in this method. The alignments of those sequences that are closely related ones are achieved in the very beginning, followed by the ones that are distantly connected. However, the method has its own limitations of an "early in the process" alignment-related errors [2]. Genetic algorithms (GAs) are algorithms based on the ideas of natural selection and genetics and on the evolution of organisms. Genetic algorithm is a product of machine learning which derives its character as a metaphor of the processes of evolution of nature. GAs represent an intelligent exploitation of a spontaneous search and heuristic process used to solve optimization-related problems [3]. In bioinformatics, comparison among the protein or other molecular sequences to identify regions of similarity is called sequence alignment. There are two different well-known techniques for sequence alignment, i.e., global alignment [every element of query sequence is aligned with same length compared sequence(s)] and local alignment (with an expectation of similarity and when the sequences to be aligned are of different lengths and are with no similarities), and sequence alignment can further be categorized as pair wise or multiple. Pair wise sequence alignment identifies similarities between two sequences at a particular time [4]. Closer to simple alignments that specifically depict a similar relation between 2 sequences), the dynamic programming approach is an employee in the sector of global alignment. In this approach, each possible and formed pair goes through a series of punctuation by a weighted sum of pairs, along with an addition of values that display similarities. Besides that, substitute processes were produced to escalate the calculative analysis, among which we are able to summit the progressive method, iterative methods as well as the hidden Markov models [5]. The series of Clustal programs are widely for their construction of trees. The vogue of the programs depends on some important factors that encompass not only the correctness of the results, but also the user friendly attribute of the programs. Although, Clustal was originally seen to be employed on a local computer, but a number of Web servers, notably at the E.B.I. (European Bioinformatics Institute) have been set up for distributed access [6]. ClustalW users are allowed to choose alignment methods which can be both fast/approximate and slow/accurate and accurate, choose initial gap openings and extension penalties, to determine the "weight matrix" which is a scoring table describing the similarity of each amino acid. The types of weight matrices for proteins used in ClustalW are: (1) PAM (Point Accepted Mutation matrix), (2) BLOSUM (BLOck Substitution matrix). ClustalW's "User-supplied" values: While using ClustalW, 2 penalties can be set by the user (though ClustalW offers its own values and there are default values, it is practicable to alter these). Two types of penalties: GOP—"Gap Opening Penalty" can be defined as the cost to open a gap in an alignment and GEP—"Gap Extension Penalty" is the cost of a gap's extension. GEP gives the cost of items in the gap. The more divergent/less closely related a sequence is, more problematic it is to align the same appropriately. In case we delay to add these most divergent sequences until the alignment process comes to its end, we may have a better opportunity to place the gaps in the process of multiple sequence alignment. Users have the liberty to choose the "cut-off identity percentage" for delaying the addition of the unwanted divergent sequences. ClustalW's default value of ignoring the addition is 40% [7]. SeaView is an editor, again employed at multiple sequence alignment that is highly skillful to add or delete one or many gaps in 1 and multiple sequences simultaneously. SeaView is an aid for intermittent users of the tree reformation as it liberates them from being defied to many technical details [8]. A progressive alignment method utilizes the "Needleman and Wunsch pairwise alignment algorithm" to build an evolutionary tree. The sequences happen to share a common ancestor, and the trees are built from "difference matrices" (derived directly from the multiple alignments). The various uncertainties and complexities concerning both the topology and branch lengths in the tree happen to be common, and huge levels of effort act as an expense towards locating the "best tree" [9]. Kalign, as accurate as the best other methods on little arrangements, are more faultless when sequences are large and are distantly related. In our comparisons, Kalign happens to be about 10 times faster than ClustalW method and depending on the alignment size, around 50 times faster than popular iterative methods. A very well suited alignment method for the increasingly significant task of aligning n numbers of sequences [10]. A software for comparative sequence analysis is ought to contain swift and speedy computational algorithms and useful statistical methods modus operandi, and have an extensive adaptable interface to allow researchers working at the front foot of sequence data generation to explore basic sequence attributes. MEGA (Molecular Evolutionary Genetics Analysis software), developed in the early 1990s, is one such method. The latest MEGA3 escalates the functionalities of MEGA2 by adding sequence data alignment as well as assembly features, along with other hard core advancements. However, MEGA3 is not intended to be a plethora of all evolutionary analysis methods. Rather, it focuses more on the exploration of sequence, constructing phylogenetic trees especially for large-scale data set [11]. Alignment Viewer is another lightweight online viewer for MSAs in a wide range that focuses on usability and performance. The architecture of Alignment Viewer allows its use without software installation and without an Internet connection. This is also one of its advantages [12]. To be able to know that the phylogenetic tree shows an accurate evolutionary relationship between species is a very crucial thing for testing. Now, selecting which sequences to align in comparing the ClustalW's original and newer version is the first step. And after the tree is constructed, there is a comparison with the known tree [13].

3   Review Methodology

Current review methodology is in accordance with the literature review that looks forward to various research questions in order to examine and identify all the high-quality research evidence necessary for the research. The aim of this research paper is to present a very fair experimental analysis of research topic by using an adaptable methodology. While going through the research papers published from the year 2004 to 2018, 31 papers were initially taken into consideration. Out of which, 12 papers appeared to be in sync with the area of concern and interest. Hence, 19 papers were excluded. Papers before 2004 are not considered due to not much work was there in this area.

4   Experimental Work and Analysis

In this section, ClustalW software for multiply aligning the protein sequences is used. Also, the aligned sequences' phylogenetic tree has also been fetched through the same software. In order to construct the phylogenetic tree, the interpreted high-level programming language Python is used. One of the key tasks in the process of carrying out the multiple sequence alignment is the construction of a phylogenetic tree.

4.1   Phylogenetic Tree

A branching pattern to reflect the process of descent with modification of a process central to the Darwin's theory of evolution ('natural selection' theory which requires inherited variation in a trait and differential survival as well as reproduction) is a phylogenetic tree (Fig. 1).

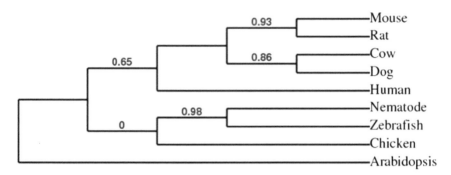

Fig. 1 Nine species aligned by the "ClustalW" program and their phylogenetic tree

Various advances in the field of genomics have been seen to enrich the range and the extent of computational methods available for assisting experts and researchers in building these trees. Among some other methods, these phylogenetic trees can be built by simply comparing genetic sequences of the required various species. The present implementations and methods to carry out multiple sequence alignment have limitations that unfortunately prevent them from constructing exact, accurate phylogenetic trees, especially whenever sequences with low similarity (divergent sequences) are contained in the dataset.

The following mentions the construction of a phylogenetic tree where we have a set of six sequences. These are the equivalent stretch taken from the equivalent gene from each of the six species and we are going to refer to these species as:

A: ATCGTGGTACTG
B: CCGGAGAACTAG
C: AACGTGCTACTG
D: ATGGTGAAAGTG
E: CCGGAAAACTTG
F: TGGCCCTGTATC

Now, the first step in this process is to align the sequences with one another.

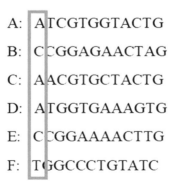

A: ATCGTGGTACTG

B: CCGGAGAACTAG

C: AACGTGCTACTG

D: ATGGTGAAAGTG

E: CCGGAAAACTTG

F: TGGCCCTGTATC

Now with the sequences aligned, we can compare each sequence with one another.

We further compare the sequences going by the order. For instance, we take sequences A and B and observe that there are in total 3 similarities and 9 differences between the two. And we create a table and set value "9" for A and B. Similarly, we take A and C and find that they comparatively have more similarities than sequences A and B, i.e., they have 10 similarities in total and only 2 that are different. We, therefore, add this value "2" for A and C in the table. We complete the table by comparing all the sequences with each other and by finding out the similarities and differences among them. And add the values simultaneously in the table and complete it for sequences starting from A to F. The desired matrix is shown in Fig. 2.

With the first table completed, we can now move to the second step which is to use this created table to identify the sequences with the fewest difference between them. We will infer that these are actually the sequences which are the most closely related sequences to one another. In our table, we observe that there are only two pairs of sequences who have the fewest differences between them, i.e., A and C and B and E. With this information, we can draw our first groupings on our phylogenetic tree (Fig. 3).

```
[Running] python "d:\dummy\aa.py"
    A    B    C    D    E    F
--  ---  ---  ---  ---  ---  ---
A        9    2    4    9    10
B             9    6    2    10
C                  5    9    10
D                       6    10
E                            10
F
```

Fig. 2 Matrix to show the distance

Fig. 3 Phylogenetic tree of the first grouping

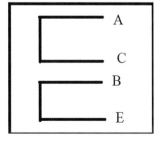

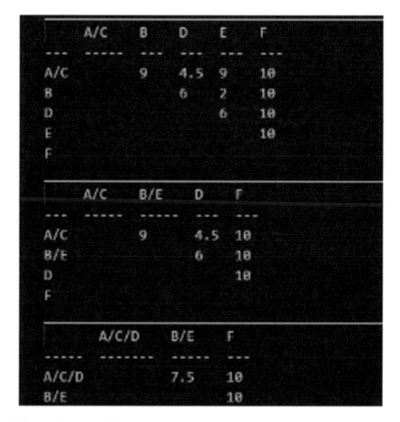

Fig. 4 Grouped sequences distance matrix

We now need to rework on our table with the grouped sequences combined together rather than combining two individual sequences on the table (e.g., A/C). Hence, we get the next three matrices in Fig. 4.

Figure 5 shows the second, third, and the fourth tables that have been constructed through repeating the process taking into consideration the similarities and differences among the sequences. Hence, the tables formed are used to create the groupings of the same protein sequences which escalate a step towards the construction of the phylogenetic or the neighbor joining tree. The final figure of the phylogenetic tree of all the six protein sequences is shown in Fig. 5.

5 Result

The phylogenetic tree in Fig. 4 that shows 10 sequences is shown in Fig. 5.

This phylogenetic tree of 10 protein sequences is obtained after the experimental work using ClustalW. This tree represents the various relationships between distantly

Phylogenetic Tree

This is a Neighbour-joining tree without distance corrections.

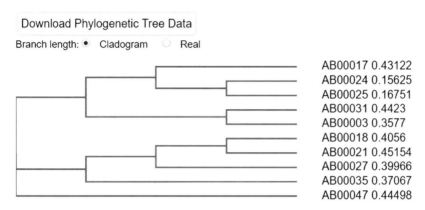

Fig. 5 Phylogenetic tree result of ClustalW

or closely related species and the final phylogenetic tree that we get by repeating the step for 6 protein sequences are shown as Fig. 6.

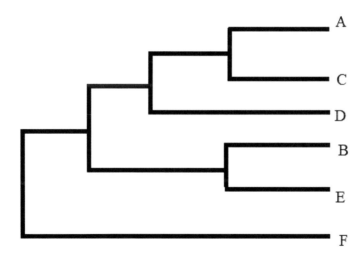

Fig. 6 Phylogenetic tree of six sequences

6   Conclusion and Future Scope

This paper presents the comparative analysis of various protein sequences and displays the multiple sequence alignment and construction of phylogenetic tree of 10 protein sequences. The step-wise analysis of six protein sequences, namely A, B, C, D, E, F, is also shown in this paper. The implementation of high-level, general purpose language, Python, can also be seen via the output tables. The crux of this paper was to generate a phylogenetic tree, a hypothesis and a structural and functional evolutionary descent of various organisms having common ancestors and multiply aligning a given set of sequences. The ClustalW series involving Clustal Omega and ClustaX will be taken into consideration in the coming future. Also, the various applications of the phylogenetic tree and the errors and approximations in the process of multiply aligning a sequence will be looked upon. We believe that there is a scope to further improve the proposed algorithms, there is a possibility to discover new weight matrix which will indeed make the results more accurate and reliable. Also, topics such as dynamic programming and pairwise alignment can be looked upon using an expanded and extended platform which could be a modified version of ClustalW itself. Furthermore, there is a scope to enhance the performance of this alignment by parallelizing it and improving the comparison methods used in ClustalW.

References

1. Vijayarani, S., & Deepa, M.S.: Protein sequence classification in data mining—a study. Int. J. Emerg. Technol. Comput. Sci. Electron. (IJETCSE) **23**(7) (2014)
2. Behera, N., Jeevitesh, M.S., Jose, J., Kant, K., Dey, A., Mazher, J.: Higher accuracy protein multiple sequence alignments by genetic algorithm. Procedia Comput. Sci. **108**, 1135–1144 (2017)
3. Rout, S.B., Dehury, S., Mishra, B.S.P.: Protein structure prediction using genetic algorithm. Int. J. Comput. Sci. Mobile Comput. **2**(6), 187–192 (2013)
4. Gupta, O.P.: Study and analysis of various bioinformatics applications using protein BLAST: an overview. Adv. Comput. Sci. Technol. **10**(8), 2587–2601 (2017)
5. Diniz, W.J.S., Canduri, F.: Bioinformatics: an overview and its applications. Genet. Mol. Res. **16**(1) (2017)
6. Chenna, R., Sugawara, H., Koike, T., Lopez, R., Gibson, T.J., Higgins, D.G., Thompson, J.D.: Multiple sequence alignment with the Clustal series of programs. Nucl. Acids Res. **31**(13), 3497–3500 (2003)
7. Luthy, R., Xenarios, I., & Bucher, P.: Protein Sci. **3**, 139–146 (1994), Russell, R.B., Barton, G.J.: Proteins **14**, 309–323 (1992)
8. Gouy, M., Guindon, S., Gascuel, O.: SeaView version 4: a multiplatform graphical user interface for sequence alignment and phylogenetic tree building. Mol. Biol. Evol. **27**(2), 221–224 (2009)
9. Feng, D.F., Doolittle, R.F.: Progressive sequence alignment as a pre requisite to correct phylogenetic trees. J. Mol. Evol. **25**(4), 351–360 (1987)
10. Lassmann, T., Sonnhammer, E.L.: Kalign—an accurate and fast multiple sequence alignment algorithm. BMC Bioinform. **6**(1), 298 (2005)
11. Kumar, S., Tamura, K., Nei, M.: MEGA3: integrated software for molecular evolutionary genetics analysis and sequence alignment. Brief. Bioinform. **5**(2), 150–163 (2004)
12. Reguant, R., Antipin, Y., Sheridan, R., Luna, A., Sander, C.: Alignment Viewer: Sequence Analysis of Large Protein Families. bioRxiv, 269720 (2018)
13. Jeanmougin, F., Thompson, J.D., Guoy, M., Higgins, D.G., Gibson, T.J.: Multiple sequence alignment with Clustal X. Trends Biochem. Sci. 403–405 (1998)

An Investigation of Simulation Tools and Techniques Based on Traffic and Network

Vijander Singh and Surendra Singh Choudhary

Abstract Researchers are using simulation tools since several years which are very important tools for all experimental fields such as computer science, mechanical engineering, electronics and electrical engineering even for medical science. Simulation has been used during the system design phase which provides a feasible, economic, easy to use and portable platform for the system design. The portable or hands-on prototype or model of any system is called simulation. This paper reconfigures as well as explores the model, which is most common not feasible in real or quite costly or unreasonable, to make it more perfect system. Working model of the system can be studied, and thereafter, characteristics related to either the real system's responses or actual system's subsystem can be derived in form of a simulator. For example, after focusing at the validity and performance of the network the outcome can be analyzed prior to implement and simulated as the real scenario. Along with this, networking technologies related to simulation reduce the cost and time of using the natural system.

Keywords VANET · DTN · Mobility · Node · Simulation · Ferry

1 Introduction

Network simulators are becoming a convenient tool for newly developed solutions of network problems. "Vehicular ad hoc networks" (VANETs) can be established when the vehicles are implemented with the devices which are capable in short-range types of wireless communication. A vehicular mobility model and a network simulation are required in accurate simulation of "VANETs" which is a challenging task. To receive exact and real-life results with cent percent accuracy, the joint utilization of

V. Singh (✉)
Manipal University, Jaipur, Rajasthan, India
e-mail: vijan2005@gmail.com

S. S. Choudhary
Sri Balaji College of Engineering & Technology, Jaipur, India
e-mail: ssc_jaipur@yahoo.com55

© Springer Nature Singapore Pte Ltd. 2020
A. K. Luhach et al. (eds.), *First International Conference on Sustainable Technologies for Computational Intelligence*, Advances in Intelligent Systems and Computing 1045,
https://doi.org/10.1007/978-981-15-0029-9_15

both simulators with powerful feedback between them is highly recommended. It is required to use two separate simulators one for generating traffic (traffic simulator) and later is for simulating the scenario (network simulator) in order to analyze and define the routings in VANETs. This paper presents various possibilities for installation and working process of both the simulators.

2 Traffic Simulator

Traffic simulation is the transportation system of mathematical modeling as a grid system of downtown, a roundabout, freeway junctions, major routes and so on. It helps in the application software of computer to provide it better plan, proper design and transportation system's operation. These models are helpful in microscopic, a macroscopic as well as in mesoscopic point of view.

There are many traffic simulators available in the field of networking for research purpose. Some of them are quite popular for the researchers like VanetMobiSim, CanuMobiSim, SUMO/MOVE, NCTUns, and TraNs. Among these, VanetMobiSim has the strongest strength because it provides support in models of macro-mobility and micro-mobility. VanetMobiSim is an expansion for the "CanuMobiSim," which is a framework with flexibility and useful for "traffic simulation."

2.1 CanuMobiSim

"CANU" is referred as "Communication in Ad hoc Networks for Ubiquitous Computing" [1] and simulator for the mobile scenario which is based on Java programming language with the graphical user interface. The project of "CanuMobiSim" was begun at the University of Stuttgart in Germany [2]. Numerous mobility models can be generated with help of this tool such as fluid traffic, pedestrian, graph walk, pedestrian, fluid walk, smooth mobility model, and mobility models based on activity. Those patterns which are drawn by "CanuMobiSim" are not generated by self but rather derived from a graph of Markov. "CanuMobiSim" does not have the capacity to create random graphs, and it also excludes simulation's obstacles of wireless network.

2.2 VanetMobiSim

"VanetMobiSim" (Vehicular Ad hoc Network Mobility Simulator) [3] is an expansion to "CanuMobiSim." Region of "CanuMobiSim" is limited at just some particular regions; details with high levels in specific scenarios were not possible to produce by "CanuMobiSim." Subsequently, "CanuMobiSim" gets expanded to accomplish realism with high level in "VanetMobiSim's" form. In the modeling of "VanetMo-biSim," "car-to-car" as well as "car-to-infrastructure" relationships are included. In this manner, the traffic signs, signal lights and macro-mobility based on activity with human mobility dynamic's support are combined. Street topologies from TIGER, GDF, arbitrary and custom topologies can be extracted. It is allowed by "VanetMo-biSim," to generate a trip on the basis of users assumption/activity based, and also, the path can be configured by users from the beginning point to the end point based on "Dijkstra's algorithm," shortest road speed or shortest density speed. A parser is being contained by the "VanetMobiSim" for extraction of the topologies between GDF, TIGER or cluster Voronoi's graphs for the use of NS.

The "Vehicular Ad hoc Network Mobility Simulator" (VanetMobiSim) is exten-sion set to "CanuMobiSim," a framework for user mobility modeling and "CANU" uses it [1]. A number of mobility models and parsers are included in the frame-work for geographic data sources which are in different patterns and in visualization module. The pluggable module concept is used in the framework so that it is easily extensible. "VanetMobiSim" provides the set of extensions, and it consists of mainly on a "vehicular spatial model" model.

Spatial elements are used in making of a "vehicular spatial model," like traffic signal lights or multi-path streets, their attributes as well as their relationship linking of these spatial elements with described vehicular areas. This model can be described in four distinct ways:

- "User-defined graph"—In it, user characterizes set of vertices and edges which is a backbone for "vehicular spatial model."
- "Geographic Data Files" (GDF)—The GDF files provide backbone data.
- "TIGER/Line Files"—It is similar as previous one, yet taking into account it is from US Census Bureau.
- "Random"—The Voronoi tessellations are used in generating the backbone, ran-domly.

After the spatial model, any of these methods required to be loaded, as all the data which describe the topology, is controlled by it. Then, vehicular specific spatial elements are added by it, like multiple-lane roads and multiple-flow roads, signage of stop and traffic signal. There are two vehicular-oriented mobility models, which are in support of mobility models of microscopic level:

- "Intelligent Driving Model with Intersection Management" ("IDM_IM"), per-fectly describes "car-to-car" and management of intersection.
- "Intelligent Driving Model with Lane Changing" ("IDM_LC"), it is an "overtaking model" and to manage lanes changes and velocity and retardation of a vehicle, it gets interacted with IDM_IM [4].

"VanetMobiSim" offers so many possibilities and features to create realistic sce-narios. Besides that, simulation scenarios for "VanetMobiSim" are defined in XML format using tags, making scenario configuration easier and in a more handy way. Definitely, "VanetMobiSim" is quite more appropriate in order to generate scenarios for VANETs than other MANETs mobility pattern generators such as CityMob [5] and BonnMotion [6].

3 Network Simulator

Network simulator is the technique in which behavior of network will be models by the program either through the calculation of interaction among the entities of a network like host router, data link, packet and so on with help of formulas based on mathematics or production network can actually capture and then play back the observations.

The main target of Network Simulator-2 (known as "NS-2") is networking research, and it is a distinct event simulator for the networking. NS gives a packet-level simulation besides so many protocols, it supports few protocols of transport, a few multicast forms, wired networking, several protocols of "ad hoc" routing as well as information broadcasting, satellites, etc. [7]. It also has the possibility to utilize mobile nodes. These nodes and their mobility can be defined either straightforwardly in the simulation file or by utilizing a trace file of mobility. "VanetMobiSim" generates the trace file, in research work. In 1989, Network Simulator starts as REAL NS's variant, and it has developed generously in the course of recent years. DARPA supported Network Simulator development with help of VINT in 1995 [8], a project of collaboration in LBNL ("Lawrence Berkeley National Laboratory"), Xerox PARC ("Xerox Palo Alto Research Center"), UCB ("University of California, Berkeley") and USC/ISINS ("University of Southern California's Information Sciences Institute").

NS-2 got built by utilizing the programming language C++, and it also gives an interface for simulation within OTcl. It has an object-oriented dialect of Tool Command Language. A network topology [9] is being described by the user with help of writing scripts of Object-oriented Tool Command Language, and then, main Network Simulator simulates the program that topology with special parameters. Moreover, NS-2 is easily extensible since the simulation kernel source code is available, which implements new routing protocols, propagation models, etc. and uses them in our simulations.

4 Languages and Files Used

The following languages and files can be used in the simulation process of research work:

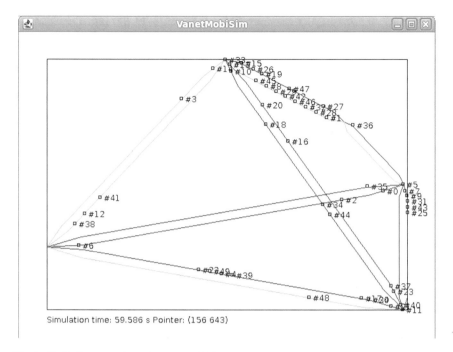

Fig. 1 XML file output through VanetMobiSim

4.1 XML and Trace File

According to the research methodology, simulators are different at the microscopic level, and this is main criteria for selecting a vehicular mobility simulation. The input to VanetMobiSim is an XML configuration file [10]. The VanetMobiSim page has explained about how to define vehicular mobility model in VanetMobiSim. In a practical vehicular movement design, it is important to characterize all the attributes like the vehicles velocity, traffic light, number of lanes, trip motion, and road topology, etc. in XML file. After defining a mobility scenario in an XML file, launching the VanetMobiSim framework is necessary in order to produce a node mobility trace file in NS-2 format. The output of XML file at particular simulation time and trace file examples is shown in Figs. 1 and 2.

4.2 TCL Language

Dr. Ousterhout J. developed "TCL" language [11] at the University of California, Berkeley, and it is a translated script language. The file is a representation of scripting which is useful in developing and coding of networking situations as per the

```
$ns_ at 7.2 "$node_(1) setdest 529.9359682970291 577.2057794302418
3.0594652"
$ns_ at 7.2 "$node_(2) setdest 224.70382527838822 334.4664168151116
0.7206531"
$ns_ at 7.2 "$node_(3) setdest 211.4378176658867 308.19806297889687
3.9231954"
$ns_ at 7.2 "$node_(4) setdest 755.0344180052855 202.27600394924556
4.3799443"
$ns_ at 7.2 "$node_(5) setdest 277.13061372576084 305.758546835616
4.3769007"
```

Fig. 2 Trace file in NS-2 format through VanetMobiSim

requirement of "vehicular ad hoc network" traffic flow on the road and is scripted for creating and connecting applicable file.

These situations are based on different parameters and their settings of generated means of transportation including mobility, safety and likewise constraints as discussed in last chapter [12]. The on-demand routing protocols require a TCL file as a input, along with particular traffic and movement file which initialize the traffic pattern and NS-2 simulate in same manner. Finally, as a resultant, it generates two files—Trace files (*.tr) as the outputs and Network Animator File (*.nam).

4.3 AWK Language

It is a tool useful in data extraction as well as reporting, it uses "data-driven scripting language," and its purpose is to produce formatted reports. In other words, AWK is an excellent filter and files of text processor. Created in 1970s in Bell Labs, and the name is derived from its authors family names "Alfred Aho, Peter Weinberger and Brian Kernighan" [13].

A file is dealt as a record of sequence, and each line is recorded by default. Every line is separated into a field's sequence, by doing this the first field in a line will be referred as the first word, the second field as the second word, etc. [14]. The AWK reads the input at a particular time line by line only. Then for each pattern in the program, a line is scanned and every matching pattern, the related action is executed.

It is easy to use AWK as compared to any other usual programming languages. It is considered as a pseudo "C" interpreter because it understands arithmetic operators in a manner as "C" does. In AWK functions for "string manipulation" are also available, using these function particular strings can be searched and output will get modified. To utilize a library function in a particular from a typed program on the command line [15], represented by the

awk [options]-f source-file

4.4 Trace and NAM File

The trace file comprises all data, for example, number of packets, which are sent, number of dropped and received packets and the sequence number, type, size of the packet and so on. This file is basically available in text format and could be called simulation's log file including all information in the format of logged in the column showing in Fig. 3, and Bentham Science does not process figures submitted in GIF format.

All the operations which are performed at the time of simulation including the positioning information and graphical information as well as the information about defined parameters are comprised in the NAM file.

NS file's Operation component call or execute the "NAM" file using built-in **nam** command. Figure 4 shows the output example of this file.

```
s -t 1.835637580 -Hs 2 -Hd -2 -Ni 2 -Nx 225.05 -Ny 334.95 -Nz 0.00 -Ne -
1.000000 -Nl RTR -Nw --- -Ma 0 -Md 0 -Ms 0 -Mt 0 -Is 2.255 -Id -1.255 -It
AODV -Il 48 -If 0 -Ii 0 -Iv 30 -P aodv -Pt 0x2 -Ph 1 -Pb 2 -Pd 9 -Pds 0 -
Ps 2 -Pss 6 -Pc REQUEST
r -t 1.836585583 -Hs 0 -Hd -2 -Ni 0 -Nx 224.43 -Ny 334.08 -Nz 0.00 -Ne -
1.000000 -Nl RTR -Nw --- -Ma 0 -Md ffffffff -Ms 2 -Mt 800 -Is 2.255 -Id -
1.255 -It AODV -Il 48 -If 0 -Ii 0 -Iv 30 -P aodv -Pt 0x2 -Ph 1 -Pb 2 -Pd
9 -Pds 0 -Ps 2 -Pss 6 -Pc REQUEST
```

Fig. 3 Trace file format

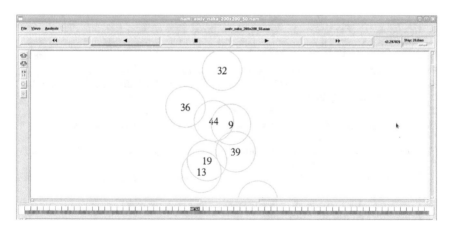

Fig. 4 "NAM" file output

5   MATLAB

MATLAB software [16] is used in high-performance numerical analysis and animation. It provides an interactive platform having several built-in functions for technical computation, graphics and visualization. The main part is that it is also possible to easily extend it with its own high-level programming language. The full form of MATLAB is MATrix LABoratory.

6   Conclusion

In this paper, basic overview of simulation tools used to analyze VANET was discussed. There are two categories of network simulators: Traffic Simulator can be used to generate traffic for the network, while Network Simulator provides the data to analyze the network using the traffic. CanuMobiSim and VanetMobiSim are the traffic simulators and NS-2 is a Network Simulator. XML and trace files are used as input to the NS-2. TCL and awk are the languages which are useful as research simulation tools.

References

1. CanuMobiSim. Available: http://canu.informatik.uni-stuttgart.de/mobisim/
2. Hassan, A.: VANET Simulation Master's Thesis in Electric Engineering, Technical Report, IDE0948, May 2009
3. VanetMobiSim-1.1: Available: http://vanet.eurecom.fr/
4. Poonia, R.C.: Integration of traffic and network simulators for vehicular ad-hoc networks. J. Inf. Optim. Sci. 1–8 (2018)
5. CityMob: Available: http://www.grc.upv.es/Software/citymob.html
6. Bonnmotion: Mobility Scenario Generator (Bonn University). Available:http://www.net.cs.uni-bonn.de/wg/cs/applications/bonnmotion/
7. NS2, DARPA Project. Available: http://nsnam.isi.edu/nsnam/index.php/User_Information/
8. VINT Project (Virtual InterNetwork Testbed). Available: http:// www.isi.edu/nsnam/vint
9. Singh, V.: Efficient routing by minimizing end to end delay in delay tolerant enabled VANETs (2015)
10. Poonia, R.C., Bhargava, D.: A review of coupling simulator for vehicular ad-hoc networks. Int. J. Technol. Comput. Sci. (2016)
11. Tool command language. Available: http://www.isi.edu/nsnam/
12. Kristiansen, S., Plagemann, T.: ns-2 distributed clients emulation: accuracy and scalability. In: Proceedings of the 2nd International Conference on Simulation Tools and Techniques. ICST (Institute for Computer Sciences, Social-Informatics and Telecommunications Engineering) (2009)

13. Singh, V., Saini, G. L.: DTN-enabled routing protocols and their potential influence on vehicular ad hoc networks. In: Soft Computing: Theories and Applications, 367–375. Springer, Singapore (2018)
14. Mefteh, W.: Simulation-based design: overview about related works. Math. Comput. Simul. (2018)
15. AWK. Available: http://www.staff.science.uu.nl/~oostr102/docs/nawk/nawk_toc.html
16. Pratap, R.: MATLAB: A Quick Introduction for Scientists and Engineers. Oxford Publication, Oxford

Sustainable Management Strategy for Software Firms to Reduce Employee Turnover Due to Freelancing

A. K. M. Wasimul Hossain, Laila Nushrat Raha, Tahmid Faiyaz, Mahady Hasan, Nuzhat Nahar and M. Rokonuzzaman

Abstract Employees tending to choose freelance rather than being employed at any software firm has become a major problem. Due to fewer professionals employed, software firms are facing some limitations that are interrupting their sustainability. Freelancers are encouraged to integrate with the software firms, but the growth of software firms tend to decline and fall into the diseconomy of scale, urging employees to be freelancers rather than an employee at any organization. In this study, the growth of the software firms and its effect on freelancers has been discussed based on data collected from primary and secondary sources. It has been observed that a healthy software firms are able to attract more employees, whereas employees are demotivated in weak software firms. Therefore, employee prefer. This paper discusses the sustainable management and economic strategies that are considered as possible ways to retain the growth of the software firms which would help them to reduce employee turnover due to freelancing.

Keywords Freelancing · Software economies of scope and scale · Sustainable growth of software firms

1 Introduction

Freelancing has greatly emerged in the outsourcing market. Day by day, the number of freelancers is rising which in turn is reducing the number of employees in the software firms. Being a freelancer, there are quite a few features that are highly attractive such as the high income of the individual, flexibility of working time, suitable work space which are some of the leading reasons for the employee turnover. In the context of

A. K. M. W. Hossain · L. N. Raha · T. Faiyaz · M. Hasan (✉) · N. Nahar
Department of Computer Science and Engineering, Independent University,
Bangladesh, Dhaka, Bangladesh
e-mail: mahady@iub.edu.bd

M. Rokonuzzaman
Department of Electrical & Computer Engineering, North South University,
Dhaka, Bangladesh

© Springer Nature Singapore Pte Ltd. 2020
A. K. Luhach et al. (eds.), *First International Conference on Sustainable Technologies for Computational Intelligence*, Advances in Intelligent Systems and Computing 1045,
https://doi.org/10.1007/978-981-15-0029-9_16

these features, individuals are being influenced to work as a solo performer rather than being an employee of a company/software firm.

The IT sector of Bangladesh is one of the fastest growing sectors of its economy. This sector has grown in demand as well as in the complexity. The continuous growth and evolution have given a chance for some people to establish their own places in this outsourcing world [1]. Bangladesh with more than 30% of citizens living below the poverty line, the country's Gross Domestic Product (GDP) per capita ranks among the lowest in the world [2]. The use of digital facilities and information and communication technology (ICT) has opened new opportunities for growth and development, which stimulate the measure of the socio-economic transformation in a remarkable way [3].

According to Bangladesh Association of Software and Information Services (BASIS), there are around 309 companies that are working closely with the development of software for the local and international markets for different ICT services. According to BASIS 2012 survey, the ICT industry has persistently grown at 20 to 30% per annum. Over 800 registered ICT companies generated total revenue of approximately $250 million. GDP growth moderated from 6.5% in Financial Year (FY) 2011 to 6.1 and 6.2% in FY2013 and FY2014, respectively [4]. In a survey of BASIS, it is observed that around 200 firms begin their journey each year, though the life of many firms lasts only 3–5 years before their downfall [5]. In [26], a guideline has been proposed for the software firms, entrepreneurs, and professionals to build a knowledgebase.

Freelancing is one of the major contributors for overall economic growth of the Bangladesh in the area of information technology/information technology enabled services (IT/ITeS) sector. According to [1], more than 50 freelancers have crossed the barrier to develop their own micro-enterprises, exporting IT products and services to the international market. The leading freelance platform in Bangladesh, Elance-oDesk, reports over 30,000 active freelancers; most of them are part-timers, with a majority being tertiary students. From 2010 to 1st quarter of 2014, a total of 435,249 jobs were awarded to Bangladeshi freelancers from Elance-oDesk market place alone [1]. It is seen that professionals are leaning towards freelancing due to some of the major reasons like low salary from software firms as of limited growth of the firms. Due to these drawbacks, employees are turning over; as a result, software firms are losing their professionals and flow of work disrupts [6].

Hence, it is important to trace out the difficulties that our software firms are facing in the market which is causing rise to freelancers. A detailed discussion on the problems of the growth of the software firms and how they are encouraging the rise of freelancers need to come into light. We have provided detail information on this issue along with some graphical representation.

2 Problem Statement

Freelancers have grown worldwide, and individuals are finding freelancing work opportunity as higher source of income than working in software firms with fixed amount of salary. Due to the diseconomy of scale, software firms are unable to pay their employees well. Paying capability of per person revenue of the firms decreases with the increasing number of team members.

It has been found that in some countries, employees working in the software firms are making more money than being a solo performer. Therefore, the problems of the software firms should be emphasized which would eventually solve the individual revenue decrement of employees working with the firms. Hence, the software firms will be making more money and as a result the lives of the employees and their families will be enhanced. In a broader sense, the society will be benefitted, and it will also create a peaceful working atmosphere.

Firms need to manage and improve their software processes, deal with rapid technological advancements and sustain their organizations through growth. Small firms are extremely responsive and flexible, but unlike large companies they do not have enough employees to develop functional specialists that would help them to perform complex tasks secondary to their products. Moreover, high finance also constraints small firms and hence cannot afford required expertise [7]. The detailed problem that software firms face is discussed below (Fig. 1).

At the beginning of any software firms, they experience economy of scale shown in Fig. 2 up to size S_1 (S_1 is a random size; in the case of Bangladesh, it is in between 20 and 30). It is because, during the starting time, the firms work on smaller projects with less number of employees. Hence, the employees can play complementary roles to each other. As software firms grow, additional professionals are recruited. When a large development team works for the same job, marginal productivity falls due to poor communication and coordination overhead, increased rework by poor representation. In Fig. 2, when S_1 has reached the marginal productivity, it starts falling due to additional recruitment. At S_2, the marginal cost is equal to the marginal

Fig. 1 Life cycle of product

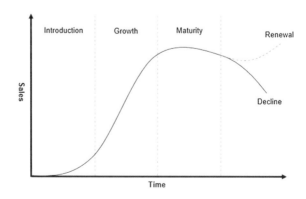

Fig. 2 Production function
of typical software firm

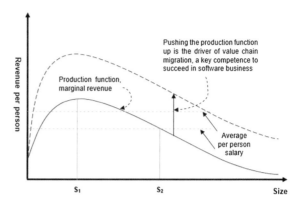

revenue which basically shows that production function has reached its growth limit
[5].

In general, the product lifecycle has an upward growth but after a certain time,
it will decline. This is the nature of products that eventually affects the economy of
scale of the software firms. Hence, production function graph also follows the same
pattern, illustrated in Fig. 2.

Not all software firms have the opportunity of creating a job. Only a handful of
small firms grow into large firms where they have mounting employee recruitment.
High growth output firms imitate high employment growth firms [8]. However, the
growth of firms depends on certain factors like the economy of scale and scope [9].
Researches have shown that reuse of code is essential as it can reduce workload and
provide efficient product [10], whereas some researches have also introduced innovation affecting the economic growth of the firm [11, 12]. Innovation is associated
with Research and Development (R&D). The more the market is explored, the better
the innovation results [13].

Bangladesh, being a developing country, is lagging in terms of some major ICT
indicators like promoting ICT products, training on ICT, etc. [14, 15]. Bangladesh's
software firms need to improve their software process to attain more stability in the
software organizations. It is identified from various studies that Bangladesh lacks
in the target set for software processes and improvements, involvement in quality
control activities and standardize business expertise practices [16]. One advantage
that the Dutch software firms in Bangladesh have received is the reduction in project
costs [17]. Comparing with India, from the IT sector, India earned US$14.7 billion in
2009–10 while Bangladesh earned US$13.85 million in 2008–09 from the IT sector
[18].

Some of the problems the freelancers face are like while withdrawing money, [19]
unstable career, improper salary scale, poor management, and requirement elicitation
[20]. These issues highlight their chronic insecurity. Since they are not traditional
employees, benefits of health insurance and retirement savings are not given to them
[21]. As employees do not stay in a company for long, the companies do not want
to spend too much on specializing them in certain departments [5]. Apart from the

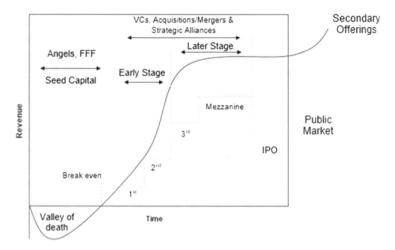

Fig. 3 Startup financing cycle

mentioned problems, freelancers in Bangladesh have to deal with some additional obstacles such as power crisis. Along with that, the Internet speed is not up to mark sometimes, which affects 40% of the business. Bangladesh Govt. still did not approve the online transaction; therefore, transferring the money sometimes is very difficult. Another problem for the freelancers is the high cost for the R&D [5]. To create useful software and to make a good profit out of it, some market analysis has to be done. Or else, the software will not produce the expected profit. For this, they have to invest some capital at the start of any project, which is termed as venture capital.

In Fig. 3, the startup financing cycle is shown. Venture capital needs to spend quite a long time to meet the break event point. Without significant financial background, venture capitals are lost in this valley of death region. This is another challenge for any startups. During this period, companies need to invest a lot but they can channel very small earning [5].

The solutions to these issues are very expensive. The individual himself cannot solve these problems, only sustainable software firms can afford to run required steps to solve these issues. In addition, to help the freelancers from these dilemmas stated earlier, the companies can take opportunity to turn the freelancers to company employees.

3   Researchable Issues

From the above discussions, it is seen that many types of research have been performed on the firms' growth and the cause of employees leaving their jobs. Some research shows that software firms' growth had been influenced by innovation where innovative ideas have been encouraged [22–24]. Innovation and R&D are interrelated; therefore, R&D has an equal importance in the growth of the software firms. Some researches were also performed trying to figure out the problems of freelancing over permanent jobs and vice versa [25]. However, employee turnover relating to the growth of the software firm has not yet been discussed in any research to the best of our knowledge. Companies are concerned with their growth, but yet no plan of strategy is made to come out of the problems. Therefore, the actual motive of this paper is to scrutinize some management strategies, which will help the software firms to maintain a sustainable position in the market reducing employee turnover and attracting freelancers to work in a better environment with a better salary and improved facilities.

4   Sustainable Management Strategies

The main concern is that software firms are unable to provide their employees with the required facilities, which in turn influences employees towards freelancing. Moreover, to reduce this influence, software firms are required to come up with some logical strategies to improve their position in the market so that they can provide better facilities.

From the previously mentioned content and graphical representation, it is seen that the growth of a software firm firmly depends on some vital factors such as economy of scope, the economy of scale, and innovation. However, when firms reach its growth limit, they have two options to follow. The first option is to improve the software production process or to access customers who are willing to pay more. With any of these options, software firms can push their production function upward. The second option is software product for the local or global market to benefit from the reuse to increase the economy of scope. When a software is reproduced, the cost of the product nearly comes to 0. With every new user, the cost of the software decreases that increases the sale.

Figure 4 shows a willingness to pay graph, Fig. 4a, which means how much the customer is willing to pay for a software, whereas in Fig. 4b, the graph shows the benefit of a consumer and producer as price and cost of a software is altered.

From Fig. 4b, at cost C_1, several the customer is n_1 whereas, at cost C_2, the number of the customer has risen to n_2. This rise in customer occurred when the cost of the product has fallen. Hence, the same product could be sold more to $n_2 - n_1$ number of customers when the cost of the product falls as well as the price. From Fig. 4b, at C_1 the price of the product will be P_1. Cost and price are at a ratio, hence influencing the cost will affect the price. When cost (C_2) is reduced, the price (P_2) also reduces which increases the number of the customer (n_2). This leads to the increase of producer surplus which means that firms are benefitted by selling their product in the market, whereas if the cost (C_3) is increased which in turns increases the price (P_3) will lead to reducing the number of customers and minimizes the producer surplus, and hence this is indicated as the red zone (Fig. 4b). Moreover, the

Fig. 4 a, b The startup
financing cycle

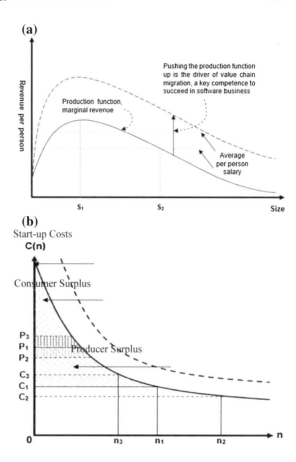

green zone indicates the firm's benefit and may lead to the economy of scale if they can maintain their zone.

With decreasing cost, revenue will increase which may lead firms to the economy of scale. To decrease the cost of the product, firms must decrease the rework and increase the reuse of the previous work. Due to scope, a lot of work for developing the new product becomes easier as only a few new additional features had to be added and the waiting time for complementary work will also be lessened. Using the scope, new products can be developed more efficiently and faster before losing their valued customers who are willing to pay for a better-improved product.

Cost of a software depends on R&D, customization and customer service. Hence, reducing cost means to reduce rework as well as waiting time and to increase reuse. To fulfill these criteria, firms should encourage unique and innovative ideas. Innovation advances when a proper R&D is done. When there is a high R&D, customization and business development per customer is low which allows reuse and reduces rework for the software. If the firm can push the graph upward as shown in Fig. 4a, then at C_1 the number of the customer will be n_3 and when this cost is reduced to C_2, the number of the customer is n_4. This upward push is due to the innovation. The blue dashed line in the graph shows the benefit of the firms, illustrated in Fig. 4a.

Though implementing these strategies all over the market will not be possible now. To implement the strategies, at first, the targeted firms are the small firms. Because with fewer employees, the implementation of these strategies will be easier. If the implementation awards a positive reply, gradually these strategies can be implemented in the large firms. Thus, firms will hold a feasible position in the market, which will stabilize their revenue, and they could increase their span of time in the market before falling into the diseconomy of scale. They could also get the scope of emerging into the global market with higher standard value. And as the revenue of the firm increases, they can provide employees with better wage and facilities. Hence, the employees will not be attracted towards freelancing.

5 Conclusion

Freelancing, being an emergent sector in the IT, individuals are leaning towards this field. With all the benefits of freelancing, there comes a time when the solo performer starts to face trouble that might be overcome if he was working in a software firm. However, the software firms are not being able to pay well to their employees as firms make less benefit out of their software and fall into the diseconomy of scale in a very short period. To help software firms retain their position, some management strategies have been suggested which if implemented can help the software firms to grow and offer a better salary to their employees. This will encourage employees to continue their complementary work rather than being a solo performer.

Implementation of these management strategies have not been performed yet because of the short span of time but when these will be executed, some limitations will be confronted. The very first limitation to be highlighted is the acceptance of the above-suggested management strategies in the software market. As these strategies are very new and fresh, firms might be reluctant in implementing the idea. As of now from research, there is a vast community of freelancers; hence they cannot be influenced towards permanent jobs very easily. But, if the strategic issues give a positive output and freelancers observe that employees in firms have a better and sustainable position than them, this will motivate freelancers to join software firms. With further investigation and implementation of the strategies for a longer period, the conclusion of the research can be generated from the result of this investigation whether more modifications in the strategy are needed or above-mentioned strategies will help to overcome the hurdles of software firms in bringing back their employees.

References

1. Bangladesh IT/ITeS Industry Development Strategy, Global Demand Assessment, Competitive Landscape, and Industry Statistics, (2015)
2. ICT Sector Study Bangladesh Bridging the Gap between Dutch and Bangladeshi ICT Sectors, Nyenrode Business Universiteit (2014)
3. Bairagi, A.K., Roy, T., Polin, A.: Socio-economic impacts of mobile phone in rural Bangladesh: a case study in Batiaghata Thana, Khulna District. Korea 71.94.7, 5–9 (2011)
4. Software & IT Service Catalogue 2014, (2014)
5. Zaman, R.: Study on software development sector of Bangladesh (2015)
6. Jimenez, C.: 10 advantages and disadvantages of becoming a freelancer. https://theamericangenius.com/entrepreneur/10-advantages-and-disadvantages-of-becoming-a-freelancer/, last accessed 2019/1/28
7. Ita, R., von Wangenheim, C.G.: Why are small software organizations different? IEEE Softw. **24**(1), 18–22 (2007)
8. Haltiwanger, J., et al.: High growth young firms: contribution to job, output and productivity growth (2016)
9. Langlois, R.N.: Scale, scope, and the reuse of knowledge. In: Economic Organization and Economic Knowledge: Essays in Honour of Brian J. Loasby, 239–254. Edward Elgar, Aldershot (1999)
10. Haefliger, S., Von Krogh, G., Spaeth, S.: Code reuse in open source software. Manage. Sci. **54**(1), 180–193 (2008)
11. Cainelli, G., Evangelista, R., Savona, M.: Innovation and economic performance in services: a firm-level analysis. Camb. J. Econ. **30**(3), 435–458 (2005)
12. Mansury, M.A., Love, J.H.: Innovation, productivity and growth in US business services: a firm-level analysis. Technovation **28**(1–2), 52–62 (2008)
13. Cohen, W.M., Klepper, S.: Firm size and the nature of innovation within industries: the case of process and product R&D. Rev. Econ. Stat., 232–243 (1996)
14. Naym, J., Hossain, M.A.: Does investment in information and communication technology lead to higher economic growth: evidence from Bangladesh. Int. J. Bus. Manag. **11**(6), 302 (2016)
15. Kabir, M.R.: Towards digital Bangladesh: scope of ICT integrated urban planning and management. J. Bangladesh Inst. Plan. ISSN 2075: 9363
16. Begum, Z., et al.: Software development standard and software engineering practice: a case study of Bangladesh (2010). arXiv preprint arXiv:1008.3321
17. Tjia, P.: The software industry in Bangladesh and its links to The Netherlands. Electron. J. Inf. Syst. Dev. Ctries. **13**(1), 1–8 (2003)
18. Khan, M.M.H.: An assessment of IT enabled services in Bangladesh: a comparative study (2012)
19. Mansur, H.T.: Freelancing in Bangladesh (2014)
20. Azad, O.: Problems of software developers in Bangladesh (2007)
21. Horowitz, S., Buchanan, S., Alexandris, M., Anteby, M., Rothman, N., Syman, S., Vural, L.: The rise of the freelance class: a new constituency of workers building a social safety net (2005)
22. Romijn, H.a.M.A.: Determinants of innovation capability in small electronics and software firms in Southeast England. Res. Policy **31**(7), 1053–1067 (2002)
23. Edison, H., Ali, N.B., Torkar, R.: Towards innovation measurement in the software industry. J. Syst. Softw. **86**(5), 1390–1407 (2013)
24. Nambisan, S.: Software firm evolution and innovation–orientation. J. Eng. Tech. Manage. **19**(2), 141–165 (2002)
25. Adey, S.: Freelance vs. Permanent employees: what to choose? http://workawesome.com/career/freelance-vs-permanent-employees/, last accessed 2019/1/28
26. Raha, L.N., Wasimul Hossain, A.K.M., Faiyaz, T., Hasan, M., Nahar, N., Rokonuzzaman, M.: A guide for building the knowledgebase for software entrepreneurs, firms, and professional students. In: 2018 IEEE 16th International Conference on Software Engineering Research, Management and Applications (SERA), pp. 165–171. IEEE, New York (2018)

Deformation Analysis of UWB Microstrip Antenna for Breast Cancer Detection

Beerpal Kaur, Lakhvinder Singh Solanki and Surinder Singh

Abstract This paper presents a novel antenna structure for body diagnosis and detection system. The antenna structure is simple to fabricate and looks like a curved structure, which is easy to place on human body's curved parts. The proposed antenna consists of rectangular patch and rectangular fractal-based defected ground. The rectangular patch has one round cut at each corner. It is fed with coaxial cable and can easily be integrated with an array of multiple elements. The antenna parameters like, S_{11}, total gain, radiation pattern, polar pattern in various planes and VSWR were evaluated for various degrees of deformation of the planner patch antenna and are compared with the conventional planner antenna. The antenna retained its various parameters within permissible limits even after deformation. The proposed antenna is suitable candidate for designing a breast cancer detection system.

Keywords Antenna · Curved · Rectangular patch · Breast cancer · On body · Detection system · Array · Electromagnetic defected ground (EBG)

1 Introduction

Cancer is a given name to any type of malignant tumor that spread rapidly to the rest of the body. Every year, more than one million people in the world are diagnosed with different types of cancers. Statistics show that breast cancer is one of the most frequent types of tumor that affect women worldwide, especially in developing countries [1]. This type of cancer overtook lung cancer which used to top the list worldwide with 1.6 million of diagnostics and death rate in 2012. In previous medical examination of year 2008, a hike of 20% in the breast cancer cases and 14% of normal cases has been observed. About 2 million women were diagnosed with breast cancer in

B. Kaur (✉) · L. S. Solanki · S. Singh
Sant Longowal Institute of Engineering and Technology, Longowal,
Distt Sangrur 148106, India
e-mail: bawabeerpal335@gmail.com

L. S. Solanki
e-mail: lakhvinder_singh_solanki@yahoo.co.uk

© Springer Nature Singapore Pte Ltd. 2020
A. K. Luhach et al. (eds.), *First International Conference on Sustainable Technologies for Computational Intelligence*, Advances in Intelligent Systems and Computing 1045, https://doi.org/10.1007/978-981-15-0029-9_17

year 2012 [2, 3]. This tumor is now considered as the major cause of death among women diagnosed with cancer in the world, with 522,000 death cases. In Tunisia, breast cancer is the most diagnosed type of cancer and represents the second cause of death after lung cancer. In 2015, 1829 cases were diagnosed with breast cancer, which represents 15% of all cancer cases and 8.5% of total death cases [4]. The early detection of cancer increases the chances of successful treatment [5, 6], which will result in the reduction of the high number of deaths caused by this disease.

Breast cancer tumor detection is a biomedical approach and an application that belongs to the microwave tomography. Many conventional techniques used for breast cancer diagnosis are X-ray, MRI, and ultrasound. These methods are not perfect to provide 100% correct diagnosis results. A false negative detection appears typical despite the fact that cancer is in body, whereas the false positive detection looks abnormal regardless of the fact that there is no cancer in the human breast [7–9]. The microwave detection of the breast cancer tumors is a non-ionizing and indeed potentially low-cost alternative. In the ultra-wide band (UWB) imaging system, a bunch of wide pulse is being transmitted from a transmitter antenna and then scattered with the different layers of the phantom. The scattered signal of different layers of breast tissue is collected by a receiver antenna surrounding by the breast [10, 11]. Microwave tomography (MWT) is the promising biomedical imaging methodology for non-invasive functional evaluation and unhealthy condition of human breast soft tissues. In microwave tomography, an array of microwave signals radiators and receivers are positioned surrounding the target human body area, and doubtful objects are detected and localized in human breast. In most of the cases, the dielectric contrast difference between the ambivalent and healthy soft tissues is not much deviated, which multiplies the microwave detection of cancer in human breast, brain, lungs, and cardiac.

Antennas play important role in the world of biomedical body sensors as microwave detector for buried objects in human body. Nowadays, microstrip antenna has achieved the extreme ease of designing and researchers prefer it due to its plentiful characteristics like low profile and low cost and many more. This antenna can be easily fabricated and conformed on cylindrical bodies [12–14]. Conformal antennas, in general, have some advantages over planner microstrip antennas such as wide angular coverage and controlled gain [15, 16]. The main drawback of microstrip antennas is their narrow bandwidth, which results in high sensitivity to any frequency change and decreases the efficiency of their overall performance [17].

Here, we use microstrip patch antenna for microwave tumors detection. Microstrip patch antenna can easily be mounted on the flat surface. It consists of flat rectangular sheet or patch of metal, mounted over a large sheet of metal called ground plane. The most commonly used employed microstrip antenna is rectangular patch antenna. Antenna bandwidth decreases as the dielectric constant of substrate increases, which increases the directivity of the antenna and therefore decreases the impedance bandwidth. Antenna emits minimal radiation compared to other systems. Rest of the paper is organized as follows: Sect. 2 describes the conventional antenna and its simulation results and discusses the microstrip antenna. Section 3 elaborates the electromagnetic band gap (EBG) structure in ground plane and its simulation results. Section 4 gives details for the antenna requirements for diagnosis. Section 5 describes the antenna deformation, design of deformed microstrip antenna, and its parametric analysis. Section 6 describes the conformed microstrip antenna for breast cancer detection array and its result and discussion.

2   Conventional Antenna

The conventional rectangular microstrip patch antenna as shown in Fig. 1 is modeled on FR4 substrate with $\varepsilon_r = 4.4$ and tan $\delta = 0.02$. The antenna dimensions are shown in Table 1.

The effect of the capacitive and inductive parts of conventional antenna are nullified and for enhancement of the bandwidth of the antenna; the corners in the lower and upper sides of the patch are rounded to inward besides adding the ground slot as shown in Fig. 2. The rounding of corners at the lower part of radiating patch increases the separation between the patch and ground planes, which adjusts their mutual capacitive coupling. Also cutting round corners in the upper corners of the patch tunes the inductive part of the antenna that neutralizes the capacitive coupling between the ground and the patch to get pure resistive input impedance. On the other side, the ground slot neutralizes the capacitive effects through the inductive nature of the patch to get nearly pure resistive input impedance. The simulation results in Fig. 3 show better impedance matching and wider bandwidth with rounding of all

Fig. 1 2D structure layout of microstrip patch antenna

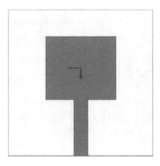

Table 1 Measurement dimensions of conventional antenna

Specifications	Measurements (mm)
Length of patch	14.5
Width of patch	15
Length of substrate	30
Width of substrate	35
Height of substrate	1.6
Feed position	−1.425, 7.5, 1.6

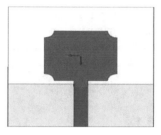

Fig. 2 2D structure layout of microstrip patch antenna with corner rounding

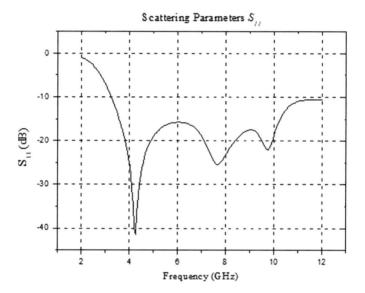

Fig. 3 The simulated results of return loss (S_{11}) of conventional antenna

the radiating patch corners, and also improvement in impedance matching has been achieved with addition of notch in the ground at middle and below the feed line [18].

2.1 The Simulation Results and Discussions for the Microstrip Antenna

The S_{11} response of the antenna spreads from 2 to 12 GHz; it shows three resonances at 4.2, 7.6, and 9.7 GHz and it has ultra wide band response from 3.3 GHz to more than 12 GHz scale limit of graph. The first resonance of 4.2 GHz reaches to S_{11} to −42 dB, second resonance to −25 dB, and last resonance −22 dB. The maximum gain (total) response of 5.4986 dB is shown for 3D pattern of total gain for modified microstrip

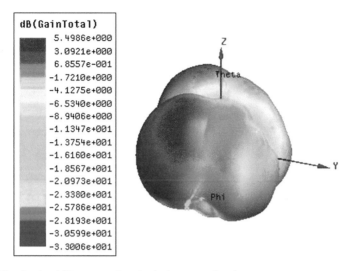

Fig. 4 The simulated 3D pattern of total gain for conventional antenna

antenna shown in Fig. 4. The response shows the bi-directional 3D radiation pattern. The 2D radiation pattern of total gain for conventional microstrip antenna in two planes at angle phi = 0°, 90° are shown in Fig. 5. The planner radiation patterns show that antenna can radiate sufficient signal in all direction as for communication with other devices at any orientation.

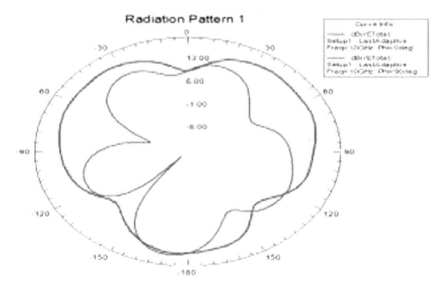

Fig. 5 2D radiation pattern of total gain for conventional antenna at phi = 0°, 90°

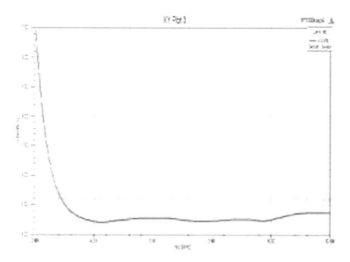

Fig. 6 VSWR of for conventional antenna

The VSWR response should be less than 2, and Fig. 6 depicts that VSWR of the modified microstrip antenna is close to 1 from 3.3 GHz to more than 12 GHz scale limit of graph, which predicts that effective working of the antenna.

3 Electromagnetic Band Gap (EBG) Structure in Ground Plane

The introduction of electromagnetic band gaps in the ground plane of the antenna structure enhances the bandwidth distributed at the operating frequency of antenna. There are various types of the EBG structures like polygonal, circular, and spiral, etc. However, it is found that circular shaped EBG exhibits higher directivity and gain as compared to others [19]. EBG structure can be with periodic metal patches placed on the dielectric layer towards ground plane or periodic slots created in the ground plane of the antenna structure. The EBG cell shape could be periodic or non-periodic [20]. In this paper, fractal rectangular-shaped cell is designed for enhanced reflection coefficient and better bandwidth. At center, square of side L_1 is etched, then Second Square with side L_2 is etched at corner of First Square and Third Square of side L_3 is etched on corner of Second Square. This process is repeated at all other corners of the First Square with side L_1. In Fig. 2, microstrip antenna is resonated at 4.2 GHz and its length and width are optimized with software HFSS. Figure 7a shows the unit cell of EBG which is based on fractal rectangular form. In first iteration, a square of $L_1 \times L_1$ mm^2 is considered and in second iteration it is cut along four corners with squares of $L_2 \times L_2$ mm^2. In third iteration, squares of $L_3 \times L_3$ mm^2 are used to cut

Fig. 7 **a** EBG unit cell
structure, **b** its equivalent

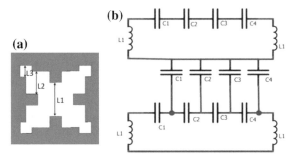

(a)

Fig. 8 Microstrip antenna
ground plane with repeated
EBG unit cell structure

the corners of second iteration square, where iteration dimensions are $L_1 = 1.5$ mm, $L_2 = 1.0$ mm, $L_3 = 0.5$ mm.

Figure 7b shows the equivalent of EBG unit cell. The EBG unit cell is periodically etched in the ground plane of microstrip antenna as shown in Fig. 8. Each EBG cell in antenna design acts a low-pass filter where the thickness of substrate layer acts as inductance and the two consecutive cells gap acts as capacitance. The fractal shape makes different gaps between the inner square and outer both squares and therefore there are dissimilar values of capacitance between the gaps.

3.1 Simulation and Result Discussion

The proposed design of microstrip patch antenna with defected ground is shown in Fig. 9, and the simulation result of the reflection coefficient S_{11} of the antenna is shown in Fig. 10. The S_{11} response of this antenna also spreads from 2 to 12 GHz, it shows three resonances at 4.16, 7.5 and 9.5 GHz, and it has ultra-wide band response from 3.3 GHz to more than 12 GHz scale limit of graph. The first resonance of 4.16 GHz reaches to S_{11} to -41 dB, second resonance to -24 dB, and last resonance -22 dB. The comparison of the modified antenna and proposed planner antenna illustrates that even resonance frequencies have shifted a very little to downside, and the resonance frequency ship is negligible. The S_{11} of the modified antenna also has negligible variation in comparison with the proposed planner antenna.

The maximum gain (total) response of 5.5919 dB is shown for 3D pattern of total gain for proposed planner microstrip antenna shown in Fig. 11. The response shows the bi-directional 3D radiation pattern. The 2D radiation pattern of total gain for modified microstrip antenna in two planes at angle phi = $0°$, $90°$ are shown

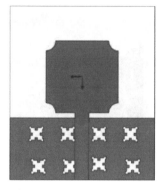

Fig. 9 2D structure layout of modified microstrip antenna with repeated EBG in ground plane

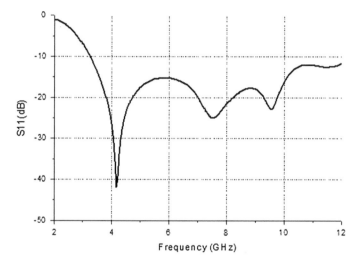

Fig. 10 Return loss (S_{11}) of modified antenna

in Fig. 13. The planner radiation patterns show that antenna can radiate sufficient signal in all direction as for communication with other devices at any orientation. The VSWR response should be less than 2, and Fig. 12 depicts that VSWR of the modified microstrip antenna is close to 1 from 3.3 GHz to more than 12 GHz scale limit of graph, which predicts that effective working of the antenna. Overall, the defected ground of the proposed microstrip antenna does not deteriorate the performance of the antenna, even after removal of the copper from ground by creating the periodical rectangular slots and shortening the ground plan toward the feed line, it strengthens the possibility to bend antenna structure over the virtual cylinder circumference.

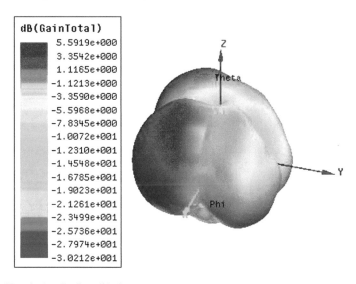

Fig. 11 3D gain (total) of modified antenna

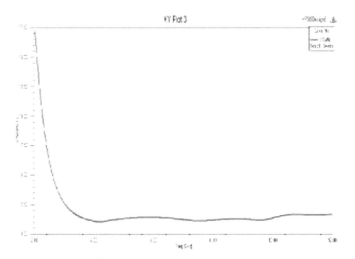

Fig. 12 VSWR of modified antenna

4 Antennas Requirements for Diagnosis

The microstrip antennas used for biomedical applications to mount on human body are the form of body-centric antennas. These antennas operate in three bands, like ISM band: 2.40–2.48 GHz, UWB band: 3–6 GHz, RFID frequencies (MHz). The body-centric communication was first planned and verified by Zimmerman [21]. The

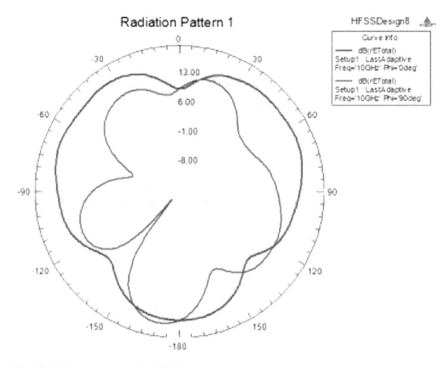

Fig. 13 Radiation pattern of modified antenna

prototype communication system operated at a low frequency of 330 kHz. After-
wards, his findings were followed by many researchers and later slightly higher
frequencies such as 10 MHz [22–24] were used. In the current scenario, ultra-high
frequency (UHF) bands such as 403 MHz, 868 MHz, and 2.45 GHz are being preferred by the applications using body-centric wireless communications [25, 26].

The breast model is imitation of a human breast. The breast model is designed as
a half sphere covered with a skin layer and outer radius of 40 mm. A fibro glandular
breast fatty tissue layer of radius 22 mm is situated inside the skin layer. The purposed
deformed antenna placed over this breast model represents a suitable candidate for
the arrays being used in research purpose applications like for breast tumor detection
in the frequency band 1 GHz to 12 GHz [27] and other body diagnosis from outer
side (Fig. 14).

5   Antenna Deformation

A mathematical model that includes the effect of curvature on fringing field and on
antenna performance is presented. Bending of the planner microstrip patch antenna
affects the performance of the patch antenna due to the effect of fringing field of

Fig. 14 Model of antenna array for tumor detection buried in breast phantom

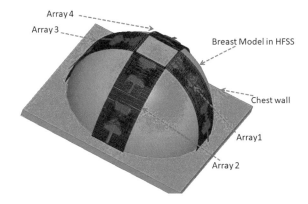

a conformed patch antenna. Effect of curvature of conformal antenna on resonant frequency by Krowne [28, 29] can be expressed as Fig. 15.

Where '2*b*' is length of the conformed patch antenna resonated at frequency $f_{r,m,n}$ (1) in planner shape, '*a*' is radius of conforming cylinder, '2*θ*' is angle bounded with width of the patch on cylinder center, is permittivity and is permeability of the substrate of antenna as is shown in Fig. 15. The cavity model is used for the analysis, which is only valid for a very thin dielectric. Also, for much small thickness than a wavelength and the radius of curvature, only *TM* modes are assumed to exist in order to calculate the radiation patterns of cylindrical-rectangular patch antenna.

Fig. 15 Geometry of microstrip patch antenna in cylindrical shape

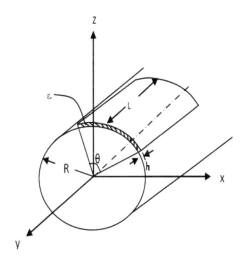

Fig. 16 2D structure layout
of proposed microstrip
antenna deformed over
cylindrical shape

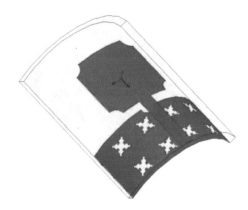

5.1 Deformed Microstrip Antenna Design

The planner microstrip antenna shown in Fig. 9 was deformed over the virtual cylinder of different radii like 15, 20, 40, 60, 1000 mm. The microstrip antenna deformed over the large radius 1000 mm was considered, as close to the planner antenna shown in Fig. 9, as the radius of the virtual cylinder was reduced to 15 mm that was considered as antenna with maximum deformation as in Fig. 16.

The deformation radius for the on-body microstrip antenna applications can vary; therefore, it is crucial to evaluate the antenna performance deterioration with deformation. In the paper, the effects of variation of the deformation radius over the antenna parameter performance were investigated [30]. The commercial electromagnetic simulator HFSS, based on the finite element method, was used to model the antenna and for it parametric analysis with variation of the deforming cylinder radius from over the range 15, 20, 40, 60, 1000 mm. The simulated reflection coefficient S_{11} of the antenna with various deformations is shown in Fig. 17.

5.2 Parametric Analysis and Discussion of Conformed Proposed Microstrip Antenna

A flat proposed microstrip antenna was deformed over a virtual non-modeled cylinder of radius (rad) using simulator HFSS, and results are shown in Fig. 17. The antenna comprises three resonances (f_1, f_2, f_3), when it is flat and as well as when it is deformed. The results indicated that the antenna produced slight deviation in the resonance frequency, when it is flat and when it is deformed over the virtual cylinder of radius (rad). The various parameters are recorded in Table 2. The flat MSA comprises first frequency resonance at 4.11 GHz ($S_{11} = -31.47$ dB), second frequency resonance at 7.3266 GHz ($S_{11} = -24.72$ dB), and third frequency resonance at 9.5879 GHz ($S_{11} = -24.95$ dB). By analyzing the data in Table 2 and Figs. 17,

Fig. 17 Frequency of
resonance, bandwidth, and
reflection coefficient S_{11}
produced by the deformed
modified microstrip antenna
over various deformations
cylinder radius (rad)

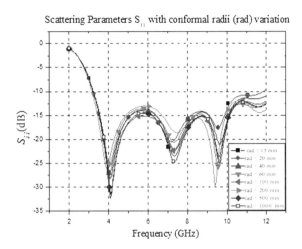

Table 2 Summary of the simulated results for the deformation of the proposed microstrip antenna over various deformation cylinder radiuses

S. No.	Deforming cylinder radius, rad (mm)	Resonance fre- quency, f_1, f_2, f_3 (GHz)	S_{11} (dB)	Bandwidth (GHz)	Gain (dB)	VSWR
1	15	4.0603, 7.1658, 9.5508	−27.3520, −22.0900, −17.7218	8.3102	3.9335	1.0975
2	20	4.0603, 7.1337, 9.5401	−28.0590, −20.3709, −17.8916	8.5667	4.2050	1.0843
3	40	4.0603, 7.2080, 9.5385	−27.5386, −19.4587, −21.0568	8.7946	5.0209	1.0932
4	60	4.1106, 7.2834, 9.6150	−30.5780, −21.8980, −22.2858	8.7710	5.1070	1.0798
5	80	4.0603, 7.3690, 9.5377	−28.0807, −19.9274, −26.1673	8.7940	5.0572	1.0869
6	1000 (close to planner MSA)	4.11, 7.3266, 9.5879	−31.47, −24.72, − 24.95	8.7700	5.0848	1.0834

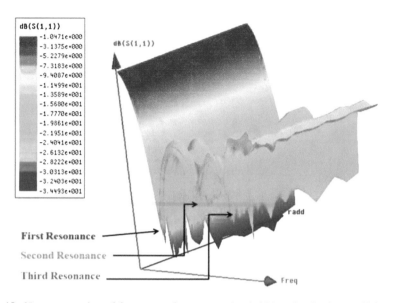

Fig. 18 3D representation of frequency of resonance, bandwidth and reflection coefficient S_{11} produced by the deformed modified microstrip antenna over various deformations cylinder radius (rad)

18, the results deduce that after deformation of the flat MSA to deformation radius 15 mm, VSWR has increased from 1.0834 to 1.0975 (1.3% increment), bandwidth has decreased from 8.77 to 8.3102 GHz (4.2% decrement), S_{11} of first resonance has reduced from −31.47 to −27.352 dB (13% reduction), S_{11} of second resonance has fallen from −24.72 to −22.09 dB (10.64% fall), S_{11} of third resonance has jumped down from −24.95 to −17.7218 dB (29% decrement), and the total gain also has fallen from 5.0848 to 3.9335 dB (22.64% reduction).

In over view, as the flat MSA is deformed from planner shape to cylindrical form, VSWR increases, and the antenna performance in total gain, bandwidth, S_{11}, is deteriorated.

In Fig. 19, the variation the all three resonances of the MSA have been represented with variations in the deformation cylinder radius (rad) from planner shape to deformed shape of radius 15 mm. The first resonance shows a fluctuation of 1.0 GHz in resonance at 900 mm deforming cylinder radius, but later resonance get conformed close to 4 GHz till deforming radius 15 mm, whereas other two resonances second and third remain confined around 7.4 GHz, 9.6 GHz, respectively; no doubt minor fluctuations at extreme deformation close to 15 mm of deforming cylinder radius have been recorded.

In Fig. 20, the variation of reflection coefficient S_{11} has been recorded over various deforming cylinder radius from planer shape of antenna to 15 mm radius. As the MSA is deformed from planer shape to 600 mm, a fall in the S_{11} has been recorded but after words with fluctuations S_{11} has decreased much up to about −27 dB. The S_{11} at other second and third resonances are almost close to each other over the full range

Fig. 19 Resonance frequency variation with variations in deformation cylindrical radius (rad) from planner shape to 15 mm

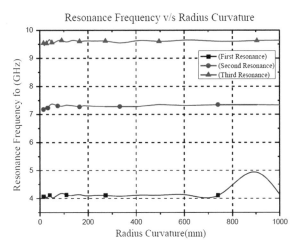

Fig. 20 S_{11} variation with variations in deformation cylindrical radius (rad) from planner shape to 15 mm

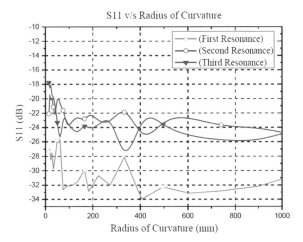

of MSA deformation. From planer shape to 100-mm deformation radius, the S_{11} at both second and third resonances were close to each other, except at 350 mm radius. Overall, the S_{11} fluctuated after 100–15 mm with extreme variations.

6 Conformed Microstrip Patch Antenna

The primary aim of the research is to design an antenna of an array for on body diagnosis like breast cancer detection and further to construct image of the target tissues places inside the human body using the confocal microwave imaging (CMI) algorithms [31].

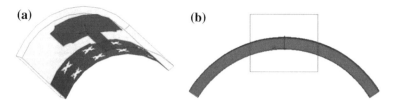

Fig. 21 **a** 3D structure layout of proposed microstrip antenna deformed over cylindrical shape, **b** front view of the deformation

The proposed antenna is able to achieve the objective to construct an array for the breast cancer detection and imaging. This type of antenna could be also part of the breast cancer diagnosis structures or products or applications. Hence, it becomes very important that antenna should have the similar shape as of the human body curvature for the ease of the female patient and sustainable structure. The antenna array could be in circular configuration around the breast as shown in Fig. 21. The array can consist 'N' equally spaced radiating elements. The radius of the breast could be around 60 mm and fabricated breast cover consisting antennas has to be placed over the human breast lying in facing upper side or may in any other position. Hence, MSA has to be deformed over the radius of around 60 mm.

6.1 Results and Discussion for Proposed Deformed Microstrip Patch Antenna

The proposed planner design of the deformed microstrip patch antenna is shown in Fig. 21, and the simulation result of the reflection coefficient S_{11} of the antenna are shown in Fig. 22. The S_{11} response of this antenna also spreads from 2 to 12 GHz, and it shows three resonances at 4.1106, 7.2834 and 9.6150 GHz, and it has ultra-wide band response from 3.325 GHz to more than 12 GHz scale limit of graph. The first resonance of 4.1106 GHz reaches to S_{11} to -30.5780 dB, second resonance to -21.8980 dB, and last resonance -22.2858 dB.

The comparison of the deformed microstrip patch antenna in Fig. 21 and planner antenna in Fig. 9 illustrates that both have three resonances (at 4.1106, 7.2834, and 9.6150 GHz) and another three (at 4.16, 7.5, 9.5 GHz), respectively. In deformed microstrip antenna it is observed that there is very little decrement of 49.4 MHz but S_{11} has also decreased by -12.424 dB. The second resonance has also decreased by 216.6 MHz, and S_{11} has decreased by -3.102 dB; the third resonance almost remains unaltered and S_{11} is not changed. In the end, it is vivid from the illustration that the performance of MSA after deformation is not deteriorated much and is acceptable for biomedical applications for the purpose of the buried unwanted tissues or tumor.

In Fig. 25, the total gain of the proposed deformed microstrip antenna is shown, and it is depicted that maximum total gain is obtained 5.1070 dB. Both major lobes

Fig. 22 Return loss (S_{11}) of planer and proposed microstrip patch antenna deformed at 60 mm radius

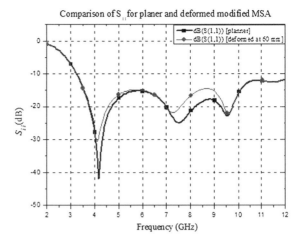

Fig. 23 VSWR of modified deformed antenna

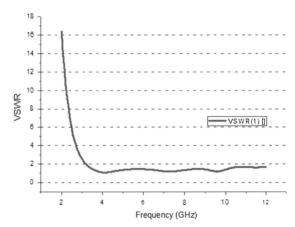

Fig. 24 Radiation pattern of deformed antenna of radius 60 mm

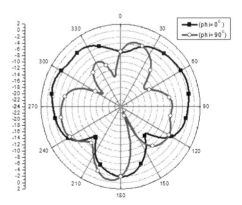

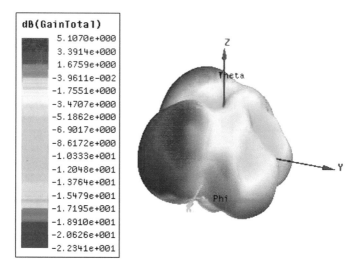

Fig. 25 Gain of deformed antenna of radius 60 mm

of the antenna are directed in both sides of *X*-axis and are symmetrical to *YZ*-plane. In Fig. 24, 2D plot of radiation pattern has been elaborated for two vertical planes *YZ* plane (phi 90°) in XZ plane (phi = 0°). The radiation pattern of the antenna shows maximum gain in direction $\theta = 260°$ of YZ plane, whereas radiation pattern in *XZ* plane $\varphi = 0°$ shows many directions for sufficient gain close to 5 dB in directions $\theta = -30°$ to 60°, 115° to 210° also from 240° to 290° the VSWR response depicted in Fig. 23 represents the faithful response of antenna from about 3.1 GHz to up to the scale limit 12 GHz. During this range, VSWR remains higher than 1 but less than 2, which is the considerable range for the practical working of the antenna.

From Figs. 22, 23, 24, and 25, the response of the antenna indicates that in comparison with the planer antenna shown in Fig. 9. After deformation, the performance of the various parameters of the antenna has not deteriorated much, and all parameters are within the considerable limit. Hence, the deformation of the antenna does not cause any negative impact over the performance of the antenna and deformed antenna could also be used for detection and diagnosis purposes for biomedical on body antenna.

In Table 3, comparison between planer and deformed antenna is tabulated. In the planer antenna, three resonances (at 4.16, 7.5 and 9.5858 GHz) and another three resonances of deformed antenna (at 4.11, 7.2834, 9.6150 GHz), respectively. There is a little change in resonances of planer and deformed antenna, and S_{11} increased in planer antenna; bandwidth of both antennas are almost same, very little change in gain and VSWR, but in the end it is vivid from the illustration that the performance of MSA after deformation is not deteriorated much and is acceptable for biomedical applications for the purpose of the buried unwanted tissues or tumor.

Table 3 Comparison of planer and deformed antenna of radius 60 mm

S. No.	Frequency (GHz)	S_{11} (dB)	Radius	BW	Gain total (dB)	VSWR
1	4.16, 7.5, 9.5858	−41.9879, −24.935, −22.7284	Planner	8.753	5.59	1.0909
2	4.1106, 7.2834, 9.6150	−30.5780, −21.8980, −22.2858	60	8.771	5.10	1.0798

7 Conclusion

This paper presents a novel antenna structure for body diagnosis and detection system. For the various degrees of deformation of planer patch antenna, the antenna parameters like VSWR, radiation pattern, S_{11} were evaluated and are compared with the conventional planner antenna. The proposed antenna deformed over various radius cylinders, while keeping the feeding method same. Various parameters of the antenna were studied, even after deformation which remained within permissible limits (like; $S_{11} < -10$ dB). The proposed antenna is a suitable candidate for designing a breast cancer detection array system, which can be directly placed on breast for diagnosis in the frequency band 1–12 GHz, and other body diagnosis from outer side.

Acknowledgements The authors acknowledge the support provided by Mr. Vijay Prashar, for the Computer Lab facility to access software HFSS also express sincere appreciation to Dr. Surinder Singh (Professor) for guiding on various aspects of ECE Deptt, SLIET, Longowal.

References

1. Zhi-gang, Y., Cun-xian, J., Cui-zhi, G., Jin-hai, T., Jin, Z., Li-yuan, L.: Risk factors related to female breast cancer in regions of Northeast China: a 1:3 matched case-control population-based study. Chin. Med. J. (Engl) **125**(5), 733–740 (2012)
2. Presse, C.D.E.: Derniéres statistiques mondiales sur le cancer En augmentation à 14 1 millions de nouveaux cas en 2012: L' augmentation marquée du cancer du sein demande des réponses En augmentation à 14 1 millions de nouveaux cas en 2012. Int. J., 2012–2014 (2013)
3. Bray, F., Ren, J.S., Masuyer, E., Ferlay, J.: Global estimates of cancer prevalence for 27 sites in the adult population in 2008. Int. J. Cancer **132**(5), 1133–1145 (2013)
4. Cancer en Tunisie Globocan 2012
5. Hossain, M.D., Mohan, A.S.: Breast cancer detection in highly dense numerical breast phantoms using time reversal. In: Proceedings of 2013 International Conference Electromagnetics in Advanced Applications ICEAA 2013, pp. 859–862 (2013)

6. Mohan, A.S., Hossain, M.D.: Breast cancer localization in three dimensions using time reversal DORT method. In: IEEE Antennas Propagation Society AP-S International Symposium, pp. 471–474 (2012)
7. Nahalingam, K., Sharma, S.K.: An investigation on microwave breast cancer detection by ultra-widebandwidth (UWB) microstrip slot antennas. In: IEEE International Symposium on Antennas and Propagation, Spokane, WA, USA, 3–8 July 2011
8. Fouad, S., Ghoname, R., Elmahdy, A., Zekry, A.: Enhancing Tumor Detection in IR-UWB Breast Cancer System. International Scholarly Research Notices, March 2017
9. Çalışkan, R., Gültekin, S.S., Uzer, D., Dündar, Ö.: A microstrip patch antenna design for breast cancer detection. Procedia Soc. Behav. Sci. **195**, 2905–2911 (2015)
10. AlShehri, S.A., Khatun, S.: UWB imaging for breast cancer detection using neural network. Prog. Electromagn. Res. C **7**, 79–93 (2009)
11. https://www.cancer.org/cancer/breast-cancer/screening-tests-and-early-detection.html, Accessed 26 Feb. 2018
12. Abd-Elrazzak, M., Al-Nomay, I.: A design of circular microstrip patch antenna for bluetooth and HIPERLAN applications. In: Proceedings of the 9th Asia-Pacific Conference on Communications (APCC '03), vol. 3, pp. 974–977, September 2003
13. Zakaria, N.A., Sulaiman, A.A., Latip, M.A.A.: Design of a circular microstrip antenna. In: Proceedings of the IEEE International RF and Microwave Conference (RFM '08), pp. 289–292, Kuala Lumpur, Malaysia, December 2008
14. Rmili, H., Miane, J.-L., Zangar, H., Olinga, T.: Design of microstrip-fed proximity-coupled conducting-polymer patch antenna. Microw. Opt. Technol. Lett. **48**(4), 655–660 (2006)
15. Wong, K.: Design of Nonplanar Microstrip Antennas and Transmission Lines. Wiley, New York (1999)
16. Josefsson, L., Persson, P.: Conformal Array Antenna Theory and Design, 1st edn. Wiley-IEEE Press (2006)
17. Elrashidi, A., Elleithy, K., Bajwa, H.: Conformal microstrip printed antenna. In: Proceedings of the ASEE Northeast Section Conference, University of Hartford, West Hartford, Conn, USA, April 2011
18. Awad, N.M., Abdelazeez, M.K.: Multislot microstrip antenna for ultra-wide band applications. J. King Saud Univ. Eng. Sci. **30**(1), 38–45 (2018)
19. Dhakad, S.K., Dwivedi, U., Baudha, S., Bhandari, T.: Performance improvement of fractal antenna with electromagnetic band gap (EBG) and defected ground
20. Mahajan, R., Vyas, R.V.: Performance improvement of microstrip antenna using fractal EBG structure and vias (2016)
21. Zimmerman, T.G.: Personal area networks: near-field intra-body communication. IBM Syst. J. **35**(3–4), 609–617 (1996)
22. Cho, N., Yoo, J., Song, S., Lee, J., Jeon, S., Yoo, H.: The human body characteristics as a signal transmission medium for intrabody communication. IEEE Trans. Microw. Theory **55**(5), 1080–1086 (2007)
23. Fujii, K., Takahashi, M., Ito, K.: Electric field distributions of wearable devices using the human body as a transmission channel. IEEE Trans. Antennas Propag. **55**(7), 2080–2087 (2007)
24. Sasaki, A., Shinagawa, M.: Principles and demonstration of intrabody communication with a sensitive electrooptic sensor. IEEE Trans. Instrum. Meas. **58**(2), 457–466 (2009)
25. Chandran, A.R., Conway, G.A., Scanlon, W.G.: Compact low-profile patch antenna for medical body area networks at 868 MHz. In: Proceedings of IEEE International Symposium Antennas Propagation Society, July 2008
26. Conway, G.A., Scanlon, W.G., Cotton, S.L.: The performance of onbody wearable antennas in a repeatable multipath environment. In: Proceedings of IEEE International Symposium Transmission Antennas Propagation Society, July 2008
27. Solanki, L.S., Singh, S., Singh, D.: Development and modeling of the dielectric properties of tissue-mimicking phantom materials for ultra-wideband microwave breast cancer detection. Optik Int. J. Light Electr. Opt. **127**, 2217–2225 (2016)

28. Krowne, C.M.: Cylindrical-rectangular microstrip antenna. IEEE Trans. Antenna Propag., pp. 194–199 (1983)
29. Wu, Q., Liu, M., Feng, Z.: A millimeter wave conformal phased microstrip antenna array on a cylindrical surface. In: IEEE International Symposium on Antennas and Propagation Society, pp. 1–4 (2008)
30. Chi, Y., Chen, F.: On-body adhesive-bandage-like antenna for wireless medical telemetry service. IEEE Trans. Antennas Propag. **62**(5), 2472–2480 (2014)
31. Khan, M.A., ul Haq, M.A.: A novel antenna array design for breast cancer detection. In: 2016 IEEE Industrial Electronics and Applications Conference (IEACon), Kota Kinabalu, 2016, pp. 354–359

Performance Analysis of Radio Over Fiber Link Using MZM External Modulator

Suresh Kumar, Deepak Sharma, Payal and Rajbir Singh

Abstract The advancements in the field of communication technology and emergence of heterogeneous multimedia applications have led to a paradigm shift from wired networking infrastructure to an optoelectronic networking interface in order to achieve higher transmission efficiency. Radio over Fiber (RoF) is an attractive technology for broadband wireless internet access scenarios as backbone data is carried between base stations using optical fiber providing the user wireless connectivity in the area of coverage. In the present work, we have given a brief description of RoF technology, various optical modulation mechanisms involved along with the recent advancements. Further, in this paper a two-channel RoF link is proposed using Mach Zehnder Modulator (MZM) as external modulator. The performance of proposed link is evaluated at transmission distance of 10, 20, and 30 km for extinction ratio of MZM varying from 10 to 30 dB using Q Factor as performance metrics.

Keywords RoF · Radio Frequency (RF) · Control Station (CS) · Voice over IP (VoIP) · Base Station (BS) · Central Office (CO) · MZM · Remote Antenna Unit (RAU)

1 Introduction

The emergence of bandwidth-hungry multimedia applications such as video streaming, cloud computing, voice, and video calling services have exerted pressure on the

S. Kumar (✉) · D. Sharma · Payal · R. Singh
Department of Electronics and Communication Engineering, University Institute of Engineering &Technology, Maharshi Dayanand University, Rohtak, Haryana, India
e-mail: sureshvashist.uiet.ece@mdurohtak.ac.in

D. Sharma
e-mail: d.29deepak@gmail.com

Payal
e-mail: payalarora325@gmail.com

R. Singh
e-mail: rajbirbits32@gmail.com

© Springer Nature Singapore Pte Ltd. 2020   229
A. K. Luhach et al. (eds.), *First International Conference on Sustainable Technologies for Computational Intelligence*, Advances in Intelligent Systems and Computing 1045,
https://doi.org/10.1007/978-981-15-0029-9_18

traditional wireless telecommunication networks. The short-range and wired access networks have evolved for seamless delivery of services but these networks suffer from electronic bandwidth bottlenecks due to scarce availability of spectrum at higher RF frequencies. Also in mountainous and hilly areas, LOS is not possible for wireless connectivity which thereby causes a connectivity bottleneck. To overcome these hindrances, advanced optical technologies that provide higher capacity and longer reach have been incorporated in backbone and backhaul networks [1]. Optical fiber is a preferred way of communication which is reliable in handling and transmitting data over hundreds of kilometers with lower bit error rate [2].

This paper presents a two-channel RoF communication system using MZM as external modulator and the performance of the proposed system has been evaluated for Q Factor for variable extinction ratios of MZM at different transmission distances. The paper is organized as follows: Sect. 2 provides a detailed study of RoF system along with dispersion models. Section 3 discusses various optical modulation techniques. The recent advancement in RoF communication is presented in Sect. 4 followed by the proposed designed RoF system in Sect. 5. Section 6 provides result and discussion and Sect. 7 concludes the paper.

2 Radio-over-Fiber (RoF)

With the growth of bandwidth-hungry applications such as online multimedia streaming and VoIP services, the requirement for network bandwidth is increasing exponentially which can be sustained by RoF system. In order to support wireless applications in RoF system, electrical signal is modulated over an RF signal followed by transmission through optical fiber. RoF system has several inherent advantages such as large bandwidth, having immunity to RF interference, low power consumption, multi-operator with multi-service operation and allocation of resources dynamic [3, 4].

Figure 1 shows the architecture of an RoF system, it includes a BS, an optical fiber and RAU. RoF technology uses OF links to dispense RF signals between central location and the RAUs.

The RF signals from a central location are transmitted over optical fiber links to the RAUs with negligible attenuation losses. The transmission between control unit and BS involves sharing of radio signals on radio carrier frequency. An electrical signal is modulated over an optical source using MZM at the transmitter side which is transmitted through an optical fiber. Optical signal is converted back into electrical domain by a photodetector at the receiver side. In order to build the baseband signal from received signal, de-modulation, and signal processing is done at Control office. Both electrical to optical (E/O) and optical to electrical (O/E) conversion operations take place in transmitter and receiver sections. Therefore, the cost of transmission equipment gets reduced thereby making RoF system more economical and efficient [5].

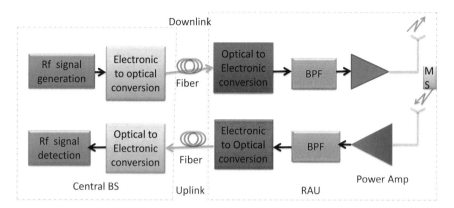

Fig. 1 Basic block diagram of RoF system

However, like traditional optical communication networks, the performance of ROF system is limited by several linear and nonlinear impairments. The inherent fading effect due to chromatic dispersion degrades the transmitted RF signal, which is analogous to multipath fading effects in wireless communication. First-order dispersion changes the phase of each sideband relative to the carrier can be expressed as:

$$\varphi = \frac{1000\pi\, L_f\ D_t(\lambda)\lambda_c^2\ f_{RF}^2}{c} \tag{1}$$

where L_f denotes the fiber link length, φ is 1st order dispersion, λ_c denotes central wavelength, f_{RF} is RF frequency, c is speed of light and D_t is the total dispersion coefficient expressed by the equation:

$$D_t(\lambda) = -\frac{\lambda_s}{c}\frac{dn}{d\lambda} - \frac{\Delta y}{2c}\left(\frac{d^2 n}{d\lambda^2}\right) - n_2\left(\frac{\Delta n}{c\lambda_s}\right) F(V) \tag{2}$$

where n_2 represents refractive-index of the cladding material, refractive-index difference is denoted by Δn, λ_s is the operating signal wavelength, $F(V)$ specifies a function for normalized frequency.

The other losses inherent to optical fiber include fiber attenuation and connector losses for a P-to-P fiber link expressed as:

$$\text{Optical Losses} = 2\big(\text{NL}_c + \text{ML}_{sp} + \alpha L_f\big)\,\text{dB} \tag{3}$$

where NL_c denotes connector loss; ML_{sp} is splicing loss, and α denotes fiber attenuation measured in dB/km.

3   Optical Modulation

In RoF communication systems, RF electrical signal is modulated over an optical carrier signal in accordance with the incoming traffic demands. This modulation process can be categorized either into direct modulation or external modulation using an external modulator. In direct modulation technique [also known as intensity modulation (IM)], the amplitude of the laser beam is directly modulated according to input RF Signal. This modulation scheme has limited bandwidth and the modulated signal suffers from pulse spreading due to which excessive chromatic dispersions and undesirable wavelength chirp arises [6].

In external modulation, an external modulator is used to modulate the optical carrier in accordance with the incoming electrical signal. The Lithium Niobate (LiNbO$_3$) MZM is the most popular modulator in optical communication system. MZM is used to control the phase and amplitude of an optical signal. It is of two types, single drive MZM and dual-drive MZM [7]. The block diagram of MZM is given in Fig. 2. The optical wave enters from the input side and then splits equally into two waveguide arms and the bias voltage is applied to these arms to induce a phase shift during the passage of signal. The optical signal traveling through both arms is recombined at the output of MZM and the phase difference between these two waves is converted into the amplitude-modulated wave.

The optical phase in each arm is controlled by changing the applied bias voltage on the electrode. An optical carrier out of one arm of the MZM is expressed as:

$$E_{\text{out}}(t) = E_0 \cos[\pi V(t)/2V_\pi] \cos(\omega_c t) \tag{4}$$

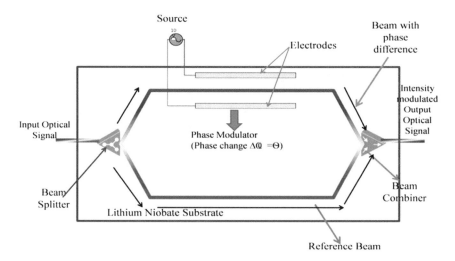

Fig. 2   Block diagram of Mach Zehnder modulator [8]

where E_0 is the amplitude of carrier, ω_c is input carrier frequency, V_π is the half-wave voltage of MZM and $V(t)$ is applied voltage.

$$V(t) = V_{\text{bias}} + V_{\text{m}} \cos(\omega_{\text{RF}} t) \tag{5}$$

where V_{bias} is the DC biased voltage, V_{m} is modulation voltage and ω_{RF} is angular frequency of driving signal.

The output optical field of MZM for an input optical power P_{in} can be expressed as:

$$E_{\text{dual}}(t) = \frac{1}{2}\left[e^{j\left(\frac{\pi V_{\text{dc1}}}{V_\pi} + \frac{\pi V_1(t)}{V_\pi}\right)} + e^{j\left(\frac{\pi V_{\text{dc2}}}{V_\pi} + \frac{\pi V_2(t)}{V_\pi}\right)}\right]\sqrt{2 P_{\text{in}}}e^{j\omega_c t} \tag{6}$$

where $E_{\text{dual}}(t)$ is the output optical field. Equation (6) can be further expressed as:

$$= \cos\left(\frac{\pi(V_{\text{dc1}} - V_{\text{dc2}})}{2V_\pi} + \frac{\pi(V_1(t) - V_2(t))}{2V_\pi}\right)e^{j\left[\frac{\pi(V_{\text{dc1}}+V_{\text{dc2}})}{2V_\pi} + \frac{\pi(V_1(t)+V_2(t))}{2V_\pi}\right]}\sqrt{2P_i}e^{j\omega_c t} \tag{7}$$

$$= \cos\left(\frac{\pi V_{\text{bias}}}{2V_\pi} + \frac{\pi V(t)}{2V_\pi}\right)e^{j\left[\frac{\pi(V_{\text{dc1}}+V_{\text{dc2}})}{2V_\pi} + \frac{\pi(V_1(t)+V_2(t))}{2V_\pi}\right]}\sqrt{2P_i}e^{j\omega_c t} \tag{8}$$

The average output optical power intensity ($P_{\text{dual}}(t)$) in terms of output optical field is given as:

$$P_{\text{dual}}(t) = |E_{\text{dual}}(t)|^2 = P_i\left[1 + \cos\left(\frac{\pi V_{\text{bias}}}{2V_\pi} + \frac{\pi V(t)}{2V_\pi}\right)\right] \tag{9}$$

The output optical field at the output of a non-ideal MZM (with a finite extinction ratio) associated with $\gamma = (\sqrt{\epsilon} - 1)/(\sqrt{\epsilon} + 1)$ is:

$$E_{\text{dual}}(t) = \frac{1}{2}\left[e^{j\left(\frac{\pi V_{\text{dc1}}}{V_\pi} + \frac{\pi V_1(t)}{V_\pi}\right)} + \gamma e^{j\left(\frac{\pi V_{\text{dc2}}}{V_\pi} + \frac{\pi V_2(t)}{V_\pi}\right)}\right]\sqrt{2P_i}\,e^{j\omega_c t} \tag{10}$$

4   Recent Advancement

The author in [9] provides a comprehensive review of RoF transmission technology, working principle, and advantages of MZM Modulator. The authors in [10] a comprehensive survey of RoF is presented for Wireless Broadband Access technologies for achieving effective delivery of wireless and baseband signal. In [11] the author proposed a distributed optical wireless network using state of art RoF technologies for effective and economical service delivery. In [12], the authors compared the performance of two WDM system based on Dual Electrode MZM and optical phase modulator for Raman crosstalk. Simulative results show that the output performance is improved using OPM based systems with considerable suppression of Raman crosstalk. In [13], the authors analyzed the performance of OFDM-RoF system for servicing higher bandwidth applications using different modulation techniques. The author concludes that OFDM-RoF system enhances bandwidth, provides lower attenuation, and lower BER using QAM and PSK. In [14] the authors analyzed the effects of four-wave mixing (FWM) crosstalk in WDM-RoF system for different fibers. Simulative results show that corning LEAF outperform other with a minimum BER of 10^{-35} and a minimum FWM optical power of -12 dBm. In [15], the author investigated the performance characteristics of optical access RoF network using soliton. Simulative results show that the total optical loss increases with increase in radio frequency and with other network parameters such as fiber link length, number of connectors, splices, and splitters. In [16], the authors analyzed the Crosstalk effects due to cross-phase modulation (XPM) in WDM-RoF system. The simulative result reveals degradation in system performance with increase in channel spacing and authors further concludes that improved performance can be achieved by optimizing by walk-off parameter, dispersion and pump, and probe wavelengths. In [17], the authors analyzed the impact of modulation index on the Stimulated Raman Scattering (SRS)-induced crosstalk. Simulative result shows that the crosstalk decreases with increasing modulation frequency and optical power. In [18], the author reviewed state of art strategies for high-performance RoF links. The authors further used OSSB + C technique for improving the modulation depth of an optical signal thereby enhanced sensitivity of RoF link. In [19], the authors presented a single channel RoF system and evaluate the performance of various modulators at variable power. The authors in [20] analyzed Digitized Radio over Fiber (DRoF) plan to achieve efficient transmission of analog signals over RoF links for sustaining higher bandwidth demands. The authors in [21] evaluated the performance of SCM-RoF system for intermodulation distortions and harmonic distortions using different modulation and amplification techniques. Simulative result shows that the use of OPM and EDFA considerably improve the performance and an improvement of 22–25 dB is found in received RF power. The authors in [22] designed WDM-RoF based Linear cell system and evaluated its performance for mitigating dispersion. The author utilizes a double-tone technology with an optical SSB modulation and for mitigating dispersion uses DSB-SC signal generation.

The earlier work reported in literature has evaluated RoF system using different modulation (direct and external modulation) and multiplexing schemes in isolations. In this work, we have evaluated the performance of external modulator MZM at variable extinction ratios over variable transmission distances using a designed a two-channel RoF system.

5 Proposed Work

We have proposed a two-channel RoF system in which MZM is used for modulation of two combined RF channels as shown in Fig. 3. The two PRBS generators act as transmitting source and generate random bitstreams. These bitstreams are digitally modulated to generate a NRZ pulse using NRZ pulse generator and then multiplexed over an RF sinusoidal source. The output of two sources is electrically combined and fed to MZM which converts the electrical signal into optical signal using a CW laser and transmits it over a fiber link.

At the receiver section, a delay interferometer is used to generate a phase delayed replica of the optical signal and corresponding electrical signal is obtained using a photodetector. The simulation parameters used are given in Table 1.

We have analyzed the performance of this proposed RoF system for Q Factor at transmission distance of 10, 20 and 30 km for variable extinction ratios of MZM. The varying extinction ratios of MZM used in simulation are 10, 15, 20, 25, 30 dB respectively. The insertion loss of MZM is kept at 2 dB; the fiber attenuation at 0.2 dB/Km and input RF frequency for channel 1 and 2 as 10 GHz and 15 GHz

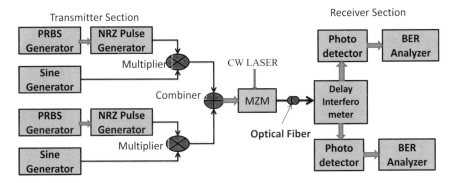

Fig. 3 Block diagram of proposed system

	Parameters	Values (units)	Fixed/variable
Table 1 The simulation parameters	Data rate	10 Gbps	Fixed
	Fiber length	10, 20, 30 km	Variable
	Modulation format	NRZ	–
	MZM extinction ratio	10, 15, 20, 25, 30 dB	Variable
	MZM insertion loss	2 dB	Fixed
	Fiber attenuation	0.2 dB/km	Varies depending on the length of fiber

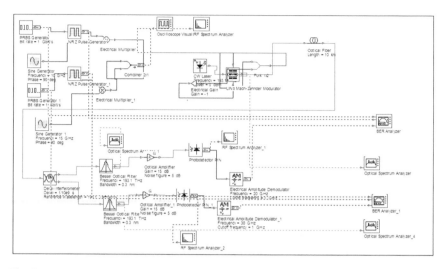

Fig. 4 Designed RoF layout

respectively. The overall fiber loss will vary depending on the length of optical fiber. Figure 4 shows the proposed layout in Optisystem simulator.

6 Result and Discussion

The performance evaluation of proposed RoF system is carried out using Optisystem taking Q factor as performance metric. The variation of Q Factor for channel 1 and channel 2 at a distance of 10 km for different extinction ratios is shown in Fig. 5.

For a fixed transmission length, the value of Q Factor increases with an increase in extinction ratio. The values of Q Factor for channel 1 are 16.8235, 17.6215, 20.6847, 22.27, and 21.6373 and for channel 2 are 18.0102, 18.6531, 19.103, 19.6365, and 19.9642 for extinction ratios of 10, 15, 20, 25, and 30 dB, respectively.

The variation of Q Factor for channel 1 and channel 2 at 20 km distance for different extinction ratios is shown in Fig. 6. The values of Q Factor for Channel 1 are 15.9966, 18.9577, 18.9293, 19.7784, and 20.6603 and for Channel 2 the values

Fig. 5 Q-factor variation with extinction ratio at 10 km fiber length

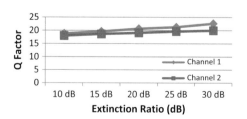

Fig. 6 *Q*-factor variation with extinction ratio at 20 km fiber length

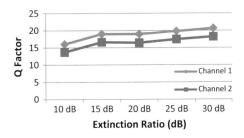

are 13.6628, 16.6026, 16.4052, 17.4336, and 18.1683 for extinction ratios of 10, 15, 20, 25, and 30 dB, respectively.

The variation of *Q* Factor for channel 1 and channel 2 at 20 km distance at different extinction ratios is shown in Fig. 7. The values of *Q* Factor for Channel 1 are 15.3417, 16.1067, 16.6323, 17.0332, 17.9332 and the values of Channel 2 are 10.7621, 12.3956, 13.7319, 15.2110, 17.2399, respectively, for extinction ratio of MZM 10, 15, 20, 25, and 30 dB. The graphical values for channel 1 and 2 at 10, 20, 30 km for variable extinction ratios are given in Table 2.

Figure 8 shows the eye diagrams of the proposed RoF system at 30 km transmission distance for 30 dB extinction ratio at 10 Gbps data rate. With an increase in fiber length, the fiber losses also increase which attenuates the signal. The distortion decreases with increasing Extinction ratio and increases with increase in fiber length which leads to spectral overlapping.

Fig. 7 *Q*-factor variation with extinction ratio at 30 km fiber length

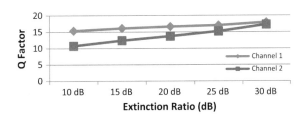

Table 2 Q-factor variation with extinction ratio at different fiber lengths

Extinction ratio (dB)	*Q* factor for channel 1			*Q* factor for channel 2		
	Transmission distance (km)			Transmission distance (km)		
	10	20	30	10	20	30
10	18.8235	15.9966	15.3417	18.0102	13.6628	10.7621
15	19.6215	18.9577	16.1067	18.6531	16.6026	12.3956
20	20.6847	18.9293	16.6323	19.103	16.4052	13.7319
25	21.27	19.7784	17.0332	19.6365	17.4336	15.2110
30	22.6373	20.6603	17.9332	19.9642	18.1683	17.2399

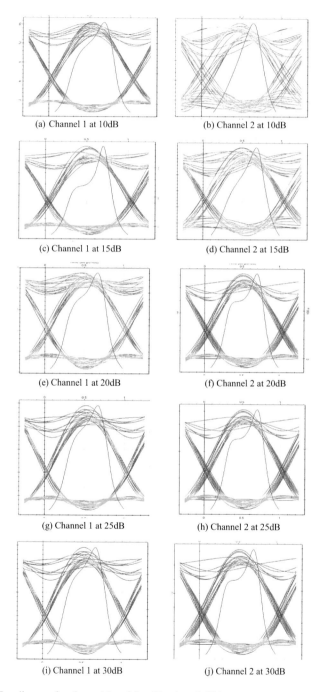

(a) Channel 1 at 10dB

(b) Channel 2 at 10dB

(c) Channel 1 at 15dB

(d) Channel 2 at 15dB

(e) Channel 1 at 20dB

(f) Channel 2 at 20dB

(g) Channel 1 at 25dB

(h) Channel 2 at 25dB

(i) Channel 1 at 30dB

(j) Channel 2 at 30dB

Fig. 8 i–j Eye diagram for channel 1 and 2 at fiber length 30 km

From Table 2 and Figs. 5, 6 and 7, it is evident that as the extinction ratio increases, the Q factor also increases. Also from graphical results, channel 1 with 10 GHz provides better Q Factor as compared to Channel 2 with 15 GHz as with increase in number of channels and RF, the different spectral components overlap with each other thereby limiting the quality of transmission. The eye diagrams for Channel 1 shows bigger eye-opening, large eye height, and width. Initially, distortion between various spectral components has been found to be less. But as the length of fiber increases, there is a slight distortion in the eye diagrams. It can be further concluded from results that Q Factor decreases with increase in transmission distance and the system works optimally up to a distance of 30 km at a data rate of 10 Gbps.

7 Conclusion

RoF technology uses the optical links to transport RF signals to the RAUs and caters for the exponentially increasing traffic demands. A two-channel RoF link with MZM as external modulator has been proposed and evaluated. The performance of the designed system has been evaluated using Q factor as a performance metric for different extinction ratios of 10, 15, 20, 25, and 30 dB at varying fiber length of 10, 20, and 30 km for each case. From simulation results and findings, it can be concluded that for the proposed RoF system, with increase in extinction ratio, the Q factor increases but decreases with increase in optical fiber length and with increase in channel RF. The proposed system works optimally even at the extinction ratio up to 30 dB for the external modulator and fiber length of 30 km.

References

1. Khosia, V.: A comprehensive review of recent advancement in optical communication networks. Int. J. Comput. Sci. Eng. **6**(9), 617–626 (2018). https://doi.org/10.26438/ijcse/v6i9.617626
2. Osman, W.M., Babiker, A., Nabi, A., Billal, K.H.: Optical fiber review. J. Electr. Electron. Syst. **7**(1), 1–4 (2018). https://doi.org/10.4172/2332-0796.1000249
3. Mitchell, J.E.: Emerging radio-over-fiber technologies and networks: challenges and issues. Invited Paper. In: Proceedings of SPIE, vol. 7234 723407-1(2014)
4. Jawad, S.S., Fyath, R.S.: Transmission performance of analog radio-over-fiber fronthaul for 5G mobile networks. Int. J. Netw. Commun. **8**(3), 81–96 (2018). https://doi.org/10.5923/j.ijnc.20180803.03
5. Singh, N., Kaur, H.: A review on radio over fiber technology with its benefits and limitations. Int. J. Innov. Res. Comput. Commun. Eng. **4**(7), 14178–14181 (2016). https://doi.org/10.15680/IJIRCCE.2016.0407139
6. Sharma, A., Rana, S.: Implementation of radio over fiber technology with different filtration techniques. Int. J. Res. Appl. Sci. Eng. Technol. (IJRASET). **5**(8), 783–789 (2017)
7. Zin, A.M., Idrus, S.M., Zulkifli, N.: The characterization of radio-over-fiber employed GPON architecture for wireless distribution network. Int. J. Mach. Learn. Comput. **1**(5), 522–527 (2011)

8. Sharma, V., Singh, A., Sharma, A.K.: Simulative investigation the impact of optical amplifica-
    tion techniques on single and dual-electrode MZM and direct modulator's in single-tone RoF
    system. Optik **122**(5), 371–374 (2011). https://doi.org/10.1016/j.ijleo.2010.02.021
9. Singh, R., Ahalawat, M., Sharma, D.: A review of radio over fiber communication system. Int.
    J. Enhanc. Res. Manag. Comput. Appl. **6**(4), 23–29 (2017)
10. Karthikeyan, R., Prakasam, S.: A survey on radio over fiber (RoF) for wireless broadband
    access technologies. Int. J. Comput. Appl. (0975–8887) **64**(12) (2013)
11. Novak, D., et al.: Radio-over-fiber technologies for emerging wireless systems. IEEE J. Quan-
    tum Electron. 0018-9197 (2015). https://doi.org/10.1109/jqe.2015.2504107
12. Kumar, S., Nain, A.: Simulative investigation of WDM RoF systems including the effect of
    the Raman crosstalk using different modulators. Telecommun. Radio Eng. **75**(14), 1243–1254
    (2016). https://doi.org/10.1615/TelecomRadEng.v75.i14.20
13. Karthikeyan, R., Prakasam, S.: A review—OFDM-RoF (Radio over Fiber) system for wireless
    network. Int. J. Res. Comput. Commun. Technol. **3**(3), 344–349 (2014)
14. Nain, A., Kumar, S.: Mitigation of FWM induced crosstalk in WDM RoF systems by employing
    different fibers. J. Optics (India) **46**(4), 492–498 (2017). https://doi.org/10.1007/s12596-017-
    0413-2
15. Mohamed, A.E., Zaki Rashed, A.N., Tabbour, M.S.F.: Transmission characteristics of radio
    over fiber (ROF) millimeter wave systems in local area optical communication networks. Int.
    J. Adv. Netw. Appl. **02**(06), 876–886 (2011)
16. Nain, A., Kumar, S., Singla, S.: Impact of XPM crosstalk on SCM-based RoF systems. J. Opt.
    Commun. **38**(3), 319–324 (2016). https://doi.org/10.1515/joc-2016-0045
17. Nain, A., Kumar, S., Singla, S.: Performance estimation of WDM radio-over-fiber links under
    the influence of SRS induced crosstalk. In: Singh, R., Choudhury, S. (eds) Proceeding of
    International Conference on Intelligent Communication, Control and Devices. Advances in
    Intelligent Systems and Computing, vol. 479. Springer, Singapore (2017)
18. Lim, C., Nirmalathas, A., Yang, Y., Novak, D.: Radio-over-fiber systems. In: Chiaroni, D. (ed.)
    Optical Transmission Systems, Switching, and Subsystems VII. Proceedings of SPIE-OSA-
    IEEE Asia Communications and Photonics, SPIE, vol. 7632, 76321S (2009)
19. Nain, A., Kumar, S.: Performance investigation of different modulation schemes in RoF sys-
    tems under the influence of self phase modulation. J. Opt. Commun. **39**(3), 343–347 (2017).
    Retrieved 15 Sept 2018, from https://doi.org/10.1515/joc-2016-0155
20. Wu, J., et al.: RF photonics: an optical microcombs' perspective. IEEE J. Sel. Top. Quantum
    Electron. **24**(4), 1–20 (July–Aug 2018). Art no. 6101020. https://doi.org/10.1109/jstqe.2018.
    2805814
21. Kumar, S., Sharma, D., Nain, A.: Evaluation of sub carrier multiplexing based RoF system
    against non-linear distortions using different modulation techniques. Int. J. Adv. Res. Com-
    put. Sci. Softw. Eng. (IJARCSSE) **7**(6), 454–461 (2017). https://doi.org/10.23956/ijarcsse/
    V7I6/019
22. Kanno, A., Dat, P.T., Yamamoto, N., Kawanishi, T.: Millimeter-wave radio-over-fiber network
    for linear cell systems. J. Light Wave Technol. **36**(2), 533–540 (2018). https://doi.org/10.1109/
    JLT.2017.2779744

A Hybrid Cryptographic Technique for File Storage Mechanism Over Cloud

Shivam Sharma, Kanav Singla, Geetanjali Rathee and Hemraj Saini

Abstract The cloud is a radical platform that conveys dynamic, virtualization asset pools and high convenience. Since distributed computing lays on web, security issues like information security, protection, secrecy, and verification are experienced. In order to get rid of these, an assortment of mechanisms and encryption algorithms are utilized in various blends. On the comparable terms, we made utilization of hybrid encryption with the utilization of crossbreed cryptographic calculations to upgrade the security of information or data file on cloud. We plan at scrutinizing different incorporation of encryption algorithms, in view of various execution constraints to reason a hybrid calculation which can anchor information more effectively on cloud.

Keywords Cloud · Hybrid cryptography · Security · Encryption techniques · Decryption techniques · File security

1 Introduction

Cloud computing was engineered to deliver computing services over the internet. Both the fully developed, ready-to-use applications, hardware resources like network, storage are provided according to user's needs. These resources are hosted at the data centers spread globally. The resources are utilized from a configurable pool of similar resources which can be relaxed and provisioned depending on their requirement [1]. According to the definition of NIST, Cloud computing is a sculpt for

S. Sharma · K. Singla · G. Rathee (✉) · H. Saini
Department of Computer Science and Engineering, Jaypee University of Information
Technology, Waknaghat, Solan, Himachal Pradesh 173 234, India
e-mail: geetanjali.rathee123@gmail.com

S. Sharma
e-mail: shivam14.cr7@gmail.com

K. Singla
e-mail: singlaheel203@gmail.com

H. Saini
e-mail: hemraj1977@yahoo.co.in

© Springer Nature Singapore Pte Ltd. 2020
A. K. Luhach et al. (eds.), *First International Conference on Sustainable Technologies for Computational Intelligence*, Advances in Intelligent Systems and Computing 1045, https://doi.org/10.1007/978-981-15-0029-9_19

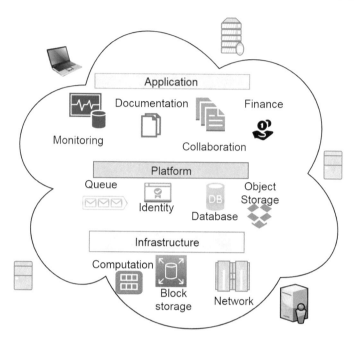

Fig. 1 Wireless mesh network

enabling suitable, on-demand system access to a mutual pool of configurable evaluating resources (e.g., storage, applications, servers, networks, and services) that can be quickly released and provisioned with minimal supervision service provider or effort interaction as depicted in Fig. 1.

- **Need for Security in Cloud Computing**

Although, Cloud computing has grown to be a popular and successful business model due to its attractive characteristics and features. In addition to the reimbursements, its features may result in severe problems specific to cloud security issues [2–4]. The entities whose concern is the security of cloud still hesitate to transfer their business to cloud. The security issues that have been governed barricade of the growth and worldwide use of cloud computing are defined as follows:

- Outsourcing: It refers to providing data to a third-party cloud hosting provider in order to support and deliver IT services that could be handed over in-house. Outsourcing means that the clients physically lose control of their tasks and data. The problem that user's data is with someone else and no longer in proximity of user is a prime security concern.
- Multi-tenancy: It implies that the cloud platform is distributed and shared, exploited by multiple clients. Furthermore, in an environment that is virtual, Different cloud user's data may reside on same physical servers in virtual environment.

A series of security problems and issues such as computation breach, data breach, and flooding attack are experienced.

- **Cryptography**

Converting useful data into unreadable text so that no one else can read it except the pre-determined user is encryption and the techniques used are called cryptographic techniques. It may be accomplished through scrambling words, using code words or using highly efficient mathematical techniques. There are different algorithms classified in two classes [5–8].

a. Symmetric Algorithms—This algorithm uses same key to decrypt and encrypt the message. For instance, if some user desires to convey a message to another and wants nobody else should read it. So, she/he can encrypt the data using a secret key that can be shared with the receiver who will decipher the data using the same key. As can be judged the key can be shared with all the users who ought to retrieve data.

b. Asymmetric Algorithms—These types of algorithms make use of two separate keys in process of decryption and encryption, respectively. Like a user ciphers, the data using one key (known as public key) and receiver decrypts the message using separate key (known as private key). Public Key—This key is nearby to all over the internet and used to cipher the secret text. Private Key—As clear form, the name it is a receiver's private key and only the receiver will be able to use this key to decrypt the required message.

We know that classic encryption schemes have long been used for security purposes and they have been successful to some extent. So why is there a need to use hybrid system? The basic answer is security. Basically, what hybrid cryptography does is that it mixes the effectiveness of symmetric encryption with easiness of public-key encryption. It enhances security level and both types of encryption efficiency. It has several advantages. The new technology was fast, dependable, and data sharing was efficient. There were certainly many benefits to it and had its limitations that were harmful to data owner. As the data was stored on outside system not within the physical reach of user there are the concerns for security of data. Data sharing may also lead to breach of sensitive data and privacy of user. There have cryptographic techniques to cipher the data and code it so that the attacker cannot understand it.

Techniques like DES, 3DES, AES, RSA, RC4 and many others have proven to be a success in hiding the data and securing it. However, everything is prone to attack and every individual encryption technique is known to be attack prone due to an increase in computational power days. Any new technique which will be developed will be known to attacks but to increase the security and efficiency hybrid encryption is used. Hybrid encryption provides effectiveness of public-key cryptography and easiness of private key cryptography. In this paper, a hybrid cryptographic mechanism is proposed by incorporating AES, DES, and RC4 techniques to implement hybrid cryptography. The remaining structure of the paper is structured as follows. Section 2 described the previous security issues proposed by different researchers. The hybrid secure mechanism is portrayed in Sect. 3. The test plan of proposed phenomenon is elaborated in Sect. 4. Further, Sect. 5 represented the performance analysis of encryption and decryption time. Finally, Sect. 6 concludes the paper.

2   Comparative Analysis of Previous Approaches

This research paper introduces cloud architecture and characteristics of cloud. Through the paper, authors have identified present security and privacy issues and discuss their proneness to attack and provide currently used defence mechanisms by analyzing their efficiency in doing the required work [9, 10]. The authors have used current strategies including Virtualization, Data centre Techniques, and MapReduce centre that helps the cloud vendor in overcoming those challenges, but each has its shortcomings and not useful. Author gives a cloud security ecosystem to model different attacks and use different mechanisms against them. Further, the paper provided the security challenges that are a big hindrance in the customer trusting the cloud technology for commercial and personal operations. The issues regarding current defence strategies have not been ratified and left open for future resolve of these issues and enhance the user trust on cloud. Various authors have proposed several cryptographic techniques in order to ensure the security overcloud. The below text describes number of security algorithms and techniques proposed by various author's.

- **Secure Data Sharing in Clouds**

In order of preserving the thoughts of challenges in cloud the researcher has designed a new SeDaSC methodology for securing the facts storage and transfer. The authors have analysed that like cl-pre-scheme, certificates-much-less encryption and el-ghamal cryptography schemes with their blessings and cons. Like the cl-pre-scheme generates a public-non-public key pair and uses bilinear method for encryption which growth the cost of encryption. The certificates much less encryption even though tries to enhance upon the cost issue, however, falls short are in trusted facts garage. On the other hand, the el-ghamal scheme uses bilinear and incremental encryption. However, complexities nonetheless exist [11]. Although, it is clear from the picture that key evaluation time changes merely with augment in file size and the encryption time too is commendable with this methodology. However, the time needs to check with file sizes of greater than 1 GB to really test [12, 13]. Additionally, the important element is that it stores key on outside server that can be underneath outside threats, so measures should be taken to make it extra comfy or limit the level of trust in the server.

- **Brief Study of Encryption Algorithms (RC4, DES, 3DES, and AES) for Information Security**

This paper gives a comparative view of different encryption algorithms on parameters like CPU usage, encryption time, ROM utilization, throughput with different file sizes, length of packet and data type [14–16].

Further, in order to ensure the privacy for cloud providers that provide the services to the clients upon request, plenty of security techniques have been proposed. Ahmad el at. [17] have proposed a game-theoretical model that assists the brokers by identifying the malicious providers that manage and create the federations. The proposed mechanism has been analyzed and simulated over a dataset through CloudSim simulator. In order to ensure security over cloud environment, Akshay et al. [18] have proposed a hybrid cryptographic mechanism that includes both asymmetric and symmetric encryption techniques. Author's have provided multilevel encryption for authenticating the clients using hash technique. The proposed technique is validated by computing the results in CloudSim simulator. From service requested clients to cloud providers, the security has been provided through various cryptographic algorithms including AES, DES, hybrid, hash or MAC algorithms. The scope of this paper is to discuss various cryptographic approaches and propose a hybrid encryption algorithm in order to ensure the security during transmission of file from one place to another. Further, author Shefali et al. [19] have used MD5 and AES encryption techniques for ensuring the security during the login and data access by the user over the cloud. The proposed mechanism need not authenticate the user upon login while during the data accessing, the AES and MD5 algorithms require user authentication.

- **Rivest–Shamir–Adleman (RSA)**

It is a public cryptosystem technique that incorporates block size encryption and variable key size. The steps involved are: Generate two distinct prime numbers, Calculate t as product of two, Now compute phi(t), Find d such that $d * e = 1$, Public Key is $(1, e)$ and Private key is $(1, d)$. It is the most apparent disadvantage is that if two numbers are of massive length then it takes more time and should be of comparable size. The below text elaborated a brief introduction of the standard algorithms.

- **Data Encryption Standard (DES)**

This encrypt algorithm is the most widely used because it works on bits. Length of message at one time to be encrypted is 64 bits and the key size of 64-bits but every $7x$ bit is a check bit which is removed in making of sub-keys. As known, it is one of the best algorithms because no such possible attack is known to crack it other than brute-force which is costly and time-consuming.

- **Triple DES**

Alias 3 DES, it is an extension of DES because it has greater key length. DES was more prone to brute attacks due to increasing computational power. Although it is better version of standard DES, but it can be breached using MITM attacks. So, to defend better against them it can be implemented using secret-key size of 112-bits.

- **Advanced Encryption Standard (AES)**

AES also recognized as Rijndael structure is an iterative algorithm rather being called a fiestel structure. It is iterative because it uses rounds to convert data to ciphertext depending on key sizes.

- **RC4**

RC4 is produced by Bokkos Rivest also called Rivest Cipher four. In this the stream figure is utilized for secret composing of the plain content. On the off chance that the basic square figures don't appear to utilize mackintosh effectively, bit-fluttering assault is doable and furthermore the stream figure assault is moreover powerless on the off chance that they're not legitimately authorized.

- **Blowfish**

Also, a symmetric cipher built as an alternate for more commercial AES and DES with varying key sizes of length from 32 to 448 bits with a block size of 64 bits. It's a 16-round cipher. It is efficient cipher and can be used commercially as alternative to AES. Most possible obvious attacks on this encryption are birthday attacks and should not be used for files of size more than 4 GB.

3 Proposed Solution

The flow graph of proposed approach where the source file is encrypted using the hybrid encryption is shown in Fig. 2. The file is selected and divided into three equal parts using the file system module. Then, each part is encrypted using the AES, DES and RC4 encryption techniques. The encrypted parts are then merged and saved into a single file which, then, can be uploaded on the cloud servers.

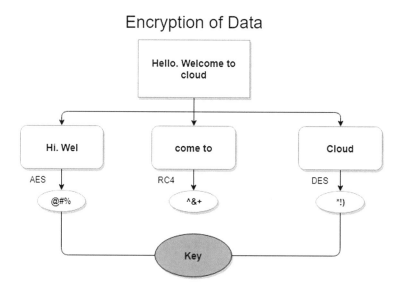

Fig. 2 Process of file encryption

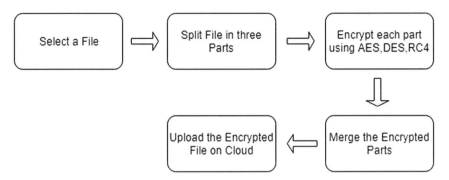

Fig. 3 Flow diagram of file encryption

File Splitting

The input file is split into three parts so that different encryption algorithms can be applied to each. Split-file module provides an efficient way to split and merge file into multiple parts.

Crypto JS

Crypto-js is a JavaScript library of crypto standards with a growing cluster of secure cryptographic algorithms implemented in JavaScript utilizing best practices and patterns. These algorithms are fast and have a simple and consistent interface. The following algorithms from module Crypto-js are used in this project. AES (Advanced Encryption Standard), DES (Data Encryption Standard), RC4 (Rivest Cipher 4), PBKDF2 (Password-Based Key Derivation Function), PKCS7 (Public Key Cryptography Standards).

File Encryption

The file is encrypted using three encryption algorithms. The steps of file encryption are shown in the given Fig. 3.

File Decryption

Further, for decryption at the receiver side as depicted in Fig. 4, the encrypted file is downloaded from the cloud servers and then split into three parts using a certain special character into three parts whereupon each part is then decrypted using the same techniques which were used for encryption, i.e., AES, DES, and RC4. The decrypted parts are merged into one and the retrieved file can then be used.

4   Test Plan

Testing will be done on different files varying from 1 MB till 30 MB.

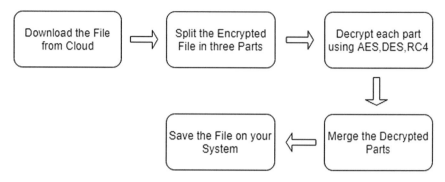

Fig. 4 Flow diagram of file decryption

Timing for encryption as well as decryption of such files using hybrid algorithm will be noted and compared with the respective encryption and decryption timings of same files using AES encryption alone. Section 5.2 offers a clear view of the time difference among the standard algorithm and the proposed algorithm. This section provides a rough idea of the comparison between the usual algorithm and the proposed algorithm. A file of size 10 MB is taken and is encrypted, firstly, AES and then using the hybrid algorithm. Also, decryption is done of the encrypted file using both, above mentioned, algorithms. The snippet of each algorithm outlines the time required to process the algorithm.

4.1 Encryption and Decryption Time for AES

A file named file.txt of size 10 MB is encrypted using the AES algorithm. Figure 5 shows how much time is required to encrypt the file while Fig. 6 depicts the decryption time.

4.2 Encryption and Decryption Time for Hybrid Algorithm

Now the file is encrypted using the hybrid algorithm (AES, DES, RC4). Figure 7 shows the time taken to encrypt the file.

Figure 8 depicts the decryption time of hybrid algorithm. The file of size 10 MB was encrypted and decrypted using the standard algorithm as well as the proposed algorithm. The timings from snippets in section show that the proposed algorithm is both comparatively faster in encryption of the file as well as decryption.

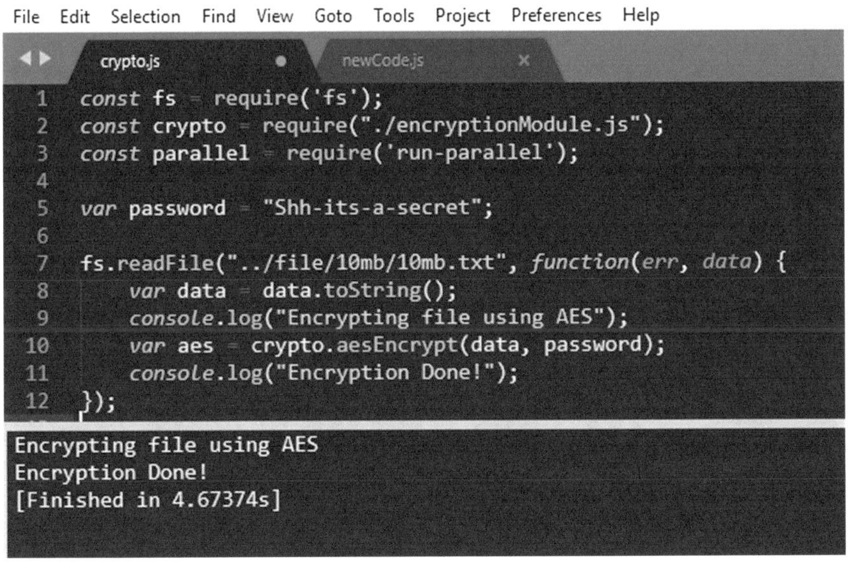

Fig. 5 AES encryption time

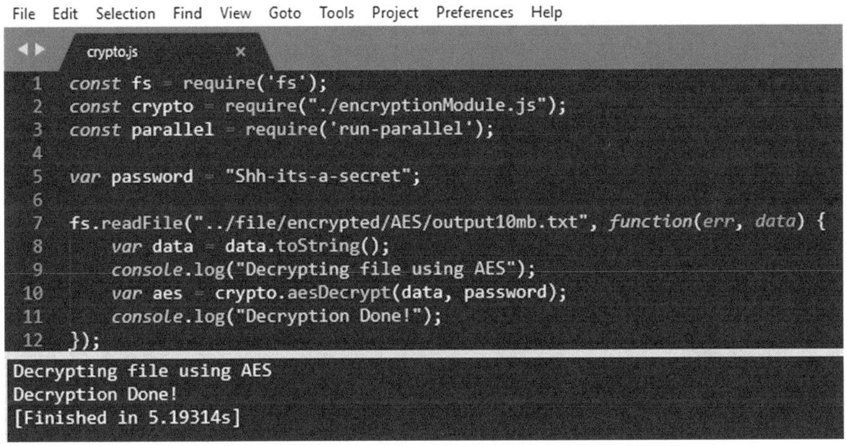

Fig. 6 AES decryption time

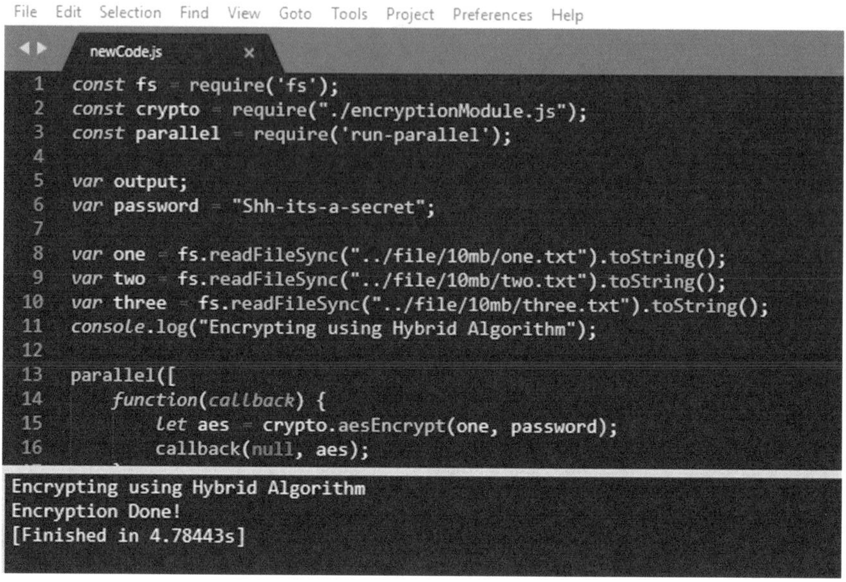

```
File   Edit  Selection  Find  View  Goto  Tools  Project  Preferences  Help

◄ ►      newCode.js          ×

1    const fs = require('fs');
2    const crypto = require("./encryptionModule.js");
3    const parallel = require('run-parallel');
4
5    var output;
6    var password = "Shh-its-a-secret";
7
8    var one = fs.readFileSync("../file/10mb/one.txt").toString();
9    var two = fs.readFileSync("../file/10mb/two.txt").toString();
10   var three = fs.readFileSync("../file/10mb/three.txt").toString();
11   console.log("Encrypting using Hybrid Algorithm");
12
13   parallel([
14       function(callback) {
15           let aes = crypto.aesEncrypt(one, password);
16           callback(null, aes);

Encrypting using Hybrid Algorithm
Encryption Done!
[Finished in 4.78443s]
```

Fig. 7 Hybrid encryption time

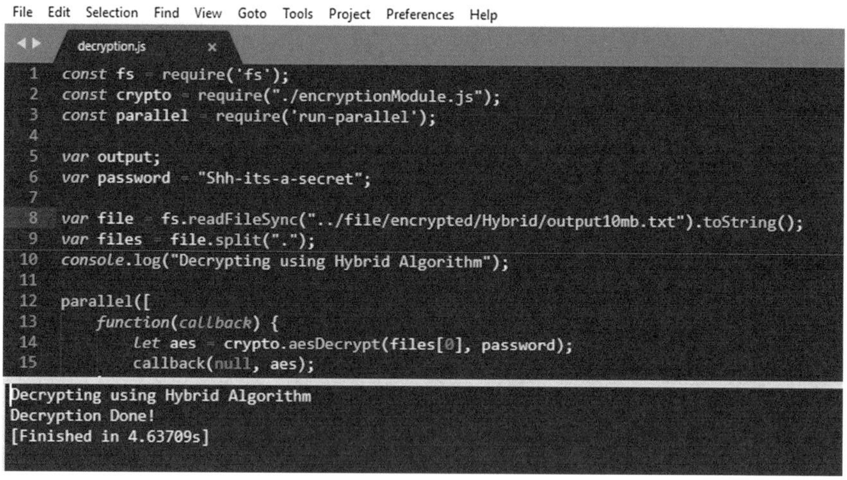

```
File   Edit  Selection  Find  View  Goto  Tools  Project  Preferences  Help

◄ ►      decryption.js          ×

1    const fs = require('fs');
2    const crypto = require("./encryptionModule.js");
3    const parallel = require('run-parallel');
4
5    var output;
6    var password = "Shh-its-a-secret";
7
8    var file = fs.readFileSync("../file/encrypted/Hybrid/output10mb.txt").toString();
9    var files = file.split(".");
10   console.log("Decrypting using Hybrid Algorithm");
11
12   parallel([
13       function(callback) {
14           let aes = crypto.aesDecrypt(files[0], password);
15           callback(null, aes);

Decrypting using Hybrid Algorithm
Decryption Done!
[Finished in 4.63709s]
```

Fig. 8 Hybrid decryption time

Table 1 Comparision of different encryption algorithms

Factors	DES	AES	RC4
Created by	IBM in 1975	Vincent Rijmen, Joan Daemen in 2001	Ron Rivest in 1987
Key length	56 bits	128, 192 or 256 bits	40–2048 bits
Round(s)	16	10, 12 or 14	1
Block size	64 bits	128 bits	2064 bits (1684 effective)
Speed	Slow	Fast	Fast
Security	Not secure enough	Excellent security	Adequate security

5 Result and Performance

In this section, various cryptographic algorithms are compared based on their key lengths, number of rounds, block sizes and other assets.

5.1 Algorithms Testing

Testing on various files will be done using AES (Standard Algorithm) and proposed algorithm (Hybrid Algorithm—AES, DES and RC4) so that the performance of the algorithm can be recorded in relation to the increasing size of the file. The files used in the tests vary from 1 MB till 30 MB. Graphs of both the algorithms along with their tables are shown below. Table 1 depicts the comparison of different encryption techniques over various parameters.

5.2 AES Testing

Encryption of file sizes ranging 1, 5, 10, 20 and 30 MB is done using AES Encryption. Encryption time of these files is calculated and plotted on a graph depicted in Fig. 9. Similarly, decryption of file sizes ranging 1, 5, 10, 20 and 30 MB is done using AES decryption algorithm. Decryption time of these files is calculated and plotted on a graph depicted Fig. 10.

5.3 Hybrid Encryption (AES, DES, RC4) Testing

Again, encryption of file sizes ranging 1, 5, 10, 20, and 30 MB is done using Hybrid Encryption. As already done on the above process, encryption time of these files is calculated and plotted on a graph depicted in Fig. 11.

Fig. 9 AES encryption time

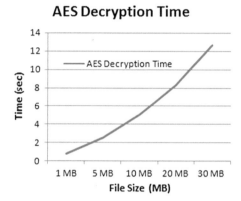

Fig. 10 AES decryption time

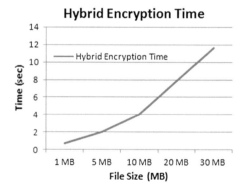

Fig. 11 Hybrid encryption time

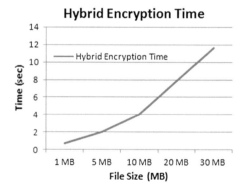

Fig. 12 Hybrid decryption time

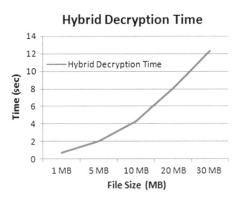

Similarly, decryption of file sizes ranging 1, 5, 10, 20, and 30 MB is done using hybrid decryption algorithm. Decryption time of these files is calculated and plotted on a graph depicted in Fig. 12.

5.4 Comparison of Encryption Algorithms Testing Data

Bih-Hwang et al. [20] have ensured the cloud data security from third-party data models. Here, the author has used cloud platform know as Heroku that is responsible for managing and integrating the services using cryptographic algorithms. Lee et al. have used AES encryption technique to provide Heroku as cloud platform and provide data security from third parties. The authors have analyzed and validated the proposed mechanism by analyzing the encryption time over various file sizes. In this paper we have used this as over base paper and compared the proposed hybrid encryption over AES with different file sizes. Comparisons of the encryption time as well as decryption time of the files ranging 1–30 MB is shown in Table 2. Hybrid algorithm needs 10–15% less time for file to be encrypted in comparison to other encryption techniques. With single encryption algorithm such kind of data security cannot be provided.

The Graphs presented in Figs. 13 and 14 depicts the encryption time and decryption time of files with various files. As depicted in Fig. 13, AES approach ensures the

Table 2 Comparision of different algorithms

File size (MB)	AES encryption	Hybrid encryption	AES decryption	Hybrid decryption
1	0.717	0.732	0.789	0.685
5	2.544	2.076	2.499	2.034
10	4.457	4.021	5.043	4.232
20	8.429	7.875	8.325	8.063
30	13.710	11.618	12.645	12.328

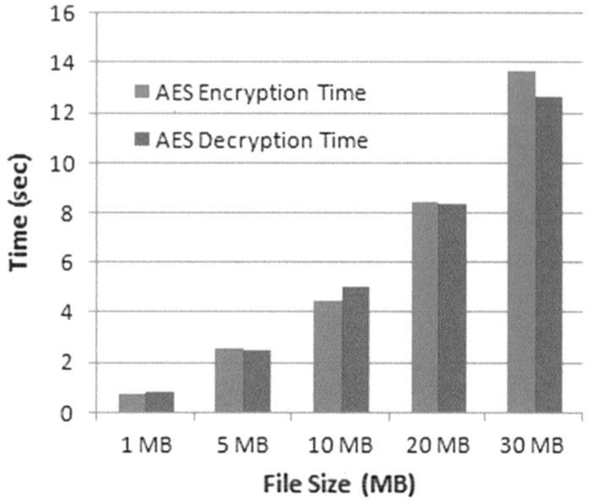

Fig. 13 Time comparison
between AES and hybrid
encryption algorithm

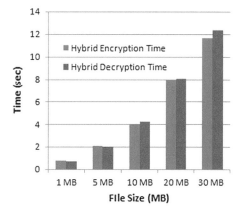

Fig. 14 Time comparison
between AES and hybrid
decryption algorithm

file security with the increase of file size, however, proposed approach gives better results as compare to AES because of hybrid encryption process. As the file size increases, proposed mechanism performs better. Similarly, the proposed algorithm performs better for the decryption process as presented in Fig. 14 with increasing file size, the decryption time taken for the hybrid algorithm gradually decreases in contrast to the AES algorithm.

6 Conclusion

By the investigation of the accompanying outcomes, we could reason that the hybrid algorithms of AES, DES, and RC4 gave us better execution time in contrast with that of AES algorithm alone. We saw that for moderately little document sizes AES individually produced better throughput in bytes per millisecond when contrasted with that of hybrid algorithm including AES, DES, and RC4. However, as the measure of record augmented, proposed algorithm indicated better outcomes. Mulling over, the substantial measure of information that business applications will in general store on the cloud, record sizes can fluctuate to extensive numbers, thus the utilization of hybrid algorithm calculation is proposed to actualize staggered security on cloud information storage. Based on the examination of execution of single encryption calculations and multiple encryption calculations, it is concluded that hybrid encryption involving multiple encryption algorithms (AES, DES, and RC4) provided much better results than single algorithm (AES) implemented alone. In the future, we plan to use certain other encryption algorithms such as Blowfish and other public-key encryption techniques like RSA to be implemented in the project.

References

1. Ali, M., Dhamotharan, R., Khan, E., Khan, S.U., Vasilakos Athanasios V., Li, K., Zomaya Albert, Y.: SeDaSC: secure data sharing in clouds. IEEE Syst. J. **11**(2), 395–404 (2017)
2. Xiao, Z., Yang, X.: Security and privacy in cloud computing. IEEE Commun. Surv. Tutor. **15**(2), 843–859 (2013)
3. Meng, S., Wang, Y., Jiao, L., Miao, Z., Sun, K.: Hierarchical evolutionary game based dynamic cloudlet selection and bandwidth allocation for mobile cloud computing environment. IET Commun. **13**(1), 16–25 (2018)
4. Esposito, C., Castiglione, A., Pop, F., Choo, K.-K. R.: Challenges of connecting edge and cloud computing: a security and forensic perspective. IEEE Cloud Comput. **4**(2), 13–17 (2017)
5. Park, J.-E., Park, Y.-H.: Fog-based file sharing for secure and efficient file management in personal area network with heterogeneous wearable devices. J. Commun. Netw. **20**(3), 279–290 (2018)
6. Zhang, H., Zhou, Z., Ye, L., Du, X.: Towards privacy preserving publishing of set-valued data on hybrid cloud. IEEE Trans. Cloud Comput. **6**(2), 316–329 (2018)
7. Thakur, J., Kumar, N.: DES, AES and Blowfish: symmetric key cryptography algorithms simulation based performance analysis. Int. J. Emerg. Technol. Adv. Eng. **1**(2), 6–12 (2011)
8. Li, J., Li, Y.K., Chen, X., Lee, P.P.C., Lou, W.: A hybrid cloud approach for secure authorized deduplication. IEEE Trans. Parallel Distrib. Syst. **26**(5), 1206–1216 (2015)
9. Seo, S.-H., Nabeel, M., Ding, X., Betrino, E.: An efficient certificateless encryption for secure data sharing in public clouds. IEEE Trans. Knowl. Data Eng. **26**(9), 2107–2119 (2014)
10. Chen, D., Li, X., Wang, L., Khan, S.U., Wang, J., Zeng, K., Cai, C.: Fast and scalable multi-way analysis of massive neural data. IEEE Trans. Comput. **64**(3), 707–719 (2015)
11. Shiu, Y.-S., Chang, S.Y., Wu, H.-C., Huang, S.C.-H., Chen, H.-H.: Physical layer security in wireless networks: a tutorial. IEEE Wirel. Commun. **18**(2), 66–74 (2011)
12. Wei, , Zhu, H., Cao, Z., Dong, X., Jia, W., Chen, Y., Vasilakos, A.V.: Security and privacy for storage and computation in cloud computing. Inf. Sci. **258**, 371–386 (2014)
13. Singh, S.P., Manini, R.: Comparison of data encryption algorithms. Int. J. Comput. Sci. Commun. **2**(1), 125–127 (2011)
14. Singhal, N., Raina, J.P.S.: Comparative analysis of AES and RC4 algorithms for better utilization. Int. J. Comput. Trends Technol. **2**(6), 177–181 (2011)
15. Sanaei, Z., Abolfazli, S., Gani, A., Buyya, R.: Heterogeneity in mobile cloud computing: taxonomy and open challenges. IEEE Commun. Surv. Tutor. **16**(1), 369–392 (2014)
16. Rong, C., Nguyen Son, T., Jaatun, M.G.: Beyond lightning: a survey on security challenges in cloud computing. Comput. Electr. Eng. **39**(1), 47–54 (2013)
17. Hammoud, A., Otrok, H., Mourad, A., Wahab, O.A., Bentahar, J.: On the detection of passive malicious providers in cloud federations. IEEE Commun. Lett. **23**(1), 64–67 (2019)

18. Arora, A., Khanna, A., Rastogi, A., Agarwal, A.: Cloud security ecosystem for data security and privacy. In: 7th IEEE International Conference on Cloud Computing, Data Science & Engineering-Confluence, pp. 288–292 (2017)
19. Ojha, S., Rajput, V.: AES and MD5 based secure authentication in cloud computing. In: IEEE International Conference on I-SMAC (IoT in Social, Mobile, Analytics and Cloud) (I-SMAC), pp. 856–860 (2017)
20. Lee, B.-H., Dewi, E.K., Wajdi, M.F.: Data security in cloud computing using AES under HEROKU cloud. In: 27th IEEE Conference on Wireless and Optical Communication Conference (WOCC), pp. 1–5 (2018)

Community Development Through Sustainable Technology—A Proposed Study with Irula Tribe of Masinagudi and Ebbanad Villages of Nilgiri District

**T. V. Rajeevan⑩, S. Rajendrakumar⑩, Thangavel Senthilkumar⑩,
S. Udhaya Kumar⑩ and P. Subramaniam⑩**

Abstract Irulars are one among six Particularly Vulnerable Tribal Groups (PVTGs) of Tamil Nadu. In the study area, Masinagudi and Ebbanad villages, the Indigenous community dwells in thick forest region of Mudumalai Wildlife Sanctuary. In comparison with other PVTGs, Irula community of these regions are behind in development due to their remote location. They have very minimum basic amenities, nutritional deficiencies, high dropout rate in schools, low literacy rate, and often migrate to other places in search of better livelihood. It resulted in the community to face the vulnerable living conditions. Unemployment, indebtedness, inadequate healthcare and poor education facility for children are increasing problems of this community. Added to that, the practice of unsustainable agriculture, chemical utilization, contaminating the wildlife habitats and nearby environment. The increasing conflicts of animal–domestic animal and animals–human shows that Irula community is at risk. The present study focuses on development of Irulas community with appropriate sustainable technologies reviewing successful intervention from different aspects.

T. V. Rajeevan
Department of Social Work, Amrita Vishwa Vidyapeetham, Coimbatore, India
e-mail: tv_rajeevan@cb.amrita.edu

S. Rajendrakumar (✉)
Centre for Sustainable Future, Department of Chemical Engineering and Materials Science, Amrita Vishwa Vidyapeetham, Coimbatore, India
e-mail: s_rajendrakumar@cb.amrita.edu

T. Senthilkumar
Department of Computer Science and Engineering, Amrita Vishwa Vidyapeetham, Coimbatore, India
e-mail: t_senthilkumar@cb.amrita.edu

S. Udhaya Kumar · P. Subramaniam
Department of Tribal Welfare, Tribal Research Centre, Udhagamandalam, Government of Tamil Nadu, India
e-mail: udhayaanthropology@gmail.com

P. Subramaniam
e-mail: trcooty@gmail.com

© Springer Nature Singapore Pte Ltd. 2020
A. K. Luhach et al. (eds.), *First International Conference on Sustainable Technologies for Computational Intelligence*, Advances in Intelligent Systems and Computing 1045, https://doi.org/10.1007/978-981-15-0029-9_20

The probability of implementation of successful intervention of appropriate sustainable technologies will be presented as framework to promote the Irula community of Masinagudi and Ebbanad village of Nilgiris district.

Keywords Technology · Sustainable · Irula · Masinagudi · Mudumalai Wildlife Sanctuary

1 Introduction

Technology is an outcome of scientific research and practical methods. It becomes indispensable for the human to survive on this earth and become cheaper, accessible to all standards of people. Technology changes human life and support in the development of the community [1]. Use of Technology for community development can transform people's lifestyle; uplifting economic conditions, educational development, improving health, medical conditions, and preservation of indigenous cultures [2]. It also provides a way for holistic development of tribal community in sustainable manner. The development customizes the community to follow democratic procedures, voluntary cooperation, self-development, and participation of the members in the community development process [3]. Innovative technological ideas, based on scientific principle can transform the tribal community by improving economic conditions, preserving indigenous, social and cultural values of the community and leads to sustainable development. These sustainable technologies can be fully exploited for developing tribe community and their population dwelling in backward regions.

In India Tribal communities live in about 15% of our country's landmass [4]. They are living in various ecological and geo-climatic conditions ranging from plains to hills, rural to urban settlements and deep of forests. The Scheduled Tribes are notified under Article-342 of the Constitution, which makes special provision for tribes or tribal communities or parts of or groups within. The Scheduled Tribes are notified in 30 States/UTs is 705 individual ethnic groups [5].

A total tribal population of the country is 10.43 crore constituting 8.6% of the total population [6]. Around 89.97% of them live in rural and 10.03% in urban regions. The state of Tamil Nadu recorded 7.95 (1.10%) lakh people are in Scheduled Tribes [7]. Irulas are one among Particularly Vulnerable Tribal Groups (PVTGs) in Tamil Nadu [8]. The name of the tribe is believed to originate from the Tamil word '*Irul*' which means dark. The tribe's name '*Irular*' means people of darkness [8]. Culturally, they are differing from other PVTGs with their own customs rites & rituals from birth to death. In Irulas, 98% are Hindu and 2% have embraced christianity but celebrating Hindu festivals [8]. Irula settlements are known as 'Ur' or 'Motta and consist of several bamboo huts and mud houses [9]. They are isolated from the mainstream due to lack of basic infrastructure facilities. Uplifting of these communities and providing good social environment is a fundamental need of these tribes population. Hence the appropriate technological methods should be identified and properly implemented for the development of these communities. Many researchers have worked on tribal development and developed several successful technological methods. Information and Communication Technology (ICT) can be a valid tool for improving livelihoods of tribal community in India [10]. Using ICTs, the socio-cultural problems were identified and methods proposed to overcome the issues of Kolam tribals [11]. Thoti tribe of Andhra Pradesh, their medical disorder, treatment methods were reported using ICTs [12]. The economic backward conditions of Koyas tribe was studied, influence of cultural and ecological factors of tribal life were, reported with measures to improve the economic conditions of tribe [13]. The study on lifestyle pattern of tribe and acceptance or rejection of government schemes was discussed with ecological and cultural barrier to implement the government schemes [14]. An assessment of food and food habits of tribal community of Bhill tribes of Rajasthan and recommended measure to improve the health conditions [15]. Impacts of migration of Banjara Tribe was addressed with a strategy to reduce the tribal migration and improve the economic conditions of this region [16].

Based on the review of successful interventions, this paper focuses on the problems of Irula community and discuss the appropriate technological framework to overcome/improve the prevailing conditions. The paper will also propose the strategy to promote the sustainable living of Irula tribes of Masinagudi and Ebbanad villages of Nilgiri District.

2   Study Area

The Nilgiris is one of the smallest districts of Tamil Nadu. It is also called as 'blue mountains' located on southwest of Coimbatore district. The average height of hills is around 1980 msl bound with Kerala and Karnataka states. For administrative purpose, the district is divided into six taluks viz. Udhagamandalam, Gudalur, Pandalur, Coonoor, Kotagiri, and Kundah [17]. Tribes like Todas, Kotas, Kurumbas, Irulas, Paniyans, and Kattunayakans are found in this district [18]. According to 2001 census, the total population of the Nilgiri District is 7.64 lakhs, out of which the total primitive tribal groups population is 28,373 constituting 4.32% of the total general population [6]. The tribal population in Nilgiri district is not evenly distributed in the six taluks. 32.08% of them are living in Kotagiri taluk; 14.33% of Gudalur, 13.16% of Udhagamandalam and 6.96% are living in Coonoor and Kundah taluks [6]. In 2011 census, the total population of the Nilgiri District came down to 7.35 lakhs but Scheduled Tribes are increased up to 32,813 constituting 4.46% of the total population [6].

Masinagudi and Ebbanad village fall under the Udhagamandalam taluk. Masinagudi is a village panchayat spread across 13,409 ha of land. Total population is 8783 where 4345 are male and 4438 are female, 1797 (889 males and 908 females) are listed as Scheduled Tribes [6]. Ebbanad village has total population of 6233 where the 3044 are male and 3189 females, 510 (251 males and 259 females) are belongs to Scheduled Tribes [6]. The area of the village is 6420.05 ha. Irulas are one of the primitive groups residing in these villages. Tribes are residing in the thick forests of Mudumalai Wildlife Sanctuary. Their settlement locally called 'Padi' [19]. Irulas settled in Siriyur, Vazhaithottam settlements of Masinagudi village, Anaikatti settlement of Ebbanad village. The exact numbers of Irulas population of these settlements are unavailable, while 46 families from Vazhaithottam, 143 and 48 families residing in Anaikatti and Siriyur settlements, respectively [20].

Irula community live on agriculture, work as labourer in tea estate, collect forest product, or on other activities. They also catch snake & rat for livelihood [8]. They also migrate to plain lands to work as labourers in agricultural field or any other job. Irulas tribes of this region are marginal farmers, holding 0.5–2.5 acre land, cultivating, ragi, cholam, millet, potato, garlic and few varieties of legumes and vegetables [21]. The staple food is rice, additionally they eat ragi, cholam, millet, and local vegetables. They also eat few tubers and green leaves collected from forest [21]. Food is prepared with local ingredients prepared by the tribes [8]. This traditionally valued food are nutrient-rich and free from synthetic ingredients. These types of food products are used by tribes for several generations and it is the identity of this indigenous group. Tribal community in these regions follows their traditional-organic cultivation practice without including chemical fertilizers and insecticides. Hence the harvested products are free from chemicals as well nutrients rich and medicinally important too [22].

3 Problems of Irula Community

The settlments of Irula community located in Mudumalai Wildlife sanctuary that is geographically inaccessible and service deficient region. Man–animal conflict is also a threat to their life and crops. Animal intervention in the habitats of the tribal settlements one of the important problem in this community during harvesting seasons [21].

People of Irula community have started migrating to cities in search of jobs which may cause cultural declination [8]. The lack of opportunity in the area and slow raise in their economic level within the specific locality have increased the tribal migration towards cities. Irula seldom socialise with the outer world and with the other communities near the localities. This makes the tribes behind in skills.

In recent years some of them have started adopting modern agriculture practice, like use of chemical fertilizers & pesticide to increase yield of crops [20]. Records from this region have shown tribal population suffering from medical illness like hypertension, diabetes, arthritis, anemia, skin infection, and vision problems [23]. These problems are severe in place where the Irular are isolated [24]. Other problems faced by the tribes are reduced birth rates, and high infant mortality rate [23].

These problems in Irulas community directly affect their existence. Migration can change their indigenous nature. The traditional-cultural values are endemic and it needs to be preserved before it gets extinct from this community.

4 Technological Interventions for Community Development

Technological application for community development is a valid method to solve the community-related issues [25]. By addressing various problems of community, by integrated sustainable approaches with proper implementation, bring a long-term solution to the community [25].

Extensive agricultural practice increases strain on natural resources. Community like Irula living in ecologically senstive area and practicing modern agriculture involving chemical can damage the adjacent natural ecosystem as well in the longterm.

Bioresources, land, water, and energy are fundamental for sustenance of agriculture [27]. In places like hill terrain and wildlife sensitive regions, the interventions should be environment friendly [26]. Harvesting runoff water by engineering methods may promoted so that harvested water can be maid use for cultivation year around. This can improve the livelihood of the farmers. A study says that supplementary irrigation system and traditional agricultural practice can increase the yield up to 12% than the normal system, further improved cropping practice increased the yield up to 45% [28]. Adoption of micro-irrigation systems such as sprinkler and drips systems possess considerable potential to improve the water use efficiency and reduce the energy consumption. The study indicates that drip irrigation technology saves water 12–84%, reduce the energy consumption by 29–45% and improves the yields by 7–98% [29]. Combination of these two methods if tried with Irula community may make their life better. The monetary expense of developing these irrigation systems can be linked with government schemes or through community contributions. Energy required to run water pumps and drier for post harvest processing can be met out from renewable resources [30].

Organic farming is sustainable option for small farmers to improve the food security and enhance the overall performance of farm and household income [31]. In traditional rain-fed agricultural systems, the organic farming increases yield significantly [32]. In the environmental dimension, use of manure made out of organic waste and residues of agriculture in farming can improve soil quality [31], soil stability, increase nutrient supply [33], soil biodiversity [34] and improve the food security in longterm [35].

In recent days, use of bioinformatics in farming is an emerging trend. Ma'ayan simple and user-friendly software for the farmers understand the different factors affects the agricultural yield. It also calculates the cost-benefits from agricultural land [37, 36]. ICT based approach in agriculture right from selection of crops to marketing of yield can give better income to the farmers as well as produced post harvest loss agricultural products [38]. A Good Agricultural Practices (GAP) and Good Handling Practices (GHP) maintain food safety. It increases the awareness to farmers to maintain the food safety and meet the food standard of India to produce the food materials with good quality and maintain food safety production system [38].

Apart from technological interventions, the farmers of Irula community need to be trained on the usage of these technological methods. There is huge demand for skill-based training to improve the farming practice. In India, Indian Council of Agriculture Research (ICAR) has been promoting new technologies for rural areas to increase productivity, reduce cost of cultivation, improve value addition, conserve resources and alternate energy generation system. This can be implemented in Irula community. One of the successfully implemented intervention is the organic ginger production in Sikkim [39]. The ginger production technology have led to the adoption of ginger as a sole crops for many farmers, and rural youth of Nandok. This also has spread to nearby villages-Thanzing, Upper Khamdong, Yangthang, Thanka and Lingtam villages of East Sikkim covering an area of 50 ha for sole crop of ginger [40]. The new interventions in agriculture has to be tested in Irula community before recommending for the whole region.

Increasing wild animal conflict is due to cultivation of crops in forest fringes, inviting wild animals. Study indicate reduction of herbivorous population forcing the predators to turn to livestock [41]. Increasing human population in wildlife regions, ecotourism can also increase, man-animals conflict. To overcome man-animal conflict the following sustainable approach can be followed, *viz.* (i) shared governance, education and awareness, (ii) knowledge on wildlife to local communities, (iii) payment of compensation by government, (iv) good land use planning, (v) protecting livestock and crops, (vi) natural barriers and deterrents (vii) growing alternate crops [41]. This can tried within Irula community in the study area.

Altered lifestyle of Irula community, with less depends on forest for food has resulted in many case of undernutrition. The issues of hypertension, diabetes, arthritis, anemia, skin infection, vision problems, are not addressed properly due to lack of medical facility [43, 44]. The poor water, sanitation and hygiene practice in tribal community is due the lack of health care facilities and services [42]. The integrated approach, that includes proper delivering of schemes like Public Distribution System, Immunization programmes, Swachh Bharat Mission and Social Forestry scheme can give better results. Linking tribal healing practice and modern medical care institutions would provide better results [42].

Tribal schools in distant areas can be improved by adopting technology. The smart classrooms increase the attentiveness of the students. The Schoolnet India Limited support in education infrastructure development and enhancing the quality of human capital of India. They developed product called 'K-Yan' (vehicle of knowledge) is a low-cost new-media product for community learning [45]. The non-formal method provides room for innovations and injects flexibility to a formal system in terms of teaching methods, content, target groups of learners [46]. Non-formal system is suitable for school dropout and working youth from tribal community. Mother tongue as medium of instruction in early stages of education place a great role in reduce a school dropout [47]. Literacy campaign, relevant study materials in local languages, appoinment of local teachers, female teacher, stipends, scholarships, residential schools, social security, and proper monitoring systems can strengthen the tribal education system. Forest Department has developed tribal schools in forest region of Top-Slip of Coimbatore district [48]. The students and teachers meet at a time which is convenient for the students and engage in the teaching–learning process.

Unemployed youth can be given non-formal technical training in electrical work, welding, carpentry, motor mechanics. These kinds of programs can bring down tendency to migrate in Irula community [8]. Government of Kerala exemplifies the skill development program in the name of "Gothrajeevika", it is a sustainable livelihood generation programme for Scheduled Tribe community. It was formulated to benefit the Adivasis in all the fourteen Districts in the State. Its chief intent is to provide skill development training and encourage the beneficiaries to venture into self-employment [49].

Empowerment of the PVTGs and bringing them to the national mainstream is one of the major policies of the Government and Non-Governmental organizations. They should take necessary steps to monitor that the government information system reaches remotest corners of the state. One of the system developed by Himachal Pradesh Government is called LOMITRA, where the readdressing the complaints of the citizens without being physically required to visit the Government office. By following a multipronged strategy would be necessary for the holistic development of the tribals.

Community development through appropriate technological options fastens the development process in any community. Many developed and developing countries have consistently invested for community development program through appropriate technology. But it needs to be streamlined based on community problems and demand [50].

5 Theoretical Framework

The Brundtland commission, 1987, definition of sustainable development is that meets the needs of the present without compromising the ability of future generation to meet their own needs. It should focus on Triple Bottom Line (TBL) means that the development should be based on social, economic and environmental values. So it denotes development on basis of economic viability, environmental sustainability, and social responsibility. Community empowerment had been proven to be a powerful approach for solving community-related problems [51]. On the process of development, the community must participate in all transformation stages and approaches should be based on Closing-Loop of resources flow. It means the entire process the natural resources should be used sustainably without exploiting environmental parameters and reuse of waste materials.

However it is very important to preserve cultural values of the Irula community. Among the Irula community inheritance of traditional knowledge in younger generation is poor due to altered lifestyle and migration for livelihood. As a result, the traditional knowledge and practice of this community is becoming extinct. In the proposed study it needs to be preserved and documented with help of ICT.

Approaches should start with complete inventory on tribal communities followed with quantification of available natural resources in these regions. Along with this inventory, the documentation of cultural and traditional values of Irula tribe are also be maintained. Based on inventory results, identification of the most vulnerable people in this community and possibility of improving them with formal and non-formal training on sustainable agricultural practices and skill development program for employability.

The design of sustainable agricultural practice should be based on the traditional agricultural system and enabling the organic farming practice. Identify suitable traditional agricultural crops which farmers can cultivate in different seasons; also identify crops which are non-palatable for the animals to be cultivated along forest fringes to reduce the animals interventions. Careful monitoring of cultivation practices should be done in the use fertilizer, insecticides and pesticides which has to be organic and prepared by the tribal community by using agro-waste materials. Marketing of agricultural products should be done by agro-co-operative society. The co-operative society will empower the tribe farmers, village leaders by direct tie-up with consumers.

Further, traditional food, food recipes should be properly documented. Food products prepared by this community can be sold by the tribe. By implementing these practices, the community member becomes a deciding authority and benefits will be received by these communities. The community can be connected to government schemes, will improve the health, education parameters, nutritional foods, hygiene practice, medical services.

6   Conclusion

Application of suitable technological advancements will improve the lives of the tribes. By adopting new agricultural technologies in a traditional system will provide scientific protection, increases the yield and co-operative based marketing will prevent the community from getting exploited by middle man, self-reliant and sustainably active. Technology can reduce not only their economic problems but also other problems related to health, education, preserving culture and inter-connected problems. It will lead the community towards holistic development.

Acknowledgements We would like to render our gratitude to all the authorities in the Ministry of Tribal Affairs, GoI for the financial support (F.No.15025/5/2017-TRI), and the management of Amrita Vishwa Vidyapeetham, Coimbatore for providing support during writing stage of this paper. We thank our project partner, Tribal Research Centre, Ooty and our project assistant Mr. P. Gunanithi who supported in field activities. Our sincere thanks to Dr. V.S. Ramchandran, Centre of Environmental Studies, Amrita Vishwa Vidyapeetham for providing valuable feedback in final review of this paper.

References

1. Malecki, E.J.: Technology and economic development: the dynamics of local, regional, and national change. University of Illinois at Urbana-Champaign's Academy for Entrepreneurial Leadership Historical Research Reference in Entrepreneurship (1997)
2. Hoadley, C.M., Kilner, P.G.: Using technology to transform communities of practice into knowledge-building communities. SIGGROUP Bull. **25**(1), 31–40 (2005)
3. Njunwa, K.M.: Community participation as a tool for development: local community's participation in primary education development in Morogoro, Tanzania—a case of Kilkaland Mindu Primary Schools. Master thesis in Development Management, University of Agder (2010)
4. MoTA.: Report of the high level committee on socio economic, health and education status of tribal communities of India. Ministry of Tribal Affairs, Government of India (2014)
5. Panduranga, R., Honnurswamy, N.: Status of scheduled tribes in India. Int. J. Soc. Sci. Humanit. Res. **2**(4), 245–252 (2014)
6. Census, The Nilgiris District: Census 2011–2018 data. Directorate of Census Operations in Tamil Nadu. https://www.census2011.co.in/census/district/31-the-nilgiris.html, last accessed 2019/02/23
7. ADTWD: Policy Note 2018–2019. Adi Dravidar and Tribal Welfare Department, Government of Tamil Nadu (2018)
8. Gnanamoorthy, K.: Present situation of Irular—a primitive tribe. IOSR J. Econ. Financ. **6**(1), 46–49 (2015)
9. Lalsangpuii: The role of chieftainessess in the Mizo resistance movement the British (1800–1900). A doctoral thesis submitted to Assam University, Assam (2014)
10. Kumar, V., Bansal, A.: Information and communication technology for improving livelihoods of tribal community in India. Int. J. Comput. Eng. Sci. **3**(5), 13–21 (2013)
11. Deogaonkar, S.G., Deogaonkar, L., Baxi: The Kolam Tribes. Concept Publishing Company, New Delhi (2003)
12. Elizabeth, A.M., Saraswathy, K.N.: Thoti Tribes of Andhra Pradesh. Abhijeet Publication, New Delhi (2004)
13. Ramaiah, P.: Tribal economy in Telangana with special reference to the Warangal district (a case study). Doctoral thesis, Department of Economics, Kakatiya University, Warangal (1980)
14. Bhowmic, P.K.: Dynamics of Tribal Development. Inter India Publication, New Delhi (1993)
15. Joshi, S., Singh, V.: Assessment of food related habits and customs of Bhill tribes of Udaipur district, Rajasthan. Food Sci. Res. J. **6**(2), 333–340 (2015)
16. Suresh Lal, B.: Rural-urban migration of tribal labour: a case study of tribal rickshaw pullers in Warangal District. Master of philosophy. Dissertation, Department of Economics, Kakatiya University, Warangal (1995)
17. Rajukkannu, K.: Anthropological of perspective tribes and inter ethnic relationship in the Nilgiris District. Int. J. Eng. Manag. Res. **5**(5), 1–7 (2007)
18. TRC, HADP.: Tribes & inter-ethnic relationship in The Nilgiri District, Tamil Nadu. Tribal Research Centre and Hill Area Development Programme, Udhagamandalam, The Nilgiris District, Tamil Nadu, (2007)
19. TRC: Scheduled tribes of Nilgiris district. Tribal Research Centre, The Nilgiris District, Udhagamandalam, Tamil Nadu (2011)
20. Rajendakumar, S., Senthil Kumar, T., Udhaya Kumar, S., Subramaniam, P., Rajeevan, T.V.: Technology inputs in promoting indigenous food recipes of Irula and Kurumba tribes and empowering disadvantaged youth of Masinagudi and Ebbanad Villages of The Nilgiris District. Progress Report-1 submitted to: Ministry of Tribal Affairs, Government of India. Submitted by Amrita Vishwa Vidyapeetham, Coimbatore and Tribal Research Centre, Ooty (2018)
21. TRC, Amrita: Technology inputs in promoting indigenous food recipes of Irulas and Kurumbas tribes and empowering disadvantaged youth of Masinagudi and Ebbanad village of The Nilgiri District. A final proposal submitted to Ministry of Tribal Affairs, Government of India, New Delhi by Tribal Research Centre, Ooty and Amrita Vishwa Vidyapeetham, Coimbatore, Tamil Nadu (2017)

22. Selvam, P., Ezhumalai, R., Vijayargavan, A., Rajasekar, A., Samydurai, P., Aravindhan, V.: Herbal healing practices of Indigenous Irular tribal peoples of Sendurai Block at Ariyalur District, Tamil Nadu. Int. J. Appl. Pure Sci. Agric. **2**(3), 196–202 (2016)
23. Kumar, N.: A study on health status and perception of Illness among Irulas—Tribal people of Nilgiri District of Tamil Nadu. Imp. J. Interdiscip. Res. **3**(3), 1208–1362 (2017)
24. Saheb, S., Yaseen., Bhanu, B., Ananda.: Health, disease and morbidity among the Irular Tribe of Tamil Nadu. Afro Asian J. Anthropol. Soc. Policy **2**(1), 17–28 (2011)
25. Sianipar, C.P.M., Widaretna, K.: NGO as Triple-Helix axis: some lessons from Nias community empowerment on cocoa production. In: Proceedings of the Triple-Helix 10th International Conference (TH-10): Emerging Triple Helix Models for Developing Countries: From Conceptualization to Implementation; 190, p. 762 (2012)
26. Mensah, A.M., Castro, L.C.: Sustainable resource use and sustainable development a contradiction? Centre for Developmental Research, University of Bonn (2004)
27. Upadhyaya, A., Sikka, A.K.: Concept of water, land and energy productivity in agriculture and pathways to improvement. Irrig. Drain. Syst. Eng. **5**(1), 1–10 (2016)
28. Birthal, P.S.: Application of frontier technologies for agricultural development. Indian J. Agric. Econ. **68**(1), 20–38 (2013)
29. Narayanamoorthy, A.: Averting water crisis by drip method of irrigation: a study of two water-intensive crops. Indian J. Agric. Econ. **58**(3), 427–437 (2003)
30. Chel, A., Kaushik, G.: Renewable energy for sustainable agriculture. Agron. Sustain. Dev. **31**, 91–118 (2011)
31. Morshedi, L., Lashgarara, F., Hosseini, F., Najafabadi, M.O.: The role of organic farming for improving food security from the perspective of far farmers. Sustainability **9**, 1–13 (2017)
32. FAO: Can organic farmers produce enough food for everybody? Food and Agriculture Organization of the United Nations. Available online: http://www.fao.org/organicag/oa-faq/oa-faq7/en/. Last accessed 2019/02/23
33. Najafabadi, O.M.A.: Gender sensitive analysis towards organic agriculture: a structural equation modeling approach. J. Agric. Environ. Ethics **27**, 225–240 (2014)
34. Azadi, H., Schoonbeek, S., Mahmoudi, H., Derudder, B., De Maeyer, P., Witlox, F.: Organic agriculture and sustainable food production system: main potentials. Agric. Ecosyst. Environ. **144**, 92–94 (2011)
35. Ghadam, Z.M.H., Bazayeh, F.A.: Weed and pest management in organic farming. J. Agric. Sustain. Dev. **52**, 34–44 (2013)
36. Silva, E.L.D.G.S., Oliveira, C.M.M., Mendes, A.B., Guerra, H.M.G.F.O.: Technology and innovation in agriculture: The Azores case study. Int. J. Interactive Mobile Technol. **11**(5), 56–66 (2017)
37. Gal, B., Gal, Y., Gelb, E.: Adoption of information technology for farm management, a case study. In: International Farm Management Congress on "Farming at the Edge", Australia (2003)
38. FAO: Success Stories on Information and Communication Technologies for Agriculture and Rural Development, 2nd edn. Food and Agricultural Organisation, Bangkok, Thailand (2015)
39. Rahman, H., Karuppaiyan, R., Kishore, K., Denzongpa, R.: Traditional practices of ginger cultivation in Northeast India. Indian Tradit. Knowl. **8**(1), 23–28 (2009)
40. ICAR, Livelihood security of tribal farmer through organic ginger cultivation in Sikkim. Indian Council for Agricultural Research (ICAR), Ministry of Agricultural and Farmers Welfare. GoI. https://icar.org.in/content/livelihood-security-tribal-farmer-through-organic-ginger-cultivation-sikkim-0. Last accessed 2019/02/23
41. FAO: Sustainable Wildlife Management and Human-Wildlife Conflict. Food and Agricultural Organisation, Kaeslin (2015)
42. Mishra, M.: Health status and disease in tribal dominated villages of Central India. Health Popul. Perspect. Issues **35**(4), 157–172 (2012)
43. Agte, V.V., Chiplonkar, S.A., Tarwadi, K.V.: Factors influencing zinc status of apparently healthy infants. J. Am. College Nutr. **24**, 334–341 (2005)
44. Mittal, P.C., Srivastava, S.: Diet, nutritional status and food related traditions of Oraon tribes of New Mal (West Bengal), India. Int. Electron. J. Rural Remote Health Res. Educ. Pract. Policy **6**, 385 (2006)

45. Shah, T.: The knowledge vehicle (K-Yan): sustainable value creation by design. In: Practice and Progress in Social Design and Sustainability, pp. 216–236. IGI Global Publishers, USA (2019)
46. Nair, P.: Whose Public Action? Analyzing Inter-Sectoral Collaboration for Service Delivery: Identification of Programmes for Study in India. International Development Department, Economic and Social Research Council, Swindon (2007)
47. Jha, J., Jhingran, D.: Elementary Education for the Poorest and Other Deprived Groups. Centre for Policy Research, New Delhi (2002)
48. TNFD: Tribal Development. Tamil Nadu Forest Department, Government of Tamil Nadu. https://www.forests.tn.gov.in/pages/view/Tribal_Development_on-going. Last accessed 2019/02/23
49. STDD, Management of Career & Higher Education Aspirations of ST Students in 2018–19. State Tribal Development Department. Government of Kerala. http://www.stdd.kerala.gov.in/training-skill-development. Last accessed 2019/02/23
50. Wicklein, R.C.: Designing for appropriate technology in developing countries. Technol. Soc. **20**, 371–375 (1998)
51. Kasmel, A., Andersen., Pernille, T.: Measurement of community empowerment in three community programs in Rapla (Estonia). Int. J. Environ. Res. Public Health **8**, 799–817 (2011)
52. Bharathi Dhevi, V.R., Bhooma Mani, B.D.: Demographic profile of selected Irular Tribes of Coimbatore District–Tamil Nadu. Int. J. Sci. Res. Publ. **4**(1), 1–8 (2014)

Hybrid Energy-Aware Routing (HEAR) Approach (Using Network Simulator)

Gaurav Vishnu Londhe and Dilendra Hiren

Abstract In connection to the research published on wireless sensor network (WSN), the main concern in data transmission is found to be the network failure. The main cause found in the earlier researches is the energy loss takes place of the nodes which helps in forming the network. The residual energy in battery terminal must be analyzed and need to be used to form a network to give better network lifetime to avoid the loss of time and energy to form a new network if current fails to transmit the message. There are several energy aware protocols available which needs to be compared and analyzed with different parameters. The total network life time can be further improved with the findings of this analysis. This analysis study involves of the comparison of the different methodologies with parameters on which the performance of the protocol is depends on. Here we compare some important parameters like distance, residual battery energy in simulations in NS2 (Network Simulator). Proposed method can give us the possibility of maximum lifetime of network and can avoid the drop in the network.

Keywords WSN · EAHR · EAR · NS2

1 Introduction

In this paper of wireless sensor network, specifically for the message to be transferred from one node to another like source to destination node (Sink node). There are various nodes are necessary to be chosen to establish the network, this can be formed by considering many different criteria's like shortest path between nodes in close vicinity, as well the nodes with maximum residual battery energy, among the nodes

G. V. Londhe
NMIMS University, Mumbai, Maharashtra 400056, India
e-mail: gaurav.londhe@nmims.edu

G. V. Londhe · D. Hiren (✉)
Pacific Academy of Higher Education and Research University (PAHER),
Udaipur, Rajasthan 313003, India
e-mail: Sigmapawan72@gmail.com

© Springer Nature Singapore Pte Ltd. 2020
A. K. Luhach et al. (eds.), *First International Conference on Sustainable Technologies for Computational Intelligence*, Advances in Intelligent Systems and Computing 1045, https://doi.org/10.1007/978-981-15-0029-9_21

in close vicinity. There are some more nodes available which have less energy who will not be considered to form the network because these will die earlier to those who has more energy and this will lead to increase the network lifetime.

Apart from this, there are few more protocols approaches which are compared and have used multiple sink nodes instead of single sink node.

In general, using network simulator, the sensing nodes are implemented with uniformity into a restricted monitored field, where the sensing nodes keep coordination among themselves in cooperation so as to make the highly precised data of given physical conditions and/or sensor properties. Every one of these imparted nodes had the capacity to gather and reroute the data items either to another nodes or returned to outer station(s).

This one of the static stations might be immovable/moving as well it can be connected to the nodes of a sensor network to the external world by using available communication setup through which a user can access earlier reported data. Mainly the lifetime of a WSN is dependent on only two parameters of physical layer, state of a channel and available energy of sensing nodes [1, 2]. We can presume that a node of very heavy communication burden will utilize maximum energy and so it will have minimum residual energy left and vice versa. Here the special nodes are considered as not active or those will be assumed to be hibernated in future. Hence to avoid it for relatively better coverage, and framing of few of the local measures, a most popular and known concept of "hot spot" in Wireless sensor nodes. This is a lifetime of network which depends on the residual energy of individual sensor nodes; to use these specific limited resources more effectively, these special routing algorithms are modified as if those are dependent on location of individual data items.

In such a task, we have many more different cases, like one or many more immovable sinks are also simulated and analyzed as well discussed the status of the energy of complete nodes within the network. The entire remaining paper is reorganized as given below. Literature survey is explained in Sect. 2; proposed methodology is also described in Sect. 4, and the results of a simulation experiment are explained in Sect. 4, whereas Sect. 5 concludes the research work.

2 Related Work

2.1 Energy-Aware Heuristic Based Protocol

Biswas [1], Muthukkumarasamy [1], Sithirasenan [1], Singh [1] had proposed the heuristic needs a routing approach to get path to proceed for source to the destination station by making use of heuristic function as well A^* search algorithm. In this the network model used with some considerations to develop the capable routing mechanism is also explained. Then, the proposed EAHR approach is suggested with depth.

In this network design, we presume that wireless networks consists of many hundreds of stationary nodes. By assuming that WSN can have below given properties:

- A unique node ID is assigned to each system node.
- Each node is aware of the location of its own position and BS as well.
- System nodes are regularly sending after some duration the information of energy to all of their neighbor nodes.
- System nodes have very limited set of not rechargeable battery supplies.
- The network contains many more static/dynamic nodes and the utilization of energy is not limited for all types of the nodes.
- System nodes have many layers of transmission of energy as well every node can modify the energy levels.

2.2 Residual Battery Energy of Entire Network Node Measurement and Utilization Method

Kumar [2] and Chaturvedi [2] had already proposed the above-given method for analysis and utilization of available system nodes with higher energy for data transmission. In such work, it is also presumed that the energy at the beginning of every node and energy and its usage rate is so taken as 1.7251 J and 576 mAh (milli Ampere Hour), respectively. To compete with such parameters, every available node must have two or more 1.5 V battery cells.

3 Comparison of Energy-Aware Routing Protocols

(1) Heuristic Function: This is the mathematical function suggests an idea how the source node is at the distant location from the destination node or the current node based on the mathematical function can be helpful to find the methodology to find the node. There are many more parameters which can represent the numeric value.

The following parameters can be taken into consideration as:

(a) Longevity Factor
(b) Distance from current node to source
(c) Euclidian distance to and from the current node so as to maintain the integrity of the special behaviors

(2) An $A*$ Heuristic Search Algorithm: This algorithm can find the best path to reach the destination from the source node with the least possible cost. This is carried out by following the heuristic function.

(3) To achieve the better throughput from the deployment of the strategies as per the proposal for the wireless sensing nodes the utilization of the batteries must

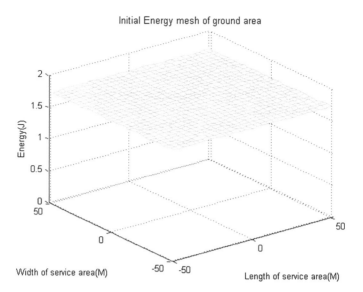

Fig. 1 An initial energy status of networks

have been done very appropriately. However, in huge networks, many more such types of sink nodes and sensor nodes must be implemented.

The energy levels of the various phases are explained with diagrammatic representation is given as follows with respective Figs. 1, 2, 3, 4 and 5, whereby the energy status of the entire network with its individual energy statuses are shown with its residual values.

(4) Yadav et al. [10] have discussed an innovative cluster-based algorithm of LEACH methodology that is proposed for maintaining the balance the entire power utilization of residual battery energy to extend the life of the network as a whole. This suggests an advantage to this architectural methodology over a non clustered architectural design which Clears the idea of the network lifetime, with its nodes are the one which expires while simulating as well another head of the cluster head need to be generated.

Leach: Low energy adaptive clustering hierarchy, this is generated by Heinzelman, Chandrakasan, and Balakrishana [1] for conservation of the cluster. In this specific clusters are made for the details of signal capacity.. The Leach methodology had different two phases like (1) setup phase and (2) steady state phase [2]. In the first phase cluster head is chosen and in an another phase, which is the head of the cluster, which is already maintained in this a data is transferred within many more nodes and an every sensing node is making a level within $0 < Ti < 1$ and which need to be tally with the earlier defined specific value (Ti). If $T(i) <$ random, at this time the node will be considered as a cluster head else the node becomes cluster member.

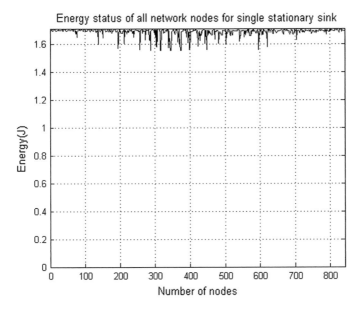

Fig. 2 Status of energy of network when only one single immovable sink is used

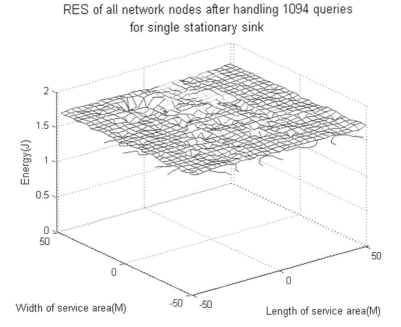

Fig. 3 Status of the residual energy of network nodes after processing

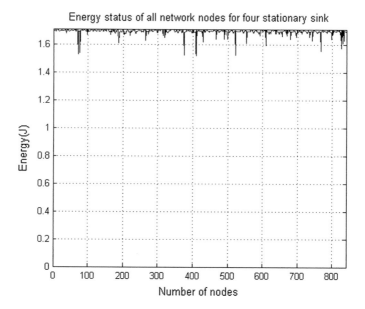

Fig. 4 Status of energy of a network when multiple sinks are deployed

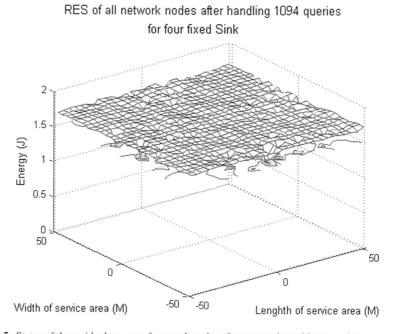

Fig. 5 Status of the residual energy of network nodes after processing with many sinks

(5) Sindhwani et al. [11] has also discussed a different version of LEACH protocol which is also called V-LEACH which proposes a less energy consumption in wireless environment. The suggested protocol implements to be improved V LEACH routing protocol, that reduces energy consumption and increases the entire network lifetime of the WSN compared to the LEACH protocol. As given in Table 1 Network Parameters.

Table 1 shows a result of the simulation of deployed network. The threshold value of the energy (0.5 J) in network is fixed and when the transmission range is to be fixed like approximately from 9.99 m and 100 nodes are to be simulated up to 10 s. These network parameters simulate the overall performance matrices.

Table 2 shows the comparison of the residual energy of various protocols like C-Leach, and Advance Leach, V-Leach with their individual node id. The resultant parameters with their individual results of the sensor nodes which has the different node id shown with the betterment of results on advance leach with respect to other leach protocols.

Table 3 describes the results and the comparison of various leach approaches w.r.t. their individual energies and lifetimes. After calculation, it is proposed that the total network lifetime can be improved after looking into the simulation of the network.

Table 1 Parameters used in network simulation

Parameter name	Values
Antenna type	Omni antenna
Base station (0, 0) m	5, 10, 35
Queue type	Drop tail
Initialize energy of the nodes	1.65 J
Transmission range	10 m
Rx energy	3.65 J
Threshold energy	0.5 J
Routing	DSDV
Channel type	Wireless channel.802_11
Simulation duration (s)	10.0

Table 2 Energy-level comparison

Node	LEACH	Advanced	C-LEACH
1	0.010	0.08	0.086
10	0.019	1.00	0.29
15	0.014	1.80	0.09
35	0.015	5.91	0.05
55	0.015	12.52	0.08
99	0.0095	28.324	0.071

Table 3 Overview of WSN protocols

Protocol	SCH-selection	Energy (J)	Lifetime
Leach	–	0.0501 J	Avg. lifetime duration on node no: 91
Leach advance	Minimum distance from Base Station	0.09 J for 100 nodes	Very high lifetime on time duration of CH 44
Leach-Cell	Residual energy	1.3 J for 100 nodes	Average lifetime from node no 2 to 5
V-Leach	Residual energy (Max)	1.5590 J for 100 nodes	Less lifetime than leach and up to 100 nodes

Table 4 mathematical representation of results with 1500 rounds

Protocol	Residual	Packet to base station	Alive node	Dead node
Leach	0.0663	27,487	3	95
Gear	5.3215	73,077	55	44
Mgear	9.2617	103,796	77	25

Table 5 Mathematical results for 2000 rounds

Protocol	Alive node	Packet to base station	Dead node	Residual energy
Leach	0	28,401	100	0.0
Gear	14	79,364	86	0.86991
Mgear	25	168,365	75	1.792

The performances of these protocols are compared with existing protocol types of the LEACH in wireless sensor network. Here we are proposing these three performances will be evaluated as shown in Tables 4 and 5.

4 Proposed Methodology

Here in the proposed methodology, the nodes with highest residual battery energy needs to be selected for establishing the network, in addition to it, we can consider the nodes with shortest distance in vicinity to the source node from which the next node to be chosen to form the network; with reference to the shortest distance algorithms, we can consider the idea for hybridizing both the concepts of energy-aware routing and shortest distance algorithms which in turn can be considered as a "Hybrid Energy-Aware Routing" (HEAR) approach.

This approach proposes that, as given in Fig. 6 the source node may have many more nodes in vicinity but only those will be chosen which have maximum residual energy and at minimum distance from the source node. All blue nodes are considered

Fig. 6 Proposed approach
of HEAR protocol for Ad
Hoc Network

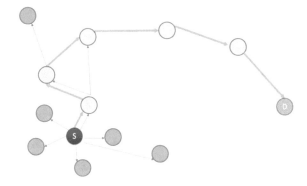

as the available nodes, the yellow nodes are the nodes with high energy and in shortest distance, so the new established network will have very less chances to get dropped in shorter period of time, which in turn increases the network lifetime and since less drops in network so the given message can be transferred in minimum time, this saves the time and the battery power needed to establish the new network to transfer the same message if the current network would have failed to submit the message.

The parameters can be used in proposed approach for the simulation are given as follows.

Parameters used in simulation

Parameter	Value
Initial battery energy of each node (B)	100 (J)
Network area	350 * 350 (m^2)
Path-loss exponent (η)	3
Data rate (r)	100 (Kbps)
Power consumption of transmitter circuit (P_t)	100 (mW)
Power consumption of receiver circuit (P_r)	100 (mW)
Maximum transmission power (P_{max})	150 (mW)
Minimum transmission power (P_{min})	15 (mW)
Maximum of transmissions in HBH system (Q_u)	7
Transmission range (d_{max})	70 (m)
Data packet size (L_d) 512 (byte)	512 (byte)
MAC ACK packet size (L_h)	240 (bit)
E2E ACK packet size (L_e)	96 (byte)
Hello packet size (L_{hello})	96 (byte)
Battery death threshold (B_{th})	0
Maximum collision probability (Pc$_{max}$)	0.3

5 Conclusion

This paper emphasizes on the residual energy, cluster heads, and the overall life-time of the existing energy-aware routing approaches and the overall comparison with their parameters and their respective values used in earlier protocols in wire-less environment. Further the proposal suggests that to increase the lifetime of the deployed network we need to consider the parameters like the distance between the high energy nodes and their high residual energy are the two parameters need to be considered to establish the new network which can be the best suitable for the uninterrupted network support and reduce the drop in network.

References

1. Biswas, K., Muthukkumarasamy, V., Sithirasenan, E., Singh, K.: An energy aware heuristic-based routing protocol in wireless sensor networks. In: 2014 17th International Conference on Computer and Information Technology (ICCIT)
2. Kumar, P., Chaturvedi, A.: Performance analysis of energy aware routing protocol for wireless sensor networks. In: 2014, 2nd International Conference on Devices, Circuits and Systems (ICDCS)
3. Al-Karaki, J.N., Kamal, A.: Routing techniques in wireless sensor networks: a survey. IEEE Wirel. Commun., pp. 1–36 (2004 Dec)
4. Chen, Y., Zhao, Q.: On the lifetime of wireless sensor networks. IEEE Commun. Lett. **9**(11), 976–979 (2005 Nov)
5. Záruba, G.V., Chaluvadi, V.K., Suleman, A.M.: LABAR: location area based ad hoc routing for GPS-scarce wide-area ad hoc networks. In: Proceedings of the First IEEE International Conference on Pervasive Computing and Communications (PerCom'03), p. 509 (2003)
6. Mir, Z.H., Ahmed Khan, S.: A zone-based location service for geocasting in mobile ad hoc networks. In: 10th Asia Pacific Conference on Communication and 5th International Symposium on Multi-dimensional Mobile Communications (2004)
7. Hou, H., Liu, X., Yu, H., Hu, H.: GLB-DMECR: Geographic location-based decentralized minimum energy consumption routing in wireless sensor networks. In: IEEE Proceedings of the Sixth International Conference on Parallel and Distributed Computing, Applications and Technologies (PDCAT'05), pp. 629–633 (2005)
8. Sha, K., Shi, W.: Modeling the lifetime of wireless sensor networks. Sens. Lett. **3**, 1–10 (2005)
9. Passino, K.M., Antsaklis, P.J.: A metric space approach to the specification of the heuristic function for the algorithm. IEEE Trans. Syst. Man Cybern. **24**(1), 159
10. Yadav, L., Sunitha, C.: Low energy adaptive clustering hierarchy in wireless sensor network (LEACH). Int. J. Comput. Sci. Inf. Technol. **5**(3) (2014). SGT Institute of Engineering & Technology, Gurgaon, Haryana-122505, India
11. Sindhwani, N., Vaid, R.: V Leach: an energy efficient communication protocol for WSN. Mech. Confab **2**(2) (2013 Feb–Mar)
12. Munir, A.: Cluster based routing protocols: a comparative study. In: Fifth International Conference on Advanced Computing & Communication Technologies (2015)
13. Singh, P.K., Prajapati, A.K., Singh, A., Singh, R.K.: Modified geographical energy-aware routing protocol in wireless sensor networks. In: International Conference on Emerging Trends in Electrical, Electronics and Sustainable Energy Systems (ICETEESES-16)
14. Goyal, R., Gupta, S., Khatri, P.: Energy aware routing protocol over leach on wireless sensor network. In: International Conference on Emerging Trends in Electrical, Electronics and Sustainable Energy Systems (ICETEESES-16)

Design and Deviation Analysis of a Semi-automated Needle Manipulation System Using Image Processing Technique

Shubham Shah, Ruby Mishra, Sourav Pramanik, Avigyan Kundu, Som Pandit and A. Mallick

Abstract Needle insertion is one of the most important parts of medical sector. For critical needle operations, precise needle insertion and removal speed control is very necessary, failing which trauma can be caused to the patients, and there will be chances of serious injuries to the internal organs. Hence, the main aim of this research work is to design and fabricate a setup that can precisely control the speed and depth of needle insertion procedure. This experimental setup contains four links and four joints having four degrees of freedom, and it was modeled in SOLIDWORKS. There are 3 revolute and 1 prismatic joint among which the revolute pairs can be used to manually position a needle and prismatic joint is automated for needle insertion. Depth and velocity of insertion of needle in the linear actuator is controlled automatically with the help of rotary encoder. The automated control system is controlled with the help of MyRIO controller. MyRIO is programmed using LabVIEW software. Six experiments were performed for each angle of insertion of the needle with respect to horizontal axis at 90°, 60°, and 25° to simulate various needle insertion medical procedures. Image processing techniques were used to analyze actual depth of insertion and angle of insertion of the needle. Basler ACA 1300 smart camera is used to capture the images and it was analyzed using MATLAB. A cost-effective alternative to test this kind of device using image processing in laboratory conditions is shown in this research. After analysis of the semi-automated needle insertion system, all the deviations from the required value were found to be minimal and within medical parameters.

Keywords Depth and velocity of insertion control · Needle manipulation · Image processing techniques · MyRIO · LabVIEW · MATLAB

S. Shah · R. Mishra (✉) · S. Pramanik · A. Kundu · S. Pandit · A. Mallick
KIIT Deemed to be University, Bhubaneswar, Odisha 751024, India
e-mail: rubymishrafme@kiit.ac.in

© Springer Nature Singapore Pte Ltd. 2020
A. K. Luhach et al. (eds.), *First International Conference on Sustainable Technologies for Computational Intelligence*, Advances in Intelligent Systems and Computing 1045, https://doi.org/10.1007/978-981-15-0029-9_22

1   Introduction

Needle insertion in the human body is one of the most vital parts of medical science. Needle insertion is used to perform some kind of specific drug insertion in the human vein or tissue aiming for the cure of some disease and also sometimes for removing some poisonous cells or tissues for curing of some deadly diseases.

   Now in the modern-age medical science, lots of difficulties and risks are found in the manual injection process. Manual injection is not sufficient to perform different critical processes, viz. for brachytherapy cancer treatment, tissue biopsies, and anesthesia drug injections. These treatment processes will fail to perform desired curing under manual injection process. According to some survey, it is found that an experienced physician does 15% error in placement of the needles for the curing of prostate cancer which was checked for 20 patients [1]. Also, for these operations, the movement of needle should be in such a way that it should not harm or penetrate any neighbor tissue or organs. Hence, in this research an attempt is made to control the speed of insertion of a medical needle by developing an initial experimental setup. This kind of system will be helpful in provincial hospitals for treatment and diagnostics. The setup is semi-automated in such a way that the physician can position a needle attached to the setup in a desired position. The needle can be inserted according to the required speed which can be controlled by a rotary encoder. This research will help in reducing the cost of such system and also increase the accuracy of a needle insertion procedure performed without causing any unintentional trauma to the body of a patient.

1.1   Literature Review

A research paper represents methods of driving hybrid positioning system using LabVIEW. The main demerits happening due to driving of pneumatic devices are removed. This system is mainly used to allow free programming during the displacement of actuator. This hybrid positioning system includes higher average speed and high accuracy. Its manufacturing cost is low [2]. Yet in another research, main priority was to analyze the variation in the accuracy of the setup during needle insertion with type of the needles at different locations. The whole project is done under CT (computed tomography). The mean accuracy was calculated by inserting the needle into liver, kidney, hip muscles. Evaluating the result, it can be concluded that the accuracy of the needle insertion does not depend upon the location as well as the type of needles [3]. An experiment for needle insertion in central venous catheterization is conducted on the static manikin. Real-time feedback is observed. As a result, the angle of the insertion, path length, and jerk are increased. The experiment can't differentiate the training groups of any 5 matrices objects, before and after the test [4]. Another research paper discusses the limitations and improvement of common practice involved in breast cancer management technologies. Newly invented needle dynamics was found to have higher rate of sampling in the case of tissue acquisition than the standard methodologies. The pneumatic needle insertion and sampling mechanism was found to be more efficient than the spring-loaded mechanism, for faster insertion and sampling procedures [5].

Research work on sensor implementation for precise positioning of the needle is carried out but the speed control of needle lacks research. Hence the primary objective of this research is to design a control system with an experimental setup that can be used to precisely position the needle and insert it at a desired speed inside the target point. Secondary objective of this research is to make the setup cost effective. Tertiary objective of this research work is to perform experiments on an inanimate object at various angles of insertion to simulate various needle insertion procedures.

2 Methodology

2.1 Designing

The setup of needle insertion device consists of 4 links including a base and 4 joints with 4 degrees of freedom. There are 3 revolute and 1 prismatic joint. The revolute joints can be positioned manually by a physician according to the requirement of the procedure at different angles. The prismatic joint is automated. The base and the revolute links are made up of wood. The wooden links are connected to the each other with the help of nut and bolts which provides the freedom to the links to rotate according to their own axis. The linear actuator, which has a prismatic joint, is connected to one of the revolute links. Hence, by the movement of the revolute links, the positioning of the linear actuator can be done. The linear actuator mainly consists of a stepper motor whose rotation is used for linear movement of the needle. A lead screw which is having a 0.5 mm pitch value is connected to the stepper motor. The needle housing is connected to this lead screw, so that with rotation of this lead screw, the needle can achieve the linear forward and backward motion. On the needle housing, there is an arrangement for connecting various different kinds of standard medical needles. Here, in this experiment two types of standard medical needles were used−one was hypodermic needle and another was the spinal needle. Front view and isometric view of the setup are shown in Fig. 1a, b. CAD model in these figures is designed in SOLIDWORKS.

2.2 Circuit Connection

The linear actuator setup which acts as end effector is connected to stepper motor driver and the motor driver is connected to MyRIO. Power supply is provided to it

(a)   (b)

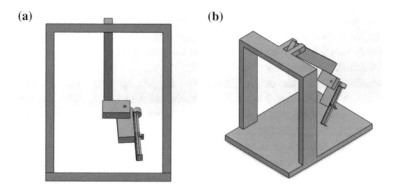

Fig. 1 **a** Front view of the setup. **b** Isometric view of the setup

with a 9 V DC power source. MyRIO works as a brain to control the automated linear actuator control system for this experimental setup. An encoder is used for controlling speed of the stepper motor and it is directly connected to MyRIO. Clockwise and anticlockwise rotation of the encoder is responsible for forward and backward motion of the needle connected to it. Increasing or decreasing speed of rotation of stepper motor changes speed of motion of needle, respectively. In this way, the speed of needle motion can be controlled precisely with the combined network of stepper motor (linear actuator), stepper motor driver, MyRIO, and rotary encoder. Connections for the stepper motor, its driver, MyRIO, and rotary encoder is shown in Fig. 2.

Fig. 2 Circuit diagram of
the setup

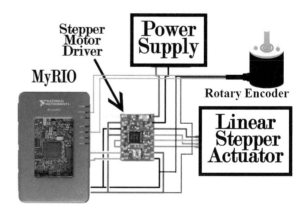

2.3   VI Setup

The programming of the MyRIO for this experiment is performed with a software, named LabVIEW (Laboratory Virtual Instrument Engineering Workbench). Virtual instrument (VI) for this experiment is shown in Fig. 3.

According to input from rotary encoder, signals are processed in MyRIO and the connected components function according to the programming. In this VI setup, different icons represent different output and input values while others are used to control different parameters of the experiment.

Counter value numerical indicator icon shows real-time current position of the encoder connected to MyRIO. Enable Boolean button helps in switching stepper motor from inactive state to active state and vice versa. Direction Boolean button helps in changing the direction of rotation of stepper motor which in turn changes the direction of the needle from backward to forward and vice versa. This button can be used by the doctors to precisely insert and remove the needle from a human body. Speedometer icon in Fig. 3 represents the speed of the stepper motor in rpm. Speed of the stepper motor is controlled by the help of rotary encoder.

Vertical indicator shows the variation of speed of needle in mm/s. Hence, controlling speed of stepper motor controls speed of needle. This kind of control on the needle insertion speed will really be very advantageous to the doctors and patients alike. Internal damage caused due to needle insertion accidents can be eliminated using this system.

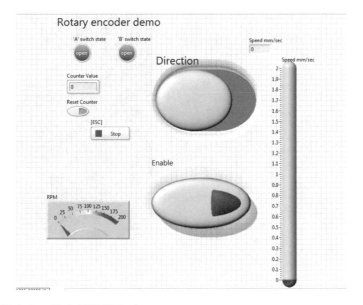

Fig. 3   VI designed in LabVIEW for the setup

2.4 Testing Equipment

Capturing of images for image processing was done using a 1278×958 resolution Basler ACA 1300 camera. This high-resolution camera is used to capture the images of the top view of the experimental setup. Width of CCD camera sensor is 3.75 μm. The focal length of the lens used is 16 mm with 1:2 aperture ratio. As the images captured are in pixels, it is necessary for analysis purposes to convert it to real-world dimensions such as millimeter. Equations (1–3) show the formulas to convert pixels into millimeter.

$$\text{Focal length of the lens}/\text{CCD sensor width} = \text{Standoff distance}/\text{Field of view} \quad (1)$$

$$\text{Size of one pixel in mm} = \text{Field of view in mm}/1278 \quad (2)$$

$$\text{Size of } n \text{ pixels in mm} = \text{Size of one pixel in mm}$$
$$\times \text{ no. of pixels found through image processing} \quad (3)$$

Figure 4 shows camera mounted on a camera mount which is connected to power supply and a laptop.

Fig. 4 Testing camera setup

Fig. 5 Complete
experimental setup for
needle insertion experiment

2.5 Complete Experimental Setup

For initiating the experiment, an apple is taken. The proper circuit connections are done and checked. A linear actuator is connected at fourth link, whose wires are connected with the stepper motor driver. MyRIO is directly connected with a rotary encoder and stepper motor driver. Then programming is done in LabVIEW software. Complete experimental setup for the experiment is shown in Fig. 5.

Before starting of the experiment, needle was setup at a desired position for penetrating the target point on the apple. Program is executed wirelessly on MyRIO. Velocity of the needle is set to required value before starting the actuator. Now, enable button is pressed to activate the actuator. Direction can be changed at any point of time using the direction button on the VI. After the experiment is performed, enable button can again be used to deactivate the linear actuator. All the VI controls and program execution are wireless and are connected over Wi-fi. The camera captures the image of the experiment, while the needle is being inserted to the desired target point.

3 Results

Using the above experimental setup, multiple experiments to analyze this novel needle insertion device were performed. A Basler Smart camera is used to analyze the performance of needle insertion system including angle of insertion, depth of insertion, and velocity of needle insertion. A target point was arbitrarily decided on another side of the apple, from that target, point of insertion on the surface of the apple was decided. This target point was kept same for all the angles of insertion while changing point of insertion according to the requirements.

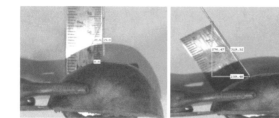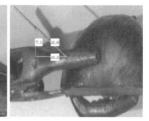

Fig. 6 Needle at 90°, 60°, and 25°, respectively, at the point of insertion

Experiments were conducted using three angles of insertions, namely 90°, 60°, and 25°. In medical procedures, if the needle is inserted at 90°, it is known as intramuscular injection process, used for local anesthesia test and allergy test. If the needle is inserted at 60°, it is known as subcutaneous injection process, used for vaccination. If the needle is inserted at 25°, it is known as intravenous injection process, used for inserting various needle inside the veins. Actual angle of insertion was measured using trigonometric relations of a right-angle triangle as obtained from image processing techniques in MATLAB. Figure 6 shows actual angle of insertions at above-mentioned angles.

To analyze the position accuracy of this semi-automated needle insertion setup, marking was made on the needle. Distance between initial and final positions of marking on the needle was measured using image processing techniques in MATLAB. Required depth of insertion was measured using a Vernier caliper and it was compared with actual depth of insertion. Figure 7 shows the needle inserted into the apple with its tip poking out of the other side at 90°, 60°, and 25°.

To analyze the velocity of insertion, a stop watch was used and time taken to insert the needle was measured in seconds. Using a rotary encoder, which was attached to MyRio, velocity was set to 1 mm/s for three experiments at each angle and it was set to 2 mm/s for other three experiments at each angle. Actual velocity of insertion was calculated by dividing the actual depth of insertion with time taken for insertion as indicated by stop watch. The data for overall experiment obtained after analyzing the images using image processing techniques in MATLAB are shown in Table 1.

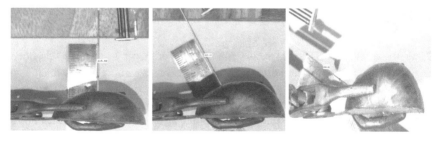

Fig. 7 Position of mark on the needle when it is completely inserted

Table 1 Data of required and actual angle, depth and velocity of insertion of a medical needle

S. No.	Required angle of insertion (°)	Actual angle of insertion (°)	Required depth of insertion (mm)	Actual depth of insertion (mm)	Required speed (mm/s)	Time taken (s)	Actual speed (mm/s)
1	90.00	89.91	25.04	25.60	1.00	25.07	1.0211
2	90.00	89.91	25.04	25.13	1.00	25.71	0.9774
3	90.00	89.91	25.04	25.54	1.00	25.74	0.9922
4	90.00	89.91	25.04	26.04	2.00	12.19	2.1361
5	90.00	89.91	25.04	25.42	2.00	12.46	2.0401
6	90.00	89.91	25.04	25.66	2.00	12.23	2.0981
7	60.00	60.02	26.50	27.15	2.00	12.45	2.1807
8	60.00	60.02	26.50	27.18	2.00	12.6	2.1571
9	60.00	60.02	26.50	27.06	2.00	11.97	2.2606
10	60.00	60.02	26.50	26.13	1.00	25.29	1.0332
11	60.00	60.02	26.50	26.21	1.00	24.73	1.0598
12	60.00	60.02	26.50	26.12	1.00	24.07	1.0851
13	25.00	25.26	28.00	28.70	2.00	14.61	1.9644
14	25.00	25.26	28.00	28.73	2.00	14.14	2.0318
15	25.00	25.26	28.00	28.72	2.00	14.35	2.0013
16	25.00	25.26	28.00	29.10	1.00	27.41	1.0616
17	25.00	25.26	28.00	28.84	1.00	27.57	1.0460
18	25.00	25.26	28.00	28.92	1.00	27.28	1.0601

From Table 1, the mean deviation in angle was found to be 0.1233°, mean deviation in depth of insertion from the required value was found to be 0.6161 mm, and the mean deviation in the velocity of insertion from the calculated values was found to be 0.0744 mm/s. According to a medical professional, deviation in insertion of the needle should be between 0.5 and 1.5 mm and deviation in velocity control is also quiet less. Hence, this setup lies well within medical parameters.

4 Conclusion

Insertion of a needle is one of the most basic and most widely used procedures but still trauma can be caused as a human cannot control the speed of the needle uniformly. Hence, a semi-automated system is designed and fabricated that can be used to smoothly control speed and depth of inserting a needle. This device has 4 links and 4 degrees of freedom. It is designed in such a way that many needle insertion procedures such as intramuscular, subcutaneous, intravenous, and intradermal can be performed. A doctor can manually adjust the setup to insert the needle at desired position and orientation using the first 3 links of the setup.

A rotary encoder with high resolution was used to control the speed of the linear actuator on which a needle was mounted. Six experiments were performed at each angle of insertion, i.e., 90°, 60°, and 25°. An apple was used as an inanimate test subject. A target point was arbitrarily selected for all the experiments, and the point of insertion was changed accordingly. To test the accuracy of angle of insertion and depth of insertion, image processing techniques were used. A smart camera was used to grab the image frame, and it was analyzed in MATLAB. After analysis, a deviation in angle of insertion and depth of insertion from the required value was found to be 0.1233°mm and 0.6161 mm, respectively. Even though the orientation of the needle was adjusted manually, the accuracy was found to be quite high. Image processing technique was a big help in adjusting the needle to its proper orientation.

Velocity of insertion was analyzed using a stop watch, and the mean deviation from the required velocity was found to be 0.0744 mm/s. As the linear actuator was of a generic variety, a little in depth of insertion and velocity was to be expected but still the deviation was well within medical parameters. This experiment showed a success of this semi-automated needle insertion setup. A novel and cost-effective alternative method is also shown in this research by using image processing techniques which can be very helpful in testing new devices in laboratory conditions.

To perfect this setup, more research is required as precise orientation control can also be integrated and the body of this setup can also be improved by making it completely automated. Hence, more attempts will be made to enhance this model so that it can be applied in provincial hospitals and help doctors and patients alike.

Acknowledgements A very special thanks to the affiliated University KIIT University of Bhubaneswar, Odisha, India, for supporting our research and providing resources for the same.

References

1. Alterovitz, R., Goldberg, K., Pouliot, J., Hsu, I.: Sensorless motion planning for medical needle insertion in deformable tissues. IEEE Trans. Inf. Technol. Biomed. **13**(2), 217–225 (2009)
2. Tan, L., Qin, X., Zhang, Q., Zhang, H., Dong, H., Guo, T., Liu, G.: Effect of vibration frequency on biopsy needle insertion force. Med. Eng. Phys. **43**, 71–76 (2017)
3. Baydere, B., Talas, S., Samur, E.: A novel highly-extensible 2-DOF pneumatic actuator for soft robotic applications. Sens. Actuators, A Phys. **281**, 84–94 (2018)
4. Hiraki, T., Matsuno, T., Kamegawa, T., Komaki, T., Sakurai, J., Matsuura, R., Yamaguchi, T., Sasaki, T., Iguchi, T., Matsui, Y., Gobara, H., Kanazawa, S.: Robotic insertion of various ablation needles under computed tomography guidance: accuracy in animal experiments. Eur. J. Radiol. **105**, 162–167 (2018)
5. Shah Shubham, K., Abhijeet, K., Mishra, R., Mohapatro, G.S.: Int. J. Mech. Prod. Eng. **4**(11), (Nov 2016)
6. Perz, P., Malujda, I., Wilczyński, D., Tarkowski, P.: Methods of controlling a hybrid positioning system using LabVIEW. Procedia Eng. **177**, 339–346 (2017)

Performance Measure of Classifier for Prediction of Healthcare Clinical Information

Santosh Kumar and T. Vijaya Shardhi

Abstract In the healthcare sector, there are various types of patient data, and that data need to be preserved for the future diagnosis of that particular patient and such a large size data can be stored using a concept of big data. In healthcare services, a huge amount of healthcare information is regularly generated at a very high speed and volume. Traditional databases are unable to handle such a huge amount of data. Every day increasing the volume of digital health care information has providing new opportunities leads to the quality of health care services and also avoid the repeated medical tests cost. If all the healthcare information is available in the form of digital, then we can use various tools and technologies to process healthcare information and generate decisions regarding the prediction of disease. Our proposed system is the automated clinical decision support system in association with a classifier. An objective of the implemented system has to predict the disease, using various classification techniques. The healthcare raw information data are stored and features are extracted that are used in further processing; based on those features, analysis is done and generates decisions on patient health information which are supplied. The proposed system is followed by a pipelined architecture and it contains the following phases: storage, feature extraction, classification, analysis, searching, and decisions. Research work emphasis on multiple classification techniques to increase the accuracy of prediction of patient health information.

Keywords Big data · CDS · EPR · Public health informatics · Classifier

S. Kumar (✉) · T. Vijaya Shardhi
Department of Computer Science & Engineering, K.L Education
Foundation, Vijayawada, AP 500002, India
e-mail: dssant@gmail.com; santosh.kumar@sitrc.org

T. Vijaya Shardhi
e-mail: saradhi1440@yahoo.com

S. Kumar
Department of Computer Engineering, Sandip Institute of Technology & Research
Centre, Nashik 422213, India

© Springer Nature Singapore Pte Ltd. 2020
A. K. Luhach et al. (eds.), *First International Conference on Sustainable Technologies for Computational Intelligence*, Advances in Intelligent Systems and Computing 1045, https://doi.org/10.1007/978-981-15-0029-9_23

1 Introduction

The increasing digitization in the various sectors has given rise to digitized data. The data in digital format are very convenient but need to be stored well. The healthcare sector also has emerged with such convenient digital data. Digitization of healthcare information is useful for all healthcare stakeholders. The health sector includes various different types of data and such a varied type of digital data needs to be stored properly. Security and privacy factors have been crucial issues for the ideal use of big data. Big data has become vital for diverse application domains and also have challenges of data security and management.

Generation of healthcare data is accelerated with a huge volume of data in every movement. It includes clinical trials of patient, past record or history, description, as well as symptoms of the disease also that patient have diagnosis information, medicines taken by the patient, doctors prescriptions, symptoms of past r disease, medical treatment was given or surgery is done, patient insurance cover etc. The large volume of data cannot be processed efficiently by using relational traditional system. The healthcare information is available or stored in different format like structured, unstructured, and semi-structured. The traditional databases are unable to manage structured, semi-structured, and unstructured data by the traditional system.

Big data analysis platform has capable to smoothly managed and processed huge volume of digital healthcare information. Because of all convenience and ease of use, if health care is merging with big data concept, it will be more efficient for process and analysis of all concerned health care. Big data techniques can be used for healthcare data storage, data retrieval, and data processing. By using big data concept for health care, all the healthcare data will be available at one place for the healthcare service providers. The advantage of the big data analytics concept is it can be handled efficiently all types of data which can be structured, unstructured, or semi-structured.

Digital health care has emerged and proved to be beneficial in all aspects of health care. Due to digital healthcare concept, it would be cheaper as all as availability of records and no need of repetitaion of medical services tand medical tests also it can be a more efficient and convenient facility for future healthcare services. The digital health care records are available in the form of electronic medical records (EMR), mobilized health records (MHR), personal health records (PHR), and electronic health records (EHR). Most of the health data is now been stored digitally and thus the available digital health data can be stored efficiently and the non-digitized data such as doctors' prescription, nurse notes, treatment past history, patient's past history, and patient's current treatment data in paper format can be converted into digital form.

The increase in digitization of data in the healthcare industry has started producing data that is of a different variety. By analyzing these digital healthcare data, there will be an improvement in disease surveillance, population health management, disease prediction, etc., in the healthcare industry. Several benefits of big data analytics in area of healthcare services include earlier detection of diseases and treatments when they are in early stages and can be controlled and treated more easily and efficiently, also patient health management by providing to the patient all required services, improving the treatment methods, etc.

Big data concept for healthcare data will be useful for the various healthcare-related decisions and also for automatic prediction of diseases based on symptoms matching with particular disease symptoms. The patient data can be stored into the system and the system will preserve the data of the patient and give the data when needed. Due to this type of data storage, all past data will be available to doctors or healthcare providers for better treatment. All the health information of patient will be available at one place and various doctors will be able to use this information for treatment of the same patient.

The traditional healthcare system includes all the healthcare data in the paperwork form that includes the handwritten prescription of the doctor; during treatment of a patient, nurse notes while the patient is admitted in a hospital, patient diagnostics, patients history in written format, etc. Due to all these written format systems, the patient always needs to carry all files and reports along with them. So to improve the healthcare system, the automated system can be developed so that all the patient data are available at one place and can be accessed by a doctor anytime. Such a system includes EHR that is electronic health Records.

In many hospitals, there are systems for the patient hospital bill, appointment management, etc. But such a system will store only a little information and it has unable to store large volume of health care information. If a system would be developed for storage of all kind of health information. It would be useful for detection of disease on past information of patient also enhance quality of health care services. Such a system can be cost-effective as unnecessary tests can be eliminated and treatment can be done by the proper way to exact disease. Early day's prediction of any kind of disease is useful in preventing major disease or defect to cause as treatment can be given earlier, before the occurrence of any disease. Such a disease can be predicted using the symptoms observed in the patient. Such a system could be beneficial to avoid treatment failure, useless test, to enhance the quality of patient's health services, which would be supported by the building of a healthy nation.

2　Literature Survey

In the field of healthcare services, data analytics play a vital role in the prediction of disease consideration with their symptoms. It would be fruitful for the society to prevention and cure of health.

An idea of big data analytics include in health care services [1]. It shows the requirement of big data analysis domain for the healthcare industry/services for clinical support services and also the limitations of traditional databases to handle a large amount of healthcare data.

Big data is used at a large scale in recent days for healthcare applications. It represents the separation of information according to the selection of features by using supervised machine learning algorithms. Time required to building the algorithm and the result is accurate. Tanagra tool [2] is used to perform data analysis, it focus on Naive Bayes, *k-NN,* and decision list algorithms to prepare the classifier.

Kharya [3] has suggested the idea of predicting the risk of breast cancer. She represents that prediction methodology for breast cancer detection by using decision trees-based prediction treatment is more suitable comparing to other algorithms.

Sellappan Palaniappan and Rafiah Awang had developed a prototype of knowledgeable heart disease prediction system (IHDPS) [3, 4] classifier with the inclusion of classification algorithms such as decision trees, Naïve Bayes, and neural network. It has applied CRISP-DM methodology to implement a framework for discovering the knowledge from data. There are six steps to perform various activities: semantic of business, to understand the data, to eliminate the unwanted values from raw data, a sampling of data, analysis, and deployment.

Adil [5] had applied the concept of big data analytics in healthcare services for prediction of multi-disease to the various areas and also they found that root causes of specific disease happen due to society, air, water, and weather. To come up with the solution for preventing global multi-disease in a particular region in the Asian continent, which would enhance the quality of health care services in terms of efficiency.

Also, identified that clinical health services have been improved, while an analysis of previous clinical data for consideration of further action to enhance the clinical trial process.

Clinical support system (CSS) works on clinical operation-oriented data such as primary information of patient as well as history of particular patient which includes diagnosis of disease, medicine, etc.

Worldwide healthcare services had the analysis of various diseases survey data with the help of big data concept to the prevention of better cure, to lead reduction rate of failure in services in the health care sector. It had been observed that collection of all types of structure, semi-structured, and unstructured medical treatment clinical services-oriented data has to analyze the huge volume of data on Hadoop architecture by using map reduce concept.

The research work emphasis on production of greatly improved in health services and gradually reduce the risk factor of clinical services in hospital.

IBM has developed health care an analytics software system [6] solution to enhance the clinical operational services, such as laboratory report, a medicine which are used for the treatment of particular disease.

Seattle's Children's Hospital has also enhanced the diagnosis and healthcare services with the help of IBM product health care, an analytics software system. It has a capability to identify common symptoms of a disease and suggest the treatment type which has required to an individual patient, to increase the accuracy of life care.

A Data Analytical framework has been developed for an analysis of the huge volume of health information, information has collected from more than thousand of a cancer patient, to analyze them and got a constructive solution for the treatment of cancer patient.

The researcher has designed a framework [7–10] for an analysis of healthcare information to deliver the constructive suggestion for treatment, to simplify the framework which is able to visualize laboratory result in the form of graphical representation that helps to show every slight change in nature of information. It has also included in the guidelines of world health organization.

A methodology is habitual to strengthen for accuracy in terms of classifying the symptoms of disease and efficient utilization of resources.

Intel had designed and developed a synergetic cloud infrastructure for analytics of health care information, it has also permit to contribute specific patient disease data so that correctly treatment of cancer is possible. Hospital medical officers can use another patient's cancer disease information and implementation and analyze the DNA of Cancer patient and own patients.

Big data healthcare analytics [7, 8, 11, 12] would provide superior treatment as well as a concrete solution for enlarged genetic diseases within a short span of time. This concept would be continued to cancer, brain tumor, autism, and a few diseases which have affected by genetic.

In the healthcare industry, the generation of multidimensional data size is accelerated in the form of structured, semi-structured, and unstructured information that might be used in clinical operational services, pharmacy, and pathology services data.

Digital health information boosts up the quality of healthcare treatment, also to accelerate the clinical operational services with respect to real-time data analysis of healthcare [13]. So that maintains the reliability of clinical care services from previous patient history. Also, uses big data analytics tools and methodology [14] to the extraction of features of healthy pattern and comparison of disease affected pattern.

Khare [3] has demonstrated a knowledgeable system for heart attack prediction concept with the integration of mining tools and method. Inclusion of mining methods or algorithms for classification of data, namely Rule Set classifiers, Decision tree algorithms like CART, ID3 C4.5, Neural Network, Neuro-fuzzy, Bayesian Network structure Discoveries. Data preprocessing uses (ODANB) and (NCC2) which are the extension of naive Bayes.

Classification techniques [3, 9, 15] are used for separate out structured and unstructured healthcare information. It is also applied to discover the knowledge of patient health information related to heart attack.

Developed a smart system that has to perform prediction of initial stages of heart disease, building the classification framework [16–18] they have uses three different data mining concept and enhance the accuracy.

The algorithm implemented to build the classifier for prediction system namely neural network, decision tree, and Naive Bayes [6].

Developed a system which can discover and extract hidden knowledge from the available data and historical heart disease databases. A system contains an extension of data mining (DMX), and SQL query has to be used for extraction of information.

There are totally fifteen medical attributes which are retrieved from the Cleveland heart disease data set. It includes total number of 909 records and those records are split into two equal data sets those are training data set and testing data set.

A medical field consists of large information but lacks in knowledge and hence there is a need to develop a smart system which will give knowledge from available information.

The system predictive system of heart disease is developed by Nishara Banu and Gomathy [14] and which may lead to control of occurrence of heart disease. They used the C4.5 Algorithm, K-means algorithm, and MAFIA algorithm. Various types of attributes are taken for examining the patient and comparing with disease's symptoms. An outcome of the system has less amount of time required to analyze, prediction of diseases and generation of result accuracy rate is more.

The accuracy of heart disease prediction has increased adopted of fuzzy intelligent mechanism for the development of the classifier model. Classification, Clustering, and Prediction model etc. early detection of various diseases which can be predicted at initial stages and controlled by giving proper treatment.

Advantages of developing automated disease prediction system are that it prepares a history database, can be used for training purpose, detects if the patient is having any disease, and is less time-consuming. Feature selection has been an emerging area of research and development in discovery of knowledge and is also broadly used in the area of retrieval of information and images, pattern classification, and many more.

Feature selection is one of the vertical which is considered for the selection of the widely usable attributes from the available information.

The aim of our work is to build the efficient classifier for clinical prediction system with integration of selection of features [19]. The drawback of big data analytics with respect to extraction of knowledge with consideration of repeatedly or irrespective features is included while the classification process is going on, and it increases the delay for completion of the task.

2.1 Problem Statement

In the healthcare sector, a very large amount of healthcare informatics data is being generated from different sources. Traditional healthcare system includes paperwork and so the patient always needs to carry files and reports. Digital data are sometimes unavailable, inadequate and are not properly utilized for patient health care effectively. Data processing for so large amount of healthcare informatics can be done using big data analytics technology [20]. A system can be made to handle data integration, generate new knowledge, translate into practice health information, and also predict the disease by giving the healthcare-related decisions automatically.

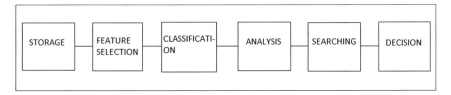

Fig. 1 Work flow of the system

3 Solving Approach and Efficiency Issues

In this section, the solving approaches and efficiency issues are studied. Figure 1 shows the steps of each and every process.

To get the solution with the help of six phases, i.e., storage, feature selection, classification, analysis, searching, and decision.

There are several responsibilities of the individual phase.

- Storage: The huge amount of healthcare data is stored for further processing in a different format. Such data may contain doctors' prescriptions, clinical test data, images, etc.
- Feature selection: From the available data, only the useful features can be extracted, in order to continue the further process. The extracted features are only taken into account for the classification phase, would be done using the principal component analysis algorithm and SIFT algorithm.
- Classification: The classification is done on the extracted features and classified according to the training data of symptoms. Classification can be done using the SVM algorithm, KNN algorithm, etc. For classifying health data, we used an ensemble learning approach which considers the results of multiple algorithms. The EDT algorithm gives the best results of classification.
- Analysis: In this phase, the classified data are analyzed. Decision tree algorithm is used for the analysis of the data. These analyzed data are then used for final decision making.
- Clinical decision support (CDS): This is the final step, and in this step, final clinical decisions are given. Decisions such as whether the patient is having a particular disease or risk of having a disease in the future.

4 Performance Analysis of System

To examine the classifier model with various parameter, selection of features as well as accuracy. Consideration of health care information and the processes that convert the data as single attribute or symptoms of diseases.

4.1 *Problem Formulation*

Mathematical model is the representation of our system in the form of computation. It is able to demonstrate every component of systems that affect shows while the calculation is going on. Also, represents the relation between various components of the system.

.It is a process to evaluate the systems to measure feasibility and sustainability of the system in the real-world application.

The function of the component would be analyzed with the help of attribute, formula, theorem, and methods. Behavior of system and subsystem would be analyzed using methods.

The system mathematically represented as, let S be the universal system such that, $S = (F, C, A, R, D, I)$.

where

I = Set of input point of the system, i.e., $I = \{I_1, I_2, I_3, \ldots, I_n\}$
O = Set of predicted output point of the system, i.e., $O = \{O_1, O_2, O_3, \ldots, O_n\}$
F = Set of selected features, i.e., $F = \{F_1, F_2, F_3, \ldots, F_n\}$
A = Set of analysis attributes, i.e., $A = \{a_1, a_2, a_3, \ldots, a_n\}$
R = Searching algorithm, i.e., $R = \{R \in 2(h_1, h_2, h_3, \ldots, h_n)\} \in R$
D = Decision or prediction set, i.e., $D = \{d_1, d_2, d_3, \ldots, d_n\}$

Function f_1: It accepts set of user data as input and performs storage onto database.

$$F_1(I) = \left[\left(\{I_{1,}, I_2, I_3, \ldots, I_n\}\right) \rightarrow S\right] \in S$$

Function f_2: Function f_2 performs reading the data from database and works as a feature selector.

$$f_2(F) = \left[\left(\{I_{1,}, I_2, I_3, \ldots, I_n\}\right) \rightarrow (F_1, F_2, F_3, \ldots, F_k)\right]$$

where $k < n$.

k = Number of selected features.

Function f_3: Function f_3 reads features and performs classifier for training data set.

$$f_3(F) = [(I_{1,}, I_2, I_3, \ldots, I_n) \rightarrow (c1, c2)] \in C$$

Function f_4 Function f_4 reads classifier and performs different analyses.

$$f_4(C) = [(c_1, c_2, c_3, c_4, \ldots, c_k) \rightarrow c_I] \in C$$

where C_1 is the best classifier with consideration of accuracy terms.

Function f_5: Function f_5 reads classifier and user data and predicts disease or risk of disease.

$$F_5(C_i, I, T) = [(C_i, (I_{1_1}, I_2, I_3, \ldots, I_n); T) \rightarrow$$
$$(d_i, d_i\ i + 1, d_i + 2, d_i + 2, \ldots, ck] \in D$$

where $k =$ no. of labels or disease.

$D_l =$ It is the highest possible predicted value.

5 Experimental Analysis

This section discusses the comparative analysis of experimental observations of the general techniques or methods and implemented system. Results are depicted in terms of parameters, viz. time, accuracy, and number of features. The experimental analysis is carried out on the Windows 7 operating system with Intel(R) Core(TM) i5 processor, 4 GB RAM. The further section describes different parameters used for automatic disease prediction system. The analysis gives a brief idea about which techniques give a better result and enhances a similar system. Another method used is graphical representation view for the analysis. On the basis of the accuracy in percentage and time in ms, the performance of the system is been measured.

To implementation of this system, we have used Eclipse. The medical images and the medical data set are stored in D drive. For testing, we have used a core-i5 system with 4 GB RAM. This system is built using Eclipse and Java language.

A portion of the data set used in the proposed approach was collected from UCI machine learning data set for the purpose of automatic prediction of patient disease. We further collected brain tumor and non-tumor images for the purpose of tumor prediction. The total time required on an average for prediction of disease is between 1 and 5 ms. The images used for testing ranges between 5 and 20 kb ranging from low-resolution pixels to high-resolution pixels.

Performance analysis measure consideration of the following factors.

- Accuracy: It measures the accuracy of the implemented system and uses various algorithms, i.e., Naive Bayes, support vector machine, KNN10, and decision tree construction (DTC) to construct various classifier data sets.
- Features/Attributes Selection: It has considered number of features selection to enhancing the accuracy of the developed system.

Figure 2 shows, measurement accuracy of constructed classifier by using different methods i.e. Navie Bayes, Support vector machine, KNN10, Decision tree construction (DTC), LR also consideration of a fixed number of the attribute to record.

KNN5 and KNN10 classification technique produce the high rate of accuracy for selection of attribute/features and support vector machine gives less i.e. is minimum accuracy, while measuring the accuracy of the constructed classifier. LR, DTC, and DTR techniques produce average accuracy everywhere in our observation subject to fixed no. of attributes selected. Overall, we observe that, KNN5 and KNN10 classification techniques perform accuracy more than 80% for fixed no of feature

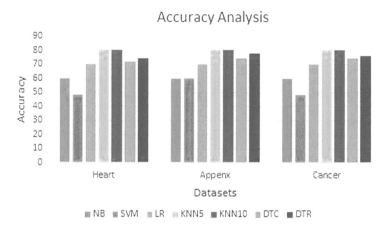

Fig. 2 Accuracy on fixed no. of attributes

or attribute. Figure 3 shows the accuracy analysis classification of medical images. Naive Bayes, Support vector machine and evolving cascade Neural Networks (CNN) techniques are used for design of medical images classifier. X-axis represent the no. of images and y-axis represent the accuracy. Medical images are used to carry out the result comparison of accuracy parameter. Designed and developed medical image classifier model has accepted only two file format of images .jpg and .png extension Image label prediction evolving cascade neural networks(CNN) has to produce the highest amount of accuracy recorded in the observation while Naive Bayes gives the minimum accuracy and support vector machine technique gives the accuracy less when no. of medical images are less if increase the no. of images as training images then it will produce the average accuracy. The experiment is carried out for multiple numbers of images and various format of images and implementation of

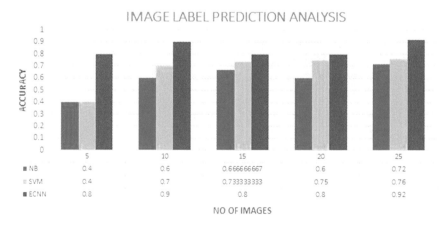

Fig. 3 Accuracy analysis on medical images

CNN algorithm for classification, it has observed that produce the best accuracy compare to NB and SVM.

Features/Attributes Selection: In this system, the training data are given and the system performs classification based on the attributes or features given. We compared the accuracy of various algorithms based on the different number of features or attributes. In Fig. 4, the attributes ranging from 1 to 10 are taken and different algorithms give result by taking a specific number of attributes. Experimental observations are compared using the constructed classifier algorithms, i.e., NB, SVM, LR, KNN5, DTC, and DTR. There is no fixed no. of the attribute are selected on Heart disease data set, consider variable no. of features is selected to measure the accuracy. Effective decision tree(EDT) shows to lead the accuracy in variable no. of the selected attribute and support vector machine shows down the accuracy label. Figure 3 shows the graphical representation of the experimental observations is compared using the constructed classifier algorithms, i.e., NB, SVM, LR, KNN5, DT, and EDT. There is fixed number of the attributes are selected on appendicitis data set (Fig. 5). As per observation support vector machine (SVM), KNN5 and effective decision tree, theses classifier technique lead the accuracy if the no selected attribute will increase then also accuracy will maintain. On the other hand, Naive Bayes, LR, and traditional

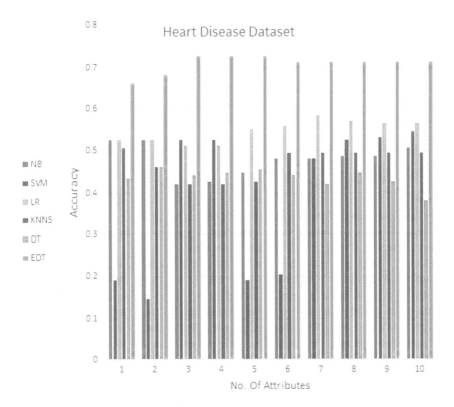

Fig. 4 Comparison using number of features taken for heart data set

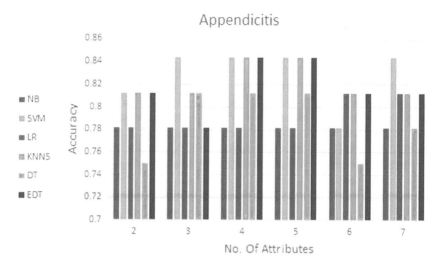

Fig. 5 Comparison of accuracy using no. of features on appendicitis data set

decision tree will maintain the accuracy of the constructed classifier for appendicitis data set. In the case of breast cancer data set, decision tree construction produces a good amount of accuracy while Navie Bayes classifier techniques go down in terms of accuracy. DTC and DTR are performed well to measure the accuracy on breast cancer data set.

We have tested our system on different data set namely heart disease data set, breast cancer data set, and appendicitis data set. Different images of a brain tumor and non-tumor are also considered. We calculate accuracy for the data sets.

6 Conclusion

Big data plays an important role in the healthcare sector and so analyzing clinical data is necessary by storing various types of digitized EPR in big data. Also, big data concepts are used for automated system that can predict the patient's health condition can be developed which will reduce healthcare cost and time needed for repetitive unnecessary clinical tests. For training automated system, various classification algorithms give varied accuracy so we have used an ensemble learning approach to increase accuracy by considering results of multiple algorithms.

In healthcare management system, big data can be used for analyzing healthcare information efficiently. Healthcare data will be easily available if all the health information is been included in big data. The big data can also be integrated with cloud technology so as health care can be provided with location independence. The system is trained based on the multiple instances for that particular disease symptoms. There is a certain disease which is very rare and can be found in one patient among many.

So, training data set is unavailable of that disease. This system can also be extended by training for such a rare disease prediction and the system can also be deployed in multiple hospitals and the system can be able for learning based on the different patients' health condition.

References

1. Kester, Q.-A.: Insight of big data analytics in healthcare industry. Int. J. Emerg. Technol. Adv. Eng. **10**(5), 61–63 (1993)
2. Rajkumar, A., Sophia Reena, G.: Diagnosis of heart disease using datamining algorithm. Glob. J. Comput. Sci. Technol., Kuala Lumpur (3 October 1995)
3. Kharya, S.: Using Data mining techniques for diagnosis and prognosis of cancer disease. Int. J. Comput. Sci. Eng. Inf. Technol. (IJCSEIT) **2**(2) (April 2012)
4. Palaniappan, S., Awang, R.: Intelligent heart disease prediction system using data mining techniques. IJCSNS Int. J. Comput. Sci. Netw. Secur. **8**(8) (August 2008)
5. Adil, A.: Analysis of multi-diseases using big data for improvement in health-care. In: IEEE UP Section Conference on Electrical Computer and Electronics (UP-CON) (2015)
6. Krishnan, S.: Application of analytics to big data in healthcare. In: 32nd Southern Biomedical Engineering Conference (2016)
7. Andrew, G., Ebenezer, J., Durga, S.: Big data analytics in healthcare: a survey. ARPN J. Eng. Appl. Sci.
8. Boneh, D., Franklin, M.: Big data in healthcare. SIAM J. Comput. **32**(3), 586–615 (2003)
9. Gangurde, H.D.: Feature selection using clustering approach for big data. Int. J. Comput. Appl. Innov. Trends Comput. Commun. Eng. (ITCCE-2014), (ISSN 0975 8887)
10. Koturwar, P., Girase, S., Mukhopadhyay, D.: A survey of classification techniques in the area of big data
11. Olaronke, I.: Big data in healthcare: prospects, challenges and resolutions. In: FTC 2016—Future Technologies Conference (2016)
12. Viceconti, M., Hunter, P., Hose, R.: Big data, big knowledge: big data for personalized health-care. IEEE J. Biomed. Health Inform. **19**(4) (2015)
13. Shobha, K., Nickolas, S.: A text information retrieval technique for big data using map reduce. Accent. Trans. Inf. Secur. **2**(7)
14. Nishara Banu, M.A., Gomathy, B: Disease predicting system using data mining techniques. Int. J. Tech. Res. Appl.
15. Shokri, R., Shmatikov, V.: Disease prediction by machine learning over big data from health care communities
16. Appari, A., Eric Johnson, M.: Information security and privacy in healthcare: current state of research. Int. J. Internet Enterp. Manag. **6**(4) (2010)
17. Somaraj, S., Hussain, M.A.: Securing medical images by image encryption using key image. Int. J. Comput. Appl. (0975 8887) **104**(3) (2014)
18. Amutha, V., Vijay Nagaraj, C.T.: Potentials and challenges of health care modeling and simulation (ICIET14)
19. Prakash, G.L., Prateek, M., Singh, I.: Computational health informatics in the big data age: a survey. Int. J. Eng. Comput. Sci. ISSN: 2319-7242
20. Puech, W.: HCAB: Healthcare analysis and data archival using big data tool

A Segmentation Method for Comprehensive Color Feature with Color-to-Grayscale Conversion Using SVD and Region-Growing Method

N. Jothiaruna and K. Joseph Abraham Sundar

Abstract Segmenting disease spots in leafs is achieved by Comprehensive Color feature (CCF), grayscale conversion and region growing method is discussed. Segmenting disease spots under real-field conditions with uneven illumination and clutter background has been a major challenge. Uneven illumination issues solves by applying Excess Red index, *H* component of HSV and grayscale conversion by using SVD, clutter background problem solves by applying region-growing method. Performance of these methods is calculated using precision, and its accurate segmentation is 89% under real-field condition.

Keywords Segmentation · Grayscale conversion · Singular value decomposition · Region growing · Disease spots

1 Introduction

In this world, agriculture plays a vital role. Major problem arises in this field like detecting diseases in plant at initial stage, and it may cause quality issues. Image processing techniques are used to solve this issue by image segmentation. Color spaces are applied in RGB image; *K* means clustering algorithm is applied to the resultant color space, for segmenting the disease spot [1]. Region growing techniques are used to segment the disease spots by examining the neighborhood pixel, if same pixels are found it forms a region. The same process is carried out until segmenting disease spots. If same regions are found, then region-merging algorithm is used to combine both regions [2]. Edge detection and morphological techniques are used

N. Jothiaruna
School of Computing, SASTRA Deemed University, Thanjavur, Tamil Nadu 613401, India
e-mail: joearuna96@gmail.com

K. Joseph Abraham Sundar (✉)
Computer Vision and Soft Computing Lab, School of Computing, SASTRA Deemed University, Thanjavur, Tamil Nadu 613401, India
e-mail: josephabrahamsundar@it.sastra.edu

© Springer Nature Singapore Pte Ltd. 2020
A. K. Luhach et al. (eds.), *First International Conference on Sustainable Technologies for Computational Intelligence*, Advances in Intelligent Systems and Computing 1045, https://doi.org/10.1007/978-981-15-0029-9_24

to segment the disease spots. Edge detection is done by Fuzzy canny method, and after detecting edges, morphological techniques are applied to segment the diseases [3]. To decolorizing an image, combination of local methods and global methods is processed to preserve color contrast of an image [4]. Adaptive K means algorithm is proposed to segment disease accurately is proposed [5]. To extract information from an image, genetic algorithm with Otsu thresholding algorithm is proposed [6] to give accurate segmentation. To optimize watermarking, genetic algorithm with singular value decomposition is proposed [7], and SVD will transform one matrix into three sub-matrices (U, S, V). An improved vegetation index, Excess Green minus Excess Red (ExGExR) was compared to the commonly used Excess Green (ExG), and the normalized difference (NDI) indices. The latter two indices used an Otsu threshold value to convert the index near-binary to a full-binary image [8]. Segmentation is done by using genetic algorithm as well as classification techniques are processed using HSV and SVM algorithms, and it identifies diseases at earlier stage [9]. RGB space is converted to HSV color space to identify greenness in an image. Removal of background pixels in an image by hue does not give accurate result when illumination is uneven [10].

2 Proposed Method: Comprehensive Color Feature Detection

Comprehensive color feature (CCF) is found by summing the Excess Red index, H component of HSV color space, and color-to-grayscale conversion. In ExR index, diseases are highlighted when illumination is uneven [11]. ExR is expressed as,

$$I_{\text{ExR}} = 1.3 I_{\text{R}} - I_{\text{G}} \tag{1}$$

where I_{R} is Red index and I_{G} is Green index taken from the input image. Hue describes the purity of the color; only hue component is extracted from HSV color space. Color-to-grayscale conversion using SVD is proposed by extracting chrominance information in an original image (Fig. 1).

CCF detection is obtained by convoluting ExR index with pillbox filter, Hue with Difference of Gaussians (DoG) and color to grayscale conversion with pillbox filter. For smoothening purpose, pillbox filter is used and DoG for blurring background of diseased leaf. General architecture for proposed method (Fig. 2).

CCF is formulated as,

$$f(I : r, \sigma_a, \sigma_b, \alpha) = \alpha(pb_{(r)} * I_{\text{ExR}}) + (\text{DoG}_{(\sigma_a, \sigma_b)} * I_H) + (pb_{(r)} * I_g) \tag{2}$$

where α is Excess Red parameter, I is an input image, $pb_{(r)}$ is an pillbox filter with radius r, I_{ExR} is excess red index (ExR), $\text{DOG}_{(\sigma_a, \sigma_b)}$ is difference of Gaussians with standard deviation a and b, I_H is H component of HSV, I_g is color-to-grayscale

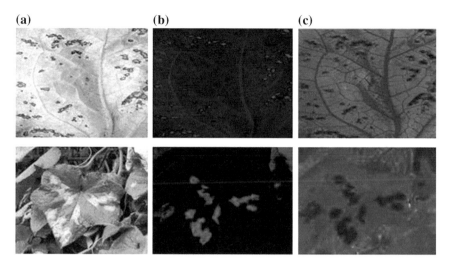

Fig. 1 **a** Input image, **b** excess red index, **c** *H* component of HSV

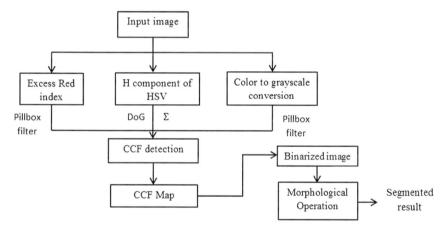

Fig. 2 General architecture

conversion and * is 2D convolution operator CCF map calculated by multiplying decrease rate β with resultant CCF process (3), and output of CCF map is binarized Fig. 3

$$CCF = -\beta(|f(I : r, \sigma_a, \sigma_b, \alpha)|) \qquad (3)$$

where β is a decrease rate and $f(I : r, \sigma_a, \sigma_b, \alpha)$ is resultant of CCF.

Color-to-grayscale conversion using SVD Using singular value decomposition (SVD), converting color to grayscale, i.e., converting three plane ($L^*a^*b^*$) from color image to one plane has grayscale image. Convert color image to $L^*a^*b^*$ color space. Take chrominance information (a^* and b^*) in color space and apply SVD

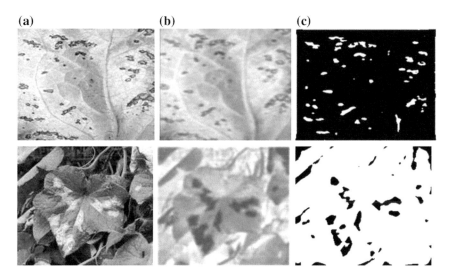

Fig. 3 **a** Input image, **b** CCF map, **c** binarized result

in chrominance matrix produce three sub-matrices (U, S, V). U defines eigenvector (a^*a^*t, b^*b^*t), V defines eigenvector (a^*ta^*, b^*tb^*), and S is a square root of both the eigenvalues $(a^*a^*t$ or $a^*ta^*)$ and (b^*b^*t, b^*tb^*). Chrominance matrix is reconstructed by multiplying r columns from U and V with singular values from r. Reconstructed weight matrix is summed with luminance information (L^*) and produces the grayscale vector (g). Convert vector form into matrix form and map the matrix form with RGB plane with matrix as luminance and chrominance as zero. Resultant grayscale image is obtained by averaging the R, G, and B planes. Architecture of color-to-grayscale conversion using SVD is Fig. 4.

Converting color-to-grayscale image using proposed method and MATLAB method using CCFR, CCPR, and E-score.

Disease spot segmentation Region-growing method is applied to resultant binarized image [2]. Clutter backgrounds are removed using region-growing method by selecting growing seeds in a binarized image. Using a threshold value, each neighborhood pixel is compared with new region and forms a loop. After region-growing process, morphological operation is applied to disease spots segmented for increase or decrease the size of holes [3] (Fig. 5).

3   Result and Discussion

Experimental setting Images are taken from standard benchmark dataset. APS Image Dataset by American Phytopathological Society, St. Paul, MN [12] and a real time images.

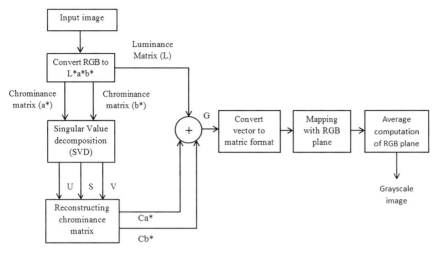

Fig. 4 Color-to-grayscale conversion using SVD

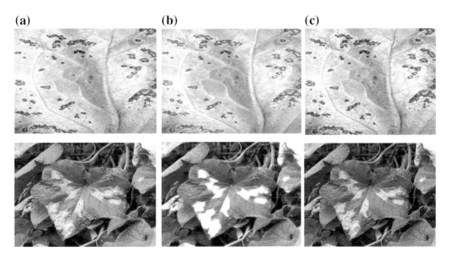

Fig. 5 **a** Color image, **b** color-to-grayscale image, **c** MATLAB grayscale image

CCF detection Excess Red parameter (α) values should be ranged from 0 to 1 for reducing the effects of uneven illumination in an infected leaf. CCF ratio is calculated for even illumination and uneven illumination. If CCF ratio is lower, its values are closest to the two regions, i.e., light affected in the leaf is very less. CCF ratio is calculated by,

$$R = \left(1/M \sum_{i=1}^{M} \text{CCF}(x, y)\right) \Big/ \left(1/N \sum_{j=1}^{N} \text{CCF}(x, y)\right) \tag{4}$$

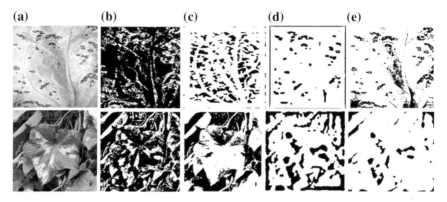

Fig. 6 **a** Input image, **b** k means, **c** Otsu, **d** CCF method, **e** proposed method

where R is CCF ratio, M is the light-affected region in the leaf, and N is the other light-affected region in the leaf. Here, the value of R increases by increasing the value of α. If the value of R is lower, it reaches the bottom. In this process, α value is assumed to be 0.01.

$$f(I : r, \sigma_a, \sigma_b, \alpha) = 0.01(pb_{(r)} * I_{\text{ExR}}) + (\text{DoG}_{((\sigma_a, \sigma_b))} * I_H) + (pb_{(r)} * I_g) \quad (5)$$

Pillbox filter with radius ($r = 3$), standard deviation value of a is 5 and b is 4. CCF map is calculated by multiplying decrease rate β is 0.5. Convoluting pillbox filter with excess red, DoG with hue and grayscale image with pillbox filter and by summing all the convoluting results, CCF is calculated.

In Fig. 6, K means clustering, Otsu method, and comprehensive color feature detection method do not give the accurate disease spot segmentation result. In proposed method, segmentation disease spot using SVD gives the accurate segmentation of disease under uneven illumination and clutter background. Its performance evaluation is calculated by using precision,

$$\text{precision} = T_P/(T_P + F_P) \quad (6)$$

where T_P is true positive, it means percentage of disease spots present in the leaf and F_P is false positive, it means percentage of misclassified disease spots in the leaf. Table 2 indicates performance evaluation of K means, Otsu, CCF method, and proposed method.

Table 1 Performance evaluation of proposed method and MATLAB function

Input image	Methods	CCPR [13]	CCFR [13]	E-score [13]
1	MatLab	94.77	93.71	94.24
	Grayscale	95.36	98.11	96.26
2	MatLab	97.63	98.47	98.05
	Grayscale	97.83	99.97	98.66

Table 2 Performance evaluation of proposed method and MATLAB function

Input image	K means	Otsu	CCF method	Proposed method
1	31.40	73.59	79.98	88.47
2	34.41	56.31	84.96	93.98

4 Conclusion

In this paper, convolution of grayscale with pillbox filter, hue with DoG and ExR with pillbox filter and by summing convolution of hue, Excess Red and grayscale conversion produces comprehensive color feature. Color-to -grayscale conversion is processed by using singular value decomposition (SVD) by incorporating chrominance value. When comparing rgb2gray MATLAB function with color to grayscale using SVD gives higher accuracy Table 1. Region-growing method segments the diseases by selecting growing seeds, and it solves the challenges of uneven illumination and clutter background. Performance evaluation is calculated by using precision Table 2. Comparing Otsu, K means and CCF method with proposed algorithm, proposed method gives accurate disease spot segmentation.

References

1. Al Bashish, D., Braik, M., Bani-Ahmad, S.: Detection and classification of leaf diseases using K-means-based segmentation. Inf. Technol. J. **10**(2), 267–275 (2011)
2. Kamdi, S., Krishna, R.K.: Image segmentation and region growing algorithm. Int. J. Comput. Technol. Electron. Eng. **2**(1) (2012)
3. Chudasama, D., Patel, T., Joshi, S., Prajapati, G.I.: Image segmentation using morphological operations. Int. J. Comput. Appl. **117**(18) (2015)
4. Zhang, X., Liu, S.: Contrast preserving image decolorization combining global features and local semantic features **34**(6–8), 1099–108 (2018)
5. Zheng, X., Lei, Q., Yao, R., Gong, Y., Yin, Q.: Image segmentation based on adaptive K-means algorithm. EURASIP J. Image Video Process. **1**(68) (2018)
6. Pruthi, J., Gupta, G.: Image segmentation using genetic algorithm and OTSU. In: Proceedings of Fifth International Conference on Soft Computing for Problem Solving. Springer, Singapore (2016)

7. Hussein, M., Bairam, M.: A survey study on singular value decomposition and genetic algorithm based on digital watermarking techniques. In: Proceedings of the First International Conference on Computational Intelligence and Informatics. Springer, Singapore (2017)
8. Meyer, G.E., Neto, J.C.: Verification of color vegetation indices for automated crop imaging applications. Comput. Electron. Agric. **63**(2), 282–93 (2008)
9. Naik, M.R., Sivappagari, C.M.: Plant leaf and disease detection by using HSV features and SVM classifier. Int. J. Eng. Sci. Comput. **3794** (2016)
10. Yang, W., et al.: Greenness identification based on HSV decision tree. Inf. Process. Agric. **2**(3–4), 149–160(2015)
11. Ma, J., Du, K., Zhang, L., Zheng, F., Chu, J., Sun, Z.: A segmentation method for greenhouse vegetable foliar disease spots images using color information and region growing. Comput. Electron. Agric. **142**, 110–7 (2017)
12. APS Image Dataset. https://imagedatabase.apsnet.org/. An Image Database and Educational Resource. American Phytopathological Society, St. Paul, MN
13. Lu, C., Xu, L., Jia, J.: Contrast preserving decolorization with perception-based quality metrics. Int. J. Comput. Vis. **110**(2), 222–39 (2014)

QSecret-Sharing Scheme (QSS)-Based Keyless: *Titanium* Secure (*Titanium*Sec) EMail Application

Deepthi Haridas, Rakesh Shukla, Hari Om Prakash, Phani Bhushan Rallapalli, Venkataraman Sarma, V. Raghu Venkatraman, Harshal Shah and Harshul Vaishnav

Abstract Securing email, i.e. "electronic mail", is a vast subject dealing with security against illegitimate usage. Standard methodology is to incorporate encryption techniques for securing email. This ultimately leads to securing the secret keys which are used for the secure communication. In order to facilitate secure communication, there arises a dependency on third-party devices as well as trusted third-party certifications. A third party is inappropriate to secure the keys storage and exchange from, whichever the case under consideration. The present work presents a novel keyless solution to secure email application. Keyless solution is based on QSecret-sharing scheme (QSS). The novelty of current work: (i) the integration of quasigroup in to secret-sharing scheme and (ii) introduces *Titanium* Secure (*Titanium*Sec) email application using indigenous QSS.

D. Haridas (✉) · R. Shukla · H. O. Prakash · P. B. Rallapalli · V. Sarma · V. R. Venkatraman
Advanced Data Processing Research Institute (ADRIN), Department of Space,
Government of India, Secunderabad, Telangana 500009, India
e-mail: deepthi@adrin.res.in; haridasdeepthi@gmail.com

R. Shukla
e-mail: rakesh@adrin.res.in

H. O. Prakash
e-mail: hariom@adrin.res.in

P. B. Rallapalli
e-mail: kalka@adrin.res.in

V. Sarma
e-mail: director@adrin.res.in

H. Shah · H. Vaishnav
Dhirubhai Ambani Institute of Information and Communication Technology (DAIICT),
Gandhi Nagar, Gujarat 382007, India
e-mail: harshal.shah031@gmail.com

H. Shah
e-mail: va.hv.02@gmail.com

© Springer Nature Singapore Pte Ltd. 2020
A. K. Luhach et al. (eds.), *First International Conference on Sustainable Technologies for Computational Intelligence*, Advances in Intelligent Systems and Computing 1045,
https://doi.org/10.1007/978-981-15-0029-9_25

Keywords Shamir's secret share (SSS) · Threshold secret share (TSS) · Quasigroup · One-way function · Verifiable secret sharing · Perfect secret sharing (PSS)

The idea of secret share communication was introduced by Shamir [1] in the year 1979 for the first time. He proposed the threshold secret-sharing (TSS) scheme (k, n), where k is the required minimum number of participants to reconstruct (out of the total set of participants involved in sharing the secret) the secret and n is the total number of participants involved in sharing the secret. Here if $k = n$, all the participants are required for secret reconstruction. For any reconstruction with less than k participants, the secret won't be revealed. Blakley [2] proposed secret-sharing scheme shared differently. Simmons [3] applied different instances of secret share with respect to different application scenarios.

There have been different variants of secret-sharing scheme over a period of time. Pedersen [4] came up with verifiable secret-sharing scheme. Blakley et al. [5] have evaluated the performance of linear secret-sharing schemes using tools of linear algebra and coding theory. Lin et al. [6] proposed a scheme where a master secret share which is issued to reconstruct different group secrets based on threshold values. Herzberg et al. [7] have periodically renewed the shares. Therefore, information gained by the adversary at any point of time is rendered worthless, because of its ever-changing share nature. Cramer et al. [8] presented a method of converting schemes of a secret in a different secret-sharing scheme using local computation with no interaction between players.

Tompa et al. [9] have proposed that Shamir's secret-sharing scheme is not secure against cheating. They proved that when they imbibe a small modification to Shamir's scheme still they could retain the security and efficiency of the original secret.

With the advent of digital security concerns, of late secret share has been used in different scenarios. Kong et al. [10] have used threshold secret-sharing mechanism to provide intrusion tolerance for mobile adhoc networks. Tian et al. [11] have used secret share-based scheme to secure data from database service provider (DSP). Lin et al. [12] have incorporated watermarking along with secret sharing for authentication. Kaya et al. [13] have proposed privacy-preserving distributed clustering protocol for horizontally partitioned data using secret-sharing scheme. Sun et al. [14] presented an efficient group key transfer protocol based on special secret-sharing scheme. Secret-sharing scheme is also used in the personal information protection for electronic identity (eID) [15] cards.

Usually if the secret is not random, then the correlation between shares generated is pretty high, due to which, it is possible that secret image could be recovered from less than k participants. In order to waive off that security concern with respect to images, Thein and Lin [16] proposed permuting the pixel order before share generation process. Thereafter in usual scenarios, the secret image is apriory Xor-ed by the permutation key matrix then subjected to share generation. Bai et al. [17] in their work they have cited that the above-mentioned scheme prevents real-time processing as the permuted image has to be obtained prior to reconstructing secret. The current work proposes that even though xoring with a random permuted key matrix would prevent real-time processing of permuted image, the secret shares generated out of the permuted matrix would be totally random hence difficult to achieve efficient compression, resulting in increased communication overhead to transfer random secret shares within the network. In the current work, the secret is first convolved (quasified) with quasigroup and then shares were generated. Sharing this quasified secret renders the following advantages over conventional secret sharing:

(i) The secret shares will not disclose any information about the secret.
(ii) Even if an adversary reconstructs the quasified secret share, without any knowledge of the base quasigroup used to quasify the secret, it is impractical (the exhaustive number of quasigroup corresponding to the order of quasigroup is a huge number) for adversary to get the original secret.
(iii) Quasigroup quasifies the secret losslessly. QSS helps in attaining good compression ratio and facilitates ease in communication.

In Thien et. al [16] scheme the size of secret share is $\frac{1}{k}$ of the total secret image size, which disqualifies this scheme as perfect secret-sharing (PSS) scheme [18, 19], whereas Shamir's secret-sharing scheme is regarded as perfect secret-sharing (PSS) scheme, as the secret cannot be revealed with any $(k-1)$ shares.

Secure email comprises securing the contents of mail through encryption techniques. Here key management plays an important role. In our previous work [20–22], keyed approach of securing mails with Blowfish encryption algorithm and latter work along with encryption, biometrics is also taken in to account. In the current work, we have introduced a keyless approach to secure email using Qsecret-sharing scheme (QSS) in *Titanium* secure email application.

The present paper works out a novel secret-sharing methodology using quasigroups as a keyless solution to protect the privacy of secure email application. At no instance, original composed image will be stored. The original composed message is stored as image at the client-side, which is converted into an intermediate quasigroup-based transformed image. Shares are constructed from an intermediate quasigroup-based transformed image. At the receiver end, they regenerate the secret image from any three or more selected shares. The secret message is unveiled at the receiver end, on reversing the quasigroup-based transformation. Hence throughout the process, the secret message is never in transit. The secret message is generated and regenerated only at the client-side, with server having no idea about the underlying secret message. The generated image and regenerated image at client end are verified by data integrity check. SHA-256 of the original image is verified with the SHA-256 of the regenerated image. Once SHA-256 is verified, then the regenerated image is displayed at the recipient's end.

The rest of the paper is structured as follows: The theory behind the Qsecret-sharing scheme is explained in Sect. 1. *Titanium* the secure email application is covered in Sect. 2, with Sect. 2.1 detailing the setup for *Titanium*Sec and Sect. 2.2 describing the components of it. Results and Discussion are contained in Sect. 3 comprising of functioning of *Titanium*Sec Sect. 3.1 and Security aspects of *Titanium*Sec. Section 4 concludes the present paper.

1 QSecret-Sharing Scheme (QSS)

The QSecret-sharing scheme (QSS) is a secret-sharing scheme based on quasigroups. Quasigroups have been used in designing many cryptographic primitives [23–26]. Quasigroups have been predominantly used for designing encryption and decryption algorithms. The present work introduces quasigroup for image exploitation. The one-way property of quasigroups makes it an ideal utility to transform an image losslessly to transformed image with pixels shuffled uniquely.

Basic Definitions

Definition 1 (*Quasigroup*) [23] Let $Q = \{a_1, a_2, \ldots, a_n\}$ be a finite set of n elements. A quasigroup $(Q, *)$ is a groupoid satisfying the law: $\forall u, v \in Q \exists$ unique solutions $x, y \in Q$ s.t $u * x = v, y * u = v$.

Definition 2 *Latin Square* [23] A Latin square L on a finite set Q of cardinality $|Q| = n$ is called an $n \times n$-matrix $L_{n \times n}$, with elements from Q such that each row and column of the matrix is a permutation of Q.

The QSecret Shaing Scheme is applied on images in the present paper. The QSecret-sharing scheme (QSS) works as follows:

* **Setup** Let the arithmetic be done on $\mathbb{Z}_{251}$. Let Q be a quasigroup s.t.
  $Q : \mathbb{Z}_{256} \times \mathbb{Z}_{256} \longrightarrow \mathbb{Z}_{256}$, defined as $Q(i, j) = i * j = k$, where i, j, and $k \in \mathbb{Z}_{256}$ and $*$ is the binary operation defining the quasigroup Q of cardinality 256. Denote the secret image as $I_{m \times n}$, where m denotes the number of scan lines of the image and n denotes the number of pix columns of the image.
* **QuasiProcessing of Secret** The present paper introduces the notion of quasitransformed images to secret share processing in the following manner:

  1. In order to avoid the problem of overshooting and undershooting in our secret image threshold, the pixels values of the secret image by setting the prime modulus as $p = 251$, i.e.

$$I(x, y) = \begin{cases} \text{pixel}_{\text{val}} & \text{if } x < 251, \\ 250 & \text{if } x \geq 251. \end{cases} \quad (1)$$

     where $0 \leq x < m$ and $0 \leq y < n$.
  2. Let $\tilde{I}$ be the quasigroup transformed secret image such that:

$$\tilde{I}(x + Q(i, j), y + j) = I(x + i, y + j) = I(x, y) * Q(i, j) \quad (2)$$

     where $0 \leq x < m, 0 \leq y < n, 0 \leq i < 256, 0 \leq j < 256$ and $0 \leq Q(i, j) < 256$.

∗ **Secret Sharing Scheme** The quasiprocessed secret is subjected to Shamir's threshold secret-sharing (k, n) scheme [1] to generate secret shares. Secret shares are generated as:

$$S_X(x, y) = s_0(x, y) + s_1 X + s_2 X^2 + s_3 X^3 + \cdots + s_{k-1} X^{k-1} (mod p) \qquad (3)$$

where $0 \le x < m$, $0 \le y < n$, and $s_0(x, y)$ is the secret, i.e. pixel value of quasiprocessed secret image $\tilde{I}_{m \times n}$ at the (x, y)th coordinates of the image. The pair of values (X_i, Y_i), where $Y_i = S_{X_i}, \forall 1 \le i \le n$ and $0 < X_1 < X_2 \cdots < X_n \le (p-1)$.

The polynomial S_X is used only to generate the shares $S_{X_1}, S_{X_2}, \ldots, S_{X_n}$ such that the value of secret if revealed to nobody. The secret won't be revealed to participants less than k secret shares. After share generation, the polynomial S_X is destroyed.

∗ **Secret Reconstruction** This takes in two steps which are listed below:

1. **Quasiprocessed Secret Reconstruction** The polynomial S_X is destroyed after the share generation. Only the quasiprocessed secret shares $S_{X_1}, S_{X_2}, \ldots,$ S_{X_n} are available with each participant. Quasiprocessed secret won't be revealed to anyone with less than k shares. If k or more than k shares are available, the quasiprocessed secret would be reconstructed using Lagrange's interpolation.

2. **Secret Reconstruction** Applying the quasigroup parastrophe to the quasiprocessed secret, the original secret image will be reconstructed back. Let $\tilde{I}$ be the quasiprocessed transformed secret image, i.e. the original secret is reconstructed using the following transformation such that:

$$\tilde{I}(x, y) * Q_{par}(i, j) = I(x, y) \qquad (4)$$

where $0 \le x < m, 0 \le y < n, 0 \le i < 256, 0 \le j < 256$ and $0 \le Q_{par}(i, j) < 256$, Q_{par} is the parastrophe of quasigroup (Fig. 1).

Use cases for *TitaniumSec* Based on the theory of QSecret-sharing scheme (QSS), *Titanium* secure email application is developed in the present work. *Titanium* is an element light in weight and has extraordinary corrosion resistance and ability to withstand extreme temperatures. *Titanium* usually used for military applications. The present work has used *Titanium* as a metaphor for our application on QSecret-sharing scheme (QSS) which secures the data of strategic users. *Titanium*Sec is client-based application for closed user group (CUG). In analogy with algebraically closed group [27], let us defined closed user group with respect to users/clients.

Definition 3 (*Closed User Group (CUG) of Users*) Let $G = \{u_i | \forall 1 \le i \le n\}$ be a finite group of "n" users, where u_i is the ith user $\forall i \in [1, n]$. If $\omega : G \times G \to [0, 1]$, such that $\Omega_{ij} = \omega(u_i, u_j)$, where $u_i, u_j \in G$ and ω (sender, receiver) as per convention ω represents the weight measure for data communicated from sender u_i to receiver u_j.

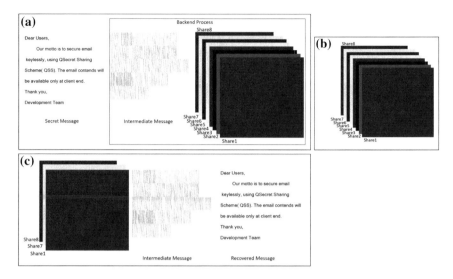

Fig. 1 **a** Processing at the sender's end, **b** data available at server end and **c** processing at the receiver's end

Suppose if $\Omega_{i-1 j} = \omega(u_j, u_i) = \Omega_{i j-1} = \omega(u_j, u_i)$, i.e. weight measure from sender u_j to receiver u_i. The different weight measures are defined as follows:

$$\Omega_{ij} = \begin{cases} 1 & \text{if } i \neq j \quad \text{and} \quad i, j \in [1, n] \\ 0 & \text{if otherwise} \end{cases}$$

That is, ω weight measure for data communicated from sender to receiver is "1" when users with in the CUG group communicate with each other, whereas when the sender and receiver are same, then data is not flowing; hence, the weight measure is allocated "0" and if the users does not belong to closed user group, then he or she cannot communicate data across; hence, weight measure is "0".

2 *Titanium* Secure (*Titanium*Sec) Email Application

The present paper incorporated the QSS-based keyless solution to *Titanium*Sec Email Application. As part of developing *Titanium*Sec email application, implemented a mail client on Windows 10PC. *Titanium*Sec App works on web browser supporting web users/clients.

2.1 *Set-up:* TitaniumSec

The algorithms are implemented on Java, C++ and JavaScript. *Titanium*Sec works on web browser supporting HTML5. QSecret-sharing scheme (3, *n*) (i.e. total *n* shares will be generated depending on the number of secret-sharing users, out of which 3 are mandatory to reconstruct back the secret) is implemented in to *Titanium* Secure Email application.

The following softwares are required for *Titanium* Secure App:

∗ **Xampp**: It is used for creating and maintaining web server. PHP is used as server-side language and database is MySQL.
∗ **Communigate Pro**: Mail server used for *Titanium* secure email application. For sending mail, simple message transfer protocol (SMTP) is used and for receiving Internet message access protocol (IMAP4)is used. An additional security measures SMTPs and IMAP4s are being used.
∗ **Eclipse Java**: Java IDE, for Java Programming.
∗ **Eclipse C++**: C++IDE, for C++ Programming.

2.2 *Components of* Titanium *Secure Email App*

TitaniumSec email application comprises of three main components: mail server, web server and storage of secret shares. The details regarding the three components are as follows:

1. **Mail Server**
   CommuniGate Pro is used as the mail server in our web application. No direct connection is allowed to email server for security reasons.
2. **Web Server and Secure Mail application**
   Apache Web server is hosted with Xampp. Additional security with secure socket layer (SSL) protocol to ensure strengthened security between client and server (over TCP/IP). All the traffic is forced to use SSL. HTML, CSS, JavaScript are used for the client-side page rendering and algorithm processing. JQuery is used for asynchronous communication with the server (from client). At the server-side, PHP is used as server-side language and for communication with MYSQL server.
3. **Storage of contents shares** Contents of mail will be stored as random shares for every mail. Thus, mails will be secured by QSS and stored at the server.

Workflow of *TitaniumSec* Email App The process of secret share generation from composed mail is shown in Fig. 2a. In Fig. 2a, at sender's end once the mail is composed and users hit send, the mail will be converted into image. The SHA-256 of the image will be computed and stored at database. Image is convolved with quasigroup of the order 256. Quasified image is split into shares. The shares reside at server/Internet. Original content is reconstructed at recipients machine only as shown in Fig. 2b. The moment mail recipient hits read mail, and the background processing

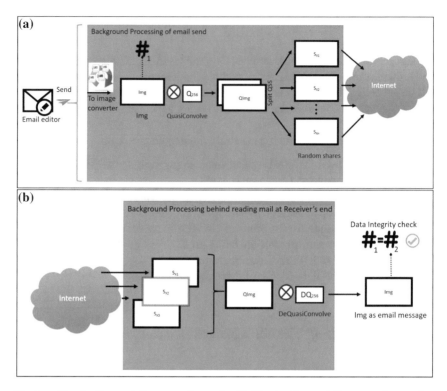

Fig. 2 **a** *Titanium*Sec work flow at senders end and **b** *Titanium*Sec work flow at receivers end

starts. The shares are fetched from the server. The quasified image is reconstructed; the quasified image is dequasified giving the back the resultant original content; SHA-256 digest is computed and checked with the SHA-256 of original content; after verification, the resultant original content is displayed at the recipient's end. Data integrity is checked for the content, as shown in Fig. 2b.

3 Results and Discussion

*Titanium*Sec email application is for closed user's group (CUG). The emails are communicated within the members of closed user group. Emails are secured without encryption keylessly. Here when the user composes the email message, it is stored as an image. Once the user composes their mail, it is converted into an image. At client-side itself, the image undergoes a lossless revamp using quasify process introduced in the current work. The quasified image is then subjected to secret-sharing methodology.

3.1 Functioning of **Titanium***Sec*

As the client sends the mail, shares are generated and pushed to the web server in SSL tunnel mode. The shares generated resides at mail server. Mail server has got no access to the mechanism of regenerating the original secret message. The recreation of secret message is available at the client-side. The process of generating the shares and recreating the secret takes place at the sender and receiver as follows:

∗ **Sender**

When the sender clicks the "Send" button, text in the body field of the mail is first converted into the image via HTML canvas module. Once the sender sends the message, the data is converted into image. At the back-end, the image is first quasified and then shares are generated from it. The back-end process is done at the client-side, such that web server cannot get access to the client's message.

∗ **Receiver**

As the receiver selects the mail to be viewed, the script at client-side request shares from the mail server, which are then processed via the algorithm used in the encryption (only one algorithm will be set as default) to reconstruct the quasified image and is further processed to construct the final image which is the image containing the message of sender.

TitaniumSec Email application takes care that the original secret message which was composed at the user end is never in transit. It is available only at the sender end and receiver end.

3.2 Security Aspects of **Titanium***Sec* **Email App**

*Titanium*Sec secures the email messages by Qsecret-sharing methodology rather than going for conventional encryption schemes. Securing mails via *Titanium*Sec are devoid of complicated key management schemes (KMS), as it being a keyless solution. The current paper presents the novel solution of quasifying the secret before subjecting it to share generation. The quasigroup used for Qsecret secret-sharing scheme is of the order 256. Quasigroups are generated dynamically at run-time. The total number of quasigroups of the order 256 are much more than $10^{58,000}$, which is more than the exhaustive number of stars in our Milkyway galaxy. Hence, the email messages will be quantum secure with future computers with Qubit Processor. The mail server has no access to share generation or reconstruction of secret, only the shares resides at mail server. In case if an adversary gaining access to the secret shares, still they will not be able to reconstruct the original secret as the secret was quasified before share generation. Direct connection is not allowed to email server, and only web browser IP is designated to communicate with the email server. Presently, CommuniGate Pro is used as mail server. It could also be implemented in Zimbra 8.0, IPLANET, MSExchange, HTML, LOTUS Domino, etc. In *Titanium*Sec, the emails are secured using perfect secret-sharing scheme (PSS) at the cost of storage space.

4   Conclusion

The present work comes up with a novel software *Titanium*Sec to secure email messages. *Titanium*Sec is a client-based application which secures the mail contents of the users using Qsecret-sharing scheme. The secret shares reside at the mail server. At the client-side when the message is to be reconstructed, the secret shares are fetched from mail server and the reconstruction is done at client-side. Data integrity of the secret message is verified and then displayed at the client end. The software solution *TitaniumSec* calls for extra storage space to provide security to the secret messages shared by the users. As every message is split in to shares, thereby storing every message leads to utilizing the storage space as multiple of total number of shares, of the secret message. Security here comes with a cost of extended storage space, which in present-day scenario is reasonable trade-off. The current application introduces quasigroups to secret sharing to encapsulate the secret, thereby improvising the secret-sharing methodology and introducing the new approach of Qsecret-sharing scheme to securing email via *Titanium*Sec. QSS scheme takes care that the secret composed message is available only to sender as well as receiver, but it is never available in transit. Therefore, this way it mimics encryption. Here QSS scheme in *Titanium* provides keyless solution to secure secret message.

References

1. Shamir, A.: How to share a secret. Commun. ACM **22**(11), 612–613 (1979)
2. Blakley, G.: Safeguarding cryptographic keys. In: Proceedings of AFIPS 1979 National Compute Conference, vol. 48, pp. 313–317. Arlington, VA (1997)
3. Simmons, G.J.: How to (really) share a secret. In: Conference on the Theory and Application of Cryptography, pp. 390–448. Springer, New York, NY (1988)
4. Pedersen, T.P.: Non interactive information theoretic secure verifiable secret sharing. In: Annual International Cryptology Conference, pp. 129–140. Springer, Berlin (1991)
5. Blakley, G.R., Kabatianskii, G.A.: Linear algebra approach to secret sharing schemes. In: Error Control, Cryptology, and Speech compression, pp. 33–40. Springer, Berlin (1994)
6. Lin, H.-Y., Yeh, Y.-S.: Dynamic multi secret sharing scheme. Int. J. Contemp. Math. Sci. **3**(1), 37–42 (2008)
7. Herzberg, A., Jarecki, S., Krawczyk, H., Yung, M.: Proactive secret sharing or How to cope with perpetual leakage. In: Annual International Cryptology Conference, pp. 339–352. Springer, Berlin (1995)
8. Cramer, R., Damgard, I., Ishai, Y.: Share conversion, pseudo random secret sharing and application to secure computation. In: Theory of Cryptography Conference, pp. 342–362. Springer, Berlin (2005)
9. Tompa, M., Woll, H.: How to share a secret with cheaters. J. Cryptol. **3**, 133–138 (1989)
10. Kong, J., Zerfos, P., Luo, H., Lu, S.W., Zhang, L.: Providing robust and ubiquitous security support for mobile adhoc networks. In: ICNP. IEEE, New York (1991)
11. Tian, X.X., Sha, C.F., Wang, X.L., Zhou, A.Y.: Privacy preserving query processing on secret share based data storage. In: International Conference on Database Systems for advanced Applications, pp. 108–122. Springer, Berlin (2011)
12. Lin, C.-C., Tsai, W.-H.: Secret sharing with steganography and authentication. J. Syst. Software **73**(3), 405–414 (2004)

13. Kaya, S.V., Pedersen, T.B., Savas, E., Saygiyn, Y.: Efficient privacy preserving distributed clustering based on secret sharing. In: Pacific-Asia Conference on Knowledge Discovery and Data Mining, pp. 280–291. Springer, Berlin (2007)
14. Sun, Y., Wen, Q., Sun, H., Li, W., Jin, Z., Zhang, H.: Authenticated group key transfer protocol based on secret sharing. Proc. Eng. **29**, 403–408 (2012)
15. Park, N., Lee, D.: Electronic identity information hiding methods using secret hsaring scheme in multimedia centric internet of things environments. Pers. Ubiquit. Comput. **22**(1), 3–10 (2018)
16. Thien, C.-C., Lin, J.-C.: Secret image sharing. Comput. Graphics **26**(5), 765–770 (2002)
17. Bai, L., Biswas, S., Ortiz, A., Dalessandro, D.: An image secret sharing method. In: 2006 Ninth International Conference on Information Fusion, pp. 1–6. IEEE, New York (2006)
18. Capocelli, R.M., De Santis, A., Gargano, L., Vaccaro, U.: On the size for shares for secret sharing schemes. J. Cryptol. **6**(3), 157–167 (1993)
19. Karnin, E., Greene, J., Hellman, M.: On secret sharing systems. IEEE Trans. Inf. Theory **29**(1), 35–41 (1983)
20. Shukla, R., Prakash, H.O., Phanibhusan, R.: OpenPGP based secure web mail. In: 2016 3rd International Conference on Computing for Sustainable Global Development (INDIACom), pp. 735–738. IEEE, New York (2016)
21. Shukla, R., Prakash, H.O., Phanibhusan, R.: Pehchanmail: IBE based E-token authenticated secure email. Int. J. Recent Technol. Eng. (IJRTE) **3** (2014)
22. Shukla, R., Prakash, H.O., Phanibhusan, R., Venkataraman, S., Vardhan, G.: Sahastradhara biometric and EToken integrated secure email system. In: 2013 15th International Conference on Advanced Computing Technologies (ICACT), pp. 1–4. IEEE, New York (2013)
23. Markoviski, S., Gligoroski, D., Andova, S.: Using quasigroups for one-one secure encoding. In: Proceedings VIII Conference Logic in Computer Science, LIRA, vol. 97, pp. 157–162 (1997)
24. Scherbacov, V.A.: Quasigroups in Cryptology. arXivpreprint. arXiv: 1007.3572 (2010)
25. Pal, S.K., Kapoor, S., Arora, A., Chaudhary, R., Khurana, J.: Design of strong crytpographic schemes based on Latin squares. J. Discrete Math. Sci. Cryptograhy **13**(3), 233–256 (2010)
26. Haridas, D., Raj, K.C.E.S., Sarma, V., Chowdhury, S.: Probabilistically generated ternary quasi-group based stream cipher. In: Progress in Intelligent Computing Techniques: Theory, Practice and Applications, pp. 158–160. Springer, Singapore (2018)
27. Scott, W.R.: Algebraically closed groups. Proc. Am. Math. Soc. **2**(1), 118–121 (1951)

Sentiment Analysis of Tweets Using Supervised Learning Algorithms

Raj P. Mehta, Meet A. Sanghvi, Darshin K. Shah and Artika Singh

Abstract The proliferation of user-generated content (UGC) on social media platforms has made user opinion tracking a strenuous job. Twitter, being a huge microblogging social network, could be used to accumulate views about politics, trends, and products, etc. Sentiment analysis is a mining technique employed to peruse opinions, emotions, and attitude of people toward any subject. This is conceptualized using digital data (text, video, audio, etc.) or psychological characteristics of humans. This procedure assists in opinion mining without having to read a plethora of tweets manually. The results could be wielded to provide an edge for businesses and governments in rolling out new entities (policies, products, topic, event). Cleaning data is an important step here, which we accomplished using regular expressions and NLTK library in Python. We implemented nine separate algorithms to classify tweets and compare their performance on cleaned data. It was observed that the convolutional neural network produces the most optimal results at 79% accuracy.

Keywords Human emotions · Machine learning · Opinion mining · Sentiment analysis (SA) · Supervised learning · Long short-term memory (LSTM) · Twitter · Naive Bayes (NB) · Tweets · Natural language processing · Multilayer perceptron (MLP) · Convolutional neural network (CNN) · Random forest · XGBoost · Max entropy · Decision tree · Support vector machine (SVM)

R. P. Mehta (✉) · M. A. Sanghvi · D. K. Shah · A. Singh
Mukesh Patel School of Technology Management & Engineering, NMIMS, Mumbai, India
e-mail: rajpareshmehta@gmail.com

M. A. Sanghvi
e-mail: meetsanghvi98@gmail.com

D. K. Shah
e-mail: dkjgraphics@gmail.com

A. Singh
e-mail: artika.singh@nmims.edu

© Springer Nature Singapore Pte Ltd. 2020
A. K. Luhach et al. (eds.), *First International Conference on Sustainable Technologies for Computational Intelligence*, Advances in Intelligent Systems and Computing 1045,
https://doi.org/10.1007/978-981-15-0029-9_26

1 Introduction

The term sentiment analysis was first observed in the year 2001 [1]. It was used to analyze the various emotions expressed in a Web page, blog, or simply an article. But in the modern era, the applications of sentiment analysis have grown largely. SA mainly targets social media networks. Sentiment analysis can be seen from various perspectives, as shown in Fig. 1.

1. **Techniques**
   (a) **Statistical Approach**: Statistical models represent each comment as a blend of hidden aspects and ratings. Assumption here is that they can be represented in multinomial distributions. Cluster head terms correspond to aspects and sentiments to ratings.
   (b) **Lexicon-Based Approach**: It concerns with the enumeration of no. of positive and negative words in a paragraph. If there are more positive words, then it is considered positive; similarly, if there are more negative words, it is considered negative. Moreover, this is done using two sub-methods, i.e., dictionary-based approach and corpus-based approach.
   (c) **Machine Learning Approach**: Machine learning is a technique which gives a computer the ability to "learn" with data, without being explicitly programmed. Machine learning for sentiment analysis includes collecting data and then applying one of the machine learning techniques to find the sentiment of the inputs. The various techniques have been described below:
      i. *Supervised Learning*: Here, the computer is given labeled training data. Each training input provided to the computer consists of an input and

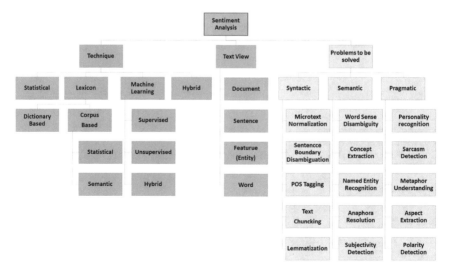

Fig. 1 Synopsis of sentiment analysis

the desired output. A supervised learning algorithm dissects the training data and produces a surmised work, which can be utilized for mapping new inputs. An ideal situation will allow for the algorithm to accurately choose class labels for unseen occurrences. For such a scenario, the algorithm should be able to generalize from the data on which it has been trained.

ii. **Unsupervised Learning**: Here, the computer is not given any labeled training data. It is the task of finding concealed structure hidden in the training data in order to capture practical insights from it. Unsupervised algorithms can prove to be more uncertain as compared to supervised learning algorithms. Even though an unsupervised learning algorithm may 'learn' to segregate between cats and dogs, it may also add unanticipated and undesired classes to deal with unwanted breeds, affecting its performance.

iii. **Hybrid Techniques**: It is a combination of supervised and unsupervised learning techniques, and the goal of this is to get the best of both, super and unsupervised learning. Usually, the training set is made up of few labeled data and plentiful of unlabeled data. The presence of labeled data helps attain a substantial improvement in learning accuracy when compared to the presence of only unlabeled data.

(d) **Hybrid Approach** The amalgamation of two approaches is suggested to obtain the best of both worlds. The lexicon and machine learning approaches are combined to improve the classification accuracy. Lexicon-based approach has high precision and low recall, whereas machine learning approach has high recall and low precision. Thus, combining them gives improved accuracy as well as recall.

2. **Text View**

This determines how the division of the input text will be carried out and fed to the algorithm. The way this is done can highly impact the performance of the algorithm, and thus, the correct choice has to be made here. The advantages and disadvantages of each technique have been discussed below:

(a) **Word-Level Sentiment Analysis**: In this text view, each word is handled individually and is mapped to a sentiment. This technique does not handle negations since each word is considered individually.
    Example 1: "He is a very good fellow." Here, good is a +ve word and others are parts of speech, which may not be given any weight.
    Example 2: "He is not a good fellow." Here, due to handling of each word individually, the algorithm may not be able to identify the sentiment.

(b) **Sentence-Level Sentiment Analysis**: In this level of the text view, distinctive levels of granularity are investigated over the input provided. Negation handling can be done since the entire sentence is considered as a single entity while evaluating the sentiment. This level of text view is closely associated with subjectivity classification that differentiates sentences expressing factual information and sentences expressing subjective views.

(c) **Feature (Entity)-Level Sentiment Analysis**: It considers different aspects of an entity, and hence, it is more accurate. Each +ve feature is assigned a +ve weight, and each −ve feature is assigned a −ve weight. After finding all features, their weights are aggregated in order to find the sentiment of the entire input. This is an excellent approach to product reviews. For example, the screen of a laptop or the quality of leather.

(d) **Document-Level Sentiment Analysis**: Here, the entire document is analyzed as a single input and a sentiment score for the entire document is provided. This is applicable when the input to be analyzed is a movie review, restaurant, etc.

3. **Preprocessing: Problems to be solved**

(a) **Micro-text Normalization**: Removal of informal text characterized by related spelling and dependence on abbreviations, emoticons, acronyms.
  • *Example*: TTYL— Talk to you later, GN—Good Night, SD—Sweet Dreams, TC—Take Care

(b) **Sentence Boundary Disambiguation**: Logical deconstructing of text into sentences.
  • *Example*: Mr. Shah—Not a sentence boundary. The weather is okay—A sentence boundary

(c) **Part-of-Speech Tagging**: Concerns with correctly categorizing each and every word that is discovered in the sentence into adverbs, adjective, noun, etc.
  • *Example*: Verbs—hear, happen, enjoy; adjective—sweet, red, good

(d) **Text Chunking (shallow parsing)**: It follows POS tagging and adds more structure to the sentence. The upshot is a syntactic substructure of words.
  • *Example*: The weather is okay—The, weather, is, okay

(e) **Lemmatization**: Converting a word to its base form.
  • *Example:* am, are, is converted to—be while car, car's, cars', cars is converted to—car.

(f) **Word Sense Disambiguation:** Identifying words having different meaning depending on its context.
  • *Example*: Fine as noun (penalty) or fine as an adjective.

(g) **Concept Extraction**: It concerns with breaking down full text into meaningful concepts, either single or multiple word expression.
  • *Example*: Pain Killer cannot be broken into Pain and Killer

(h) **Named Entity Recognition**: It concerns with locating and classifying entities into person, organization, location, etc.

(i) **Anaphora Resolution**: The pointing back of a reference to a previous point in the text.
  • *Example:* John travelled around France twice.

(j) **Subjectivity Detection**: Removal of factual or neutral content and retain subjective content.
   • *Example*: The weather is good—Removal of: The, is

(k) **Personality Recognition**: Personality is a combination of an individual's behavior, emotion, motivation, and thought-pattern characteristics.

(l) **Sarcasm Detection**: In this, there exist a target person or an entity against whom the utterance is directed. These kinds of sentences can mislead the polarity and is important to identify.
   • *Example*: Light travels faster than sound. This is why some people appear bright until you hear them speak.

(m) **Metaphor Understanding**: They are usually complex substitutes bearing similar meaning or ideas but conveyed indirectly.

(n) **Aspect Extraction**: They are specific features in an entity. Therefore, it is important to identify it because a commentator may express contradictory polarities for aspects in the same comment.
   • *Example*: My phone's screen is beautiful, but its performance is even better

(o) **Polarity Detection**: It is the most popular sentiment analysis. Some research works have even used "sentiment analysis" and "polarity detection" interchangeably. Its an NLP task that aims to categorize sentences as positive, negative, or neutral.
   • *Example*: It made her sad (negative). It made him happy (positive).

2   Literature Survey

1. **Application of Machine Learning Techniques to Sentiment Analysis**: The author of this paper [2] lays out details regarding the procedure of sentiment analysis of data collected from Twitter using machine learning and its text analysis framework using Apache Spark. Apache Spark is universally known for its fast cluster computing system which, in memory, is approximately 100 times faster than Hadoop MR and approximately 10 times quicker on disk. Proposed framework uses Naïve Bayes and decision trees algorithms. It could be theorized from the results given by the paper that:

   (a) **Multinomial Naïve Bayes** does not perform well with the small training dataset as compared to decision tree.
   (b) **Decision tree** requires longer time to be trained.
   (c) **The proposed framework** is domain independent and can operate on datasets with varying sizes.

2. **Sentiment Analysis Is a Big Suitcase**: The author in [3] has recognized some big issues that machine learning suffers from:

   (a) **Dependency**: It demands a lot of training data and is domain-dependent.
   (b) **Consistency**: Difference in training data gives different results.

(c) **Transparency**: The underlying algorithms are uninterruptible and hence no transparency.

The hurdles in natural language processing and machine learning could be resolved by devising a top-down and bottom-up approach. A Top-Down approach to take advantage of models as semantic network and conceptual dependency representations to encode the actual meaning. A bottom-up approach to use sub-symbolic techniques like deep neural networks and multiple kernels attempting to deduce syntactic patterns from the given data. The author addresses this problem by dividing the problem domain into three layers, namely syntactic layer, semantic layer, and pragmatics layer. This is inspired by NLP curves' paradigm [3].

(a) **Syntactic Layer**: This layer aims to preprocess text and convert it into simple, computer understandable English language. Infected verbs and nouns are normalized here, and general sentence structure is put forth.
(b) **Semantic Layer**: This layer deconstructs the output given by syntactic layer to obtain concepts and resolve any references given. It also aims at filtering neutral, unbiased data from the input to ameliorate the accuracy of final results.
(c) **Pragmatics Layer**: This layer aims at distilling the actual meaning from outputs of syntactic and semantic layer. It can perform interactions with the user to infer meanings and understand the metaphor, etc.

3. **Review of Sentiment Semantic Analysis Technology and Progress**:
Concepts of sentiment computing with its essential feature vector is jotted down here. Several vital issues that hazard the sentiment analysis process are also presented. Subjective and objective contents are categorized using an algorithm. The algorithm employs either single-modal analysis (text, image, audio, video) or multimodal analysis. Emotion detection and fusion strategy are summarized in this paper. The author in [4] uses wearable devices to determine sentiment characteristics under different pressure levels. He is the first one to propose "sentiment computing."

(a) **Single Modal** Carriers of SA are complexion of our skin, its conductance, our heartbeat, along with gestures, texts, and images to name a few.
    i. *Sentiment Computing and Measurement of Text*: Some famous lexicons used nowadays for English are GI, BL Lexicon, MPQA Lexicon. Universally used evaluation dictionaries are HowNet lexicon and NTU. There exist some specialized lexicons that can extract more desirable features than existing lexicons using Dirichlet distribution and PMI.
    ii. *Sentiment Computing and Measurement of Image*: It is the most natural method to depict sentiment. Color, texture, shape, and contour could be used to show emotion, where color and texture are the principal components. The classical methods for feature extraction here include

      gray-level co-occurrence matrix, multiple reference frame theory, local binary patterns, fractal theory, wavelet theory, Tamura feature, etc.

  iii. *Some Classifications under this topic*: Facial expression, gesture recognition, ordinary image processing.

  iv. *Sentiment Computing and Measurement of Audio*: Audio can accommodate several characteristics for depicting sentiment, like the quality of sound, rhymes, spectrum, even when ignoring the semantic features. Speaking person's tone, intensity, and speed are greatly aroused by situations, and hence, it gives clear-cut insight into the person's sentiment. Deep learning algorithms like CNN and LSTM are applied best here.

(b) **Multimodal** The author goes on to explain bimodal and multimodal systems, wherein a lot of information is automated to evaluate sentiments lying under the hood. Basically, mixtures of single-modal systems fused together give multimodal systems. Decision and feature fusion layers are the types identified by the author.

4. **A Parallel Semantic Sentiment Analysis**:
   Tweets are analyzed in parallel using WordNet and AFFIN, wherein synonymy plays a crucial role. HDFS and MapReduce are used for distributed and parallel processing. The author strives to combine a work using semantic similarity, opinion mining, big data, sentiment analysis, and information retrieval.
   The authors in [5] propose a dictionary-based approach where the input is classified into three categories: motivated, demotivated, and neutral. The cynosure here is to enhance the AFINN dictionary that contains weight value for each word ranging from -5 to 5 (mildly negative and strongly positive, etc.), with the semantics. Semantics that is employed here is synonymy which is sourced from WordNet. WordNet is apprehended as a semantic network where each node in it is a concept. The flowchart shown in Fig. 2 illustrates that the words not found in AFINN dictionary are searched in WordNet for synonyms and the same process of finding sentiment value is repeated. A threshold value is then set to classify the tweet as aforementioned.

5. **Twitter Sentiment Analysis using Various Classification Algorithms**: Here, the authors have used the Sanders Twitter dataset and applied various algorithms to conduct sentiment analysis. Preprocessing of tweets has been done, wherein tangential data like stop words have been removed. This is attained by using the NLTK library. WEKA data mining tool has been used for executing the classification algorithms. Emoticons, opinion lexicons, punctuations, and unigrams have also been evaluated as features. The classifiers mentioned below have been used for performing sentiment analysis are Naïve Bayes, Bayes nets, discriminative multinomial Naïve Bayes, sequential minimal optimization, hyperpipes, and random forest. The following performance metrics have been used for comparing results of the classifiers: accuracy, F-score, false alarm rate, recall (TPR), FNR, FPR, and precision. A graph juxtaposing precision and recall have been provided by the authors in Fig. 3. It can be perceived that discriminative multi-

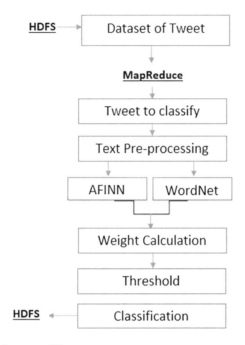

Fig. 2 Flowchart of the system [5]

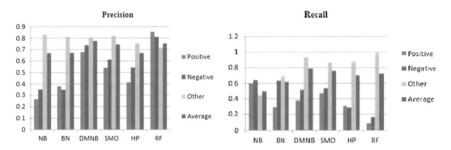

Fig. 3 Precision and recall comparison [6]

nomial Naïve Bayes and sequential minimal optimization provide a balance of performance and results [6].

6. **A Brief Review on Sentiment Analysis:** In the age of big data, companies can clinch market share by knowing precisely what their customers' likes and dislikes are. This problem of knowing the feelings of people toward certain products is known as sentiment analysis. A succinct review on SA is given in this paper. Several techniques have been compared. Textual info can be chiefly divided into two types [7], namely:

   (a) **Facts**: subjective information

(b) **Opinions**: represents the sentiment of people about the product. This is the genre of data that has to be appraised.

Finally, the authors have suggested a solution of their own based on Naïve Bayes classifier. It works by identifying keywords in each document and assigning higher weights to the opinion classes that have more number of keywords. Finally, the aggregated weightage is applied on each opinion class and then a decision is made [7].

7. **Sentiment Analysis of Text Using Deep Convolutional Neural Networks**: The utilization of convolutional neural networks (CNN) has been propounded by the authors. This machine learning algorithm is a feedforward NN, inspired by the formation of an animal's visual cortex. The training dataset is converted to vector form using word2vec, which is then provided as input to the algorithm. Figure 4 below shows a pictorial representation of the algorithm. Subsampling is used for minimizing the data that has to be processed, and convolutions extract features from the subsampled data. After performing multiple convolutions, the data becomes too huge, and thus, subsampling has to be done again to ensure the data can be efficiently handled. Fully connected operation ensures that each neuron/node is connected to all other nodes in its adjacent layer. Proposed technique generates an accuracy of 80.69% [8].

8. **Analyzing Twitter Sentiments Through Big Data**: There are four steps in the architecture devised by the authors in paper [9].

   (a) **Data collection**
   (b) **Data cleaning And preprocessing**
   (c) **Data analysis**
   (d) **Results**

In the data collection step, authors collected tweets from twitter API tweepy on reviews expressed by customers of "Bharti Airtel." The total number of tweets were 80,000 using "#Airtel" tag. After collection, the data extracted was cleaned and preprocessed.

   (a) **Removal of any unwanted notations or symbols**

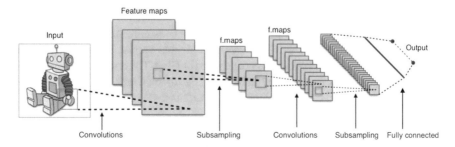

Fig. 4 Typical CNN architecture [8]

(b) **Removal of repetitive words**
(c) **Spell check to correct misspelt words**

The data analysis step was performed on an Apache Hadoop: a framework which has Hadoop Distributed File System. It has a NameNode/DataNode combination of various nodes. MapReduce deals with parallel processing of queries across these nodes. The dataset was divided into a 70–30 ratio for training and testing the tweets. It successfully calculated the polarity of the tweets with results being: 20% negative, 38% neutral, and 40% positive tweets.

9. **Twitter Sentiment Analysis using Deep Learning Methods**: The authors [10] performed SA using deep learning methods. The data is unstructured and was found on the Internet via various microblogging Web sites. The first step is performed to pull out meaningful and useful textual data and information from a vast collection of the data. It is also known as text mining and three steps, namely:

   (a) **Information retrieval**
   (b) **Information extraction**
   (c) **Data mining**

The next step is text cleaning and preprocessing. Here, the retrieved information is cleansed to remove the trivial and unfitting data. It has many different steps like stemming and lemmatization, conversion to lowercase, removal of numbers and stop words, elimination of white spaces and other characters such as '@, RT, links, punctuations'. The deep learning architecture used in the paper has three hidden layers with feedforward neural network system. Feedforward NN connotes a network where no cycles exist and the movement of information is only in one direction. The architecture also used a ReLU and Sigmoid function with about 100 neurons as input. Each hidden layer has its own functionality. One layer filters the words, another does it based on sentences, and the third does it based on popularity. ReLU stands for rectified linear unit, and it is an activation function and usually used to speed up the training process of a network.

10. **Sentiment Analysis of Facebook Posts: the Uber Case**:
    The first two steps are much alike with respect to other papers [9, 10]. An analysis is performed on the cleaned and preprocessed word features of textual data. Facts which are objective are removed while subjective opinions are retained. User's subjective opinions are classified by taking into account the context, level, subjectivity, orientation, and strength of sentiment. The results of the classification are represented in a graphical manner by using pie, bar graphs, line graph, etc. The authors performed SA on reviews of people for Uber. The database was made by collecting reviews from Facebook by using Facebook Graph API. ProSuite—an online commercial software along with a lexicon-based approach—was used.

3   Dataset

The dataset employed here is taken from Kaggle competition dataset [12]. It contains approximately 400,000 comments that includes **99,989** train and **299,989** test data. The training data contains **56,457** positive and **43,532** negative comments. A number of preprocessing steps like word validation, handling emojis/punctuations, handling URL, user Mention, retweets, stemming were performed on it (Figs. 5 and 6).

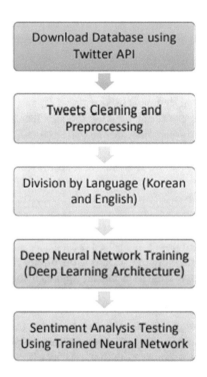

Fig. 5   Proposed method for sentiment analysis [10]

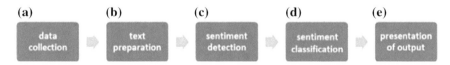

Fig. 6   Stages of sentiment analysis process [11]

4 Experimental Analysis

This section is devoted to the analysis performed while experimenting with the dataset. Algorithms that are used to train the system are as follows:

1. **Naive Bayes (NB)**: They are a collection of probabilistic classifiers that are built on the Bayes theorem that assumes strong autonomy among the features [13]. Multinomial Naive Bayes from sklearn library in Python is used.
2. **Support Vector Machine (SVM)**: It is a supervised learning technique that produces a mapping function to generate an output based on the input from a collection of labeled training data [14]. SVM from sklearn library in Python is used.
3. **Decision Tree (DT)**: These are built by recursive partitioning of the dataset by a univariate split. Pruning generally follows this step [15]. Decision tree classifier from sklearn library in Python is used.
4. **Max Entropy**: This algorithm supports the idea that the most uniform model that satisfy any given constraint must be preferred [16]. MaxEnt classifier from NLTK library is used.
5. **XGBoost**: It is a scalable and open-source machine learning setup for tree boosting. It can do parallel computing and can take many different types of input [17]. XGBoost classifier from XGBoost library in Python is used.
6. **Random Forest**: These are classifiers that employ ensemble learning methods. They function by generating a multitude of decision trees [16]. Random forest classifier from sklearn library in Python is used.
7. **Convolutional Neural Network (CNN)**: Convolutional neural networks (CNN) wield layers with convolving filters that are applied to local features in the dataset [18]. CNN classifier from Keras library in Python is used.
8. **Long Short-Term Memory (LSTM)**: An RNN architecture is designed to overcome error back-flow problems. It is used to bridge time intervals in excess of 1000 steps even in case of noisy input sequences, without loss of short time lag capabilities [19]. Keras library in Python provides it.
9. **Multilayer Perceptron (MLP)**: It extends the perceptron with hidden layers, that is, layers of processing elements which are not directly connected to the external world [20]. Keras library in Python provides it.

   The comparison of all the algorithms after training, testing, and retrieving the accuracy has been shown in Table 1.

   From Fig. 7, it is evident that CNN and SVM give the highest accuracy when applied to the preprocessed dataset. CNN uses the Global Vectors for Word Representation (GloVe) for Twitter [21]. Naive Bayes outputs acceptable accuracy since independent features are used here.

Table 1 Comparison of different algorithms

Features	Naïve Bayes	Max entropy	XGBoost	SVM	Decision tree	Random forest	MLP	CNN	LSTM
Based on	Bayes theorem	Feature based classifier	Ensemble of weak prediction models	Non-probabilistic binary classifier	Predictive modeling approach	Ensemble of weak prediction models	Feedforward ANN that generates a set of outputs from a set of inputs	Uses a variation of MLP and has convolution, pooling and F-C layers	Special type of RNN, capable of learning long-term dependencies
Complexity	Very low	High	High	Low	Low	Low	High	Very high	Very high
Performance	High	Medium	High	Medium	Very low	High	Medium	Very high	High
Accuracy	Good	High	High	Good	Good	High	Medium	High	Very high
Memory and processing requirement	Very low	High	High	High	High	High	High	Very high	Very high
Other applications	Spam detection, document classification	Diagnostic tests in pathology laboratories	Identification of dissatisfied bank customers	Face detection, classification of images	Customer relationship management	Rotating machinery faults	Autonomous driving, game playing, cancer detection	Image classification, action, voice and emotion recognition	Robot control, time series prediction, speech recognition
Time required for training classifier	Very low	Medium	High	Medium	Medium	High	Very high (directly proportional to no. of layers)	Very high (directly proportional to no. of layers)	Very high (directly proportional to no. of layers)

ACCURACY %

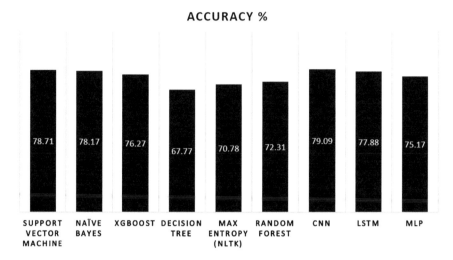

Fig. 7 Accuracy comparison of implemented algorithms

5 Conclusion

Recent advances in machine learning have ameliorated the field of NLP to make great progress. The extended list of stumbling blocks to be addressed while building a remarkable SA system have been laid out and a summary of several implementations of SA systems has been provided in this paper. It is hoped that this paper will serve as a basis for understanding various techniques that can be used to simulate the manner in which humans interpret and comprehend natural language. It could be reasonably theorized that real-world problems such as processing of natural language by machines, analysis of human behaviors through texts can be resolved by single or parallel implementations of artificial intelligence algorithms. Some algorithms indeed need vast amounts of data, whereas some are immensely complicated and power intensive. Preprocessing is one of the most important step, which was implemented using regular expressions and NTLK library, without which the accuracy degrades. After implementing multiple machine learning algorithms, it is observed that the CNN, along with GloVe embeddings, gives the best accuracy with 79% accuracy. Support vector machine also manages to give similar accuracy.

6 Future Work

In the future, implementing advanced machine learning algorithms on the same dataset using multiple performance measures will help us to compare their performance and to obtain better results. After implementing the algorithms on a powerful standalone computer, it becomes necessary to make do the same on other online platforms that can benefit from our solution, for example, to identify negative comments on social media. Corporates demand that their core model must be portable and scalable, and at the same time, this makes it pellucid that cloud platforms be explored. Machine learning algorithms are too complex and need great amounts of processing time even on powerful computers; hence, research for optimizing their performance cost and time needs to be done. After text sentiment analysis, there is facial emotion recognition to be explored that integrates various features of texts and images (or video) considered all together. Instead of just using a single lexicon (Language—e.g., English), various other lexicons could be employed to augment the performance of the sentiment extraction. Though there is a lack of dataset/available lexicons in regional languages, text sentiment analysis for it can be implemented, by first focusing on dataset creation. This can be done using softwares like TXM.

References

1. Pang, B., Lee, L.: Opinion mining and sentiment analysis. Found. Trends Inf. Retr. **2**(1–2), 1–135 (2008)
2. Jain, A.P., Dandannavar, P.: Application of machine learning techniques to sentiment analysis. In: 2nd International Conference on Applied and Theoretical Computing and Communication Technology (iCATccT) (2016), pp. 628–632
3. Cambria, E., Poria, S., Gelbukh, A., Thelwall, M.: Sentiment analysis is a big suitcase. IEEE Comput. Soc., pp. 74–80 (November/December 2017)
4. Wang, Y., Rao, Y., Wu, L.: A review of sentiment semantic analysis technology and progress. In: 2017 13th International Conference on Computational Intelligence and Security, pp. 452–455
5. Youness, M., Mohammed, E., Jamma, B.: A parallel semantic sentiment analysis (2017)
6. Deshwal, A., Sharma, S.K.: Twitter sentiment analysis using various classification algorithms. In: 2016 5th International Conference on Reliability, Infocom Technologies and Optimization (ICRITO) (Trends and Future Directions), Sep. 7–9, 2016, AIIT, pp. 251–257. Amity University Uttar Pradesh, Noida, India
7. Raghuvanshi, N., Patil, J.M.: A brief review on sentiment analysis. In: International Conference on Electrical, Electronics, and Optimization Techniques (ICEEOT) (2016), pp. 2827–2831
8. Chachra, A., Mehndiratta, P., Gupta, M.: Sentiment analysis of text using deep convolution neural networks. In: Proceedings of 2017 Tenth International Conference on Contemporary Computing (IC3), 10–12 August 2017. Noida, India (2017)
9. Kumar, M., Bala, A.: Analyzing twitter sentiments through big data. In: 2016 International Conference on Computing for Sustainable Global Development (INDIACom)
10. Ramadhani, A.M., Goo, H.S.: Twitter sentiment analysis using deep learning methods. In: 2017 7th International Annual Engineering Seminar (InAES), Yogyakarta, Indonesia
11. Baj-Rogowska, A.: Sentiment analysis of facebook posts: the uber case. In: The 8th IEEE International Conference on Intelligent Computing and Information Systems (ICICIS 2017)
12. Kaggle Page For Twitter Sentiment Analysis Dataset Datahttps: www.kaggle.com/c/twitter-sentiment-analysis2/data, 22 Dec 2018
13. Wikipedia Page for Naive Bayes Classifier [Online], Available: https://en.wikipedia.org/wiki/Naive_Bayes_classifier, 22 Dec 2018
14. Wang, L. ed.: Support Vector Machines: Theory and Applications (Vol. 177). Springer Science & Business Media, Berlin (2005)
15. Kohavi, R.: Scaling up the accuracy of Naive-Bayes classifiers: a decision-tree hybrid. In: KDD, vol. 96, pp. 202–207(August 1996)

16. Gupte, A., Joshi, S., Gadgul, P., Kadam, A., Gupte, A.: Comparative study of classification algorithms used in sentiment analysis. Int. J. Comput. Sci. Inf. Technol. **5**(5), 6261–6264 (2014)
17. Chen, T., He, T., Benesty, M.: Xgboost: extreme gradient boosting. R package version 0.4-2, pp. 1–4 (2015)
18. Kim, Y.: Convolutional neural networks for sentence classification. arXiv preprint arXiv:1408.5882 (2014)
19. Hochreiter, S., Schmidhuber, J.: Long short-term memory. Neural Comput. **9**(8), 1735–1780 (1997)
20. Principe, J.C., Euliano, N.R., Lefebvre, W.C.: Sky Software HelptoWord B. (Chap. 3). Multilayer Perceptrons. Available online at http://www.cnel.ufl.edu/courses/EEL6814/chapter3.pdf. Accessed Nov 2018
21. Pennington, J., Socher, R., Manning, C.: Glove: Global vectors for word representation. In: Proceedings of the 2014 Conference on Empirical Methods in Natural Language Processing (EMNLP), pp. 1532–1543 (2014)

A Comparative Analysis of Artificial Neural Network and Support Vector Regression for River Suspended Sediment Load Prediction

Barenya Bikash Hazarika, Deepak Gupta, Ashu and Mohanadhas Berlin

Abstract The artificial neural network (ANN) model and support vector regression (SVR) model have gained tremendous popularity among the researchers during the past couple of decades. Both of the models are very powerful in prediction and have several applications in different fields, which also include suspended sediment load prediction. In this work, the predictive capability of ANN and SVR model is investigated to estimate the daily suspended sediment load (SSL) in Tawang Chu River, Jang of Arunachal Pradesh, India. The performance of the models is evaluated using three quality measuring parameters, i.e., mean squared error (MSE), root-mean-square error (RMSE), and mean absolute error (MAE). From the experimental results, one can conclude that the predictive capability of SVR is better compared to ANN in terms of all of the quality measuring parameters.

Keywords Artificial neural network · Support vector regression · Suspended sediment load · Prediction

1 Introduction

Sediment transportation plays a vital role in the environmental changes in the rivers. They also act as contamination in water due to the presence of silt and clay which induce into the decrease of water quality. Mainly, the transportation of these sediments happens in suspension form in nonlinear behavior and could not be predicted

B. B. Hazarika · D. Gupta (✉) · Ashu · M. Berlin
National Institute of Technology Arunachal Pradesh, Yupia, India
e-mail: deepak@nitap.ac.in

B. B. Hazarika
e-mail: barenya1431@gmail.com

Ashu
e-mail: ashujarwal1996@gmail.com

M. Berlin
e-mail: berlin1982@gmail.com

© Springer Nature Singapore Pte Ltd. 2020
A. K. Luhach et al. (eds.), *First International Conference on Sustainable Technologies for Computational Intelligence*, Advances in Intelligent Systems and Computing 1045, https://doi.org/10.1007/978-981-15-0029-9_27

easily with classical like Einstein's model or the traditional physical models. Models like multilayer perceptron, sediment rating curve, linear regression methods, ANN are used to predict the suspended sediment loads in rivers or in any water system. With time, the application of neural networks for prediction of various problems took a leap as they could tackle the nonlinear behavior of these problems without expressing their inner architecture. Hence, different ANN models were used to predict the different environmental problems like soil erosion, sediment prediction, pollutant transport. But, with the development of this ANN technology, another technology developed by [1], which is known as support vector machine (SVM). It works for the regression and classification problems based on structural risk minimization (SRM). As compared to ANN, the support vector regression (SVR) attempts to minimize the prediction error. The SVR models put an upper bound on the risk, and hence, it has a greater ability to generalize the problem. The civic projects, like building dams and the floods, depend on sediments in a way. Further, the industrial waste or chemicals are dumped into the river; they are carried by the river along with sediments in which these contaminated materials may settle. This may cause danger to aquatic life as well as contaminating the present water. Further, these sediments will be transported from one place to another in a nonlinear way in rivers; hence, sediment measurement is needed to forecast the utility of dam building at the place. Similarly, it is the case with flood, i.e., if sediment load could be predicted before time at a particular place, the probability of flood also might be predicted. Sediment loads consist of two loads, namely bed load and suspension load. Suspension loads are normally following the nonlinear behavior. This complex nonlinear nature needs advance soft computing models for accurate prediction.

There are many studies available for the prediction of sediment loads. An ANN model was developed by [2], for prediction of the sediment load in the river. For prediction accuracy, the ANN performance is compared with other traditionally used sediment discharge formulas. It was found that ANN predictions are much accurate than the other traditionally used sediment discharge formulas. Tayfur [3] applied the fuzzy logic approach to predict the sediment loads in bare soil surfaces at different rainfall intensities. Results showed that at very high intensities fuzzy approach performed better. Raghuwanshi et al. [4] compared linear regression model and ANN for the sediment load prediction in a watershed called Nangwan in India. They have used seven years data, out of which five years were applied as a testing data and two years as training data. The results show that ANN out-performed the linear regression model in prediction accuracy. Cigizoglu [5] used two ANN techniques, namely radial basis function (RBF) and backpropagation, to predict the total daily suspended sediment load (SSL) for the data belongs to Juniata River in the USA. It was found that ANN performs better than conventional multi-linear regression technique. Misra et al. [6] used the SVM model to predict the monthly, weekly, and daily sediment yield in a watershed in India. Multilayer feedforward neural network (MFNN) is compared with SVR model by [7] for the prediction of rainfall–runoff. A form of regression trees is used by [8] for the estimation of the suspended sediment near-bed concentration in New Zealand. Genetic programming, ANN, adaptive neuro-fuzzy inference system (ANFIS), SVM models are compared for the prediction of suspended sediment load by [9]. Using ANN and SVM, the suspended sediment load was predicted by [10]. Adib and Mahmoodi [11] used an ANN-based genetic algorithm (GA) and Markov chain hybrid model on the Marun River in the southwest of Iran. The different parameters of ANN are optimized using GA. The researchers found that GA can minimize the normalized mean square error (NMSE) to 80%, but it does not increase the correlation coefficient (R) significantly. Khan et al. [12] used three different types of ANN models to estimate the monthly mean SSL of Ramganga River in India. They found that there was an insufficiency in rainfall values, but the water discharge values enhance the performance of the three algorithms.

In this work, SSL prediction is done using two computing models such as—ANN and SVR. Both models are applied with one-year data of sediment load in Tawang Chu River provided by NHPC Limited. It could be said that SSL acts as a suitable index to estimate the future condition of river for the possible occurrence of flood or other activities.

Normally, ANN is easy to train than SVR technique for the collected data but this is applicable for large set of data, whereas there is less data available, and the training time does not count for comparison of performance of both models in their prediction accuracy, and hence, for performance comparison, only errors are compared.

The main objectives of this research are:

1. To explore the ability of SVR and ANN as a regression model on a river dataset.
2. To apply both models on Tawang Chu River dataset to predict the sediment load concentration on a river.
3. To discuss the comparative performance of ANN and SVR based on different measurements.

In Sect. 2, the basic details of artificial neural network and their general architecture are discussed. Also, we have introduced the formulation of support vector regression in details. Section 3 has the description of the data. In Sect. 4, numerical experiment is discussed as the comparative performance between the two models, viz. ANN and SVR. Finally, we have concluded our work and suggest the future work in Sect. 5.

2 Models

2.1 Artificial Neural Network (ANN)

In 1943, [13] desired to understand the functioning of brain and if in some way it could be emulated in practical, hence giving birth to the idea of ANN. Then nearly for four decades, the subject took the silent development with a very slow progress and less keen eyes toward it. With the emergence and development of more powerful computing tools, ANN saw its renaissance when the need of auto-associable networks arose [14] and a great growth after rediscovery of backpropagation algorithm [15]. After this, ANN found its application in several areas of study such as biomedical science, electrical engineering, neurophysiology, image processing, robotics, image processing, and hydrology. In hydrology-related areas, ANN saw its successful application in stream flow forecasting, water quality management, rainfall–runoff modeling, water reservoir operations, sediment prediction, and many others.

The ANN could be said as a parallel distributed processing system for the available information which resembles a human brain in its performance characteristics. Mainly, ANN does not show linear nature in its performance but shows nonlinear nature in the variables [2]. ANN model could be called as mathematical model of neurons working in human cognition. It consists of units or elements or nodes where the information processing is done—in fact they work as neurons in brain, hence, also sometimes known as neurons. Now between these neurons or nodes, the signals are passed through defined connection links which are associated with some weight (for each connection link) determining the strength of connection link. Hence, to determine the output signal or output for ANN, a nonlinear transformation known as activation function is applied by each neuron or node to the input available.

ANN architecture consists of three layers, viz. input layer, output layer, and hidden layer, where each layer has neurons or information processing elements in it. Now, to process the information it is needed to be stored; hence, these neurons also have their local memory. If flow of information is similar to feedforward, the neurons are connected with unidirectional links that carry the information. As an activation function for these neurons, the sigmoid function could be used.

To train the ANN, an output vector or output value is generated where the error between the target value and output value is minimized. This is done through iterative adjustment of neurons and by optimizing connection strengths or connection weights of neurons and their threshold value. The main aim of training the algorithm is to diminish the error function as much as possible so that target value could be attained satisfactorily. Hence, only the input of the input layer does not change, whereas the inputs of the other layers are adjusted to minimize the error depending on its prior layer. After ANN is trained, it is considered that ANN will perform satisfactorily on the test data. The architecture of ANN could be given as shown in Fig. 1.

The input vector represents all the inputs given to the neurons in the input layer could be given as $(x_1, x_2, \ldots, x_n)$. In this architecture, output is scaled and normalized, i.e., only output value is considered which are normalized to fall between the value 0 and 1. As it is known, generally output is in the form of a vector $(y_1, y_2, \ldots, y_m)$ like input vector depicting the value of the neurons in the output layer. The output produced by ANN could be denoted as y while target could be denoted as t. Hence, for each n sample or pattern, average error square could be given as:

$$X_p = \frac{1}{2} \sum_k (t_p - y_p)^2 \tag{1}$$

Fig. 1 ANN architecture

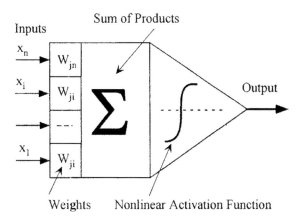

According to the delta rule, the weight change in the pair from jth node to kth nodes can be written as:

$$W_{kj} = -\partial \varepsilon X / \partial W_{kj} = \varepsilon\, \partial_k\, y_j \tag{2}$$

where $\partial_k = -(\partial X / \partial o_k)[f'(\text{net}_k)]$, ε is learning rate, and $f'(\text{net}_k) = \partial o_k / \partial \text{net}_k$. Similarly, the weight change in the pair from jth to ith nodes can be set as

$$W_{ji} = -\partial \varepsilon X / \partial W_{ji} = \varepsilon \partial_k y_j \tag{3}$$

where $\partial_j = -(\partial X / \partial o_j)[f'(\text{net}_j)]$ and $f'(net_j) = \partial o_j / \partial \text{net}_j$.

The sigmoid function represents y_j and the threshold value θ_j.

$$o_j = \frac{1}{1 + \exp\left[-\left(\sum_i W_{ji} y_i - \theta_j\right)\right]} \tag{4}$$

Hence, the following expressions may be given, for hidden and output layer, respectively:

$$\delta_{pk} = \alpha y_{pj}(1 - y_{pj}) \sum_k \delta_{pk} w_{kj} \tag{5}$$

$$\delta_{pk} = \alpha(t_{pk} - y_{pk}) y_{pk}(1 - y_{pk}) \tag{6}$$

2.2 Support Vector Regression (SVR)

Support vector machines (SVM) are used for both classification and regression. SVM is successfully applied in regression, i.e., function approximation problem. By introducing a novel ε-insensitive loss function, [16] has proposed support vector regression (SVR) [17].

Suppose that a set of input training samples $\{(u_1, v_1), (u_2, v_2), \ldots, (u_m, v_m)\}$ is given where for any input feature vector $u_i = (u_{i1}, u_{i2}, \ldots, u_{in})^t \in R^n$, its observed output be $y_i \in R$ for $i = 1, 2, \ldots, m$ where m corresponding to input training samples. The regression function is given by:

$$f(u) = u^t w + b \tag{7}$$

where $w \in R^m$ and $b \in R$ are the variables.

The formulation of SVR is given as:

$$\min_{w,b,\xi,\xi^*} \frac{1}{2} w^t w + C(e^t \xi + e^t \xi^*)$$

subject to

$$
\begin{aligned}
& v_i - (u_i^t w + b) \leq \varepsilon + \xi_i, \\
& u_i^t w + b - v_i \leq \varepsilon + \xi_i^* \\
& \text{and } \xi_i \geq 0, \xi_i^* \geq 0 \text{ for } i = 1, 2, \ldots, m.
\end{aligned}
\tag{8}
$$

where $\varepsilon > 0$ is the input parameter, $C > 0$ is the regularization parameter and $\xi = (\xi_1, \ldots, \xi_m)^t, \xi^* = (\xi_1^*, \ldots, \xi_m^*)^t$ are the vectors of slack variables.

Vapnik has introduced the ε-insensitive error function as

$$
L|f(u) - v| = \begin{cases} 0 & |f(u) - v| \leq \varepsilon \\ |f(u) - v| \leq \varepsilon & \text{otherwise} \end{cases}
\tag{9}
$$

By introducing the Lagrangian multiplier $\ell = (\ell_1, \ldots, \ell_m)^t$ and $\ell^* = (\ell_1^*, \ldots, \ell_m^*)^t$, the solution of primal problem (8) is obtained by solving its dual form as a quadratic programming problem (QPP):

$$\min_{\delta, \delta^*} \frac{1}{2} \sum_{i,j=1}^{m} (\ell_i - \ell_i^*)^t u_i^t u_j (\ell_j - \ell_j^*) + \varepsilon \sum_{i=1}^{m} (\ell_i + \ell_i^*) - \sum_{i=1}^{m} v_i (\ell_i - \ell_i^*)$$

subject to

$$\sum_{i=1}^{m} (\ell_i - \ell_i^*) = 0, \quad 0 \leq \ell, \ell^* \leq Ce, \tag{10}$$

where $\varphi(.)$ is the mapping function that maps the input examples into a higher dimensional feature space. Here, the kernel function is given by $k(u_i, u_j) = \varphi(u_i)^t \varphi(u_j)$. $\varphi(.)$ is nonlinear transformation function. The decision function $f(.)$ for nonlinear case is given as,

$$f(u) = \sum_{i=1}^{m} (\ell_{1i} - \ell_{2i}) k(u, u_i) + b \qquad (11)$$

3 Data Description

In this work, Tawang Chu River dataset of sediments load concentration is used which is collected from NHPC Limited, Tawang Basin Project that contains the data from January 1, 2012 to December 31, 2012, and further apply on both the models to predict the sediments load concentration in the river. Here, the window size is fixed to 5, i.e., estimation is obtained with respect to the sediment load concentration of the previous five days. Hence, the dataset contains 6 numbers of attributes and 360 numbers of samples. ANN and SVR methods are applied on this dataset. The minimum mean square error (MSE), root-mean-square error (RMSE), and mean absolute error (MAE) are used to measure the errors of the two models.

4 Numerical Experiments

Usually, backpropagation algorithm is used with feedforward neural network to train the ANN model. In this, the errors generated at output layer between target value and output value are backpropagated to the ANN, i.e., to the hidden layer and accordingly the neurons and their weights are adjusted to minimize the error with different weights and again produce output value which will again backpropagated if the desired error is not achieved. In training, the ANN optimization is achieved through reducing the quadratic error between the target data and the computed data or produced output values. It could be said that ANN training is as easy as compared to the SVR as ANN works on empirical risk minimization principle, whereas SVR works on SRM principle.

The performance of SVR and ANN is compared on the basis of different quality measure parameters. These parameters are:

- Mean absolute error (MAE)
- Mean squared error (MSE)
- Root-mean-square error (RMSE).

Here, $Q_{(s.pred)}$ and $Q_{(s.obs)}$ denote the predicted and observed values, respectively, and N denotes the total quantity of targets or observed values used to calculate the error. In the start, or ideally it is known that the error values for MAE and RMSE should be equal to zero. The errors MAE and RMSE have indistinguishable unit and need similar scale from research center and hence are benefit to use as checking parameters with each other. The MSE, RMSE, MAE could be calculated as:

$$\text{MSE} = \left[\frac{1}{N} \sum_{N=1}^{N} (Q_{(s.obs)} - Q_{(s.pred)})^2 \right] \tag{12}$$

$$\text{RMSE} = \left[\frac{1}{N} \sum_{N=1}^{N} (Q_{(s.obs)} - Q_{(s.pred)})^2 \right]^{1/2} \tag{13}$$

$$\text{MAE} = \frac{1}{N} \sum_{N=1}^{N} \left| Q_{(s.obs)} - Q_{(s.pred)} \right| \tag{14}$$

4.1 Result of ANN

ANN model is successfully implemented for prediction of sediments load concentration. ANN model was performed on 365 input data. In this paper, 60% of the data are used for model training, 5% data and 35% data are implemented for model validation and testing, respectively. Further for ANN model, the number of nodes has taken in the hidden layer as 10. For training the data, we use scaled conjugate gradient algorithm. The ANN produces the errors and depicted in Table 1. In Fig. 2, blue line shows the original data of sediment loads, whereas red line shows the predicted data of sediments loads.

4.2 Result of SVR

SVR models are characterized by decreasing the target function error between the observed and the anticipated suspended load values in the test period for each input

Table 1 Performance comparison of artificial neural network and support vector regression in terms of different quality measurements

Quality measures	ANN	SVR
MSE	0.0481	**0.0093**
RMSE	0.2193	**0.0967**
MAE	0.0477	**0.0306**

Bold type shows the best results

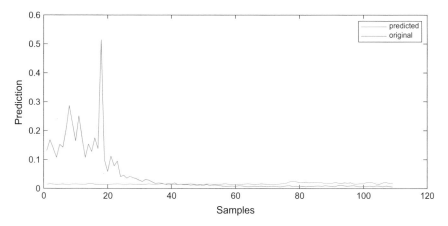

Fig. 2 Graph between the original data and predicted data using ANN model

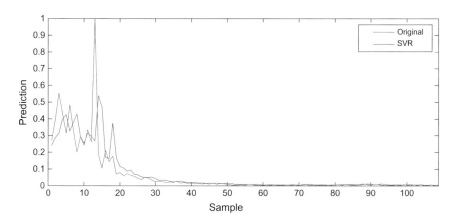

Fig. 3 Graph between the original data and predicted data using SVR model

combination. SVR model was performed based on 360 input data. Among every one of the information, 65% information is utilized for preparing and remaining information are utilized enemy testing the model. The SVR produces the errors shown in Table 1. In Fig. 3, blue line shows the original data of sediment loads, whereas red line shows the predicted data of sediments loads.

5 Conclusion and Future Work

In this work, the prediction capability of two models, i.e., ANN and SVR, for sediment load prediction was performed. Theoretically, ANN requires less training time as compared to SVR. But, for small dataset, it does not make much difference in training time of both models. Hence, in results of performance of both the models we focused only on error determined based on the quality measures like RMSE, MSE, and MAE. The best fit model is based on the error value obtained with the procedure presented in this study. Model which has smaller error values is the best model for prediction. Above figures and table illustrate the accuracy of each model. Prediction by ANN and SVR is compared with each other to check the performance. It could be derived from the above tables that RMSE, MSE, and MAE of SVR model is less than the ANN model which justify its generalization ability. It could be seen from the numerical results that SVR estimates closely to observed values. This shows that the SVR is capable of good prediction of the sediment load. The outcomes indicate that the SVR model gives better result as compare to ANN. Since, in this work, we have considered the dataset of only one year which is limited, we can take more data and apply other prediction models to achieve the better prediction of sediment load concentrations in future.

Acknowledgements This work was fully supported by the Science and Engineering Research Board, Government of India (SERB), under early career research award ECR/2016/001464. We are also thankful to NHPC LIMITED, Tawang Basin Project to provide this data.

References

1. Vapnik, V.: The Nature of Statistical Learning Theory. Springer Science & Business Media, Berlin (2013)
2. Nagy, H.M., Watanabe, K.A., Hirano, M.: Prediction of sediment load concentration in rivers using artificial neural network model. J. Hydraul. Eng. **128**(6), 588–595 (2002)
3. Tayfur, G.: Artificial neural networks for sheet sediment transport. Hydrol. Sci. J. **47**(6), 879–892 (2002)
4. Raghuwanshi, N.S., Singh, R., Reddy, L.S.: Runoff and sediment yield modeling using artificial neural networks: Upper Siwane River, India. J. Hydrol. Eng. **11**(1), 71–79 (2006)
5. Cigizoglu, H.K.: Estimation and forecasting of daily suspended sediment data by multi-layer perceptrons. Adv. Water Resour. **27**(2), 185–195 (2004)
6. Misra, D., Oommen, T., Agarwal, A., Mishra, S.K., Thompson, A.M.: Application and analysis of support vector machine based simulation for runoff and sediment yield. Biosys. Eng. **103**(4), 527–535 (2009)
7. Botsis, D., Latinopulos, P., Diamantaras, K.: Rainfall-runoff modeling using support vector regression and artificial neural networks. In: 12th International Conference on Environmental Science and Technology (CEST2011) on 2011, pp. 8–10
8. Oehler, F., Coco, G., Green, M.O., Bryan, K.R.: A data-driven approach to predict suspended-sediment reference concentration under non-breaking waves. Cont. Shelf Res. **46**, 96–106 (2012)
9. Kisi, O., Dailr, A.H., Cimen, M., Shiri, J.: Suspended sediment modeling using genetic programming and soft computing techniques. J. Hydrol. **450**, 48–58 (2012)
10. Lafdani, E.K., Nia, A.M., Ahmadi, A.: Daily suspended sediment load prediction using artificial neural networks and support vector machines. J. Hydrol. **478**, 50–62 (2013)
11. Adib, A., Mahmoodi, A.: Prediction of suspended sediment load using ANN GA conjunction model with Markov chain approach at flood conditions. KSCE J. Civil Eng. **21**(1), 447–457 (2017)

12. Khan, M.Y., Hasan, F, Tian, F.: Estimation of suspended sediment load using three neural network algorithms in Ramganga River catchment of Ganga Basin, India. Sustainable Water Resources Management, pp. 1–7 (2018)
13. McCulloch, W.S., Pitts, W.: A logical calculus of the ideas immanent in nervous activity. Bull. Math. Biophys. **5**(4), 115–133 (1943)
14. Hopfield, J.J.: Neural networks and physical systems with emergent collective computational abilities. Proc. Natl. Acad. Sci. **79**(8), 54–58 (1982). (Springer)
15. Rumelhart, D.E., Hinton, G.E., Williams, R.J.: Learning internal representations by error propagation. California Univ. San Diego La Jolla Inst. for Cognitive Science on 1985. La Jolla
16. Vapnik, V., Mukherjee, S.: Support vector method for multivariate density estimation. In: Advances in Neural Information Processing Systems on 2000, pp. 659–665. MIT Press, Cambridge
17. Sharma, N., Zakaullah, M., Tiwari, H., Kumar, D.: Runoff and sediment yield modeling using ANN and support vector machines: a case study from Nepal watershed. Model. Earth Syst. Environ. **1**(3), 23 (2015)

Optimum Parallelism in Spark Framework on Hadoop YARN for Maximum Cluster Resource Utilization

P. S. Janardhanan and Philip Samuel

Abstract Spark is widely used as a distributed computing framework for in-memory parallel processing. It implements distributed computing by splitting the jobs into tasks and deploying them on executors on the nodes of a cluster. Executors are JVMs with dedicated allocation of CPU cores and memory. The number of tasks depends on the partitions of input data. Depending on the number of CPU cores allocated to executors, one or more cores get allocated to one task. Tasks run as independent threads on executors hosted on JVMs dedicated exclusively to the executor. One or more executors are deployed on the nodes of the cluster depending on the resource availability. The performance advantage provided by distributed computing on Spark framework depends on the level of parallelism configured at 3 levels, namely node level, executor level, and task level. The parallelism at each of these levels should be configured to fully utilize the available computing resources. This paper recommends optimum parallelism configuration for Apache Spark framework deployed on Hadoop YARN cluster. The recommendations are based on the results of the experiments conducted to evaluate the dependency of parallelism at each of these levels on the performance of Spark applications. For the purpose of the evaluation, a CPU-intensive job and an I/O-intensive job are used. The performance is measured by varying the parallelism at each of the 3 levels. The results presented in this paper help Spark users in selecting optimum parallelism at each of these levels for achieving maximum performance for Spark jobs by maximum resource utilization.

Keywords Distributed computing · Apache Spark · Hadoop YARN · SparkBench · Spark configuration · Multi-level parallelism · Resource optimization

P. S. Janardhanan (✉)
SunTec Business Solutions Pvt Ltd, Thejaswini, TechnoPark, Trivandrum, India
e-mail: janardhananps@suntecgroup.com

P. Samuel
Department of Computer Science, Cochin University of Science & Technology,
Kochi, India
e-mail: philips@cusat.ac.in

© Springer Nature Singapore Pte Ltd. 2020   351
A. K. Luhach et al. (eds.), *First International Conference on Sustainable Technologies for Computational Intelligence*, Advances in Intelligent Systems and Computing 1045,
https://doi.org/10.1007/978-981-15-0029-9_28

1 Introduction

The Spark computing framework provides a parallel run time environment that hides complexities of fault-tolerance and cluster computing. In general, Spark framework is ideal for iterative computations on large volumes of data, and workloads can be deployed for applications like machine learning, stream processing, graph processing, and SQL query processing [1]. Spark core can be deployed on different types of clustering platforms. The performance of these applications depends on the level of execution parallelism configured for the Spark and the clustering platform. The execution parallelism can be configured in 3 levels, namely node level, executor level, and task level. A study is done to determine the impact of parallelism at all these levels on the performance of the applications. This paper recommends ideal parallelism configuration at each of the levels for optimum resource utilization. These recommendations help in configuring optimal parallelism for achieving maximum performance from available computing resources. Also, the recommendations become useful in hardware sizing of the nodes of the cluster and in determining the number of nodes of the cluster to guarantee SLA levels in the completion of time-bound applications running on the Spark framework.

2 Levels of Parallelism

The basic unit of execution in Spark is called task and the number of tasks depends on the number of partitions of the input data. Tasks run as independent threads on executors hosted on JVMs, dedicated exclusively for each executor. One or more executors get deployed on the nodes of the cluster depending on the resource allocation. The performance advantage provided by distributed computing on Spark framework depends on the level of parallelism configured on the clustering framework on which Spark is deployed. The multiple levels of parallelism existing in spark framework are shown in Fig. 1. They are node level (P1), executor level (P2), and task level (P3).

2.1 Node Level

Spark framework can be deployed on clustering platforms like Standalone, Hadoop YARN, Mesos, Kubernetes, etc. The performance of Spark applications depends on the capabilities of the clustering platform in managing distributed scheduling and execution of parallelism on the nodes [2]. Some of these clustering platforms insist on identical hardware configuration for all the nodes. Hadoop YARN cluster can be configured on nodes with heterogeneous hardware resources.

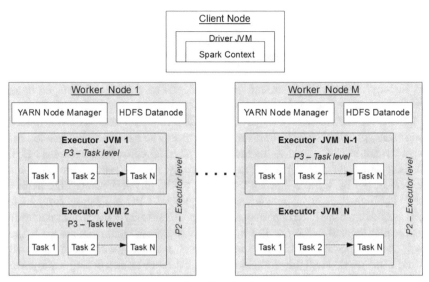

Fig. 1 Levels of parallelism in Spark framework

Hadoop YARN allows Spark applications to co-exist with other workloads on the same cluster [2]. The Spark cluster nodes are classified as client nodes and worker nodes. There needs to be at least one client node on which the Spark Context will be running. Spark jobs get launched from Spark Contexts. Spark can be configured to run in cluster mode also. In this mode, the Spark Context runs on one of the cluster nodes, which can also acts as a worker node. The node-level parallelism depends on the number of nodes of the cluster. The clustering framework introduces overheads and imposes upper limits on the number of nodes that can be supported [3]. Hence, it is required to select an appropriate clustering framework to configure maximum node-level parallelism for the application [2]. Increasing node-level parallelism helps in improving the performance of I/O-intensive applications since multiple disks will be spinning in parallel when the job is in execution. We need to select a clustering platform like Hadoop, which allows online addition and removal of nodes for elastic scaling of node-level parallelism without disrupting running applications. On Hadoop cluster, the names of the nodes are to be entered in the file */etc/hadoop/slaves*. And for the Spark environment, the names of the nodes are to be entered in the configuration file */conf/slaves*.

2.2 Executor Level

Spark is a flexible, fault-tolerant, and scalable distributed computing platform with concise, powerful APIs and higher-order tools, and programming languages [1]. On Hadoop clusters, Spark executors get deployed on containers managed by the resource manager known as YARN [4]. Spark executors have dedicated allocation of CPU cores and memory. For partitioning CPU as a resource, YARN introduced a new concept called *vcores*, short for virtual cores. The general recommendation is to set the number of *vcores* to the number of physical cores on the node for CPU-intensive tasks. But for I/O bound tasks, generally, seen in Hadoop environments, the *vcores* can be set to a value 4 times that of the physical cores. This helps in achieving higher levels of parallelism and improved CPU utilization. Support for CPU as a resource has been implemented in *CapacityScheduler* by the introduction of *DominantResourceCalculator*. In Hadoop YARN cluster, containers are predefined partitions of resources; CPU cores and memory. Containers get created dynamically as and when needed and they transition through different states during deployment. In the case of Hadoop cluster, the resource allocation for the executors is limited by the resources allocated to the containers on which they get deployed. The executor process supports multiple tasks getting executed concurrently as multiple threads of the JVM. Tasks slow down due to excessive garbage collection, when executors are configured with high memory. When executors are configured with minimum number of *vcores*, the benefits of running multiple tasks in a single JVM will be lost. One or more executors get deployed on the nodes of the cluster depending on the number of containers configured on the nodes of the Hadoop YARN cluster. Hence, the total number of containers configured on the nodes determines the execution parallelism at executor level on the cluster [4].

Container resource allocation is defined in the hadoop configuration file */etc/hadoop/yarn-site.xml*. The parameter *yarn.scheduler.maximum-allocation-vcores* defines the maximum *vcores* that can be allocated to one container. The parameter *yarn.scheduler.maximum-allocation-mb* defines the maximum size of memory in MB that can be allocated to one container. Spark executor resource allocation is defined in */conf/spark-defaults.conf*. The parameter *spark.executor.memory* defines the memory allocation for the executor in GB. And the parameter *spark.executor.cores* defines the cores allocated to an executor. The parameter *spark.executor.instances* defines the total number of executors in the spark environment and *spark.default.parallelism* defines the number of parallel executions. For maximum utilization, the resource allocations configured for Hadoop YARN containers and Spark executors should be identical.

2.3 Task Level

Partitioning of input data can have major impact on performance in Spark since the number of partitions decides the job granularity. The number of tasks launched from an executor depends on the number of CPU cores allocated to the executor. Hence, the number of partitions correlates to the number of CPU cores defined in the executors. Spark manages distributed data using Resilient Distributed Datasets (RDD). Partitioning of RDDs is a key property to parallelize a Spark application on a cluster. RDDs produced by textFile or Hadoop methods have their partitions determined, by default, by the number of blocks on the HDFS file system. Partitions for RDDs produced by parallelize method come from the parameter given by the user, or the configuration parameter *spark.default.parallelism* if none is given. Spark default parallelism depends on the cluster manager. In Mesos fine-grained mode, the value is 8. On all other cluster managers, it is equal to the total number of cores on all executor nodes. The main goal is to run enough concurrent tasks so that the data destined for each task fits in the memory of the executor available for that task. If there are fewer concurrent tasks than the cores available to run them in, the stage will not be taking advantage of all the CPU cores available. A small number of tasks, also mean that more memory pressure, are placed on any aggregation operations that occur in each task. The rule of thumb is, too many tasks is usually better than too few. The Spark configuration parameter *spark.task.cpus* specifies the number of cores to allocate for each task.

3 Problem Statement

Distributed computing and parallel execution are key factors helping Spark applications to achieve high performance. The goal is to run enough concurrent tasks to make use of all the cores allocated to all executors on all nodes of the cluster. If there are fewer tasks than the cores allocated to the executors, it will not be taking advantage of all the cores available on the executors. The performance advantage of Spark applications depends on the level of parallelism introduced in the execution of jobs. The parallelism can be introduced in 3 levels as discussed in Sect. 2. It is always a dilemma to decide on the extent of parallelism to be configured at each of these levels to fully utilize the resources. There exist no guidelines on defining parallelism for Spark applications to achieve maximum performance with minimum resource allocation. Hence, it becomes difficult to choose the right number of nodes for a Spark cluster, the right number of concurrent executions on each of the executors, and the right number of executors on each of the nodes of the cluster.

4 Test Environment

4.1 Hardware

The experiments were performed on 6 node cluster with identical hardware configurations. The hardware details of the nodes are:

Processor: Intel(R) Core $^{(TM)}$ i5-3570 CPU @ 3.40 GHz.
Cache size: 6144 KB.
Number of cores: Physical 4, Logical 16
Memory available: 16 GB
Physical storage: 512 GB with 7200 RPM
Network Card: Full duplex network card with 1000 Mbps speed.

4.2 Software

Apache Hadoop Release 2.8.0   Configured with Capacity Scheduler
Apache Spark Release 2.2.1   Configured with KryoSerializer in Client mode.
SparkBench Release 2.1.1   Disabled dynamic allocation of executors
Java Version   1.8.

5 SparkBench for Performance Evaluation

SparkBench is an open-source benchmarking suite for Spark distributed computing framework [5]. SparkBench originally began as a benchmarking suite to get timing numbers on very specific algorithms, mostly in the machine-learning domain. Since then, it has morphed into a highly configurable and flexible framework suitable for many use cases.

SparkBench can be used for various scenarios: performance comparison, cluster provisioning, in-depth study of Spark, etc. The SparkBench data generator automatically generates input data sets with desired sizes. SparkBench is designed to provide quantitative comparison for different platforms and hardware cluster setups. It enables in-depth study of performance implication of Spark system in various aspects like workload characterization, parameter impact, scalability, etc. It also provides insight and guidance for cluster sizing and provisioning. If a user aims to provision a Spark cluster for usage, SparkBench will help in finding out the performance and help them identifying resource bottlenecks affecting the performance.

SparkBench has representative workloads and can be easily extended with new workloads. Workloads are available for typical applications like machine learning, graph processing, streaming, and SQL query applications. It allows users to explore different parameter configurations easily and study the impact on application performance. Typical use cases of SparkBench include regression testing with changes to Spark, comparing performance of different hardware, and Spark tuning options, simulating multiple notebook users hitting a cluster at the same time, comparing parameters of a machine-learning algorithm on the same set of data, providing insight into bottlenecks through use of compute-intensive- and I/O-intensive workloads and benchmarking.

6 Performance Evaluation

The aim of the evaluation is to determine the performance impact of parallelism at the node level, executor level, and task level. The test cluster consists of Hadoop YARN cluster configured with one Spark client and 5 worker nodes. The executors are deployed on the worker nodes. One CPU-intensive application (K-means cluster) and one I/O-intensive application (SQL query) are used for the evaluation. The machine-learning application K-means clustering is an unsupervised iterative algorithm that groups input data in a predefined number of K clusters. Each cluster has a centroid, which is a cluster center. It is a highly iterative machine-learning algorithm that measures the distance (between a vector and centroids) as the nearest mean. Iterations are repeated till they converge within a specified number of steps. For the purpose of evaluation, an input file is created on HDFS with the data generator provided in the tool with 40 million rows and 24 columns. The evaluation is done by setting a value of 10 to K, the number of clusters in the input data.

The SQL query workload called *CSV* versus *Parquet* is used as the I/O-intensive workload. This is an experimental setup for benchmarking the performance of some simple SQL queries over the same data set stored in CSV and Parquet formats on the HDFS file system. For this study, an input file is created with the data generator workload called *data-generation-kmeans* with 1 million rows with 36 columns and stored in CSV format. In this case, only one spark-submit is needed. Within that spark-submit, several workload-suites get run serially. The first workload in that suite picks up the data set in CSV format and writes it out as Parquet. The second workload runs an SQL query on the data stored in both the formats.

The performance evaluation is done by running these workloads by varying the parallelism at each of these levels. When performance benchmark experiments are done at one level, the parallelism at the other two levels is kept constant.

6.1 Parallelism at Node Level

The purpose of the evaluation is to determine the impact of parallelism at node level on the performance of Spark applications. One executor with 12 GB memory and 5 cores is configured on each of the nodes of the cluster. The number of tasks launched on an executor depends on the number of cores, and the experiments are repeated with cores of 5 and 10. The jobs are deployed on clusters by varying the nodes from 2 to 5 and the execution time is noted. The results are shown in Fig. 2a, b. For the CPU-intensive job (K-means), the execution time comes down with a number of nodes. For I/O-intensive jobs (SQL query) also, the execution time reduces with the number of nodes, and the reduction in execution time is mainly attributed to the parallel I/O happening on multiple nodes. In both the cases of CPU-intensive jobs and I/O-intensive jobs, the reduction in execution time is exponential. When the number of nodes increases beyond 4, the reduction in execution time is marginal and there will not be adequate returns on investments on additional nodes.

6.2 Parallelism at Executor Level

The aim of the experiment is to determine the impact of executor parallelism on the performance of the Spark applications. Executors are configured with 4 GB memory and 4 cores each. This enables one node to run up to 3 executors concurrently without resource congestion. The total number of executors is varied from 1 to 15 on a cluster in which the number of nodes is varied from 1 to 5. Spark can be deployed on a Hadoop cluster with any number of nodes and data nodes of HDFS will be equal to the number of nodes of the Hadoop cluster. On an N node Hadoop cluster, the input data for the Spark applications get stored on HDFS with N Datanodes. Benchmarking is done with Hadoop clusters in which the number of nodes of the Hadoop cluster is equal to the number of nodes of the Spark cluster. The performance benchmarking is done with the K-means clustering and SQL query workloads. The execution times are graphically represented in Fig. 3a, b.

From Fig. 3a, it is seen that for CPU-intensive K-means clustering workload, the execution time reduces when three executors are deployed on one node since the computing resources are fully utilized. The benchmark test gives minimum execution time with 12 executors running on 4 nodes. Hence, there is no advantage on investing on computing resources beyond 4 nodes on a Hadoop YARN cluster.

For the I/O-intensive SQL query workload, the execution time reduces marginally beyond one node as seen in Fig. 3b And beyond 6 executors deployed on 2 nodes, the reduction in execution time is ignorable. On a single node cluster with 3 executors, the execution time increases with the number of executors. This is because of the I/O congestion at JVM level.

The application performance depends on the CPU power available on the nodes of the cluster and parallelism in disk I/O when the input data is read. Hence, for Spark workloads, there is no advantage in parallelizing the I/O beyond a certain limit without taking into consideration of the computing power available on the nodes of the cluster. The optimum level of I/O parallelism depends on the nature of the application. The number of nodes configured in an HDFS cluster used for deploying Spark workloads should be equal to the number of nodes of the Spark cluster.

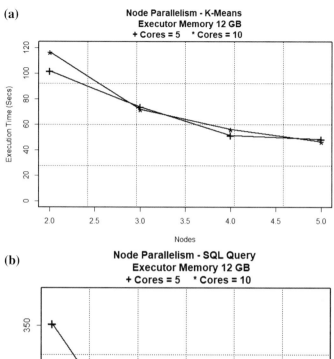

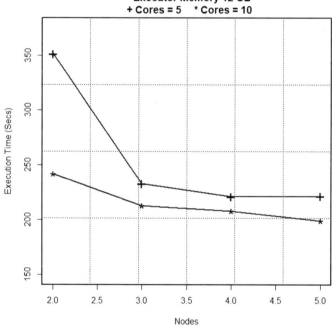

Fig. 2 **a** Impact of node parallelism on CPU-intensive workload. **b** Impact of node parallelism on I/O-intensive workload

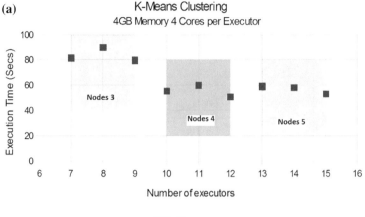

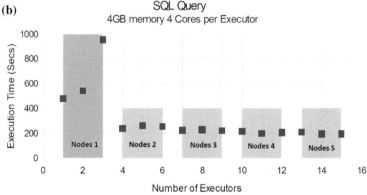

Fig. 3 a Impact of executor parallelism on CPU-intensive workload. **b** Impact of executor parallelism on I/O-intensive workload

6.3   Parallelism at Task Level

A benchmark test is conducted to evaluate the dependency of task-level parallelism on the performance of applications. For the experimental study, an executor with memory of 12 GB is configured on each node of the cluster. CPU cores are allocated to the executors from 1 to 14. There will be 4 executors running on the 4 worker nodes of the cluster. Hence, the total number of cores varies from 4 to 56. This means there will be 4–56 tasks running in parallel on the cluster. With each of the task allocations, the execution time of test workloads is measured. In each case, the experiment is repeated 3 times and the average time is taken. For benchmarking using K-means clustering, an input file with 40,000 K rows and 24 rows is created using the data generator program provided in SparkBench suite. The K-means workload is invoked with $K = 10$, the number of clusters in the input data. Figure 4a, b shows the impact of task-level parallelism on performance. From Fig. 4a, b, it can be inferred that the execution time does not reduce beyond a limit by increasing the task-level

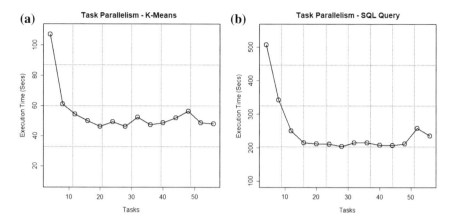

Fig. 4 **a** Impact of task-level parallelism on CPU-intensive workload. **b** Impact of task-level parallelism on I/O-intensive workload

parallelism in the case of both CPU-intensive and I/O-intensive jobs. It is seen that the performance is limited by the concurrent I/O overheads introduced at the JVM level. This limit is seen at a task level of 20. With 4 executors running in parallel, each executor will be running 5 tasks. Hence, there is no performance improvement if more than 5 tasks are configured to run on one executor.

Cluster resources will not be fully utilized unless the level of parallelism for the job is high enough. Spark automatically sets the number of partitions of an input file according to its size and for distributed shuffles. By default, Spark creates one partition for each block of the file in HDFS. Now, if we have a cluster with 16 cores, the total number of tasks is 35. Then, the tasks get executed in groups of 16, 16, and 3. If processing one group takes 4 min, it takes a total of 12 min to complete all the tasks of the job. While processing the last group, 13 CPU cores will be idle. The number of idle cores can increase with the total number of cores if the number of tasks is less. So, it is a requirement to increase the number of tasks by increasing the number of partitions of RDDs. To overcome this problem, we should make RDDs with the number of partitions equal to multiple of the number of cores in the cluster. When dividing the RDDs into partitions, as a rule of thumb, all tasks should take at least 100 ms to execute. If the tasks take considerably longer than that, keep increasing the level of parallelism until performance stops improving.

7 Recommendations

For Spark deployments on Hadoop YARN cluster, there exist a limit beyond which the application performance does not increase significantly with increasing the number of nodes. This limit depends on the nature of the application and differs for CPU-intensive- and I/O-intensive applications. The execution time decays exponentially and the decrease in execution time beyond 4 nodes is found to be marginal and does not yield desired return on investments. The performance gain of I/O-intensive applications is attributed to concurrent I/O on multiple nodes and nodes with multiple hard disks would give better performance.

The resource allocation for YARN containers and Spark executors must be matching. Container resource allocation should be done in such a way that the entire computing resources available on the server are fully utilized. The configuration of Spark default configuration should be same as the total number of executors configured on the nodes of the cluster. For I/O-intensive jobs, more than 12 cores per executor can lead to bad I/O throughput at JVM level.

RDD partitions decide the total number of tasks created for Spark applications. The Spark configuration allows us to configure multiple cores to each task on executors. The main goal in Spark configuration is to run enough concurrent tasks so that the data destined for each task fits in the RDD memory available on the executors. If there are fewer tasks than the cores available to run them, each stage will not be taking advantage of all the CPU cores. The rule of thumb is, too many tasks is usually better than too few. It is recommended to select the number of RDD partitions as an exact multiple of the total number of tasks configured on executors on the cluster. This ensures that all cores on all executors are fully used in all stages of the computations. The JVM introduces concurrent I/O bottleneck for each executor and there will not be significant performance improvement if more than 5 tasks are configured to run on one executor.

8   Conclusion

The results of the study are summarized in this Section. For Spark cluster computing framework on Hadoop cluster, there is an upper limit for the number of nodes beyond which the performance improvement is marginal. This limit depends on the nature of the application and differs for CPU-intensive- and I/O-intensive applications. When the cluster is configured, the number of nodes is to be selected within this limit to conserve investments in hardware resources. It is also seen that the execution time varies non-linearly with increasing parallelism at executor level. For CPU-intensive applications, overheads in OS level scheduling and garbage collection impact the performance with parallelism at executor level. For I/O-intensive applications, the performance improvement results from multiple concurrent I/O happening at each JVM with increasing number of executors. For task-level parallelism, the limits are imposed at the JVM level and it is seen that the ideal number of task parallelism is seen when 5 tasks run on one executor. Beyond this limit, the performance improvement is marginal. The limits seen at each of the levels of parallelism has mutual dependency on parallelism at node, executor, and task levels. When resources are allocated to the Spark cluster, all these limiting parameters are to be taken into consideration. In Spark cluster computing framework, it is recommended to configure an HDFS cluster with the same number of nodes as that of the Spark cluster.

References

1. Zaharia, M., Chowdhury, M., Franklin, M.J., Shenker, S., Stoica, I.: Spark Cluster Computing with Working Sets, HotCloud 2010 (2010)
2. Mane, D.: How to Plan Capacity for Hadoop Cluster, Hadoop Magazine, April (2014)
3. Janardhanan, P.S., Samuel, P.: Analysis and modeling of resource management overhead in Hadoop YARN Clusters. In: IEEE DataCom 2017, The 3rd IEEE International Conference on Big Data Intelligence and Computing Orlando, Florida, USA (2017)
4. Janardhanan, P.S., Samuel, P.: Study of execution parallelism by resource partitioning in Hadoop YARN. In: ICACCI'17—6th International Conference on Advances in Computing, Communications and Informatics, Manipal University, Karnataka, India (2017)
5. Li, M., Tan, J., Wang, Y., Zhang, L., Salapura, V.: SPARKBENCH a comprehensive benchmarking suite for in memory data analytic platform spark, IBM TJ Watson Research Center. In: CF'15 Proceedings of the 12th ACM International Conference on Computing Frontiers, Article No. 53 Ischia, Italy (2015)

Novel Approach of Intrusion Detection Classification Deeplearning Using SVM

Pritesh Nagar, Hemant Kumar Menaria and Manish Tiwari

Abstract The main objective of intrusion detection systems (IDS) is to discover the dynamic and the virulent form of network traffic that simply changes according to the characteristics of the network. The IDS methodology represents a prominent developing area in the field of computer network technology and its security. Different form of IDS has been developed working on distinctive approaches. One such kind of approach where it is used is the machine learning mechanism. In the proposed methodology, an experiment is applied on the data set named as KDD-99, including its subclasses such as denial of service (DOS), other types of attacks and the class without any form of attack. Depending upon the machine learning algorithms various distinct forms of IDS have been developed which further checks the optimization-based potential features in connection with the neural network classifier for the various forms of IDS-based attacks. This approach provides a comparative study between the ANN and the optimizer-based ANN technology. The experimental analysis shows the convolution neural network with SVM show effective analysis providing accurate forms of IDS, thereby improving its detection based on individual class along with maintaining its results fundamentally.

Keywords Intrusion detection system · Denial of service · Artificial neural network

1 Introduction

In the present scenario, the use of Internet is growing at a large pace with highly developed and emerging forms of ever-growing network and its connectivity, but the use of Internet poses a great threat to cyber security. In order to maintain the high level of security, there is an important need to overcome the cyber threats posing problems to various organizations, companies, and firms. One of the major challenges among the cyber security is to maintain the integrity of the intrusion detection system (IDS),

P. Nagar (✉) · H. K. Menaria · M. Tiwari
Geetanjali Institute of Technical Studies, Udaipur, India
e-mail: priteshnagar1983@gmail.com

© Springer Nature Singapore Pte Ltd. 2020
A. K. Luhach et al. (eds.), *First International Conference on Sustainable Technologies for Computational Intelligence*, Advances in Intelligent Systems and Computing 1045, https://doi.org/10.1007/978-981-15-0029-9_29

thereby protecting it from major forms of attacks and to conquer the various form of risks of the intruded system [1]. The main function of the IDS is to identify a more precise form of intrusion. The illegal hackers of the security have found a large number of ways to break the security of the system whether it is a cloud network or the wireless-based network. Many researches have been performed by the technologists to curb the security threats from distinct forms of intrusions done to the cloud computing systems and the wireless system. So, the main objective of IDS is to protect the information whether it is governmental, public or private entity [2]. The use of IDS is mainly required in detecting the false and the poor detection rates. Whenever an attack is observed by the system or a harmful activity is done to the system, it automatically generates an alarm resulting in a false-positive alarm [1]. The research mainly focusses upon the enhanced capabilities of the intrusion detecting system and thereby reduces the occurrence of the false type alarms.

1.1 Intrusion Detection System

The term intrusion detection system, i.e. IDS is a developing area having various forms of application in the computer technology and its inter-linked networks. Some of the important forms of IDS which identifies the traffic-data and its changing activities by using an algorithm (single class). But some of the single-class algorithms are not able to fetch a good detection rate and does not provide a low occurrence of the false alarms. So, the working methodology is based on using an intelligent hybrid technology comprising of different technology comprising of different sets of classifiers, which are helpful in enhancing the productivity of the system in an intelligent way. In IDS intelligent based mechanism, various forms of data-mining approaches such as genetic algorithms, classification, decision trees, artificial neural networks, and clustering have been used in the mining of data for the development in the field of IDS also the SVM, i.e. support vector machines technology provides the best technique for classification of the clean as well as the intrusive form of data [3]. The SVM technology deals with high-class accuracy in detecting the data intrusions. To avoid redundancy, inadequacy and the noisy data forms, there is an urgent need to go for selection, i.e. feature-based [4, 5]. The basic operation of an intruder is to search the faulty operative conditions in the network or the systems. So, an intruder helps to find out the best-optimized solutions to identify the intrusions in the data. The main requirement of the IDS is not only to encounter the intruders in the data path, but also to supervise the intruders of the data. The most important security aspects of an intrusion detection system consist of maintaining the following conditions:

- *Confidentiality*: Only an authorized user can detect the system.
- *Availability*: Here, the computer technology provides various forms of resources and the access to the legal users of the system without disturbing the working operation of the system.
- Integrity: The information must be protected from any kind of malicious act.

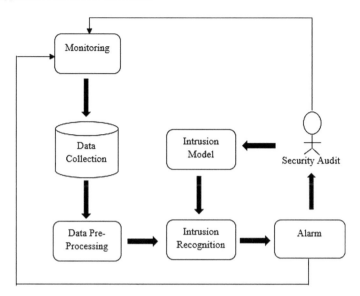

Fig. 1 Basic structure of IDS

The process of intrusion detection system popularly started its operation in 1990. The process of IDS acts as a security alarm where it provides an alarming state in case of any kind of violation in the form of messages, emails or audio-video [6] (Fig. 1).

The IDS is designed as a tool for securing the system from various types of malwares or intrusions interrupting the working if the system [2]. The main function of IDS is to inspect the various types of attacks done on the system, thereby providing a defence mechanism to fight against these attacks in such a way that it also provides information about the intrusions. So, IDS provides a mechanism that deals with the safety of current network security system [7].

1.2 IDS: Architecture

The architecture of IDS comprises of its unique core element, i.e. sensor popularly known as the analyzing engine to pin-point the intrusions occurring in the system. The sensor consists of a mechanism that helps in detecting the intrusions. In Fig. 2, the sensor gets the data (raw) from the given sources as shown which consists of the audit trails, knowledge-based data, and syslog. The 'syslog' includes the authority to the particular system or the system file configuration [7].

The sensor consists of a component known as event generator which performs the data collection shown in Fig. 2. It detects the way of collecting the data. The event generator consists of network, operating system, and the network applications where

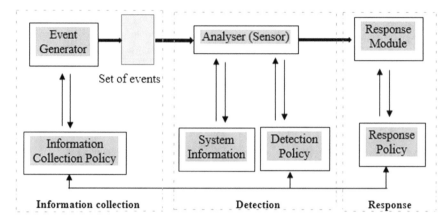

Fig. 2 IDS components

it generates a set of events including audit (log) of the system or the packets of the network. This form of set events also involves the policy of information collection, i.e. in or out of the system. Sometimes, it is not necessary to store the data as it reaches simply to the analyser. So, basically, the key role of the sensor is to extract or filter the data and remove the unwanted form of the data that is achieved from the event data set system [6, 8]. Additionally, the database holds the configurational parameters of IDS that includes its mode of communication methods based on the response module. The sensor itself contains its own data observing all the historical multiplex forms of intrusions. Practically, the IDS may follow a structure based on an 'agent' principle where small modules (autonomous) are designed on 'per-host' basis approach.

The agent mainly monitors and filters the activities scheduled within the area, i.e. fully protected and further starting its initial analysis by undertaking a response action [3]. When a suspicious act or event is detected, an agent issues an alarm. These can be shifted or cloned on another system. The system further may include the transceivers monitoring all the operations effected by agents of another host, i.e. specific. The results fetched by the transceivers are provided to a single unique monitor where the monitor can coordinate the distributed form of information. In addition, some filters are used for aggregation and the selection purpose [9, 10].

1.3 IDS: Classification and Types

There are various categories of IDS based on structure or detection. The IDS are classified based on characteristics as represented in Fig. 3.

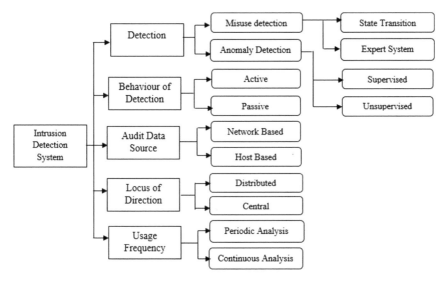

Fig. 3 Classification of IDS based on its characteristics

1.3.1 Based on Structure

The process of IDS is divided into three of its important categories based on its framework. These are host-based IDS, application-based IDS, and network-based IDS.

1. *Host-based intrusion detection system [HIDS]*: The type of detection that is placed in the computer server represents the host of the system usually called as HIDS. As the name suggests a mechanism that helps in analyzing the stored and the system files and further tells about the changes or the deletions done by the attacker in the system files. The HIDS simply detects the part, i.e. not detected by the NIDS mechanism. These are more liable to the attacks that are direct in nature and are inclined to attacks based on DOS, i.e. denial of service.

2. *Application-based IDS*: The application-based IDS is another development of HIDS, which monitors the different types of events such as the inspection of the files, checking the abnormal functions like exceeded permission, void-file execution, etc. It helps in analyzing the communication between the user and the application and monitors the traffic of the network, i.e. encrypted [11].

3. *Network-based intrusion detection system [NIDS]*: The Network-based IDS represents a passive network analyzing the traffic related to the network and for finding out the evidence of various forms of attacks. When a NIDS detects an attack, it provides an instant report to the administrator. It basically checks the types of attack that are incoming and outgoing networks and is usually placed inside the router. But the NIDS is unable to find out the encrypted source of information and is not able to distinguish some forms of attacks. There is no effect of system-failure over the NIDS. The main function after the installation

process is to identify and match the signatures present in the database with the attacking form of signatures.

2   Related Work

This section of literature survey represents the most important section of the thesis. The research study includes the extractions of various articles, books, journals, and research papers from various distinct publications at the national and the international levels. Modi et al. [7] conducted a survey on different intrusions that affected the integrity of cloud-resources, confidentiality, availability, and the services linked. The proposals of subsuming the intrusion prevention systems (IPS) and intrusion detection systems (IDS) in cloud technology are examined. The researcher's recommended the positioning of IDS/IPS in clouded environment to acquire the needed security in the next generation future-based network developments. Kamarudin et al. [9] proposed their study on technology of network security that has become a supreme method for the protection of information or the data. With the excessive growth of Internet technology, various forms of attack cases are observed in a day-to-day life. It includes performance analysis based on machine learning algorithm known as decision tree (J48) where a comparison has been done with two of the other machine learning algorithms named as the neural networks (NN) and the support vector machines (SVM's). These algorithms were tested on the strategy of false alarm rate, detection rate, accuracy, and accuracy of four classes of attacks. From the experimental analysis, it was observed that the decision tree (J48) algorithm performed well as compared to the other two machine learning algorithms. Elshoush et al. [1] focused on proper prevention of attacks that were linked to the computer-based systems. As the motive of complete prevention of attacks is not possible, so the process of using the intrusion detection systems (IDSs) play a crucial role to overcome the harm that is done to the operating systems. Two most important forms-sof methods based on intrusion detection were used, the first one was misuse-based detection and the second was the anomaly based detection. A CIIDS, i.e. collaborative intelligent intrusion detection system was proposed to examine both the methods, as the individually obtained results from both the methods resulted in less form of accuracy. Specifically, there are two major challenges in CIIDSs research strategy. Both of them were reviewed and highlighted. The two challenges were the architecture of CIIDSs and alert-correlation algorithms. Further, it concluded, the occasion for the integrated-solution to large-scale CIIDS. Mohammed and Sulaiman [3] conducted a study on using smart and intelligent form of data-mining methods in order to observe the intrusion occurring in the local type of networks. This paper suggested an improved strategy IDS that combines the expert systems, the processes of data mining as carried out in WEKA. The classification generally consists of a principle based on detection as well as it includes some of the conditions of WEKA such as data-mining open-source processes. The combined methodology gives an improved performance of IDS-based systems and helps to maintain the detection in its more effective form. The result was based on evaluating a new design produced a better form of detection based on efficiency. So, the study presented a good approach to analyse the experiments on behalf of intrusion detection. Vinchurkar et al. [8] conducted a research on intrusion detection systems that consisted of high-level security of networks, and thus provides the system dealing with security of network and the intrusion-based attacks. The ideal features of IDS include a monitoring activity of network and the threats. The intrusion detection system is generally classified on the basis of the model and the data-source. But some of IDS techniques are more challenging in nature. The anomaly based IDS can be detected easily using various anomaly detection techniques. The process of dimension reduction is based on the analysis of principle component. The problem of construction classifier can be identified using a support vector machine methodology. Nadiammai et al. [6] focused upon the security issue of the networks and various developments in applications running on distinct platforms capturing an attention towards security of the network. This type of paradigm exploited the vulnerabilities of security that are technically difficult and expensive to solve. Hence, intrusion is used as a key to compromise the integrity, availability, and confidentiality of a computer resource. Four issues such as classification of data, high-level of human interaction, lack of labelled data, and effectiveness of distributed denial of service attack are being solved using the proposed algorithms like EDADT algorithm, hybrid IDS model, semi-supervised approach and varying HOPERAA algorithm, respectively. Our proposed algorithm has been tested using KDD cup data set. All the proposed algorithm shows better accuracy and reduced false alarm rate when compared with existing algorithms. Agrawal et al. [4] worked on the need of the present world dealing with huge amounts of data, i.e. transferred and stored from location or another. When the data gets transferred or is stored somewhere then it gets exposed to many forms of attack. However, many types of techniques and the detection mechanism have been developed to overcome the problem of data risk. Thus, to examine the data and to identify the kind of attack done on the data various mining techniques of data have emerged to make it free from any kind of loss related activity. The process of anomaly detection uses mining techniques of data to recognize the hidden behaviour inside the whole set of data, which might increase the chance of being attacked in an easy way. The use of hybrid-based approaches has also been made to judge the form of attacks whether known or unknown in nature. This research has reviewed data-mining techniques for anomaly detection to provide better understanding among the existing techniques that may help the interested researchers to work in future. Jabez et al. [12] proposed a study based on intrusion detection system (IDS) that presents an application of a software monitoring the activities of the system network generating reports to the management system. The main objective of this study was based on IDS detection and the prevention systems (IDPS). So, this work proposed a new strategy known as outlier-detection which used a data set measured by the neighbourhood outlier factor (NOF). Here, a model, i.e. trained consists of large data sets with storage environment of distributed type to improve the IDS-based performance of the system. The experimental analysis proved that the proposed strategy detects the malicious content in a very effective way. Jabbar et al. [5] proposed the research based on the intrusion detection system to notify and identify the type of activities or normal users or the hackers performing malicious operations. In this paper, a model has been designed for intrusion detection system (IDS) using a classifier based on random forest, where the random forest (RF) denoted an ensemble classifier and that performed very well as compared to the other classifiers that worked traditionally for an effective classification of different forms of attacks. Gupta et al. [2] proposed work on IDS to pin-point the harmful malicious attacking activities done to the system. But this kind of framework carried a disadvantage of the number of false-positive and alert rates, which resulted in decreased proficiency of the IDS systems. This research has complied in many other researches and provides a single-metric technique to the false-positive alarms in the process of IDS which would further help the future researchers to extract and gain the knowledge to propose the further studies. He et al. [11] proposed the method of fuzziness based on the technique of instance selection for a huge amount of data sets in order to increase the supervised learning algorithm-based efficiency. It did so by improving the design shortcomings of intrusion detection system (IDS). The methodology proposed was dependent over a new type of single-layer feed-forward neural network (SLFN) known as random weight neural network (RWNN). Siddiqui and Farooqui [10] proposed a work using the IDS tool for anomaly detection that provides network security to the system. The IDS represents a method to detect the processes of cyber-attacks and this process of detection is based on the amount of distinct forms of intrusive activities occurring in the operation of the system as the detection of an intrusion denotes a very complicated process. So, this paper conducted a study based on classifier named cascaded support vector or it might be called as an improved version of ensemble classifier using a function, i.e. kernel function. This kernel function represents a Gaussian function. A neural network technique has been used for collecting its features based on different and distinct forms of attacks and this algorithm is more effective than the earlier method used. Wang et al. [13] proposed a methodology that focused on the fact that the security of the network has been increased at a very large pace for all the organizations, firms, and the most important is the security of an individual. The main aim was to obtain high-quality improvement in detection for the trained data set. As the ratio of marginal density denotes a powerful classifier, i.e. univariate in its nature, the study adopted for the obtaining the results is based on framing an IDS based on SVM method entailing its augmented features. Uniquely, a method has been implemented based on logarithmic values of the ratios of the marginal density in order to obtain a good quality of its transformed features, which improved the rate of detection based on SVM model. The set of data named NSL-KDD is basically used for the proposed method and the experimental results showed that the results are far much better than the existing forms or methods specifically targeting the rate of accuracy, its training speed, and the false alarm rate.

3 The Proposed Method

In this section, we discussed the proposed approach and the methodology used to achieve the results.

3.1 Proposed Technique

The proposed technique involves the following steps:

The author has proposed a hybrid model which consists of SVM, i.e. support vector machine combined different classification-algorithm to mitigate the rates of the false-positive alarms. To obtain the pre-thesis objective, a methodology has been proposed which is further divided into three types of phases.

Phase 1: Collection and preprocessing

- Data set collection.
- Extraction of features through a data, i.e.'tcpdump'.
- Converting the obtained features into binary representation.
- Preparation of the input for its classification.

Phase 2: Classification

- To find the best classifier from the available classifier.
- To test and train the tool of classification by the data set-partitioning process.

Phase 3: Result analysis

- To compare the obtained results with their existing work.

The proposed working methodology is designed as in Fig. 4.

3.2 Proposed Flowchart

The proposed steps of flow chart are given as below:

1. *KDD-99 Data set*: This is a type of data set used for the (Third International Knowledge Discovery and Data Mining Tools) competition, held in conjunction with KDD-99 (The Fifth International Conference on Knowledge Discovery and Data Mining). The main task was to build a network based on intrusion detection and to predict a model, i.e. capable of discriminating a good or a bad form of data set. This form of data set maintains a standard including a wide variety of network-based intrusions.
2. *Label features*: A label helps in providing a complete information regarding the set of data.

Fig. 4 Proposed
methodology

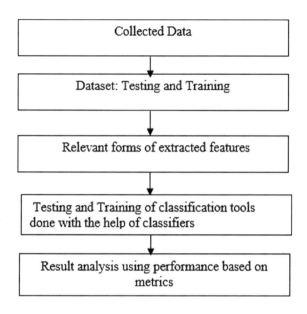

3. *Input in PSO*: Each of the particles has its velocity and position to search for better solution. So, the velocity and position are the inputs used in PSO.
4. *Initialize particles*: The PSO-based technique is initialized with a population of random solution.
5. *Update fitness function*: It helps in judging the individual solutions based on how well they can handle the problem.
6. *Optimize objective form*: Here, the objective is optimized.
7. *Initialize chromosomes*: The process is initialized by building a population of chromosomes which is a set of possible solutions to the optimization problem.
8. *Check the convergence*: These type of methods helps in testing the conditional-convergence, absolute-convergence, interval of convergence or divergence of an infinite series.
9. *Cross over*: A point or place of crossing from one side to the other.
10. *Roulette selection*: It is a method used in genetic algorithms for selection of potentially useful solutions for the purpose of recombination.
11. *Optimize features*: This type of method achieves the best designing technique.
12. *Neural networks:* It represents biologically inspired information processing system.
13. *Test model*: It performs a system or software system.
14. *Precision and recall accuracy:* The precision is a good measure that determines the cost of false-positive is high (Fig. 5).

C. *Algorithm*

Following are the algorithms that are used in the proposed work.

Fig. 5 Proposed flowchart

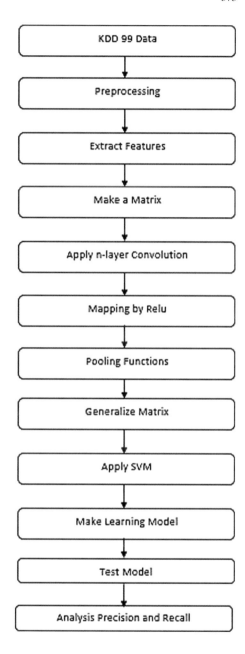

Algorithm

Step 1: Input bugs in the form of KDD-99l data.
Step 2: Preprocessing of data test to remove the noisy data.
Step 3: Extract the bigrams and make the matrix.
Step 4: Apply n-layer convolution and mapping by Relu.
Step 5: Pooling of function and generalize the matrix.
Step 6: Apply the SVM

With optimization model $min_{\omega,\xi,Q} P(\omega,\xi)$ we describe the model of SVM classification.

$$min_{\omega,\xi,Q} P(\omega,\xi_r) = \frac{1}{2}\omega^g\omega + \frac{1}{2}\gamma\sum_{r=1}^{n}\xi_r^2$$

$$s_r[\omega^t\phi(u_r) + Q = 1 - \xi_r, r = 1, 2, \ldots, n$$

$$\xi = (\xi_1, \xi_2, \ldots, \xi_n)$$

Where
$\xi_r \leftarrow$ Slack variable
$Q \leftarrow$ Offset
$\omega \leftarrow$ Support vector
$\gamma \leftarrow$ Classification parameter for balancing the model complexity and fitness error.

Then, describing the classification decision function:

$$F(z_r) = sgn\left(\sum_{r=1}^{n}\alpha_r s_r L(q, q_r) + Q\right)$$

Step 7: Make learning and testing model.
Step 8: Analyze the precision and recall.

4 Result Analysis

4.1 Experimental Setup

(a) *Data set description*: The experiments as discussed above are mainly executed with the help of KDD-99 data set having around 41 sets of feature sets. Such type of features is mainly used for the process of optimization and further it involves the mechanism of learning and currently, these are basically used for analyzing in terms of various kinds of attacks. This work is comprised of evaluating the rate of accuracy in IDS methodology. In the part of analysis, the data is taken based on the number of intrusions in the process. The attacks are divided into four of its classes: (1) Probe, (2) Dos, (3) U2R, and (4) R2L. In our methodology, we use three of the following categories: (1) other attack which mainly consists of probe, U2R and R2L (2) Normal attacks, and (3) DoS-attack. Further, the evaluation of precision, accuracy, F-measure, and recall is done in several cases.

Figure 6 shows the analysis in respect of precision, accuracy, f-measure, recall, etc. This figure involves the demonstration of efficiency analysis from all four types of algorithm. Analysis demonstrates that ANN with both SPO and GA give better result in terms of all the four parameters, i.e. precision, accuracy, recall, and f-measure. In Fig. 7, represents the parameters investigation of various kind of classifiers and the proposed approach. In investigation parameters like exactness, review, precision, and f-measure fluctuate as indicated by classifier yet one examination clear about proposed methodology (ANN with GA in neural system) demonstrate huge enhance all parameters. In the event that examination just proposed methodology, review demonstrate huge enhancement then different parameters so it will clear sign of decreasing false negative rate so assaults distinguishing proof is successful in proposed approach in view of enhance weight given by ANN_GA approach.

Figures 7 and 8 parameters investigation of various classifier and proposed approach. In investigation parameters like exactness, review, precision, and f-measure fluctuate as indicated by classifier yet one examination clear about proposed approach (PSO with GA in neural system) demonstrate huge enhance all parameters. In the proposed methodology review demonstrate huge enhancement then different parameters so it will clear sign of decreasing false negative rate so assaults distinguishing proof

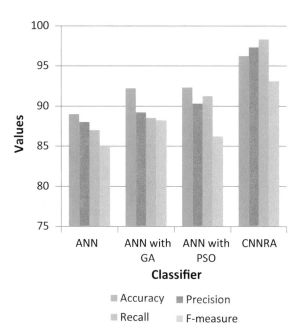

Fig. 6 Simulated graph of comparison parameters

Fig. 7 Analysis of ANN

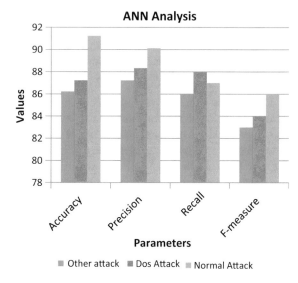

Fig. 8 Analysis with ANN and ANN_GA

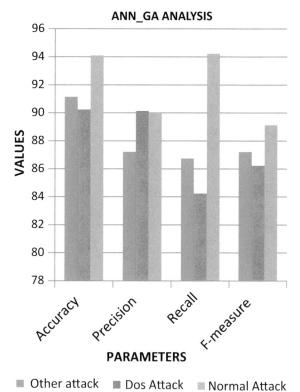

is successful in proposed approach in view of enhance weight given by PSO_GA approach CNNRA.

In Fig. 8, profundity investigation of every one of the three classes in ANN and ANN_GA. In this examination, we endeavour to indicate what the criticalness of our methodology is. This exchange we precede in perception (3) moreover. So first point which examination by typical class n which no assault working and in the two cases ANN and ANN with GA perform well contrast with other parameter like exactness, review, and f-measure yet ANN_GA still preferable precision over ANN, so include weighted by enhancement by one way or another perform in view of decreasing covering data learning. On the off-chance that examination through DOS assaults it additionally indicate higher exactness in ANN with GA. So we can close feature advance weight is better methodology so by what means can enhance improvement these perception talk about in next standard (Fig. 9).

At last from the whole analysis it can be concluded that algorithm CNNRA gives better result for all the attacks we examined in our work. In Fig. 10, examination proceed from perception (2) and attempt to discovering centrality of streamlining enhancement impact on various classes' recognition by characterization. On the off-chance that investigation the both chart demonstrate the compelling review; however, for ordinary class so decrease the false-positive rate this enhancement occurring with all classes like DOS assault and different assaults yet the viable outcome appear in other assault which increment fundamentally in proposed approach. So PSO stream-lining is great, however, PSO with GA more enhance in other assault and ordinary class.

Fig. 9 Analysis of ANN_PSO

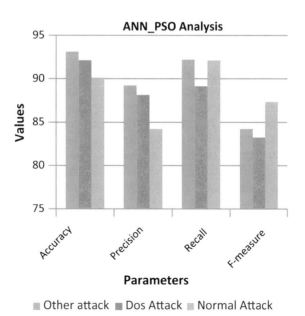

Fig. 10 Analysis with
ANN_PSO and CNNRA

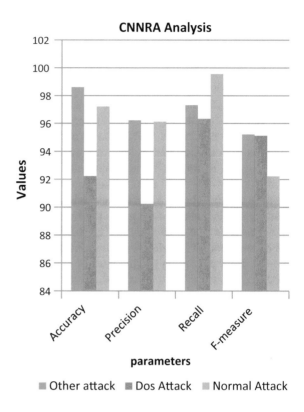

5 Conclusion

The present scenario experiences various forms of developments and a huge growth in advanced processing technologies consisting of connectivity among different networks, but methodology is vulnerable by the activities of the intruders or the attackers of the system. These specifically smart attackers interrupt the operation with new and fascinating methods of data-breaching among large networks. Though there are various forms of available intrusion of intrusion detection systems that can detect the intrusions occurring in the network, i.e. based on the false-positive detection rate and the alert rates, but with the detection rate of intrusions, they also have a high false-positive rate resulting in an adequate system comprising of low accuracy level of the system and are generally more prone to different kinds of attack. This usually helps the intruder to enter into the system and perform a pre-planned attack. So, this pre-thesis will propose a hybrid approach to reduce the false-positive alarms. The experimental analysis consists of a specified particular form of data set and the process of feature-based selection will be done to improve the analysis. These features obtained will be used for the classification-tool training and testing the performance of the system. Finally, the result obtained will be compared with the results that already exist.

References

1. Elshoush, H.T., Osman, I.M.: Alert correlation in collaborative intelligent intrusion detection systems—A survey. Appl. Soft Comput. **11**(7), 4349–4365 (2011)
2. Gupta, N., Srivastava, K., Sharma, A.: Reducing false positive in intrusion detection system: A survey. Int. J. Comput. Sci. Inf. Technol. **7**(3), 1600–1603 (2016)
3. Mohammed, M.N., Sulaiman, N.: Intrusion detection system based on SVM for WLAN. Proc. Technol. **1**, 313–317 (2012)
4. Agrawal, S., Agrawal, J.: Survey on anomaly detection using data mining techniques. Proc. Comput. Sci. **60**, 708–713 (2015)
5. Farnaaz, N., Jabbar, M.A.: Random forest modeling for network intrusion detection system. Proc. Comput. Sci. **89**, 213–217 (2016)
6. Nadiammai, G.V., Hemalatha, M.: Effective approach toward intrusion detection system using data mining techniques. Egypt. Inform. J. **15**(1), 37–50 (2014)
7. Modi, C., Patel, D., Borisaniya, B., Patel, H., Patel, A., Rajarajan, M.: A survey of intrusion detection techniques in cloud. J. Network Comput. Appl. **36**(1), 42–57 (2013)
8. Vinchurkar, D.P., Reshamwala, A.: A review of intrusion detection system using neural network and machine learning (2012)
9. Jalil, K.A., Kamarudin, M.H., Masrek, M.N.: Comparison of machine learning algorithms performance in detecting network intrusion. In: 2010 International Conference on Networking and Information Technology (ICNIT), pp. 221–226. IEEE, New York (2010)
10. Siddiqui, A.K., Farooqui, T.: Improved ensemble technique based on support vector machine and neural network for intrusion detection system. Int. J. Online Sci. **3**(11) (2017)
11. Ashfaq, R.A.R., He, Y., Chen, D.: Toward an efficient fuzziness based instance selection methodology for intrusion detection system. Int. J. Mach. Learn. Cybernet. **8**(6), 1767–1776 (2017)
12. Jabez, J., Muthukumar, B.: Intrusion Detection System (IDS): Anomaly detection using outlier detection approach. Proc. Comput. Sci. **48**, 338–346 (2015). ISSN 1877-0509. http://dx.doi.org/10.1016/j.procs.2015.04.191
13. Wang, H., Jie, G., Wang, S.: An effective intrusion detection framework based on SVM with feature augmentation. Knowl.-Based Syst. **136**, 130–139 (2017)

Authentication Protocol with Privacy Preservation for Handover in Wireless Mesh Networks

Amit Kumar Roy and Ajoy Kumar Khan

Abstract As we know that accessing the Internet through a mobile device is growing rapidly worldwide. Therefore, a reliable and efficient authentication scheme for fast handover is essential to avoid temporary disconnection, authentication delay and non-registered access to the Internet. The handover took place when mobile device left its home mesh access point (HMAP) that covers a limited area within which it can get access to the Internet and joins the new MAP. To allow Internet access to the mobile device, mutual authentication took place between the mesh access point (MAP) and a client/user. Therefore, we proposed an efficient handoff authentication protocol along with preserving the privacy of the transfer ticket between clients and MAPs which is not considered in the existing method. The experiment result proves that our protocol achieves efficient handoff authentication with the privacy of the transfer ticket.

Keywords Wireless mesh networks · Login authentication · Handover authentication · Tickets · Security requirements

1 Introduction

For accessing the Internet, wireless mesh networks (WMNs) become a next-generation network for everyone because the technology provides quick and efficient network deployment [1]. Due to its dynamic nature, WMN could easily be self-organized and self-healing while offering mesh connectivity among them automatically [2]. The architecture of WMNs is categorized into three tiers-mesh routers (MR), mesh clients (MC), and Internet gateways (IGWs) [3]. Presently, for accessing the Internet, mobile is employed frequently because of its small size and easy to carry anywhere. Therefore, mobile changes its place from one MAP to another

A. K. Roy (✉) · A. K. Khan
Assam University, Silchar, Assam, India
e-mail: amitkroy12@gmail.com

A. K. Khan
e-mail: ajoyiitg@gmail.com

© Springer Nature Singapore Pte Ltd. 2020
A. K. Luhach et al. (eds.), *First International Conference on Sustainable Technologies for Computational Intelligence*, Advances in Intelligent Systems and Computing 1045,
https://doi.org/10.1007/978-981-15-0029-9_30

383

MAP as a result of which handoff took place. During handoff, there is a possibility of temporary disconnection, delay, and illegal use of the Internet when a mobile leaves its coverage area and enters to a new coverage area under the new MAP [4]. This handoff process leads to authentication delay.

Enormous work had been done in the past for efficient handoff authentication for WMN. Among them, ticket-based authentication method had taken a large step toward the establishment of a trust relationship during a handoff operation. A single authentication process during handover via ticket-based allows neighboring MAPs to share a pre-stored symmetric key [5–7]. A group key authentication process during handover allows the authentication server (AS) to divide the network into multiple groups [8–10]. A broadcast authentication process during handover allows AS to authenticate the clients while maintaining every MAP [11, 12]. Therefore, to minimize latency during the handoff process, enormous protocols had been developed during recent years for handoff authentication [13–17]. Our proposed work provides efficient and fast handoff as well as providing privacy for the clients. Our work had been compared with [18, 19].

2 Related Work

To minimize the handoff latency several existing works had done based on the protocol known as Kerberos [10, 20]. This section overviews some relevant work that had been done in the past.

A ticket-based authentication protocol during handoff had been proposed for WMNs. The protocol provides efficient handoff by forming a temporary connection between mesh entities. However, the proposed work suffers from long latency as the process of authentication was done in a multi-hop fashion with AS during handoff. Moreover, as the users are restricted to limited power supply, the pre-distribution of transfer tickets to mesh clients was a burden which consumed their power very fast [19]. For wireless network, a re-authentication method had proposed during handover. However, the method suffered from long latency as multi-hop authentication is required from the AAA server [12]. In 2013, an authentication method had been proposed to overcome from multi-hop authentication. In their scheme, authentication was done within one hop between the client and requesting MAP. However, the protocol comes up with certain disadvantages. Sharing transfer ticket in plaintext form and redistributing the similar ticket among other MAPs could violate the security [18]. In 2010, for WLAN, a handover authentication method had been proposed. However, in the proposed protocol, the mobile clients suffered from extra burden during pre-storing the set of tickets as mobile users have limited storage and restricted bandwidth [11]. Kassab et al. proposed a proactive authentication method for securing fast handover in WLANs. However, in their proposed work, the AS had to distribute the PMK every time with an increase in mobile clients throughout the network with limited bandwidth [21]. In 2008, for wireless mesh networks, a secured authentication protocol via ticket had been proposed. However, their work neglects the authentication process among the user and MAP [22]. A privacy-preserving method for EAP-based wireless networks was proposed by Jing et al. However, the protocol also suffered from the extra burden for clients while informing the home MAP to which foreign MAP it wanted to associate, as a client was restricted to limited bandwidth and power [5]. For 802.16j network, a multi-hop authentication method had been proposed. However, for wireless mesh networks, their protocol may suffer from long handoff latency due to multi-hop authentication as the number of relay stations was not fixed between the clients and AS [6]. In 2013, for WiMAX networks, an authentication on grouping method had proposed. The work was fully dependent on domain-based with a fixed number of BSs. Therefore, it was not secured and not scalable for large wireless mesh networks [23].

3   The Trust Model

Trust model employs a TA who issues tickets for mesh entities to authenticate each other during the roaming process. Ticket agent acts as a centralized authority in mesh network. Types of trust relationship built between mesh entities are as follows.

1. TA-MAPs: Trust is established between TA and MAP on a request of MAP ticket.
2. TA–Users: Trust is established between TA and users on a request of user ticket.
3. MAP–User: Trust is established between MAP and user through their respective tickets.
4. MAP_1-MAP_2: Trust is established between any two neighboring MAPs through a digital certificate. This trust permits a user to associate with the neighboring MAPs.

3.1   Types of Tickets Issued for Mutual Authentication Among Mesh Entities

1. User tickets (T_C): It established a trust relationship between user and MAP. T_C allows the user to prove its legality to MAP.

   Elements within T_C are shown below

$$T_C = \left\{ I_C, I_A, \tau_{exp}, P_C, Sig_A \right\} \tag{1}$$

where

I_C Users identity.
I_A TA identity.
τ_{exp} expiry time of T_C.

P_C Users' public key.

Sig_A Digital signature of TA in mesh entities ticket ensures its legality during LAP.

2. MAP tickets (T_M): It established a trust relationship between MAP and user. T_M allows the MAP to prove its legality to user.

   Elements within T_M are shown below

$$T_M = \{I_M,\ I_A, \tau_{\exp}, P_M, \text{Sig}_A\} \tag{2}$$

where,

I_M MAPs identity.
I_A TA identity.
$\tau_{\exp}$ expiry time of T_M.
P_M MAPs' public key.
Sig_A Digital signature of TA.

3. Transfer tickets (θ_C): It builds trust between a user and the foreign MAP (e.g., MAP_2). This transfer ticket is generated by a home MAP (e.g., MAP_1) after mutual trust is established between user and MAP. θ_C allows the user to prove its legality to MAP_2.

   Transfer ticket θ_C contains the following elements

$$\theta_C = \{I_C, I_M, I_A, \tau_{\exp}, V_{K_{MAC}}(I_C,\ I_M,\ I_A,\ \tau_{\exp})\} \tag{3}$$

where

I_C Users' identity owning θ_C.
I_M MAPs' identity owning θ_C.
I_A TA identity.
$\tau_{\exp}$ Expiry time of θ_C. Requires fresh θ_C request from MAP (Table 1).

4   Analysis of Existing Protocol

The protocol is based on two different authentication methods: First, login authentication protocol (LAP) for mutual authentication and secondly, handover authentication protocol (HAP) during migration process. Both the protocols depends on three types of keys—a pairwise master key, a pairwise transient key, and a group transient key for authentication between mesh entities. As mesh users are restricted to power constraint, the exchange of these keys should be minimized. We analyzed the protocols in detail and found that the protocol suffers from various security threats. Firstly, θ_C are transmitted in a plaintext format. Therefore, user could easily obtain similar

Table 1 Notation used

Notation	Description
T_A	Ticket Agent
T_C	Ticket of User
T_M	Ticket of MAP
Θ_C	Transfer ticket
τ_{exp}	Expiry time of ticket
I_x	Identity(ID) of entity x
P_x	Public key of entity x
$E_{pub_x}(m)$	Message encryption via public key of x
$D_{pub_x}(m)$	Message decryption via public key of x
$E_K(m)$	Message encryption via key K
$D_K(m)$	Message decryption via key K
K_{MAC}	Key employed to form message authentication code
Sig_A	Digital Signature of T_A

MAC value and misuse it. Secondly, θ_C in the plaintext format allows an intruder to tamper the integrity of the transfer ticket θ_C which involves I_C, where I_C is rarely changed. Thirdly, as a user want to migrate to only one MAP (i.e., MAP$_2$), the home MAP$_1$ redistributes the similar ticket among other MAPs (i.e., MAP$_3$ and MAP$_4$) which could violate the security [18].

5   The Proposed Protocol

To overcome against the problem analyzed in the above section, we proposed a protocol that provides an efficient authentication among the client/user and MAPs along with preservation of transfer ticket θ_C. Furthermore, the proposed work restricts the distribution of the same ticket by MAP$_1$ to all its neighboring MAPs. In our proposed protocol, we have considered a change in elements of the existing transfer ticket θ_C.

A transfer ticket θ_C contains the following elements

$$\theta_C = \left\{ I_C,\ I_M,\ I_A,\ N_1,\ \tau_{exp} \right\} \tag{4}$$

where elements are described below

I_C Identity of the user owning θ_C.
I_M Identity of the MAP issuing θ_C.
I_A Identity of TA.
N_1 nonce 1 to prevent from a replay attack.

The proposed protocol in this paper had been compared with [18, 19] and overcome the drawbacks found in their work.

5.1   Login Authentication Protocol

Assuming the TA had issued the user and the MAP_1 the user ticket and the MAP ticket, respectively. Thereafter, both parties share their tickets for mutual authentication.

1. $C \rightarrow MAP_1 : I_C$
2. $MAP_1 \rightarrow C : T_{M1}$
3. $C \rightarrow MAP_1 : E_{P_{M1}}(T_C, N_{C1})$
4. $MAP_1 \rightarrow C : E_{P_C}(N_{M1}, N_{C1})$
5. $C \rightarrow MAP : V_{K_{MAC}}(N_{M1})$
6. $MAP \rightarrow C : V_{K_{MAC}}(\theta_C)$

1. User C broadcast request information along with its identity to MAP_1 for accessing the Internet.
2. On receiving the request information, MAP_1 sends T_{M1} to user C. On the acceptance of T_{M1}, the user verifies the T_{M1} via signature (Sig_A) and expiry time τ_{exp} present in T_{M1}.
3. Ticket T_{M1} after successfully verified by C, the public key P_{M1} of MAP_1 is extracted from T_{M1}. Further, the ticket T_C with nonce N_{C1} is encrypted by C using P_{M1} and forwarded to MAP_1. On the arrival of $E_{P_{M1}}(T_C, N_{C1})$, MAP_1 decrypts the information and verifies the elements of T_C.
4. On successful verification, the public key P_C is extracted by MAP_1 from T_C. Further, two nonces N_{M1} and N_{C1} are encrypted by MAP_1 using P_C and forwarded to C. Later, both parties compute $K_{MAC} = N_{C1} \parallel N_{M1}$. On the acceptance of $E_{P_C}(N_{M1}, N_{C1})$, user C uses a private key to decrypt the information to extract N_{M1} and N_{C1}. Thereafter, the user verifies the nonce N_{C1} whether it matches with the nonce issued by the user before. If the nonce is similar, user computes the $K_{MAC} = N_{C1} \parallel N_{M1}$. The nonce N_{C1} and N_{M1} are secured via public-key cryptography.
5. To complete the authentication protocol for accessing the mesh network, the user sends $V_{K_{MAC}}(N_{M1})$ to MAP_1. On the acceptance of the information, MAP_1 checks the accuracy of $V_{K_{MAC}}(N_{M1})$ via MAC key and then verifies the nonce N_{M1} whether it matches with the nonce issued by the MAP_1 before. If N_{M1} matches, then the MAP_1 has successfully authenticated C. Finally, the authentication protocol for accessing the mesh network is completed.
6. After successful authentication, MAP_1 sends $V_{K_{MAC}}(\theta_C)$ to C. On the acceptance of $V_{K_{MAC}}(\theta_C)$, the user checks the accuracy of $V_{K_{MAC}}(\theta_C)$ using the MAC key. If the accuracy of the $V_{K_{MAC}}(\theta_C)$ fails, user C rejects this information.

5.2 Handover Authentication Protocol

For users to move from MAP_1 to foreign MAP, an efficient handoff with privacy preservation is proposed. User C informs the MAP_1 to which foreign MAP_x it wants to join. MAP_1 then encrypts the MAC key (i.e., $K_{MAC} = N_{C1} \parallel N_{M1}$) via P_x of its neighboring MAP_x and transmits the encrypted MAC key to MAP_x. We assumed public key of MAP's is known to each other. Then, MAP_1 sends the encrypted MAC key to relevant MAP_x ($x = 2$) to whom the C wants to join. Upon receiving the encrypted MAC key, MAP_x decrypts it with its private key. User C during migration from MAP_1 to MAP_x performs the following handover authentication process:

1. $C \rightarrow MAP_x : V_{K_{MAC}}(\theta_C, N_c)$
2. $MAP_x \rightarrow C : V_{K_{MAC}}(N_c)$

1. A user transmits the θ_C and a fresh nonce N_c as $V_{K_{MAC}}(\theta_C, N_c)$ to MAP_x. On receiving the $V_{K_{MAC}}(\theta_C, N_c)$, MAP_x checks the accuracy of $V_{K_{MAC}}(\theta_C, N_c)$ via K_{MAC} issued from the MAP_1. If the accuracy is passed, MAP_x authenticates the validity of the information. Then, MAP_x checks the elements of θ_C. If everything goes well, then MAP_x sends $V_{K_{MAC}}(N_c)$ to C for the confirmation of acceptance.
2. On the acceptance of the information, the user checks the accuracy of $V_{K_{MAC}}(N_c)$ via K_{MAC} and authenticates the nonce N_c if it is similar to the nonce issued by a user before. If the accuracy is successful, user C accepts the handover authentication procedure (Figs. 1 and 2).

6 Security Analysis of Proposed Protocol

Based on some security requirements and threats, we analyzed our proposed method.

1. Mutual authentication: Mutual authentication took place between user/client and MAP_1 during the login process to verify the legality of each other. Their legality can be proved via their tickets exchanged between client and MAP_1. The digital signature of the TA ensures the authentication of MAP_1 and user. Further, the user encrypts its information via $E_{P_{M1}}$ as $E_{P_{M1}}(T_C, N_{C1})$. As the information is encrypted via public key of MAP_1, only MAP_1 could decrypt and extract this information. Therefore, public-key cryptography restricts an adversary to generate a correct reply message.
2. Privacy preservation: Existing method neglects the security consideration for a transfer ticket and sends it in a plaintext format between mesh entities during LAP and HAP [18]. Therefore, the integrity of the transfer ticket could be tampered easily by intruders. To overcome the existing problem, we proposed the privacy of the transfer ticket, i.e., $V_{K_{MAC}}(\theta_C)$ during LAP and $V_{K_{MAC}}(\theta_C, N_c)$ during HAP

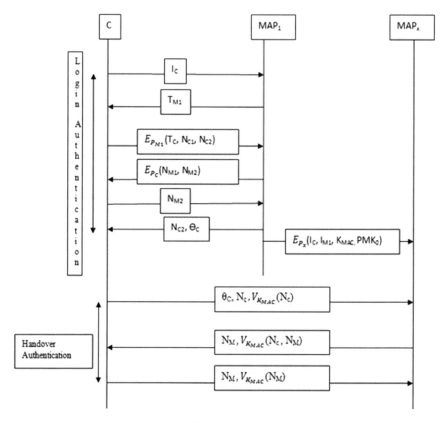

Fig. 1 Existing authentication method [18]

in our proposed work. Therefore, it resists against tampering of transfer ticket by an attacker.

3. Forgery attack resistance: Digital signature of TA present in mesh entities ticket ensures its legality during LAP. During HAP, as the foreign MAP received the legal information of θ_C beforehand from home MAP_1 from which C wants to migrate, C cannot be tampered the θ_C. Therefore, HAP in our proposed protocol prevents any kind of attacks from intruders.

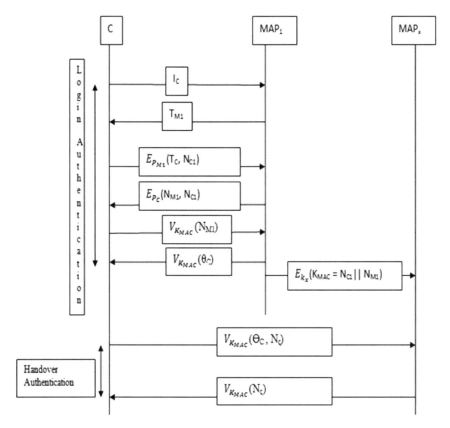

Fig. 2 Proposed authentication protocol

7 Experimental Results

7.1 Performance Metrics

The performance metrics of our proposed work is based on authentication delay (latency), computation cost, and communication cost. The simulation is done using NS-3.21.

1. Authentication delay is the time required to send the authentication request and receive the acceptance confirmation between a client and MAP. It is computed as the addition of computation cost and communication cost mentioned in the bottom row of Table 2.

2. Computation cost occurs due to delay in processing caused by various security operations as shown in column 1 of Table 2 during login (LAP) and handover (HAP).

Table 2 Performance comparison

Operation	Algorithm	Time (ms)	[18] Login	[18] Handover	[19] Login	[19] Handover	Proposed Login	Proposed Handover
$E_{pub_x}(m)$	RSA	1.42	2	0	2	0	2	0
$D_{pub_x}(m)$	RSA	33.3	2	0	2	0	2	0
TA_{SigA}	ECDSA	11.6	0	0	1	0	0	0
Verification of TA_{SigA}	ECDSA	17.2	2	0	3	0	2	0
E_K	AES	2.1	0	0	0	1	0	0
D_K	AES	2.2	0	0	0	1	0	0
MAC	HMAC	0.015	2	7	0	4	6	4
Hash	SHA-2	0.009	0	0	4	1	0	0
Total computational cost (ms)			103.8	0.105	132.67	4.369	103.9	0.06
Number of messages			6	3	9	5	6	2
Authentication latency (ms)			$103.87 + 6d$	$0.105 + 3d$	$132.676 + 9dh$	$4.369 + 5d$	$103.93 + 6d$	$0.06 + 2d$

3. The communication cost is computed with respect to overall information transmitted between a user and MAP during login process and handover process. In column 6, bottom row of Table 2 the symbol d denotes the average delay transmission over single hop and h is the multiple hops between mesh entities. Since only [19] requires multi-hops communication during login process, notation h is applicable to only [19].

7.2   Result Analysis

(1) Comparison based on total computational cost during login (LAP) is shown in row 11, column 4, 6, and 8 of Table 2.
(2) Comparison based on total computational cost during handover (HAP) is shown in row 11, column 5, 7, and 9 of Table 2.
(3) Comparison based on total communication cost during login (LAP) is shown in row 12, column 4, 6, and 8 of Table 2.
(4) Comparison based on total communication cost during handover (HAP) is shown in row 12, column 5, 7, and 9 of Table 2.

7.3   Simulation Results

(1) Figure 3 shows the authentication delay of the proposed protocol versus Li et al. protocol versus Xu et al. protocol with different network size, i.e., with 10, 20, 30, and 40 nodes in the network during LAP.

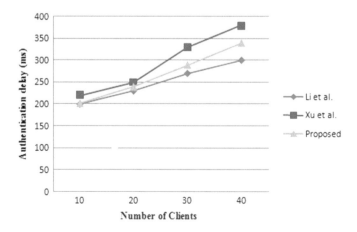

Fig. 3 Authentication delay during login with 10, 20, 30 and 40 clients

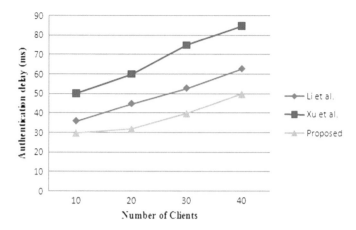

Fig. 4 Authentication delay during handover with 10, 20, 30 and 40 clients

(2)   Figure 4 shows the authentication delay of the proposed protocol versus Li et al. protocol versus Xu et al. protocol with different network size, i.e., with 10, 20, 30, and 40 nodes in the network during HAP.

8   Conclusion

The proposed protocol in this paper achieves an efficient handoff authentication for wireless mesh networks (WMNs). The proposed work offers security along with privacy preservation of transfer ticket. Proposed protocol also includes preservation from forgery attack. The simulation is done in Network Simulator 3 (NS3). The experimental result shows that our proposed protocol performed well compared to existing protocols. The comparison is mainly based on three performance metrics, authentication delay (latency), computation cost, and communication cost. In future, we could further extend our work to offer more advanced protocols to meet all the advanced future needs.

References

1. Fu, Y., He, J., Wang, R., Li, G.: Mutual authentication in wireless mesh networks. In: 2008 IEEE International Conference on Communications, pp. 1690–1694. IEEE, Beijing, China (2008)
2. Akyildiz, I.F., Wang, X.: A survey on wireless mesh networks. IEEE Commun. Mag. **43**(9), S23–S30 (2005)

3. Seyedzadegan, M., Othman, M., Ali, B.M., Subramaniam, S.: Wireless mesh networks: WMN overview, WMN architecture. In: International Conference on Communication Engineering and Networks IPCSIT 19, p. 2. Singapore (2011)

4. He, D., Chan, S., Guizani, M.: Handover authentication for mobile networks: Security and efficiency aspects. IEEE Network **29**(3), 96–103 (2015)

5. Jing, Q., Zhang, Y., Fu, A., Liu, X.: A privacy preserving handover authentication scheme for EAP-based wireless networks. In: Global Telecommunications Conference (GLOBECOM 2011), pp. 1–6. IEEE, New York, USA (2011)

6. Tie, L., Yi, Y.: Extended security analysis of multi-hop ticket based handover authentication protocol in the 802.16 j network. In: 8th International Conference on IEEE Wireless Communications, Networking and Mobile Computing (WiCOM), pp. 1–10. Shanghai, China (2012)

7. Yang, X., Huang, X., Han, J., Su, C.: Improved handover authentication and key pre-distribution for wireless mesh networks. Concurrency Comput.: Pract. Exp. **28**(10), 2978–2990 (2016)

8. Du, W., Deng, J., Han, Y.S., Varshney, P.K., Katz, J., Khalili, A.: A pairwise key predistribution scheme for wireless sensor networks. ACM Trans. Inf. Syst. Secur. (TISSEC) **8**(2), 228–258 (2005)

9. Fu, A., Zhang, Y., Zhu, Z., Liu, X.: A fast handover authentication mechanism based on ticket for IEEE 802.16m. IEEE Commun. Lett. **14**(12), 1134–1136 (2010)

10. Kohl, J., Neuman, C.: The Kerberos network authentication service (V5). No. RFC 1510 (1993)

11. Li, G., Chen, X., Ma, J.: A ticket-based re-authentication scheme for fast handover in wireless local area networks. In: 6th International Conference on Wireless Communications Networking and Mobile Computing (WiCOM), pp. 1–4. IEEE, Chengdu, China (2010)

12. Li, G., Ma, J., Jiang, Q., Chen, X.: A novel re-authentication scheme based on tickets in wireless local area networks. J. Parallel Distrib. Comput. **71**(7), 906–914 (2011)

13. Blom, R.: An optimal class of symmetric key generation systems. In: Workshop on the Theory and Application of Cryptographic Techniques, pp. 335–338. Springer, Berlin, Heidelberg (1984)

14. Khedr, W.I., Abdalla, M.I., Elsheikh, A.A.: Enhanced inter-access service network handover authentication scheme for IEEE 802.16m network. IET Inf. Secur. **9**(6), 334–343 (2015)

15. Chaudhry, S.A., Farash, M.S., Naqvi, H., Islam, S.H., Shon, T.: A robust and efficient privacy aware handover authentication scheme for wireless networks. Wireless Pers. Commun. **93**(2), 311–335 (2017)

16. Wang, K., Wang, Y., Zeng, D., Guo, S.: An SDN-based architecture for next-generation wireless networks. IEEE Wirel. Commun. **24**(1), 25–31 (2017)

17. Srivatsa, A.M., Xie, J.: A performance study of mobile handoff delay in IEEE 802.11-based wireless mesh networks. In: 2008 IEEE International Conference on Communications, pp. 2485–2489. IEEE, Beijing, China (2008)

18. Li, C., Nguyen, U.T., Nguyen, H.L., Huda, N.: Efficient authentication for fast handover in wireless mesh networks. Comput. Secur. **37**, 124–142 (2013)

19. Xu, L., He, Y., Chen, X., Huang, X.: Ticket-based handoff authentication for wireless mesh networks. Comput. Netw. **73**, 185–194 (2014)

20. Pirzada, A.A., McDonald, C.: Kerberos assisted authentication in mobile ad-hoc networks. In: Proceedings of the 27th Australasian conference on Computer science, vol. 26, pp. 41–46. Australian Computer Society, Dunedin, New Zealand (2004)

21. Kassab, M., Bonnin, J.M., Guillouard, K.: Securing fast handover in WLANs: A ticket based proactive authentication scheme. In: Globecom Workshops, pp. 1–6. IEEE, New York (2007)

22. Qazi, S., Mu, Y., Susilo, W.: Securing wireless mesh networks with ticket-based authentication. In: 2008 2nd International Conference on Signal Processing and Communication Systems, pp. 1–10. IEEE, Australia (2008)

23. Fu, A., Zhang, G., Zhang, Y., Zhu, Z.: GHAP: An efficient group-based handover authentication mechanism for IEEE 802.16m networks. Wireless Pers. Commun. **70**(4), 1793–1810 (2013)

An Efficient Distributed Approach to Construct a Minimum Spanning Tree in Cognitive Radio Network

Deepak Rohilla, Mahendra Kumar Murmu and Shashidhar Kulkarni

Abstract The increasing interest in cognitive radio ad hoc network (CRAHN) has driven study and development for latest approaches. The minimum spanning trees are advantageous for data broadcasting and disseminating. This article presents an associate in nursing algorithm rule for the construction of minimum spanning tree (MST) in cognitive radio network (CRN). For communication network, MST square measure is used for important network tasks like broadcast, leader election, and synchronization. We tend to be commenced our message and time restriction-based cost or weight estimation efficient distributed algorithm for construction of a minimum spanning tree. Proposed algorithm describes including facilitates of state diagram illustration. The verification demonstration of the proposed algorithm is additionally enclosed.

Keywords Distributed algorithm · Cognitive radio networks · Minimum spanning tree

1 Introduction to CRN

Cognitive radio networks (CRNs) are a rising area of research in the wireless communication environment. A cognitive radio is a creative communication system that senses its around and accommodate its internal seats for the statistical change within incoming radio frequencies signals by building comparable changes in a conversed operating parameter in a real time.

The major functionalities of the cognitive radio networks are:

D. Rohilla · M. K. Murmu (✉) · S. Kulkarni
Department of Computer Engineering, NIT Kurukshetra, Haryana 136119, India
e-mail: mkmurmunitkkr@gmail.com

D. Rohilla
e-mail: acerohilla1995@gmail.com

S. Kulkarni
e-mail: shashidhar.kulkarni@dell.com

© Springer Nature Singapore Pte Ltd. 2020
A. K. Luhach et al. (eds.), *First International Conference on Sustainable Technologies for Computational Intelligence*, Advances in Intelligent Systems and Computing 1045,
https://doi.org/10.1007/978-981-15-0029-9_31

1.1 Spectrum Sensing (SS)

Identify the un-used spectrum and sharing them whereas not destructive interference with completely different-different user. It is a vital demand of the psychological feature radio networks to sense the spectrum hole and the detection of primary users. The foremost productive approach to detect spectrum is to detect prime users in the environment. Spectrum sensing techniques are categorized into three.

Transmitter detection: psychological feature radio having an aptitude to see if a proof of a primary transmitter is domestically gifted in very sure spectrum, different approaches present:

- Matched Filter detection
- Energy detection
- Cyclo-stationary feature detection.

Cooperative detection: it tends to a spectrum sensing strategy wherever data from different psychological feature radio user are integrated for the detection of primary user.

Interference-based detection.

1.2 Spectrum Management (SM)

In CRN, the un-used spectrum band may send over wide frequency ranges having licensed and unlicensed band both. The unlicensed band detected through spectrum sensing. Since cognitive radio network decides the most effective spectrum band that satisfy the quality of services requirement over available spectrum band. Spectrum management function is much needed for cognitive radio networks.

1.3 Spectrum Mobility (SMO)

Cognitive radio network objective is to use the spectrums in a very dynamic manner to permit the CR to operate within the finest of available frequencies band. The concept of the optimum available frequency band is for communication purpose.

1.4 Spectrum Sharing (SSH)

Spectrum usage faces an open challenge in cognitive radio network, i.e., spectrum sharing, sensing of spectrum, allocation of spectrum, transmitter–receiver handshake, access of spectrum, and the mobility of spectrum are the major steps for the process of spectrum sharing.

The nodes present in a cognitive radio network (CRN) are classified into two components: licensed user or primary user or PU and unlicensed users (secondary users or SU). The primary user node holds a license of a spectrum, the secondary users always cooperates the environment while came in influence on other using a white hole or free spectrum of primary users. The network formed by the secondary user (SU) node is temporarily available.

In the cognitive radio network, secondary user node is set up with the following characteristics: reliability, intelligence, learning, efficiency, and robustness for working with the heterogeneous channels. The communication channels are available to the SU node depending on the activity of the primary user in environment. Channels detected by every secondary user node are known as LCS or local channel set, and at each position of secondary user node, the availability of the local channel sets are different. In LCS, a few channels could have selected for the connection of secondary user nodes. The set of these channel set is termed as common control channel (CCC). Therefore, CCC is the subset of LCS.

The common control channel creates a link (logical link) with secondary node in the cognitive radio networks, and by using logical link two SU can interact with each other; hence, sometimes logical link also called interaction graph (or edge). For the creation of a MST, a collection of edges required, still a challenge in a cognitive radio network. Minimum weight spanning tree is a graph that connects all secondary user node (or vertex) in such a way that obtained tree is acyclic in the environment.

2 The Minimum Spanning Tree Construction Problems in CRN

There have a few distinct threats faced while designing distributed algorithm for the construction of minimum spanning tree in a cognitive radio network. The threats are:

2.1 Multiple Channel Access

SU node or unlicensed node may or may not work with single or multiple channels. In any time, unlicensed node coordinate with only one channel known as common control channel. For the construction of MST in the cognitive radio network, the channel must be available for a sufficiently long time to the unlicensed nodes for communication.

2.2   Common Control Channel

To find the neighbors node in the network environment, the common control channel is taken into the action and edge is from among the secondary user. If multiple access channels are available between the nodes while the construction of a minimum spanning tree in CRN, lowest cost associated channel has the lowest priority.

2.3   Mobility

The cognitive radio network may be divided into two parts in terms of mobility (a) the spectrum mobility and (b) the node mobility. At any point of time, SU nodes can join or may leave the networks like an ad hoc manner for having the property of mobility. PU interference in the network should affect the spectrum mobility.

2.4   Primary User Interface

It seems difficult to forecast the activities of a primary user in a cognitive radio network. The afflicted areas of spectrum are not considered as the part of a MST in cognitive radio network.

2.5   Neighbor Discovery

For the construction of a spanning tree in a cognitive radio network, neighbor node is appreciable if connected with a minimum edge. From the available channels, it is an open challenge to select a unique channel with minimum cost edges for the construction of a tree which is should be minimum.

3   Motivation of MST Creation Approach and Discussion of Proposed Basic Idea

The algorithms for the construction of MST are strongly analyzed in the asynchronous model of mobile ad hoc networks (MANETs). A famous distributed algorithm for the construction of minimum spanning tree for the asynchronous model was proposed by authors of [1]. When it achieves the ability, level grows up and it switches to the next level for each iteration. For the growing of spanning tree fragment identity is used, fragment identity is the user. To switch from present level to next level, more waiting time is required which is a highly disadvantageous condition for the cognitive radio network. Further, the proposed work in [2] is efficient for the construction of a MST for the asynchronous networks. Network considered as fully connected and we are assuming that not necessary for achieving ability on every level complexity is reduced. For the improvement of a minimum spanning tree, root id is used. The proposed work in [3, 4] describes a distributed algorithm having a multiprocessor co-operation while mobility of nodes excepted in the network. Further, a simple linear time algorithm [5] is proposed which is difficult to use in cognitive radio environment directly while having different-different working characteristics. Because of functionality differences between CRN and mobile ad hoc network, algorithm for construction of the distributed minimum spanning tree in MANETs is not directly applicable to cognitive radio network. Cognitive radio networks had unique challenges in access capabilities for singlechannel or multichannel. The distributed minimum spanning tree algorithm (ad hoc network) required some modification to make it flexible in CRN. The agreement of this paper is determined with [6] due to different reasons that are as follows.

- For the construction of a minimum spanning tree in CRN, participation between the CR nodes is much required.
- The cognitive radio node can sense any channel in environment, which should be a part of a minimum spanning tree edges.
- The SU node is unlicensed node and cognitive radio network is the opportunistic type, so balance growth is not reasonable for CRN.

4 Organization of Model

Let us assume n secondary user nodes and k channels are present in cognitive radio network. In the communication weighted graph G having $(V*, E*, LCS)$, $V*$ represent the total number of connected node, $E*$ represents the total number of connected edges, and LCS represents the total number of channel sensed by each node. Each edge associated with unique cost. The connected node shows the relation graph between nodes and edges, known as fragment. In the network model, N number of vertices connected with $N - 1$ edges. Radio node sensed the channel called global channel set (GCS), while LCS belongs to GCS. SU nodes are connected through a link called an edge, which are used to build the sub-graph of the network. Sub tree should be represented by a unique identity called root id by the root process. Between any SU nodes, there may exist more than one CCC and one CCC should be associated with the nodes in MST at any time. While, there can exists more than one link between two SU or unlicensed node, but there one unique path exists from source node to destination node and final set of fragments is known as a minimum spanning tree of a cognitive radio network. In this proposed algorithm, we neglect the activities of primary user. We have used different state of nodes. The explanations are as following.

4.1 States of Root Process

4.1.1 Active_State

The root process observes to every LCS and outgoing edges to find the new sub tree.

4.1.2 Ready_State

After the confirmation of request message from all neighbors, root process change their state to ready_state, in order to participate in the graph.

4.1.3 Wait_State

If root process receives the reject message form the neighbor's node then it resides in wait_state with the same channel.

4.1.4 Lock_State

If root process receives the accept message then it resides in lock_state.

4.2 Messages Types

4.2.1 Request_Message (ID, LCS, RTT)

Root process sends this message to all neighbor's node to create the relation graph.

4.2.2 Reply_Message (ID, LCS, RTT)

Receiver used this message to allocate the sub-tree identification.

4.2.3 Update_Message (ID, LCS)

This message used by the root process to all neighbor's node to update about relation graph.

4.2.4 Accept/Reject_Message (ID, Accept/Reject)

Receiver used this message for the acceptance or rejection of the update message and sends to the sender.

5 Concept of the Proposed Algorithm

Minimum spanning tree is created by the collection of relation graph and an undirected communication graph. In the minimum spanning tree, weights are distributed among many relation graphs. The relation graph contains secondary user node, edge, and local channel set, termed as the sub tree of minimum spanning tree. Root process should represent the sub tree uniquely. Root process can represent a sub tree or may represent simple node. Root process is identified by a unique root id. The group of sub trees are merged based on a relation order in increasing order to from next sub tree for the construction of a MST in a cognitive radio networks. This algorithm explained in detail using state diagram as shown in Fig. 1. In start, we assumed that all the root processes are in *active_state.* In active_state, root process of a sub tree sensed the LCS and send *request_message* to all neighbor's node and wait for the acknowledgement from the other root processes of the sub tree. If there is at least one CCC, then receiver node can receive the *request_message*. If the root processes of sub tree receive the request_message, then they reply to sender using *reply_message*. After receiving the reply_message, root process changes their state to ready_state and calculate the RTT (round-trip time) for each node. Now, root process compares their id with each node. If sender id is less than receiver node, RTT is minimum and if the destination node is present with only one of the two LCS then the CCC is taken from that LCS else if destination node is present in both LCS then take minimum cost and add to CCC.

Root process sends update_message to all neighbor's node.

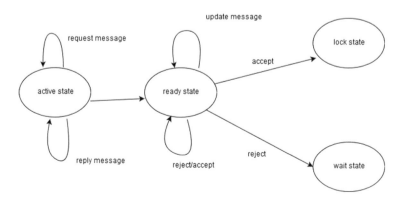

Fig. 1 State diagram of state of node

If the message is accepted by all neighbor's node then node sends connect_message to notify all others in the constructed tree about newly root else receiver send reject message using reject_message to sender. If there are no further outgoing edges, then the algorithm terminates. So, the process for the construction of MST is completed, and resultant tree is the finalized MST of the cognitive radio network. This finalize minimum spanning tree may be used for leader election, broadcasting, and many more in the cognitive radio network.

6 Proposed Algorithm

Steps for the creation of minimum spanning tree are as below:

1. Assume the root process of sub trees are in *active_state* and sensed LCS.
2. The root process sends a *request_message* (ID, LCS, RTT) to its neighbor's node.
3. Receiver node generates *reply_message* (ID, LCS, RTT) after receiving the *request_message* (ID, LCS, RTT).
4. After receiving the reply_message (ID, LCS, RTT), root process change its state from *active_state* to *ready_state* and calculate the RTT (round-trip time) of each neighbor's node.
5. Now, root process starts comparison with receiver, suppose the root (a) and root (b) are sender and receiver, respectively.

   5.1 If (ID root (a)) > (ID root (b)), RTT is minimum and if destination node is present in only one of the two LCS then the CCC takes from that LCS.
   5.2 Else, if destination node is present in both LCS, then take minimum cost and add to CCC.

6. Root process send *update_message* (ID, CCC) to all neighbor's node.
7. If the *update_message* is accepted by all neighbors, then root node sends *connect_message* to inform others in the constructed tree for new root.
8. Else receiver sends rejection message using *reject_message* to sender.
9. Repeat steps from 2.
10. If there are no further outgoing edges, then root process should terminate.

Proof 1 Cycle is not present in the constructed graph.

Proof Assuming a cycle present in the MST of CRN.

If a cycle is present in the MST, then there should be a path from a node to itself while not traversing any edges doubly. But, as the flow of the algorithmic rule is formed, graph should exists just unique path from a node to different node and node keep on passing a message to successive nodes, which indicates to their CCC. In the initial network graph, message might return to starting node via distinctive path.

Although, those edges once execution of algorithm are changed to wait_state. However, message cannot come back to the starting node unless it enters to a wait_state, that is not allowable for the message after the completion of algorithm. So, by contradiction, we have proven that there is no cycle present in proposed algorithm.

Proof 2 Terminated algorithm having n vertices cover $n - 1$ edges, final tree is a minimum spanning tree.

Proof For MST to be accurate, we are assuming the number of edges present in the network should equal to the number of node within the same network, that is, $n - 1$ edge states $n - 1$ node present within network. According to the definition of MST, we had got ascertained that $n - 1$ edges cover path between nodes. Progress of algorithm results in MST and a unique path may present between nodes. However, we may not have same number of edge and node unless cycle can present in MST of network while, in Proof 1 we had proved that a result to our algorithm, there should not present any cycle within network. So, with the help of contradiction, we have proved that n vertices for $n - 1$ edge present in network.

7 Implementation

We assume a CRN as shown in Fig. 2, where distinct cost is distributed to all secondary user nodes. SU1 is assumed to be a root process and executed all the steps of the proposed algorithm. SU1 send a request_message (ID, LCS, RTT) to their neighbor's nodes SU2 and SU3. The request_message received by neighbor's nodes (SU2 and SU3) and a reply_message (ID, LCS, RTT) is generated to sender. Now, root process (SU1) calculates the round-trip time of each neighbor's nodes and starts the comparison. The minimum round-trip time of the neighbor's node is selected and checked whether the node is present in only one of the two LCSs. If node present in

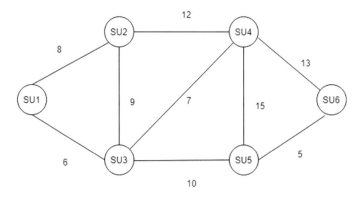

Fig. 2 CRN

only one of the two LCSs, then it is selected as the part of minimum spanning tree. SU3 have the minimum round-trip time so it is selected as the part of the minimum spanning tree. We mark the channel between SU1 and SU3 as red.

Now, the proposed algorithm is executed to both SU2 and SU3 simultaneously and the channel between the SU1 and SU2, SU1 and SU3, SU3 and SU4, and SU3 and SU5 selected as the part of minimum spanning tree and mark red as shown in Fig. 3. The channel of SU2 and SU3 not selected because of the formation of cycle. Repeat the same procedure on SU4 and SU5.

The channel between SU5 and SU6 selected and mark as red. Channel SU4 and SU5 and SU4 and SU6 are not selected because the round-trip time is high, node is present in both two LCS and it should form the cycle in the network. Now, the proposed algorithm executed on SU6, but it does not make any modification in the network (Fig. 4).

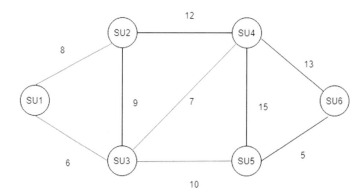

Fig. 3 .

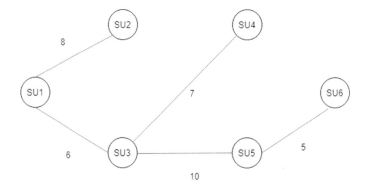

Fig. 4 Resultant network is formed as the minimum spanning tree of CRN

8 Conclusion

We had proposed an efficient distributed algorithm having a mechanism of passing a message for the construction of a minimum spanning tree in CRN. The proof of algorithm is also included. Our future plan is to simulate this proposed algorithm in network simulator. Our proposed efficient MST algorithm is straightforward and distributed in nature for a cognitive radio network.

References

1. Murmu, M.K., Firoz, A.M., Meena, S., Jain, S.: A distributed minimum spanning tree for cognitive radio networks. IMCIP Proc. Comput. Sci. **89**, 162–169 (2016). (Elsevier)
2. Singh, G., Kumar, N., Verma, A.K.: Ant colony algorithms in MANETs: A review. J. Netw. Comput. Appl. **6**, 1964–1972 (2012). (Elsevier)
3. Sun, X., Chang, C., Su, H., Rong, C.: Novel degree constrained minimum spanning tree algorithm based on an improved multicolony ant algorithm. Math. Probl. Eng. Article ID 601782 (2015) (Hindwai Publishing)
4. Ibanez, M.L., Stutzle, T., Dorigo, M.: Ant colony optimization: A component-wise overview. IRIDIA Technical Report Series (2015)
5. Qiang, H.Z., Kai, N., Tao, Q., Tao, S., Jun, X.W., Li, G., Ru, L.J.: A bio-inspired approach for cognitive radio networks. Int. J. Chin. Sci. Bull. Theor. Wirel. Networks **57**(28), 3723–3730 (2012). (Springer)
6. Song, Z., Shen, B., Zhou, Z.: Improved ant routing algorithm in cognitive radio networks. In: IEEE Conference (2009)
7. Alam, S.S., Marcenaro, L., Regazzoni, C.: Opportunistic spectrum sensing and transmissions. In: Cognitive Radio and Interference Management: Technology and Strategy: Technology and Strategy (Chap. 1) (2012)
8. Ducatelle, F., Caro, G.D., Gambardella, L.M.: Using ant agents to combine reactive and proactive strategies for routing in mobile ad hoc networks. J. Comp. Intel. Appl. **5**(169), 1–15 (2005). (Researchgate)
9. Fernandez-Marquez, J.L., Serugendo, G.D.M., Montagna, S.: BIO-CORE: Bio-inspired self-organising mechanisms core. In: Bio-Inspired Models of Networks, Information, and Computing Systems on LNICST, vol. 103, pp. 59–72 (2012)
10. Dorigo, M., Di Caro, G., Gambardella, L.M.: Ant algorithms for discrete optimization. IRIDIA Artif. Life **5**(2), 137–172 (1999) (MIT Library)
11. Mao, X., Ji, H.: Biologically-inspire distributed spectrum access for cognitive radio network. In: IEEE Conference. Beijing University of Posts and Telecommunications (2010)
12. Qiang, H.Z., Kai, N., Tao, Q., Tao, S., Jun, X.W., Li, G., Ru, L.J.: A bio-inspired approach for cognitive radio networks. Int. J. Chin. Sci. Bull. Theor. Wireless Netw. **57**(28)

Path Planning of Unmanned Aerial Vehicles: Current State and Future Challenges

Aditi Zear and Virender Ranga

Abstract Path planning is the preliminary requirement of unmanned aerial vehicles (UAVs) for their autonomous functions. This paper discusses the significant usage of UAVs in distinct applications and the need for path planning in order to increase their service rate in different applications. UAV's path planning can be either start to goal or coverage path planning. Path planning techniques can be generally categorized as roadmap/skeleton, approximate/exact cell decomposition, potential field, sampling-based, and bio-inspired/machine learning-based methods. These methods are briefly discussed in this paper. Finally, the present state of the art of UAV's path planning using these techniques is discussed. This paper will be a seed source for the researchers who are actively working on UAVs to implement efficient path planning techniques according to their usability in different applications.

Keywords Multi-UAV · Optimal path planning · Collision avoidance · Adhoc networks

1 Introduction

Recently due to expeditious growth in robotics, unmanned aerial vehicles (UAVs) are observed as another aviation technology in the domain of networking [1–3]. UAVs are gaining recognition because of their numerous advantages such as affordable development and maintenance cost, reduced operational and structural complexities, flexible deployment, reproducible architecture and less constrained flight [4, 5]. UAVs are categorized into three main groups which are rotary-wing, fixed-wing, and hybrid layout UAVs [6–8]. Fixed-wing UAVs have advantages such as high speed and capability to fly for longer distances. In spite of these benefits, they require mechan-

A. Zear (✉) · V. Ranga
Department of Computer Engineering, NIT Kurukshetra, Haryana, India
e-mail: aditizear93@gmail.com

V. Ranga
e-mail: virender.ranga@nitkkr.ac.in

© Springer Nature Singapore Pte Ltd. 2020
A. K. Luhach et al. (eds.), *First International Conference on Sustainable Technologies for Computational Intelligence*, Advances in Intelligent Systems and Computing 1045,
https://doi.org/10.1007/978-981-15-0029-9_32

ical systems, e.g., landing gears for vertical takeoff and landing (VTOL) [8–10]. More applications use rotary-wing UAVs because of their better maneuverability with VTOL ability. Hybrid layout UAVs have combined properties of fixed-wing and rotary-wing UAVs. However, they have complicated mechanisms during the flight in order to change from rotary-wing to fixed-wing [1, 10–14]. Unmanned aerial vehicles contribute tremendous benefits in different applications such as area surveillance, post-disaster management, aerial photography, delivering packages, managing agriculture, weather observation, target identification and destruction in military, forest fire detection, planetary exploration, reconnaissance. The utilization of UAVs in different applications will be improved significantly if they can operate autonomously through an environment different from the human operator's envision [13–15]. UAV's motion planning is one of the considerable areas where such autonomy could benefit from different applications [16]. Generally, the task of motion planning includes two levels: path planning and trajectory planning. Path planning generates path or curve needs to be traced by vehicle in its configuration based on environment description and start–goal points as inputs [16–19]. The trajectory consists of time-parameterized equation of motion depends on dynamic attributes of mobile agent (UAV or vehicle) like its position, velocity, and acceleration [18–20]. In accordance with different applications, majority of path planning methods are categorized into coverage path planning (CPP) and start to goal path planning (SGPP) [21–25]. The start to goal path planning determines path between fixed start to goal points together with avoiding obstacles [24–26]. Mostly proposed methods that consider start to goal path planning problem are random sampling search algorithms, probabilistic roadmap, and bio-inspired algorithms [26–29]. In CPP, a path that crosses over all the points of region of interest is determined in addition to bypassing obstacles. Maximum CPP algorithms reorganize the target region into subregions or cells in order to simplify the path planning [24–28]. The classification of CPP approaches can be done in accordance with exploited decomposition methods [16–20]. Various reported CPP algorithms are trapezoidal decomposition, visibility graph, topological coverage algorithm, etc.

1.1 Path Planning Techniques

There are various path planning approaches reported in the literature, However, the known path planning approaches can be generally divided into five categories: roadmap/skeleton, approximate/exact cell decomposition, potential field, sampling-based planning, and bio-inspired/machine learning-based methods [19–22]. Figure 1 provides the summarized details of path planning techniques reported under these categories [1] and a brief discussion about these path planning approaches is given below:

1. **Roadmap/Skeleton method**: In roadmap/skeleton-based methods, the path planning problem is reduced into the graph search problem by fitting graph

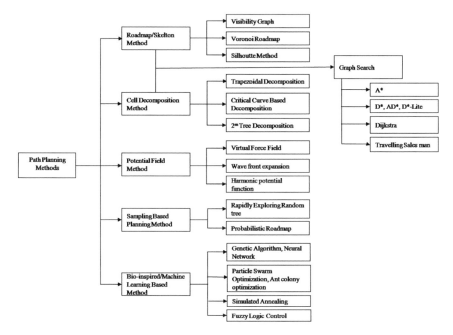

Fig. 1 Path planning techniques

also called roadmap/skeleton to space [7–11]. Here the free space is represented in the form of one-dimensional path. The actual path can be identified by moving flying robot on points on the roadmap from start to goal configuration and connecting the points as paths on roadmap. The algorithm formed on roadmap should follow some topographical properties in order to be complete. Various well-known roadmap approaches are the visibility graph, Voronoi roadmap, silhouette method [23–27].

2. **Cell Decomposition Method**: In cell decomposition approach planning, surface is depicted in the form of union of cells or fixed grid. These are further modeled using undirected graph which usually encompasses the solution path. Trapezoidal decomposition, 2^m tree decomposition (quadtree or octtree representation), and critical curve-based decomposition are some familiar cell decomposition approaches [22–27].

Search algorithms such as traveling salesman A*, D*, Dijkstra, D*-Lite, and AD* are used in various previously reported path planning techniques along with roadmap/skeleton and cell decomposition methods [15–19].

3. **Potential Field Method**: Potential field method works by assigning a potential function to the workspace. The mobile agent simulates by responding to the forces because of potential fields. The flying UAV gets attracted by goal configuration and at the same time obstacles and other uncertainties repel the flying robot. Potential field, including gradient descent also called virtual force field,

harmonic potential function, and wavefront expansion are famous potential field methods [23–29].

4. **Sampling-Based Planning Method**: In sampling-based approach, path planning is done by using sampling configuration also called C-space and try to acquire connectivity by this sampling of C-space [20–24] . It can be considered as black box where the feasible collision-free path is returned between start and goal state. The global plans are specified using high-level planner and low-level controller plans the path execution. This approach has advantage of providing fast suboptimal solutions for complex problems. Probabilistic roadmap (PRM) and rapidly exploring random tree (RRT) are famous sampling-based planning methods [29].

5. **Bio-inspired/Machine Learning-based Methods**: Different nature-inspired algorithms can excellently manage complex unstructured as well as NP-hard problems [20–24]. These techniques can be applied in an unknown environment and knowledge about the environment can be acquired by the flying UAV while navigating. Various previously reported machine learning and bio-inspired-based approaches used for path planning are genetic algorithm (GA), neural network (NN), particle swarm optimization (PSO) [7–12], simulated annealing, ant colony optimization (ACO), fuzzy logic control etc. [24–29].

These machine learning techniques can also be combined with above four path planning techniques in previously communicated research papers in order to have better results. These hybrid algorithms can be used to overcome the drawbacks of individual algorithm.

2 Literature Review

Review of previously reported start to goal and coverage path planning techniques is presented below. Table 1 summarizes these reported path planning techniques and some details about their objectives, environment (Envir), optimization parameters, algorithms, and performance measures.

2.1 *Start to Goal Path Planning*

Cheng et al. [10] proposed dynamic path planning strategy for UAV using improved PSO by dynamically changing the inertia weights results in order to improve local and global search ability of PSO. However, it can still fall into local optimum. Therefore, chaos optimization search theory is combined with PSO to improve its search performance. Kalman filtering algorithm is used to estimate the track of moving obstacles. These combined algorithms effectively perform UAV real-time path planning and provide superior results as compared to immune PSO and standard PSO algorithm. Liu et al. [16] presented scheme for local collision avoidance in combi-

Table 1 Analysis of path planning techniques

References	Type	Objectives	Algorithm	Optimization	Environment	Performance measure
Sánchez-García et al. [7]	Coverage	Trajectory generation to explore disaster scenario area, converge UAVs to victim groups	d-PSO-U	Maximize coverage and minimize convergence time	3D	Percentage victims discovered, percentage of UAV's coverage, UAV's convergence time, No. of connections and time of connection between node and UAV, comparison with lawnmower algorithm
Loscri et al. [9]	Coverage	Coverage of dynamic events in zones of interest	VFA-C, PSO-S and VFA-D	Minimum travel distance and maximum node coverage	2D	Distance travelled, coverage, No. of iterations, reactivity, comparison with VFA-c
Cheng et al. [10]	Start to goal	UAV path planning, collision avoidance	Improved PSO, chaos optimization, Kalman filtering	Alleviate curse of dimensionality, maximize local and global search ability of PSO	2D	Collision avoidance at different angles between UAV and moving obstacle
Yoon et al. [12]	Coverage	Path planning with traversal time and delivery deadline constraints, delivering delay sensitive information at disaster scenario areas	Distributed path planning using grid points, Task Naïve, K-means and balanced K-means clustering for task division	Maximize No. of relevant nodes to deliver data and minimize travel time	2D	Percentage of nodes successfully receive data within deadline, comparison with random, nearest and mTSP-GA, Percentage of grid points covered by UAVs with different clustering algorithms
Xiao et al. [13]	Start to goal and coverage	Path planning for complicated scenarios including obstacles, combination of start to goal and coverage path planning	Virtual regional field	Maximize coverage ratio and by pass obstacles	2D	Coverage rate in particular time duration, obstacle avoidance
Liu et al. [16]	Start to goal	Global path planning with collision avoidance, consideration of aviation traffic rules for decision making, height and altitude control system design	RRT, fuzzy logic, EKF, PID control	Minimize cross track error and bypass obstacles	2D	Computational complexity, obstacle avoidance at different coordinates, capability of cross track error prediction

(continued)

Table 1 (continued)

References	Type	Objectives	Algorithm	Optimization	Environment	Performance measure
Salamat et al. [17]	Start to goal	Real-time trajectory generation with obstacle avoidance	Quintic B-splines, PSO	Optimization of violation function with obstacle avoidance, Euler angle, torque, position, velocity, acceleration constraints	2D	Execution time, path length, Comparison with A*, RRT*, GA in terms of execution time and path length
Altmann et al. [18]	Start to goal	Path planning along with obstacle avoidance	Cell decomposition, 3D Field D*, linear interpolation	Minimize turning radius, flight altitude and maximize flight path angle	3D	Computational time, cruising altitude. No. of node expansions in the initial planning phase and node expansion in re-planning phase and post-processing phase, path length
Copot et al. [19]	Start to goal	Path planning along with collision avoidance	Visibility graph, A*	Minimize path length	2D	Path length, obstacle avoidance
Wang et al. [21]	Coverage	Trajectory planning for efficient data collection, energy conservation	Cell decomposition, MIP-OTP	Minimize path length and maximize packet receiving ratio	2D	Comparison with TSP solution, path length, No. of sensor nodes covered, packet receiving ratio
Arantes et al. [22]	Start to goal	Path planning with uncertainties	Multi-population GA, visibility graph	Minimize computational time and avoid obstacles	2D	Comparison with CPLEX and CSA, computational time
Roberge et al. [23]	Start to goal	Trajectory generation in 3D complex environment, comparison of GA and PSO	GA, PSO, cell decomposition, parallel implementation	Minimize computation time, travel distance, Altitude	3D and 2D	Computation time, path length
Pehlivanoglu et al. [24]	Start to goal	Autonomous path planning	Vibrational GA, terrain models	Minimize path length, turn angle and execution time	3D	Execution time, path length, turn angle
Kothari et al. [26]	Start to goal	Trajectory generation in dynamic environment, collision avoidance	CC-RRT + Heuristics (Exploration and optimization)	Minimize exploration time and path cost during tree expansion	2D	Path length, computational time
Perez-Carabaza et al. [28]	Coverage	Trajectory generation for lost target	ACO, Node and TIME encodings, Adhoc heuristics	Minimize search time	3D	Computation time, comparison with other optimization method

nation with global path planning. The global path planning is done using RRT. The local collision avoidance mechanism using fuzzy logic modifies planned global path in case of any obstacles. Extended Kalman filter (EKF) is employed for cross-track error estimation. The proportional integral derivative (PID) control method is added to develop fundamental altitude controller to stabilize the UAV's altitude. PID in combination with fuzzy logic control (FLC) method regulates baseline controller to enhance the working capabilities of UAV in different conditions. Simulation studies show that this scheme is effective in both static and dynamic environments. The method proposed in [17] aimed to generate realistic stochastic trajectories for UAVs. Quintic B-splines are implemented for trajectory generation.

The B-splines provide smooth trajectories by using second-order derivative of acceleration and Euler angles. PSO with multiple objective optimization function is used for the parameter adjustment of B-splines. This approach optimizes the violation function with obstacle avoidance, Euler angle, torque, position, velocity, and acceleration as constraints. Simulation results show that PSO generates minimum length trajectory as compared to GA, RRT*, and A* algorithm. Altmann et al. [18] applied heuristic incremental interpolation-based search approach for fixed-wing UAV operation in large environment. The path planning approach is based on 3D Field D*. The algorithm is being modified to account limited path angles of fixed-wing aircraft and for minimum turning radius. Inclusive of this a method is proposed to control the selected minimum cruising altitude. Furthermore, post-processing algorithm is used for path improvement and removes irrelevant path points. Simulation results demonstrate that the algorithm produces shorter paths. Copot et al. [19] aimed on generating collision-free trajectory for mobile agent between start and goal state. A* algorithm along with extra heuristic function is used for shorter trajectories and for faster convergence to solution. The optimal path is obtained by considering two components of multiple objective function: the safer path and shortest path. Here grid graph is preferred, and experimental platform is designed to simulate quadrotor motion in virtual environment. Experimental results show that this algorithm can generate possible paths that can be implemented successfully in real life. Arantes et al. [22] proposed a hybrid path planning method in non-convex regions with uncertainties for unmanned aerial vehicles. In this hybrid path planning method, visibility graph is combined with multi-population GA. The performance of hybrid genetic algorithm (HGA) is tested for 50 maps. The computational results of HGA algorithm are compared with heuristic customized solution approach (CSA) and exact CPLEX methods. The method proposed in [23] worked on two deterministic algorithms, i.e., PSO and GA, for fixed-wing UAV's path planning considering its dynamic properties in 3D environment. The paths created consists of circular arcs, vertical helices, line segments, and their characteristics are represented using multi-objective cost function developed in this approach. The execution time is reduced by using parallel programming "single program multiple data" paradigm. Simulation results show that by using parallel implementation apparently a linear speedup of 7.3 is achieved on 8 cores and both algorithms have 10 s execution time. The comparison between these two algorithms shows that by using same encoding technique GA produces better trajectories than PSO. Pehlivanoglu et al. [24] emphasized on creating algorithm for effective path planning using autonomous UAV. Bezier curves are used for representing smooth and global continuous path based on the basis of control points coordinates. These Bezier curves are coupled with vibrational genetic algorithm (VGA), and the main emphasis is on the vibrational mutation (VM) operator to efficiently escape from local optimum in order to obtain global optimum. The proposed approach is tested in various terrains such as Bezier surface, spatial database-based, and trigonometric terrain model. Simulation results describe the efficiency of VM technique for low population rates. It also constructs feasible smooth optimal paths in less execution time. In [26], the method developed a real-time path planning strategy for single UAV system functioning in an unpredictable environment. In this strategy, chance constraint formulation is incorporated with RRT to discover probabilistically robust paths in the uncertain environments. The chance-constrained formulation along with number of heuristics is used manage unpredictable dynamic obstacles. The relevance of generated trajectories is determined using chance constraints at every time step. The proposed algorithm provides quality solution as the mission progresses by avoiding unpredictable obstacles on known paths.

2.2   Coverage Path Planning

Sánchez-García et al. [7] worked on deploying UAV network autonomously to provide communication in disaster scenario. A variant of well-known PSO algorithm called d-PSO is used for such task. The second main goal deals with converging UAVs to several victims groups that are being discovered during exploration. Nodes in the UAV network follow delay-tolerant network (DTN) approach. The proposed algorithm is simulated for different set of inertia values, local and neighbor best to identify the finest PSO parameters combination. The algorithm discovers maximum victims when compared with lawnmower algorithm. However d-PSO requires extra time to sweep the scenario area. This approach is faster in discovering 25, 50, and 75% of victims than lawnmower algorithm and faster connections to victim groups. The method discussed in [9] focused on covering specific regions of interest that changes dynamically. The mobility of UAV swarms is achieved by using two different algorithms. The first algorithm PSO-serialized (PSO-S) is the combination of local PSO and distributed VFA (VFA-D) in which flying UAV's update their velocity and trajectory based on local information they receive from their neighbors. The second algorithm uses VFA-D to cover zones of interest. The proposed algorithms performed better as compared to VFA-C in terms of reactivity, travelled distance, and coverage. The method in [12] considered the path planning problem with the help of multiple UAVs for delivering delay-sensitive data in disaster scenario. The main focus is to generate optimal paths for multi-UAVs to service maximum nodes within certain packet deadline and reduce the travel time for visiting over virtual grid topology. Task division mechanism is employed in this approach to distribute unvisited virtual grid points among UAVs. Simulation studies show that mixer of path planning with task division algorithm provides better node service rate within deadline and total travel time in comparison with baseline counterpart algorithms.

Wang et al. [21] investigated to find the shortest possible trajectory and also ensure communication constraints for wireless sensor nodes (motes). This problem is considered as mixed integer programming (MIP) problem which is NP-hard problem, and it is impractical to identify its exact solution. Therefore, heuristic algorithm to deal with OTP problem is proposed in which the coordinates of motes and the packet size are considered. The trajectory is generated by connecting waypoints from the waypoints set. The waypoints set are obtained using four major steps: initialization, rotation, optimization, and smooth. Simulation studies demonstrate that this trajectory planning approach for OTP problem generates trajectories with reduced path length along with satisfying communication constraints of motes. Carabaza et al. [28] proposed a new scheme using ACO to reduce the search time of UAV swarms in order to obtain high-level superior straight-segmented trajectories. These high-quality trajectories can be achieved by combining the information from the pheromone trails left by previous better trajectories and benefits of new heuristics especially created for minimum time search (MTS) problem. Two variants of this scheme aided by NODE and TIME encodings for pheromone table are presented in this paper. The first variant resembles the original ACO, and the second variant is similar to the codification employed in alternative search techniques. The capabilities of the approach are analyzed using six search scenarios and also compared with various existing MTS solutions statistically by using mean computation time. The comparative analysis of this new approach with other search algorithms shows that it is quicker than other approaches and obtains similar results in lesser computational time. In [13], the method focused on combining SGPP and CPP problem for complicated scenarios including obstacles. The algorithm specializes in CPP problem but if the target region is single, then it is converted to start to goal problem. The algorithm is based on updating virtual regional field and its gradients continuously. Simulation studies are performed on real data that shows that the algorithm successfully avoids obstacles. It provides high coverage rate in given time duration for target regions. This regional field-based approach provides options for cheaper hardware.

3 Conclusion

This paper discussed UAVs as other aviation technology gaining interest in various applications and the requirement of UAV's path planning in order to increase its usage in these applications. Various path planning techniques are briefly explained in this paper. Further, the applicability of these techniques in previous communicated research papers of path planning is discussed in tabular format. After surveying different proposed path planning techniques, following points are concluded by us (1) In some occasions, intelligent learning-based algorithms yield poor solutions. Thus, the algorithm would require subsequent re-execution. Therefore, the path planner sub-module of UAV control system should include such decision-making module in order to handle such cases. (2) Path planning can be made more realistic by also considering other objectives such as bypassing forbidden areas, energy consumption, moving targets along with reaching goal positions while avoiding uncertainties.

Acknowledgements This research work is being supported by SERB-DST, Government of India.

References

1. Goerzen, C., Kong, Z., Mettler, B.: A survey of motion planning algorithms from the perspective of autonomous UAV guidance. J. Intell. Rob. Syst. **57**(1–4), 65 (2010)
2. Goddemeier, N., Daniel, K., Wietfeld, C.: Role-based connectivity management with realistic air-to-ground channels for cooperative UAVs. IEEE J. Sel. Areas Commun. **30**(5), 951–963 (2012)
3. Yang, L., Qi, J., Song, D., Xiao, J., Han, J., Xia, Y.: Survey of robot 3D path planning algorithms. J. Control Sci. Eng. **2016**, 5 (2016)
4. Debnath, S.S.K., Omar, R., Latip, N.B.A.: A review on energy efficient path planning algorithms for unmanned air vehicles. In: Computational Science and Technology, pp. 523–532 . Springer, Berlin (2019)
5. Jawhar, I., Mohamed, N., Al-Jaroodi, J., Agrawal, D.P., Zhang, S.: Communication and networking of uav-based systems: Classification and associated architectures. J. Netw. Comput. Appl. **84**, 93–108 (2017)
6. Bekmezci, I., Sahingoz, O.K., Temel, Ş.: Flying Ad-Hoc networks (fanets): A survey. Ad Hoc Netw. **11**(3), 1254–1270 (2013)
7. Sánchez-García, J., Reina, D., Toral, S.: A distributed PSO-based exploration algorithm for a UAV network assisting a disaster scenario. Future Gener. Comput. Syst. **90**, 129–148 (2019)
8. Tuna, G., Nefzi, B., Conte, G.: Unmanned aerial vehicle-aided communications system for disaster recovery. J. Netw. Comput. Appl. **41**, 27–36 (2014)
9. Loscri, V., Natalizio, E., Mitton, N.: Performance evaluation of novel distributed coverage techniques for swarms of flying robots. In: 2014 IEEE Wireless Communications and Networking Conference (WCNC), pp. 3278–3283. IEEE, New York (2014)
10. Cheng, Z., Wang, E., Tang, Y., Wang, Y.: Real-time path planning strategy for UAV based on improved particle swarm optimization. J. Comput. **9**(1), 209–215 (2014)
11. Li, J., Chen, J., Wang, P., Li, C.: Sensor-oriented path planning for multiregion surveillance with a single lightweight UAV SAR. Sensors **18**(2), 548 (2018)
12. Yoon, J., Jin, Y., Batsoyol, N., Lee, H.: Adaptive path planning of UAVs for delivering delay-sensitive information to Ad-Hoc nodes. In: 2017 IEEE Wireless Communications and Networking Conference (WCNC), pp. 1–6. IEEE, New York (2017)
13. Xiao, Z., Zhu, B., Wang, Y., Miao, P.: Low-complexity path planning algorithm for unmanned aerial vehicles in complicated scenarios. IEEE Access **6**, 57049–57055 (2018)
14. Ghorbel, M.B., Rodriguez-Duarte, D., Ghazzai, H., Hossain, M.J., Menouar, H.: Energy efficient data collection for wireless sensors using drones. In: 2018 IEEE 87th Vehicular Technology Conference (VTC Spring), pp. 1–5. IEEE, New York (2018)
15. Wang, Z., Li, Y., Li, W.: An approximation path planning algorithm for fixed-wing UAVs in stationary obstacle environment. In: Proceedings of the 33rd Chinese Control Conference, pp. 664–669. IEEE, New York (2014)
16. Liu, Z., Zhang, Y., Yuan, C., Ciarletta, L., Theilliol, D.: Collision avoidance and path following control of unmanned aerial vehicle in hazardous environment. J. Intell. Rob. Syst. 1–18 (2018)
17. Salamat, B., Tonello, A.: Stochastic trajectory generation using particle swarm optimization for quadrotor unmanned aerial vehicles (UAVs). Aerospace **4**(2), 27 (2017)
18. Altmann, A., Niendorf, M., Bednar, M., Reichel, R.: Improved 3D interpolation-based path planning for a fixed-wing unmanned aircraft. J. Intell. Rob. Syst. **76**(1), 185–197 (2014)

19. Copot, C., Hernandez, A., Mac, T.T., De Keyse, R.: Collision-free path planning in indoor environment using a quadrotor. In: 2016 21st International Conference on Methods and Models in Automation and Robotics (MMAR), pp. 351–356. IEEE, New York (2016)
20. Mengying, Z., Hua, W., Feng, C.: Online path planning algorithms for unmanned air vehicle. In: 2017 IEEE International Conference on Unmanned Systems (ICUS), pp. 116–119. IEEE, New York (2017)
21. Wang, Q., Chang, X.: The optimal trajectory planning for UAV in UAV-aided networks. In: International Conference on Cloud Computing and Security, pp. 192–204. Springer, Berlin (2016)
22. Arantes, M.S., Arantes, J.S., Toledo, C.F.M., Williams, B.C.: A hybrid multi-population genetic algorithm for UAV path planning. In: Proceedings of the Genetic and Evolutionary Computation Conference 2016, pp. 853–860. ACM, New York (2016)
23. Roberge, V., Tarbouchi, M., Labonté, G.: Comparison of parallel genetic algorithm and particle swarm optimization for real-time UAV path planning. IEEE Trans. Ind. Inform. **9**(1), 132–141 (2013)
24. Pehlivanoglu, Y.V., Baysal, O., Hacioglu, A.: Path planning for autonomous UAV via vibrational genetic algorithm. Aircraft Eng. Aerosp. Technol.: Int. J. **79**(4), 352–359 (2007)
25. Kroumov, V., Yu, J., Shibayama, K.: 3D path planning for mobile robots using simulated annealing neural network. Int. J. Innovative Comput. Inf. Control **6**(7), 2885–2899 (2010)
26. Kothari, M., Postlethwaite, I.: A probabilistically robust path planning algorithm for UAVs using rapidly-exploring random trees. J. Intell. Rob. Syst. **71**(2), 231–253 (2013)
27. Kim, S., Oh, H., Suk, J., Tsourdos, A.: Coordinated trajectory planning for efficient communication relay using multiple UAVs. Control Eng. Practice **29**, 42–49 (2014)
28. Perez-Carabaza, S., Besada-Portas, E., Lopez-Orozco, J.A., Jesus, M.: Ant colony optimization for multi-UAV minimum time search in uncertain domains. Appl. Soft Comput. **62**, 789–806 (2018)
29. Elbanhawi, M., Simic, M.: Sampling-based robot motion planning: A review. IEEE Access **2**, 56–77 (2014)

HCCD: Haar-Based Cascade Classifier for Crack Detection on a Propeller Blade

R. Saveeth and S. Uma Maheswari

Abstract Crack detection in aircraft components is an important assessment because even a small unnoticed crack tends to critical crack length. Aviation demands reliability, and therefore, periodical inspection of cracks in aircraft parts like engine turbine blade, aircraft skin, rivets, wing spar, bulk fuselage, and airwings has to be detected in a fixed interval, but it requires human effort and expert's knowledge. Features are extracted using extended Haar-like features and it has been given as input to cascade classifier to classify cracks and non-cracks images of a propeller blade. The supervised learning algorithm is developed and trained by a set of positive and negative images. The experimental results validate the test images by the cascading classifier to locate cracks.

Keywords Propeller blade · Crack detection · Cascading classifier · Haar-like features · Supervised learning

1 Introduction

A propeller blade in aircraft is one of the highly stressed components, consists of two or more blades connected by a hub. Thrust is a force generated by the propeller blade to move the aircraft through air. Centrifugal force acting on the hub which will pull the blades which tends to damage the propeller that will leads to occurrence of crack. The number of blades is indirectly proportional to workload acted on each blade to generate thrust which will be a main parameter to lift the aircraft. Dynamic

R. Saveeth (✉)
Department of Computer Science and Engineering, Coimbatore Institute of Technology,
Coimbatore, Tamil Nadu, India
e-mail: saveeth.r@cit.edu.in

S. Uma Maheswari
Department of Electronics and Communication Engineering,
Coimbatore Institute of Technology, Coimbatore, Tamil Nadu, India
e-mail: umamaheswari@cit.edu.in

© Springer Nature Singapore Pte Ltd. 2020
A. K. Luhach et al. (eds.), *First International Conference on Sustainable Technologies for Computational Intelligence*, Advances in Intelligent Systems and Computing 1045, https://doi.org/10.1007/978-981-15-0029-9_33

Fig. 1 Causes of cracks

force and moment act constantly on the propeller which cause nicks that leads to cracks on the propeller.

Figure 1 shows the factors which cause cracks are (i) corrosion, (ii) fatigue load, (iii) welding effects and (iv) physical and environmental factors. Corrosion is a significant factor which will loosen the structural integrity of the components. It occurs when the aircraft is exposed to bad weather conditions constantly. The area around fastener holes is easily prone to cracks due to heavy load applied during assembly; cracks at the edges of the wing found to be significant than around fastener holes. The necessity of crack inspection not only depends on the ageing of aircraft, but of how many cycles the aircraft is in operation. During welding process, improper polishing may tend to surface cracks on the propeller. Internal cracks may occur at the rib of the wings which run through from back to the front.

Painting on the surface of the propeller will prevent erosion and also this will help to find out the crack. Aircraft components are made up of aluminium alloy, and on compromising the composite materials to reduce the cost, it will affect the structural integrity of the aircraft structure. Ground strikes, unintended stress acting on the propeller exposing to overspeed condition, object strikes and lightning strikes cause micro-cracks on the propeller blade and most of the mechanical damage may be seen in sharp edges. Periodical manual inspection is required to ensure the health conditions of the aircraft components.

Federal Aviation Administration (FAA) had a survey and gave the report like most of the cracks are in tip region of the blade. The small unnoticed cracks might land up in some disastrous event; therefore, propeller blade should be inspected periodically with the fixed time interval. In aviation, reliability is a major concern, and even 99.9% reliability is not an acceptable one because it involves human loss.

In this paper, Sect. 2 discusses about the need for research. Section 3 describes the previous works related to crack detection and Sect. 4 covers the methodology of the proposed work and experimental results are discussed in Sect. 5. Finally, Sect. 6 presents conclusion and references.

2 Motivation

Aviation is highly associated with risks, and therefore, safety is a major concern. Periodical inspection should be on highly risk parts, like all rotating parts which will be subjected to dynamic loads. During take-off, the propeller blade will absorb the vibration caused by the engine power pulse and also vibration caused by online stream, and this will be a centrifugal force pulling the blade from the hub which may cause damage to the blade. Forced landing may cause dent or nicks in the blade. Federal Aviation Administration (FAA) has stated that 'nicks are the cracks starter'. When nicks and dints subjected to stress, then that will tend to fatigue cracking and land up in the failure of the blade.

In 21 August 1995, Empresa Brasileira de Aeronautica S. A. (Embraer) EMB-120RT, an airplane met with an accident during emergency landing because of the loss of propeller blade from the left engine. This results in the death of three passengers and four passengers' fatal injuries. The probable reason for the loss of propeller is due to lack of propeller maintenance. The propeller blade detached from the hub due to fatigue cracking from multiple corrosion pits which was neglected by the maintenance authority.

The blades separated from the hub might occur either during flight or ground operation and the loss of one or more blades will cause imbalance. The imbalance scenario will cause over-vibration which makes the engine to shut down. This will lead to the occurrence of some unhealthy event, and it may be a catastrophic disaster.

3   Related Works

This section discussed the previous work related to crack detection on various textures. Traditional methods like edge detection and morphological operations are explored to detect cracks. In recent times, deep learning is an emerging field in learning features to detect objects. Deep learning techniques are used to detect crack in concrete structure [1–3], metallic structure [4], road pavement [5] and wooden structure [6]. Tong et al. investigate crack detection on concrete bridge by transforming input to binarized image [7]. Hutchinson et al. [8] used wavelet transformation and Bayesian theorem to locate cracks at reduced computational time on the concrete surface. Yamaguchi et al. [9] employ percolation-based image processing method to detect cracks on the concrete surface. Vision-based crack detection methods have been used to review steel surface during production phase for metallic surface [10]. In paper [11], image and object approaches are combined to classify crack and non-crack images. Adhikari et al. [12] proposed a model which uses neural network and 3D visualization for identifying cracks.

Yang et al. [13] proposed an image analysis method to detect hairline cracks in concrete surface. Iyer et al. [7] used curvature evaluation and mathematical morphology to detect crack-like patterns in bottom surface of concrete bridges. Salmon et al. [14] used Gabor filtering to detect crack automatically from the digital images. Multidirectional crack detection can be achieved by Gabor filtering. After the completion of filtering process, cracks aligned to different directions are detected. Fan et al. [15] use magneto-optic imager to detect cracks in the rivet side in the aircraft skin. He classifies the detected rivet into good rivet and cracked rivet. Motion-based filtering is used to separate the defect signals from the background noise. Xu et al.

[16] proposed a crack detection method to identify cracks nearby fastener holes using X-ray imaging. Aircraft wing X-ray images are trained to fix the SMART threshold to identify the fastener from any incoming wing image.

From the above study, it is observed that cracks can be detected in any texture and powerful approaches had been used to detect crack. Based on the study, cascading classifier has been proposed to classify crack and non-crack images in propeller blade.

4 Methodology

The research paper presents the study of detecting cracks in propeller blade by analysing the images. The Haar-like features are calculated by taking contrast variance between the two or three adjacent group pixels. The supervised learning algorithm is used for training the set of crack and non-crack images. Cascading classifier is used to detect the existence of cracks in the test image.

4.1 The Crack Detection Framework

The Overall proposed architecture is shown in Fig. 2. The crack detection framework consists of two phases namely crack detection framework development and crack detection framework deployment.

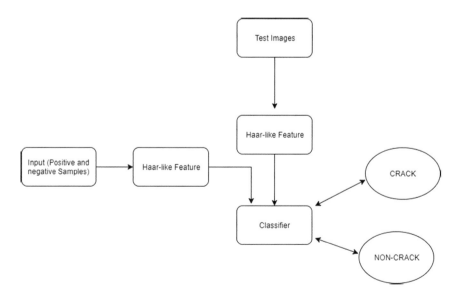

Fig. 2 Schematic diagram of crack detection framework

In the first stage, the Haar-like features are obtained and cascade classifier is employed to train the images. The following algorithm explains the steps in Phase 1.

4.1.1 Algorithm for HCCD

Step 1: Creating training dataset

(i) Collect propeller blade images
(ii) Manually extract crack regions from the collected crack images as positive and collect a set of non-crack images while its possible background as negative samples for training.

Step 2: Feature extraction using Haar-like features

(i) Haar-like feature $H = \sum i(x', y') - \sum i(x', y')$

   $H \rightarrow$ Value of Haar-like feature.
Step 3: Cascading classifier

(i) Set number of stages as n.
(ii) Extracted Features will be given as input to the cascade classifier
(iii) Train the cascading classifier.

Step 4: Classification of the training images

(i) Classify the training image set
(ii) Select the better Haar feature based on the computational result
(iii) Apply the trained cascading classifier to develop the crack detection model.

4.2 Haar-like Features

Harr-like features are used for object recognition. The significant advantages of Harr-like Features consider over pixel values are

(1) There is no limitation for the training samples.
(2) Object detection is faster.
(3) Harr-like features are used to compute the difference between white and black rectangles which are vigorous to noises and lighting variations.

Fig. 3 **a** Edge feature, **b** line feature and **c** surround feature

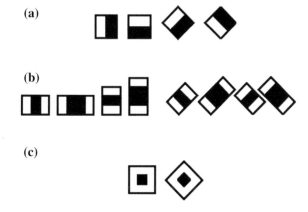

Three types of Harr features shown in Fig. 3 are (a) edge feature (b) line feature and (c) surround feature.

Rectangular two-dimensional image features can be computed rapidly using an intermediate representation called the integral image (in Fig. 3). In greyscale images, the difference between white and grey rectangles in pixels is computed in Eq. (1)

$$H = \sum i(x', y') - \sum i(x', y')$$ (1)

$(x', y') \in$ grey region $(x', y') \in$ White region

$H \rightarrow$ Haar-like feature value.

$I(x', y') \rightarrow$ pixel value of the coordinate (x', y') in an image.

Harr-like features can be computed at high speed by introducing the integral image.

The summation of the pixel value in a selected rectangular region will give the Harr-like features, e.g. to compute the shaded rectangular region. The integral image is calculated as shown in Eq. (2) and illustrated in Fig. 4.

$$\text{Sum} = I(C) + I(A) - I(B) - I(D)$$ (2)

where points A, B, C, D belong to the integral image.

To improve the performance of object recognition, Lienhart and Maydt [17] used extended Haar-like features (in Fig. 3) which is the tilted Haar-like feature. It is used to increase the dimensionality of the set of feature in order to improve the object recognition.

Fig. 4 Integral image I

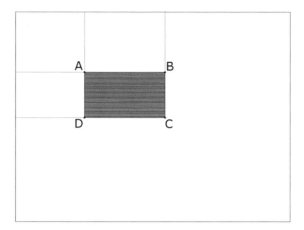

4.3 The Cascading Classifier

The cascading classifier is like ensemble learning which is based on concatenation of several classifiers as shown in Fig. 5. The output of the one classifier will be given as the input of the next cascading classifiers. It is trained by a set of images with positive sample and arbitrary negative samples. After the completion of training, it can be applied to a region of an image and the object can be recognized. If an object is in need to detect in an image, the search window will scan the image in a horizontal manner to locate the object. Therefore, it is used for object recognition or object tracking.

It comprises of cascading classifiers of multistage based on Haar-like features. At the initial stage, extracted features will be given as input to the cascading classifier, and subsequent classifiers will be enhanced by taking more features into account. If a positive image (image containing cracks) is taken for training, it can pass through

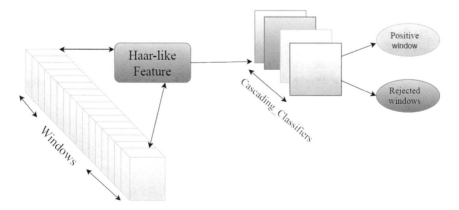

Fig. 5 Cascading classifier

all the stages of classifiers. If former stage classifiers determine a window containing cracks, then only subsequent classifiers will be activated. The cascading classifiers will filter out the negative image (image without cracks) at a great rate; thus, the cascading classifiers reduce the computational time.

4.4 Adaboost

Adaboost cascading classifier (adaptive classifier) is deployed to strengthen the weak classifiers. The main objective of this classifier is to minimize the false positive rate. This classifier uses many learning algorithms to improve the performance and it will tweak the weak classifiers which are misclassified by previous classifiers. It is sensitive to noise and outliers. The individual learners may be weak, but the performance of each one is better than the other by random guessing, and at last, the final model can be proven to converge as a strong classifier as shown in Eq. (3). Adaboost classifier is a method of training a boosted classifier

$$FT(x) = \sum_{t=1}^{T} ft(x) \tag{3}$$

where *ft* is a weak learner and *x* is the input and returns a value which is the class of the object. The weak classifier output predicts the class of the object and absolute value gives betterment in the classification.

Each weak learner produces an output $h(x_i)$ (Eq. in 4) for each training sample, and for every iteration, a weak learner is selected and coefficient αt is assigned to reduce the sum training error Et of the resulting t-stage boosted classifier.

$$E_t = \sum_i E\big[F_{t-1}(x_i) + \alpha\, t\, h(x_i)\big] \tag{4}$$

$F_{t-1}(x_i)$ is the boosted classifier that has been built up to the previous stage of $Ft(x_i)$ classifier.

$E(F)$ is the error function. $f_t(x) = \alpha t\, h(x)$ is the weak learner that will be considered for the addition to the final classifier.

At each iteration, a weight w_t is assigned to each sample in the training set equal to the current error $E(F_{t-1}(x_i))$ on that sample.

5   Overall Architecture of HCCD MODEL

In the detection process, testing images can be classified as either crack or non-crack images. The testing image can be scanned by the detection window of suitable

size. The pre-defined window size Δ will scan the image horizontally and vertically. Cracks' size is of random manner so carefully scale the detection window to
cover different crack size. The proposed method considers the features of cascading
classifier to fix the size of the detection window.

Multiple windows were used to detect the existence of cracks and to avoid multiple
detections; a post-processing is employed to combine two overlapping directions.

6   Experimental Results

The cascading classifier is developed based on Haar-like features and it is improved by
incorporating extended Haar-like features. The images were trained by the extended
cascading classifier and test images were scanned by the appropriate window size.
The extended cascading classifier provides a better performance than the cascading
classifier.

6.1   Training Samples

Manually collect images of crack and extract the regions of crack and make it as
positive sample. Collect images without cracks and train it as a negative sample
(Fig. 6).

6.2   Feature Extraction

The image size will be transformed into greyscale images. Extract features using
extended Haar-like features. There are two features namely black and white which
helps out to detect the crack region from the background. The black colour is identified as crack region.

6.3   Development of the Cascading Classifier

Cascading classifier comprises of number of stages and used to detect the crack region
in the propeller blade based on Haar-like features. At the initial stage, classifier is
based upon few sets of Haar-like features to achieve pre-defined false alarm rate.
The classifiers will be enhanced by taking more number of Haar-like features in the
subsequent stages. If the negative sample is fed, it will not be passed through the
next stages since only if the former stage detects cracks, the subsequent stages will
be activated, thereby reducing the computational time.

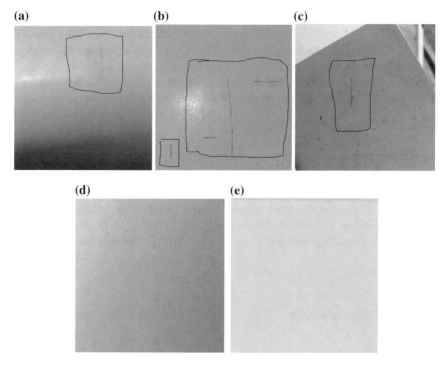

Fig. 6 Training samples. **a** Positive samples and **b** negative samples

6.4 Validation Results

In this work, 60 images were taken, of these 30 images were for training. The training images consist of crack and non-crack. The count of 30 images was taken for testing and got the output. Cracks are detected and its locations are marked by a blue rectangle (Fig. 7).

7 Conclusions

In this research, cascading classifier has been deployed to depict the location of cracks in the images. The extended Haar-like features are used to extract the features and this will be given as input features to the cascading classifier for classification. The scalable window is used to screen the test images and fix out the regions of crack. The cascading classifier will reject the negative sample images thus increasing the processing speed.

The proposed work was validated by the images of the propeller blade. The accuracy can further be improved by incorporating additional Haar-like features and

Detected Crack in image   **Detected Crack in image**

Fig. 7 Crack detection

improving cascading classifier. This research will be a beneficial one to assess the lifetime of the components, thereby easing the work of manual monitoring.

8 Future Work

This work has to be extended to test with a larger set of dataset and also to incorporate deep learning techniques to classify the images with crack and non-crack. In addition have to calculate the number of cracks in each images and also the significant of cracks can be classified by labelling each crack individually.

References

1. Abdel-Qader, O.A., Kelly, M.E.: Analysis of edge detection techniques for crack identification in bridges. J. Comput. Civil Eng. **17**(4), 255–263 (2003)
2. Fujita, Y., Hamamoto, Y.: A robust automatic crack detection method from noisy concrete surfaces. Mach. Vis. Appl. **22**(2), 245–254 (2010)
3. Jahanshahi, M.R., Masri, S.F.: Adaptive vision-based crack detection using 3D scene reconstruction for condition assessment of structures. Autom. Constr. **22**, 567–576 (2012)
4. Chen, F.-C., Jahanshahi, M.R., Wu, R.-T., Joffe, C.: A texture-based video processing methodology using Bayesian data fusion for autonomous crack detection on metallic surfaces. Comput.-Aided Civil Infrastruct. Eng. **32**(4), 271–287 (2017)
5. Zhang, L., Yang, F., Zhang, Y.D., Zhu, Y.J.: Road crack detection using deep convolutional neural network. In: Proceedings of the IEEE International Conference on Image Processing, pp. 3708–3712 (Sep 2016)
6. Meinlschmidt, P.: Wilhelm-klauditz-institut (wki), fraunhoferinstitute for wood research, braunsch weig. thermo-graphic detection of defects in wood and wood-based materials. In: 14th international Symposium of nondestructive testing of wood, Hannover, Germany, May 2–4 (2005)

7. Tong, X., Guo, J., Ling, Y., Yin, Z.: A new image-based method for concrete bridge bottom crack detection. In: 2011 International Conference on Image Analysis and Signal Processing, pp. 568–571 (2011)
8. Hutchinson, T.C., Chen, Z.: Improved image analysis for evaluating concrete damage. J. Comput. Civ. Eng. **20**(3), 210–216 (2006)
9. Yamaguchi, T., Nakamura, S., Hashimoto, S.: An efficient crack detection method using percolation-based image processing. In: 2008 3rd IEEE Conference on Industrial Electronics and Applications, pp. 1875–1880 (2008)
10. Choi, D.-C., Jeon, Y.-J., Lee, S.J., Yun, J.P., Kim, S.W.: Algorithm for detecting seam cracks in steel plates using a Gabor filter combination method. Appl. Opt. **53**(22), 4865–4872 (2014)
11. Choudhary, G.K., Dey, S.: Crack detection in concrete surfaces using image processing, fuzzy logic, and neural networks. In: Fifth International Conference on Advanced Computational Intelligence (ICACI). IEEE, New York (2012)
12. Adhikari, R.S., Moselhi, O., Bagchi, A.: Image-based retrieval of concrete crack properties for bridge inspection. Autom. Constr. **39**, 180–194 (2014)
13. Yang, Y.-S., Yang, C.-M., Huang, C.-W.: Thin crack observation in a reinforced concrete bridge pier test using image processing and analysis. Adv. Eng. Softw. **83**, 99–108 (2015)
14. Salman, M., Mathavan, S., Kamal, K., Rahman, M.: Pavement crack detection using the Gabor filter. In: Proceedings of 16th International IEEE Annual Conference on Intelligent Transportation Systems, pp. 2093–2044 (2013)
15. Fan, Y., Deng, Y., Zeng, Z., Udpa, L., Shih, W., Fitzpatrick, G.: Aging aircraft rivet site inspection using magneto-optic imaging: Automation and real-time image processing. In: 9th Joint FAA/DoD/NASA Aging Aircraft Conference (2006)
16. Xu, J., Liu, T., Yin, X.M., Wong, B.S., Hassan, S.B.: Automatic X-ray crack inspection for aircraft wing fastener holes. In: 2nd International Symposium on NDT in Aerospace 2010–Mo. 5. A. 4
17. Lienhart, R., Maydt, J.: An extended set of Haar-like features for rapid object detection. In: Proceedings. International Conference on Image Processing, vol. 1, pp. I-900–I-903 (2002)
18. Chen, F.-C., Jahanshahi, M.R.: NB-CNN: Deep learning-based crack detection using convolutional neural network and Naïve Bayes data fusion. IEEE Trans. Ind. Electron. **65**(5) (May 2018)
19. Iyer, S., Sinha, S.K.: A robust approach for automatic detection and segmentation of cracks in underground pipeline images. Image Vis. Comput. **23**(10), 931–933 (2005)
20. Qiao, W., Lu, D.: A survey on wind turbine condition monitoring and fault diagnosis—Part I: Components and subsystems. IEEE Trans. Ind. Electron. **62**(10), 6536–6545 (2015)
21. Wu, X.-Y., Xu, K., Xu, J.-W.: Application of undecimated wavelet transform to surface defect detection of hot rolled steel plates. Proc. Congr. Image Signal Process. **4**, 528–532 (2008)
22. Neogi, N., Mohanta, D.K., Dutta, P.K.: Review of vision-based steel surface inspection systems. EURASIPJ. Image Video Process. **2014**(1), 1–19 (2014)
23. Viola, P., Jones, M.: Rapid object detection using a boosted cascade of simple features. In: Proceedings of the 2001 IEEE Computer Society Conference on Computer Vision and Pattern Recognition. CVPR 2001, vol. 1, pp. I-511–I-518 (2001)
24. Lee, J.-K., Park, J.-Y., Oh, K.-Y., Ju, S.-H., Lee, J.-S.: Transformation algorithm of wind turbine blade moment signals for blade condition monitoring. Renew. Energy **79**, 209–218 (2015)
25. Musolino, A., Raugi, M., Tucci, M., Turcu, F.: Feasibility of defect detection in concrete structures via ultrasonic investigation. In: Progress in Electromagnetics Research Symposium 2007, Prague, Czech Republic, August 27–30
26. Xu, W., Tang, Z., Zhou, J., Ding, J.: Pavement crack detection based on saliency and statistical features. In: 20th IEEE International Conference on Image Processing (ICIP), pp. 4093–4097 (2013)
27. Prasanna, P., Dana, K., Gucunski, N., Basily, B., La, H., Lim, R., Parvardeh, H.: Automated crack detection on concrete bridges. IEEE Trans. Autom. Sci. Eng. **13**(2), 591–599 (2016)
28. Fitzpatrick, G.L., et al.: Magneto-optic/eddy current imaging of subsurface corrosion and fatigue cracks in aging aircraft. Rev. Prog. Quant. Nondestr. Eval. **15** (1996)

Placement of Routers and Cloud Components in Cloud-Integrated Wireless-Optical Broadband Access Network

Sangita Solanki, Raksha Upadhyay and Uma Rathore Bhatt

Abstract Cloud-integrated wireless-optical broadband access network (CIW) is a hybrid network which combines the advantages of wireless access network, cloud computing and optical access network. They build a new advantage which provides flexibility, scalability and easy to deployment. It is an infrastructure as a service which provides to facilitate different cloud services with their access network. In this paper, we have presented deployment of routers, cloud components and users in cloud-integrated wireless-optical broadband access network (CIW) using mixed-integer linear programming (MILP) model. Further, we have proposed cloud component connectivity routing algorithm (CCCRA), which provides connectivity between routers and cloud components, using the advantages of cloud service for the end users. Simulation results show different configurations for different number of routers and cloud components illustrating the connectivity between cloud components and routers. This work may be extended to evaluate survivability, energy utilization, etc. of the network.

Keywords Cloud component · CCCRA · Manhattan distance · MILP

1 Introduction

Fiber-wireless (Fi-Wi) is a service-centric network architecture that creates flexible network. Therefore, it enables end user Internet access in any manner [1]. Cloud integrated with wireless-optical broadband access network (WOBAN) enables WOBAN to use cloud services in the network. It provides local allocation of traffic, offloading

S. Solanki (✉) · R. Upadhyay · U. R. Bhatt
Department of Electronics & Telecommunication Engineering, Institute of
Engineering & Technology DAVV, Indore, India
e-mail: ssolanki@ietdavv.edu.in

R. Upadhyay
e-mail: rupadhyay@ietdavv.edu.in

U. R. Bhatt
e-mail: umarathore@rediffmail.com

© Springer Nature Singapore Pte Ltd. 2020
A. K. Luhach et al. (eds.), *First International Conference on Sustainable Technologies for Computational Intelligence*, Advances in Intelligent Systems and Computing 1045, https://doi.org/10.1007/978-981-15-0029-9_34

backhauls computation decreases and increases the offered load at front end [2]. WOBAN is a promising architecture for future access network [3]. Currently, cloud-integrated wireless-optical broadband access network (CIW) has got significant research attention as they can support broadband access network with other feature like survivability, scalability and reliability, etc. It is the combination of cloud computing, optical access network and wireless access network. Wireless access network provides flexibility and easy networking. Optical network facilities higher capacity, low losses and high bandwidth, while cloud components provide greater flexibility, efficiency, local support and survivability.

Figure 1 shows a typical architecture of cloud-integrated WOBAN, which can be classified into two sections wireless front end and optical back end. The wireless front end consists of routers and cloud components (storage or server) and end users. At the front end, the wireless routers constitute wireless mesh networks (WMNs), and cloud components are connected with wireless routers providing facilities associated with server and database. The optical back end consists of optical line terminal (OLT), optical network units (ONUs) and splitter [4]. At the optical back end, the OLT connects with splitter and splitter connects through ONUs to backbone. In WMNs, wireless access from OLT to every client may not be possible because of limited spectrum. Furthermore, most of the traffic go through one of the few gateways and that create bottleneck traffic nearby gateways, to address this limitation, we have

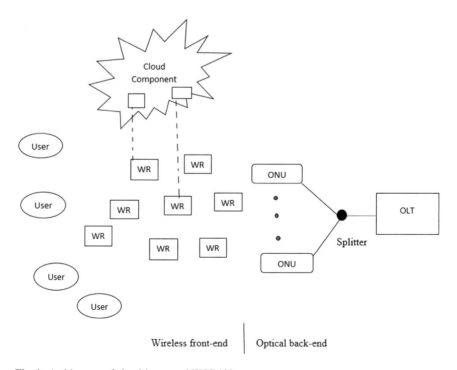

Fig. 1 Architecture of cloud-integrated WOBAN

proposed placement of routers and cloud components in CIW. In this manner, cloud provides services for end users with better coverage capacity and energy efficiency.

The rest of the paper is organized as follows. Section 2, summarizes literature survey. Section 3 describes MILP-based problem formulation, variable, notation, constraints and objective. Section 4 describes proposed algorithm. Section 5 describes performance evaluation. Section 6 concludes this work.

2   Literature Survey

Wireless-optical broadband access network (WOBAN) is a promising solution for Internet access. We study various issues based on WOBAN such as architecture, placement of equipment, fault tolerance, routing algorithms and energy efficiency. The review associated with placement of equipment and quality of service has been presented in hybrid wireless-optical broadband access network (WOBAN) [3]. In [4], authors have included energy-efficient mechanisms such as ONU sleep mechanisms, power save mode and cooperative-based energy saving on Fi-Wi access network. It describes placement algorithm like random, deterministic, greedy and simulated annealing approaches in WOBAN. It also discussed network connectivity-based routing algorithms like minimum hop shortest path and delay aware routing algorithm. In [5], authors have proposed advanced architecture for passive optical network (PON) which supports Fi-Wi network. In this paper, fully distributed ring-based WDM-PON architecture is proposed that can be utilized for the support of next-generation mobile infrastructure. In [6], authors have proposed photonics aided millimeter-wave (mm-wave) technology which integrated with fiber wireless network. This paper summarizes the integration of various multi-dimensional multiplexing techniques and radio-frequency transparent photonic demodulation technology for fiber-wireless-fiber network to achieve large capacity and long-distance fiber wireless transmission system. In [7], authors have proposed cost metric (euclidean distance) for the placement of multiple ONUs in a WOBAN. The objective of this paper is to minimize the average Euclidean distance between the ONUs and nodes. A mixed-integer programming (MIP) model has been proposed for optimum placement of base station and ONUs in a hybrid wireless-optical broadband access network (WOBAN) [8]. They apply lagrangian relaxation (LR) method, primal algorithm, upper bound (UB) and lower bound (LB) of MIP, in order to achieve feasible solution. A mixed-integer linear programming (MILP) has been proposed for cloud-integrated WOBAN (CIW) [9]. These papers create infrastructure platform for different cloud services within access network. They also presented energy-saving routing referred to as, green routing for CIW (GRC). It provides self-manage activation of the network equipment's like ONUs and cloud components. The result showed GRC achieves energy saving with low average packet delay. In [10], authors proposed a novel design for WOBAN, called integrate cloud over WOBAN (COW). It facilitates offload traffic from wireless link, reduce bottleneck near gateway and reduce delay. They propose mathematical formulation as a "mixed-integer linear programming (MILP)."

Yu et al. [11] proposed an approach for the planning of survivable cloud-integrated WOBAN. They apply two approaches, namely integer linear programming (ILP) and heuristic solution to obtain maximal coverage under survivability constraint for end users. In this paper, author assumes when a distribution fiber fails, data transmit to its backup ONU through wireless routers with the help of node disjoint wireless path. The result shows that relationship between network coverage and cost for different number of wireless routers that give maximal coverage for end user with network survivability. Yu et al. [12] presented an integer linear programming (ILP) solution called survivability strategies against multi-fiber failure (SSMF). Author discussed two problems: first, how to deal when multiple distribution fiber failure and coverage and capacity; second, how to locate router and ONU in dynamical manner, in order to achieved maximal coverage at certain number of users. The result and analysis show the configuration of ONUs, routers and also coverage in different ways.

In above literature, the challenges such as survivability and scalability of cloud-integrated WOBAN (CIW) are not discussed. It is also found that not much work has been done to deploy cloud components and router in front-end cloud-integrated WOBAN (CIW) using MILP. We propose Manhattan distance (MD) for better connectivity between cloud components to routers that give service for end users. This paper considers the placement of cloud component and router in CIW. In the future, we will extend our research work in terms of failure of components at the front-end cloud-integrated WOBAN that will be very useful for quantifying survivability in CIW.

3   MILP-Based Problem Formulation

In this section, we present mathematical formulation for placement of routers and cloud components using mixed-integer linear programming in the cloud-integrated WOBAN (CIW). First of all, we define notation and variable for this [13].

3.1   Notations

The notation used for mixed-integer linear programming model is as follows.

- S = Set of routers.
- K = Set of cloud components.
- B = Set of location for end users, where $|B|$ is the total number of end users.
- M = Area of the grid.
- Φ = Set of router and cloud components.
- E_n = End user density where $n \in B$.
- $d_{j,y}$ = Distance from site j to site y.
- R = Distance for interconnection of router or cloud components.

- $\tau_n = \{j, y \in M \,|\, d_{j,y} \leq R, j \neq j\}$ show cloud component or router cover end users.
- $\alpha_j = \{j, y \in M \,|\, d_{j,l} \leq R, j \neq l\}$ set of router or cloud components placed by site j.

3.2 Variables

- $x_n = \begin{cases} 1, & \text{if end users exist in site } n \\ 0, & \text{otherwise} \end{cases}$

- $y_\theta = \begin{cases} 1, & \text{if cloud component or router placed exist} \\ 0, & \text{otherwise} \end{cases}$

- $\mu_i = \begin{cases} 1, & \text{if a router exist in site } i \\ 0, & \text{otherwise} \end{cases}$

- $z_j = \begin{cases} 1, & |\text{if a cloud components exist in site } j \\ 0, & |\text{otherwise} \end{cases}$

- $\rho_r = \begin{cases} 1, & \text{if an cc or a router is placed at location } r \\ & \text{placed at location } r \\ 0, & \text{otherwise} \end{cases}$

- $\omega_{r,y} = \begin{cases} 1, & \text{if the router site } r \text{ is connected to cc at} \\ & \text{site } y \text{ with the available wireless path} \\ & \text{between them} \\ 0, & \text{otherwise} \end{cases}$

- $\sigma_{j,l}^{x,y} = \begin{cases} 1, & \text{if the wireless path from the cc at site } x \\ & \text{to the cc at site } y \text{ traverses the wireless} \\ & \text{link from cc or router site } j \text{ to cc or router site } l \\ 0, & \text{otherwise} \end{cases}$

- $\omega_{j,l}^{r,y} = \begin{cases} 1, & \text{if the wireless path from the router at site } r \\ & \text{to the cc at site } y \text{ traverses the wireless} \\ & \text{link from cc or router site } j \text{ to cc or router site } l \\ 0, & \text{otherwise} \end{cases}$

3.3 MILP Formulation

We formulate MILP for optimal placement of routers and cloud components in CIW.

Objective: Our objective is to maximize the number of covered end users by MILP-based CCCRA approach.

$$\text{Maximize} \sum_{n \in B} E_n x_n \tag{1}$$

(1) Constraint of end user's coverage:

$$x_n \leq \sum_{\theta \in \tau_n} y_\theta \ \forall n \in B \tag{2}$$

$$x_n \geq \sum_{\theta \in \tau_n} y_\theta / |B|, \ \forall n \in B \tag{3}$$

Constraints on cloud components and router, which subject are described below

(2) Constraints of cloud components:

$$\sum_{j \in K} (C^c) \cdot z_j \tag{4}$$

$$\sum_{j \in K} (C^c) = 1 \forall j \in K \tag{5}$$

$$\sum_{l \in \alpha_j} \sigma_{j,l}^{x,y} - \sum_{l \in \alpha_j} \sigma_{l,j}^{x,y} = \begin{cases} \sigma_{x,y}. & \text{if } j = x \\ -\sigma_{x,y}. & \text{if } j = y \\ 0. & \text{otherwise} \end{cases} \tag{6}$$

$$\sum_{j \in \varphi} \sum_{l \in \alpha_j} \sigma_{j,l}^{x,y} \leq H \ \ \forall x, \ y \in M, \ j \in \varphi, \ x \neq y \tag{7}$$

Equation (4) shows the required number of cloud components which will be placed at the grid area sites. Equation (5) indicates the binary variable for cloud components if cloud components placed in grid area site that mean 1 is there, otherwise 0. Equation (6) enforces the flow conversions of the wireless link between cloud component to router or router to cloud component. Equation (7) shows that number of hops for wireless routing path. Each placed cloud component in the network should be connected to at least one router through wireless path between them as shown in Eq. (8).

$$\sum_{y \in M} \omega_{r,y} = p_r \ \ \forall r \in \{\varphi - M\} \tag{8}$$

(3) Constraints of routers:

$$\sum_{i \in S} C^r \mu_i \tag{9}$$

$$\sum_{i \in S} C^r = 1 \ \forall i \in S \tag{10}$$

$$\sum_{l \in \alpha_j} \omega_{j,l}^{ry} - \sum_{l \in \alpha_j} \omega_{l,j}^{ry} = \begin{cases} \omega_{r,y}. & \text{if } j = r \\ -\omega_{r,y}. & \text{if } j = y \\ 0 & \text{otherwise} \end{cases} \tag{11}$$

$$\sum_{j \in \varphi} \sum_{l \in \alpha_j} \omega_{j,l}^{r,y} \leq N \ \ \forall r \in \{\varphi - M\}, y \in M, j \in \varphi \tag{12}$$

Equation (9) indicates the placement of router in grid area. Equation (10) shows the binary variable for placement of router in the network. If 1 is there that mean router is present in the network otherwise 0. Equation (11) defines the flow conversion from cloud components to routers or router to cloud components. Equation (12) shows that number of hops for wireless routing path.

4   Proposed Algorithm

In this section, we described our proposed algorithm for placement of routers and cloud components to provide connectivity to end users. First, we initialize the parameters, then deployed routers, cloud components and clients in random manner by mixed-integer linear programming (MILP). We proposed cloud components connectivity routing algorithm. In this algorithm, we find out the shortest distance by Manhattan algorithm because it provides fast calculation of all the parameter. Manhattan distance algorithm is the sum of horizontal and vertical distances between points on a grid which is given by Eq. (13). It checks connectivity between cloud components and routers, as it allows four direction movements on a square grid area [14]. Thus, users can achieve services.

$$d(x, y) = \sum_{i=1}^{n} |x_i - y_i| \tag{13}$$

Algorithm: Cloud component connectivity routing algorithm

1 Start

2. // **Step 1 Initialize input parameter**

3. No. of clients, routers and cloud components

4. // **Step 2 Deployment of router, cloud component & end user**

5. For n $\in$ B do

6. Compute according to eq. (1), (2) & (3)

7. For i $\in$ S do

8.   Compute according to eq. (4), (5)

9.   for j $\in$ K do

10.   Compute according to eq. (9), (10)

11.   plot (B, S, K)

12. end for

13.   end for

14.   end for

15. // **Step 3 Generate objective function vector**

16. determine distance between cc-router

17. for ii $\in$ 1: S

18.   for jj $\in$ 1: W

19.   Compute according to eq.   (6), (11)

20.   end for

21.   end for

22. update objectives

23. // **Step 4 Each cloud component connected through at least one router**

24.   for ii $\in$ 1:S do

25.   Compute according to eq. (8)

26.   end for

27. // **Step 5 Connectivity between router to cloud component**

28. for ii $\in$ 1: S-------------- do

29.   Compute according to eq. (7)

30.   jj = router associate with ii

31.   Compute according to eq. (12)

32. if d<R

33. plots (y direction cc to router or router to cc)

34. else

35. plots (x direction cc to router or router to cc)

36. end if

37.   end for

Output: Connectivity between cloud components and routers.

4.1   Algorithm Description

Step 1 Initialize input parameter: S, K and B, where S is the number of routers, K is the number of cloud components and B is the number of clients or end users.

Step 2 Deployment of routers, cloud components and end users: Number of routers, cloud components and end users are placed in network area using mathematical Eqs. (1–10).

Step 3 Generate objective function vector: Determine the distance between cloud component to router using Eqs. (6 and 11) and generated objectives.

Step 4 Each cloud component connected through at least one router: This section calls integer linear programming function. Input parameters are taken from step 1, 2, 3. The output of these functions is optimal solution.

Step 5 Connectivity between router to cloud components: We use Manhattan distance for finding shortest path to check network connectivity between router to cloud components using Eqs. (7 and 12).

5   Performance Evaluation

In this section, we deploy the cloud-integrated WOBAN in grid area. In which routers, cloud components and clients are placed. We compute the connectivity of router with cloud component in terms of hop counts using CCCRA. Initially, we consider CIW with 4 cloud components and 1200 clients, where the routers various as 10, 20, 30. These configurations are shown in Figs. 2, 3 and 4, respectively. Summary of these configurations is shown in Table 1.

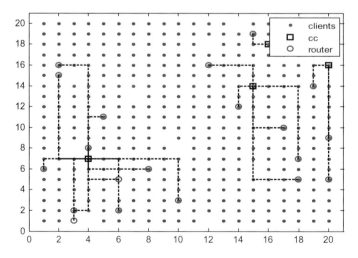

Fig. 2 Placement of routers, CC and clients for CCCRA ($R = 10$, CC $= 4$ and Clients $= 1200$)

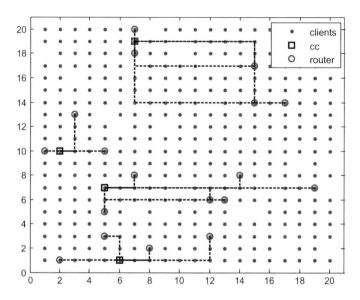

Fig. 3 Placement of routers, CC and clients for CCCRA ($R = 20$, CC $= 4$ and Clients $= 1200$)

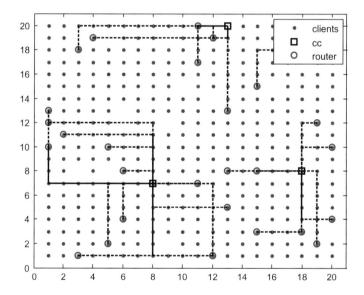

Fig. 4 Placement of routers, CC and clients for CCCRA ($R = 30$, CC $= 4$ and Clients $= 1200$)

Table 1 Simulation results

CC = 4 and Clients = 1200								
R (10)			R (20)			R (30)		
H1	H2	H3	H1	H2	H3	H1	H2	H3
6	4	0	12	8	0	14	8	2

It is clear from table that as we vary the number of routers from 10 to 30, single-hop connectivity increases from 6 to 14. These results are shown in Fig. 5.

Now we consider CIW with 50 routers and 1200 clients, where the cloud components varies as 2, 3, 4. Again, we compute the connectivity of router with cloud component in terms of hop counts using CCCRA. These configurations are shown in Figs. 6, 7 and 8, respectively. Summary of these configurations is shown in Table 2.

It is clear from the table that as we vary the number of cloud components from 2 to 4, single-hop connectivity increases from 11 to 21. These results are shown Fig. 9.

Fig. 5 No. of hop count versus no. of router

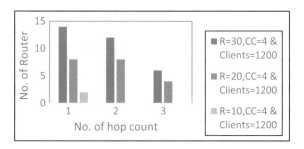

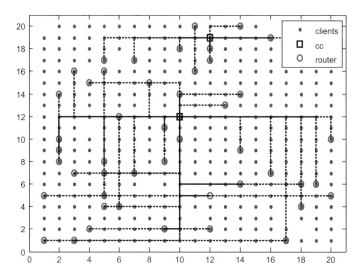

Fig. 6 Placement of routers, CC and clients for CCCRA ($R = 50$, CC = 2 and Clients = 1200)

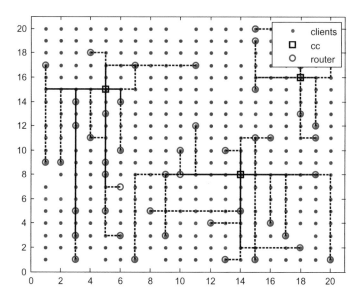

Fig. 7 Placement of routers, CC and clients for CCCRA ($R = 50$, $CC = 3$ and Clients = 1200)

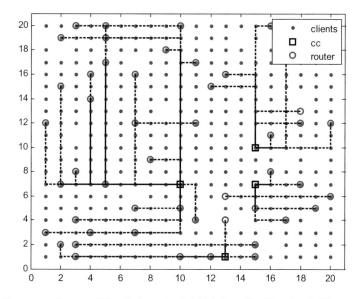

Fig. 8 Placement of routers, CC and clients for CCCRA ($R = 50$, $CC = 4$ and Clients = 1200)

Table 2 Simulation results

$R = 50$ and Clients $= 1200$

CC (2)							CC (3)					CC (4)				
H1	H2	H3	H4	H5	H6	H7	H1	H2	H3	H4	H5	H1	H2	H3	H4	H5
11	13	9	9	4	2	2	15	15	10	6	4	21	19	11	5	5

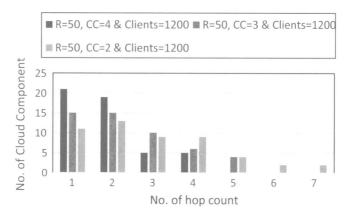

Fig. 9 No. of hop count versus no. of cloud component

This analysis may be very useful in providing the survivability to CIW, because in case of failure of components in CIW through CC, router, and whole network is assuring connectivity with cloud.

6 Conclusion

In WMNs, wireless access from OLT to every client may not be possible because of limited spectrum. Moreover, because most of the traffic go through one of the few gateways and that create bottleneck traffic nearby gateways. To address these limitations, we have proposed MILP for placement of routers and cloud components. We further proposed cloud component connectivity algorithm for achieving connectivity solution, which provides services for the end users. These end users may be mobile node, desktop, cloud component and resource constraint smart devices [15]. In the future, this work can be used to enhance survivability and energy efficiency at the front end and back end of the network.

References

1. Maier, M., Ghazisaidi, N., Reisslein, M.: The audacity of fiber-wireless (FiWi) networks. In: Proceedings, ICST International Conference on Access Networks (Access Nets), Las Vegas, NV, USA, pp. 1–10, Oct 2008
2. Maier, M., Rimal, B.P: Invited paper: The audacity of fiber wireless (FiWi) networks: revisited for clouds and cloudlets. China Communication, Aug 2015
3. Sarkar, S., Dixit, S., Mukherjee, B.: Hybrid wireless-optical broadband-access network (WOBAN): a review of relevant challenges. J. Lightwave Technol. **25**(11), 3329–3340 (2007)

4. Mishra, V., Upadhyay, R., Bhatt, U.R.: A review of recent energy efficient mechanisms for fiber wireless (FiWi) access network. In: Advances in Intelligent Systems and Computing, vol. 563 (2018)
5. Ellinas, G., Vlachos, K., Christodoulou, C., Ali, M.: Advanced architecture for PON supporting Fi-Wi convergences. In: Fiber-Wireless Convergence in Next-Generation Communication Network, pp. 235–263, Sept 2017
6. Yu, J., Li, X., Zhou, W.: Tutorial: Broadband fiber-wireless integration for 5G+ communication. APL Photonic **3**(11) (2018)
7. Sarkar, S., Dixit, S., Mukherjee, B.: Optimum placement of multiple optical network units (ONUs) in optical-wireless hybrid access networks. In: IEEE/OSA Optical Fiber Communications (OFC), Anaheim, California, March 2006
8. Sarkar, S., Dixit, S., Mukherjee, B.: A mixed integer programming model for optimum placement of base stations and optical network units in a hybrid wireless-optical broadband access network (WOBAN). In: IEEE Wireless Communications and Networking Conference (WCNC), Hong Kong, March 2007
9. Reaz, A., Ramamurthi, V., Tornatore, M., Mukherjee, B.: Cloud-Integrated WOBAN: an offloading-enabled architecture for service-oriented access networks. Comput. Netw. **68**, 5–19 (2014)
10. Reaz, A., Ramamurthi, V., Tornatore, M.: Cloud-over-WOBAN (COW): an offloading-enabled access network design. In: IEEE ICC Communication Society (2011)
11. Yu, Y., Liu, Y., Zhou, Y., Han, P.: Planning of survivable cloud-integrated wireless-optical broadband access network against distribution fiber failure. Opt. Switch. Netw. **14**(3), 217–225 (2014)
12. Yu, Y., Liu, Y., Zhou, Y., Han, P.: Survivable deployment of cloud-integrated fiber-wireless networks against multi-fiber failure. Photon. Netw. Commun. **31**, 559–567 (2015)
13. Yu, Y., Ranaweera, C., Lim, C., Wong, E., Guo, L., Liu, Y., Nirmalathas, A.: Hybrid fiber-wireless network: an optimization framework for survivability deployment. J. Opt. Commun. Netw. **9**, 466–478 (2017)
14. Sharmal, S.K., Kumar, S.: Comparative analysis of Manhattan and Euclidean distance metrics using A* algorithm. J. Res. Eng. Appl. Sci. **1**(4), 196–198 (2016)
15. Soni, A., Upadhyay, R., Kumar, A.: Wireless physical layer key generation with improved bit disagreement for the internet of things using moving window averaging. Phys. Commun. **33**, 249–258 (2019)

Optimal Variation of Bridge Column Cross Section of Muscat Expressway, Oman

Himanshu Gaur, Ram Kishore Manchiryal and Biswajit Acharya

Abstract In this chapter, an attempt is made to find optimal variation of cross sections of structures when subjected to different loading conditions such as axial, bending and torsion or coupled loading such as axial with bending or axial with torsional loading, etc. Variation of important structures such as variation of a bridge column section along the length, horizontal variation of bridge deck and variation of a chimney subjected to horizontal loading, etc. can be optimized with this methodology. In this chapter, we focus on optimal variation of bridge piers of Muscat Expressway by considering axial loading only. Variation of column section is first of all expressed/assumed mathematically which represents different possible variation pattern of section along the length such as linear, quadratic and cubic variation and so on. This assumed mathematical expression of sectional area is then plugged into the differential equation of equilibrium of the column which is derived by variational principle. Based on the solution found such as deformation of the column, optimal variation of the column section can be concluded. It is found that linear variation is the best variation when only axial load is considered. Square variation is the second best shape to be adopted for design of the columns when only axial loading is acting. The study also laid guidelines to the architectural community to choose the best possible variation of a structural element when aesthetics is a concern.

Keywords Optimization · Bridge piers · Optimum variation of cross section

H. Gaur (✉)
Bauhaus–Universitat Weimar, Weimar, Germany
e-mail: himanshugaur82@gmail.com

H. Gaur · R. K. Manchiryal
Middle East College, Muscat, Oman

B. Acharya
University College of Engineering, Rajasthan Technical University, Kota, India

© Springer Nature Singapore Pte Ltd. 2020
A. K. Luhach et al. (eds.), *First International Conference on Sustainable Technologies for Computational Intelligence*, Advances in Intelligent Systems and Computing 1045, https://doi.org/10.1007/978-981-15-0029-9_35

1 Introduction

From the last four decades, there have been considerable improvements in the optimization problems where its development goes slowly from shape or size optimization to topology optimization. Development of our method starts after the work done by Y. M. Xie et al. in 1993 where a procedure for structural optimization is proposed, named evolutionary structural optimization (ESO) [1–3]. It works on the principle that after finite element analysis, by gradually removing inefficient elements in the design domain of the structure leads to the optimal structural shape/topology of the structure. Further development of ESO method is bidirectional evolutionary structural optimization (BESO) where the material is removed as well as added in the design domain of the structure [4, 5].

Optimization of structures can be done based on constraints applied to stress, stiffness, displacement, energy and frequency, etc. Recently, few research papers [6–9] show optimization based on fatigue criteria in aerospace and mechanical engineering where any mechanical part is optimized for specific boundary and loading conditions. Metals behave equally for the tensile as well as for the compressive loading. This is not the case for the reinforced concrete structures.

History of the development of topological optimization for reinforced concrete structures goes back to W. Ritter, in 1899 [10], who attempted first to determine the reasonable distribution of reinforcement in the concrete structures, so-called D-regions. D-regions are complex-shaped structural members such as corbels, walls with openings and deep beams with openings which do not satisfy the Saint-Venant's Principle.

In 1991, one of the pioneering works in this direction was done by Schlaich and Schafer [11]. It is popularly known as the strut-and-tie model. This method is developed based on the directions of principal stresses in the finite element analysis. Compressive regions are represented as struts, and tensile regions are represented as ties in the continuum structure. According to the directions of strut and ties, a feasible direction of lying reinforcement and concrete can be determined for complex geometrical problems. Later in 2006, Leu et al. [12] extended the application of this method with conjunction of evolutionary structural optimization (ESO) method [1–3].

In 2013, Luo and Kang [13] did the layout design of reinforced concrete structures with Drucker–Prager yield constraints. The methodology was proposed to be useful for the important reinforced concrete structures where there is high crack control requirement such as nuclear reactor vessel or offshore oil platforms. From the numerical results, it was proven that the methodology makes the best use of compressive strength of concrete and tensile strength of steel.

From the last ten years, the development of topology optimization shows some improvement in Structural Engineering. Recently, few research articles implement its use in the design of concrete structures as well [12–21]. Topological optimization also attracts the architectural community where the interest is growing with the aesthetic view of the geometry which topology optimization gives to the structure [22–24].

Fig. 1 Continuously varying cross sections of bridge piers used in Muscat Expressway, Muscat, Oman

In this chapter, we intend to study the variation of column cross sections of piers of Muscat Expressway bridges in Muscat, Oman. Usually, these variations are provided for aesthetic and better appearances. Figure 1 shows a few of these column sections as well as deck sections, which are varying continuously throughout the span. In this chapter, we discover optimal variations of these sections. In this study, only bridge piers are considered when subjected to axial loading only.

2 Methodology Adopted for Optimization

For finding the optimal variation, we assume a bar (Fig. 2) of constant cross section A_0 till length l_1 and after that, continuously varying with the following relation till further length l_2.

$$A_x = A_0\left(1 + \frac{x - l_1}{P}\right)^n l_1 \leq x \geq l_2 \tag{1}$$

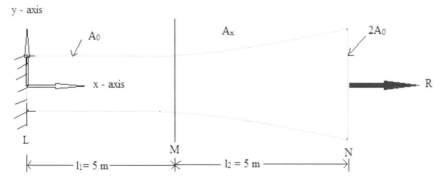

Fig. 2 Cross section of the bar is shown

In this expression, constant **P** and **n** can be suitably selected to define the variation of cross section. Arbitrary value of **P** can be selected to adjust the length with the desired increase in cross section, and the parameter **n** is selected to define the variation of cross section. The value of **n** zero means the straight bar, one means liner variation and two means quadratic variation and so on.

This cross-sectional assumption is plugged into the differential equation of equilibrium which is found by the variational principle. It is to be noted that we limit our study for the axial loading only in this chapter.

3 Variational Principle-Differential Equation of Equilibrium for Axial Loading

When the bar is loaded with load '**R**' at free end, the total potential of the bar can be stated as

$$\Pi = \text{Internal work} - \text{External work}$$

Hence,

$$\Pi = \int_0^l \frac{1}{2} E A \left(\frac{du}{dx} \right)^2 dx - Ru|_{x=10}\ 0 \le x \ge l_1 + l_2 \tag{2}$$

With essential boundary condition,

$$u(0) = 0 \tag{3}$$

where **u** is the displacement field, **A** is cross-sectional area and **E** is elasticity of the material used. Here, it is worthy to note that we are neglecting anybody forces of the element.

By invoking stationarity condition $\delta\Pi = 0$ and integrating by parts, we get

$$\frac{d}{dx} \left(E A \frac{du}{dx} \right) = 0 \tag{4}$$

And the natural boundary condition,

$$E A \frac{du}{dx}|_{x=l_1+l_2} = R \tag{5}$$

Here, $\delta\Pi$ is the small variation in total potential with respect to the variation in **u** [1].

Table 1 Numerical values of constant **P** for getting double area at free end

S. No.	Type of variation of cross section (n)	P	x-limit (m)
1	Straight bar ($n = 0$)	–	5–10
2	Linear variation ($n = 1$)	5	5–10
3	Square variation ($n = 2$)	12.071	5–10
4	Cubic variation ($n = 3$)	19.23661	5–10
5	Quartic variation ($n = 4$)	26.426	5–10
6	Quantic variation ($n = 5$)	33.625	5–10

4 Cross-Sectional Variation of the Bar

Let us assume initial cross section of the bar is A_0, after 5 m of length it becomes double of the section as shown in Fig. 2.

We vary the cross section with the relation expressed in Eq. (1) by adjusting the values of **P**. For $A = 2A_0$ at the free end, the following values of **P** are calculated (Table 1).

5 Calculation For Deformation Function for Different Variations

Now, let us assume first the linear variation in cross section. That is, $n = 1$ in Eq. (1). Hence,

$$A_x = A_0\left(1 + \frac{x - l_1}{P}\right)^1 \quad l_1 \leq x \geq l_2 \tag{6}$$

Integrating Eq. (4), with the boundary condition in Eq. (5), we have

$$EA\frac{du}{dx} = R$$

with Eq. (6)

$$EA_0\left(1 + \frac{x - l_1}{P}\right)\frac{du}{dx} = R$$

Solving the differential equation with a condition $u(l_1) = \frac{R}{EA}l_1$ (straight bar length l_1), we get the displacement field as

$$x = \frac{Rl_1}{EA} + \frac{PR}{EA}\log_e\left(1 + \frac{x - l_1}{P}\right) \tag{7}$$

In the same way, other displacement fields are evaluated for different values of **n** and are shown in Table 2.

Assuming $R = 5000$ kN, $A_0 = 25$ cm^2, $l_1 = 5$ m, $l_2 = 5$ m, modulus of elasticity of steel $= 200$ GPa, the following numerical deformations are found:

These variations can also be read graphically in Fig. 3.

Figure 3 shows the deformation of 10-m-long bar, sectional variation of which changes after 5 m length as shown in Fig. 2. It can be clearly observed in Fig. 3 and the numerical values of Table 3 that linear variation of the straight bar is the best section when only axial loading is acting on the bar. Square variation is the second best section for designing. Other higher-order variations are not suitable for design.

Table 2 Displacement fields for different values of **n**

S. No.	Cross-sectional area	Displacement function (u)	P	x-limit
1	Straight bar, $n = 0$	$= \frac{R}{EA}x$		l_1–l_2
2	Linear variation of cross section, $n = 1$	$= \frac{Rl_1}{EA} + \frac{PR}{EA}\log_e\left(1 + \frac{x-l_1}{P}\right)$	5	l_1–l_2
3	Square variation of cross section, $n = 2$	$= \frac{Rl_1}{EA} + \frac{PR}{EA} - \frac{PR}{EA\left(1+\frac{x-l_1}{P}\right)}$	12.071	l_1–l_2
4	Cubic variation, $n = 3$	$= \frac{Rl_1}{EA} + \frac{PR}{EA} - \frac{PR}{EA\left(1+\frac{x-l_1}{P}\right)^2}$	19.23661	l_1–l_2
5	Quartic variation, $n = 4$	$= \frac{Rl_1}{EA} + \frac{PR}{EA} - \frac{PR}{EA\left(1+\frac{x-l_1}{P}\right)^3}$	26.426	l_1–l_2
6	Quantic variation, $n = 5$	$= \frac{Rl_1}{EA} + \frac{PR}{EA} - \frac{PR}{EA\left(1+\frac{x-l_1}{P}\right)^4}$	33.625	l_1–l_2

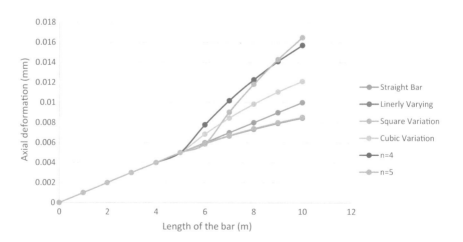

Fig. 3 Axial deformation of bars along the length

Table 3 Axial deformation of bar (mm) for different values of **n**

x	Straight bar	Linear variation	Square variation	Cubic variation	Quartic variation	Quantic variation
0	0	0	0	0	0	0
1	0.001	0.001	0.001	0.001	0.001	0.001
2	0.002	0.002	0.002	0.002	0.002	0.002
3	0.003	0.003	0.003	0.003	0.003	0.003
4	0.004	0.004	0.004	0.004	0.004	0.004
5	0.005	0.005	0.004999	0.005000001	0.005	0.005
6	0.006	0.005911608	0.005922495	0.006854195	0.007786	0.005879
7	0.007	0.006682361	0.006714728	0.008452676	0.010195	0.009042
8	0.008	0.007350018	0.007401827	0.00984039	0.012286	0.011832
9	0.009	0.007938933	0.008003418	0.011052826	0.014112	0.0143
10	0.01	*0.008465736*	**0.008534528**	0.012118305	0.015713	0.016491

Bold value indicates the least value of axial deformation of the bar for n=2 and italic value shows the second least value of axial deformation of the bar (n=1)

6 Conclusions

With these results, it can be concluded that the linear variation of the column section is the best selection for the stiffest design of the column. Second best selection is square variation which gives the best design if has to be adopted for aesthetic or architectural point of view of design.

References

1. Xie, Y.M., Steven, G.P.: A simple evolutionary procedure for structural optimization. Comput. Struct. **49**(5), 885–896 (1993)
2. Xie, Y.M., Steven, G.P.: Evolutionary Structural Optimization. Springer, London (1997)
3. Huang, X., Xie, Y.M.: Evolutionary Topology Optimization of Continuum Structures. Wiley, Chichester (2010)
4. Querin, O.M., Steven, G.P., Xie, Y.M.: Evolutionary structural optimisation (ESO) using a bidirectional algorithm. Eng. Comput. **15**(8), 1031–1048 (1998)
5. Yang, X.Y., Xie, Y.M., Steven, G.P., Querin, O.M.: Bidirectional evolutionary method for stiffness optimization. Am. Inst. Aeronaut. Astronaut. J. **37**(11), 1483–1488 (1999)
6. Holmberg, E., Torstenfelt, B., Klarbring, A.: Fatigue constrained topology optimization. Struct. Multidisc. Optim. (2014)
7. Jeong, S.H., Choi, D.-H., Yoon, G.H.: Topology optimization method for dynamic fatigue constrained problem. In: 10th World Congress on Structural and Multidiscipline Optimization, Orlando, Florida, USA, 19–24 May 2013
8. Belyaev, A., Maag, V., Speckert, M., Obermayr, M., Küfer, K.-H.: Multi-criteria optimization of test rig loading programs in fatigue life determination. Eng. Struct. **101**, 16–23 (2015)
9. Gaur, H., Murari, K., Acharya, B.: Optimization of sectional dimensions of I-section flange beams and recommendations for IS 808: 1989. Open J. Civ. Eng. **6**, 295–313 (2016)

10. Ritter, W.: The Hennebique construction and methods. Schweizerische Bauzeitung 33/34 (1899)
11. Schlaich, J., Schafer, K.: Design and detailing of structural concrete using strut-and-tie models. Struct. Eng. **69**, 113–125 (1991)
12. Leu, L.-J., Huang, C.-W., Chen, C.-S., Liao, Y.-P.: Strut-and-tie design methodology for three-dimensional reinforced concrete structures. J. Struct. Eng. ASCE **132**, 929–938 (2006)
13. Luo, Yangjun, Kang, Zhan: Layout design of reinforced concrete structures using two-material topology optimization with Drucker-Prager yield constraints. Struct. Multidisc. Optim. **47**, 49–110 (2013)
14. Amir, O., Bogomolny, M.: Topology optimization for conceptual design of reinforced concrete structures. In: 9th World Congress on Structural and Multidiscipline Optimization, Shizuoka, Japan, 13–17 June 2010
15. Liang, Q.Q., Xie, Y.M., Steven, G.P.: Topology optimization of strut-and-tie models in reinforced concrete structures using an evolutionary procedure. ACI Struct. J. (2000) (March–April)
16. Amir, Oded: A topology optimization procedure for reinforced concrete structures. Comput. Struct. **114–115**, 46–58 (2013)
17. Bogomolny, M., Amir, O.: Conceptual design of reinforced concrete structures using topology optimization with elastoplastic material modeling. Int. J. Numer. Methods Eng. (2012)
18. Bruggi, M.: Finite element analysis of no–tension structures as a topology optimization problem. Struct. Multidisc. Optim. **50**, 957–973 (2014)
19. Liu, S., Qiao, H.: Topology optimization of continuum structures with different tensile and compressive properties in bridge layout design. Struct. Multidisc Optim **43**, 369–380 (2011)
20. Kingman, J.J., Tsavdaridis, K.D., Toropov, V.V.: Applications of topology optimization in structural engineering: high- rise buildings and steel components. Jordan J. Civ. Eng. **9**(3) (2015)
21. Warshawsky, B.L.: Practical Application of Topology Optimization to the Design of Large Wind Turbine Towers. University of Iowa (2015)
22. Mullins, M., Kirkegaard, P.H., Jessen, R.Z., Klitgaard, J.: A topology optimization approach to learning in architectural design. In: eCAADe, vol. 23, session 3, Digital Design Education
23. Rosko, P.: Design optimization in statics of structures and aesthetics of structures. In: Proceedings of International Conference on Architectural, Structural and Civil Engineering, Antalya, Turkey, pp. 1–4, 7–8 Sept 2015
24. Picellia, R, Townsenda, S., Bramptonb, C., Noratoc, J., Kimad, H.A.: Stress-based shape and topology optimization with the level set method. Comput. Methods Appl. Mech. Eng. **329**, 1–23

Analysis of Customer Churn Prediction in Telecom Sector Using CART Algorithm

Sandeep Rai, Nikita Khandelwal and Rajesh Boghey

Abstract Predicting client churn in telecommunication industries becomes the most significant topic for analysis in recent years. Because its helps in detecting which customer are likely to change or cancel their subscription to a service. Analysis of information that is extracted from telecommunication companies will help to seek out the explanations of client churn and also uses the knowledge to retain the purchasers. Thus, predicting churn is extremely necessary for telecommunication firms to retain their customers. During this paper, we have designed the classification model using call tree, evaluated the performance measures, and compared its performance with logistic regression model.

Keywords Classification · Churn prediction · Telecom data · Logistic regression model · Customer retention · CART algorithm

1 Introduction

Data mining strategies lie at the intersection of computing, statistics, and machine learning info systems. Data processing techniques help in building the prediction models to get future developments and actions permitting the organizations to require good selections derived from the data from knowledge [1].

Churn prediction is associate application of client performance in data processing. Churn [2] could be a key issue sweet-faced through associate enterprise associated denoted the value of extending a replacement client is almost five times more than the value of maintaining an recent client. As a result of the fight of the enterprise market

S. Rai (✉) · N. Khandelwal · R. Boghey
Department of Computer Science and Engineering, Technocrats Institute of Technology
(Excellence), Bhopal, India
e-mail: sandtec@gmail.com

N. Khandelwal
e-mail: nikitakhandelwal0000@gmail.com

R. Boghey
e-mail: rajeshboghey@gmail.com

© Springer Nature Singapore Pte Ltd. 2020
A. K. Luhach et al. (eds.), *First International Conference on Sustainable Technologies for Computational Intelligence*, Advances in Intelligent Systems and Computing 1045,
https://doi.org/10.1007/978-981-15-0029-9_36

declined through churn, churn prediction [3] is dispensed through data processing to boost the client maintenance. Corporations establish that the customers' organization is not passionless to maneuver close to a contender through churn prediction. After that, appropriate advertising operations' square measure would not preserve and hold the shoppers [4]. Churn prediction permits corporations to boost the potency of client retention operations and to reduce the prices joined with churn.

Churn prediction method on medium business is a vital analysis space for the popularity of the faults. Client churn [5] is characterized because the loss of shoppers as they leave to their competitors. It is key issue in medium industries as taking new customer's [6]. In telecommunication business, the churn is additionally known as client attrition or subscriber churning. It is termed because the development [1] of loss of a client. The method of movement from one supplier to a different is occurring due to the higher rates or services or different blessings that the contender company provides whereas language up.

Within the business setting, churn indicates client migration and loss of import. Churn rate is calculated because the share of shoppers' organization finishes reference to the organization or with customers receiving their services. In associations, there square measure giant demand that predicts the shoppers to take care of them promptly by reducing the prices and risks. It additionally will increase the potency and fight. They are employed in market advanced analytics tools and applications designed to spot the massive quantity of information within the teams. It additionally creates prediction derived from the knowledge earned by examining and exploring the information.

2 Related Work

In our work, we have pointed out some previous work that is from the start of the information mining [2, 7] that is employed to get new knowledge's from the databases will serving to numerous issues and helps the business for his or her solutions. Telecommunication companies improve their revenue by retentive their customers' client churn in telecommunication sector to depart a one subscription and be a part of the opposite subscription. This paper predicting the client churn by victimization numerous R packages [8] and that they created a classification model and that they train by giving him a dataset and when coaching they will classify the records into churn or non-churn then they visualize the result with the assistance to visualization techniques [9]. Regression model live used and this models initial train on coaching information; at that time, they will take a look at the model on take a look at information to work out the performance measure of the classification model and obtain the assorted parameters.

Shopper [10] value of each shopper totally fully altogether totally different, so misclassification value of each sample. Associate inclination tends to propose the partition cost-sensitive CART [11] model throughout this paper. The experiment supported the desired information [12], and it is showed that the maneuver not alone obtains associate honest classification performance and, however, additionally reduces the complete misclassification costs effectively. As we [7] have a tendency to tend to any or all perceive that nearly all profits of business return from the consumer directly, telecommunication [13] firms are not exceptions. Firms' organization has heaps of shoppers and maintains a good client relationship [5]. Will get substantial benefits. Studies have shown that, [14] inside the telecommunication business, the worth of effort new shopper is five to six times over mindful existing customers. Moreover, recent [15] customers can generate higher profits. For this reason, students have studied and resolved consumer churn prediction draw back from completely different views.

3 Problem Formulation

The problem is followed as:

Telecommunication corporations are presently not capable to predict corporations initiated churn.
How will telecommunication corporations use data processing techniques and model selection techniques for churn prediction?
Client retention focuses on the subject of client churn, whereby churn pronounces earnings of shoppers, and management of churn designates efforts a business makes to spot and management the client churn drawback.

3.1 Proposed Work

We proposed a machine learning classification techniques based on CART algorithm [9] through which we can build the churn classification model for prediction. In this, we use R programming which is an open-source language used for machine learning for building models [12].

3.2 Proposed Methodology

1. *Dataset*: A medium dataset is taken for predicting churn that we tend to taken is in .arff format. The dataset consists of 20,000 observations (lines, rows) over 12 variables (fields, columns) reading features of customers of a mobile phone provider.
2. *Data Preparation*: We can name each attributes. These attributes are similar to column names in which each attribute is a collection of similar type of record values.

3. *Data Preprocessing*: In these, we can change the data type according to our need such as attributes consist value like true or false value, so we can transform into 0 or 1 (numeric) data for our algorithm needs.
4. *Data Extraction*: We have got worked with numerical and categorical values, and from the 12 attributes, the feature selection process selects 12 attributes which is best for tree construction.
5. *Decision*: Based on these 12 attributes, the tree model generates a decision tree. Now, we have predicted value comes from decision tree model and the actual values, so we can compare the predicted value with the actual values and create a confusion matrix, and from these matrix values, we can calculate the various performance measures like accuracy, specificity, and sensitivity.

3.3 Algorithm

Input:

Data partition, D, that may be a set of coaching tuples and their connected category labels;

Attribute_list;

Attribute_selection_method, to determine the splitting criterion that "best" partitions.

Output: A decision tree (Figs. 1 and 2).

PHASE 1—CREATING A ROOT NODE

i.   *Built a root node N*
ii. *If tuples in D are all of the similar class, C then*
iii. *Return N as a child node label with the class C;*
iv. *If attribute list is empty then*
v.   *Return N as a child node label with the majority class in D*

PHASE 2—ATTRIBUTE SELECTION

vi.   *Apply attribute_selection_method(D, attribute_list) to discover the "best "splitting _criterion attribute;*
vii. *Label node N with splitting _criterion;*
viii. *Update the attribute_list*

PHASE 3—SPLIT THE TREE

ix. *for each outcome j of splitting_criterion*
     //partition the tuples and produce subtrees for each partition
x.   *Based on splitting_criterion attribute*
xi. *Split the tree into two part*
xii. *Attach a child labeled with the common class D in node N:*

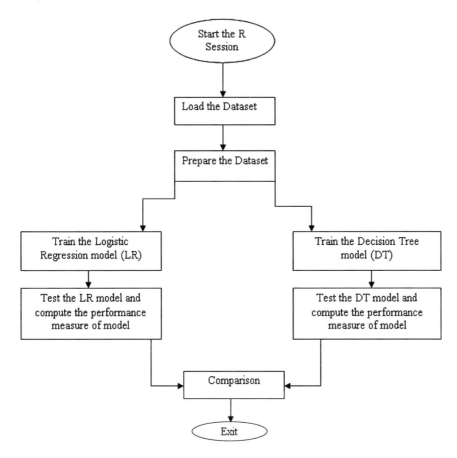

Fig. 1 Flowchart

xiii.   *else attach the node give back by Generate_decision_tree(Dj:attribute_list)to*
        node N:
        end for
xiv.   *return N,*

4   Simulation and Results

4.1   Simulation

The dataset consists of 20,000 observation (lines, rows) over 12 variables (fields, columns) reading features of customers of a mobile phone provider, including the

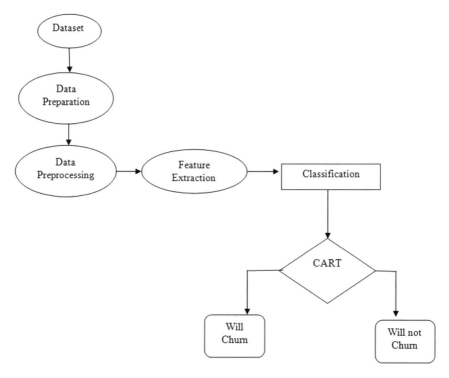

Fig. 2 Churn prediction framework

class variable LEAVE represent whether e customer decided to quit the service or not.

4.2 Result Analysis

Logistic Regression Model

Let us use the logistic regression for better understanding of which variables are important in the dataset. We will use the bootstrapping method to train the model this time. We can preprocess the data for transform dataset for logistic regression model, and then, we are using `glmnet` package. The performance measure are shown in Fig. 3.

Decision Tree Model

Now, we will try a decision tree model. We will use repeated cross validation to train the model with 10 folds and 10 repeats. This means that our model will essentially be training on 100 random subsections of our data. After the model gets trained, we can

```
> data.glm <- train(leave ~ .,
+                     data = data,
+                     method = "glm",
+                     trControl = fitControl)
>
> data.glm
Generalized Linear Model

20000 samples
   11 predictor
    2 classes: 'LEAVE', 'STAY'

No pre-processing
Resampling: Cross-Validated (10 fold, repeated 10 times)
Summary of sample sizes: 18000, 18000, 18000, 18000, 18000, 18000, ...
Resampling results:

  Accuracy   Kappa
  0.6402703  0.2799667

> |
```

Fig. 3 Performance measure of LRM

compute the performance of the model and we get the performance measure which is shown in Fig. 4.

The decision tree model performs superior to logistic regression. And the comparison of both the models based on different parameters is shown below.

Here in Fig. 5, accuracy comparison graph shows that the decision tree models' resultant value is 70.25 and logistic regression models' resultant value is 64.02 that shows that the accuracy of decision tree is better than the logistic regression.

Here also in Fig. 6, Kappa value comparison graph shows that the decision tree models' resultant value is 40.63 and logistic regression models' resultant value is 27.99 that shows that the Kappa value of decision tree is better perform than the logistic regression.

Now, the comparison graphs show that the decision tree model performs higher than the logistic regression model on categorical information.

5   Conclusion and Future Scope

Analysis of information that is extracted from medium firms will help to seek out the explanations of client churn and furthermore utilize the data to hold the client. So predicting churn is very essential for telecom organizations to hold their client. From the result analysis, we say that decision tree algorithm provides better result as compared with the logistic regression model. By comparing the accuracy value based on the criteria, the decision tree model gives better result than logistic regression model (LRM).

```
  RStudio

File   Edit   Code   View   Plots   Session   Build   Debug   Profile   Tools   Help

  ○   ·   ○        ·                 ≡    ↦ Go to file function          ⋯  · Addins  ·

  Source                                                                                   ⊟⊏

  Console    Terminal ×                                                                    ⊐⊟

     C:/Users/abhishek/Desktop/report new/nikita tit kerborus/new update/

        2 classes: 'LEAVE', 'STAY'

No pre-processing
Resampling: Cross-Validated (10 fold, repeated 10 times)
Summary of sample sizes: 17999, 18000, 18000, 18000, 18000, 18000, ...
Resampling results across tuning parameters:

  cp      Accuracy   Kappa
  0.001   0.7001852  0.4011120
  0.003   0.7005351  0.4019931
  0.005   0.7025251  0.4063841
  0.007   0.7025251  0.4063841
  0.009   0.6981052  0.3976516
  0.011   0.6920452  0.3861127
  0.013   0.6906752  0.3832570
  0.015   0.6861552  0.3729942
  0.017   0.6856501  0.3710794
  0.019   0.6861252  0.3718709
  0.021   0.6861252  0.3718709
  0.023   0.6861252  0.3718709
  0.025   0.6861252  0.3718709
  0.027   0.6861252  0.3718709
  0.029   0.6861252  0.3718709
  0.031   0.6861252  0.3718709
  0.033   0.6831754  0.3657166
  0.035   0.6732903  0.3448552
  0.037   0.6691954  0.3359207
  0.039   0.6690804  0.3356024
  0.041   0.6691254  0.3356520
  0.043   0.6691254  0.3356520
  0.045   0.6691254  0.3356520
  0.047   0.6691254  0.3356520
  0.049   0.6691254  0.3356520

Accuracy was used to select the optimal model using the largest value.
The final value used for the model was cp = 0.007.
> |
```

Fig. 4 Performance measure of decision tree

Fig. 5 Accuracy
comparison

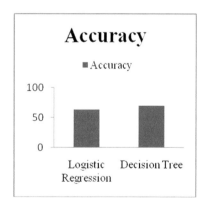

Fig. 6 Kappa comparison

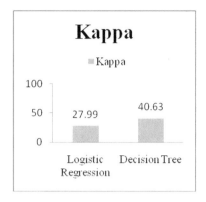

This may encourage unequivocal appropriate intersession techniques to use at any given time. Additionally, there is still space for the development of prediction rates. Furthermore, churn datasets suffer from category imbalance. As expressed antecedent, this makes it onerous for a few data processing techniques to acknowledge the minority category instances though they will reach high overall accuracy. To resolve this issue its required additional analysis. These are a number of the problems; we have a tendency to want to deal with in our future analysis.

References

1. Jadhav, R.J., Pawar, U.T.: Churn prediction in telecommunication using data mining technology. In (IJACSA) International Journal of Advanced Computer Science and Applications, vol. 2, No. 2, Feb 2011
2. Li, P., Li, S., Bi, T., Liu, Y.: Telecom customer churn prediction method based on cluster stratified sampling logistic regression. In: IEEE (2015)
3. Dalvi, P.K., Khandge, S.K., Deomore, A., Bankar, A., Kanade, V.A.: Analysis of customer churn prediction in telecom industry using decision trees and logistic regression. In: 2016 Symposium on Colossal Data Analysis and Networking (CDAN). IEEE (2016)
4. Kamalraj, N., Malathi, A.: A survey on churn prediction techniques in communication sector. Int. J. Comput. Appl. (0975 – 8887) **64**(5), 39–42 (2013)
5. Wang, C., Li, R., Wang, P., Chen, Z.: Partition cost-sensitive CART based on customer value for Telecom customer churn prediction. In: Proceedings of the 36th Chinese Control Conference. 2017. IEEE (2017)
6. Dahiya, K., Bhatia, S.: Customer churn analysis in telecom industry. In: IEEE (2015). 978-1-4673-7231-2/15
7. Wu, X., Meng, S.: E-commerce customer churn prediction based on improved SMOTE and AdaBoost. In: 2016 13th International Conference on Service Systems and Service Management (ICSSSM), pp. 1–5. IEEE (2016)
8. Backiel, A., Verbinnen, Y., Baesens, B., Claeskens, G.: Combining local and social network classifiers to improve churn prediction. In: 2015 IEEE/ACM International Conference on Advances in Social Networks Analysis and Mining (ASONAM), pp. 651–658. IEEE (2015)
9. Ning, L., Lin, H., Jie, L., Zhang, G.: A customer churn prediction model in telecom industry using boosting. IEEE Trans. Industr. Inf. **10**(2), 1659–1665 (2014)

10. Yihui, Q., Chiyu, Z.: Research of indicator system in customer churn prediction for telecom industry. In: 2016 11th International Conference on Computer Science & Education (ICCSE), pp. 123–130. IEEE (2016)
11. Shen, Q., Li, H., Liao, Q., Zhang, W., Kalilou, K.: Improving churn prediction in telecommunications using complementary fusion of multilayer features based on factorization and construction. In: The 26th Chinese Control and Decision Conference (2014 CCDC), pp. 2250–2255. IEEE (2014)
12. Dahiya, K., Talwar, K.: Customer churn prediction in telecommunication industries using data mining techniques—a review. Int. J. Adv. Res. Comput. Sci. Softw. Eng. 5(4), 417–433 (2015)
13. Maldonado, S., Flores, Á., Verbraken, T., Baesens, B., Weber, R.: Profit-based feature selection using support vector machines—general framework and an application for customer retention. Appl. Soft Comput. 35, 740–748 (2015)
14. Backiel, A., Baesens, B., Claeskens, G.: Predicting time-to-churn of prepaid mobile telephone customers using social network analysis. J. Oper. Res. Soc. 1–11 (2016)
15. Ganesh Sundarkumar, G., Ravi, V., Siddeshwar, V.: One-class support vector machine based undersampling: application to churn prediction and insurance fraud detection. In: 2015 IEEE International Conference on Computational Intelligence and Computing Research (ICCIC), pp. 1–7. IEEE (2015)

Fluctuating Time Quantum Round Robin (FTQRR) CPU Scheduling Algorithm

Chhaya Gupta and Kirti Sharma

Abstract In multitasking world of today, CPU scheduling is the only process to determine how different requests can be serviced which are in the ready queue waiting for execution. CPU scheduling algorithms make sure to increase the CPU utilization by decreasing "average turnaround time (TAT)," "average waiting time (WT)," and "context switches" among different processes. Round robin scheduling algorithm is the most efficient algorithm which is used to increase the efficiency of CPU. This paper is proposing a new fluctuating time quantum round robin algorithm (FTQRR) using attributes of "Dynamic Time Quantum based Round Robin" (DTQRR) (Berhanu et al. in Int J Comput Appl 167(13):48–55, 2017 [1]) and "Improved Round Robin with Varying Time Quantum" (IRRVQ) (Kumar Mishra and Rashid in Int J Comput Sci Eng Appl 4(4):1–8, 2014 [2]) such that "average turnaround time, average waiting time, and context switches" among various process will be minimized.

Keywords Round robin scheduling algorithm · Average waiting time · Average turnaround time · Context switch · Fluctuating time quantum

1 Introduction

CPU scheduling means process assigned to CPU for a particular time of interval [3]. In operating system, scheduling is the only procedure by which processes are assigned different resources like processor, access to input–output devices, and many other resources. In order to perform multitasking and multiplexing, the need of scheduling algorithms arises [4]. Processes are switched among waiting queue and ready

C. Gupta (✉) · K. Sharma
Vivekananda School of Information Technology,
Vivekananda Institute of Professional Studies, New Delhi, India
e-mail: Chhayagupta.spm@gmail.com

K. Sharma
e-mail: kirtisharmaa.11@gmail.com

© Springer Nature Singapore Pte Ltd. 2020
A. K. Luhach et al. (eds.), *First International Conference on Sustainable Technologies for Computational Intelligence*, Advances in Intelligent Systems and Computing 1045,
https://doi.org/10.1007/978-981-15-0029-9_37

queue, so the processor efficiency can be maximized [5]. CPU scheduling algorithms enhance CPU performance by decreasing "average turnaround time, average waiting time, and context switches" among the processes which are waiting in the ready queue for execution. In operating system, it is important to allocate process to CPU for a specific time in order to avoid starvation of processes in CPU and this is implemented with the help of scheduler. In round robin algorithm, the processes are grouped as per their "arrival times" in a queue which is known as ready queue. If all the processes are arriving at the same time, then they follow first come first order in the queue where processes are waiting for execution. Each process has their own processing time which is the time required by a process for being executed by the CPU. It is also called as execution time of a process. A time quantum [5] is specified for each process. Time quantum is the time measured in ms (milliseconds) for which a process can continue to be executed in CPU other than its burst time. Once the time quantum is finished, the process which was executed by the CPU is taken back, and if its burst time is zero, then its execution is complete, but if burst time is not equal to zero, then it is put at the back of queue and it waits for its execution turn. The succeeding process is executed which is picked from the beginning of the ready queue, and time quantum is allocated to it for execution. This process keeps on repeating itself. If processing time interval of any procedure is smaller than the time quantum specified by round robin, then process is executed completely, but if processing time is not smaller than the "time quantum," then process executes for the specified amount of time, and then, it is taken back from processor and is put back in ready queue.

1.1 Performance Criteria for CPU Scheduling

Various criterions are used to measure the performance and efficiency of CPU. CPU must be busy all the time to optimize the systems' performance. Processor is allocated to a process from ready queue, whenever processor is free [1]. Different criterions are as follows:

Burst Time—The time required by processes to complete its execution.
Arrival Time—When process is loaded from input queue to ready queue, the time is known as arrival time.
Waiting Time—The time when a process is waiting in ready queue for execution by CPU.
Completion Time—Time consumed by process to finish its execution when CPU is allocated to it, and it is known as completion time.
Turnaround Time—The difference between process's completion time and arrival time gives turnaround time.
Context Switch—When CPU sits idle, it switches to another process and executes it, and this is known as context switching.
Throughput—It refers to the total count of processes executed per time frame.

In the proposed fluctuating time quantum round robin (FTQRR) algorithm, the parameters which are used for performance evaluation are "turnaround time, waiting time, and context switches." The important objective is to decrease "average waiting time, average turnaround time, and count of context switches" using this proposed CPU scheduling algorithm.

1.2 Different Schedulers

The process during its life moves from one scheduling queue to another. Processes from these queues must be selected by the operating system in some order to perform scheduling. It is the scheduler's job to do this selection process. Operating system has three types of schedulers which help in maximizing the CPU utilization. Firstly, there is "long-term scheduler" (job scheduler). This scheduler determines which job will go to ready queue first. Secondly, there is "short-term scheduler" (CPU scheduler), and this allocates processor to the processes which are waiting in the queue for their execution. And thirdly, we have midterm scheduler which removes the process form RAM and puts it in secondary memory when a process has some I/O work to be completed. Switching a process from main to memory to secondary memory is known as swapping out, and when we put any process in main memory from secondary storage for execution, it is known as swapping in. This happens when a process is both CPU and I/O bound. Short-term scheduler works really fast in order to make the best utilization of CPU and makes sure that CPU never sits idle. Long-term scheduler is responsible for multiprogramming.

1.3 Various Scheduling Algorithms

Different scheduling algorithms are proposed to enhance CPU performance and minimize idle time for CPU, hence increasing multiprogramming and multitasking. The processes have various states, and there are various queues which are shown in Fig. 1.

The new processes created are put into *new pool*. The long-term scheduler determines which procedure to put in the "*ready queue*." Now, the processes are waiting for CPU in the "*ready queue*" and "short-term scheduler" determines which process to be given the CPU first and to change the state of a process from "*ready* to *running*." If process is only CPU bound, then it finishes its execution and is terminated, but if it is I/O bound and it requires to execute some I/O action, then the process's state is changed from "*running* to *waiting*," and the CPU is taken back from the process and is given to some other processes, and the earlier process is kept in "*waiting queue*." This is done by midterm scheduler. The process which is waiting is put back in "*ready queue*" on the completion of I/O action, and it now again waits for its turn to be allocated for CPU. The processes which will be assigned the CPU by "short-term

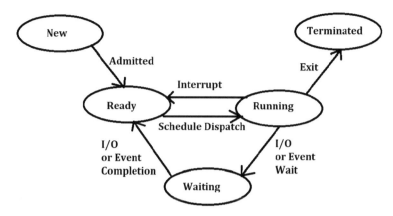

Fig. 1 Process state diagram

scheduler" are selected on the basis of some CPU scheduling algorithms. Few of them are listed below:

A. First Come First Serve (FCFS)

- Process that requests first will get CPU first.
- This algorithm works on non-preemptive mode.
- It is quite simple to apply this algorithm as compared to other algorithms.
- Disadvantage is that it may lead to Convoy effect.

B. Shortest Job First (SJF)

- It works on both preemptive and non-preemptive modes.
- Process with minimum execution time is selected next for the execution.
- By executing the shorter processes, more processes can be completed in stipulated time.
- It is difficult to know in advance that how much time a process needs to complete its execution.

C. Priority-Based Scheduling

- It is a non-preemptive CPU scheduling algorithm.
- It is depending on the CPU burst time, and priorities are assigned to different processes.
- It is simple and user friendly.
- It may lead to starvation.

D. Shortest Remaining Time First (SRTF)

- It is preemptive mode algorithm which is based on SJF algorithm.
- The number of context switching is very high.
- Similar to SJF, one should know the burst time beforehand.

E. Round Robin (RR) Scheduling

- It is preemptive algorithm.
- Here, we define a fix amount of time given to all jobs that is called as time quantum (TQ).
- Each task or job is executed for one time quantum, and then, it is preempted for next process.
- The preempted job or task or process is put at end of "ready queue" and waits for its execution.

2 Related Work

Research scholars have proposed many CPU scheduling algorithms for improvement of system performance. Few of them are reviewed in this paper. Time quantum is kept fixed for the processes which are waiting in ready queue, and then, shortest job first scheduling algorithm is chosen to pick next processes for execution in "An improved Round Robin Scheduling Algorithm for CPU Scheduling" [6]. In "An Improved Round Robin (IRR) CPU Scheduling Algorithm" [6], processor is assigned to the first job present in queue (where all other processes are waiting for execution) for one time quantum. When it is done, the abided processes's burst time is checked with given time quantum, and if this is smaller than specified time quantum, then processor is given to this same job again and it finishes its execution hence reducing waiting time of actively running process. "An Improved Round Robin CPU Scheduling algorithm with Varying Time Quantum (IRRVQ)" [2] uses attributes of shortest job first and round robin scheduling algorithms. In this ALGO, all processes are grouped in increasing order of their processing time and the processor is assigned to first job. Time quantum has been kept same as processing time of initial process in "ready queue." In [1], "Dynamic Time Quantum-based Round Robin CPU Scheduling Algorithm" processes are grouped (ascending order) as per processing times and TQ is made same as processing time of first job. Once a cycle of execution is completed, first process is terminated from the "ready queue," and the second process in "ready queue" is given CPU, but the TQ is kept same as processing time of initial process only. Now after execution of process for amount time quantum specified, abided "burst time" of executing process is compared with specified TQ, if it is found to be smaller than time quantum, then CPU is given to this process only and if it is found to be greater than process is put at the back of "ready queue." "A New Round Robin based Scheduling Algorithm for Operating Systems: Dynamic Quantum Using the Mean Average" [7] has been proposed which uses varying time quants which is same as mean processing time of processes after every cycle of execution. "An Additional Improvement in Round Robin CPU Scheduling Algorithm" [8] is an advancement which is proposed in round robin and "Improved round robin scheduling algorithms" [6]. "A New Improved Round Robin (NIRR) CPU Scheduling Algorithm" [9] is an improvement for IRR [6].

3 Proposed FTQRR CPU Scheduling Algorithm

The proposed fluctuating time quantum round robin (FTQRR) algorithm takes advantage of various features of proposed approaches like IRR [6], IRRVQ [2], and DTQRR [1]. The proposed algorithm combines features of all these approaches to reduce various scheduling criteria like "turnaround time, waiting time, and number of context switches" among various jobs and processor so that it can be utilized efficiently.

First of all, all the processes which are waiting in queue where all the processes are waiting for processor are grouped in increasing order of their execution times. Time quant is made equal to processing time of the first process waiting in queue where all the processes are waiting for execution. The processor is designated to first process for the amount of time quant assigned. When process is done with execution part, the time quant is compared with execution time of second job or task, and if burst time is lacking than the specified amount of time, then there is no change in TQ and processor is assigned to the same procedure, but if specified time quantum is less than the processing time, then time quant is made same as processing time of next process in queue and then processor is given to next process and when it finishes the execution the amount of time(TQ) is again compared to the execution time of next process and the process keeps on going on for all the processes residing in the queue where all the other processes are residing for execution. This technique encompasses features of round robin as well as FCFS CPU scheduling algorithms.

The following is proposed FTQRR scheduling algorithm:

Step 1: Create ready queue RQUEUE for processes ready for execution.
Step 2: DO steps 3 –11 WHILE queue RQUEUE becomes empty.
Step 3: Processes are sorted in non-decreasing order in RQUEUE according to CPU execution time.
Step 4: Set time quantum (TQ) same as processing time of first job in RQUEUE.
Step 5: CPU is assigned to first procedure in RQUEUE for one time interval. Remove currently running process from RQUEUE as it has finished its execution. If the arrival time of all processes is zero, then continue with steps 6 and 7 for each process in RQUEUE. If arrival time is different, then follow steps 8, 9, and 10.
Step 6: Select next process from RQUEUE, and check if TQ < processing time of process; if TQ is less, then assign TQ = Burst time of process and processor is given to it. If TQ ≥ burst time of process, then don't make any changes in TQ and assign the processor to the process.
Step 7: Remove the process from RQUEUE when they are finished with their execution with CPU.
Step 8: If arrival time is nonzero, then set TQ = processing time of first process, and till process is running arrange remaining processes in increasing order of processing time which are coming in RQUEUE while process is executing.
Step 9: Select next process from RQUEUE and check if TQ < processing time of process; if it is found to be less, assign TQ = execution time of process, and then, allocate processor to this process. If TQ ≥ burst time of process, then don't make

any changes in TQ and assign the CPU to the process and again repeat step 8 while the process is using the CPU for execution.

Step 10: Remove the process from RQUEUE when they are finished with their execution with CPU.

Step 11: Repeat steps 6–10 for each process in RQUEUE depending upon value of arrival time.

3.1 Illustration of Proposed Algorithm

For illustration purpose, we have taken five processes in ready queue as P1, P2, P3, P4, and P5. The processes are coming at "0" time with processing time equal to 10, 1, 2, 1, and 5, respectively. According to the proposed algorithm, the processes P1, P2, P3, P4, and P5 are categorized into increasing order of their execution times in the queue, where all the processes are waiting for execution which provides sequences as P2, P4, P3, P5, and P1. The value of time quant is set as processing time of first process present in queue, i.e., TQ = 1. Now, processor is assigned to the next procedure P2 in waiting in queue for time quantum of 1 ms. Process P2 has finished its execution and its remaining execution time = 0; therefore, it is terminated. Time quantum is differentiated with different burst times of second process. If TQ is < than burst time of second process, then TQ is changed to the processing time of second process, and if TQ ≥ burst time of second process, then there is no change in the time quantum. In our illustration example, the processing time of second process P4 is 1 which is same as TQ; therefore, there is no change in TQ and CPU is allocated to P4. When P4 is completely executed, it is also terminated from the "ready queue." Now, next job or task, i.e., P3, has burst time equal to 2, as 2 is > TQ or TQ < burst time of process P3; therefore, the value of TQ is changed to 2 now and CPU is allocated to P3. When it is done with its completion, it is also terminated from the queue. The next procedure residing in queue where all processes wait for processor is P5 which is having burst time = 5. As TQ = 2 and it is less than burst time of process P5; therefore, we make change in the TQ, and now, the value of TQ is 5. CPU is allocated to P5, and when it is done with its execution, it is detached from queue where all the processes are waiting for their execution. The last task left in ready queue is P1 having execution time = 10 and as this value is greater than the present time quantum value, and hence, the value of time quantum = 10 and CPU is assigned to this process as it has finished the execution.

Now, the waiting time for the processes is 9, 0, 2, 1, and 4, respectively. The average waiting time (AWT) = 3.2 ms. The average turnaround time (ATAT) = 7 ms, and the number of context switches (CS) is 4.

The AWT = 4.2 ms and ATAT = 8 ms with DTQRR [1] CPU scheduling algorithm and CS is 13.

The AWT becomes 3.8 ms, and ATAT becomes 7.6 ms with 9 context switches in IRRVQ [2] when we have same above defined processes.

When round robin algorithm is applied on the above, the average waiting time is equal to 5.8 ms and average turnaround time is equal to 9.6 ms with 9 context switches with time quantum set to 2 ms, and if we set time quantum to 5 ms, then the AWT is 7.4 ms and ATAT is 11.2 ms.

The above illustration proves that the proposed FTQRR CPU scheduling algorithm gives better results and better performance than DTQRR, IRRVQ, and RR CPU scheduling.

4 Experimental Analysis

4.1 Assumptions

For the purpose of evaluation for the proposed FTQRR algorithm, we assume that all the processes are having equal priority. The processes in ready queue and their respective burst times are known beforehand. CPU overhead has not been taken into account for arranging the processes. Every procedure in ready queue is assumed to be processor bound only, and none of them is input–output bound. The burst time and time quantum are measured in milliseconds (ms).

4.2 Experiments Performed

In this paper, two different cases have been taken into account for performance evaluation of our proposed FTQRR algorithm. In first case that is Case I, CPU burst time is in variant order and processes arrival time is assumed to be zero. In Case II, again the CPU burst times are in random order and processes arrival time are assumed to be nonzero. CPU burst time in ascending or descending order has not been considered.

Case I—When Arrival Time is ZERO

Arrival time for processes is zero, and CPU burst times are taken in variant order. In the queue where all the processes wait for execution, there five processes; namely, P1, P2, P3, P4, and P5 are present which are shown in Table 1.

The results after comparing various results from RR, IRRVQ, DTQRR, and proposed FTQRR are shown in Table 2. Figures 2 and 3 shows axis charts representation of RR with time quantum 5 and 10, respectively. Figure 4 shows the axis charts of IRRVQ with dynamic time quantum. Figure 5 shows the axis chart for DTQRR with dynamic time quantum. Figure 6 shows the axis chart representation for our proposed FTQRR with dynamic time quantum.

Table 1 Processes with arrival and burst times (Case I)

Processes	Arrival time (AT)	Execution time (ET)
P1	0	10
P2	0	1
P3	0	2
P4	0	1
P5	0	5

Table 2 Comparison of RR, IRRVQ, DTQRR, and FTQRR (Case I)

Algorithms	Time quantum [TQ (ms)]	Average waiting times of various algorithms(ms)	Average turnaround times of various algorithms(ms)	Context switches
Round robin	2, 5	5.8, 7.4	9.6, 11.2	9, 5
IRRVQ	1, 1, 3, 5	4.2	7.8	9
DTQRR	1, 4, 5	3.4	7.2	6
FTQRR	1, 2, 5, 10	3.2	7	4

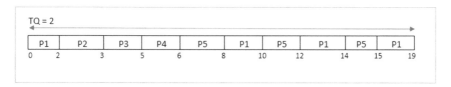

Fig. 2 Axis chart for RR when TQ = 2 (Case I)

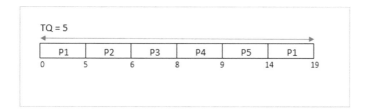

Fig. 3 Axis chart for RR when TQ = 5 (Case I)

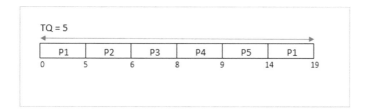

Fig. 4 Axis chart for IRRVQ with dynamic TQ (Case I)

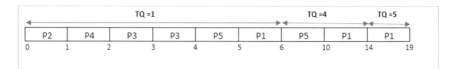

Fig. 5 Axis chart for DTQRR with dynamic TQ (Case I)

Fig. 6 Axis chart representation of FTQRR with dynamic TQ (Case I)

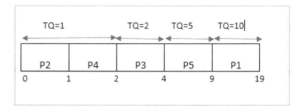

Case II—When Arrival Time is NONZERO

In this case, processes are not arriving at the same time, and therefore, various different arrival times have been taken for different processes, and hence, arrival time is considered to be nonzero. CPU burst time also taken randomly. The experiment is performed on a set of following processes P1, P2, P3, P4, and P5 which are shown as Table 3. Here, the time quant is same as processing time of first process which will be arriving in ready queue. Till the process is running, other processes will be added in the ready queue and they will be categorized in increasing order of processing time. TQ is made same to next process's execution time in ready queue, and rest of the process remains the same.

The compared results are shown in Table 4. Figures 7 and 8 show the axis charts of RR with TQ equals to 5 and 10 ms, respectively. Figure 9 shows axis chart of IRRVQ. Figure 10 shows axis chart of DTQRR. Figure 11 shows axis chart of proposed FTQRR algorithm with dynamic time quantum.

Table 3 Processes with arrival and burst times (Case II)

Processes	Arrival time (AT)	Execution time (ET)
P1	0	9
P2	3	20
P3	5	6
P4	10	28
P5	14	15

Table 4 Comparison of various algorithms RR, IRRVQ, DTQRR, and FTQRR (Case II)

Algorithms	Time quantum (ms)	Average waiting time of various algorithms (ms)	Average turnaround time of various algorithms (ms)	Context switches
Round robin	5, 10	29.4, 25.4	45, 41	14, 8
IRRVQ	9, 6, 5, 8	22.8	38.4	7
DTQRR	9, 11	21.6	37.2	6
FTQRR	9, 6, 15, 20, 28	14.4	30	4

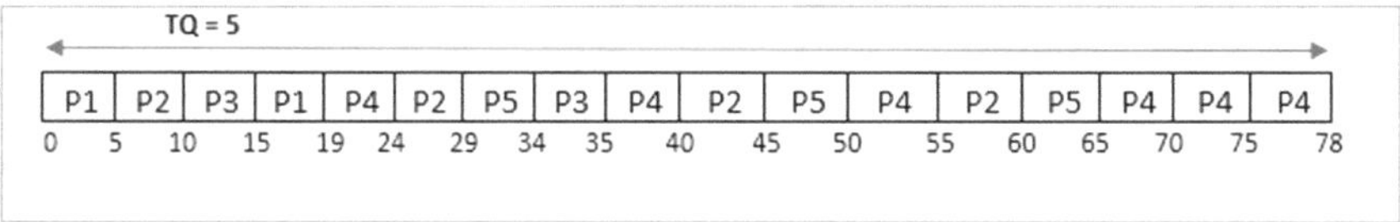

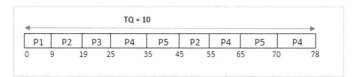

Fig. 7 Axis chart for RR with TQ = 5 (Case II)

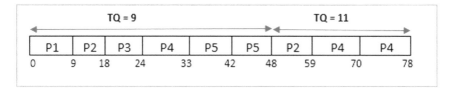

Fig. 8 Axis chart for RR with TQ = 10 (Case II)

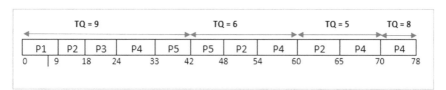

Fig. 9 Axis chart for IRRVQ with dynamic TQ (Case II)

Fig. 10 Axis chart for DTQRR with dynamic TQ (Case II)

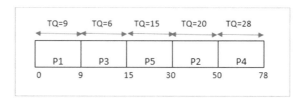

Fig. 11 Axis chart for FTQRR with dynamic TQ (Case II)

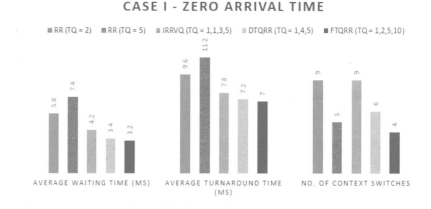

Fig. 12 Bar chart showing analysis of various scheduling heads for Case I

5 Result Analysis

In this section, this paper is giving a comparative analysis of the results obtained by performing the above experiments with the help of bar charts. Figure 12 gives the analysis of result for experiments performed for Case I that is when arrival time is zero, and Fig. 13 gives the analysis of result for experiments performed for Case II that is when arrival time is nonzero. The analysis clearly shows that proposed FTQRR algorithm in this paper is having superior results and is good for reducing the various times and scheduling criteria when compared with different algorithms discussed above.

6 Conclusion

This paper proposes new scheduling algorithm. Experiments are performed which shows that proposed algorithm is giving better results when compared with different already existing and proposed algorithms. The results clearly showcased that

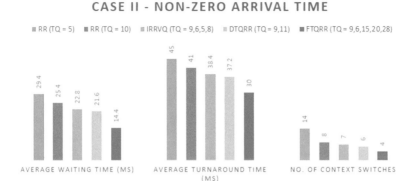

Fig. 13 Bar chart showing analysis of various scheduling heads for Case II

proposed FTQRR algorithm is giving healthier results when compared with RR, IRRVQ, and DTQRR CPU scheduling algorithms and is helpful in reducing the various scheduling heads among processes residing in ready queue. This proposed FTQRR algorithm can be used to refine the efficiency of operating systems.

References

1. Berhanu, Y., Alemu, A., Kumar, M.: Dynamic time quantum based round robin CPU scheduling algorithm. Int. J. Comput. Appl. **167**(13), 48–55 (2017)
2. Kumar Mishra, M., Rashid, F.: An improved round robin CPU scheduling algorithm with varying time quantum. Int. J. Comput. Sci. Eng. Appl. **4**(4), 1–8 (2014)
3. Imran, Q.: CPU scheduling algorithms: a survey. Int. J. Adv. Netw. Appl. **1973**, 1968–1973 (2014)
4. Stallings, W.: Operating Systems Internal and Design Principles (2006)
5. Silberschatz, A., Galvin, P.B., Greg, G.: Operating System Concepts (2009)
6. Yadav, R.K., Mishra, M.K., Navin, P., Himanshu, S.: An improved round robin CPU scheduling algorithm for CPU scheduling. J. Glob. Res. Comput. Sci. **3**(6), 64–69 (2012)
7. Noon, A., Kalakech, A., Kadry, S.: A new round robin based scheduling algorithm for operating systems: dynamic quantum using the mean average. J. Comput. Sci. **8**(3), 224–229 (2011)
8. Abdulrahim, A., Aliyu, S., Mustapha, A.M., Abdullahi, S.E.: An Additional Improvement in Round Robin (AAIRR) CPU scheduling algorithm. Int. J. Adv. Res. Comput. Sci. Softw. Eng. **4**(2), 601–610 (2014)
9. Abdulrahim, A., Abdullahi, S.E., Sahalu, J.B.: A New Improved Round Robin (NIRR) CPU scheduling algorithm. Int. J. Comput. Appl. **90**(4), 27–33 (2014)

Implementation of a Markov Model for the Analysis of Parkinson's Disease

K. M. Mancy, G. Suresh and C. Vijayalakshmi

Abstract In this paper, Parkinson's disease which is the most common form of neurodegenerative disorder is analyzed. As research advances more than quite a long while, it is apparent that it has turned out to be tricky to follow the growth of Parkinson's disease. The mathematical model has been implemented for this disease. The progressive nature of this study is focused through Markov chain process. The long-run fraction method is used to calculate the individuals residing in each state. Succeeding, the paper portrayed the traits of finite-state Markov models in nonstop time frequently used to show the course of illness for assessing change rates and probabilities. Our results encourage the statistical approach to analyze the severity of patient health conditions to determine the steady-state distribution of the Markov model. It is expected that the theoretical analysis and experimental results of the study aid to improve hypothetical methods and interventions for the neurodegenerative disorder diseases and to control the risk in future.

Keywords Mental health disorders · Parkinson's disease · Markov processes · TPM · Steady-state distribution · Neurodegenerative disorder

1 Introduction

In spite of extensive studies over the past few decades, mental disorders are severe diseases that affect more number of people worldwide every year. According to World Health Organization (WHO), mental disorder is generally associated with distress or disability which expresses major mental health issues [1]. Mental health disorders

K. M. Mancy · C. Vijayalakshmi (✉)
Mathematics Division, SAS, Vellore Institute of Technology, Chennai, India
e-mail: vijayalakshmi.c@vit.ac.in

K. M. Mancy
e-mail: mancykm.2017@vitstudent.ac.in

G. Suresh
Department of Mathematics, VV College of Engineering, Tirunelveli, Tamil Nadu, India
e-mail: shurace_g@yahoo.co.in

© Springer Nature Singapore Pte Ltd. 2020
A. K. Luhach et al. (eds.), *First International Conference on Sustainable Technologies for Computational Intelligence*, Advances in Intelligent Systems and Computing 1045, https://doi.org/10.1007/978-981-15-0029-9_38

481

are of various types and one among them is neurodegenerative disorders, which seeks the most significant attention among all medical diseases and deferred recognition heightens the danger of the patients encountering genuine psychological well-being issues. These are medical conditions that disrupt a person's thinking, feeling, mood, and ability to relate to others and daily functioning. In general, neurodegeneration refers to the progressive damage of nerve cells and is the umbrella term for the progressive loss of structure or function of neurons. Many neurodegenerative diseases are caused by genetic mutations, most of which are located in completely unrelated genes and cause a loss of cognitive abilities such as memory and decision making.

It is seen that neurodegenerative maladies (Alzheimer's, Parkinson's, and Huntington's) have numerous likenesses at cell and atomic dimension as they convey parallel components including protein collection and incorporation body arrangement brought about by protein misfolding [2]. Albeit each of the three of the ailments show with various clinical highlights, the ailment forms at the cell level seem, by all accounts, to be fundamentally the same as. For example, Alzheimer's ailment harms diverse pieces of the cerebrum and leads to dynamic loss of memory; Parkinson's malady influences the basal ganglia of the mind, exhausting it of dopamine prompts firmness, inflexibility, and tremors in the significant muscles of the body. Huntington's infection is a dynamic hereditary confusion that influences significant muscles of the body prompting extreme engine limitation and in the long-run demise.

In this paper, the major contribution of the work is focused about the Markov chain analysis of Parkinson's disease (PD), the second most common neurodegenerative disorder effects from the death of dopamine generating cells in the considerable nigra, an area of the midbrain; the reason of cell death is unknown. The target of the investigation is sorted out as pursues: Sect. 2 gives an audit of past works. Section 3 portrays the nature, causes and factors, discovery stages and treatment about Parkinson's sickness. In Sect. 4, the disease progression model is described and provides a way to analyze the severity of patient's health conditions. Section 5 describes the attributes of finite-state Markov model used for the estimation of transition probabilities. In Sect. 6, a novel statistical approach is proposed based on finite-state Markov chains to determine the steady-state distribution of the Markov model. The numerical illustration is presented in Sect. 7 and the results are discussed succeeding.

2 Review of Literature

A several of mediation thinks about have been completed to investigate the location of Parkinson sickness. Ongoing examinations edify that Parkinson's is perpetual and gradually dynamic, implying that side effects proceed and compound over a time of years. Parkinson's is not viewed as a lethal sickness. In [3], two techniques proposed for recognizing sound controls and patients determined to have Parkinson's ailment by methods for recorded smooth interest eye developments. The results showed that the two parameters are effectively separated in terms of the considered metric; however, they are demonstrative of the capability of the exhibited strategies as diagnosing or organizing apparatuses for Parkinson's illness. The work in [4] reports that the obsessive gathering of Aβ, tau, and α-synuclein is at the core of Alzheimer's and Parkinson's sicknesses. Extracellular stores of Aβ and intraneuronal tau incorporations characterize Alzheimer's illness, though intracellular considerations of α-synuclein make up the Lewy pathology of Parkinson's infection. Most instances of malady are sporadic; however, some are acquired in a prevailing way. It is demonstrated that the transformations every now and again happen in the qualities encoding Aβ, tau, and α-synuclein and overexpression of these freak proteins can offer ascent to ailment-related phenotypes.

In [5], the PD is detected by using hidden Markov models and is trained iteratively using common algorithms such as Viterbi and Baum-Welch used for Parkinsonian gait. It is demonstrated that the proposed technique decreases picture highlight from the two-dimensional plane to a one-dimensional vector. The work in [6] provided a predictive system to follow the growth of Parkinson's disease and proposed an expert system of finding the progression type through various state transitions. Geman and Costin [7] applied Markov chain process to the patients with Parkinson's disease in order to predict the evolution of the disease over time and the study improved the quality of patient's life in terminal stage. In [2], the study established the phylogenetic analysis revealing insight on evolutionary relationship caused by protein misfolding of neurodegenerative diseases. The investigation demonstrated that the protein successions of neurodegenerative infections had high grouping likeness and personality to one another as portrayed by the developmental tree. Oskooyee et al. [1] analyzed and modeled the Major Depressive Disorder (MDD) using Markov chain models. The investigation is actualized under a clinical situation demonstrates the precision of the proposed technique used to analyze nervousness and schizophrenia also. In [8], this investigation went for surveying the execution of various approaches for walk stages discovery in patients with Parkinson's malady in OFF and in ON levodopa through a relative examination. The aftereffects of the present examination recommend that all strategies researched kept up a high exactness in stride identification both in ON and in OFF conditions. The best execution was accomplished by HMM that demonstrated the most astounding precision both amid a relative typical walk under levodopa impacts and amid Parkinsonian step variations from the norm. In [9], the stochastic model can be designed for human gestation period. The Markov and parametric models are used to design the human fertility. The human pregnancy can be evaluated and analyzed by using stochastics. Keeping the above papers as base, even more stochastic models can be designed for human gestation period.

3 Parkinson Disease

PARKI, caused by mutations in the SNCA gene, which codes for the protein alpha-synuclein. SNCA gene is analyzed by different NCBI tools and bioinformatics softwares such as GenScan, BLAST, FASTA, and PhyloDraw. The instrument by which the mind cells in Parkinson's are lost may comprise of a strange amassing of the protein alpha-synuclein bound to ubiquitin in the harmed cells [10]. This protein gathering shapes proteinaceous cytoplasmic considerations called Lewy bodies and the most recent research on pathogenesis of infection has demonstrated that the passing of dopaminergic neurons by alpha-synuclein is because of an impact in the hardware that vehicles proteins between two noteworthy cell organelles—the endoplasmic reticulum and the Golgi mechanical assembly (Figs. 1 and 2).

The two major reasons of Parkinson's disease are—hereditary causes and environmental causes [6]. Today, with existing complexities in medical research, it is found that several gene mutations that can cause the disease directly and the mutations involving genetic factor that play a vital part in dopamine cell functions. The two primary categories of genes are casual genes and associated genes. Here, alpha-synuclein is a casual gene that a person inherits and develops Parkinson's disease, whereas related genetic factor does not cause the disease on its own, but the chances of developing the disease will get increased.

When compared to the genetic factors, the researchers have identified several environmental factors such as rural living, well water, manganese, and usage of pesticides, prolonged occupational exposure to chemicals may be the hazard of affecting

Fig. 1 Alpha-synuclein protein structure, a pathological hallmark of Parkinson disease

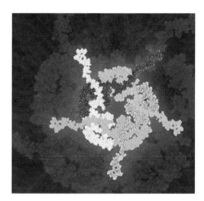

Fig. 2 A Lewy body (stained brown) in a brain cell of the substantial nigra

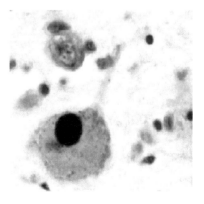

Fig. 3 Report of Da Tscan
test for Parkinson

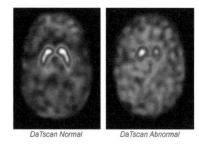

DaTscan Normal   DaTscan Abnormal

Parkinson's disease. The person affected by Parkinson's disease has various types of symptoms which may vary from individuals to individuals. The most general symptom between the people with the disease is motor symptoms who exhibit a lack of sense of smell, personal strength, personality traits being shy, rigidity and postural instability, causing tremors and mood disorders, post-traumatic stress disorders, etc. There is no well-known method of identifying Parkinson's disease before the beginning of symptoms. The researchers found that the use of smell testing is one of the processes of identifying Parkinson's disease in at danger zone, that a person who carries an unusual genetic factor transformation identifies the objective of the disease. One of the predictive tests, known as Da Tscan (shown in Fig. 3), is used as a technique to improve the asymptomatic Parkinson's disease in at danger zone or possibly even avoid the beginning symptoms of Parkinson's disease [11]. It does not necessarily mean that a person will develop the disease but facilitates the researchers to better understand the diagnosis process of the disease.

4 Disease Progression

In general, it describes the change of disease status over time as function of disease process and treatment effects. A few people with Parkinson's find that side effects, for example, discouragement or weakness meddle more with day by day life than do issues with development. Based on the movement symptoms, the most commonly used rating scale (to measure the severity of the disease) based on motor and non-motor symptoms are such as:

Scales	Stages and Range of Measurements		
Hoehn and Yahr	1 and 2 represent early stage	2 and 3 represent mid stage	4 and 5 represent advanced stage
United Parkinson's Disease Rating Scale (UPDRS)	It is more far reaching than the Hoehn and Yahr scale, which centers around development manifestations. The scale incorporates psychological troubles, capacity to complete everyday exercises, and treatment entanglements.		

(continued)

(continued)

Scales	Stages and Range of Measurements
Schwab and England Activities of Daily Living Scale (SE)	It is applied to estimate the incapacity and the scale is explicitly framed for patients with Parkinson's disease to implement daily actions in terms of rapidity and individuality estimated on 11-point index varying from 0 to 100%

As this disease is chronic, slowly progressive and worsens over several years, any medical therapies involve three phases for treatment. The primary stage is the *detection phase* which is mostly used and the most beneficial at the early stage of mental issues. It instructs individuals to be aware of early symptoms of neurodegenerative disorders with plans on seeking help for medical treatment. Succeedingly, *in-hospital treatment* is the second phase, where individuals might be admitted to the hospital to receive regular and continuous treatment would be a better choice. Based on motor and non-motor symptoms of the disease, this system evaluates progressive of the disease shown in Fig. 4. The final stage is the *patient follow-up* after the patient received treatment in the hospital.

Parkinson's disease is not considered a fatal disease, and the way it progresses is different from everyone based on the above levels. If the patient has mild symptoms, then systematic workout improves and continues mobility, variety of indication and stability and also decreases depression. Further, the individual may not interfere with second and third levels, i.e., the disease may not progress to the advanced stage.

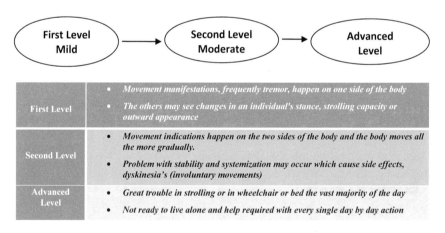

Fig. 4 Disease progressive system of Parkinson's disease

5   The Finite-State Markov Model

Numerous perpetual infections have a characteristic translation as far as arranged movement and hence Markov models are used with growing frequency to characterize the progression of disease [12]. In general, Markov models take a shot at the supposition that the future state is controlled by an irregular procedure subordinate just on the present condition of the system. Before actualizing a Markov demonstrate, it is important to decide the rundown of every single imaginable express that the model takes. Henceforth, the least complex kind of Markov display is the finite-state Markov chain which models the quantity of conceivable states is limited, and change starting with one state then onto the next happens at predefined focuses in time [13]. Such models are often a good alternative for modeling many medical problems. Based on Markov assumption, the statistical approach is proposed for finite states which evolve over time. It means that the patient stays or moves from one state to another state for a particular period of time or may continue in the same state for different consecutive phases. In this section, the necessary attributes of finite-state Markov model such as states, stages, and transition probability matrix are described for modeling the disease [14].

5.1   States

In medicinal portfolio, states frequently speak to different dimensions of ailment movement and might be characterized as either transient or engrossing. A transient is impermanent, and once entered it will be abandoned with sureness and though, a retaining state is one from which there is no kept running off, i.e., once entered, no different states are conceivable. Along these lines, the conditions of a Markov show portray the total arrangement of fundamentally unrelated elective conditions under which a system works.

5.2   Stages

The phases of a Markov model are the focuses in time when the system is observed and information is gathered. The phases in a Markov show contrast from malady arranges in the restorative writing, which are like the conditions of a Markov model. For example, the primary perception stage may demonstrate a seriousness of PD later, at that point a patient may stay in a similar state and once an engrossing state is entered, each progressive stage will show the patient stay in a similar state for significant lot.

5.3 Transition Probability Matrix

All in all, the change probabilities are capacities dependent on introductory and last states, yet in addition the season of progress also. At the point when the one-advance progress probabilities are autonomous of the time variable, n, at that point the Markov chain has stationary change probabilities. By defn, If $P_{ij}^{n,n+1} = P_{ij}$ is independent of n, and P_{ij} is the temporary likelihood that the state esteem experiences a progress from I to j in one preliminary, then it is normal to arrange the numbers P_{ij} in the form of rectangular array of matrix, refers to $P = |P_{ij}|$, known as the Markov matrix or transition probability matrix (TPM) of the process.

$$P = \begin{bmatrix} P_{11} & P_{12} & P_{13} \\ P_{21} & P_{22} & P_{23} \\ P_{31} & P_{32} & P_{33} \end{bmatrix}$$

6 The Proposed Method

In this section, a novel statistical approach is proposed to analyze the severity of Parkinson's disease based on finite-state Markov chain. The proposed approach known as finite-state Markov decision process (FSMDP) and the conceptual framework of the method are presented to analyze the steady-state distribution of the homogeneous Markov chain whose initial state distribution is known. The crucial aim of the proposed method is to determine in the long run, the probability that a patient stays in the same state for quite a few successive stages. The novel scheme is depicted in the form of flow chart (shown in Fig. 5) which summarizes the nature of disease apparently.

Figure 5 gives an idea of how the Markov chain theory actually works in the proposed system. In this method, a state transition from t_k to t_{k+1} may represent one of three possibilities—disease progression, patient recovery or health conditions unchanged. Also, the computation of steady-state transition probabilities determines the probability that how long a patient may remain in same state or remain unchanged. The description about state transitions and corresponding decision premises is described in the succeeding section with data illustration.

7 Numerical Calculations

Consider a numerical data which illustrates the analysis of finite-state Markov chains to determine the steady-state distribution of the model. Irrespective of sex and age

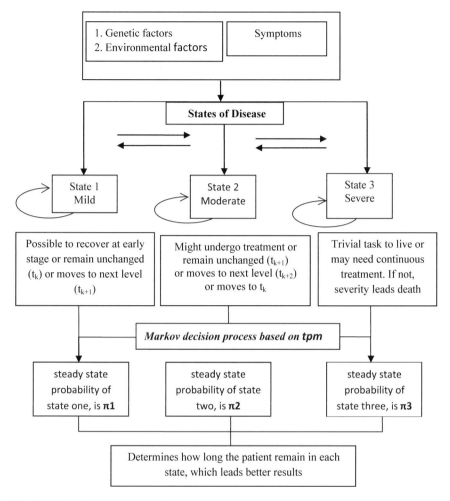

Fig. 5 Schematic representation of the proposed finite-sate Markov decision approach

factors, it is assumed that disease in a certain locality can be modeled as the homogeneous Markov chain whose transition probability matrix is shown below.

Mild Moderate Severe

$$P = \begin{matrix} \text{Mild} \\ \text{Moderate} \\ \text{Severe} \end{matrix} \begin{pmatrix} P_{11} & P_{12} & 0 \\ P_{21} & P_{22} & P_{23} \\ 0 & P_{32} & P_{33} \end{pmatrix} = \begin{pmatrix} 2/3 & 1/3 & 0 \\ 1/3 & 1/3 & 1/3 \\ 0 & 1/2 & 1/2 \end{pmatrix}$$

It represents the finite state transition probabilities between three possible states such as mild, moderate, and severe. At the point when the Markov chain is homogeneous, the one-advance change likelihood is indicated by P_{ij}. The network $P =$

P_{ij} is called one-advance progress likelihood lattice. Since the transition probability matrix of the above Markov chain is a stochastic matrix, and since all $P_{ij}^s \geq 0$, the sum of all elements of any row of the transition matrix is one. The description of the state transition probabilities is shown below.

If the initial state distribution for the above data is assumed as $P^{(0)} = 0.7, 0.2, 0.1$, then the probability distribution is given by,

$P^{(0)}$	$(0.7, 0.2, 0.1)$
$P^{(1)}$	$P^{(0)} * P = (0.7, 0.2, 0.1) * P = (0.72, 0.195, 0.085)$
$P^{(2)}$	$P^{(1)} * P = (0.72, 0.195, 0.085) * P = (0.7245, 0.1920, 0.0835)$

Hence the state transition diagram (Fig. 6) for the above transition probability matrix is shown.

To find in the long run, $\lim_{n \to \infty} P^{(n)}$, if $\pi = \pi_1 \pi_2 \pi_3$ is the steady-state distribution of the Markov chain then by the property of π, it is known that

$$\pi P = \pi \tag{1}$$

$$\pi_1 + \pi_2 + \pi_3 = 1 \tag{2}$$

$$(6.1) \Rightarrow \pi_1 \pi_2 \pi_3 \begin{bmatrix} 0.66 & 0.33 & 0 \\ 0.33 & 0.33 & 0.33 \\ 0 & 0.50 & 0.50 \end{bmatrix} = (\pi_1 \pi_2 \pi_3) \tag{3}$$

$$\Rightarrow 0.66\,\pi_1 + 0.33\,\pi_2 = \pi_1 \tag{4}$$

$$0.33\,\pi_1 + 0.33\,\pi_2 + 0.33\,\pi_3 = \pi_2 \tag{5}$$

$$0.50\,\pi_2 + 0.50\,\pi_3 = \pi_3 \tag{6}$$

By property of π satisfying the condition (2) and solving the above equations. The estimated values are

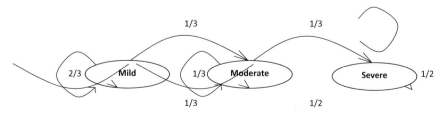

Fig. 6 State transition diagram for the proposed model

$$\Pi_1 = 0.325$$
$$\Pi_2 = 0.336$$
$$\Pi_3 = 0.336$$

Therefore, the steady-state distribution of the chain is, $\pi = \pi_1 \, \pi_2 \, \pi_3$

i.e., $\quad \pi = (0.325, 0.336, 0.336)$

It may be concluded that in the long run, the probability that a patient will remain in three states such as mild, moderate, and severe will be 32%, 34%, and 34%, respectively. The description of the results is discussed in the next section.

8 Results and Discussions

The paper addressed the issue of mental health disorders (MHD), the fourth biggest weight of malady, by the World Health Organization. The long-haul investigation of scatters indications required for exact finding is not feasible for some cases. Just like chronic diseases, mental health disorders are treatable, but the association with neurodegenerative diseases has turned out more critical to follow the disease progression. With the challenging task, the work focused the implementation of finite-state Markov analysis for Parkinson's disease (Sects. 4 and 5). With the context of Markov chain theory, the one-step transition matrix is constructed and shown in the form of transition diagram shown in Fig. 6. The description of transitions from one state to another state is presented in Table 1 with corresponding state transition probabilities.

Based on the proposed statistical approach, a numerical data is presented with illustration. From the given data, it is evident that a patient may progress over three possible states. The primitive objective of the paper is to analyze the steady-state transition probability for the proposed model. With the initial state distribution $P^{(0)}$, the two-step transition matrices such as $P^{(1)}$ and $P^{(2)}$ were determined. In order to find the long run, the steady-state probabilities such as $\pi_1 \, \pi_{2,}$ and π_3 were computed based on the steady-state equations. From the computed steady-state probabilities $\pi_1 = 32\%$, $\pi_2 = 34\%$, and $\pi_3 = 34\%$, it is clear that any individuals exposing the symptoms of Parkinson's disease might have a chance to remain the same state in the long run, as mild, moderate, and severe, respectively.

9 Conclusions

The paper focused a theoretical view of Parkinson's disease, one of the major neurodegenerative diseases in medical arena. Today, with existing research complexities, there is presently no remedy accessible to treat the disease. Such diseases are

Table 1 State transitions based on finite states

State transition probabilities	Description of states
Mild to mild (P_{11})	Patient remains in the same state and possible to recover in the early stage itself
Mild to moderate (P_{12})	Disease progresses slowly and may not progress to the advanced stage
Mild to severe (P_{13})	Patient may not develop advanced stage directly from mild level under one-step transition
Moderate to mild (P_{21})	Patient is under treatment phase which may notice changes in health conditions and possibly becomes mild state
Moderate to moderate (P_{22})	Patient remains in the same state and possibly move toward advanced level or mild level
Moderate to severe (P_{23})	Disease may progress to the advanced stage and patient needs effective treatment from hospital
Severe to mild (P_{31})	Disease in the advanced stage may not possible to recover easily
Severe to moderate (P_{32})	Based on age factor and improved test treatments, the patient possibly moves to the moderate level, but not completely recovered
Severe to severe (P_{33})	The main known risk factor is age. If the patient remains in severe state for long run, life is more challengeable

incurable, resulting in progressive degeneration or death of neuron cells. Hence, the analysis of disease progression is inevitable and hence the model is proposed based on finite-state Markov chains. Based on Markov chain theory, the study analyzes the progressive features of Parkinson's disease under three states. In addition, the computation of steady-state probabilities is found to be very low in percentage for long run, it is evident that the patient remaining in the mild state is slightly lower than the other two states. Hence, it is evident that a disease is slowly progressive, which means that the symptoms continue and worsen over a period of years.

The overall conclusion about Parkinson's disease based on recent studies reports that the use of pesticides may add to an expanded hazard for the illness by causing alpha-synuclein protein to misfold. Examination into new, viable medications for Parkinson's ailment has turned out to be troublesome, in all probability since what really causes the dopamine—creating cells to cease to exist is not known. The animating data is that ongoing advances and disclosures in science, including the distinguishing proof of qualities explicit to Parkinson's illness, have started innovative work into new treatment approaches.

References

1. Oskooyee, K.S., Rahmani, A.M., Riahi Kashani, M.M.: Predicting the severity of major depression disorder with the Markov chain model. In: International Conference on Bioscience, Biochemistry and Bioinformatics, vol. 5, pp. 30–34. IACSIT Press, Singapore (2011)
2. Hussain, B., Khalid, H., Nadeem, S., Sultana, T., Aslam, S.: Phylogenetic and chronological analysis of proteins causing Alzheimer's, Parkinson's and Huntington's diseases. Int. J. Bioautom. **16**(3), 165–178 (2012)
3. Jansson, D., Medvedev, A., Axelson, H., Nyhlom, D.: Stochastic anomaly detection in eye-tracking data for quantification of motor symptoms in Parkinson disease. Adv. Exp. Med. Biol. **823**, 63–82 (2015)
4. Goedert, M.: Alzheimer's and Parkinson's diseases: the prion concept in relation to assembled Aβ, tau, and α-synuclein. The Science **349**, 557–664 (2015)
5. Shaw, L.: HMM based Parkinson's detection by analyzing symbolic postural gait image sequences. Int. J. Tech. Res. Appl. **2**(4), 211–216 (2014)
6. Poongodai, A., Bhuvaneswari, S.: Prognostic system for Parkinson disease: an overview. Int. J. Adv. Res. Comput. Commun. Eng. **3**(8), 7825–7828 (2014)
7. Geman, O., Costin, H.: Parkinson's disease prediction based on multistate Markov models. Int. J. Comput. Commun. **8**(4), 523–537 (2013)
8. Mileti, I., Germanotta, M., Alcaro, S., Pacilli, A., Imbimbo, I., Petracca, M., Erra, C., Di Sipio, E., Aprile, I., Rossi, S., Bentivoglio, A.R., Padua, L., Palermo, E.: Gait partitioning methods in Parkinson's disease patients with motor fluctuations: a comparative analysis. In: 2017 IEEE International Symposium on Medical Measurements and Applications (MeMeA) (2017). 978-1-5090-2985-3
9. Mancy, K.M., Vijayalakshmi, C.: Design of stochastic model for human gestation period in genetics—a review. Int. J. Pharm. Res. **10**(3) (2018)
10. Parkinson's Disease Foundation Website (online). http://www.pdf.org/. Accessed on May 2013
11. Sriram, T.V.S., Rao, M.V., Satya Narayana, G.V., Kaladhar, D.S.V.G.K., Pandu Ranga Vital, T.: Intelligent Parkinson disease prediction using machine learning algorithms. **3**(3) (2013). ISSN: 2277 – 3754
12. Jackson, C.H., Sharples, L.D., Thompson, S.G.: Multistate Markov models for disease progression with classification error. The Statistician **52**, 193–209 (2003)
13. Taylor, H.M., Karlin, S.: An introduction to stochastic modeling, 3rd edn. Academic Press, New York (1998)
14. Daniel Mullins, C., Weisman, E.S.: A simplified approach to teaching Markov models. Am. J. Pharm. Educ. **60**, 42–47 (1996)
15. Haji Ghassemi, N., Hannink, J., Martindale, C.F., Gaßner, H., Müller, M., Klucken, J., Eskofier, B.M.: Segmentation of gait sequences in sensor-based movement analysis: a comparison of methods in Parkinson's disease. Sensors **18**, 145 (2018)
16. San-Segundo, R., Navarro-Hellín, H., Torres-Sánchez, R., Hodgins, J., De la Torre, F.: Increasing robustness in the detection of freezing of gait in Parkinson's disease. Electronics **8**(2), 119 (2019)
17. Alavijeh, A.H.P., Raykov, Y.P., Badawy, R., Jensen, J.R., Christensen, M.G., Little, M.A.: Quality control of voice recordings in remote Parkinson's disease monitoring using the infinite hidden Markov model. In: IEEE International Conference on Acoustics, Speech and Signal Processing (ICASSP) IEEE (2019) (Accepted/In press)

Multi-font Devanagari Text Recognition Using LSTM Neural Networks

Teja Kundaikar and Jyoti D. Pawar

Abstract Current research in OCR is focusing on the effect of multi-font and multi-size text on OCR accuracy. To the best of our knowledge, no study has been carried out to study the effect of multi-fonts and multi-size text on the accuracy of Devanagari OCRs. The most popular Devanagari OCRs in the market today are Tesseract OCR, Indsenz OCR and eAksharayan OCR. In this research work, we have studied the effect of font styles, namely Nakula, Baloo, Dekko, Biryani and Aparajita on these three OCRs. It has been observed that the accuracy of the Devanagari OCRs is dependent on the type of font style in text document images. Hence, we have proposed a multi-font Devanagari OCR (MFD_OCR), text line recognition model using long short-term memory (LSTM) neural networks. We have created training dataset *Multi_Font_Train*, which consists of text document images and its corresponding text file. This consists of each text line in five different font styles, namely Nakula, Baloo, Dekko, Biryani and Aparajita. The test dataset is created using the text from benchmark dataset [1] for each of the font styles as mentioned above, and they are named as *BMT_Nakula*, *BMT_Baloo*, *BMT_Dekko*, *BMT_Biryani* and *BMT_Aparajita* test dataset. On the evaluation of all OCRs, the MFD_OCR showed consistent accuracy across all these test datasets. It obtained comparatively good accuracy for *BMT_Dekko* and *BMT_Biryani* test datasets. On performing detailed error analysis, we noticed that compared to other Devanagari OCRs, the MFD_OCR has consistent, insertion and deletion type of errors, across all test dataset for each font style. The deletion errors are negligible, ranging from 0.8 to 1.4%.

Keywords Multi-font · OCR · Devanagari

T. Kundaikar (✉) · J. D. Pawar
Department of Computer Science and Technology, Goa University, Taleigao, India
e-mail: dcst.teja@unigoa.ac.in

J. D. Pawar
e-mail: jdp@unioa.ac.in

© Springer Nature Singapore Pte Ltd. 2020
A. K. Luhach et al. (eds.), *First International Conference on Sustainable Technologies for Computational Intelligence*, Advances in Intelligent Systems and Computing 1045, https://doi.org/10.1007/978-981-15-0029-9_39

1 Introduction

India is country with 22 constitutionally recognized languages and 12 scripts. Devanagari script is officially used by 10 Indian languages. Historically, lot of Indian literature, religious and scientific books are written in this script. Most of these work is not converted into digital form. In case this literature can be made available in digital form, it will enrich the research field of natural language processing for Indian languages. Optical character recognition (OCR) is a research field in which the text is recognized from scanned documents. There are two methods to recognize text from the scanned document. The first is segmentation-based and the second is segmentation-free approach. In the former approach, the characters are extracted from the input document using segmentation techniques, and further, these extracted characters are recognized using neural network techniques. In the segmentation-free approach, deep learning techniques are used to recognize the text in each line. Using these techniques a lot of work is done in OCR for Roman script, and it shows impressive accuracy on variety of documents. Similar techniques have been used to create OCR for Devanagari script, but the accuracy of this OCR is low as compared to OCR for the Roman script. There are three competitive OCRs for Devanagari script, namely Tesseract OCR [2] open-source tool, Indsenz OCR [3] commercial product and eAksharayan OCR [4] freely available on TDIL [5].

Document in the Roman script is written in different font styles like Times New Roman, Calibri and same is true for document written in Devanagari text. However, we use different font style to represent text for various purposes such as invitation cards, formal documents, notices. On using a different font, the look of text differs as shown in Fig. 1.

We were interested to know the performance of OCR on the scanned document having text in Devanagari script in different font styles. We have tested the Devanagari OCRs with the Devanagari text document images in different font styles and observed that there is a difference in the OCR's accuracy for each of the font styles. In this paper, we have worked on multi-font Devanagari OCR. The content of paper is organized into five sections. Section 1 consists of Introduction and followed by Sect. 2, which gives an idea about related work. Section 3 describes the effect of text document images in different font style on accuracy of Devanagari OCRs. And Sect. 4 discusses creating training and test dataset in form of text document images in different font styles, model to recognize text using LSTM and the result. Section 5 concludes the paper.

Fig. 1 Text in different types of fonts style for Devanagari script

वह कथा या कविता बदल जाती है
वह कथा या कविता बदल जाती है
वह कथा या कविता बदल जाती है
वह कथा या कविता बदल जाती है
वह कथा या कविता बदल जाती है

2 Related Work

The first efforts towards the recognition of Devanagari characters in printed documents started in 1970 using segmentation-based approaches. Researchers at Indian Institute of Technology, Kanpur, India, developed a syntactic pattern analysis system for the Devanagari script [6]. In the 1990s, Palit and Chaudhuri [7], Pal and Chaudhuri [8] developed the first complete end-to-end OCR system for Devanagari. Although in the 1990s OCR for Devanagari was contained only at the research level, in the early 2000s it took a major leap when Center for Development of Advance Computing (CDAC) India released the first commercial Hindi OCR called "Chitrankan" [9]. Recently, segmentation-free approach was used for word classification in printed Devanagari documents by [10, 11]. BLSTM-based transcription scheme is used to achieve text recognition for the seven Indian languages [12].

Bidirectional long short-term memory (BLSTM) architecture with connectionist temporal classification (CTC) output layer was employed to recognize printed Urdu text. They have evaluated BLSTM networks for two cases: one ignoring the character's shape variations and the second by considering the variations. The recognition error rate at character level for the first case is 5.15% and for the second is 13.6% [13]. Deep convolutional neural network has been used for handwritten Devanagari text [14]. The iterative procedure using segmentation-free OCR was able to reduce the initial character error of about 23% (obtained from segmentation-based OCR) to less than 7% in few iterations [15]. There has been a lot of work for different scripts and little work for text document images in multi-font for same scripts. We have attempted to create multi-font Devanagari OCR (MFD_OCR) for Devanagari text using LSTM neural networks.

3 Effect of Devanagari Text Document in Different Fonts Styles on Devanagari OCRs

3.1 Devanagari Font Styles—An Overview

Font classification is based on different properties such as internal representation, glyph width, installation of fonts, physical existence, encoding and readability. The internal representation consists of either bitmap or vector fonts. In bitmap font, glyphs are represented as a matrix of dots and vector is represented as curves. The vector fonts consist of true type, open type, meta font and postscript. The Devanagari font design model [16] can visually describe any given Devanagari typeface, and it is this model which has been used as a base for developing the tool. The facets of the model consist of tool, grey value, contrast, axis, vertical terminal, built , horizontal terminal, angular terminal, basic hand, turns, counters, inclination, vertical stem, curves, C to C joinery, end knot, beginning loop, end loop, neck join, vertical proportion, horizontal proposition and neck proportion [16].

Fig. 2 Text in Nakula font
style

वह कथा या कविता बदल जाती है

Fig. 3 Text in Baloo font
style

वह कथा या कविता बदल जाती है

Fig. 4 Text in Dekko font
style

बह कथा या कविता बदल जाती हे

Fig. 5 Text in Biryani font
style

वह कथा या कविता बदल जाती है

Fig. 6 Text in Aparajita font
style

वह कथा या कविता बदल जाती है

The style of fonts differs based on the facets chosen to design font style. There are more than 300 fonts for Devanagari script. We have considered different properties such as vertical terminal with serif, horizontal and left swoosh, grey value with light and dark. Considering these properties of the font, we have chosen fonts, namely Nakula, Baloo, Dekko, Biryani and Aparajita, which has regular style. The description for each of it is given below.

Nakula Font: The Nakula Devanagari font, which has been developed by IMRC, India, for the University of Cambridge. The font is Unicode compliant, and the font is TrueType. It contains all the conjuncts and other ligatures (including Vedic accents) likely to be needed by Sanskritists. It follows the rounded glyphs and little thin/thick variation as shown in Fig. 2.

Baloo Font: The Baloo font is Unicode compliant and libre licensed font. It has a distinctive heavy spurless design as shown in Fig. 3 [17].

Dekko Font: The inter-letter spacing of this font is wider, allowing for it to be used at smaller sizes on screens. This fonts stroke contrast has thick horizontals. The pen angles traditionally associated with Devanagari has some diagonal stress, but here the script uses a vertical stress as shown in Fig. 4 [18].

Biryani Font: Biryani is a libre font development project. Its fonts are designed in a monolinear, geometric sans serif style. Like several early geometric sans typefaces from the last century, Biryani font characters have a strong flavour to them. Figure 5 shows Biryani font in regular font style [19].

Aparajita Font: The Aparajita font is high OpenType font and is Unicode-encoded. Figure 6 [20] shows Aparajita font in regular font style.

3.2 Performance of OCRs on Devanagari Text Document Images in Aparajita and Nakula Font Styles

We have used the text from [1] to create text document images in Aparajita and Nakula font style. The same is used to check the performance of the OCRs on text document in different font style.

Table 1 Performance of OCR's character accuracy for text in Aparajita and Nakula font style

Font	OCR		
	Tesseract (%)	Indsenz (%)	eAksharayan (%)
Nakula	**90.14**	81.74	70.14
Aparajita	**77.27**	33.21	22.20

Bold indicates the highest accuracy for each of the fonts

Table 2 Error analysis of OCRs for text document image in Aparajita font styles

Type of error	OCR		
	Tesseract (%)	Indsenz (%)	eAksharayan (%)
Insertion	9.01	5.17	15.07
Substitution	9.54	41.52	55.32
Deletion	3.37	19.34	7.39

Table 3 Error analysis of OCRs for text document image in Nakula font styles

Type of error	OCR		
	Tesseract (%)	Indsenz (%)	eAksharayan (%)
Insertion	4.27	5.59	10.58
Substitution	3.53	7.49	33.19
Deletion	1.86	4.20	8.09

We have studied the performance of the three most popular Devanagari OCRs, namely Tesseract, Indsenz and eAksharayan OCR on text document image in different font styles, namely Nakula and Aparajita. The performance of OCRs is shown in Table 1. The results showed that OCR accuracy is dependent on the text document image in different font styles. We were interested to know the reason for the errors in the OCR, for which it does not achieve 100% accuracy. This made us look at the types of errors it generates. There are three types of errors OCR produces, i.e. insertion, deletion and substitution errors. These number of errors can be understood with the help of OCR evaluation tool [21]. Table 2 shows the percentage of each type of errors for text document images in Aparajita font style, and Table 3 shows the percentage of each type of errors for document images in Nakula font style. We can observe from this that insertion and deletion errors are different for these two font styles. This shows that the type of OCR error differs for text document in different font styles. As there are three types of errors in OCR, i.e. insertion, substitution and deletion. It is difficult to handle the insertion and deletion type of error in post-processing of OCR output. In this work, we have attempted to minimize insertion and deletion type of errors and to create OCR which shows consistent accuracy across all font styles.

4 Proposed Multi-font Devanagari OCR (MFD_OCR) Using LSTM Neural Network

4.1 Data Creation of Text Document Image for Different Font Styles

We have identified this above-mentioned five different fonts to create training dataset using text from corpus available at Centre for Indian Language Technology [22]. We created the images for each line in text corpus for each of the font styles using python packages. The training dataset is created for five fonts, namely Nakula, Baloo, Dekko, Biryani and Aparajita. It consists of an image file of each line in text document images and its corresponding text. Training dataset consists of total 500 K images and the same number of text files. This training dataset is named as *Multi_Font_Train* dataset.

To create test data, we have used text from benchmark dataset IIITH [1], which consists of 100 image files and corresponding text files . The test data is created using following steps:

1. We created text files named as *Text_Files_TestDataset* for each of the text line in these 100 text files of benchmark dataset. There are total 2812 text files in *Text_Files_TestDataset*.
2. For each of the file in *Text_Files_TestDataset*, we have created corresponding text document images for Nakula font style using python script, which take input as text and font style and output is same text as document image in ".png" format. This is named as *BMT_Nakula* dataset, which as 2812 files in ".png" format.
3. Like step 2, we have created test dataset for Baloo font style named as *BMT_Baloo* dataset consisting of 2812 files in ".png" format.
4. Similarly, we have created test dataset for Dekko, Biryani and Aparajita font style named as *BMT_Dekko*, *BMT_Biryani* and *BMT_Aparajita*, respectively.

4.2 Multi-font Devanagari OCR: Text Recognition Using LSTM Neural Networks

OCRopus [23] is an open-source document analysis and recognition tool, and we have used it to create the multi-font Devanagari OCR. Recurrent neural networks (RNNs) are used for recognition, especially 1D LSTM architecture. In this architecture, a window of height equals to that of the input image named as "depth of the sequence" and width equal to 1 pixel called a "frame" and is moved over the text line. These 1D sequences (frames) are then fed to the input layer of the LSTM network. To achieve normalization, it is important to make the depth of sequence equal for all input text line images. We have used an LSTM recurrent neural net to learn this sequence mapping of Devanagari text in different font styles. Our model has 48 inputs, 200 nodes in a hidden layer and 249 outputs. The inputs to the network are columns of pixels. The columns in the image are fed into the network, one at a time, from left to right. The outputs are scores for each possible letter. Learning rate and momentum were set to a standard value which is 0.0001 and 0.9, respectively. It took 800 K iteration, to generate a model with stabilized accuracy.

4.3 Experimental Results

Accuracy of Tesseract, Indsenz, eAksharayan and MFD_OCR on text document images in different font styles.

We have used an abbreviation to each of the OCR as mention here, T_OCR for Tesseract OCR, I_OCR for Indsenz OCR and E_OCR for eAksharayan OCR. ISRI analytic tool is used for the evaluation of OCRs [21]. We have evaluated the performance of the proposed MFD_OCR along with T_OCR, I_OCR and E_OCR with the created test dataset for all five font styles. The performance of OCRs for each of the test dataset is shown in Table 4.

The best character accuracy for *BMT_Nakula* test dataset is shown by T_OCR. Again the accuracy for Text in BMT_Baloo test dataset is greater for T_OCR compared to all other OCRs. The proposed MFD_OCR performs better for *BMT_Dekko* and *BMT_Biryani* test dataset compared to all other OCRs. Finally, Tesseract OCR is more accurate for *BMT_Aparajita* test dataset. The Indsenz and eAksharayan OCRs produce very low accuracy for *BMT_Aparajita* test dataset. It has been observed that each OCR accuracy varies for text document image in different font styles, but MFD_OCR obtained consistent accuracy across all the text document in different font styles.

Error analysis for T_OCR, I_OCR, E_OCR and MFD_OCR for each of the test dataset.

We have also performed the error analysis of the OCRs with the help of OCR evaluation tool [21]. The error in OCR is due to insertion, substitution and deletion of the character. We have discussed these type of errors for each of the test dataset.

Table 4 OCRs character accuracy on test dataset

Dataset	OCR			
	T_OCR (%)	I_OCR (%)	E_OCR (%)	MFD_OCR (%)
BMT_Nakula	**90.14**	81.74	70.14	89.23
BMT_Baloo	**89.64**	78.15	62.93	88.49
BMT_Dekko	77.87	72.34	47.51	**89.03**
BMT_Biryani	88.56	77.28	59.41	**88.89**
BMT_Aparajita	**77.27**	33.21	22.20	69.87

Bold indicates the highest character accuracy for each of the test datasets

Table 5 Percentage of errors of OCRs for *BMT_Nakula* test dataset

Type of error	OCR			
	T_OCR (%)	I_OCR (%)	E_OCR (%)	MFD_OCR (%)
Insertion	**4.27**	5.59	10.58	4.91
Substitution	**3.53**	7.49	33.19	5.55
Deletion	1.86	4.20	8.09	**0.86**

Bold indicates the lowest error rate

Table 6 Percentage of errors of OCRs for *BMT_Baloo* test dataset

Type of error	OCR			
	T_OCR (%)	I_OCR (%)	E_OCR (%)	MFD_OCR (%)
Insertion	4.98	7.78	20.19	**4.58**
Substitution	**2.72**	8.46	14.24	5.39
Deletion	2.10	3.90	2.62	**0.90**

Bold indicates the lowest error rate

Table 7 Percentage of errors of OCRs for *BMT_Dekko* test dataset

Type of error	OCR			
	T_OCR (%)	I_OCR (%)	E_OCR (%)	MFD_OCR (%)
Insertion	6.05	7.07	6.50	**2.23**
Substitution	7.71	12.60	30.35	**6.79**
Deletion	7.86	7.19	15.62	**1.49**

Bold indicates the lowest error rate

Table 5 shows the percentage of each error type for above-mentioned OCRs for *BMT_Nakula* test dataset. The insertion and substitution types of error are less in T_OCR for *BMT_Nakula* dataset. The MFD_OCR has less deletion type of errors, and insertion type of error is nearly the same as that of T_OCR.

Table 6 shows the percentage of errors for each of the OCR for *BMT_Baloo* dataset. The insertion and deletion type of errors are lowest in MFD_OCR for the *BMT_Baloo* dataset. The T_OCR has less substitution type of errors.

Table 7 shows the percentage of errors for each of the OCR for *BMT_Dekko* dataset. The insertion, substitution and deletion type of error are less in MFD_OCR for *BMT_Dekko* dataset.

Table 8 shows that the percentage of errors for each of the OCR for *BMT_Biryani* test dataset. The T_OCR has less substitution and insertion type of errors. The MFD_OCR has less deletion type of errors and insertion, and substitution type of error is nearly the same as that of T_OCR.

Table 9 shows that the percentage of errors for each of the OCR for *BMT_Aparajita* test dataset. The insertion type of error is less in I_OCR for *BMT_Aparajita* dataset. The MFD_OCR has less deletion type of errors, and The T_OCR has less substitution type of errors.

Table 8 Percentage of errors of OCRs for *BMT_Biryani* test dataset

Type of error	OCR			
	T_OCR (%)	I_OCR (%)	E_OCR (%)	MFD_OCR (%)
Insertion	**4.29**	5.63	4.56	4.42
Substitution	**4.09**	10.75	23.37	4.94
Deletion	2.54	5.22	12.65	**1.00**

Bold indicates the lowest error rate

Table 9 Percentage of errors of OCRs for *BMT_Aparajita* test dataset

Type of error	OCR			
	T_OCR (%)	I_OCR (%)	E_OCR (%)	MFD_OCR (%)
Insertion	9.01	**5.17**	15.07	6.24
Substitution	**9.54**	41.52	55.32	22.61
Deletion	3.37	19.34	7.39	**1.40**

Bold indicates the lowest error rate

Percentage of types of errors for each of the OCRs for text document image in different font styles.

Here we have shown the plot of the percentage of each type of errors for all OCRs across all the test datasets as shown below. Figure 7 is a plot of the percentage of deletion type of errors for each of the test dataset versus OCRs.

It is observed from Fig. 7 that the proposed MFD_OCR gives fewer deletion errors for all types of font, which has been tested with a particular type of dataset.

Figure 8 is a plot of the percentage of insertion type of errors for each of the test dataset versus OCRs. Figure 8 shows MFD_OCR has consistent insertion type of error for all fonts, which has been tested with a particular type of dataset.

Figure 9 shows that the substitution types of error are also consistent for all fonts in MFD_OCR except the text document in Aparajita font style.

Fig. 7 Deletion type of errors of OCRs for text document in different fonts

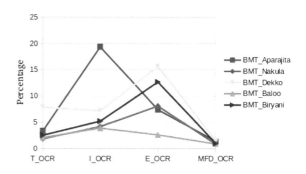

Fig. 8 Insertion type of errors of OCRs for text document in different fonts

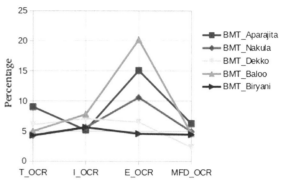

Fig. 9 Substitution type of errors of OCRs for text document in different fonts

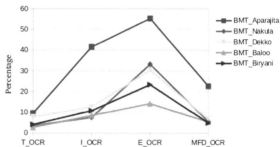

5   Conclusion and Future Scope

In this paper, initially, we carried out the study of the effect of text in different fonts style on Devanagari OCRs. We also analysed the percentage of different types of error for which the accuracy of OCRs was low. We observed that the most popular OCR's accuracy is different for text in different font style. The Tesseract OCR obtained good accuracy among all OCRs but showed the highest insertion and deletion type of errors compared to the substitution type of errors. In post-processing of OCR output, it is difficult to handle insertion and deletion type of errors.

We implemented multi-font Devanagari OCR(MFD_OCR) using LSTM neural networks and *Multi_Font_Train* dataset. On the evaluation of MFD_OCR, it showed consistent accuracy, across all the text document images in different font style. It obtained comparatively good accuracy for text document images in Dekko and Biryani font style. On performing detailed error analysis, it was noticed that compared to other Devanagari OCRs, the insertion and deletion type of errors are consistent for MFD_OCR across all text document images in different font style. The deletion errors are negligible, ranging from 0.8 to 1.4%. The Tesseract OCR has the lowest substitution type of errors compared to all OCRs. The accuracy of Tesseract is highest for three test datasets, namely *BMT_Nakula*, *BMT_Baloo* and *BMT_Aparajita*. Using the proposed MFD_OCR, we were able to minimize the insertion and deletion type of errors. These type of errors are difficult to handle in the post-processing stage. We are currently working on the possibility of further improving the accuracy by understanding the substitution types of error.

References

1. Mathew, M., Singh, A.K., Jawahar, C.: Multilingual OCR for indic scripts. In: 2016 12th IAPR Workshop on Document Analysis Systems (DAS), pp. 186–191 (2016)
2. Source OCR: Tesseract OCR [Computer software manual]. Retrieved from https://github.com/tesseract-ocr/. Accessed 14 Aug 2018
3. Hellwig, O.: Indsenz ocr [Computer software manual]. Retrieved from http://www.indsenz.com/int/index.php. Accessed 14 Aug 2018
4. tdil: eaksharaya [Computer software manual] (2018). Retrieved from http://tdil-dc.in/eocr/index.html. Accessed 14 Aug 2018
5. Technology development for Indian languages [Computer software manual]. Retrieved from http://tdil.meity.gov.in/. Accessed 26 Jan 2019
6. Sinha, R.: A syntactic pattern analysis system and its application to Devanagari script recognition. Electrical Engineering Department (1973)
7. Palit, S., Chaudhuri, B.: A feature-based scheme for the machine recognition of printed Devanagari script. In: Das, P.P., Chatterjee, B.N. (eds.) Pattern Recognition, Image Processing and Computer Vision, pp. 163–168. Narosa Publishing House, New Delhi, India
8. Pal, U., Chaudhuri, B.: Printed Devanagari script ocr system. VIVEK-BOMBAY **10**, 12–24 (1997)
9. Pal, U., Chaudhuri, B.: Indian script character recognition: a survey. Pattern Recogn. **37**(9), 1887–1899 (2004)
10. Karayil, T., Ul-Hasan, A., Breuel, T.M.: A segmentation-free approach for printed Devanagari script recognition. In: 2015 13th International Conference on Document Analysis and Recognition (ICDAR), pp. 946–950 (2015)
11. Sankaran, N., Jawahar, C.: Recognition of printed Devanagari text using BLSTM neural network. In: ICPR, vol. 12, pp. 322–325 (2012)
12. Krishnan, P., Sankaran, N., Singh, A.K., Jawahar, C.: Towards a robust OCR system for indic scripts. In: 2014 11th IAPR International Workshop on Document Analysis Systems (DAS), pp. 141–145 (2014)
13. Ul-Hasan, A., Ahmed, S.B., Rashid, F., Shafait, F., Breuel, T.M.: Offline printed Urdu Nastaleeq script recognition with bidirectional LSTM networks. In: 2013 12th international conference on Document Analysis and Recognition (ICDAR), pp. 1061–1065 (2013)
14. Acharya, S., Pant, A.K., Gyawali, P.K.: Deep learning based large scale handwritten Devanagari character recognition. In: 2015 9th International Conference on Software, Knowledge, Information Management and Applications (SKIMA), pp. 1–6 (2015)
15. Ul-Hasan, A., Bukhari, S.S., Dengel, A.: Ocroract: A sequence learning OCR system trained on isolated characters. In: DAS, pp. 174–179 (2016)
16. Dalvi, G.: Terminology of Devanagari typefaces [Computer software manual]. Retrieved from http://dsource.in/tool/devft/en/terminology.php. Accessed 20 Jan 2019
17. Type, D.E.: Baloo fonts [Computer software manual]. Retrieved from https://fonts.google.com/specimen/Baloo. Accessed 24 Jan 2019
18. Type, D.S.: Dekko fonts [Computer software manual]. Retrieved from https://fonts.google.com/specimen/Dekko. Accessed 24 Jan 2019
19. Font Biryani, G.: Biryani fonts [Computer software manual]. Retrieved from https://fonts.google.com/specimen/Biryani. Accessed 24 Jan 2019
20. Microsoft Corporation: Aparajita fonts [Computer software manual]. Retrieved from https://catalog.monotype.com/font/microsoft-corporation/aparajita/regular. Accessed 24 Jan 2019

21. Rice, S.V., Nartker, T.A.: The ISRI analytic tools for OCR evaluation. UNLV/Information Science Research Institute, TR-96-02 (1996)
22. cfilt.iitb: Word frequency for Hindi [Computer software manual]. Retrieved from http://www.cfilt.iitb.ac.in/Downloads.html. Accessed 14 Aug 2018
23. OCRopy: Open source document analysis and OCR system. [Computer software manual]. Retrieved from https://github.com/tmbdev/ocropy. Accessed 14 Jan 2019

Experimental Evaluation of CNN Architecture for Speech Recognition

Md Amaan Haque, Abhishek Verma, John Sahaya Rani Alex and Nithya Venkatesan

Abstract In recent days, deep learning has been widely used in signal and information processing. Among the deep learning algorithms, Convolution Neural Network (CNN) has been widely used for image recognition and classification because of its architecture, high accuracy and efficiency. This paper proposes a method that uses the CNN on audio samples rather than on the image samples in which the CNN method is usually used to train the model. The one-dimensional audio samples are converted into two-dimensional data that consists of matrix of Mel-Frequency Cepstral Coefficients (MFCCs) that are extracted from the audio samples and the number of windows used in the extraction. This proposed CNN model has been evaluated on the TIDIGITS corpus dataset. The paper analyzes different convolution layer architectures with different number of feature maps in each architecture. The three-layer convolution architecture was found to have the highest accuracy of 97.46% among the other discussed architectures.

Keywords Convolution Neural Networks (CNN) · Deep Neural Networks (DNN) · Kernel · Mel-Frequency Cepstral Coefficients (MFCC) · Speech Recognition (SR)

1 Introduction

Nowadays, speech recognition (SR) has become a fundamental feature of almost every smart machine. SR systems try to recognize human voice input and respond appropriately based upon the application. SR is used in applications such as password verification, personal devices, gaming and security. Systems use different algorithms to recognize speech. The most common ones are Hidden Markov Model (HMM), Deep Neural Network (DNN) [1] and Recurrent Neural Networks (RNNs) [2]. However, this paper proposes modified approach to apply Convolutional Neural Network (CNN) algorithm for speech recognition.

M. A. Haque · A. Verma · J. S. R. Alex (✉) · N. Venkatesan
Vellore Institute of Technology, Chennai 600127, India
e-mail: jsranialex@vit.ac.in

© Springer Nature Singapore Pte Ltd. 2020
A. K. Luhach et al. (eds.), *First International Conference on Sustainable Technologies for Computational Intelligence*, Advances in Intelligent Systems and Computing 1045, https://doi.org/10.1007/978-981-15-0029-9_40

Our aim is to modify the one-dimensional speech features extracted the from speech data and apply the state of the art CNN to train our machine in order to see the classification accuracy of CNN's for SR. CNNs have been used previously for the purpose of speech recognition. Kim [3] trained CNN on top of a pre-trained word vector for the classification of sentences. Sainath and Parada [4] compared CNN with for small footprint keyword spotting by limiting the number of parameters and multiplies. The automatic speech recognition (ASR) system in [5] shows that the relationship between the raw speech signal and the phones can be directly modeled and ASR systems competitive to standard approach can be built. But normalizations such as cepstral mean normalization were not performed on the extracted MFCC. Zhao et al. proposed CNN which was based on time-frequency kernel [6]. In contrast to the above works, we have applied 2D CNN by modifying the dimension of extracted speech features.

This paper is organized as follows: Sect. 2 describes the system design. Section 2.1 gives a theoretical overview of CNN architecture which has been used for our research. Section 2.2 gives a brief workflow of the proposed system. Experimental setup has been described in Sect. 3. Classification performance of CNN on our modified speech data is analyzed in Sect. 4. Finally, based on the results, a brief conclusion is made in Sect. 5.

2 System Design

The Convolutional Neural Networks (CNN) is a deep learning algorithm that is predominantly used for image identification [7] and classification/categorization. The architecture of CNN is shown in Fig. 1. CNNs are giving unprecedented results with high accuracy in the above-mentioned field. Due to the reason that CNN works best on a 2D array, not much work has been done to apply this algorithm in speech recognition with audio signals being the 1D data. The speech signal is framed with window size of 20 ms with an overlap of 10 m. Thirty-nine MFCC speech features are extracted for a frame of 20 ms. The frames are stacked vertically and the corresponding MFCC speech features are kept as rows.

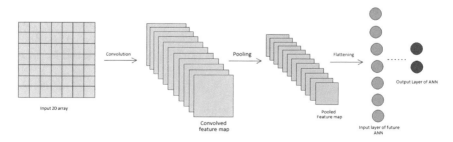

Fig. 1 CNN architecture

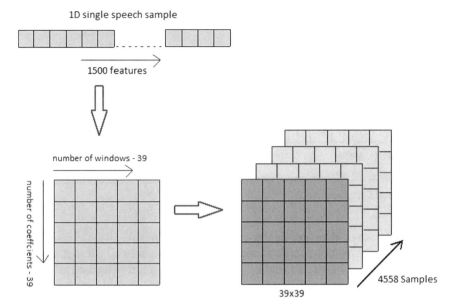

Fig. 2 1D speech feature to 2D array

In this way, each 1D audio sample was converted into 2D array of speech features and this is shown in Fig. 2.

The CNN was applied on 2D array and prediction accuracy was obtained.

2.1 CNN Overview

A CNN is a modified version of DNN where the input data undergoes several processes like convolution, max pooling [8], flattening and full connection in ordered fashion before being fed into the deep neural network.

- Convolution
  A convolution is a combined integration of two functions which is shown in Eq. 1. It tells us how one function modifies the shape of the other function.

$$(f * g)(t) \overset{\text{def}}{=} \int_{-\infty}^{\infty} f(\tau)g(t - \tau)\mathrm{d}\tau \tag{1}$$

In the case of images, the first function is the image itself, and the second function is the feature detector/kernel [9]. The feature detector convolves over the whole image stride by stride and outputs the convolved feature map containing the most relevant/dominant information, thus reducing the dimension of the input image as

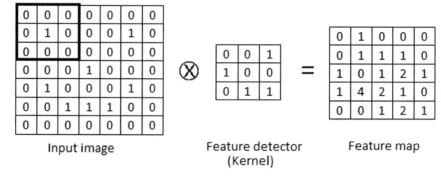

Input image   Feature detector   Feature map
                     (Kernel)

Fig. 3 Typical CNN convolution

well. This convolution process creates feature maps where each individual feature map is a convolved result of different individual feature detector which is shown in Fig. 3. Rectified linear unit (ReLu) activation function is then applied to increase the nonlinearity in the resulting feature maps in order to distinguish adjacent pixels of the maps more accurately.

- Pooling

  It is applied on the feature maps after convolution operation.

  Even similar input data (images or 2D array vary), i.e., belonging to the same class vary widely in shape, texture, etc. For example, not all dogs look same, there may be different breeds of dogs which look completely different from each other, take German Shepherd and a Pug in this case, they look entirely different but come into the same category of 'Dogs'. Let's take one more scenario, this time the three input pictures are of the same dog namely German Shepherd but in different orientations, i.e., in one on them the German Shepherd is standing and in the other, it is in the sitting position or maybe the third picture is flipped upside down. The point here is, to a computer, these three pictures are completely different. Therefore, in order to predict all of them under the same category, pooling is applied. Pooling digs deeper into the feature maps and pools out the most relevant dominant features though fewer in numbers. For example, no matter which dog it is, it'll definitely have two frontal eyes, a tail, etc., so these dominant features will be selected/pooled out by the pooling layer. Thus, pooling accounts for special and texture distortions, preserve the feature and further reduces the size of the feature maps by as much as 75%, thus reducing number of parameters going to the neural network and in turn prevents overfitting.

- Flattening

  The pooled feature map needs to go as an input in Artificial Neural Network (ANN) for further processing. For this, the 2D feature map is flattened, i.e., converted into 1D column by taking row by row values of the feature map. This results in dimensional column of feature map which acts as an input to the ANN.

Fig. 4 Flowchart of the proposed system

- Full Connection
  The flattened layer, i.e., input layer to the ANN is fully connected with the subsequent layer of the ANN as the combination of input features in the input layer and its attributes in the next layer in ANN results in much better accuracy.

2.2 Proposed System

CNN's are widely used for image recognition application due to its ability to work efficiently on 2D array. Images are nothing but two-dimensional arrays of numeric pixel values, where each pixel in the image corresponds to a numeric value ranging from 0 to 255, where 0 corresponds to black and 255 corresponds to white color. So basically, when CNN is applied on any image/picture, in actual case CNN is being applied on a two-dimensional array of numbers which forms the image. This idea is used to apply CNN algorithm on TIDIGITS database [10] containing 4558 utterances in total where each sample contains 1D array of 1500 feature vectors extracted using MFCC technique. The system flow of the proposed system is shown in Fig. 4.

Each one-dimensional sample was converted in 2D 39 × 39 array where the number of columns, i.e., 39 is obtained by dividing the total duration (1500 ms) by width of the window taken, i.e., 25 ms with 15 ms overlap and the number of rows, i.e., 39 are the total MFC coefficients comprising of 13 static, 13 delta and 13 acceleration coefficients. Thus, in this way, 4558 two-dimensional arrays were obtained on which 2D CNN can be applied.

3 Experimental Setup

TIDIGITS database having in total 4558 samples containing utterances of digits 'zero' to 'nine' is taken as the dataset where each digit is pronounced couple of times by the speaker. The train and test split is made as 75% and 25%. In total, 1500 features are extracted using MFCC technique for each sample. Each one-dimensional MFCC feature was converted in 2D 39 × 39 array where the number of columns, i.e., 39 is the total MFC coefficients comprising of 13 static, 13 delta and 13 acceleration coefficients of one frame of speech signal which is of 25 ms duration. Since speech signal is divided into multiple frames with 25 ms duration, speech features of multiple frames are stacked which form 2D array as that of an image. Thus, in this way, for each speech sample, 2D array is obtained, hence 4558 two-dimensional arrays were obtained for the TIDIGITS corpus which is then can be applied to 2D CNN. CNN model is trained on the Intel processor, having 1.70 GHz CPU with 4 GB RAM. Jupyter notebook is used as a working environment. Numpy and Keras library is used with TensorFlow framework backend. The training and testing split done with scikit-learn library. The proposed CNN model is trained on 50 epochs.

4   Results and Discussions

The CNN model trained on the modified dataset of speech samples gave 97.46% accurate predictions on the test set with the maximum validation accuracy of 98.07%. The CNN model was trained with three different convolution layer architectures. The first architecture consists of only one convolution layer with 64 feature maps and one pooling layer after the convolution layer. This architecture gave an accuracy of 95.16% on the training dataset. The second architecture consists of two convolution layers, the first layer with 64 feature maps and then the second layer with 64 feature maps and has a pooling layer after each convolution layer. This architecture showed an accuracy of 96.05% on the training dataset. The third architecture with three convolution layers has feature maps increasing twice the number of feature maps in the previous layer, the first layer has 64 feature maps that doubles to 128 feature map in the second layer and then becomes 256 feature maps in the final layer. Each convolution layer is followed by a max pooling layer. This architecture with three convolution layers shows an accuracy of 97.46% which is the highest among the three convolution architectures discussed in this paper. The model was trained with 50 epochs and different dimensions of feature maps. Table 1 lists the various parameters of the proposed CNN architecture of speech recognition (Fig. 5).

Table 1 CNN architecture parameters

Parameters	Values
Total samples	4558
Training set samples	3419
Test set samples	1139
Modified array dimensions	39×39
Epochs	50
DNN fully conn layer activation function	ReLu
Number of CNN layers	3
Max CNN accuracy	97.46%

Accuracy(%)

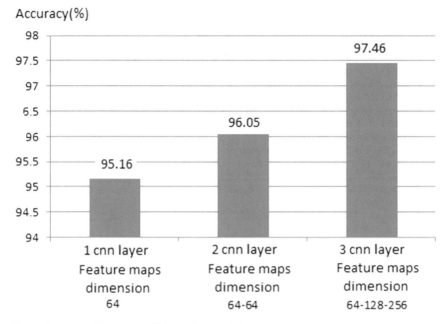

Fig. 5 Accuracy of proposed CNN speech recognition system

5 Conclusion

The 2D CNN architecture takes 2D data as input such as image. In this paper, one-dimensional MFCC speech features were converted into two-dimensional format and applied as input to the CNN architecture. A one-layer CNN was able to give the classification accuracy as good as 95.16%. Further, CNN layers were added to the network which resulted in much higher classification accuracy. A three-layer CNN was able to give the best classification results of 97.46% on the modified speech features. In future, the MFCC features could be reduced using dimensionality reduction methods and the transformed features could be evaluated with the three-layer CNN architecture.

References

1. Hinton, G., et al.: Deep neural networks for acoustic modeling in speech recognition: the shared views of four research groups. IEEE Sig. Process. Mag. **29**(6), 82–97 (2012)
2. Graves, A., Mohamed, A., Hinton, G.: Speech recognition with deep recurrent neural networks. In: 2013 IEEE International Conference on Acoustics, Speech and Signal Processing (ICASSP). IEEE (2013)
3. Kim, Y.: Convolutional neural networks for sentence classification(2014). arXiv preprint arXiv: 1408.5882

4. Sainath, T.N., Parada, C.: Convolutional neural networks for small-footprint keyword spotting. In: Sixteenth Annual Conference of the International Speech Communication Association (2015)

5. Palaz, D., Magimai-Doss, M., Collobert, R.: Analysis of cnn-based speech recognition system using raw speech as input. In: Sixteenth Annual Conference of the International Speech Communication Association (2015)

6. Zhao, T., Zhao, Y., Chen, X.:. Time-frequency kernel-based CNN for speech recognition. In: Sixteenth Annual Conference of the International Speech Communication Association (2015)

7. Krizhevsky, A., Sutskever, I., Hinton, G.E.: Imagenet classification with deep convolutional neural networks. In: Advances in Neural Information Processing Systems (2012)

8. Tolias, G., Sicre, R., Jégou, H: Particular object retrieval with integral max-pooling of CNN activations (2015). arXiv preprint arXiv:1511.05879

9. Li, S., et al.: Shape driven kernel adaptation in convolutional neural network for robust facial traits recognition. In: Proceedings of the IEEE Conference on Computer Vision and Pattern Recognition (2015)

10. Leonard, G., Doddington, G.: TIDIGITS LDC93S10. Web Download. Linguistic Data Consortium, Philadelphia (1993)

Design and Simulation of a Single-Phase SHEPWM Inverter-Fed Induction Motor Drive Using Generalized Hopfield Neural Network

Velide Srinath, ManMohan Agarwal and Devendra Kumar Chaturvedi

Abstract This paper discusses the design of a single-phase $(1 - \phi)$ inverter-fed induction motor drive which uses GHNN controlled selective harmonic elimination pulse width modulation (SHEPWM) technique. SHEPWM is a popular switching technique which reduces the lower-order harmonics by proper selection of the switching angles. The initial switching angles for modulation index (M) varying from 0.1 to 0.9 with an interval of 0.1 are derived by solving five nonlinear algebraic transcendental equations using genetic algorithm technique. In this paper, a generalized Hopfield neural network (GHNN) is used for reduction of the fifth-, seventh-, eleventh-, and thirteenth-order harmonics in the output voltage of the inverter-fed $1 - \phi$ induction motor for any modulation index which changes continuously. A set of five ordinary differential equations (ODEs) describing the behavior of GHNN is obtained from the energy function which is formulated using the solution set of nonlinear algebraic transcendental equations. In some application, the modulation index (M) changes continuously due to various factors and the switching angles to eliminate the low-order harmonics are determined by solving the above-obtained ODEs using Runge–Kutta fourth-order method. The performance of the proposed inverter drive was simulated for both $1 - \phi$ capacitor start and $1 - \phi$ capacitor start and run induction motors (IM), for varying modulation index, and with the sudden change in the load torque, with and without a filter are presented. The proposed SHEPWM using GHNN for induction motor drive has been simulated using MATLAB/Simulink with satisfactory results.

Keywords Generalized Hopfield neural network (GHNN) · Induction motor · Modulation index · MATLAB/Simulink · RK4 · SHEPWM

V. Srinath (✉) · M. Agarwal · D. K. Chaturvedi
Faculty of Engineering, Dayalbagh Educational Institute, Agra, India
e-mail: srinathlal@gmail.com

M. Agarwal
e-mail: a.manmohan@yahoo.co.in

D. K. Chaturvedi
e-mail: dkc.foe@gmail.com

© Springer Nature Singapore Pte Ltd. 2020
A. K. Luhach et al. (eds.), *First International Conference on Sustainable Technologies for Computational Intelligence*, Advances in Intelligent Systems and Computing 1045, https://doi.org/10.1007/978-981-15-0029-9_41

1 Introduction

The economic growth and development of any state or country will be depended on the rapid industrialization and reliable power supply. Due to the modernization of human lifestyle and continuous increase in the demand for electrical energy for growing industrial and household application of electrical motors forces to increase the power generation capacity especially in countries like India. In recent years, power electronic devices are playing a vital role in industrial drives and control systems as they have several technical advantages compared to other conventional devices and are also easily available in the market at a low cost. In inverter fed, electric drive systems harmonics are the major issue for voltage changing and produce fluctuating torque, creating losses as well as surges in power supply. The PWM inverter output has a several unwanted harmonics which can be eliminated by several techniques proposed by scientist and engineering in [1] a review of bio-inspired HE technique are described in detail.

In [2–4], single-phase AC voltage controller is used for various circuit configurations of R–L load and induction motor. A two-phase motor is considered as a single-phase induction motor, this gives us an opportunity to use a single-leg voltage source inverter [5–7]. In [8] DTC of single-phase symmetrical and asymmetrical induction motor for various load conditions, it was observed that it is possible to obtain a balanced operation of machine irrespective of stator asymmetry. In recent years [9–16], extensive research is performed on various drives like BLDC, switched reluctance motor, SRM-driven water pumping system controlled by different converter topologies and strategies. A literature survey of various existing converter topologies, which have been proposed for adjustable speed single-phase induction motor drives, is discussed in [17]. In [18–22], single-phase induction motor drives for various applications, speed control, water pumping, and vector method control is investigated. In [23], single-phase step-down cyclo-converter is used to control the speed of induction motor.

In this paper, single-phase selective harmonics elimination PWM inverter-fed induction motor is simulated using MATLAB/Simulink. SHEPWM inverter has gained popularity since 1970 and has been a subject of research for both academics and industry. Obtaining the pattern of the switching angles in the case of SHEPWM involves the solution of a set of nonlinear transcendental equations. These equations can be solved by various optimization techniques like GA, PSO, BEE, etc.; recent year's researcher used ANN for determining the switching instants. In [24–28], design and analysis of SHEPWM inverter using GHNN for continuously varying modulation are discussed but for only R and R–L load. In this paper, SHEPWM inverter using GHNN with filter and without a filter for the induction motor is investigated.

This paper is organized as follows. The application of GHNN for single-phase SHEPWM is discussed in Sect. 2, and simulation results and performance of single-phase induction motor without and with filter are discussed in Sect. 3. In Sect. 4, conclusion, future scope, and reference are mentioned at the end.

2 Implementation of SHEPWM Using GHNN

The Simulink diagram of the proposed inverter-fed induction motor using GHNN is given in Fig. 1. The PWM generator is connected to two different $1 - \phi$ inverter circuits which are further fed to two different single-phase induction motor, i.e., one is a capacitor start and the other is a capacitor start and run single-phase induction motors. In [29], a novel approach for solving a set of nonlinear algebraic equations using GHNN-type architecture to optimize an energy function has been proposed and stability analysis for a dynamic model has been carried out. In this paper, using the above technique, five algebraic transcendental SHEPWM nonlinear equations are solved to reduce the fifth-, seventh-, eleventh-, and thirteenth-order harmonics and the higher-order harmonics can be filtered using a simple RC circuit. The desired fundamental voltage is retained in that inverter which is fed to a $1 - \phi$ IM using MATLAB and Simulink. In Fig. 1, the switching angles for any given modulation index (M) are determined by the GHNN block which reduces the unwanted harmonics; using these angles, the PWM block will generate the gating singles which are applied to the power switches of H-bridge inverter whose output is further fed to $1 - \phi$ IM.

Figure 2 shows the unipolar voltage waveform of a standard $1 - \phi$ H-bridge inverter.

Equation (1) is determined by solving the waveform of Fig. 2 with the help of Fourier series to represent the periodic waveform as a sum of sine and cosine functions.

$$v(\omega t) = \frac{4V_{dc}}{\pi} \left\{ \sum_{n=1,3,5,\ldots}^{\infty} \frac{\sin(n\omega t)}{n} \times (\cos(n\alpha_1) - \cos(n\alpha_2)) \right.$$

$$\left. + \cos(n\alpha_3) - \cos(n\alpha_4) + \cos(n\alpha_5)) \right\} \tag{1}$$

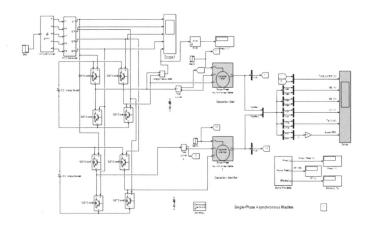

Fig. 1 Simulink model of proposed GHNN controlled inverter-fed induction motor

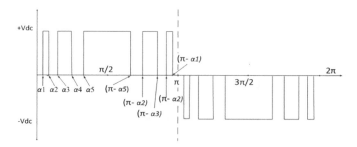

Fig. 2 The output voltage waveform of single-phase H-inverter

The problem here is to reduce the lower-order (i.e., fifth, seventh, eleventh, and thirteenth) harmonics by calculating the unknown switching angles α_1, α_2, α_3, α_4, and α_5 with the help of the five transcendental nonlinear algebraic equations from (2) to (6) for any given desired fundamental voltage (V_1) and M, where $M = V_1/(4V_{dc}/\pi)$.

$$\cos\alpha_1 - \cos\alpha_2 + \cos\alpha_3 - \cos\alpha_4 + \cos\alpha_5 = M \tag{2}$$

$$\cos 5\alpha_1 - \cos 5\alpha_2 + \cos 5\alpha_3 - \cos 5\alpha_4 + \cos 5\alpha_5 = 0 \tag{3}$$

$$\cos 7\alpha_1 - \cos 7\alpha_2 + \cos 7\alpha_3 - \cos 7\alpha_4 + \cos 7\alpha_5 = 0 \tag{4}$$

$$\cos 11\alpha_1 - \cos 11\alpha_2 + \cos 11\alpha_3 - \cos 11\alpha_4 + \cos 11\alpha_5 = 0 \tag{5}$$

$$\cos 13\alpha_1 - \cos 13\alpha_2 + \cos 13\alpha_3 - \cos 13\alpha_4 + \cos 13\alpha_5 = 0 \tag{6}$$

The energy function (7) is formulated using five nonlinear equations from (2) to (6)

$$
\begin{aligned}
E = -0.5\big\{ & (\cos\alpha_1 - \cos\alpha_2 + \cos\alpha_3 - \cos\alpha_4 + \cos\alpha_5 - M)^2 \\
& + (\cos 5\alpha_1 - \cos 5\alpha_2 + \cos 5\alpha_3 - \cos 5\alpha_4 + \cos 5\alpha_5)^2 \\
& + (\cos 7\alpha_1 - \cos 7\alpha_2 + \cos 7\alpha_3 - \cos 7\alpha_4 + \cos 7\alpha_5)^2 \\
& + (\cos 11\alpha_1 - \cos 11\alpha_2 + \cos 11\alpha_3 - \cos 11\alpha_4 + \cos 11\alpha_5)^2 \\
& + (\cos 13\alpha_1 - \cos 13\alpha_2 + \cos 13\alpha_3 - \cos 13\alpha_4 + \cos 13\alpha_5)^2 \big\}
\end{aligned}
\tag{7}
$$

The differential equations (DE) from (8) to (12) are obtained using energy function (7) which governs the network dynamics for implementing GHNN. These five first-order DEs are solved by applying Runga–Kutta-order (RK4) method and the flowchart of this algorithm is shown in Fig. 3.

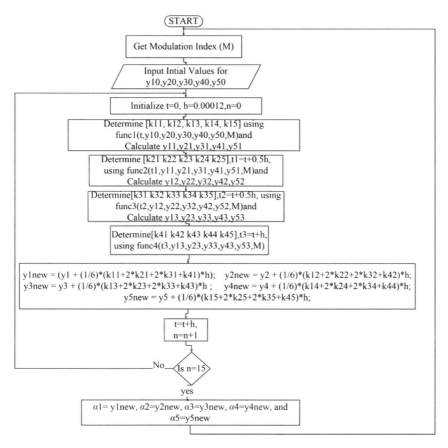

Fig. 3 Flowchart to determine switching angles using RK4 method

$$\frac{d\alpha_1}{dt} = -\frac{\partial E}{\partial \alpha_1} \tag{8}$$

$$\frac{d\alpha_2}{dt} = -\frac{\partial E}{\partial \alpha_2} \tag{9}$$

$$\frac{d\alpha_3}{dt} = -\frac{\partial E}{\partial \alpha_3} \tag{10}$$

$$\frac{d\alpha_4}{dt} = -\frac{\partial E}{\partial \alpha_4} \tag{11}$$

$$\frac{d\alpha_5}{dt} = -\frac{\partial E}{\partial \alpha_5} \tag{12}$$

Table 1 shows the results of initial input values (y10, y20, y30, y40, y50) of switching angles obtained using MATLAB GA toolbox. The energy function (7)

Table 1 Initial switching angles obtained from GA toolbox

Modulation index	α_1 in radians	α_2 in radians	α_3 in radians	α_4 in radians	α_5 in radians
0.1	0.8497	0.8833	1.192	1.248	1.543
0.2	0.8593	0.8836	1.1919	1.248	1.543
0.3	0.8432	0.8919	1.161	1.27	1.514
0.4	0.8243	0.8974	1.131	1.289	1.4179
0.5	0.8156	0.8962	1.114	1.299	1.426
0.6	0.7974	0.8812	1.034	1.301	1.347
0.7	0.6048	0.6458	0.8645	1.044	1.114
0.8	0.4441	0.5226	0.6992	0.8662	0.9897
0.9	0.3362	0.416	0.5793	0.8407	0.9194
1	0.2473	0.3399	0.492	0.7436	0.7823

along with the given constraints is optimized in such a way that the lower-order harmonics are eliminated.

3 Simulation Results and Discussion

3.1 Generation of Gate and Carrier Signal

The block diagram of GHNN inverter fed to $1 - \phi$ IM is shown in Fig. 4 which is simulated using MATLAB/Simulink. The modulation index (M) is accepted from 0.05 to 1 with a resolution of 0.01 and fed to the GHNN block where the five nonlinear algebraic transcendental equations from (2) to (6) are solved using RK4 and determine the five unknowns α_1, α_2, α_3, α_4, and α_5. These switching angles which are in radians are given to the PWM generator.

In this PWM generator, two carrier triangular signals of magnitude 1.57 with a frequency of 50 Hz are generated as shown in Fig. 5a, c. The triangular carrier signal and the switching angles from GHNN block are compared using a comparator which produces the gate signals as shown in Fig. 5b, d.

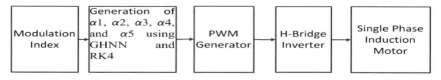

Fig. 4 Block diagram to implement GHNN inverter-fed induction motor

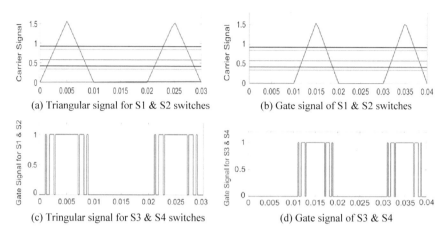

Fig. 5 Gate and carrier signal of H-bridge inverter

The gate signal generated from the PWM generator is fed to the opposite arms (S1, S3) and (S2, S4) of the H-bridge inverter shown in Fig. 5. A single-phase induction motor is connected to this inverter.

3.2 Output Voltage and Current Waveforms

Figure 6 shows output voltage waveform of single-phase inverter for a modulation index m = 0.865. Figure 7 shows the harmonic spectrum of the output voltage shown in Fig. 6.

Figure 8 shows the variation in modulation index at time 1, 2, and 3 s and the corresponding variation in the output voltage waveforms is shown in Fig. 9a, b. Figure 10a, b shows the RMS voltages fed to induction motor with and without RC filter.

Figures 11 and 12 show the stator current in the main and auxiliary winding. Figure 13 shows the voltage across the capacitor. Figures 14 and 15 show the speed and torque variations for with and without a filter.

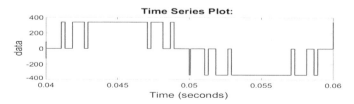

Fig. 6 Voltage waveform for a modulation index $m = 0.865$

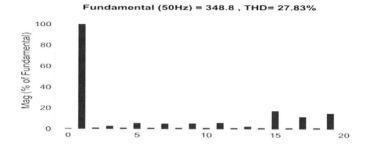

Fig. 7 Harmonic spectrum of output voltage using FFT

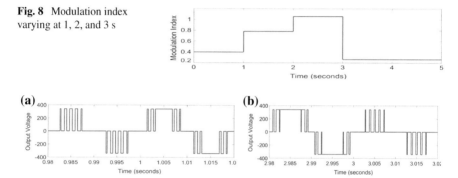

Fig. 8 Modulation index varying at 1, 2, and 3 s

Fig. 9 Output voltage for varying modulation index **a** 0.41 to 0.785, **b** 1.06 to 0.267

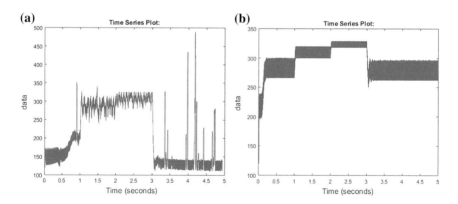

Fig. 10 RMS voltages for changing modulation index **a** without a filter, **b** with filter

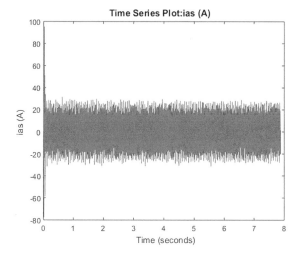

Fig. 11 Main winding stator current ia

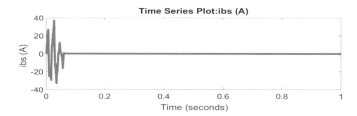

Fig. 12 Auxiliary winding stator current ib

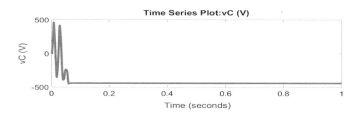

Fig. 13 Capacitor voltage

4 Future Scope of the Proposed Work

This work can be extended for a solar controlled induction motor. For changing irradiation and temperature using a suitable MPPT method, different modulation index can be calculated and fed to the proposed GHNN inverter-fed induction motor for better performance.

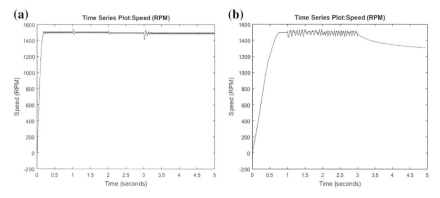

Fig. 14 Speed variation **a** with RC filter, **b** without filter

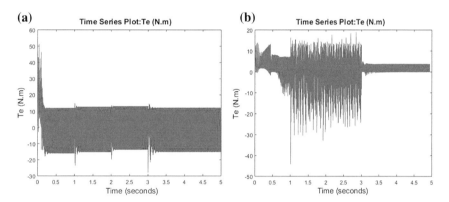

Fig. 15 Torque variation **a** with RC filter, **b** without the filter

5 Conclusion

In this paper with the use of GHNN and RK4, the switching angle for SHEPWM inverter which minimizes the fifth-, seventh-, eleventh-, and thirteenth-order harmonic in the output voltage is fed to the single-phase induction motor. The higher-order harmonics can be easily removed by use of a simple RC filter of the very low rating. The various performances result with and without a filter and analyzed. The proposed SHEPWM can also be applied to the renewable energy system and other low and medium-high power applications.

References

1. Memon, M., Mekhilef, S., Mubin, M., Aamir, M.: Selective harmonic elimination in inverters using bio-inspired intelligent algorithms for renewable energy conversion applications: a review. Renew. Sustain. Energy Rev. **82**, 2235–2253 (2018)
2. Shepherd, W.: Thyristor control of AC Circuits, 1st edn. Book, Lockwood staples Ltd., England (1975)
3. Dubey, G.K., Doradla, S.R., Hoshi, A., Sinha, R.M.: Thyristorized Power Controllers. Book, Wiley, India (1986)
4. Hamed, S., Chalmers, B.: Analysis of variable-voltage thyristor controlled induction motors. IEEE Proc. **137**(3), Pt. B, 184–193 (1990)
5. Popescu, M., Demeter, E., Micu, D., Navrapescu, V., Jokinen, T.: Analysis of a voltage regulator for a two-phase induction motor drive. In: Proceedings of IEMD—International Conference on Electric Machines and Drives, Seattle, USA, pp. 658–660 (1999)
6. Klíma, J.: Analytical model of induction motor fed from four-switch space vector PWM inverter. Time domain analysis. Acta Tech. CSAV **44**, 393–410 (1999)
7. Chomát, M., Lipo, T.A.: Adjustable-speed single-phase IM drive with reduced number of switches. In: Conference Record IEEE-IAS Annual Meeting, pp. 1800–1806 (2001)
8. Dangeam, S., Kinnares, V.: A direct torque control for three-leg voltage source inverter fed asymmetrical single-phase induction motor. In: 2015 18th International Conference on Electrical Machines and Systems (ICEMS), Pattaya, pp. 1569–1574 (2015)
9. Narayana, V., Kumar Mishra, A., Singh, B.: Development of low-cost PV array-fed SRM drive-based water pumping system utilizing CSC converter. IET Power Electron. **10**(2), 156–168 (2017)
10. Singh, B., Kumar, R.: Simple brushless DC motor drive for solar photovoltaic array fed water pumping system. IET Power Electron. **9**(7), 1487–1495 (2016)
11. Kumar, R., Singh, B.: BLDC motor-driven solar PV array-fed water pumping system employing zeta converter. IEEE Trans. Ind. Appl. **52**(3), 2315–2322 (2016)
12. Singh, B., Kumar, R.: Solar PV array fed brushless DC motor driven water pump. In: IEEE 6th International Conference on Power Systems (ICPS), New Delhi, pp. 1–5 (2016)
13. Kumar, R., Singh, B.: Solar PV-battery based hybrid water pumping system using BLDC motor drive. In: 2016 IEEE 1st International Conference on Power Electronics, Intelligent Control and Energy Systems (ICPEICES), Delhi, pp. 1–6 (2016)
14. Mishra, A.K., Singh, B.: Solar energized SRM driven water pumping utilizing modified Landsman converter. In: 2016 IEEE 1st International Conference on Power Electronics, Intelligent Control and Energy Systems (ICPEICES), Delhi, pp. 1–6 (2016)
15. Singh, B., Kumar, R.: Solar photovoltaic array fed water pump driven by brushless DC motor using Landsman converter. IET Renew. Power Gener. **10**(4), 474–484 (2016)
16. Singh, B., Mishra, A.K., Kumar, R.: Solar powered water pumping system employing switched reluctance motor drive. IEEE Trans. Ind. Appl. **52**(5), 3949–3957 (2016)
17. Ba-Thunya, A.S., Khopkar, R., Kexin, W., Toliyat, H.A.: Single phase induction motor drives-a literature survey. In: IEEE International Electric Machines and Drives Conference IEMDC, pp. 911–916 (2001)
18. Oladepo, O., Adegboyega, G.A.: MATLAB simulation of single-phase SCR controller for single phase induction motor. Int. J. Electron. Electr. Eng. **5**(2), 135–142 (2012)
19. Leicht, A., Makowski, K.: Analysis of a single phase capacitor induction motor operating of two power line frequencies. Arch. Electr. Eng. **61**(2), 251–266 (2012)
20. Khader, S.H.: Modeling and simulation of single phase double capacitors induction motor. In: Proceedings of the 2nd WSEAS International Conference on Biomedical Electronics and Biomedical Informatics, pp 21–27 (2009)
21. Lee, K.J., Kim, H.G., Lee, D.K., Chun, T.W., Nho, E.C.: High-performance drive of single-phase induction motor. In: Proceedings of IEEE International Symposium on Industrial Electronics, vol. 2, pp. 983–988 (2001)

22. Abdel-Rahim, N.M.B., Shaltout, A.: High performance single-phase induction motor drive system. In: Proceedings of IEEE Eleventh International Middle East Power Systems Conference, vol. 1, pp. 385–392 (2006)
23. Epemu, A.M., Enalume, K.O.: Speed control of a single phase induction motor using step-down cycloconverter. Int. J. Ind. Manuf. Syst. Eng. **3**(1), 6–10 (2018)
24. Balasubramonian, M., Rajamani, V.: Design and real-time implementation of SHEPWM in single-phase inverter using generalized Hopfield neural network. IEEE Trans. Ind. Electron. **61**(11), 6327–6336 (2014)
25. Anand, B., Balasubramonian, M.: Generalized Hopfield neural network based SHEPWM in a single phase inverter. Int. J. Adv. Inf. Sci. Technol. (IJAIST) **22**(22) (2014)
26. Karunanidhi, A.: Artificial neural network based SHEPWM controller for single phase voltage source inverter. Int. J. Appl. Eng. Res. (2015)
27. Patil, D.R., Guruprasad, K.: Selective harmonic elimination PWM using generalized Hopfield neural network for multilevel inverters. Int. Res. J. Eng. Technol. (IRJET) **4**(6) (2017)
28. Ghonge, P.A., Agiwale, P.L.: GHNN based selective harmonic elimination implementation for single phase inverter. Int. J. Eng. Res. Technol. (IJERT) **8**(2) (2019)
29. Mishra, D., Kalra, P.K.: Modified Hopfield neural network approach for solving nonlinear algebraic equations. Eng. Lett. **14**(1), 135–142 (2007)

A Novel Mechanism for Host-Based Intrusion Detection System

Ch. Gayathri Harshitha, M. Kameswara Rao and P. Neelesh Kumar

Abstract Today, the world is progressively associated with the Internet so that the attackers and hackers are having a high opportunity to enter PCs and networks. In today's world hackers are using different types of attacks for receiving valuable information. It is important to recognize these attacks ahead of time to secure end clients and the system effects. Intrusion detection system (IDS) has been generally conveyed in PCs and systems to recognize the variety of attacks. In this paper, the basic observation is on log monitoring in host-based intrusion detection systems. In this paper, host-based intrusion detection is achieved using OSSEC tool. By using the OSSEC, the system is capable of detecting the malicious logs which run in the background from the system.

Keywords Intrusion detection system · Host-based · OSSEC · Malicious logs · Log monitoring

1 Introduction

By observing the drastic increase in cybercrime in present scenarios from worldwide, it is important to take safety efforts which guarantee the integrity, availability, and reliability of computer systems to protect clients from the intruders. In this paper, a tool is implemented in the systems such as firewalls, virtual private networks, and authentication mechanism to protect an organization from malicious attacks. However, this mechanism finds vulnerabilities as well, and cybercriminals who intrude

Ch. Gayathri Harshitha (✉) · M. Kameswara Rao · P. Neelesh Kumar
Department of Electronics & Computer Engineering, KLEF,
Vaddeswaram, Guntur, India
e-mail: gayathrichittelu@gmail.com

M. Kameswara Rao
e-mail: Kamesh.manchiraju@kluniversity.in

P. Neelesh Kumar
e-mail: nani.pokala@gmail.com

© Springer Nature Singapore Pte Ltd. 2020
A. K. Luhach et al. (eds.), *First International Conference on Sustainable Technologies for Computational Intelligence*, Advances in Intelligent Systems and Computing 1045,
https://doi.org/10.1007/978-981-15-0029-9_42

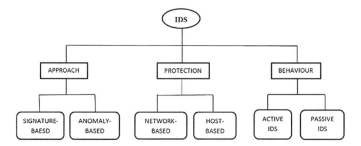

Fig. 1 IDS classification

into the organizations through computer systems. In this mechanism, security mechanisms which can monitor the system real time, detect attacks, and take countermeasure. Accordingly, by these implications, intrusion detection systems started to play a vigorous role in computer security.

Intrusion detection system monitors network traffic for suspicious action and cautions the server when such action is found. These IDSs are accomplished to take actions when any of the malicious activity or bizarre traffic is detected such as it blocks traffic sent from suspicious IP addresses. IDSs are alarm systems that are designed to detect unauthorized access which enters into the system without proper authorization (Fig. 1).

1.1 Approach

Signature-based IDS

Signature-based IDS performs analysis and comparison between server data and client data with the signature of known attacks. There are many attacks which have their own signatures. In case of similarity in between signatures, an alert will be delivered to the server. It has more precision and standard cautions assured to the client.

Anomaly-based IDS

AIDS has the ability to detect new and unique attacks. It comprises of a statistical model of ordinary system traffic which comprises of the data transmission, protocols, ports, and the devices. It frequently monitors the network traffic, and comparison is done with the statistical model. If there should arise an occurrence of any disturbance, the server will get alarms.

1.2 Behavior

Active IDS

AIDS is also known as intrusion detection and prevention system (IDPS). AIDS is configured to habitually block disbelieved assaults with no intercession required by a server. It provides quick healing activity because of an assault.

Passive IDS

PIDS configured just to monitor and break down the system traffic action and alert organization systems to potential vulnerabilities and assaults. PIDS cannot perform any defending or counteractive actions in response to attack its own.

1.3 Protection

Network-based IDS

The network-based IDS comprises of network-based applications (like sensors) that identify intrusions by looking at system traffic and monitor different hosts. Network-based applications capture overall network traffic and analyze the contents from the individual packets to find malicious traffic. Some of the tools of NIDS are Snort, Suricata, Fail2Ban, Sagan, OpenWIPS–NG, and Bro IDS.

Host-based IDS

Host-based IDS involves small programs (such as agents) that are to be introduced on individual systems which are to be checked. HIDS does not monitor the entire network; it can only monitor the individual host systems on which the agents are installed. Agents on a host which recognizes interruptions by the examination of system calls, application logs, document system alterations (binaries, secret key records, limit databases, access control records, etc. and other host works out).

1.4 Features

- IDSs monitor and analyze the system activities of the user.
- It looks at the documents which are available in the system and different setups and the working system.
- It surveys the integrity of the system and information records and leads the investigation of examples dependent on known assaults.
- It distinguishes misrepresentation in the system setup and alarms the system.

2 HIDS

A host-based IDS is a system that monitors a PC system on which it is introduced to distinguish an intrusion and additionally abuse and it reacts by logging the action and telling the approved client. HIDS agent monitors and breaks down in the case of anything or anybody, regardless of whether interior or outside, has maintained a strategic distance from the systems security rules (Fig. 2).

2.1 OSSEC

OSSEC (Open-Source Security) tool performs analysis of system logs, to check the integrity, Windows registry monitoring, and detection of the rootkit, time-based alerting, and send active responses.

   The OSSEC tool is used for the results because of the following:

- Open-Source log analysis
- Easy installation
- Easily customized (rules and configuration in XML format)
- Scalable (client/server design)
- Multi-platform (Windows, Solaris, Linux, *BSD, and so on.)
- By default secure (need to make the declaration/private key for SSL)
- Accompanies numerous decoders/rules which investigation our logs: (telnetd, Su, Sudo, vsftpd, Postfix, Apache, Syslog, and so forth.)

Architecture

OSSEC uses client/server design. It has a server for system observing and getting data from the agents.

Fig. 2 Packets transfer from the Internet to the system

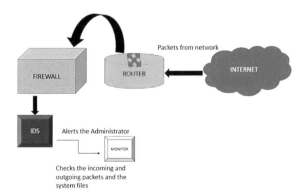

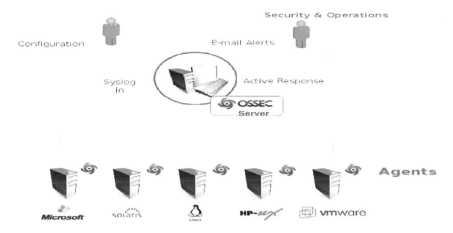

Fig. 3 Sending alerts from server to agents

Server

It saves the record integrity by checking the databases, log documents, events, and system checking gets to. Every rule, decoders, and significant arrangement choices are put away midway in the server, making it simple to control even a substantial number of agents.

Agents

The agent contains a small program, or a group of programs, which are monitored by installing on the systems. For the analysis and association, the agent will gather information and forward it to the server (Fig. 3).

2.2 OSSEC Flow Diagram

The communication between agents and server occurs on port 1514/udp on secure mode (Fig. 4).

File Integrity Monitoring (FIM)

FIM files change the server's file system. System logs can be changed in the monitored file, and then, it should be sent to be centralized logging for examining security incidents. It will send alerts directly if any irregularities occurred in the system and can be customized the alert settings.

Benefits of FIM

- Real-Time Alerting
- Active Response

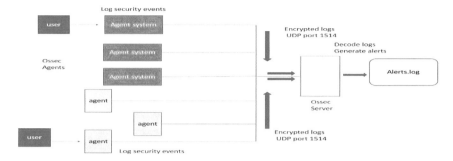

Fig. 4 Identifying logs and alerting the system

- Centralized Management Server
- File Integrity Checking
- Log Monitoring

Installation of OSSEC Agent in Linux

1. OSSEC agent should be downloaded and command should be entered: Ossec-hids-3.1.0.tar.gz
2. It will be saved into a registry called ossec-hids-3.1.0. Go to that registry.
3. Click on the OSSEC to get started the process of installation. Choose agent mode while the installation is in process on the server machine and end hosts.
4. Configuration path should set as/var/ossec.
5. The server IP address should be entered in the OSSEC agent.
6. In the client mode, integrity check and rootkit detection, and active response features should be enabled.
7. Press the enter button to start the installation process in the OSSEC server.

After starting the process, it demonstrates the begin/stop contents and setup way for OSSEC. Press the enter button to finish the installation procedure in the OSSEC server.

- manage_agents should run on the OSSEC server.
- Add an agent.
- Key should be extracted for the agent.
- Copy that key to the agent.
- manage_agents should run on the agent.
- Copy the key from the manager.
- Click on the option as restart which manages OSSEC processes.
- The agent should get started [19] (Fig. 5).

Fig. 5 Log monitoring

Algorithm for Log Monitoring

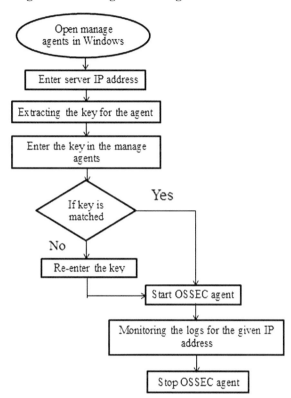

3   Results and Discussion

See Figs. 6, 7, and 8.

4   Conclusion

This paper concludes that host-based system for intrusion detection is one of the strategies where it can avoid from the intruder from entering into the system and it tends to be perceived by checking the logs of the system with the OSSEC tool. This can identify the intruders which enter without any access by examining log files and can send alerts with help of system logs to the server. By using this, safety measures can be taken in the organization by which can avoid the attacks that cause to the organization within the organization limits. We can identify the logs of every system by just getting the system IP address, and the client can be able to identify the logs by having the system IP address and MD5 hash key generated by the OSSEC server.

Fig. 6 Server agent

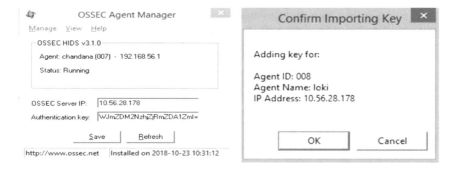

Fig. 7 Client agent

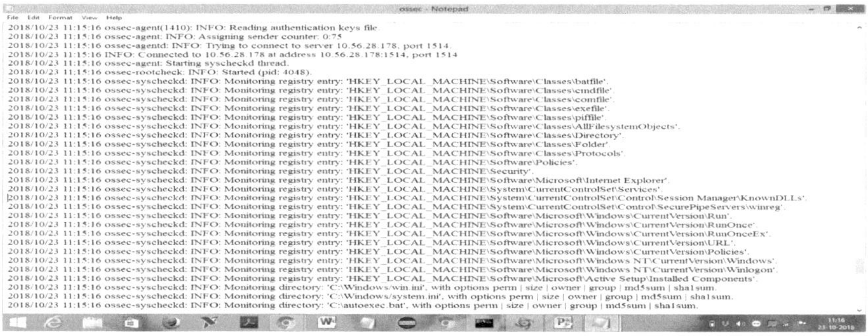

Fig. 8 Log running in the system

By using this mechanism, system logs cannot be changed and are difficult to crack the key which is generated by the OSSEC server. The server has an option as maild from which logs can be mailed to the head of that organization, i.e., admin. This can prevent the intruder not to enter into the network through this HIDS (host-based intrusion detection system)—log monitoring.

References

1. Korba, A.A., Nafaa, M., Ghamri-Doudane, Y.: Anomaly-based intrusion detection system for ad hoc networks. 978-1-5090-4671-3/16/$31.00 ©2016 IEEE
2. Shaikh, A.A., Qi, H., Jiang, W., Tahir, M.: A novel HIDS and log collection based system for digital forensics in cloud environment. In: 2017 3rd IEEE International Conference on Computer and Communications
3. Jacoby, G.A., Davis, N.J.: Mobile Host-Based Intrusion Detection and Attack Identification. IEEE Wirel. Commun. **14**, 53–60 (2007). 1536-1284/07/$20.00 ©2007 IEEE
4. Suda, H., Natsui, M., Hanyu, T.: Systematic intrusion detection technique for an in-vehicle network based on time-series feature extraction. In: 2018 IEEE 48th International Symposium on Multiple-Valued Logic

5. https://searchsecurity.techtarget.com/definition/intrusion-detection-system
6. Garcia, K.A., Monroy, R., Trejo, L.A., Mex-Perera, C., Aguirre, E.: Analyzing log files for post-mortem intrusion detection. IEEE Trans. Syst. Man Cybern. Part C Appl. Rev. **42**(6) 1690–1704 (2012)
7. Lydia Catherine, F., Pathak, R., Vaidehi, V.: Efficient host based intrusion detection system using Partial Decision Tree and Correlation feature selection algorithm. In: 2014 International Conference on Recent Trends in Information Technology
8. Nobakht, M., Sivaraman, V., Boreli, R.: A host-based intrusion detection and mitigation framework for smart home IoT using OpenFlow. In: 2016 11th International Conference on Availability, Reliability and Security
9. Zhu, M., Huang, Z.: Intrusion detection system based on data mining for host log. 978-1-4673-8979-2/17/$31.00 ©2017 IEEE
10. Lounis, O., Malika, B.: A new vision for intrusion detection system in information systems. In: Science and Information Conference (2015)
11. Raghavender, K.V., Premchand, P.: Host based intrusion detection system-file integrity check
12. Nema, S., Raghuwanshi, S.S.: An innovative method to improve security in cloud: using LDAP and OSSEC. Int. J. Innov. Res. Comput. Commun. Eng. **2**(11) (2014)
13. Ambati, S.B., Vidyarthi, D.: A brief study and comparison of, open source intrusion detection system tools. Int. J. Adv. Comput. Eng. Netw. ISSN: 2320-2106
14. Badgujar, T., More, P.: An intrusion detection system implementing host based attacks using layered framework. In: IEEE Sponsored 2nd International Conference on Innovations in Information, Embedded and Communication systems (ICIIECS) (2015)
15. Mishra, V.P., Shukla, B.: Development of simulator for intrusion detection system to detect and alarm the DDoS attacks. 978-1-5386-0514-1/17/$31.00 ©2017 IEEE
16. https://ossec-docs.readthedocs.io/en/latest/index.html
17. https://www.elprocus.com/basic-intrusion-detection-system/
18. http://parthicloud.com/ossec-agent-installation-in-linux-step-by-step/

Free-Space Optics: A Shifting Paradigm in Optical Communication Systems in Difficult Terrains

Payal and Suresh Kumar

Abstract Free-space optical (FSO) communication is a high-speed technology that can be used to promote rapid deployment of ubiquitous wireless service at the geographical locations such as hilly areas, where radio frequency (RF) technology is inaccessible and laying of optical fibers is not physically and economically viable. FSO has gained significant importance due to several advantages such as extremely high bandwidth, unlicensed spectrum allocation, reduced power consumption about half of the RF, ease of deployment, improved channel security, and reduced size which is one-tenth of the diameter of RF antenna. However, along with many advantages of FSO, atmosphere poses a serious limitation on its performance causing absorption, scattering, scintillations, and atmospheric turbulences. This paper presents a comprehensive review of FSO technology. FSO basics along with its advantages over RF are described so that readers can easily get the concept of shifting from RF to optical communication. Further, the paper also highlights the challenges faced by FSO, various models to characterize atmospheric turbulence fluctuations, issues, and methods to overcome them.

Keywords On–off keying (OOK) · Subcarrier intensity modulation (SIM) · Pulse position modulation (PPM) · FSO · Optical fiber cable (OFC) · Line of sight (LOS) · RF

1 Introduction

The advancement in the field of information and communication and the emergence of high-speed multimedia applications, live streaming, video conferencing, etc. have led to the exponential increase in bandwidth requirements for efficient service delivery. These rising bandwidth-hungry multimedia and data services require a paradigm shift from RF to optical communication due to RF spectrum congestion. FSO is a

Payal (✉) · S. Kumar
University Institute of Engineering and Technology, Maharshi Dayanand University,
Rohtak, Haryana, India
e-mail: payalarora325@gmail.com

© Springer Nature Singapore Pte Ltd. 2020
A. K. Luhach et al. (eds.), *First International Conference on Sustainable Technologies for Computational Intelligence*, Advances in Intelligent Systems and Computing 1045, https://doi.org/10.1007/978-981-15-0029-9_43

high-speed communication technology that adapts the LOS principle and uses optical signal as carrier in free space medium to transfer information between transmitter and receiver separated by certain distance. FSO link is a wireless link between transmitter and receiver in optical communication system. It works in a similar manner as OFC networks with a difference that optical beams are sent through free air instead of optical fiber cables [1, 2]. This wireless technology uses highly directional narrow beam of light which leads to inherent security with efficient transmission and a major portion of transmitted optical power is collected by the receiver [3].

In order to achieve optimum performance from a practical FSO link, some major challenges are required to be overcome at the transmitter side such as transmitting wavelengths, choice of modulation formats [4, 5], optical source with suitable output power, and type of antenna [6]. Similarly, at the receiver, the antenna, receiver sensitivity, and losses incurred in the free space are equally important [7].

The structure of paper is as follows: Sect. 1 gives an overview of FSO followed by its advantages over RF communication in Sect. 2. The application areas of FSO are described in Sect. 3. Section 4 gives a detailed explanation of atmospheric and geometric attenuation affecting the performance of FSO link. The various models to characterize atmospheric turbulence fluctuations are given in Sect. 5. Section 6 is concerned with the several issues in FSO along with the methods used to overcome them and Sect. 7 concludes the paper.

2 Why FSO Over RF Communication?

The optical communication is better than RF communication but for difficult terrains such as at hilly areas at the remote location (mountains and wider river banks), the laying of optical fibers is not economically and physically feasible as the physical distance is more than the aerial distance. This problem can be solved by using FSO by implementing it in the last mile to carry larger amount of data.

FSO differs from RF due to a large difference in wavelength. From several transmission windows in the wavelength range of 0.7–10 μm, there are various atmospheric transmission windows but FSO systems are designed to operate in "0.78–0.85 μm and 1.52–1.6 μm" ranges due to various advantages of using in particular wavelength region [8]. The RF transmission window lies within ("30 mm to 3 m") which is much larger than their optical counterparts. This gives rise to an interesting comparison between FSO and RF given as:

1. **Narrower divergence**: In optical carriers, the beam divergence is much narrower than that of RF carrier, which results in an increased received signal strength at the receiver for a given transmitter power. This divergence is directly proportional to carrier wavelength and diameter of aperture and can be represented as λ/D_R where λ and D_R denote carrier wavelength and diameter of aperture, respectively.

2. **High Modulation Bandwidth**: The information carrying capacity of the communication system increases with increase in carrier frequency. The acceptable

bandwidth in RF communication is around one-fifth of the carrier frequency. However, in optical domain, the acceptable bandwidth is up to the order of 100 THz even if the bandwidth taken is 1% of the carrier frequency ($\approx 10^{16}$ Hz). The usable bandwidth is in THz which is almost 10^5 times that of an RF carrier [9, 10].

3 **Low power and mass requirement**: The received optical intensity is more due to narrow beam divergence for a given transmitter power level. Thus, smaller λ of optical carrier enabled to design an FSO system with an antenna smaller than RF system for achieving the same gain.

4 **High directivity**: Smaller antennas operating at shorter optical wavelengths provide very high value of directivity. The antenna directivity very much relates to its gain. The optical to RF antenna directivity ratio is expressed as:

$$\frac{\text{Gain}_{(\text{optical})}}{\text{Gain}_{(\text{RF})}} = \frac{4\pi/\theta^2_{\text{div(optical)}}}{4\pi/\theta^2_{\text{div(RF)}}} \tag{1}$$

where $\theta_{\text{div(optical)}}$ and $\theta_{\text{div(RF)}}$ denote the beam divergence of optical and RF carrier

5 **Unlicensed spectrum**: In RF system due to spectral congestion, the adjacent carriers interfere which arises the requirement of spectrum licensing where as the optical system does not require spectrum licensing which reduces the deployment time and cost of initial setup [11].

6 **High Security**: In FSO system, interception and detection become difficult due to highly directional and narrow beam divergence. Unlike RF, FSO signal prevents eavesdropping as it cannot pass through walls.

For optical communications, fiber is the most consistent means but the laying of fiber is economically prohibitive due to digging, associated costs, and delays. The optical fiber once used cannot be redeployed and is difficult to recover the investment in a reasonable time frame. FSO is an optical communication at the speed of light as light travels faster through air than glass. The RF technology provides longer reach as compared to FSO but requires enormous capital investments for acquiring spectrum license which is not economic for service providers looking to optical network extension. The current RF bandwidth ceiling is 622 Mbps and hence cannot scale up to optical capacities of 2.5 Gbps. The coaxial (wire) infrastructure is not a feasible alternative for solving the connectivity bottleneck although it is available everywhere. The biggest issue is of bandwidth scalability.

FSO is the most optimum and viable solution for connectivity bottleneck due to its speed of deployment, bandwidth scalability, redeployment, portability, and lesser installation cost.

3 FSO Application Areas

FSO communication link provides many services described as follows:

1. **Last Mile Access**: The cost of laying of fibers in difficult terrain (i.e., river, mountain peaks, valleys, etc.) may not be economically viable for end users. FSO is a high-speed link which can solve this issue by deploying it in backhaul networks for last mile connectivity.
2. **Backup link**: FSO may be used in redundant links for fiber back up in place of a second fiber link in cases of communication breakdown or unavailability of the optical fiber link.
3. **Multicampus connectivity**: FSO is relevant for interconnecting LAN segments to connect two buildings or campus networks due to its easily installable feature.
4. **Cellular communications Backhaul**: FSO can carry cellular traffic at higher speeds from antenna towers back to the public switched telephone network.
5. **Military Access**: FSO being a secure and undetectable system is suitable for military applications as it can connect inaccessible remote areas safely in cost-effective manner.
6 **Difficult terrains**: The adverse weather conditions like heavy fog, dry snow, wet snow, etc. at hilly areas cannot be ignored. FSO becomes an attractive link to carry data in difficult terrains such as hilly areas, across a river, rail tracks, and very busy streets.

4 Challenges in FSO Communication

The FSO link performance is mainly affected by two types of attenuation: Geometric attenuation and Atmospheric attenuation [12]. Figure 1 depicts the challenges in FSO

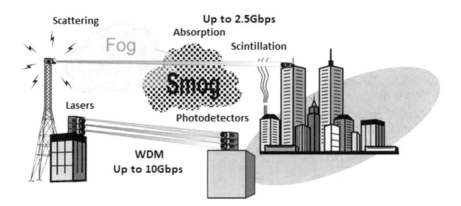

Fig. 1 Challenges in FSO communication

communication involving the geometric losses and environmental factors responsible for limiting its performance.

Geometric attenuation is due to misalignment between antennas at the transmitter and receiver which get disturbed due to vibrations, base motion, wind effect, etc. Building sway also referred to as base motion is classified into three categories namely: low-, moderate-, and high-frequency base motion. Low-frequency base motion has periods from minutes to months and is dominated by seasonal and diurnal temperature variations such as twists and bends in buildings due to thermal gradients. Moderate-frequency base motion includes wind-induced building motion and has periods of seconds [13]. High-frequency base motion has periods of less than one second which includes motion of large machineries such as large fans, human activities such as shutting of doors, walking. Due to narrowness of the transmitter beam and receiver field of view, base motion can disrupt the communication by affecting FSO transceiver's alignment.

The optical transmitter transmits highly directional and narrow light beam that must be imposed on the receiver's aperture area of the communication link. For FSO link to perform, it is necessary to maintain alignment between both ends of the link. However, with increase in transmission distance, the beam spread increases to a size larger than the receiver aperture diameter. This leads to reduced power at the receiver. Geometric losses are expressed by Eq. 2:

$$\text{Geometric loss (dB)} = 10 \times \log \left\{ \frac{R_d(\text{m})}{T_d(\text{m}) + L(\text{km}) \times D(\text{mrad})} \right\}^2 \qquad (2)$$

where D denotes beam divergence in (mrad), T_d and R_d denote the transmitter and receiver aperture diameter, respectively in (m), L is transmission range in (km).

The second type of attenuation is atmospheric attenuation which occurs due to bad weather conditions such as rain, snow, fog, etc. It includes absorption, scattering, scintillation, and atmospheric turbulences which reduces fidelity of the FSO link.

Absorption is a wavelength-dependent phenomenon which occurs due to interaction between the propagating photons and atmospheric molecules along the propagation path [14]. The wavelength range to be chosen for FSO should have minimum absorption which is referred as atmospheric transmission window. The aerosol or molecular absorption attenuation should be less than 0.2 dB/km. Table 1 shows the numeric data related to molecular absorption for clear weather conditions [15].

Table 1 Various molecular absorption at characteristic wavelengths

S. No.	Wavelength (nm)	Absorption (dB/km)
1	1550	0.01
2	850	0.41
3	690	0.01
4	550	0.13

Like absorption, scattering losses are strongly wavelength dependent. It refers to redirection or redistribution of light and is due to light interaction with atomic and molecular components present within the medium. The absorption and scattering losses in the atmospheric channel explained by Beer's law expressed as the attenuation coefficient [16]:

$$\tau_a = e^{-(\beta_{absp}+\beta_{scatt})R} \tag{3}$$

Here, R denotes the atmospheric path length, β_{absp} and β_{scatt} denote the absorption and scattering coefficients, respectively [17, 18]. The scattering phenomenon is dependent upon the characteristic parameter x_0 given by "$x_0 = 2\pi r/\lambda$" where λ is the wavelength of optical signal and r is the size of aerosol particle throughout propagation. If $x_0 \ll 1$, the Rayleigh scattering is produced as the size of atmospheric particles is smaller than the optical wavelength. This scattering is generally experienced at shorter wavelengths, i.e., around visible and UV range and neglected near infrared range. Air molecules and haze contribute to this scattering [19]. Mie scattering is produced when $x_0 \approx 1$ and fog, aerosol particles are its major contributors. When $x_0 \gg 1$, i.e., the atmospheric particles exceed the size of optical wavelength, the scattering is called nonselective or geometric scattering [20].

Fog causes absorption and mie scattering and hence acts as a major contributor in atmospheric attenuation thereby reducing the optical signal power. The fog attenuation is inversely proportional to wavelength. It is difficult to describe fog by its physical means so we use dense or thin fog to give an explanation of fog characteristics. Advection and radiation are the two types of fog [21]. The advection fog is created by combination of environmental packets with distinct temperatures or densities. The attenuation due to advection fog is expressed as [22]:

$$\sigma_{adv} = \frac{0.11478\lambda + 3.8367}{V} \tag{4}$$

Here, V and λ denote the visibility in km and wavelength in nm, respectively. Radiation fog generally arises at night or at the day end when temperature reaches dew point causing condensation and obstructs visibility. The attenuation by radiation fog is given by the relation:

$$\sigma_{rad} = \frac{0.18126\lambda^2 + 0.13709\lambda + 3.7502}{V} \tag{5}$$

where λ denotes the wavelength in micrometers. Due to the complex physical characteristics of fog such as nonavailability of particle distribution and size of particle, empirical models are used to predict the fog induced attenuation of the optical signal. Kim and Kruse are the most widely used models for this purpose. Equation 6 given below expresses the attenuation coefficient:

$$\beta_a(\lambda) = \left(\frac{3.91}{V}\right)\left(\frac{\lambda}{550}\right)^{-q} \qquad (6)$$

where q denotes the size distribution. In Kim model [23], the value of q is taken as ("0 for $V < 0.5$ km, V-0.5 for 0.5 km $< V <$1 km, $0.16V + 0.34$ for 1 km $< V < 6$ km, 1.3 for 6 km $< V < 50$ km and 1.6 for $V > 50$ km"). The values of q for Kruse model are ("$0.585V^{1.3}$ for $V < 6$ km, 1.3 for 6 km $< V < 50$ km, and 1.6 for $V > 50$ km"). Grabner and Kvicera [24] have experimentally demonstrated the relationship of fog attenuation and atmospheric visibility at 830 and 1550 nm using four different models (Kruse, Kim and Advection and Convection Al Naboulsi models). The Kim and Al Naboulsi models are not much different in results at 830 nm with Kruse model with worst results. At 1550 nm, models other than the Kruse give the better measured data than at the former wavelength.

Ijaz et al. [25] compared the performance of FSO link under controlled laboratory fog conditions at 830 and 1550 nm. It has been verified that fog attenuation shows wavelength dependency with considerably lower optical loss of 7 dB at 1550 nm than 27 dB at 830 nm. It was concluded that selection of 1550 nm wavelength is favorable in dense and moderate fog conditions. Andrej et al. [26] investigated the performance of FSO link for different fog densities using OFDM with 512 subcarriers and 16-QAM. In dense fog, the modulation has minimum effect on the transmission quality. For moderate fog conditions, the performance level of both modulations is very similar. OFDM suppresses multipath signal propagation better than 16-QAM. Ali and Ali [27] investigated the performance of fog models in FSO link at four different wavelengths (650, 850, 950, and 1550 nm) using 16-PPM and NRZ-OOK modulation techniques. A significant increase in distance is achieved at 1550 nm with 16-PPM. The maximum data transmission with 16-PPM reached to 1.3 km and 1.25 km for Kruse and Kim models, respectively and 0.8 km for Al Naboulsi models, while up to 1.13 and 1.19 km for Kruse and Kim model and 0.7 km with NRZ-OOK.

The attenuation due to rain is wavelength-independent and increases with rainfall rate [28]. The quantity of rain decides the visibility of FSO system. It can produce fluctuation effects in laser delivery. Heavy rain can restrict the passage of beam due to absorption, scattering, and reflection of the optical beam [29]. The effect of rain is not much pronounced as it exhibits nonselective scattering. The attenuation for light to heavy rain ("2.5–25 mm/h") lies within 1–10 dB/km for wavelengths around 850 and 1500 nm [30, 31]. ITU-R has proposed empirical models for prediction of attenuation due to rains in FSO communication system [32]. The rain attenuation in FSO link is expressed by [33]:

$$\alpha_{\text{rain}} = k_1 R^{k_2} \qquad (7)$$

where R denotes the rate of rainfall in "mm/h" and model parameters k_1 and k_2 depends on various parameters such as rain temperature and raindrop size. Table 2 depicts the proposed models of rain attenuation by ITU-R for FSO communication.

The performance of a region-specific FSO system under heavy rainfall attenuation is analyzed for different number of optical beams [34]. A significant improvement

Table 2 Proposed models of rain attenuation by ITU-R for FSO communication

Model	Origin	k_1	k_2
Japan	Japan	1.58	0.63
Carbonneau	France	1.076	0.67

is achieved using four-beam FSO system in terms of link distance, received optical power, and geometrical losses than the one-, two-, and three-beam FSO system.

The attenuation due to snow (30–350 dB/km) lies within the values of rain and fog attenuation. The path of laser beam is blocked by high-density snowflakes in heavy snowfall. The snow attenuation is classified into dry and wet snow attenuation given by the following equation:

$$\alpha_{\text{snow}} = a S^b \tag{8}$$

The value of the constants (a and b) varies for dry and wet snow. For dry snow, the values of a and b are $5.42 \times 10^{-5} + 5.49$ and 1.38, respectively and for wet snow, $1.02 \times 10^{-4} + 3.78$ and 0.72, respectively.

The various air packets with different refractive indices (eddies) in different atmospheric layers formed due to temperature inhomogeneities also cause atmospheric turbulence thereby limiting FSO communication performance. These eddies create intensity (scintillation) and phase fluctuations in the received signal by redistributing signal energies by constructive or destructive interference during propagation [35]. The scintillation is given by:

$$\sigma_I^2 = \frac{\langle I^2 \rangle - \langle I \rangle^2}{\langle I \rangle^2} \tag{9}$$

where I denote ensemble average of irradiance. σ_I^2 is scintillation index which is a function of refractive index structure parameter C_n^2 also called as turbulence strength and is the main considerable parameter for defining the turbulence. The turbulence strength varies with the time, season, altitude, and location of the day. The value of C_n^2 for weak turbulence is "10^{-17} m$^{-3/2}$" and can be up to "10^{-13} m$^{-3/2}$" or greater in case of strong turbulence.

The particles of the atmospheric channel in combination with several natural phenomenons degrade the quality of received signal. When the sun rays strike the Earth's surface, the air near the surface gets warmer than at higher altitudes. The mixing of warm and cool air changes the characteristics of channel leading to atmospheric turbulence. Figure 2 shows atmospheric channel with different eddies. The smallest eddies are called interior size I_0 which is of the order of few millimeters, while the largest eddies are called outer size L_0 of turbulence which is of few meters order.

The random refractive index fluctuations in atmospheric channel are expressed by [36]:

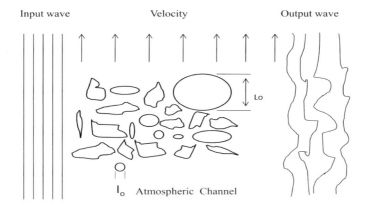

Fig. 2 Atmospheric channel with turbulence eddies

$$n = 1 + \frac{77.6(1 + 7.52 \times 10^{-3}\lambda^{-2})P}{T} \times 10^{-6} \tag{10}$$

where n denotes the refractive index, P denotes the atmospheric pressure in millibars, and T denotes the temperature in kelvin.

5 Modeling of Atmospheric Turbulence Fluctuations

Several statistical models are used to analyze the FSO channels based on the intensity fluctuations. The log-normal distribution model is used in weak turbulences and Gamma–Gamma model is used for moderate to strong turbulences.

Log-Normal Model: In the log-normal model, the irradiance fluctuations at the receiver are expressed by the following PDF:

$$P_l(I) = \frac{1}{I\sqrt{2\pi\sigma_p}}\exp\left[\frac{-\left[\ln\left(\frac{I}{I_0}\right) + \frac{\sigma_p^2}{2}\right]^2}{2\sigma_p^2}\right] \tag{11}$$

where I and I_0 are the irradiance with and without turbulence and σ_p^2 denotes the Rytov variance expressed by Eq. 12:

$$\sigma_p^2 = 1.23C_n^2 k^{\frac{7}{6}} L^{\frac{11}{6}} \tag{12}$$

Here, C_n^2 and k denote turbulence strength and the optical wave number. The Rytov variance represents the scintillation index calculated for a plane wave using the Kolmogorov spectrum which is defined as:

$$\phi_n(\chi) = C_n^2 \chi^{11/3} \tag{13}$$

where χ denotes the scalar spatial wave number. For probability error calculation, average atmospheric turbulence is assumed over 'K' log-normal random variables which are given as:

$$z = \frac{(I_1 + I_2 + I_3 + \cdots I_k)}{I_0} = \exp(w) \tag{14}$$

The PDF of z is expressed as:

$$P(z) = \frac{1}{z\sqrt{2\pi\sigma_z^2}} \exp\left[-\frac{(\ln z - w_z)^2}{2\sigma_z^2}\right] \tag{15}$$

where w_z is the mean and σ_z^2 is the new log intensity variance and both are expressed by Eqs. 16 and 17, respectively.

$$w_z = \ln(K) - \frac{1}{2}\ln\left[1 + \frac{\exp(\sigma_p^2) - 1}{K}\right] \tag{16}$$

$$\sigma_z^2 = \ln\left[1 + \frac{\exp(\sigma_p^2) - 1}{K}\right] \tag{17}$$

The log-normal distribution is for weak turbulences with scintillation index less than 0.75.

Gamma–Gamma Model: For moderate to strong turbulences, Gamma–Gamma model is used and its PDF is expressed as [37]:

$$P_h(I) = \frac{2(\alpha\beta)^{(\alpha+\beta)/2}}{\Gamma(\alpha)\Gamma(\beta)} I^{[((\alpha+\beta)/2)-1]} K(\alpha - \beta)\left(2\sqrt{\alpha\beta I}\right) \tag{18}$$

Here, $\Gamma(.)$ is Gamma functions, and α and β denote the small and large scale eddies of the scattering environment to decide the turbulence level and the Bessel function with order of $(\alpha - \beta)$ is given by $K(\alpha - \beta)$. The values of α and β are given as:

$$\alpha = \left[\exp\left(\frac{0.49\sigma_p^2}{\left(1 + 0.56\sigma_p^{12/5}\right)^{7/6}}\right) - 1\right]^{-1} \tag{19}$$

$$\beta = \left[\exp\left(\frac{0.51\sigma_p^2}{\left(1 + 0.69\sigma_p^{12/5}\right)^{5/6}}\right) - 1\right]^{-1} \tag{20}$$

Negative Exponential Model: With increase in link length, the scattering due to several unrelated particles increases thereby making irradiance fluctuations stronger.

In strong turbulences, the irradiance due to fluctuations of the optical beam follows a negative exponential statistics given by [38]:

$$f(I) = I/I_0 \exp(-I/I_0), \quad I \geq 0 \tag{21}$$

where I and I_0 denote the irradiance and mean received irradiance, respectively.

6 Issues in FSO and Possible Methods to Overcome Them

The main reason for quality deterioration in FSO is its vulnerability toward various atmospheric turbulences due to adverse weather conditions. Besides these, the transmitter power and receiver sensitivity are also the important parameters. At the transmitting end, the size, power, and beam quality of the optical source are the limiting factors which determine the laser intensity and beam spread. The choice of modulation format and operating wavelength also poses a challenge in selection of transmitter parameters in FSO. For effective reception at the receiver, choice of various parameters such as type of detector, sensitivity, responsivity, and effective area of detector has to be selected in an optimized manner for efficient working.

Several techniques such as aperture averaging, diversity techniques, modulation and coding techniques, etc. are used to mitigate these atmospheric turbulences for improving the fidelity of FSO under adverse weather conditions. The aperture averaging technique reduces the atmospheric scintillations and fluctuations due to small size turbulence cells by increasing the receiver aperture size and thus improves BER performance. However, large aperture area improves system performance but at the cost of increase in noise due to background radiance [39, 40]. Therefore, choice of optimum aperture diameter is essential for efficient and optimized working of FSO system.

Diversity techniques can be implemented in frequency, time, and spatial domain for mitigation of atmospheric turbulences. In diversity techniques, a single-large receiver aperture is replaced with an array of receivers with smaller aperture diameters. These array of smaller aperture receivers creates flexibility such that multiple and mutually uncorrelated copies of signals can be transmitted in spatial, time, and frequency domain improving the availability of link and BER performance [41].

The diversity techniques also include [(i) equal gain, (ii) maximum ratio (iii) selection linear] combining techniques. The BER is improved by maximum ratio combining but is more complex as compared to equal gain combining technique. However, extra care and pre-study of the FSO medium can be helpful in guiding what parameters should be considered before system setup.

At the transmitting end, modulation formats along with different amplification mechanism should be incorporated to increase transmission efficiency [42–44]. OFDM involves use of orthogonal low rate subcarriers to support higher data rate and provides flexibility in resource utilization and enhances efficiency of transmitter section [45].

The choice of modulation and coding techniques generally depends on bandwidth and optical power efficiency. It is easier to implement power-efficient modulation schemes. For low data rates, they are effective in mitigating the turbulence effects. However, during high turbulences, it limits the maximum propagation distance. There are several other modulation techniques such as OOK, PPM and SIM are used in FSO communication system for enhancing efficiency. OOK is the simplest and most commonly used modulation format. However, it is difficult to set the receiver threshold value according to the turbulence level which makes receiver designing a complicated task. PPM, on the other hand, excellent in power efficiency but poor in bandwidth efficiency, does not require any adaptive threshold. SIM overcomes the drawbacks of both OOK and PPM and provides high spectral efficiency.

7 Conclusion

FSO has gained significant importance in geographical locations such as hilly areas, where RF technology is inaccessible and laying of optical fibers is not physically and economically viable. However, along with many advantages of FSO, atmosphere poses a serious limitation on its performance causing absorption, scattering, scintillations, and atmospheric turbulences. The paper presents a broad review of FSO technology. Further, the paper also discusses the challenges faced by FSO, various models to characterize atmospheric turbulence fluctuations, issues, and methods to overcome them. This paper will help the researchers to comprehend all the related aspects of FSO and to follow their research in an effective manner.

References

1. Neha, Kumar, S.: Free space optical communication: a review. Int. J. Electr. Electr. Comput. Syst. (IJEECS) **5**(9), 4–8 (2016). ISSN 2348-117X
2. Chan, V.W.S.: Free-space optical communications. J. Lightwave Tech. **24**(12), 4750–4762 (2006)
3. Popoola, W.O., Ghassemlooy, Z., Lee, C.G., Boucouvalas, A.C.: Scintillation effect on intensity modulated laser communication systems—a laboratory demonstration. J. Opt. Laser Technol. **42**(4), 682–692 (2009)
4. Zhong, W.D., Fu, S., Lin, C.: Performance comparison of different modulation formats over free space optical (FSO) turbulence link with space diversity reception technique. IEEE photon. J. **1**(6), 277–285 (2009)
5. Rajbhandari, S., et al.: On the study of the FSO link performance under controlled Turbulence and Fog atmospheric condition. In: 11th International Conference on Telecommunications, ConTEL, G Raz, Austria, pp. 223–226 (2011)
6. Borah, D.K., Voelz, D.G.: Pointing error effects on free space optical communication links in the presence of atmospheric turbulence. J. Lightwave Technol. **27**(18), 3965–3973 (2009)

7. Kumar, S., Rathee, S., Arora, P., Sharma, D.: A comprehensive review on fiber bragg grating and photodetector in optical communication networks. J. Opt. Commun. DG Gruyter. **0**(0), aop (2019) https://doi.org/10.1515/joc-2018-0205
8. Bloom, S., Korevaar, E., Schuster, J., Willebrand, H.: Understanding the performance of free-space optics [Invited]. J. Opt. Netw. **2**(6), 178–200 (2003)
9. Williams, W.D., et al.: RF and optical communications: a comparison of high data rate returns from deep space in the 2020 timeframe. In: Technical Report: NASA/TM-2007-214459 (2007)
10. Neha, Kumar, S.: Role of modulators in free space optical communication. Int. J. Eng. Technol. Manage. Appl. Sci. (IJETMAS) **4**(9), 92–96 (2016). ISSN 2349-4476
11. Henniger, H., Wilfert, O.: An introduction to free-space optical communications. J. Radio Eng. **19**(2), 203–212 (2010)
12. Hansel, G., Kube, E.: Simulation in the design process of free space optical transmission systems. In: Proceedings of the 6th Workshop, Optics in Computing Technology, Paderborn (Germany), pp. 45–53 (2003)
13. Payal, Kumar, S.: Nonlinear impairments in fiber optic communication systems: analytical review. In: Futuristic Trends in Network and Communication Engineering. FTNCT-2018. Communications in Computer and Information Science, Springer, Singapore, vol. 958, pp. 28–44 (2019) https://doi.org/10.1007/978-981-13-3804-5_3
14. Popoola, W., Ghassemlooy, Z., Awan, M.S., Leitgeb Piteti E.: Atmospheric Channel Effects on terrestrial free space optical communication link. In: ECAI 2009—International Conference 3rd edn, pp. 17–23 (2009)
15. Rouissat, M., Borsali, A.R., Chiak-Bled, M.E.: Free space optical channel characterization and modeling with focus on Algeria weather conditions. Int. J. Comp. Netw. Inf. Secur. **3**, 17–23 (2012)
16. Achour, M.: Free-space optics wavelength selection: 10µ versus shorter wavelengths. In: Proceedings of SPIE (SPIE, Bellingham, WA), vol. 5160, pp. 234–246 (2003) https://doi.org/10.1117/12.502483
17. Awan, M.S., Horwath, L.S., Muhammad, S.S., Leitgeb, E., Nadeem, F., Khan, M.S.: Characterization of fog and snow attenuations for free-space optical propagation. J. Commun. **4**(8), 533–545 (2009)
18. Sandalidis, H.G., Tsiftsis, T.A., Karagiannidis, G.K., Uysal, M.: BER performance of FSO links over strong atmospheric turbulence channels with pointing errors. IEEE Commun. Lett. **12**(1), 44–46 (2008)
19. Willebrand, H., Ghuman, B.S.: Free Space Optics: Enabling Optical Connectivity in Today's Networks. Sams Publishing (2002)
20. Mahalati, R.N., Kahn, J.M.: Effect of fog on free-space optical links employing imaging receivers. Opt. Exp. **20**(2), 1649–1661 (2012)
21. Wainright, E., Refai, H.H., Sluss, J.J.: Wavelength diversity in free-space optics to alleviate fog effects. In: Proceedings of SPIE, vol. 5712, pp. 110–118 (2005)
22. Al Naboulsi, M., Sizun, H., De Fornel, F.: Fog attenuation prediction for optical and infrared waves. Opt. Eng. **43**(2), 319–329 (2004)
23. Kim, I.I., McArthur, B., Korevaar, E.: Comparison of laser beam propagation at 785 nm and 1550 nm in fog and haze for optical wireless communications. In: Proceedings of SPIE, vol. 4214, Boston, MA, USA (2001)
24. Grabner, M., Kvicera, V.: Fog attenuation dependence on atmospheric visibility at two wavelengths for FSO link planning. In: 2010 Loughborough Antennas & Propagation Conference, Loughborough, pp. 193–196 (2010)
25. Ijaz, M., Ghassemlooy, Z., Rajbhandari, S., Le Minh, H., Perez, J., Gholami, A.: Comparison of 830 nm and 1550 nm based free space optical communications link under controlled fog conditions. In: 8th International Symposium on Communication Systems, Networks & Digital Signal Processing (CSNDSP), Poznan, pp. 1–5 (2012)
26. Andrej, L., et al.: Features and range of the FSO by use of the OFDM and QAM modulation in different atmospheric conditions. In: Proceedings of SPIE, vol. 9103, Wireless Sensing, Localization, and Processing IX, 91030O, pp. 1–10 (2014)

27. Ali, M., Ali, A.: Performance analysis of fog effect on free space optical communication system. IOSR J. Appl. Phys. **7**(2), 16–24 (2015)
28. Fadhil, H.A., et al.: Optimization of free space optics parameters: an optimum solution for bad weather conditions. Optik **124**(19), 3969–3973 (2013)
29. Rahman, A.K., Anuar, M.S., Aljunid, S.A., Junita, M.N.: Study of rain attenuation consequence in free space optic transmission. In: Proceedings of the 2nd Malaysia Conference on Photonics Telecommunication Technologies (NCTT-MCP '08), pp. 64–70, IEEE, Putrajaya, Malaysia (2008)
30. Suriza, A.Z., Rafiqul, I.M., Wajdi, A.K., Naji, A.W.: Proposed parameters of specific rain attenuation prediction for free space optics link operating in tropical region. J. Atmos. Solar Terres. Phys. **94**, 93–99 (2013)
31. Vavoulas, A., Sandalidis, H.G., Varoutas, D.: Weather effects on FSO network connectivity. J. Opt. Commun. Netw. **4**(10), 734–740 (2012)
32. Crane, R.K., Robinson, P.C.: ACTS propagation experiment: rain-rate distribution observations and prediction model comparisons. Proc. IEEE **86**(6), 946–958 (1997)
33. Al Naboulsi, M., Sizun H., De Fornel F.: Propagation of Optical and Infrared Waves in the Atmosphere. http://www.ursi.org/proceedings/procga05/pdf/F01P.7(01729).pdf
34. Al-Gailani, S.A., Mohammad, A.B., Shaddad, R.Q.: Enhancement of free space optical link in heavy rain attenuation using multiple beam concept. Opt. Int. J. Light Electron Opt. **124**(21), 4798–4801 (2013)
35. Andrews, L.C., Phillips, R.L., Hopen, C.Y.: Laser Beam Scintillation with Applications. SPIE Press (2001)
36. Ghassemlooy, Z., Popoola, W.O.: Terrestrial free-space optical communications. Optical Communications Research Group, NCR lab, Northumria University, Newcastle upon Tyne, **5**(7), 195–212 (2014)
37. Andrews, L.: Field Guide to Atmospheric Optics. SPIE Press (2004)
38. Popoola, W.O., Ghassemlooy, Z.: BPSK subcarrier intensity modulated free-space optical communications in atmospheric turbulence. J. Lightwave Technol. **27**(8), 967–973 (2009)
39. Perlot, N., Fritzsche, D.: Aperture-averaging, theory and measurements. In: Proceedings of SPIE, Free-Space Laser Communication Technologies XVI, vol. 5338, pp. 233–242 (2004)
40. Wasiczko, L.M., Davis, C.C.: Aperture averaging of optical scintillations in the atmosphere: experimental results. In: Proceedings SPIE, Atmospheric Propagation II, vol. 5793, pp. 197–208 (2005)
41. Zhu, X., Kahn, J.M.: Maximum-likelihood spatial-diversity reception on correlated turbulent free-space optical channels. In: IEEE Conference Global Communication, San Francisco, CA, vol. 2, pp. 1237–1241 (2000)
42. Payal, Kumar, S., Sharma, D.: Performance analysis of NRZ and RZ modulation schemes in optical fiber link using EDFA. Int. J. Adv. Res. Comput. Sci. Softw. Eng. (IJARCSSE) **7**(8), 161–168 (2017). https://doi.org/10.23956/ijarcsse/v7i8/0102
43. Payal, Sharma, D., Kumar, S.: Analyzing EDFA performance using different pumping techniques. Int. J. Comput. Sci. Eng. **6**(5), 195–202 (2018). https://doi.org/10.26438/ijcse/v6i5. 195202
44. Deepti, Payal, Kumar, S.: Performance evaluation of proposed WDM optical link using EDFA and FBG combination. J. Opt. Commun. DG Gruyter, **0**(0), aop (2018). https://doi.org/10.1515/joc-2018-0044
45. Sharma, D., Kumar, S.: Design and evaluation of OFDM based optical communication network. J. Eng. Appl. Sci. **12**(Special Issue 2), 6227–6233 (2017). https://doi.org/10.3923/jeasci.2017. 6227.6233

Classification of Plants Using Convolutional Neural Network

Gurinder Saini, Aditya Khamparia and Ashish Kumar Luhach

Abstract Plant leaf classification plays a foremost role in botanical research, Ayurveda, agriculture practices, medicine and drug, weed detection, and many more areas. It is a technique by which a plant leaf is categorized based on its different morphological structures. The aim of this paper is to offer a deep learning technique for plant leaf classification with the help of deep Convolutional Neural Network as a substitute of conventional classification methods like k-nearest neighbor, probabilistic neural network, support vector machine, genetic algorithm, and principal component analysis, which all need feature extraction and are time-consuming. CNN has been used since it has outperformed in various image recognition challenges and feature extraction is performed automatically, which takes less time. This work uses a 5000 leaf images of two plant species. 4000 images are used for training and 1000 for testing purpose. CNN is trained in such a way that it can classify the species and predict the class for a new leaf image. The proposed model is run for different epochs and results were recorded. It was observed that the CNN model performed effectively in distinguishing the plant leaf images and achieved 99.96% training accuracy and 99.90% testing accuracy.

Keywords Plant leaf classification · Convolutional Neural Network (CNN) · Deep learning · Machine learning

G. Saini · A. Khamparia (✉)
School of Computer Science and Engineering,
Lovely Professional University, Jalandhar, Punjab, India
e-mail: aditya.khamparia88@gmail.com

G. Saini
e-mail: gurindersaini25@gmail.com

A. K. Luhach
The PNG University of Technology, Lae, Papua New Guinea
e-mail: ashishluhach@acm.org

© Springer Nature Singapore Pte Ltd. 2020   551
A. K. Luhach et al. (eds.), *First International Conference on Sustainable Technologies for Computational Intelligence*, Advances in Intelligent Systems and Computing 1045,
https://doi.org/10.1007/978-981-15-0029-9_44

1   Introduction

Plant identification or classification has an advantageous perspective in agricultural practices and medicine. It also plays a significant role in biological diversity research. Plant leaf classification has many applications such as botanical studies, and in tea, cotton, and other industries [1]. It is also helpful in weed detection in the plants. Plants serve an important role in the protection of the environment also. But it is an essential and very tough job to diagnose a variety of plant species that exist on our planet. Many plant varieties carry important evidence for the development of human civilization. The crucial state is that a variety of plants are at the danger of extinction due to an increase in pollution and deforestation. So, it is a basic necessity to propose an automatic technique for plant species identification and to build a database of those species.

Leaf recognition by a computer system is a chief step in plant classification. Plants are generally recognized based on their flowers and fruits. But, due to their three-dimensional nature, it becomes relatively complex. If we use flowers and fruits for classification purpose, it requires morphological characters such as the number of stamens present in flower and how many ovaries are found in fruits. So, it becomes time-consuming as well as requires the supervision of trained botanists. It has other pitfalls such as the absence of required morphological characters information and use of botanical terms which can only be understood by botanical experts. There is another option to use leaves for plant identification as they can be obtained without any difficulty and are present in all the seasons, while flowers can only be found in blooming season. The leaf shape is a characteristic feature for the identification of many plants' species visually. Plant leaves also have plane nature and thus they are suitable for image processing. In our research work, we have used color leaf images of plants.

The present paper proposes a deep learning CNN model for classification of plant leaves because it has outperformed in various image recognition challenges and can easily handle a large dataset efficiently [2]. Researchers have already used various conventional effective classification methods such as the k-nearest neighbor, support vector machine, genetic algorithm, probabilistic neural network, and principal component analysis for plant leaf classification [3].

The rest paper structure is arranged as described: Sect. 2 provides a summary of the related works, Sect. 3 gives information about the dataset used in the experimentation, Sect. 4 presents proposed CNN model and workflow diagram; discussion of the results is demonstrated in Sect. 5, and lastly, Sect. 6 delivers the conclusion and the future scope.

2 Related Work

In 2006, Dyrmann et al. [4] proposed a novel technique using CNN through which the species of plant can be recognized from color images. There are six various datasets used to gather images used in the dataset here. In terms of resolution, lighting and types of soil, the properties of these images are different. With respect to the stabilization and illumination of the camera, the images gathered under controlled conditions are included here. Simulations are performed which show around 86% of classification accuracy for the proposed technique (Fig. 1).

In 2017, Barré et al. [5] proposed a deep learning model named LeafNet through which the discriminative features from leaf images can be learned. Further, the species of plants are identified by applying a classifier. It is seen through the comparative analysis of results that the feature representation of leaf images can be improved as compared to that achieved by traditional techniques by applying the proposed technique. Lee et al. [6] proposed a novel approach in which CNN is applied to learn important leaf features directly from the representations provided from input provided. Further, based on deconvolutional network, the intuition of selected features was provided. The hierarchical botanical definitions of leaf types are very appropriate as per these findings. The new hybrid feature extraction models were provided in this paper through which the discriminative power of plant classification systems was enhanced. Affonso et al. [7] done a comprehensive assessment of a Convolutional Neural Network (CNN) in comparison to texture descriptors to categorize wood board samples. The author performed experiments using a dataset acquired from a real-world problem to evaluate the classification performance of this technique. The author also compared CNN with other Machine Learning (ML) techniques for the problem. The key variation of CNN is that it can separate features from the inputted image, whereas the conventional classification techniques require a set of features explaining an image, as input.

In 2018, Ji et al. [8] proposed an innovative 3D CNN-based approach through which the crops can be classified automatically from the remote sensing images given as input. For improving the accuracy to a required threshold value such that higher efficiency can be achieved, an active learning mechanism is introduced within the CNN model. In comparison to other existing approaches, comparative analysis

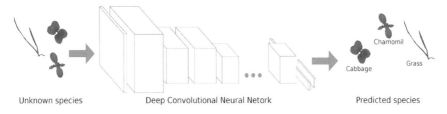

Fig. 1 Prediction of plant species using deep CNN [4]

is performed which show that the dynamics of crop growth can be categorized efficiently as compared to existing traditional techniques. 3D CNN performs better than 2D CNN. SVM achieved high accuracy than other methods such as KNN and PCA. Zhu et al. [9] proposed an enhanced deep Convolutional Neural Network through which the benefits of Inception V2 were considered along with BN. Thus, within the region proposal networks, multiscale image features were provided by this approach. Thus, the leaf species within complex backgrounds are identified at a better accuracy level by the proposed approach in comparison to other traditional approaches. Xu et al. [10] proposed a novel CNN through which the complex network topology adjacent matrix can be reformatted within an image. In order to extract the relevant features and classify them, a CNN of around 10 layers that include several components is designed here. The target features can be extracted and around 95% of accuracy can be achieved in feature classification as per the experimental results achieved by applying the proposed CNN model. Hamuda et al. [11] proposed a new technique which used a Kalman filtering along with Hungarian algorithm for crop detection using image processing which was able to reach 99.34% classification accuracy.

3 Dataset

PlantVillage dataset [20], a freely available dataset, was used for the experimental study. It comprises around 50,000 images of healthy and unhealthy leaves of different plant species [12]. But we have selected only two plant species forming two classes viz. apple and grape. The reduced dataset has 5000 images. There are 4000 images in the training set, whereas 1000 images in the test set. Two types of plant leaves are taken into account for this study, namely 'apple' and 'grape' whose description is shown in Table 1.

The preprocessing of images is done before they are inputted to the CNN, such as rescaling, shearing, zooming, and horizontal flipping. Figure 2 illustrates an arbitrarily chosen sample of plant leaf images from the dataset.

Table 1 Dataset description

Class	Training set	Test set
Apple	2000	500
Grape	2000	500

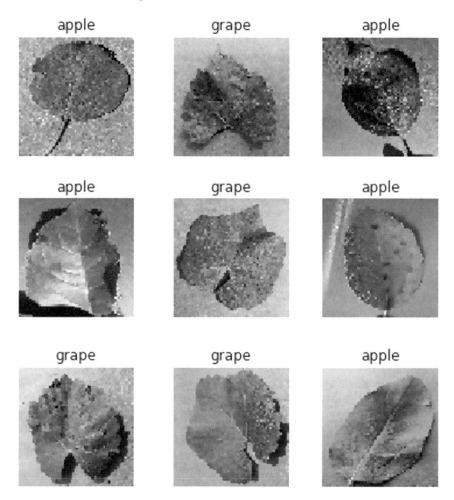

Fig. 2 An arbitrarily chosen sample of leaf images

4 Convolutional Neural Network Architecture and Workflow Diagram

4.1 Convolutional Neural Network Architecture

Convolutional Neural Network (CNN or ConvNet) is a prominent deep learning architecture mainly used for image classification It was inspired by biological process of the visual cortex system of animals [13]. In 1959, it was found by Hubel and Wiesel that cat and monkey visual cortex contains neurons that individually respond to small regions of the visual field [14]. In 1980, a neocognitron model was proposed based on CNN [15]. A Convolutional Neural Network (CNN) is a sequence of layers, and each layer of a CNN converts one set of points to another over a differentiable limit [16]. A CNN consists of some important layers such as convolutional layer, subsampling/downsampling layer and fully connected layer, and may include some supplementary layers [17]. CNN applications include text detection and recognition, image classification, natural language processing, object detection, object tracking, recommender systems, medical image analysis, and many more.

The basic units of CNN architecture are enlightened as follows:

(a) **Convolutional Layer**

A convolutional layer is the essential layer of CNN architecture that performs most of the convolution operation [18]. In the convolution operation, we calculate the element-wise product of weight matrix (also known as kernel or filter) and image pixels. It helps to get the important features from the input image like edges of an object in the image.

(b) **Activation Functions**

The outcome of convolution operation is surpassed through an activation function which is used to add nonlinearity to the network. A rectified linear unit (ReLU) has been used as an activation function internally in the convolutional layers. The ReLU values can be computed as $R(z) = \max(0, z)$. It is easy to calculate and provide result much faster as compared to the sigmoidal function which involves calculating exponential terms [19]. But at the output layer, sigmoidal function has been used.

(c) **Max-Pooling Layer**

It is a subsampling/downsampling layer which is usually added after the convolutional layer. It helps to reduce the redundancy present in the input features and also keep a check on overfitting. Max-pooling is used in the CNN architecture which selects the maximum value and pool into an output matrix.

(d) **Fully connected layer**

This is a flattening layer which results into a feature vector and the neurons have full connections with neurons in the preceding layer. Then the feature vectors are passed through multiple dense layers.

(e) **Output layer**

The output layer is responsible for producing probabilities of each class. We have used sigmoidal activation function to find the final class of given input image.

Figure 3 shows the CNN architecture being used for plant leaf classification. Conv2d_1 is the first-convolutional layer applied on input image which is followed by a max-pooling layer, i.e., MaxPool2d_1. Again, a convolution and a max-pooling layer are added. Finally, a flattening layer, i.e., Flatten_1 and two dense layers, Dense_1 and Dense_2, are added. Table 2 represents the generated Convolutional Neural Network model summary. It is showing different layers used in the model as well as the output shape obtained from different layers. The total parameters are 813,217 which all are trainable.

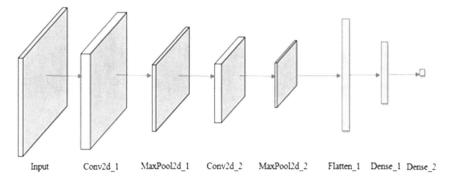

Fig. 3 CNN architecture

Table 2 CNN model summary

Layer (Type)	Number of filters used	Size of each filter	Output shape	Number of parameters
Input_1 (Input layers)			(None, 64 × 64 × 3)	0
conv2d_1 (Conv2D)	32	3 × 3	(None, 62 × 62 × 32)	896
max-pooling2d_1 (MaxPooling2D)		2 × 2	(None, 31 × 31 × 32)	0
conv2d_2 (Conv2D)	32	3 × 3	(None, 29 × 29 × 32)	9248
max-pooling2d_2 (MaxPooling2D)		2 × 2	(None, 14 × 14 × 32)	0
flatten_1 (Flatten)			(None, 6272)	0
dense_1 (Dense)			(None, 128)	802,944
dense_2 (Dense)			(None, 1)	129
Total parameters			813,217	
Total trainable parameters			813,217	
Total nontrainable parameters			0	

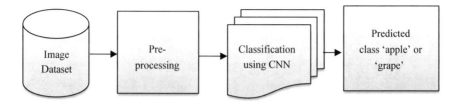

Fig. 4 Proposed workflow

4.2 Workflow Diagram

The complete workflow is described below as shown in Fig. 4:

(a) Initially, the image dataset is preprocessed to reduce the dimensions to 64 * 64.
(b) After that, input image is subjected to CNN as shown in Fig. 3. It will be initially convolved with a 3 * 3 filter size with 32 filters, in the first-convolutional layer. This is done to extract useful features. ReLU activation function is internally applied.
(c) Then a max-pooling layer with pool-size 2 * 2 is added to further reduce the dimensions.
(d) Again, a second-convolutional layer with 32 filters and 3 * 3 filter size is applied followed with a max-pooling layer of 2 * 2 pool-size.
(e) Then a flattening layer is supplied to obtain a linear vector having full connections.
(f) After that, two dense layers are applied, first employs ReLU and second uses a sigmoidal activation function.
(g) After the input image is passed through all the CNN layers, classification is done and the model will be able to predict the class of a different image other than used for training.

5 Results and Discussions

The proposed work provides a way to classify leaves of different plants such as apple and grape using a Convolutional Neural Network. The implementation of CNN is done in python language with the help of Spyder Anaconda tool. The training set comprises 4000 images; half of them are of apple leaves and half of them are grape leaves. 1000 images are used for testing, 500 for apple, and 500 for grape leaves. The python code was executed for a various round of epochs to record corresponding accuracy and losses. Table 3 illustrates the training as well as testing accuracy for the several epochs such as 2, 5, and 10. The best training accuracy 99.96 which is obtained in the 10th epoch. The best testing accuracy is 99.90 when the number of epochs was 2 and 10.

Table 3 Training and testing accuracy for the different number of epochs

Number of epochs	Convolution filter size	Training accuracy	Testing accuracy
2	3 * 3	99.71	99.90
5	3 * 3	99.94	99.80
10	3 * 3	99.96	99.90

Figures 5, 6, and 7 show the loss and accuracy curve on the proposed CNN model for 2, 5, and 10 epochs, respectively. In the case of 2 epoch, the accuracy is increasing linearly while losses are decreasing as shown in Fig. 5. When the number of epochs was increased to 5, there was a sudden increase in training accuracy and after that, it increased slowly as shown in Fig. 6. In the case of 10 epochs as in Fig. 7, the training accuracy escalated instantly till second epoch and then, it increased slowly, but the testing accuracy showed some variation. Hence, from these experimental executions, it can be perceived that the classification accuracy fluctuates with the number of epochs performed.

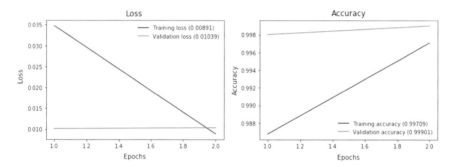

Fig. 5 Model loss and accuracy curve for 2 epochs

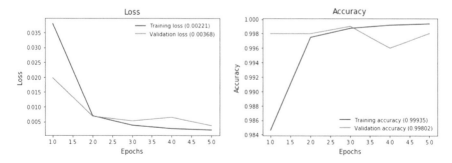

Fig. 6 Model loss and accuracy curve for 5 epochs

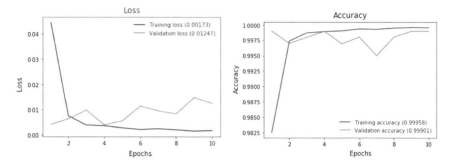

Fig. 7 Model loss and accuracy curve for 10 epochs

6 Conclusions

In this paper, a deep learning model of Convolutional Neural Network (CNN) is used to classify and predict the plant leaf diseases which provide a way of sustainable computing. This network has achieved effective results with best training accuracy 99.96% and best testing accuracy as 99.90%, which is far better than the conventional techniques. We have considered only two classes of the plant species; however, this problem can be stretched for other plant species to cover a wide range of common plants. Also, we noticed that deep learning techniques are able to handle a large dataset efficiently and there is no need for manual feature extraction, which is not possible with the conventional techniques. But the main challenge is that you need high-speed GPUs for processing such large data. Deep learning is an evolving discipline, so it can provide more possibilities. It is also possible to use a hybrid model like CNN in combination with other techniques, which can further reduce the processing.

References

1. Rahmani, M.E., Amine, A., Hamou, M.R.: Plant leaves classification. Alldata **82**(c), 75–80 (2015)
2. Liu, W., Wang, Z., Liu, X., Zeng, N., Liu, Y., Alsaadi, F.E.: A survey of deep neural network architectures and their applications. Neurocomputing **234**, 11–26 (2017)
3. Kaur, L.: A review on plant leaf classification and segmentation. Int. J. Eng. Comput. Sci. **5**(8), 17658–17661 (2016)
4. Dyrmann, M., Karstoft, H., Midtiby, H.S.: Plant species classification using deep convolutional neural network. Biosyst. Eng. **151**(2005), 72–80 (2016)
5. Barré, P., Stöver, B.C., Müller, K.F., Steinhage, V.: LeafNet: a computer vision system for automatic plant species identification. Ecol. Inform. **40**(May), 50–56 (2017)
6. Lee, S.H., Chan, C.S., Mayo, S.J., Remagnino, P.: How deep learning extracts and learns leaf features for plant classification. Pattern Recognit. **71**, 1–13 (2017)
7. Affonso, C., Rossi, A.L.D., Vieira, F.H.A., de Carvalho, A.C.P., de L.F.: Deep learning for biological image classification. Expert Syst. Appl. **85**, 114–122 (2017)

8. Ji, S., Zhang, C., Xu, A., Shi, Y., Duan, Y.: 3D convolutional neural networks for crop classification with multi-temporal remote sensing images. Remote Sens. **10**(1), 75 (2018)
9. Zhu, X., Zhu, M., Ren, H.: ScienceDirect method of plant leaf recognition based on improved deep convolutional neural network Action editor: Ali Minai. Cognitive Syst. Res. **52**, 223–233 (2018)
10. Xu, Y., Chi, Y., Tian, Y.: Deep Convolutional Neural Networks for Feature Extraction of Images Generated from Complex Networks (2018)
11. Hamuda, E., Mc Ginley, B., Glavin, M., Jones, E.: Improved image processing-based crop detection using Kalman filtering and the Hungarian algorithm. Comput. Electron. Agric. **148**, 37–44 (2018)
12. Khamparia, A., Saini, G., Gupta, D., Khanna, A., Tiwari, S., de Albuquerque, V.H.C.: Seasonal crops disease prediction and classification using deep convolutional encoder network. Circ. Syst. Signal Process. **32**, 1–19 (2019)
13. Aloysius, N., Geetha, M.: A review on deep convolutional neural networks. In: International Conference on Communication and Signal Processing (ICCSP), pp. 0588–0592 (2017)
14. Zhang, Z., Xing, F., Su, H., Shi, X., Yang, L.: Recent advances in the applications of convolutional neural networks to medical image contour detection (2017)
15. Al-Saffar, A.A.M., Tao, H., Talab, M.A.: Review of deep convolution neural network in image classification. In: International Conference on Radar, Antenna, Microwave, Electronics, and Telecommunications (ICRAMET), pp. 26–31 (2017)
16. Gu, J., Wang, Z., Kuen, J., Ma, L., Shahroudy, A., Shuai, B., Liu, T., Wang, X., Wang, G., Cai, J., Chen, T.: Recent advances in convolutional neural networks. Pattern Recognit. (2017)
17. Guo, Y., Liu, Y., Oerlemans, A., Lao, S., Wu, S., Lew, M.S.: Deep learning for visual understanding: a review. Neurocomputing **187**, 27–48 (2016)
18. Bhandare, A., Bhide, M., Gokhale, P., Chandavarkar, R.: Applications of convolutional. Neural Netw. **7**(5), 2206–2215 (2016)
19. Qian, S., Liu, H., Liu, C., Wu, S., Wong, H.S.: Adaptive activation functions in convolutional neural networks. Neurocomputing **272**, 204–212 (2018)
20. PlantVillage Disease Classification Challenge. https://www.crowdai.org/challenges/plantvillage-disease-classification-challenge/dataset_files. Last accessed 25 July 2018

The Internet of Drone Things (IoDT): Future Envision of Smart Drones

Anand Nayyar, Bao-Le Nguyen and Nhu Gia Nguyen

Abstract The Internet of Drone Things (IoDT) is envisioned as Future direction of Drones backend via Internet of Things, Smart Computer vision, Cloud Computing, advanced wireless communication, big data, and high-end security techniques. The utilization of drones is increasing in diverse fields from Agriculture to Industry, from Government to private organizations and from Smart Cities to Rural area monitoring. With IoDT based implementations, all the existing sectors will become intelligent and smart for performing Monitoring, surveillance, search and rescue and more. In this paper, we present, a conceptual presentation of new terminology, i.e., Internet of Drone Things (IoDT), along with its related technologies, applications, security issues and real-time implementation of IoDT by taking case studies of Agriculture and Smart Cities.

Keywords Robotics · Internet of robotics things (IoDT) · Smart cities · Future farms · Internet of things (IoT) · Drones · Smart agriculture

1 Introduction

Unmanned Aerial Vehicle (UAV) or Unmanned Aircraft System [1, 2] was designed and developed in the twentieth century and has gained lots of attention in recent times, due to their implementations in diverse fields because of their flexible and dynamic capability to carry out real-time operations and reduced acquisition costs. Since, their development, tremendous technological advancements are made in UAV systems like Weight reduction, improvised battery life, Smart sensor integration, 4K camera's

A. Nayyar (✉) · N. G. Nguyen
Graduate School, Duy Tan University, Da Nang, Vietnam
e-mail: anandnayyar@duytan.edu.vn

N. G. Nguyen
e-mail: nguyengianhu@duytan.edu.vn

B.-L. Nguyen
Provost, Duy Tan University, Da Nang, Vietnam
e-mail: baole@duytan.edu.vn

© Springer Nature Singapore Pte Ltd. 2020
A. K. Luhach et al. (eds.), *First International Conference on Sustainable Technologies for Computational Intelligence*, Advances in Intelligent Systems and Computing 1045, https://doi.org/10.1007/978-981-15-0029-9_45

integration, and above all implementations of machine learning and deep learning techniques to bring a high degree of autonomy in operations. The word "Unmanned Aircraft System" is adopted by regulators of Federal Aviation Administration (FAA) under various terms like "Flying Robot", "Pilotless Aircraft", "Remote Plane", but the main word "Drone" was proposed by military. The FAA Modernization and Reform act of 2012 has defined "Unmanned Aircraft" as "An Aircraft operated autonomously without any human intervention within or on aircraft" [3, 4].

An Unmanned Aerial Vehicle (UAV) can be defined as *"Smart Aircraft System that can be remotely controlled or using on-board computer system"* to perform monitoring, capturing video or other real-time operations. UAV is a part Unmanned Aerial System (UAS), whose main components are—Controller, Communication system, Sensors, Camera and other Mechanical components. The adoption of UAV is increasing across end-users across the globe in diverse fields like Agriculture, Entertainment and media, Oil & Gas, Industries especially E-Commerce, Hospitals and many more.

According to ResearchandMarkets.com [5], *"The UAV market has reached $8.02 billion in 2016 with a strong share of small UAVs"*. According to Business Intelligence [6], *"The sale of drones will cross $12 billion in 2021, which is up by a CAGR of 7.6% from $8 billion in 2016.* The growth of drones will occur in three main categories— Consumer Drones—expected to touch 29 million units by 2021; Enterprise drones will touch 8 million in 2021; Government drones' adoption is increasing 50% every year. The following Fig. 1 gives a clear picture of Drones adoption in varied industries (Source: PvC).

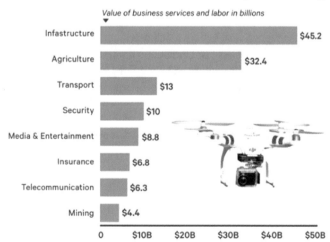

Fig. 1 Drone industry (*Source* PvC) [https://www.businessinsider.com/author/skye-gould]

The term "Internet of Things" is growing continuously with advancements in diverse fields like Wireless Sensor Networks, Machine-to-Machine (M2M) communication, ubiquitous computing [7]. IoT revolves around the ability to sense and understand things and act accordingly with regard to environmental requirements. The development of Internet and Smart Sensor Systems has brought revolution in Pervasive computing. Internet of Things consists of "Things" [8, 9] communicating to one another, Machine-to-Machine (M2M) communication as well as end user-to-computer communications. The primary objective of IoT is to uniquely identify and address devices to identify and exchange information with or without or with very less human intervention.

IoT applications typically comprise of:

- Sensor/Actuator/Thermostat embedded into the device to sense environmental conditions.
- Smart Wireless connectivity between sensor and cloud infrastructure.
- Data Collection and Analytics by smart integration of Cloud, Fog, Big Data, and Data centers.

Unmanned Aerial Systems (UAS) are already finding their way with tight integration with IoT and with strong collaboration with IoT, UAS systems can become more advanced for legacy applications and open opportunity for a plethora of new real-time applications. With IoT, UAV's/UAS/Drones can be deployed at challenging locations, can carry flexible and diverse payloads, mission re-programming in real-time and will be able to measure just about anything, anywhere.

Organization of Paper

Section 2 reviews associated terminologies—Unmanned Aerial Vehicles (UAV)/Unmanned Aerial Systems (UAS) along with its categories, Flying Adhoc Networks, and Drones. Section 3 highlights the concept of "Internet of Drone Things (IoDT)"—Introduction, Technologies powering IoDT, UAV's v/s IoDT, Applications and Security challenges. Section 4 visualizes the implementation of IoDT in Agriculture and Smart Cities. The paper concludes with future scope in Sect. 5.

2 Terminologies

IoT (*Technology considered as a fusion of Smart Sensors, Embedded Systems, Cloud Computing, Big Data, Machine Learning, Deep Learning*) in the recent times, has introduced its wide adoption in the area of Nano Technology, Medical, Agriculture, Robotics, Industry and its booming with leaps and bounds and the near future is certain towards its adoption in the area of Drones. This section outlines terminologies associated with Internet of Drone Things.

Table 1 Classification of UAVs

Category	Type of size	Max take-off weight	Operating altitude (ft)	Airspeed (Knots)
Group 1	Small	0–20	Less than 1200	Less than 100
Group 2	Medium	21–55	Less than 3500	Less than 250
Group 3	Large	Less than 1320	Less than 18,000	Less than 250
Group 4	Larger	Greater than 1320	Less than 18,000	Variable
Group 5	Largest	Greater than 1320	Less than 18,000	Variable

2.1 Unmanned Aircraft/Aerial Systems (UAS)/Drones

The term Unmanned Aircraft System (UAS) or Unmanned Aerial Vehicles (UAV) are used interchangeably. The proliferation of Unmanned Aircraft Systems (UAS) [10–12] in diverse applications like Agriculture, Military, Industrial Monitoring has led to significant technological advancements to integrate UAS into new areas and markets especially in civil domains. UAS, also termed as "Drones" are deployed in lots of real-time operations like wildlife monitoring and management, pollutant studies, weather monitoring, military observations, emergency response, Photography, Linear infrastructure monitoring like Railways, Pipelines, etc. In the near future, UAS deployments are envisioned towards tons of civil, commercial and scientific applications. As per FAA, USA—Till 2017, 777,000 UAS are registered and the number is expected to increase every year by 23%.

Unmanned Aerial System (UAS) has three main components—(1): Autonomous or Human-operated control system—located mostly on the ground or embedded in controller chip of drone to keep it airborne; (2): Unmanned Aerial Vehicle (UAV); (3): Communication, Command, and Control (C3) system to link all sorts of communications.

The following Table 1 elaborates UAVs classification according to the US Department of Defense [13] [Source: http://www.dtic.mil/docs/citations/ADA518437].

The following Table 2 enlists the classifications of UAV's as per Technical Characteristics.

2.2 Flying Adhoc Network

The critical challenge, until today since the evolution of UAV/UAS and even Drones is "Efficient Communication". Typically, communication in any UAS/UAV falls under three main categories [14]—(1) IMC (Internal Machine Communications)—communication between inbuilt components or modules of UAV like camera, sensor

Table 2 Technical characteristics of UAV's

Types of UAV's	Technical characteristics
Very small UAV	Size-30–50 cm; flapping or rotary wings-Insect Size; Small size, lightweight and utilize conventional aircraft configuration; Range-5 km; Flight Time-20 to 45 Min; Examples: US Aurora Flight Sciences Skates, Israeli IAI Malat Mosquito
Small UAV	Size-50 cm to 2 m; Fixed-wing; Launched via Air throw, Range-30 to 400 m above ground level; Flight speed-10 to 50 m/s; Examples: RQ-11 Raven, Turkish Bayraktar
Medium UAV	Wingspan-5 to 10 m; Payload support-100 to 200 kg; Altitude: 10,000 to 30,000 ft; Flight Time: 24–48 h; Range-100 to 700 Nautical Miles; Speed: Mach 0.83; Examples: US Boeing Eye, RQ-2 Pioneer, RQ-5A Hunter.
Large UAV	Wingspan-35 to 55 ft; Weight support up to 2250 lbs; Control-Pilotless (on-board), three personnel on ground (Pilot, Sensor Operator, Intelligence Analyst); Range: 1000–1500 kms; Examples: US General Atomics Predator A and B

communication, etc. (2) Machine-to-Machine Communication (M2M)—Communication with other UAV's comprising a network in specific regions. (3) Machine-to-Infrastructure Communication—Communication between UAV with Communication Network (Ground Network/Satellite Network or combination of Both).

To facilitate information communication, existing communication system in UAV is direct ground-to-UAV communication and mostly uses unlicensed spectrum of 2.4 GHz. However, this approach has several limitations in terms of security, interference vulnerability, limited data rate, LoS operation and utilization of expensive hardware to perform long-distance communication. Sometimes, reliability of communication is not there in challenging environmental, operational scenarios like High Altitude terrains, Dense Forests, etc. and sometimes mission-critical applications also require dynamic topology which becomes difficult for the UAV to manage. So, efficient reliable communication is required for break-free communication. UAV's can be deployed in groups forming a reliable ad hoc network termed as "Flying Adhoc Network" [15]. Adhoc Networks operate according to implementation, utilization, communication and above all mission types. For Mobile Nodes based communications—MANETs is deployed; For Vehicular based communication—VANETs is deployed; For Underwater/SEA communications—UANETS/SEANETS are deployed; In case of Drones/Flying UAV's—FANETs is best. The following Fig. 2 highlights the concept of FANETs.

Considering FANETs, UAV communications will face less issues because of the following advantages:

- High degree of Mobility is facilitated in FANETs as compared to MANETs which facilitates communication to operate without any overhead and information loss during topology change.
- Peer-to-Peer connectivity support for coordination and collaboration of UAVs operational in network to gather environmental data and transmit back to the

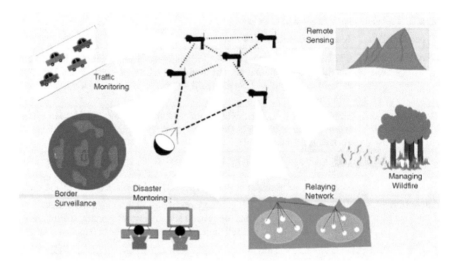

Fig. 2 FANETs (Flying Adhoc Network)

control center, almost like WSN. FANETs support multiple connections among UAVs.

- Communication Range in FANETs is more as compared to other ad hoc networks to link UAVs to ground stations. This requires more sophisticated communication hardware to maintain radio links.
- FANETs supports heterogeneous sensors as every UAV require different methodologies for distribution of data.
- FANETs supports complex environmental operations, as with higher node mobility, random block links are there in UAVs to support varied communication paths.

2.3 Drones

A Drone is termed as "Unmanned Aerial Vehicle" or a Flying robot [16] that can be controlled remotely or fly autonomously via software-controlled flight plans installed in microcontrollers of drone, working in a coordinated manner with sensors, communication systems, and GPS technology. The word and the concept of Drone originated from World War I and II for military purpose via Hot air balloon marked as first drone. Drones, till today, are remarkably improvised and used in varied configurations depending on the platform and the type of mission assigned. Today, Drones are classified as MAV (Micro or Miniature Air Vehicles), NAV (Nano Air Vehicles), VTOL (Vertical Take-off and Landing), LASE (Low Altitude, Short Endurance), LASE Close, LALE (Low Altitude, Long Endurance), MALE (Medium Altitude, Long Endurance) and HALE (High Altitude, Long Endurance).

Table 3 Types of Drones

Type of drone	Description
Quadcopters	Usually, have 4 rotors arranged in a square pattern and two pairs of identical fixed pitch propellers—Two clockwise and two counterclockwise. Mostly used in Research, Military, Photography, Sports
GPS drones	Linked to satellite via GPS to locate flight path, equipped with Autopilot system including camera and other aerial options including collision avoidance technology. Mostly used in Aerial Photogrammetry, 3D mapping, Multispectral imaging and Film making
RTF (Ready to Fly) drones	Plug and Play drones, already assembled and just ready to fly out of the box. Mostly used by beginners, kids, and hobbyists
Helicopter drones	Single rotor drone used by military for spying and observations
Delivery drones	Used by E-Commerce companies to deliver goods to clients. Customized with Small Basket to attach goods carrier. Small in size as compared to Big Drones and equipped with Smart Sensors and GPS tracking functionality
Racing drones	Small quadcopter basically designed to compete in the First-Person View race. Inexpensive and built with smart camera, Long battery and have fixed wings

The following Table 3 enlists some Real-time Operational drones.

3 Internet of Drone Things (IoDT)

The usage of drones in diverse fields is shining and progressing and in the near future, it can be skyrocketing. According to Gartner, 3 million drones were purchased in 2017 year itself by both end-users and industrial enterprises for multi-functions. It is also expected that world market of drones will touch $11.2 by 2020. Drone Technology together with fusion of Cloud Computing, Fog Computing, Smart cum Intelligent Sensors [17], Big Data and above all Internet of Things (IoT) will shape a next-generation future termed as "Internet of Drone Things (IoDT)" [18]. The term "Internet of Drone Things" will facelift the existing drone technology by broadening the scope towards new missions, new applications, faster connectivity, improved collaboration and coordination with other objects operational in network and high-end data processing via machine learning techniques.

Internet of Drone Things (IoDT) will make the drone technology applicable to those challenging environments like Landslides terrain, rural area monitoring, underwater monitoring, and even underground coal and oil gas mines, etc. where monitoring right now is almost impossible. IoT, as of today, is no longer confined to connect smart homes, cities to improvise the quality of an individual's life, but with the implementation of drone technology—it can enhance and transform the drone utilization in air, ground and even underwater. The objective of this research paper is to present and highlight the novel evolving concept in the near future with respect to drones, i.e., Internet of Drone Things (IoDT).

3.1 Technologies Shaping Internet of Drone Things (IoDT)

IoT and Drones technology integration has created various use-cases for organizations right from security to logistics, from agriculture to industry the drones and IoT offer interactivity and ubiquitous connectivity. To shape up Internet of Drone Things (IoDT), the following technologies are key players:

- **Cloud Computing** [19]—Cloud Computing is the key player in the concept of IoDT. Cloud computing heavily relies on resource sharing, which is the primary requirement of any IoT platform. Its primary characteristic is, it is location independent, the end-users can access all types of cloud services from any location and with any sort of device, just connected to the Internet as we talk of any IoT platform. Cloud computing offers rapid elasticity and dynamic scalability of resources so with a fusion of Cloud and IoT, IoDT has huge scope of diverse applications-based implementations in real-world. Two main approaches for Cloud Computing in IoDT are Cloud-based IoT and IoT-centric cloud. With regard to IoDT, as all communication will be done via IoT infrastructure, so Quality of Service (QoS) becomes a primary concern. So, to facilitate, break-free IoDT operations, service providers provide cloud computing services with best in QoS based on the Internet to gather information from the clients on a regular basis. With IoT-Centric cloud, it becomes easy to share the acquired data to the ground stations by eliminating latency, high traffic load, less hop count and above all less overhead.
- **Fog Computing** [20]—With fog computing, drones can become more efficient and process necessary information in an intelligent manner. With fog computing, Internet of Drone Things (IoDT), can handle the issue of best Internet connectivity, QoS requirements and real-time processing in remote areas with limited connectivity. Fog Computing is another crucial aspect along with Cloud for IoDT. With fog computing, drone applications can perform computations of approximate solutions locally and also rely on ground stations for more computational-intensive tasks to improvise the solution accuracy and reliability. So, with Fog Computing, IoDT will be benefited in terms of: Low latency, performing tasks, services and operations in real-time without any loss at limited or no connectivity areas; large number of drones can be integrated in the network; Heterogeneity and above all performing varied streaming's like—Adaptive Video Streaming, Parallel successive-refinement based streaming and live Camera Feeds.
- **Sensors and IoT Protocols**—Smart Sensors are becoming smaller and intelligent data by day and can be integrated into controller chips without any hiccup. Various IoT communication protocols like Ultra-Wideband, BLE, and protocols like

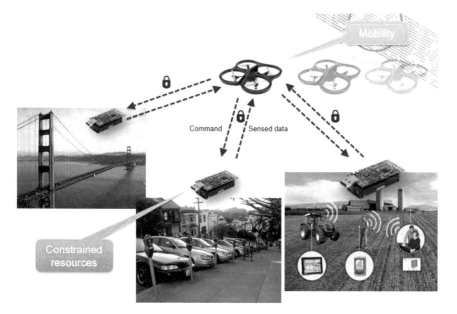

Fig. 3 IoDT scenario

ZWave or ZigBee will empower IoDT with prompt communication and reliable transmission (Fig. 3).

3.2 Comparison of UAV's with Internet of Drone Things (IoDT)

UAV's or Drone technology has seen several transformations in the past few years. Considering the path-breaking improvement in drone technology, currently, seven generations have gone underway.

The following points highlight the Generations of Drone Technology—[21].

- Generation 1: Basic RF-based drones. Small UAV's and Kid Drones toys.
- Generation 2: Static Design, Camera Integration, Video/Still Photography, Entire manual Pilot Control.
- Generation 3: Static Design, 2-AXIS Gimbal, HD Video Recording, Flying Assistance.
- Generation 4: Transformative Designs, 3-AXIS Gimbal, Full HD-1080 p Video Recording, Autopilot Integration, Safety Directions, and Collision Avoidance technology.
- Generation 5: 360° Gimbals, 4K Video Recording, Gesture-based Photos, Smart Navigation, and Flight controls.

Table 4 Comparison between UAVs and IoDT

Parameter	UAV's	IoDT (Internet of Drone Things)
Technology	Flying Adhoc Network (FANET)	IoT + Cloud computing + Smart sensors + Fog computing + Drones
Real-time complex operations	Only possible in those areas having strong connectivity	With cloud and fog computing, real-time complex operations are possible in complex environments with limited or no connectivity
Support for latest and well-updated Information	Limited	Fully updated at short intervals of time
Cloud-based information processing	Not supported	Highly supported
Operating conditions	Ground, air	Ground, air, underwater
Flying assistance and safety precautions	Not very advanced	Fully advanced with safety precautions as per FAA and autonomous control

- Generation 6: Autonomous controls, Safety Landing, Collision Avoidance, Micro Drones, Face and Object Recognition, Machine Learning Integration, Safety, and Regulatory Standards-based design.
- Generation 7: Automated Safety Modes, Enhanced Pilot Controls, Auto Action, Autonomous Flight, Smart Battery Monitoring and Landing Procedures and Integration of IoT and Cloud.

Internet of Drone Things (IoDT) is the next wave of advancement of UAVs to meet key features. IoDT links to Cloud Computing, real-time streaming, up-to-date information and intelligence response to dynamically changing conditions. With IoDT, complex missions can be carried out with ease, enhancing reliability and applicability.

The following Table 4 enlists key differences between UAV and Internet of Drone Things (IoDT).

3.3 Applications of Internet of Drone Things (IoDT)

As, stated above, IoT is gaining strong attention towards rapid innovations and implementations in diversified industries. The concept of Internet of Drone Things (IoDT) can also be visualized as an area with endless possibilities. In this section of the paper, we discuss various applications highlighting the real-time implementations of IoDT [22, 23].

a. **Agriculture**

With IoT, Agriculture field is already revolutionizing and some farmers are already making use of drones to monitor crop production and field conditions.

- Smart and Advanced Monitoring: With IoDT, farmers can get better precise information anytime and anywhere with regard to soil quality, crop maturity, soil moisture levels, evaporation level and with that farmer can make water and fertilizer adjustments, making the crop yield better. IoDT drones are also equipped with advanced cameras (4K/Full HD) with thermal sensors to detect cooler, best watered and under-watered regions of farms.
- Threats Identification: Crop yield is vulnerable from various threats like Wild animal, fire. With IoDT, farmers get early alerts of all sorts of threats and even with precision location to detect and correct the problems at the earliest point of time.
- Ease of Utilization and GIS Mapping: Traditional UAV's are very difficult to operate and setup, but IoDT has started Piloting feature with regard to collision detection, autonomous flying so IoDT makes the work very easy and with GIS Mapping, IoDT acts as a strong key player for farmers to boost yields, cut costs and enhance profits.

b. **Undersea Operations**

With IoDT undersea drones, marine robots can bring revolution to shipping industry, fisheries, undersea researchers and even military.

- Undersea Aquatic Life Monitoring: With undersea Aerial IoDT based drones, the best undersea photography and aquatic life monitoring can be done with high quality and ease and IoDT based drones can go more down the sea level as compared to UAV drones. So, high-end scientific researchers can explore unexplored areas of underwater for marine research. And even other observatory tasks in terms of predator nets damage, mooring line inspection and other submerged infrastructure can be done without any hiccup.
- Undersea Infrastructure Maintenance: With undersea IoDT drones, local municipal corporations and other service companies and other service providers can use a remote camera to perform regular inspection checks with regard to drainage pipes, sewerage pipes and even water tanks to detect any sort of damage and to ensure normal functioning.
- Search and Recovery: With advanced IoDT drones-based cameras, location and retrieval of drowned persons and searching victim people in case of any water emergency situations will become easy.

c. **Delivery and Healthcare**

With IoDT, Delivery and Healthcare industry can gain huge advantages in terms of quick response and control.

- Delivery: With IoDT drones, a drone can transport the goods ranging from Simple Pizza to Luxury items at ease. Even the local letters can reach the residential homes at short span of time. An Online E-commerce company like AMAZON.COM has already deployed more than 300 IoDT based drones for package delivery in the U.S.
- With IoDT, lots of victims can be saved and emergency medical supplies can reach the destination quickly. On the identification of the pinpoint location of the victim, IoDT drone can carry medical supply and even with Flying Ambulance, the patient can reach the hospital quickly.

d. **Military**

Military is one of the crucial areas for deployment of IoDT. As till today, the most early and widespread use of drones is done by Military of various counties.

- Border Surveillance: Considering the cross-border threats and to combat with all sorts of illegal trespassing and intrusions, countries are defending their lines of border with varied smart drones. With IoDT, military can fly the drone not only for a prolonged period of time, but can also get pinpoint location of that area at the earliest, where any intrusion in progress and with IoDT, militaries all over the world can even monitor those areas, where connectivity is either less or even almost nothing as IoDT based drones have direct satellite-based linkages.
- Early Threat detection: With IoDT based drones monitoring the sensitive areas, the drones can give early warning with regard to enemy vehicles like Fighter Jets or Missiles entering in one's border.
- Safety of War Personnel in Combat Zones: During War times, IoDT based drones, can give pinpoint location of wounded soldiers and most soldiers can be made safe during complex missions and above all, sensitive areas monitoring can be performed with ease to gain military intelligence data to frame strategic operations.

3.4 Security Issues

Internet of Drone Things (IoDT), as mentioned above, will not only bring revolution to varied fields, but will make the existing fields work with more precision, intelligence and integrate next-generation smart technologies. But, as IoDT is new technology, and new technology always comes with some Challenges and Shortcomings, that must be solved [24].

The primary issue connected with regard to IoDT is with regard to Security and handling Big Data in an efficient manner. With IoDT, tons of drones will work in cooperative manner collecting data, and IoDT can bring tons of unstructured data at Cloud servers. The data is required to be processed in real-time and with Big Data clustering and Mining techniques which require lots of servers and technologies, making IoDT implementation an expensive affair.

Considering the real-time applications, depending on the applicability scenario, sometimes fleet of drones is required. As with the increase in number of devices, addressing and device management by considering Quality of Service (QoS) parameter will become a critical issue. So, this can be overcome, by deploying devices in an ad-hoc manner to join and leave the network at any point of time. So, all applications must be tailored/customized from scratch to enable extensible services and operations.

Another significant challenge is to deal with connectivity. In order to make IoDT run in an efficient manner, preferably 4G/5G Connectivity is required, and if the connectivity breaks, connection will be lost, which seriously impacts the IoDT operational performance and can compromise the sensitive tasks and even data security.

In this section, of research paper, we highlight some of the security attacks which can pose a serious threat to IoDT at different layers.

The following Table 5 highlights some of the security issues associated with IoDT.

4 Internet of Drone Things (IoDT)—Real-Time Implementations

In this section of research paper, we present, some real-time implementations of Internet of Drone Things (IoDT) in two scenarios—Agriculture transforming into Smart Agriculture and Smart Cities real-time monitoring

4.1 Smart Agriculture

IoT has already strong roots in the field of agriculture, but still Agriculture farms have certain issues with regard to monitoring of large area, because of limited or no connectivity. Farmers require a precision solution to monitor crop fields, agriculture fields—Moisture level, soil quality, pesticides quality, animals' detection in farms, cattle monitoring and other agricultural equipment's [26, 27]. No doubt, WSN and WBAN can give prompt solution, but sensors deployment is limited, the data acquired is often delayed and sometimes not of real-time and above all sensors are prone to frequent failures because of life and energy is a big problem to combat.

Solution via IoDT

So, to combat the issue of precision Agriculture Farm Monitoring, Cattle Monitoring, Agri Bots Data Acquisition, and Weather Monitoring IoDT come to rescue. The following Fig. 4 highlights Smart Agriculture Farm Monitoring via IoDT.

With implementation of IoDT, farm monitoring can come to ease, and all the potential threats like Animal Intrusion, Farm Border can be seen easily, which is highly expensive with regard to Sensor Deployment or Remote HD camera's installation.

Table 5 Security Issues associated with IoDT

Layer	Security threats associated
Physical layer	Jamming Attack—Interferences to Radio Signals making connectivity issues for Drones to operate efficiently and also impact energy consumption.
	Tampering—Intruder can break in the nodes sensitive data and access confidential information
	GPS Spoofing—Utilization of Fake GPS signal rather than genuine one
Data link layer	Collision Attacks—Making the Information collide, when two or more drones operate on the same frequencies, making a network unreliable to operate
	Unfairness Attack—Attacker can overall degrade the entire system, causing other drone nodes to miss their MAC protocol-based transmission deadline in real-time
Network layer	Selective Forwarding Attack—Malicious nodes causing all types packet disruption/dropping/corruption passing through them, making IoDT overall unreliable
	Other attacks associated like any other Ad Hoc Network causing packet loss are: Sinkhole, Wormhole, HELLO FLOOD, Sybil attack which impacts the Packet Delivery Ratio, throughput and creates routing overhead
	To combat and counterfeit these attacks, Broadcast Storm Problem (BSP) techniques are required [25]
Transport layer	Flooding Attack—Causing extreme packet flow leading to overall network congestion
	Desynchronization Attack—Malicious node pushing messages, by conveying sequence numbers to more than one node operational in the network
	Hijacking Attacks—Making some nodes unoperational, causing disruption in the network
	Man-in-Middle Attack—Attacker can deploy a malicious drone in-between operational network, causing information leakage, disruption, and other security attacks
Perception layer	Heterogeneity of Drones and other IoDT-based things non-compatibility to operate. In addition to other issues like—Limited power, less computational capacity and limited storage facility, overall act as backdoors for IoDT
Application layer	Misc Attacks—Denial of Service Attacks, Privacy Attacks, Malicious Access Points, Unlawful attacks, Intrusion attacks

With IoDT, Cattle monitoring can be done at precision level. All the cattle's can be monitored by sensor deployment on every animal. With the field surveillance from drones, the cattle-raising area can be monitored to take timely actions. Any unusual activity, like cattle health, cattle going out of farms can be known at timely manner to trigger required action.

IoDT drones can acquire the live data from sensors at regular intervals with regard to Agri bots cultivating the farms, sea sensors to detect water quality levels and even weather monitoring can be done [26–28].

So, IoDT can not only transform Agriculture, but also makes complex problems solved for farmers for better crop yields.

Fig. 4 Smart agriculture solutions based on IoDT

4.2 Smart Cities Monitoring

To transform the legacy cities to Smart Cities, various technologies like Cloud Computing, Internet of Things (IoT), Big Data play a significant role. But IoT based infrastructure can be easily deployed in urban areas, as no issues with regard to connectivity but in rural areas, smart cities facilitations are a big drawback. In addition to this, Governments and IT organizations are taking less interests in deployment of these due to lack of demand and high expenditure to deploy latest technologies [29].

But with IoDT implementation in Smart Cities, new ventures and opportunities will be facilitated in terms of crime detection, search and rescue, people/object tracking, smart monitoring, etc. So, IoDT has huge potential to transform existing smart cities architecture to next level.

Existing Problems in Smart Cities

Considering the existing smart cities, the main issue is the monitoring coverage to all the city ends as all areas don't facilitate smart internet connectivity and don't update/synchronize the data at regular intervals so efficient monitoring is not facilitated.

In addition to this, emergency response systems are not highly reliable as existing network infrastructure of Camera Deployments and Sensors is not available across nook and corner of the city and moreover, it is also expensive to have entire city coverage due to high infrastructure cost of Sensors and HD cameras. And sometimes,

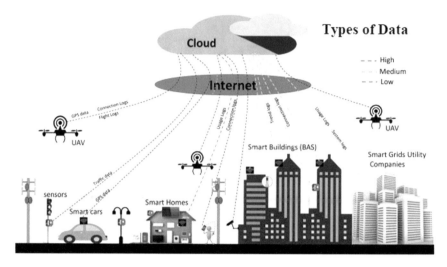

Fig. 5 Smart cities solutions and transformation via IoDT

Mobile technologies coverage in terms of GSM, 4G is also not fully available which limits emergency response teams to take rapid action.

So, IoDT can be a one-stop solution to combat the existing problems of smart cities.

Solutions via IoDT

The following Fig. 5 elaborates the solution proposed via IoDT in smart cities.

With IoDT, the monitoring of entire smart city is covered and every nook and corner can be monitored and updated at short intervals of time, making not only monitoring easy but also to provide live traffic feed and even weather monitoring.

With IoDT, the city's electric power companies can perform electricity bills reading automation via Smart Grids, as manual reading is time-consuming and cumbersome task. With every residential house, equipped with RFID tags based for Smart Reading and IoDT drones capturing the data and providing it to the company's back-end data centers, the electricity billing can become very easy.

IoDT can transform Smart Buildings and Smart Homes by facilitating smart Mobile Coverage and emergency teams with response towards any sort of critical situation. Natural disasters, Environmental Monitoring like Gases, Pollution levels can be easily monitored and in case of any demolition, necessary steps can be taken care of.

5   Conclusion and Future Scope

Internet of Things (IoT) in the near future can transform everything via drones. In this paper, the conceptual foundation of new terminology, i.e., Internet of Drone Things (IoDT) is outlined. IoDT is just a fusion of Cloud, Big Data, Smart Sensors and IoT to bring revolution in existing IoT infrastructures and transform various sectors. In the lines of IoT, IoDT has advanced capabilities of performing complex missions, advanced monitoring and rapid response times for necessary actions. IoDT can be considered as network of Drones, operating in cooperative and coordinated manner to give optimized, safe and reliable cum secure results.

In this paper, we address the common terminologies associated with IoDT, technologies empowering IoDT, real-time applications, comparison with UAV and serious security threats acting as a significant research area for researchers to make IoDT secure for implementations in near future. In addition to this, the research paper highlights two important live case studies with regard to IoDT implementations in Smart Agriculture and Smart Cities.

Future Scope

In the near future, we propose a novel framework for implementing IoDT in Government sector especially Energy, Pipeline based monitoring and how IoDT can transform legacy industries to Smart industries. The future research direction is strongly focused on proposing real-time operational fleet of 10+ drones forming IoDT network in a city doing Traffic and Weather monitoring.

References

1. Heermann, P.D.: Unmanned Aerial Systems (No. SAND2015-9558PE). Sandia National Lab. (SNL-NM), Albuquerque, NM (United States) (2015)
2. Cavoukian, A.: Privacy and Drones: Unmanned Aerial Vehicles, pp. 1–30. Information and Privacy Commissioner of Ontario, Ontario, Canada (2012)
3. Colomina, I., Molina, P.: Unmanned aerial systems for photogrammetry and remote sensing: a review. ISPRS J. Photogrammetry Remote Sens. **92**, 79–97 (2014)
4. FAA Draft Advisory Circular: Unmanned air Vehicle Design Criteria, Section 6.j, 15 July 1994
5. https://www.researchandmarkets.com/reports/4455437/global-unmanned-aerial-vehicle-uav-market-value. Accessed on 5 Nov 2018
6. https://www.businessinsider.com/commercial-uav-market-analysis-2017-8. Accessed on 5 Nov 2018
7. Ray, P.P.: A survey on Internet of Things architectures. J. King Saud Univ. Comput. Inf. Sci. **30**(3), 291–319 (2018)
8. Silva, B.N., Khan, M., Han, K.: Internet of things: a comprehensive review of enabling technologies, architecture, and challenges. IETE Tech. Rev. **35**(2), 205–220 (2018)
9. Solanki, A., Nayyar, A.: Green internet of things (G-IoT): ICT technologies, principles, applications, projects, and challenges. In: Handbook of Research on Big Data and the IoT, pp. 379–405. IGI Global
10. Hackenberg, D.L., Linkel, J., Wolfe, R., Noble, J.: Unmanned Aircraft Systems Demand and Economic Benefit Forecast Study (2018)

11. Oh, P., Piegl, L.A.: Unmanned Aircraft Systems. In Valavanis, K. P. (ed.). Springer (2013)
12. Frew, E.W., Brown, T.X.: Networking issues for small unmanned aircraft systems. J. Intell. Rob. Syst. **54**(1–3), 21–37 (2009)
13. http://www.dtic.mil/docs/citations/ADA518437. Accessed on 5 Nov 2018
14. Campion, M., Ranganathan, P., Faruque, S.: A review and future directions of UAV swarm communication architectures. In: 2018 IEEE International Conference on Electro/Information Technology (EIT), pp. 0903–0908. IEEE (2018)
15. Nayyar, A.: Flying adhoc network (FANETs): simulation based performance comparison of routing protocols: AODV, DSDV, DSR, OLSR, AOMDV and HWMP. In: 2018 International Conference on Advances in Big Data, Computing and Data Communication Systems (icABCD), pp. 1–9. IEEE (2018)
16. Batth, R.S., Nayyar, A., Nagpal, A.: Internet of robotic things: driving intelligent robotics of future-concept, architecture, applications and technologies. In: 2018 4th International Conference on Computing Sciences (ICCS), pp. 151–160. IEEE (2018)
17. Nayyar, A., Puri, V., Le, D.N.: A comprehensive review of semiconductor-type gas sensors for environmental monitoring. Rev. Comput. Eng. Res. **3**(3), 55–64 (2016)
18. Gharibi, M., Boutaba, R., Waslander, S.L.: Internet of drones. IEEE Access **4**, 1148–1162 (2016)
19. Biswas, A. R., Giaffreda, R.: IoT and cloud convergence: opportunities and challenges. In: 2014 IEEE World Forum on Internet of Things (WF-IoT), pp. 375–376. IEEE (2014)
20. Chiang, M., Zhang, T.: Fog and IoT: an overview of research opportunities. IEEE Internet Things J. **3**(6), 854–864 (2016)
21. https://airdronecraze.com/drone-tech/. Accessed on 5 Nov 2018
22. Hassanalian, M., Abdelkefi, A.: Classifications, applications, and design challenges of drones: a review. Prog. Aerosp. Sci. **91**, 99–131 (2017)
23. Pigatto, D.F., Rodrigues, M., de Carvalho Fontes, J.V., Pinto, A.S.R., Smith, J., Branco, K.R.L.J.C.: The internet of flying things. In: Internet of Things A to Z: Technologies and Applications, pp. 529–562 (2018)
24. Lin, C., He, D., Kumar, N., Choo, K.K.R., Vinel, A., Huang, X.: Security and privacy for the internet of drones: challenges and solutions. IEEE Commun. Mag. **56**(1), 64–69 (2018)
25. Srilakshmi, A., Rakkini, J., Sekar, K.R., Manikandan, R.: A comparative study on internet of things (IoT) and its applications in smart agriculture. Pharmacognosy J. **10**(2) (2018)
26. Puri, V., Nayyar, A., Raja, L.: Agriculture drones: a modern breakthrough in precision agriculture. J. Stat. Manage. Syst. **20**(4), 507–518 (2017)
27. Swapna1, B., Manivannan, S.: Analysis: Smart Agriculture And Landslides Monitoring System Using Internet of Things (IOT) (2018)
28. Rathore, M.M., Paul, A., Hong, W.H., Seo, H., Awan, I., Saeed, S.: Exploiting IoT and big data analytics: defining smart digital city using real-time urban data. Sustain. Cities Soc. **40**, 600–610 (2018)
29. de Melo Pires, R., Arnosti, S.Z., Pinto, A.S.R., Branco, K.R.: Experimenting broadcast storm mitigation techniques in FANETs. In: 2016 49th Hawaii International Conference on System Sciences (HICSS), pp. 5868–5877. IEEE (2016)

Solving Grid Scheduling Problems Using Selective Breeding Algorithm

P. Sriramya and R. A. Karthika

Abstract Grid scheduling is characterized as the way toward settling on planning choices including resources over multiple administrative domains. This procedure can search through different administrative areas to utilize a particular machine or scheduling one job for exhausting various resources at a particular node or multiple nodes. From a grid point of view, a job is anything that needs a resource. The primary objective of grid is to give service with high dependability and minimal effort for substantial volumes of clients and support teamwork. In this paper, we consider a directed acyclic graph (DAG) with nodes and edges where the nodes are considered the task and the edges specify the order of execution of the tasks as a grid. This kind of problem is called the precedence-constrained problem. The selective breeding algorithm is an efficient algorithm to solve NP-hard problems. One such example of NP-hard problem is the precedence-constrained problem. So we consider SBA algorithm to solve precedence-constrained problems and found optimal solution of 13 units when compared with the traditional methods of 23 units. And it is also proved that the amount of waiting time is reduced greatly when compared to the traditional methods. So by implementing SBA for the grid scheduling problem more time is saved and is proved to be efficient.

Keywords Selective breeding algorithm · Precedence-constrained problem · Grid scheduling · Directed acyclic graph

1 Introduction

Grid scheduling is characterized as the way toward settling on planning choices including a resource over multiple administrative domains. This procedure can search

P. Sriramya
Saveetha School of Engineering, Saveetha Institute of Medical and Technical Sciences, Chennai, India

R. A. Karthika (✉)
Vels Institute of Science Technology & Advanced Studies, Chennai, India
e-mail: karthika.se@velsuniv.ac.in

© Springer Nature Singapore Pte Ltd. 2020
A. K. Luhach et al. (eds.), *First International Conference on Sustainable Technologies for Computational Intelligence*, Advances in Intelligent Systems and Computing 1045, https://doi.org/10.1007/978-981-15-0029-9_46

through different administrative areas to utilize an individual machine or scheduling a particular job for using numerous resources at a particular node or multiple nodes. From a grid point of view, a job is anything that needs a resource. The primary objective of grid is to give service with high dependability and minimal effort for substantial capacities of clients and likewise upkeep teamwork. The utmost significant problem in grid computing is control in resource managing, dependability, and security. We need to improve the efficiency of grid scheduling. To upsurge the efficiency of grid an appropriate and valued scheduling is required.

Generally, a grid scheduler obtains request from the user's grid and then utilizes the information acquired from the source modules. It chooses the grid service information for the needs as well as based on the objective function and resource competence, which could be predicted. And then generates a mapping among the source requests. There are two approaches used in the list scheduling: static and dynamic. In static list, scheduling the list of nodes is statically constructed before task allocation begins, and the sequencing in the list is not modified during the other operations. In dynamic list, scheduling priorities are updated after each allocation and the list of ready tasks is consequently rearranged. Dynamic list scheduling algorithms can potentially generate better schedules, in particular if they have to operate in a grid environment, which is extremely dynamic.

Evolutionary computation is one of the growing soft computation research disciplines that are inspired by principles of natural evolution. Evolutionary algorithms adopt these mechanisms of natural evolution in a simplified way and progressively breed better solutions to a wide variety of complex optimization and design problems like grid scheduling problems.

Breeders of animals and plants in today's world are looking to produce organisms that will possess desirable characteristics, such as high crop yields, resistance to disease, high growth rate, and many other phonotypical characteristics that will benefit the organism and species in the long term. This practice of selecting parents is so-called artificial selection or selective breeding.

It is based on the concept of genetics. The selection of parents is based on the non-random method. Here, the parents who have the preferred properties are selected to yield the hybrid in which at least one will show all of those required features. This concept of selective breeding is used to develop the proposed selective breeding algorithm. This new evolutionary algorithm consists of three main mechanisms namely:

- Breeding mechanism
- Fusion process
- Inbreeding depression.

2 Related Work

Foster et al. [1] developed a grid computing infrastructure for both hardware and software that delivered access to high-level computational abilities into dependable, reliable, prevalent, and reasonably priced offers. Grid is a pooled environment, which is executed with the formation of long-term services and standards. Those services are used to generate and segment the resources that are to be allocated.

Kesselman et al. [2] introduced an opportunistic algorithm, which takes into account the dynamic characteristics of grid environments without the necessity to inquiry the remote sites. The opportunistic algorithm profits from the dynamic characteristics of the grid environment. If a site ensues to accomplish poorly, then the number of jobs allotted to this site decreases. Likewise, if a site process jobs rapidly, then extra jobs are scheduled to that site.

Jin et al. [3] illustrated the fundamental objective of grid by furnishing administrations with high dependability and smallest amount for expansive dimensions of clients and support team could work and the greatest imperative concern in grid computing are the resource managing and control, dependability, and security. Nowadays expanded productivity of grid is an essential dispute. In order to build the effectiveness of a network, an appropriate and beneficial scheduling is required. But, the vibrant idea of grid resource and the requests of various clients were initiating intricacy in scheduling framework.

Yagoubi et al. [4] insisted on the load balancing for an immense number of frameworks, which is a critical issue to be solved keeping in mind, the end goal to empower the productive utilization of parallel computer frameworks. This issue can be contrasted with issues emerging in regular work distribution process like that of scheduling all tasks expected to develop an expansive building. The basic goal of a load balancing, comprises principally, in enhancing the normal reaction time of utilizations, which regularly implies keeping up the workload relatively comparable all in all assets of a systems.

Lina et al. [5] proposed a heuristic task scheduling algorithm fulfilled load balancing on grid condition. The algorithm plans undertakings by utilizing mean load in view of assignment predictive execution time as heuristic data, to get an underlying booking technique. At that point optimal scheduling system is accomplished by choosing two machines fulfilled condition to change their loads through reassigning their assignments under the heuristic of their mean load.

Hoseini et al. [6] depicted that a mixture heuristic to take care of parallel machines assignment planning problem. Genetic algorithm (GA) and ant colony optimization (ACO) share information structures and develop in parallel with a specific end goal to enhance the execution of the calculation. In ACO algorithm, for balancing job scheduling, grids are presented in which, the grid is viewed as a heterogeneous multi-processor and its motivation is to create a decent errand booking calculation to relegate employments to assets in both figuring grid and information grid.

Ramya et al. [7] introduced an algorithm named optimized hierarchical load balancing algorithm (OHLBA), which is used for job scheduling and load balancing. This recommended technique is to powerfully make an ideal schedule to finish the employments inside least makespan. The foremost involvements are to provide stability to the system load and to lessen the makespan of the jobs. GridSim was used to scrutinize the performance of OHLBA algorithm with another algorithm in relationship to makespan and its proficiency. Tentative outcomes show that the suggested algorithm can accomplish well in a grid environment.

Keerthika et al. [8] proposed a new bacteria schedule algorithm (BSA) that considers client fulfillment along with adaptation to internal failure. The fundamental commitment of this work incorporates accomplishing client fulfillment with fault tolerance to non-critical failure and limiting the makespan of employments. The execution of this proposed calculation is assessed utilizing GridSim in light of makespan and number of employments finished effectively inside client due date.

Piyush Chauhan and Nitin propose a new algorithm to solve grid scheduling problem where the resources are attached to aP2P networks. The authors propose a fully decentralized P2P grid scheduling algorithm, which solves the problem in a decentralized environment [9]. The algorithm takes precise decisions in terms of cost and time taken.

Zhu et al. [10] proposed server on time (SOT) algorithm to have an efficient scheduling for solving optical grid applications when tasks are more important. The authors focus mainly on task scheduling algorithm for dynamic multi DAG. The results proved in the paper say that it takes less time and utilizes the resources effectively when the DAG size is more.

SBA is a genetic-based algorithm based on a non-random method. It is proposed by Sriramya [11]. In SBA parents with preferred possessions are carefully chosen to yield the hybrid in which at least one would express all of those characteristics. SBA is applied for solving the one-dimensional bin-packing problem [12] with precedence constraints. It is also applied for many task scheduling problems.

3   Grid Scheduling Based on DAG

Scheduling of precedence-constrained task graphs is an NP-complete problem in its common form. According to Kwok et al. [13] the problem is NP-complete even in two simple cases:

1. Scheduling unit-time tasks to an arbitrary number of processors
2. Scheduling one or two time unit tasks to two processors.

The problem of scheduling in parallel applications is formed by multiple interacting jobs in a dynamic environment formed by a set of heterogeneous resources and an un-uniform interconnected network. In this context, a parallel application is commonly exhibited by a precedence-constrained task graph, which is a directed acyclic graph (DAG) with node and edge weights, i.e., the nodes symbolize the tasks and the directed edges signify the execution dependencies as well as the extent of communication. In this model, a task could not start execution before all of its parents have completed their execution and direct all of the messages to the machine allocated to that task. Figure 1 shows an example of precedence-constrained task graph where task n cannot start before tasks n1, n to n edge labels are data transfer times

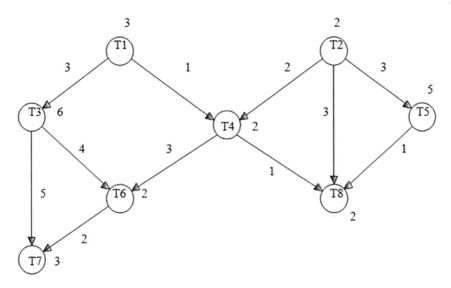

Fig. 1 Example of DAG

and node labels are computation times. The scheduling objective is to minimize the program completion time.

A DAG is a tuple $G = (V, E)$ where $V = \{n, j = 1\ldots v\}$ is the set of nodes with $|V| = v$, E is the set of communication edges and $|E| = e$. Nodes and edges are labeled as follows:

The weight c, c is the communication cost incurred along the edge e_{ij} $i, j = (V_i, V)$ E, which becomes zero if both nodes are mapped to the same processor
The weight Cij
T of the node V, V is the computation cost (or the expected execution time) of the task represented by the node V_i.

A task is a set of instructions that should be accomplished sequentially within the same processor without preemption. A task could not start execution before all of its parents have completed their execution and directed all of those messages to the machine assigned to that task. Two tasks are termed independent if there are no dependence paths among them. The width of a DAG is the size of the utmost set of independent tasks. A node is a source node if it has no incoming edges and it is an exit node if it has no outgoing edges.

Each edge (V_i, Vj) in the DAG having weight D, which actually displays the time necessary to transferring data that should be transferred to run Vj from V_i and likewise note that before the end of V_i those data would not be movable to Vj. Every single node in a DAG is an input node and an output node that has scheduling problems starting with resource distribution to the input node V_i, ends on accomplished resource distribution to the output node Vj, and run it.

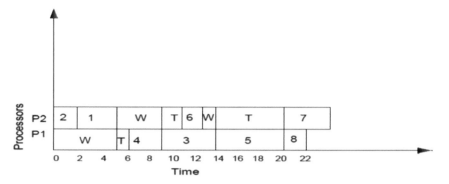

Fig. 2 Gantt chart for allocation of tasks using traditional method

Figure 1 shows DAG representation of selected grid for the grid scheduling problem. In the assumed grid, T is number of tasks arrived for scheduling the grid. C is the data transmission cost matrix among the depending nodes to facilitate C ($T * T_{ij}$) which signifies the cost of data transfer from V_i to Vj. D_{T*T} matrix is the dependence matrix. There are two processors were in the tasks are supposed to be scheduled. We represent our results using a Gantt chart. The Gantt chart defines the processor assignment, the starting and the completion times for each task.

Based on the Gantt charts the maximum execution time taken to process all eight tasks is totally 23 units including the wait time of the processors in the case of traditional methods. This is shown in the Gantt chart in Fig. 2.

4 Proposed SBA for Solving Grid Scheduling Based on DAG

The cost and dependence matrix for grid is based on dependence among tasks, execution time, and data transfer time for tasks. We create the matrix C_{T*T}, D_{T*T} based on the DAG precedence values.

Arrived matrices C_{8*8} and D_{8*8} are as follows:

$$
C_{8*8} = \begin{array}{c} \\ T1 \\ T2 \\ T3 \\ T4 \\ T5 \\ T6 \\ T7 \\ T8 \end{array}
\begin{array}{c} T1\ T2\ T3\ T4\ T5\ T6\ T7\ T8 \\
\begin{pmatrix}
0 & 0 & 3 & 1 & 0 & 0 & 0 & 0 \\
0 & 0 & 0 & 2 & 3 & 0 & 0 & 0 \\
0 & 0 & 0 & 0 & 0 & 4 & 5 & 0 \\
0 & 0 & 0 & 0 & 0 & 3 & 0 & 1 \\
0 & 0 & 0 & 0 & 0 & 0 & 0 & 1 \\
0 & 0 & 0 & 0 & 0 & 0 & 2 & 0 \\
0 & 0 & 0 & 0 & 0 & 0 & 0 & 0 \\
0 & 0 & 0 & 0 & 0 & 0 & 0 & 0
\end{pmatrix}
\end{array}
$$

Data transmission cost matrix in the example desired graph

$$
D_{8*8} = \begin{array}{c} \\ T1 \\ T2 \\ T3 \\ T4 \\ T5 \\ T6 \\ T7 \\ T8 \end{array}
\begin{array}{cccccccc}
T1 & T2 & T3 & T4 & T5 & T6 & T7 & T8 \\
\left(\begin{array}{cccccccc}
0 & 0 & 0 & 0 & 0 & 0 & 0 & 0 \\
0 & 0 & 0 & 0 & 0 & 0 & 0 & 0 \\
0 & 0 & 1 & 0 & 0 & 1 & 1 & 0 \\
1 & 1 & 0 & 0 & 0 & 1 & 0 & 1 \\
0 & 1 & 0 & 0 & 0 & 0 & 0 & 1 \\
1 & 1 & 1 & 1 & 0 & 0 & 1 & 0 \\
1 & 1 & 1 & 1 & 0 & 1 & 0 & 0 \\
1 & 1 & 0 & 1 & 1 & 0 & 0 & 0
\end{array}\right)
\end{array}
$$

Dependency among tasks matrix in the sample desired graph.

5 Solving Grid Scheduling Problem Using SBA

Using the above cost matrix and dependency matrix as input to the SBA algorithm the precedence-constrained problem is being solved. The steps of the algorithm with the implementation of problem are given as follows:

Step 1: Initial population is

$$
\begin{array}{cc}
1 - 3 - 6 - 7 & 2 - 4 - 5 - 8 \\
1 - 3 - 7 - 6 & 2 - 5 - 4 - 8 \\
1 - 4 - 6 - 7 & 2 - 8 - 4 - 5 \\
1 - 4 - 3 - 7 & 2 - 8 - 4 - 5 \\
1 - 4 - 3 - 6 & 2 - 4 - 8 - 5
\end{array}
$$

Step 2: Calculation of objective function and breeding factor for each sequence

(i)
$$
\begin{aligned}
\text{Breeding factor} &= 1/\text{objective function} \\
&= 1/\text{Makespan}
\end{aligned}
$$
(ii) Organized sequences in descending order related to its breeding factor.

Step 3: Divide the population set into two sets and form di-series sets

The presiding set R contains five task sequences namely R1, R3, R4, and R5. The abandon set r contains five task sequences namely r, r2, r3, r4, and r5. Form the di-series set such as {R1r1, R2r2, R3r3, R4r4, and R5r5}.

$$R-\text{presiding set} \qquad r-\text{abandon set}$$

Makespan   **Makespan**

$$2 - 4 - 5 - 8 => 17 \quad 1 - 3 - 6 - 7 => 19$$
$$2 - 5 - 4 - 8 => 15 \quad 1 - 3 - 7 - 6 => 18$$
$$2 - 8 - 4 - 5 => 14 \quad 1 - 4 - 6 - 7 => 16$$
$$2 - 8 - 5 - 4 => 15 \quad 1 - 4 - 3 - 7 => 18$$
$$2 - 4 - 8 - 5 => 14 \quad 1 - 4 - 3 - 6 => 17$$

Step 4: Breeding process

Potential breeds for one set combination by considering two di-series namely R1r1 and R2r2.

	R2	r2
R1	R1R2	R1r2
r1	r1R2	r1r2

Find all the possible breeds (di-series sets).

$$\mathbf{R1R2} => 2 - 4 - 5 - 8 \,\&\, 2 - 5 - 4 - 8$$
$$\mathbf{R1r2} => 2 - 4 - 5 - 8 \,\&\, 1 - 3 - 7 - 6$$
$$\mathbf{r1R2} => 1 - 3 - 6 - 7 \,\&\, 2 - 5 - 4 - 8$$
$$\mathbf{r1r2} => 1 - 3 - 6 - 7 \,\&\, 1 - 3 - 7 - 6$$

Similarly form di-series sets for all possible combinations.

Step 5: Fusion process

Fusion process means that selecting the points where we need to interchange the value in the series of values. The number of fusion points is found by the following formula:

$$\text{Number of fusion points} = \text{length of the given series}/2.$$

Fusion points are chosen randomly. At the fusion points we interchange the values. Consider one set of di-series: Fusion points are 2 and 4

$$2 - 4 - 5 - 8 \qquad 2 - \mathbf{3} - 5 - \mathbf{7}$$
$$1 - 3 - 6 - 7 \qquad 1 - \mathbf{4} - 6 - \mathbf{8}$$

Original sequence After Fusion

Step 6: Inbreeding depression process

In SBA, there are chances of risk that we may lose some valid values for the pool of values which cannot be retrieved back. This process is called inbreeding depression. To overcome this disadvantage we add some 10% of newly generated values for each iteration. Because of this we are able to search from a newer set of pooled values.

Step 7: Sort the final sequences based on breeding factor

First ten sequences are used for next iteration which is given below.

Makespan		Makespan	
$2 - 4 - 5 - 8$ =>	13	$3 - 8 - 2 - 5$ =>	18
$1 - 3 - 6 - 7$ =>	13	$3 - 4 - 5 - 7$ =>	19
$2 - 4 - 8 - 5$ =>	14	$2 - 6 - 8 - 4$ =>	19
$1 - 5 - 3 - 7$ =>	14	$1 - 5 - 2 - 4$ =>	19
$1 - 3 - 7 - 6$ =>	17	$1 - 4 - 6 - 8$ =>	20

After completing 2 iterations, the result is:

$$2 - 4 - 5 - 8- > \text{Completion time(Makespan)} = 13$$
$$1 - 3 - 6 - 7- > \text{Completion time (Makespan} = 13$$

The SBA algorithm is implemented in C language on personal computer Pentium IV 2.4 GHz.

6 Conclusion and Discussion

The input for the SBA algorithm is taken from the two matrices C_{8*8} and D_{8*8} generated. If we use the traditional method to schedule the tasks then the tasks T1, T2, T6, and T7 are allotted to processor 1 and tasks T3, T4, T5, and T8 are allotted to processor 2, which takes totally 23units to complete all eight tasks [14]. Figure 2 gives the results obtained by traditional method using Gantt chart. By using SBA algorithm, the tasks T2, T4, T5, and T8 are allotted to the processor P1 and T1, T3, T6, and T7 to the processor P2 which takes totally 13 units to complete. Figure 3 gives the optimized results obtained by SBA using a Gantt chart. The graph comparison for the completion of all the tasks using the traditional method and proposed SBA are given in Fig. 4. When comparing the Gantt chart of traditional method and SBA there is less amount of waiting time in SBA method. So this proves that SBA algorithm is an efficient algorithm to solve grid scheduling problems. As a future work, the same algorithm can be used to solve fully decentralized P2P grid scheduling problem.

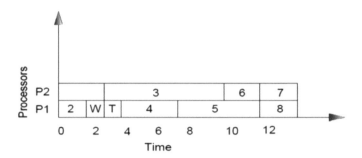

Fig. 3 Optimal task scheduling proposed by SBA algorithm

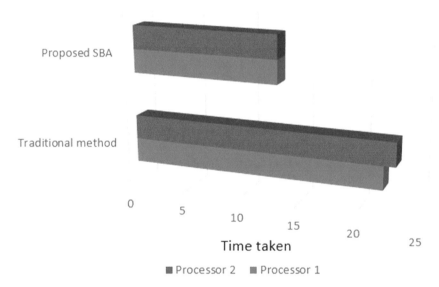

Fig. 4 Graph comparison of time taken for SBA algorithm and traditional method for allocation of tasks

References

1. Foster, I., Kesselman, C.: Computational Grids Blueprint for a New Computing Infrastructure, pp. 15–52. Morgan Kaufmann, San Francisco, CA (1999)
2. Foster, I., Kesselman, C., Salisbury, C., Tuecke, S.: The data grid: towards an architecture for the distributed management and analysis of large scientific data sets. J. Netw. Comput. Appl. **23**(3), 187–200 (2001)
3. Jin, H., Zheng, R., Zhang, Q., Li, Y.: Components and workflow based grid programming environment for integrated image-processing applications. In: Concurrency and Computation: Practice and Experience, vol. 18, no. 14. Wiley Ltd., pp. 1857–1869 (2006)
4. Yagoubi, B., Meddeber, M.: A load balancing model for grid environment. In: Proceeding of 22nd international symposium on computer and information sciences (ISCISC 2007), pp. 1–7 (2007)

5. Ni, L., Zhang, J., Yan, C., Jiang. C.: Heuristic algorithm for task scheduling based on mean load. In: First international conference on semantics, knowledge and grid (SKG'05), Beijing, China 27–29 Nov 2009
6. Shah, H: A low-complexity task scheduling algorithm for heterogeneous computing systems. In: Asia Modeling Symposium (AMS'09), Indonesia (2009)
7. Ramya, R., Thomas, S.: An optimal job scheduling algorithm in computational grids. In: The international conference on communication, computing and information technology (ICC-CMIT), pp. 12–16 (2012)
8. Keerthika, P., Kasthuri, N.: An efficient grid scheduling algorithm with fault tolerance and user satisfaction. Math. Prob. Eng. **2013**(Article ID 340294), 1–9 (2013)
9. Chauhan, P., Nitin: Decentralized scheduling algorithm for DAG based tasks on P2P grid. J. Eng. **2014**(Article ID 202843), 14 pages (2014)
10. Zhu, L., Su, Z., Guo, W., Jina, Y., Suna, W., Hua, W.: Dynamic multi DAG scheduling algorithm for optical grid environment. In: Network Architectures, Management, and Applications V, Proc. of SPIE, vol. 6784, 67841F (2007)
11. Sriramya, P., Parvathavarthini, B., Balamurugan, T.: A novel evolutionary selective breeding algorithm and its application. Asian J. Sci. Res. **6**, 107–114 (2013)
12. Sriramya, P., Parvathavarthini, B.: Performance analysis of selective breeding algorithm on one dimensional bin packing problems. J. Inst. Eng. **93**, 255–258 (2013)
13. Galinier, P., Hao, J.-K.: Hybrid evolutionary algorithms for graph coloring. J. Comb. Optim. **3**, 379–397 (1999)
14. Bidgoli, A.M., Nezad, Z.M.: A new scheduling algorithm design for grid computing tasks. In: 5 Symposium on Advances in science and Technology, Iran, pp. 21–30 (2011)

Correntropy-Induced Metric-Based Variable Step Size Normalized Least Mean Square Algorithm in Sparse System Identification

Rajni Yadav, Sonia Jain and C. S. Rai

Abstract This paper incorporates a correntropy-induced metric (CIM)-based sparsity constraint into variable step size normalized least mean square (VSSNLMS) algorithm for identification of sparse system corrupted by additive noise. The proposed CIM-VSSNLMS algorithm makes a good trade-off between convergence characteristics, filter stability, and steady state error. The proposed (CIM-VSSNLMS) algorithm incorporates a variable step size that accelerates the convergence characteristics and lowers the normalized misalignment (NMSA) error with respect to CIM-NLMS with constant value of step size. An expression for variant step size is derived under the stability condition of proposed algorithm. Finally, the implementation of the proposed algorithm is carried out in MATLAB software to manifest the improved estimated behavior in the identification of sparse system.

Keywords Gaussian input · Sparse system · Stability · Correntropy · Convergence rate · Normalized misalignment

1 Introduction

The wideband communications systems consist of dense wireless channels i.e. having few negligible coefficients and most of the coefficients are non-zero [1]. Sparse channel is often observed when the signal is transmitted from source location to desired destination using broadband wireless channel. In sparse channel, the essential significant coefficients are low in number and rest of the coefficients are either redundant or zeros. Some other examples of sparse channel are: (i) echo channel with having low active time span, (ii) high definition digital TV (HDTV), (iii) and under

R. Yadav (✉) · S. Jain
Maharaja Agrasen Institute of Technology, Guru Gobind Singh
Indraprastha University, New Delhi, Delhi 110078, India
e-mail: radhagulati1986@gmail.com

C. S. Rai
University School of Information, Communication and Technology,
Guru Gobind Singh Indraprastha University, New Delhi, Delhi 110078, India

© Springer Nature Singapore Pte Ltd. 2020
A. K. Luhach et al. (eds.), *First International Conference on Sustainable Technologies for Computational Intelligence*, Advances in Intelligent Systems and Computing 1045,
https://doi.org/10.1007/978-981-15-0029-9_47

water acoustic channel [2–4]. In all these examples of sparse channel, the number of significant coefficients is negligible. The usual method adopted so far was normalized least mean square algorithm (NLMS) and was extensively used for system identification as it exhibits simplicity in performing the computational analysis. [5]. But NLMS algorithm possesses fixed step size for every coefficient in each iteration. This makes it difficult to manage a good trade-off between convergence characteristics and normalized misalignment (NMSA) [6]. Therefore, to compensate for the limitation of fixed step size, an algorithm named variable step size normalized least mean square (VSSNLMS) was developed which incorporated the different values of step sizes for each coefficient. This accelerated the convergence characteristics and minimized NMSA [7].

The following work has been carried out considering the sparse system. Algorithms considering the elements of sparse coefficients have been implemented based on least absolutely shrinkage and selection operator (LASSO) [8] and compressive sensing (CS) [9]. These take in account the prior information of the system. The most known sparse penalties are zero attraction (ZA) and reweighted zero attraction (RZA) [10–15]. However, the implementation of ℓ_1—norm penalty-based sparsity aware VSSNLMS algorithm does not meet with the performance ℓ_0—norm penalty-based sparsity aware VSSNLMS algorithm. The ℓ_0—norm penalty improves the estimation behavior by providing a high-intensity zero attraction on smaller coefficients of system. But the sparse solution based on ℓ_0—norm minimization problem is NP-hard. So various approximation of ℓ_0—norm are used in the past. Nowadays, several researches are done on correntropy-induced metric (CIM)-based sparsity penalty [16–18]. CIM-based sparsity penalty closely resembles ℓ_0—norm penalty. The proposed work combines CIM with VSSNLMS algorithm for sparse system identification.

2 Review of Correntropy-Induced Metric (CIM)

The correntropy is used to evaluate resemblances between two random variables $C = [c_1, c_2, \ldots, c_L]$ and $D = [d_1, d_2, \ldots, d_L]$ in kernel space defined as [16]:

$$V(C, D) = E[\kappa_\sigma(C, D)] = \int \kappa_\sigma(c, d) \mathrm{d}F_{C,D}(c, d) \qquad (1)$$

Here, κ_σ is the most popular Gaussian kernel defined as:

$$\kappa_\sigma(c, d) = \frac{1}{\sqrt{2\pi}\sigma} \mathrm{e}^{-\frac{e^2}{2\sigma^2}} \qquad (2)$$

where e is the error between random variable c and d.

Practically, joint distribution function $F_{C,D}(c, d)$ of random variable C & D is unknown for calculating correntropy measure. Hence, correntropy is estimated using available length of data samples and is defined as:

$$\widehat{V}(C, D) = \frac{1}{L} \sum_{i=1}^{L} \kappa_\sigma (c_i, d_i) \tag{3}$$

where L is number of data samples of stochastic variables C & D.

CIM does not exhibit linearity and therefore is defined as:

$$\mathrm{CIM}(C, D) = \sqrt{\left(\kappa(0) - \widehat{V}(C, D) \right)} \quad \text{and}$$

$$\mathrm{CIM}^2(C, D) = \left(\kappa(0) - \widehat{V}(C, D) \right) \tag{4}$$

Here $\kappa(0) = \frac{1}{\sqrt{2\pi}\sigma}$

The CIM constraint leads to a better approximation to l0-norm than other previous smoothing continuous approximations [18]. CIM-based sparsity constraint is defined as:

$$\|A\|_0 \sim \mathrm{CIM}^2(C, 0) = \frac{\kappa(0)}{L} \sum_{i=1}^{L} \left(1 - \frac{1}{\sqrt{2\pi}\sigma} e^{-\frac{a_i^2}{2\sigma^2}} \right) \tag{5}$$

From Eq. (5), it is clear that for any $c_i \neq 0$ and $|c_i| > \sigma$, the solution reaches very close to ℓ_0—norm for $\sigma \to 0$. Thus, CIM closely resembles ℓ_0—norm sparsity constraint and can be used for developing algorithms in sparse system identification.

3   Proposed CIM-VSSNLMS

The proposed CIM-VSSNLMS incorporates CIM-based sparsity constraint into the cost function of the usual VSSNLMS algorithm.

The reference output $d(k)$ of the system to be identified is given:

$$\mathbf{d}(k) = \mathbf{w_0}^\mathrm{T} \mathbf{u}(k) + \mathbf{z}(k) \tag{6}$$

Here, $\mathbf{z}(k)$ is channel noise and $\mathbf{u}(k) = [u(k)\, u(k-1)\, u(k-2) \ldots u(k-M+1)]^\mathrm{T}$ is Gaussian input signal vector with variance σ_u^2.

$\mathbf{w}_0$ is unknown M-dimensional system coefficient vector which is sparse, and $\mathbf{z}(k)$ is measurement noise

The output, $\mathbf{y}(k)$ of adaptive filter is given as:

$$\mathbf{y}(k) = \hat{\mathbf{w}}^\mathrm{T}(k)\mathbf{u}(k) \tag{7}$$

Let $e(k)$ is the error between reference output and estimated output of adaptive filter given by,

$$e(k) = \mathbf{d}(k) - \mathbf{y}(k) \tag{8}$$

$\hat{\boldsymbol{w}}(k) = \left[\hat{w}_0(k), \hat{w}_1(k), \hat{w}_2(k), \ldots, \hat{w}_{M-1}(k)\right]^{\mathrm{T}}$ represents estimated system coefficients vector of adaptive filter of length M.

Thus, the objective function of CIM-VSSNLMS can be written as:

$$G_{\text{CIM-VSSNLMS}}(k) = |e(k)|^2 + \lambda_{\text{VCIM}}\left(\text{CIM}^2(\hat{\mathbf{w}}(k+1), 0)\right) \tag{9}$$

Here, λ_{VCIM} is a balancing factor for sparsity constraint.

Taking gradient of Eq. (9) both sides, we get

$$\frac{\partial G_{\text{CIM-VSSNLMS}}(k)}{\partial \hat{w}(k+1)} = 2 * (e(k))\frac{\partial e(k)}{\partial \hat{w}(k+1)} + \lambda_{\text{VCIM}}\frac{\hat{\boldsymbol{w}}(k)}{\sqrt{2\pi}\sigma^3}\mathrm{e}^{-\frac{\hat{w}^2(k)}{2\sigma^2}}$$

$$= -2 * e(k)u(k) + \lambda_{\text{VCIM}}\frac{\hat{\boldsymbol{w}}(k)}{\sqrt{2\pi}\sigma^3}\mathrm{e}^{-\frac{\hat{w}^2(k)}{2\sigma^2}} \tag{10}$$

By gradient descent method, the updating equation of CIM-VSSNLMS becomes:
So new updating equation of CIM-VSSNLMS becomes

$$\hat{\boldsymbol{w}}(k+1) = \hat{\boldsymbol{w}}(k) + \frac{\mu(k+1)}{\boldsymbol{u}^{\mathrm{T}}(k)\boldsymbol{u}(k)}\boldsymbol{u}(k)e(k) - \rho_{\text{VCIM}}\frac{\hat{\boldsymbol{w}}(k)}{\sqrt{2\pi}\sigma^3}\mathrm{e}^{-\frac{\hat{w}^2(k)}{2\sigma^2}} \tag{11}$$

The CIM-VSSNLMS algorithm solves the scaling input problem by adding variable step size in each iteration. This is achieved by providing variable step size in such a fashion that the larger step size is initially utilized for accelerating the convergence rate and thereafter a smaller step size is used to minimize NMSA error.

Where the parameter $\mu(k+1)$ denotes variable step size which depends on estimation error and input signal-to-noise ratio. The parameter $\rho_{\text{VCIM}} = \mu(k+1)\lambda_{\text{VCIM}}$ is used to exploit the sparsity of system.

In order to calculate an optimal step size, we have assumed input $\mathbf{u}(k)$ and channel noise $\mathbf{z}(k)$, and output $\mathbf{d}(k)$ as mutually independent. For simplicity, we have considered ρ_{VCIM} as constant instead of variant.

We can write the weight misalignment vector $\mathbf{s}(k)$ of adaptive filter as:

$$\mathbf{s}(k) = \mathbf{w_0} - \hat{\boldsymbol{w}}(k+1) \tag{12}$$

Hence

$$\mathbf{s}(k+1) = \mathbf{s}(k) - \frac{\mu(k+1)}{\boldsymbol{u}^{\mathrm{T}}(k)\boldsymbol{u}(k)}e(k)\boldsymbol{u}(k) - \rho_{\text{VCIM}}F_\sigma(\hat{\boldsymbol{w}}(k)) \tag{13}$$

and

$$s(k+1)s^{\mathrm{T}}(k+1)$$

$$= \left[s(k) - \frac{\mu(k+1)}{\boldsymbol{u}^{\mathrm{T}}(k)\boldsymbol{u}(k)} e(k)\boldsymbol{u}(k) \right.$$

$$\left. - \rho_{\mathrm{VCIM}} F_\sigma(\hat{\boldsymbol{w}}(k)) \right] \left[s(k) - \frac{\mu(k+1)}{\boldsymbol{u}^{\mathrm{T}}(k)\boldsymbol{u}(k)} e(k)\boldsymbol{u}(k) - \rho_{\mathrm{VCIM}} F_\sigma(\hat{\boldsymbol{w}}(k)) \right]^{\mathrm{T}}$$

$$= s(k)s^{\mathrm{T}}(k) + \frac{\mu^2(k+1)}{\boldsymbol{u}^{\mathrm{T}}(k)\boldsymbol{u}(k)} e^2(k) + \rho_{\mathrm{VCIM}}^2 F_\sigma(\hat{\boldsymbol{w}}(k)) F_\sigma^{\mathrm{T}}(\hat{\boldsymbol{w}}(k))$$

$$- 2\mu(k+1)\frac{e(k)s^{\mathrm{T}}(k)\boldsymbol{u}(k)}{\boldsymbol{u}^{\mathrm{T}}(k)\boldsymbol{u}(k)} + 2\mu(k+1)\frac{e(k)\boldsymbol{u}(k)F_\sigma^{\mathrm{T}}(\hat{\boldsymbol{w}}(k))}{\boldsymbol{u}^{\mathrm{T}}(k)\boldsymbol{u}(k)}$$

$$- 2\rho_{\mathrm{VCIM}} s^{\mathrm{T}}(k) F_\sigma(\hat{\boldsymbol{w}}(k)) \tag{14}$$

Taking Expectation on both sides in Eq. (14), we get mean square deviation (MSD) as

$$E\left[s(k+1)s^{\mathrm{T}}(k+1)\right] = E\left[s(k)s^{\mathrm{T}}(k)\right] + \mu^2(k+1)E\left[\frac{1}{\boldsymbol{u}^{\mathrm{T}}(k)\boldsymbol{u}(k)} e^2(k)\right]$$

$$+ \rho_{\mathrm{VCIM}}^2 E\left[F_\sigma(\hat{\boldsymbol{w}}(k)) F_\sigma^{\mathrm{T}}(\hat{\boldsymbol{w}}(k))\right]$$

$$- 2\mu(k+1)E\left[\frac{e(k)s^{\mathrm{T}}(k)\boldsymbol{u}(k)}{\boldsymbol{u}^{\mathrm{T}}(k)\boldsymbol{u}(k)}\right]$$

$$+ 2\rho_{\mathrm{VCIM}}\mu(k+1)E\left[\frac{e(k)\boldsymbol{u}(k)F_\sigma^{\mathrm{T}}(\hat{\boldsymbol{w}}(k))}{\boldsymbol{u}^{\mathrm{T}}(k)\boldsymbol{u}(k)}\right]$$

$$- 2\rho_{\mathrm{VCIM}} E\left[s^{\mathrm{T}}(k) F_\sigma(\hat{\boldsymbol{w}}(k))\right] \tag{15}$$

Based on above assumptions, fifth term in Eq. (15) become zero
We get the following result

$$\boldsymbol{D}(k+1) = \boldsymbol{D}(k) + \mu^2(k+1)E\left[\frac{1}{\boldsymbol{u}^{\mathrm{T}}(k)\boldsymbol{u}(k)} e^2(k)\right]$$

$$+ \rho_{\mathrm{VCIM}}^2 E\left[F_\sigma(\hat{\boldsymbol{w}}(k)) F_\sigma^{\mathrm{T}}(\hat{\boldsymbol{w}}(k))\right]$$

$$- 2\mu(k+1)E\left[\frac{e(k)s^{\mathrm{T}}(k)u(k)}{\boldsymbol{u}^{\mathrm{T}}(k)\boldsymbol{u}(k)}\right] - 2\rho_{\mathrm{VCIM}} E\left[s^{\mathrm{T}}(k) F_\sigma(\hat{\boldsymbol{w}}(k))\right] \tag{16}$$

According to Eq. (16), mean square deviation (MSD) depends on parameter $\mu(k+1)$ and ρ_{VCIM}. The value of ρ_{VCIM} is dependent on system sparsity and additive noise, so its best fitted value cannot be directly obtained. In proposed CIM-VSSNLMS algorithm, the value of ρ_{VCIM} is taken same as in CIM-NLMS to make a fair comparison to obtain the best fitted step size. In above Eq. (16), second and fourth terms are functions of step size, so the MSD defined by above Eq. (16) can be written as sum of two terms.

$$E\left[\mathbf{s}(k+1)\mathbf{s}^\mathrm{T}(k+1)\right] = J_0 - J(\mu) \tag{17}$$

where

$$J_0 = \mathbf{D}(k) + \rho_{\mathrm{VCIM}}^2 E\left[F_\sigma^2(\hat{\boldsymbol{w}}(k))\right] - 2\rho_{\mathrm{VCIM}}\mathbf{E}\left[\mathbf{s}^\mathrm{T}(k)F_\sigma(\hat{\boldsymbol{w}}(k))\right] \tag{18}$$

$$J(\mu) = -\mu^2(k+1)E\left[\frac{1}{\boldsymbol{u}^\mathrm{T}(k)\boldsymbol{u}(k)}e^2(k)\right]$$
$$+ 2\mu(k+1)E\left[\frac{e(k)\mathbf{s}^\mathrm{T}(k)\boldsymbol{u}(k)}{\boldsymbol{u}^\mathrm{T}(k)\boldsymbol{u}(k)}\right] \tag{19}$$

Because the function $F_\sigma(\hat{\boldsymbol{w}}(k))$ is bounded, therefore J_0 is also bounded [19].

When ρ_{VCIM} is fixed, then finding the optimum value of step size becomes our optimization problem. So the step size μ should be chosen such that $J(\mu(k+1))$ is maximized, and evidently therefore the MSD will decrease maximally from previous iteration to next iteration.

$$\mu_0(k+1) = \arg \max_{\mu_0(k+1)} J(\mu(k+1)) \tag{20}$$

Taking the gradient of Eq. (19) with respect to $\mu(k+1)$ and moving in positive direction, we get

$$2\mu_0(k+1)E\left[\frac{1}{\boldsymbol{u}^\mathrm{T}(k)\boldsymbol{u}(k)}e^2(k)\right] = 2E\left[\frac{e(k)\mathbf{s}^\mathrm{T}(k)\boldsymbol{u}(k)}{\boldsymbol{u}^\mathrm{T}(k)\boldsymbol{u}(k)}\right]$$
$$\mu_0(k+1) = \frac{E\left[\frac{e(k)\mathbf{s}^\mathrm{T}(k)\boldsymbol{u}(k)}{\boldsymbol{u}^\mathrm{T}(k)\boldsymbol{u}(k)}\right]}{E\left[\frac{1}{\boldsymbol{u}^\mathrm{T}(k)\boldsymbol{u}(k)}e^2(k)\right]} \tag{21}$$

Considering estimation error $e(k) = z(k) + \mathbf{s}^\mathrm{T}(k)\mathbf{u}(k)$ and assumption (1), we can write numerator of Eq. (21) as

$$E\left[\frac{e(k)\mathbf{s}^\mathrm{T}(k)\boldsymbol{u}(k)}{\boldsymbol{u}^\mathrm{T}(k)\boldsymbol{u}(k)}\right] = E\left[\frac{z(k)\mathbf{s}^\mathrm{T}(k)\boldsymbol{u}(k) + \mathbf{s}^\mathrm{T}(k)\boldsymbol{u}(k)\boldsymbol{u}^\mathrm{T}(k)\mathbf{s}(k)}{\boldsymbol{u}^\mathrm{T}(k)\boldsymbol{u}(k)}\right]$$
$$= E\left[\frac{\mathbf{s}^\mathrm{T}(k)\boldsymbol{u}(k)\boldsymbol{u}^\mathrm{T}(k)\mathbf{s}(k)}{\boldsymbol{u}^\mathrm{T}(k)\boldsymbol{u}(k)}\right] \tag{22}$$

Let

$$\mathbf{r}(k) = \left[\left[\frac{\mathbf{s}^\mathrm{T}(k)\boldsymbol{u}(k)\boldsymbol{u}^\mathrm{T}(k)\mathbf{s}(k)}{\boldsymbol{u}^\mathrm{T}(k)\boldsymbol{u}(k)}\right]\right] \tag{23}$$

And

$$
\begin{aligned}
||\mathbf{r}(k)||^2 &= \left[\frac{\mathbf{s}^{\mathrm{T}}(k)\mathbf{u}(k)\mathbf{u}^{\mathrm{T}}(k)\mathbf{s}(k)}{\mathbf{u}^{\mathrm{T}}(k)\mathbf{u}(k)}\right]\left[\frac{\mathbf{s}^{\mathrm{T}}(k)\mathbf{u}(k)\mathbf{u}^{\mathrm{T}}(k)\mathbf{s}(k)}{\mathbf{u}^{\mathrm{T}}(k)\mathbf{u}(k)}\right]^{\mathrm{T}} \\
&= \left[\frac{\mathbf{s}^{\mathrm{T}}(k)\mathbf{u}(k)\mathbf{u}^{\mathrm{T}}(k)\mathbf{s}(k)}{\mathbf{u}^{\mathrm{T}}(k)\mathbf{u}(k)}\right]
\end{aligned}
\tag{24}
$$

Similarly, the denominator of Eq. (21) can be written as

$$
\begin{aligned}
E\left[\frac{1}{\mathbf{u}^{\mathrm{T}}(k)\mathbf{u}(k)}e^2(k)\right] &= E\left[\frac{\{z(k) + \mathbf{s}^{\mathrm{T}}(k)\mathbf{u}(k)\}\{z(k) + \mathbf{s}^{\mathrm{T}}(k)\mathbf{u}(k)\}^{\mathrm{T}}}{\mathbf{u}^{\mathrm{T}}(k)\mathbf{u}(k)}\right] \\
&= \left[\frac{\{z(k)\mathbf{s}^{\mathrm{T}}(k)u(k) + \mathbf{s}^{\mathrm{T}}(k)\mathbf{u}(k)z(k)\mathbf{s}(k)\} + \{z(k)z^{\mathrm{T}}(k) + \mathbf{s}^{\mathrm{T}}(k)u(k)\mathbf{u}^{\mathrm{T}}(k)\mathbf{s}(k)\}}{\mathbf{u}^{\mathrm{T}}(k)\mathbf{u}(k)}\right]
\end{aligned}
\tag{25}
$$

Using similar approach as in Eq. (24), we can write Eq. (25) as

$$
E\left[\frac{1}{\mathbf{u}^{\mathrm{T}}(k)\mathbf{u}(k)}e^2(k)\right] = \sigma_z^2 \mathrm{tr}\left\{E\left[\frac{1}{\mathbf{u}^{\mathrm{T}}(k)\mathbf{u}(k)}\right]\right\} + ||r(k)||^2
\tag{26}
$$

where $\sigma_z^2 = E\left[z^{\mathrm{T}}(n)z(n)\right]$ is additive noise power
Hence

$$
\begin{aligned}
\mu_0(k+1) &= \frac{E\left[\frac{e(k)\mathbf{s}^{\mathrm{T}}(k)u(k)}{\mathbf{u}^{\mathrm{T}}(k)u(k)}\right]}{E\left[\frac{1}{\mathbf{u}^{\mathrm{T}}(k)u(k)}e^2(n)\right]} \\
&= \left[\frac{||r(k)||^2}{||r(k)||^2 + \sigma_z^2 \mathrm{tr}\left\{E\left[\frac{1}{\mathbf{u}^{\mathrm{T}}(k)u(k)}\right]\right\}}\right]
\end{aligned}
\tag{27}
$$

From Eq. (17), $r(k)$ depends on $\mathbf{w}_0$ which is unknown; this causes a great difficulty in calculating $\mu_0(k+1)$.

When additive noise $z(k) = 0$,

$$
\begin{aligned}
r(k) &= \frac{\mathbf{s}^{\mathrm{T}}(k)\mathbf{u}(k)\mathbf{u}^{\mathrm{T}}(k)}{\mathbf{u}^{\mathrm{T}}(k)\mathbf{u}(k)} = \frac{(\mathbf{w}_0 - \hat{\mathbf{w}}(k+1))^{\mathrm{T}}\mathbf{u}(k)\mathbf{u}^{\mathrm{T}}(k)}{\mathbf{u}^{\mathrm{T}}(k)\mathbf{u}(k)} \\
r(k) &= \frac{e(k)\mathbf{u}^{\mathrm{T}}(k)}{\mathbf{u}^{\mathrm{T}}(k)\mathbf{u}(k)}
\end{aligned}
\tag{28}
$$

From above Eq. (28), we propose to calculate the $r(k)$ by time averaging as

$$\hat{r}(k) = \alpha\hat{r}(k-1) + (1-\alpha)\frac{\boldsymbol{u}^{\mathrm{T}}(k)e(k)}{\boldsymbol{u}^{\mathrm{T}}(k)\boldsymbol{u}(k)} \tag{29}$$

where α is smoothing constant to control the step size $0 \le \alpha < 1$.

$$\mu(k+1) = \left[\frac{\left|\left|\hat{r}(k)\right|\right|^2}{\left|\left|\hat{r}(k)\right|\right|^2 + \sigma_z^2\mathrm{tr}\left\{E\left[\frac{1}{\boldsymbol{u}^{\mathrm{T}}(k)\boldsymbol{u}(k)}\right]\right\}}\right] = \frac{\left|\left|\hat{r}(k)\right|\right|^2}{\left|\left|\hat{r}(k)\right|\right|^2 + Q} \tag{30}$$

$$\sigma_z^2\mathrm{tr}\left\{\mathrm{E}\left[\frac{1}{\boldsymbol{u}^{\mathrm{T}}(k)\boldsymbol{u}(k)}\right]\right\} = \left[\frac{\sigma_z^2}{(\sigma_u^2)}\right] \tag{31}$$

Here the parameter, Q is a positive value related to inverse of signal-to-noise ratio (SNR).

The step size $\mu(k+1)$ will be confined in range $0 < \mu(k) < 2$ for convergence of proposed algorithm.

Hence

$$\mu(k) = \mu_{\max}\frac{\left|\left|\hat{r}(k)\right|\right|^2}{\left|\left|\hat{r}(k)\right|\right|^2 + Q} \tag{32}$$

where $\mu_{\max}$ is maximal step size and is such that $0 < \mu_{\max} < 2$.

When $\left|\left|\hat{r}(k)\right|\right|^2$ is very large, then $\mu(k) \to \mu_{\max}$.

When $\left|\left|\hat{r}(k)\right|\right|^2$ is small, then $\mu(k)$ is small. Thus, the step size is affected by weight misalignment vector error directly. As the stability of the system is maintained through out the simulation, the performance of adaptive system is better even in the case of smaller step size. When step size rises, convergence rate of algorithm also rises but at low computational cost. Thus, CIM-VSSNLMS algorithm obtains an optimal step size by decreasing the step size to maintain stability of sparse system and making a positive increment in the step size to accelerate the convergence behavior.

4 Simulation Results

In this section, we have simulated the proposed algorithm in MATLAB software. For simulations, we have assumed the equal number of coefficients for the system to be recognised and adaptive filter. The input signal $\mathbf{u}(k)$ and measurement noise $\mathbf{z}(k)$ are considered white Gaussian sequence having zero mean.

We have taken normalized misalignment (NMSA) for evaluating the performance of the proposed algorithm, where normalized misalignment is defined as:

$$\mathrm{NMSA} = 10 * \log_{10} E\left[\frac{\left\|w_0 - \hat{w}(k)\right\|^2}{\left\|w_0\right\|^2}\right] \tag{33}$$

We have performed four experiments in MATLAB to evaluate the performance of the proposed CIM-VSSNLMS algorithm. In the first experiment, we have compared the performance of proposed algorithm with ZA-NLMS, RZA-NLMS, CIM-NLMS, ZA-VSSNLMS, and RZA-VSSNLMS algorithms for different sparsity levels, K. We have considered an unknown system with 64 coefficients with different sparsity $K = \{2,4,8\}$, where K signifies the number of nonzero coefficients of sparse system. We have taken $\rho_{CIM} = \rho_{ZA} = \rho_{RZA} = \rho_{VCIM} = \rho_{VZA} = \rho_{VRZA} = \rho = 5 \times 10^{-5}$, where ρ_{CIM}, ρ_{ZA}, ρ_{RZA}, ρ_{VCIM}, ρ_{VZA}, and ρ_{VRZA} correspond to sparsity constant for CIM-NLMS, ZA-NLMS, RZA-NLMS, CIM-VSSNLMS, ZA-VSSNLMS, and RZA-VSSNLMS, respectively. In all the simulation, we have taken SNR equal to 20 dB. We have taken $\mu_{max} = 0.5$ for VSSNLMS algorithm and its variants. We have taken $\mu = 0.5$ for NLMS algorithm and its variants. The CIM kernel width, $\sigma = 0.01$, is adopted. The average of simulation outcomes of 10 Monte Carle iterations is obtained. Figures 1, 2, and 3 show the performance of proposed algorithm for different K. It is clear that the results of the proposed CIM-VSSNLMS algorithm are better than other algorithms in every case of sparsity level. Even when the system is becoming denser, the proposed algorithm has better performance than others.

In second experiment, we have evaluated the effect of SNR on the implementation of proposed algorithm. We have considered $K = 4$. We have taken SNR = {5 dB, 10 dB, 20 dB, 30 dB) and corresponding parameter, $C = \{1 \times 10^{-4}, 0.5 \times 10^{-4}, 1 \times 10^{-5}, 1 \times 10^{-6}\}$, respectively. All other parameters are taken same as in first experiment. From Fig. 4, we observe that as SNR increases, the performance of proposed algorithm is getting better.

In next experiment, we have evaluated the implementation of the algorithm proposed for different values of CIM kernel width, σ. We have taken $\sigma = \{0.001, 0.01, 0.03, 0.08, 0.1,$ and $0.5\}$. From Fig. 5, it is clear that when we take the value of σ lies between 0.001 and 0.01 the implementation results of the algorithm proposed is good,

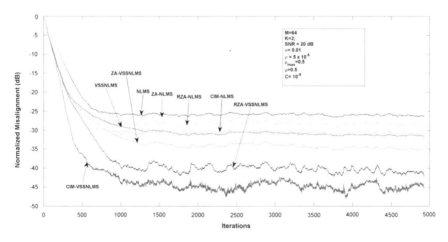

Fig. 1 Implementation of the proposed algorithm for sparsity level, $K = 2$

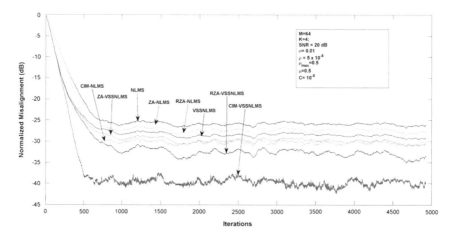

Fig. 2 Implementation of the proposed algorithm for sparsity level, $K = 4$

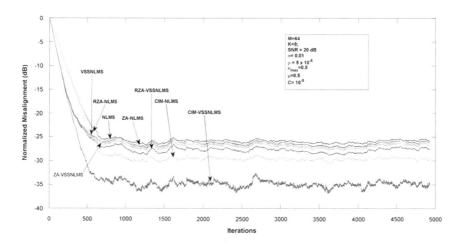

Fig. 3 Implementation of the proposed algorithm for sparsity level, $K = 8$

but as we increase the value of σ above 0.01, the performance starts deteriorating. So we have taken $\sigma = 0.01$ in above simulations.

In the final experiment, the implementation of the proposed algorithm for different sparsity constant, $\rho = \{5 \times 10^{-6}, 8 \times 10^{-5}, 5 \times 10^{-5}, 8 \times 10^{-4}, 5 \times 10^{-4}, 5 \times 10^{-3}\}$ is carried out. The parameter ρ should be adopted very carefully for better performance. All the other parameters are same as in the first experiment. From Fig. 6, it is clear that when the value of ρ starts decreasing from 5×10^{-3} to 5×10^{-5}, the performance improves but when it goes below 5×10^{-5}, the performance begins to deteriorate. Hence, the sparsity constant $\rho = 5 \times 10^{-5}$ is utilized in the simulations.

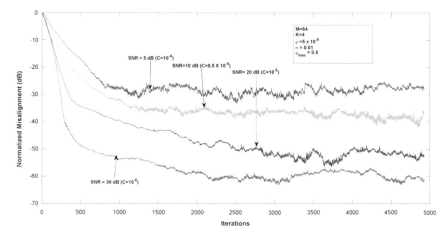

Fig. 4 Effect of SNR on the implementation of the proposed algorithm

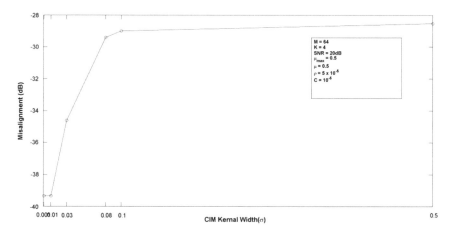

Fig. 5 Effect of CIM kernel width, σ on the implementation of the proposed algorithm

5 Conclusion

Considering the previous ℓ_1—norm penalties based sparsity aware ZA-VSSNLMS and RZA-VSSNLMS, here we have proposed CIM-VSSNLMS algorithm with more efficient sparse penalty. We observed that this improves the estimation behavior by providing a high intensity of zero attraction. The proposed algorithm has performed well for all values of sparsity levels, K. We have also evaluated the effect of SNR, CIM kernel width (σ), and sparsity constant (ρ) on the implementation of the proposed algorithm. The proposed algorithm has lower normalized misalignment and higher convergence rate than all other existing algorithms.

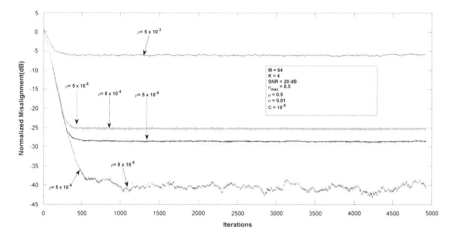

Fig. 6 Effect of sparsity constant, ρ on the implementation of the proposed algorithm

6 Future Scope

In the proposed algorithm, we have not considered the input noise. In future, we will develop the CIM-VSSSNLMS algorithm considering an input noise. Now in the proposed work, the linear constraint is not considered for constrained adaptive applications. In future, a linear constraint can be added further for adaptive beam forming and linear phase system applications.

References

1. Adachi, F., Kudoh, E.: New direction of broadband wireless technology. Wirel. Commun. Mob. Comput. **7**(8), 969–983 (2007)
2. Krishna, V.V., Rayala, J., Slade, B.: Algorithmic and implementation aspects of echo cancellation in packet voice networks. In: Proceedings of 36th Asilomar Conference Signals, Systems Computers, vol. 2, pp. 1252–1257 (2002)
3. Schreiber, W.: Advanced television systems for terrestrial broadcasting. Proc IEEE **83**(6), 958–981 (1995)
4. Berger, C.R., Zhou, S., Preisig, J.C., Willett, P.: Sparse channel estimation for multicarrier underwater acoustic communication: from subspace methods to compressed sensing. IEEE Trans. Sig. Process. **58**(3), 1708–1721 (2010)
5. Diniz, P.S.R.: Adaptive Filtering: Algorithms and Practical Implementation. Springer, USA. https://doi.org/10.1007/978-1-4614-4106-9 (2013)
6. Gui, G., Kumagai, S., Mehbodniya, A., Adachi, F.: Variable is good: adaptive sparse channel estimation using VSS-ZA-NLMS algorithm. In: International Conference on Wireless Communications and Signal Processing, Hangzhou, pp. 1–5 (2013)
7. Shin, H., Sayad, A.H., Song, W.-J.: Variable step-size NLMS and affine projection algorithms. IEEE Sig. Process. Lett. **11**(2), 132–135 (2004)
8. Tibshirani, R.: Regression shrinkage and selection via the lasso. J. R. Stat. Soc. Ser. B. Methodol. **58**(1), 267–288 (1996)

9. Donoho, D.L.: Compressed sensing. IEEE Trans. Inf. Theory **52**(4), 1289–1306 (2006)
10. Chen, Y., Gu, Y., Hero, A.: Sparse LMS for system identification. In: IEEE International Conference on Acoustics, Speech and Signal Processing, pp. 3125–3128 (2009)
11. Wang, Y., Li, Y., Jin, Z.: An improved reweighted zero-attracting NLMS algorithm for broadband sparse channel estimation. In: IEEE International Conference on Electronic Information and Communication Technology (ICEICT), Harbin, pp. 208–213 (2016)
12. Yan, Z., Yang, F., Yang, J.: Block sparse reweighted zero-attracting normalised least mean square algorithm for system identification. IEEE Electron. Lett. **53**(14), 899–900 (2017)
13. Chen, J., Richard, C., Song, Y., Brie, D.: Transient performance analysis of zero-attracting LMS. IEEE Sig. Process. Lett. **23**(12), 1786–1790 (2016)
14. Gui, G., Peng, W., Xu, L., Liu, B., Adachi, F.: Variable-step-size based sparse adaptive filtering algorithm for channel estimation in broadband wireless communication systems. EURASIP J. Wirel. Commun. Networking (2014)
15. Gwadabe, T.R., Aliyu, M.L., Alkassim, M.A., Salman, M.S., Haddad, H.: A new sparse leaky LMS type algorithm. In: 22nd Signal Processing and Communications Applications Conference (SIU), Trabzon, pp. 144–147 (2014)
16. Liu, W., Pokharel, P., Principe, J.: Correntropy: Properties, and applications in non-Gaussian signal processing. IEEE Trans. Sig. Process. **55**(11), 5286–5298 (2007)
17. Ma, W., Qu, H., Gui, G., Xu, L., Zhao, J., Chen, B.: Maximum correntropy criterion based sparse adaptive filtering algorithms for robust channel estimation under non-Gaussian environments. J. Frankl. Inst. **352**(7), 2708–2727 (2015)
18. Wang, Y., Li, Y., Albu, F., Yang, R.: Sparse channel estimation using correntropy induced metric criterion based SM-NLMS algorithm. In: IEEE Wireless Communications and Networking Conference (WCNC), San Francisco, CA, pp. 1–6 (2017)
19. Wang, Y., Li, Y., Albu, F., Yang, R.: Convergence analysis of a correntropy induced metric constrained mixture error criterion algorithm. In: 9th International Conference on Wireless Communications and Signal Processing (WCSP), Nanjing, pp. 1–5 (2017)

Identification of Suitable Basis Wavelet Function for Epileptic Seizure Detection Using EEG Signals

H. Anila Glory, C. Vigneswaran and V. S. Shankar Sriram ⓘ

Abstract Selection of suitable order of Daubechies (DB) wavelet for the decomposition of Electroencephalogram (EEG) signals to detect epileptic seizures is quite challenging, as experimentation is time-consuming. In existing methods, the selection of basis wavelet function for decomposition of EEG signals is carried out by considering the literature or by trial and error method. There is a very little significant literature which discusses the comparative analysis for the identification of suitable basis wavelet function (mother wavelet). However, the existing methods often fail to provide proper justification for selecting the mother wavelets. Hence, this research work addresses the fore-mentioned setback by identifying the suitable basis wavelet function based on wavelet selection methods for epileptic seizure detection. Further, the entropy-based features are extracted and classified using SVM, DT, ANN, and KNN with five complex cases (University of Bonn, Germany EEG dataset): A-B-C-D-E, AB-CD-E, C-D-E, AB-C-D, and ABCD-E. From the entropy analysis, it is evident that while extracting entropy-based features, tenth order of Daubechies wavelet (DB10) is found to be the most suitable basis wavelet function for the accurate detection of Epileptic Seizures. The performance metrics confirm the suitability of the identified basis wavelet function in terms of sensitivity, specificity and classification accuracy.

Keywords Electroencephalogram (EEG) · Discrete wavelet transform (DWT) · Daubechies (DB) · Basis wavelet function (BWF)

1 Introduction

Unusual neural activity which causes seizure results in epilepsy [1]. A report presented by World Health Organization (WHO) states that about 50 million people are suffering from epilepsy, out of which 75% of epileptic patients are residing in low and

H. Anila Glory · C. Vigneswaran · V. S. Shankar Sriram (✉)
School of Computing, Centre for Information Super Highway (CISH), SASTRA Deemed University, Thanjavur, Tamil Nadu 613401, India
e-mail: sriram@it.sastra.edu

© Springer Nature Singapore Pte Ltd. 2020
A. K. Luhach et al. (eds.), *First International Conference on Sustainable Technologies for Computational Intelligence*, Advances in Intelligent Systems and Computing 1045, https://doi.org/10.1007/978-981-15-0029-9_48

middle-income countries [2]. For the proper diagnosis of epilepsy, many electronic devices are used to monitor the activity of the human brain. Among the available brain imaging techniques [3] EEG is cost-effective and records the electrical activity directly, therefore it can be used in low and middle-income countries. EEG records the electrical signals from the brain cells which help in determining the brain activity, picked up by electrodes attached to the scalp [4].

The diagnosis of epilepsy by mere visual inspection is tedious [5] and results in low accuracy. There arises the need for automation, to ease the diagnosis process and provide accurate results. Thus, the concept of automation is applied in medical domain for the precise diagnosis of epileptic seizures. The recent research works related to this study are tabulated in Table 1. Hence, utilizing the concept of Machine Learning, which enables a system to act without being explicitly programmed; this paper presents an epileptic seizure detection model which accurately classifies EEG signals as healthy, ictal and interictal [6]. In order to design an efficient epileptic seizure detection model, the identification of suitable basis wavelet function is inevitable as it directly influences the overall performance of the model [7]. However, from the literature, it has been inferred that none of the existing work provides a proper explanation for the identification of suitable BWF to detect epileptic seizures. Thus, in this paper, an attempt is made to identify appropriate BWF by employing wavelet selection methods and the justifications are provided. The aim of this work and steps involved for choosing the DB wavelets (DB2, DB4, DB6, DB8, DB10, and DB12) are highlighted.

(i)   For the selection of suitable basis wavelet function, wavelet selection methods are employed and the outcomes are interpreted
(ii) The extraction of relevant features from the decomposed EEG signal is mandatory as analysis with a larger volume of data becomes tedious. Since the EEG signals are non-stationary, entropy-based features are extracted
(iii) The extracted feature vectors are classified using the state-of-the-art classifiers (SVM, Decision Tree, ANN, KNN) and the results are compared
(iv) The aim of this paper is to identify the suitable basis wavelet function for epileptic seizure detection that provides highly discriminative features which in turn improves the accuracy rate.

The rest of this paper is structured as follows: A detailed insight into the proposed work is provided in Sect. 2. In Sect. 3, the experimental setup and discussions regarding our idea are presented and Sect. 4 holds the conclusions

2   Proposed Method

This section provides an insight into different BWF of DB family that is suitable for epileptic seizure detection. The mother wavelets (Fig. 1; DB2, DB4, DB6, DB8, DB10, and DB12) are applied on the EEG signal to decompose the signal into signal components (frequency sub-bands) and obtain the amplitudes of the same in five

Table 1 Related works

Authors	Techniques	Evaluation criteria
Hadi Ratham Al Ghayab et al. (2019)* [8]	*Feature Extraction*: Tunable-Q-Factor Wavelet Transform, statistical features *Classification*: KNN, SVM	1. Sensitivity 2. Specificity 3. Classification accuracy
Manish Sharma et al. (2018)* [9]	*Feature Extraction*: *Wavelet* Decomposition, Kraskov entropy-based features *Classification*: SVM	1. Sensitivity 2. Specificity 3. Classification accuracy 4. AUC
Md Mursalin et al. (2017)* [5]	*Feature Extraction*: Time and Frequency-domain features (DWT), Entropy-based features *Feature Selection*: ICFS *Classification*: Random Forest	1. Specificity 2. Classification accuracy 3. Sensitivity
Ozan Kocaagli et al. (2017)* [3]	*Feature Extraction*: Discrete *Wavelet* Transform *Feature Selection*: Fuzzy Relations *Classification*: ANN by Cross-Entropy	1. Cross entropy 2. Mean square error 3. Akaike info criterion 4. Corrected AIC 5. Bayesian info criterion
Manish Sharma et al. (2017)* [4]	*Feature Extraction*: Analytic Time-Frequency Flexible *Wavelet* Transform *Feature Selection*: Students t-test *Classification*: LS SVM	1. Specificity 2. Classification accuracy 3. Sensitivity
Mingyang Li et al. (2017)* [2]	*Feature Extraction*: *DWT* and Hilbert transform *Classification*: Neural Network Ensemble	1. Specificity 2. Classification accuracy 3. Sensitivity
Ahnaf Rashik Hassan et al. (2016)* [10]	*Feature Extraction*: Tunable-Q-Factor *Wavelet* Transform *Feature Selection*: Statistical hypothesis testing *Classification*: Bootstrap aggregating	1. Specificity 2. Classification accuracy 3. Sensitivity
Mingyang Li et al. (2016)* [1]	*Feature Extraction*: DD- *DWT Classification*: GA Optimised SVM	1. Classification accuracy 2. Sensitivity 3. PPV 4. Specificity
Piyush Swami et al. (2016)* [11]	*Feature Extraction*: Multi-resolution analysis using Dual-Tree Complex *Wavelet* Transform *Classification*: GRNN	1. Specificity 2. Classification accuracy 3. Sensitivity 4. Computation time
Thiago M. Nunes et al. (2014)* [12]	*Feature Extraction*: *DWT* *Feature Selection*: CFSS, Relief, InfoGain *Classification*: OPF, Bayesian, Multilayer ANN, SVM	1. Classification accuracy 2. Sensitivity 3. PPV 4. F-measure

*University of Bonn, Germany EEG Benchmark dataset for Epileptic Seizure Detection

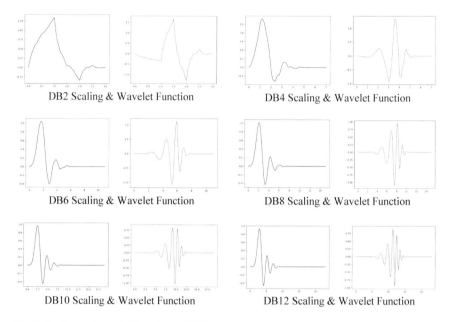

DB2 Scaling & Wavelet Function   DB4 Scaling & Wavelet Function

DB6 Scaling & Wavelet Function   DB8 Scaling & Wavelet Function

DB10 Scaling & Wavelet Function   DB12 Scaling & Wavelet Function

Fig. 1 Scaling and wavelet functions of DB

biological EEG sub-bands namely, gamma, beta, alpha, delta, and theta [5]. From the decomposed EEG signals, entropy-based features are extracted, which forms a feature vector. To identify the appropriate BWF of DB family, minimum entropy criterion [13] and reconstruction criterion [14] are employed. Further to confirm the suitability of the BWF for epileptic seizure detection, the performance of each BWF is evaluated with respect to sensitivity, specificity and classification accuracy using state-of-the-art classifiers: KNN, DT, SVM, ANN. The flow diagram of the proposed method is depicted in Fig. 3.

2.1 EEG Signal Decomposition

Signal decomposition is the process of breaking the raw signal into signal components through which the signals can be examined in a better way. Since EEG signals are nonstationary and more complex [1], DWT is adopted for EEG signal decomposition. Among the several wavelet families, DB wavelet is extensively used for EEG signal processing due to the smoothing feature and shape of this wavelet and its ability to deal with signal discontinuities [7]. The DWT of the EEG signal is given in Eq. (1),

$$\text{DWT}(i, k) = \frac{1}{\sqrt{|2^i|}} \int\limits_{-\infty}^{\infty} x(t) \varphi\left(\frac{t - 2^{ik}}{2^i}\right) \mathrm{d}t \qquad (1)$$

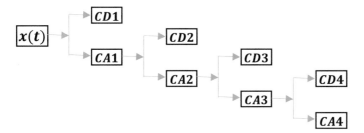

Fig. 2 Wavelet decomposition diagram

where 2^{ik} replaces the translation parameter b and 2^i replaces the scaling parameter a for the given wavelet function $\varphi(t)$. Figure 2 shows the 4th level signal decomposition, as the frequency of the brain signals is divided into five sub-bands. In the four levels, CD1-CD4 represents the detailed coefficients and CA4 denotes the approximate coefficient, the decomposed signal can be reconstructed by the summation of detailed coefficients and the last approximate coefficient [11].

2.2 Wavelet Selection Methods

Selection of the suitable basis wavelet function for the EEG signal decomposition is crucial as it determines the classification results significantly. Selection of appropriate BWF is influenced by its characteristic properties such as orthogonality, compact support, vanishing moments, regularity and symmetry. In order to identify the appropriate BWF various approaches have been proposed which are discussed as follows:

i.   *Variance Method* [15]: The wavelet packet coefficients whose sum of variance is maximum could be selected as a mother wavelet
ii. *Correlation Coefficient (CC) Method* [15]: It measures the degree of correlation among the signal and wavelet packet coefficients. The function that maximizes CC can be chosen as a basis wavelet function
iii. *Reconstruction Criterion* [14]: It measures the Root Mean Square Difference (RMSD) between the original EEG signal and reconstructed EEG signal. The function with minimum RMSD can be chosen as a basis wavelet function
iv. *Energy to Entropy ratio criterion* [13]: This criterion shows the uniqueness of the EEG signal that measures the ratio between energy and entropy values. The function that maximizes this ratio is chosen as the mother wavelet

The wavelet selection methods (i) and (ii) are apt for Wavelet Packet Transform [15], whereas (iii) and (iv) are appropriate for DWT. Since the EEG signals are decomposed using DWT, this paper employs Reconstruction criterion and Energy to Entropy ratio criterion for the selection of suitable BWF.

Reconstruction criterion

In this method, [14] the original signal is reconstructed from the corresponding decomposed wavelet coefficients and the root mean square difference between the original signal (x_i) and the reconstructed signal (x_j) has been computed using Eq. (2). The basis wavelet function with the least RMSD (cost-function) can be chosen as a basis wavelet function to represent the signal.

$$\varepsilon_{\text{diff}} = \sqrt{\frac{1}{N} \sum_{i=1}^{N} (x_i - x_j)^2} \tag{2}$$

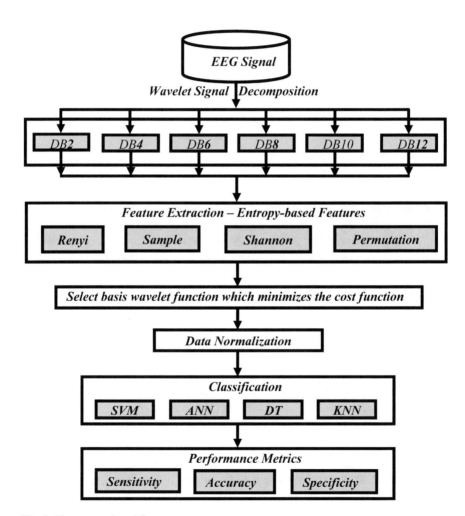

Fig. 3 The proposed workflow

Energy to Entropy ratio criterion

In general, the term energy is used to characterize the signal. On the other hand, entropy is the measure of randomness of a signal. As the EEG signals are non-stationary and aperiodic, entropy and energy values of a signal have been framed as a proper selection criterion for choosing the BWF. By considering the energy and entropy values, three basic selection criteria are formulated [13] namely (i) Maximum energy criterion, (ii) Minimum Shannon entropy criterion, and (iii) Maximum energy to Shannon entropy Ratio Criterion. Of the all selection criteria, [13] concluded that minimizing the Shannon entropy is a well-suited cost-function, to identify the optimum BWF. Hence, in this work for n different orders of BWF, Minimum Shannon Entropy criterion Eq. (3) has been employed.

$$\text{Cost\_function} = \text{Min}(\text{Shann}_{\text{Ent}}(\text{DBn})) \tag{3}$$

The suitable mother wavelet can represent the signal in a better way. Nevertheless, the cost-function to be considered has to quantify the number of coefficients necessary to represent the signal more accurately and should show the additive property. Shannon entropy satisfies the above criteria thus making it better cost-function to select the most suitable BWF. Subsequently, Renyi entropy [16] generalizes Shannon entropy therefore, it can also be considered as an appropriate cost-function Eq. (4).

$$\text{Cost\_function} = \text{Min}\big(\text{Renyi}_{Ent}(\text{DBn})\big) \tag{4}$$

2.3 Feature Extraction

Feature extraction is carried out to transform a set of data points into significant features that help in discriminating the class labels accurately. In this work, the entropy-based features are extracted namely, Sample, Shannon, Renyi and Permutation entropy. Those features are discussed as follows,

Sample Entropy [5]: It quantifies the uniformity of physiological time-series signals that diagnose victims' state. It is a negative natural logarithm which estimates the conditional probability of length m. Sample Entropy can be calculated using Eq. (5)

$$\text{Sample}_{\text{Ent}} = -\log \frac{X}{Y} \tag{5}$$

where X represents the number of vector pairs of dimension $m + 1$ and Y denotes the number of vector pairs of dimension m. It is relatively consistent across diverse pattern dimensions.

Renyi Entropy [16]: A generalization of Shannon entropy which measures the spectral complexity of the signal. Renyi's definition Eq. (6) is parameterized by a single parameter α, if α approaches unity, then it is said to be Shannon entropy.

$$\text{Renyi}_{\text{Ent}} = \frac{1}{1 - \alpha} \log \sum\nolimits_{i=1}^{m} (\text{sp}_i)^{\alpha} \tag{6}$$

where α represents the order and sp_i denotes the spectral power of the signal. The significant advantage of Renyi Entropy is though the variables are rescaled the entropy value varies as an additive constant.

Shannon Entropy [17]: It estimates the average least number of bits required to translate a string of symbols, based on the occurrence of the symbols. Let x is a random variable having finitely many values, and p_j is a probability of an event. Then the Shannon Entropy of x is given by Eq. (7),

$$\text{Shann}_{\text{Ent}} = - \sum_{k=1}^{x} p_j \log_2 p_j \tag{7}$$

Permutation Entropy [16]: It measures arbitrary time-series based on the study of permutation patterns. Besides, permutation entropy is utilized to determine the embedding arguments among time-series Eq. (8).

$$\text{Permutation}_{\text{Ent}} = - \sum_{k=1}^{m} p_k \log_2 p_k \tag{8}$$

where p_k denotes the relative frequencies of the signal pattern.

2.4 Data Normalization

The range of features extracted from the data is standardized by the data normalization process, thus minimizing the computational complexities. In this approach, we have adopted the Min-Max Normalization [18] which scales the data within the range [0,1] by linearly transforming the high-valued features to low-valued features Eq. (9).

$$\text{Min\_Max\_Norm} = \frac{d - d_{\min}}{d_{\max} - d_{\min}} \tag{9}$$

2.5 Classification

Support Vector Machine (SVM) [1]: It is a kernel-based technique which falls in the supervised learning category which is stated by a separating hyperplane. SVM is used for both classification and regression task. It maps the data in a high-dimensional feature space, accepting attributes as input and classifies it based on the decision class tags with well-suited hyperplane that best separates the tags. With respect to hyperplane, the classification is done by maximizing the margin between the classes

Decision Tree (DT) [19]: DT belongs to the family of supervised learning algorithms and is built top-down from a root node. The leaf node denotes a classification or decision. DT can have as many levels as possible. They do not require any preceding assumptions about the category of probability distributions fulfilled by the class and other features. The main advantages of DT are, it has higher interpretability and the rules can be easily compressed with Boolean logic.

Artificial Neural Network (ANN) [3]: It is a brain-inspired computational model that mimics the network structure of the biological neural network. It is a nonlinear statistical data modelling technique where the unidentified nonlinear relationships among inputs and outputs are modelled. It has the ability to model the data with high volatility and non-constant variance and can learn hidden relationships between them.

K-Nearest Neighbour (KNN) [20]: It is a data classification technique that identifies what group a sample is in, by observing the samples around it. KNN algorithm stores all the available trails and classifies new trails by means of similarity measure such as distance metrics. Trails are categorized based on majority voting decision logic. The data for the KNN algorithm can be any measurement scale from ordinal, nominal to quantitative scale.

3 Experimental Setup and Discussions

This section provides an insight into the experimental setup and other observations for the identification of suitable BWF. To assess the performance of different basis wavelet functions, the following metrics are considered: sensitivity, specificity, and classification accuracy [8] and to validate the work, Stratified K-fold cross-validation ($K = 10$) has been exploited. Further to achieve the generalized and unbiased result, the classification process is repeated 20 times and the average of it has been taken into account. The proposed approach has been carried out using a laptop with Intel ® Core™ i5 processor, a physical memory of 4 GB RAM running on Windows 7 operating system and implemented using Python. The decomposition of the EEG signal has been carried out via PyWavelets-1.0.1, and the entropy-based features from the decomposed EEG signal are extracted by Pyentrp package. The extracted features are finally classified by the fore-mentioned classifiers using Scikit-learn-0.20 library.

Table 2 Dataset summary

Sets	A	B	C	D	E
Subjects	Five healthy volunteers		Five epileptic patients		
State	Eyes open	Eyes closed	Interictal[a]	Interictal[a]	Ictal[b]

[a]Epileptic patient's recordings at the seizure-free interval
[b]Epileptic patient's recordings during seizure activity

3.1 Dataset Description

To evaluate the performance, we have used EEG dataset which is made publicly available by Andrzejak et al. [21] the University of Bonn, Germany which comprises of five different classes (Table 2). Each class consists of 100 samples taken from five healthy volunteers and five epileptic patients taken with the sampling rate of 173.61 Hz for the duration of 23.6 s. In this work, five complex cases are identified: C1: A-B-C-D-E, C2: AB-CD-E, C3: C-D-E, C4: AB-C-D, and C5: ABCD-E which highlights the clinical relevance and complexity [4].

3.2 Discussions

The entire process can be divided into two phases, (i) identification of suitable BWF (ii) feature extraction and classification. In the first phase using wavelet selection methods, the suitable mother wavelet for the decomposition of EEG signals are identified.

Table 3 shows the comparison of RMSD values for the chosen BWF of DB family. From the reconstruction criterion, it is inferred that DB6 has the least RMSD, which means DB6 is more suitable mother wavelet that preserves the statistical properties of the decomposed EEG signal. However, the EEG signals are non-stationary, features extracted based on its statistical properties are not sufficient to discriminate the EEG signals effectively.

Further, (Table 4) it has been observed that based on the entropy criterion, DB10 is the most appropriate BWF than DB6. For both Renyi and Shannon entropy criterion, DB10 has the minimum entropy value. Though there is a minuscule change in the values of DB10 and DB12, the entropy-based features extracted using DB10 are highly discriminative, it is evident from the classification results obtained using the state-of-the-art classifiers. The reason behind is, if the order of the Daubechies wavelet increases, the frequency resolution of a signal is improved, on the other

Table 3 Comparison of BWF in terms of RMSD between the original and reconstructed signal

BWF	DB2	DB4	DB6	DB8	DB10	DB12
RMSD	2.69E−12	2.42E−12	**2.30E−12**	2.98E−12	2.79E−12	3.60E−12

Table 4 Comparison of BWF in terms of Minimum Entropy Criterion

BWF	DB2	DB4	DB6	DB8	DB10	DB12
Renyi Ent	32.5801	31.7230	31.5071	31.4400	**31.4210**	31.4242
Shann Ent	33.6976	32.8415	32.6162	32.5406	**32.5114**	32.5137

hand, the time resolution of the signal is reduced. Therefore it has been analyzed that some of the distinctive features have been left out which deteriorates the performance of the epileptic seizure detection model with respect to sensitivity, specificity, and classification accuracy.

In the second phase, entropy-based features, namely Shannon, Renyi, Sample and Permutation entropies are extracted from the decomposed EEG signals. The extracted features are compared with respect to classification accuracy, sensitivity and specificity. Table 5 shows the classification results of the chosen BWF of DB wavelets which are classified using SVM, DT, ANN, and KNN. From the table, we inferred that DB10 provides better results for all the five different cases with an accuracy rate of 88.26 (KNN) for C1; 98.11 (KNN) for C2; 86.47 (KNN) for C3; 89.75 (KNN) for C4 and 99.21 (KNN) for C5. Similarly, in terms of sensitivity and specificity (Fig. 4) DB10 clearly outperforms. Though in some cases there are minuscule differences between DB10 and DB12, an ample difference would be observed for larger datasets. Thus, DB10 is the most appropriate basis wavelet function for epileptic seizure detection.

4　Conclusions

EEG recordings play a vital role in the diagnosis and treatment of neurological disorders. Researchers prefer DB wavelets for decomposing the EEG signals, owing credits to its predominant usage in biomedical signal processing. The selection of the appropriate BWF is a formidable task, which can be done by trial and error method and is time-consuming. This work aims to provide a solution to this limitation by identifying the most suitable mother wavelet for epileptic seizure detection. In this research work, the decomposition of EEG signals is done choosing different basis wavelet functions and entropy-based features are extracted and classified using state-of-the-art classifiers. Further, for the identification of suitable BWF, wavelet selection methods are employed. From the results obtained using University of Bonn EEG dataset, it was found that DB10 satisfies the minimum entropy criterion and is superior to other basis wavelet functions for the accurate detection of epileptic seizures using EEG signals. The performance metrics ratifies that DB10 dominates other BWFs in terms of sensitivity, specificity, and accuracy. It is worthy to state that, the usage of DB10 for EEG signal decomposition can improve the performance of Epileptic Seizure detection model.

Table 5 Comparison of BWF in terms of classification accuracy (%)

Cases	C1: A-B-C-D-E				C2: AB-CD-E				C3: C-D-E			
Classifiers	SVM	DT	ANN	KNN	SVM	DT	ANN	KNN	SVM	DT	ANN	KNN
DB2	69.50	77.56	79.25	83.66	92.30	93.01	95.48	96.94	67.90	71.50	75.33	77.98
DB4	72.52	81.97	73.58	85.02	91.62	90.29	95.84	97.01	69.83	69.52	75.93	80.25
DB6	75.00	79.16	80.14	86.47	93.70	93.64	96.61	97.38	69.80	71.80	75.47	80.33
DB8	75.83	79.29	81.33	87.01	93.75	94.56	96.58	97.93	70.67	71.85	73.47	79.78
DB10	**77.75**	**83.01**	**82.86**	**88.26**	**95.82**	**95.93**	**97.61**	**98.11**	**71.47**	**78.47**	**77.67**	**86.47**
DB12	76.55	81.42	81.41	87.30	95.58	94.70	96.42	98.04	69.87	73.20	77.43	83.33

Cases	C4: AB-C-D				C5: ABCD-E			
Classifiers	SVM	DT	ANN	KNN	SVM	DT	ANN	KNN
DB2	73.51	77.46	79.75	84.40	97.66	96.11	96.60	98.23
DB4	73.56	79.38	76.25	85.26	97.09	95.62	97.40	97.86
DB6	73.66	78.18	77.75	85.83	97.81	96.23	98.20	97.90
DB8	74.33	81.65	80.25	86.16	97.54	96.32	97.60	98.74
DB10	**75.70**	**84.70**	**82.50**	**89.75**	**98.02**	**97.26**	**98.80**	**99.21**
DB12	73.39	81.78	77.75	89.00	97.73	96.28	98.36	98.67

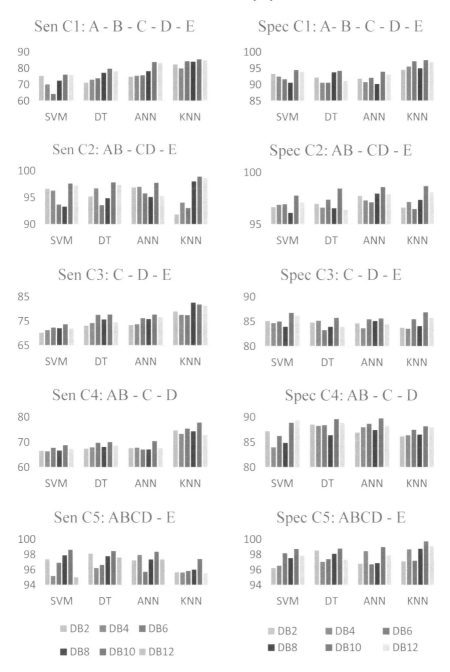

Fig. 4 Comparison of BWF in terms of sensitivity (Sen), and specificity (Spec) (%)

Acknowledgements This work was supported by the IBM Shared University Research Grant and the Department of Science and Technology, India through Fund for Improvement of S&T Infrastructure (FIST) Programme (SR/FST/ETI-349/2013).

References

1. Li, M., Chen, W., Zhang, T.: Automatic epilepsy detection using wavelet-based nonlinear analysis and optimized SVM. Biocybernetics Biomed. Eng. **36**(4), 708–718 (2016)
2. WHO Homepage: https://www.who.int/news-room/fact-sheets/detail/epilepsy. Last accessed 7 Feb 2019
3. Kocadagli, O., Langari, R.: Classification of EEG signals for epileptic seizures using hybrid artificial neural networks based wavelet transforms and fuzzy relations. Expert Syst. Appl. **88**, 419–434 (2017)
4. Sharma, M., Pachori, R.B., Acharya, U.Rajendra: A new approach to characterize epileptic seizures using analytic time-frequency flexible wavelet transform and fractal dimension. Pattern Recogn. Lett. **94**, 172–179 (2017)
5. Mursalin, M., Zhang, Y., Chen, Y., Chawla, N.V.: Automated epileptic seizure detection using improved correlation-based feature selection with random forest classifier. Neurocomputing **241**, 204–214 (2017)
6. Li, M., Chen, W., Zhang, T.: Classification of epilepsy EEG signals using DWT-based envelope analysis and neural network ensemble. Biomed. Sig. Process. Control **31**, 357–365 (2017)
7. Gajic, D., Djurovic, Z., Gligorijevic, J., Di Gennaro, S., Savic-Gajic, I.: Detection of epileptiform activity in EEG signals based on time-frequency and non-linear analysis. Front. Comput. Neurosci. **9**(38), 1–16 (2015)
8. Al Ghayab, H.R., Li, Y., Siuly, S., Abdulla, S.: A feature extraction technique based on tunable Q-factor wavelet transform for brain signal classification. J. Neurosci. Methods **312**, 43–52 (2019)
9. Sharma, M., Bhurane, A.A., Acharya, U.Rajendra: MMSFL-OWFB: a novel class of orthogonal wavelet filters for epileptic seizure detection. Knowl.-Based Syst. **160**, 265–277 (2018)
10. Hassan, A.R., Siuly, S., Zhang, Y.: Epileptic seizure detection in EEG signals using tunable-Q factor wavelet transform and bootstrap aggregating. Comput. Methods Programs Biomed. **137**, 247–259 (2016)
11. Swami, P., Gandhi, T.K., Panigrahi, B.K., Tripathi, M., Anand, S.: A novel robust diagnostic model to detect seizures in electroencephalography. Expert Syst. Appl. **56**, 116–130 (2016)
12. Nunes, T.M., Coelho, A.L.V., Lima, C.A.M., Papa, J.P., De Albuquerque, V.H.C.: EEG signal classification for epilepsy diagnosis via optimum path forest—a systematic assessment. Neurocomputing **136**, 103–123 (2014)
13. Zaeri, R., Ghanbarzadeh, A., Attaran, B., Moradi, S.: Artificial neural network based fault diagnostics of rolling element bearings using continuous wavelet transform. In: The 2nd International Conference on Control, Instrumentation and Automation, Shiraz, Iran, pp. 753–758 (2011)
14. Megahed, A.I., Moussa, A.M., Elrefaie, H.B., Marghany, Y.M.: Selection of a suitable mother wavelet for analyzing power system fault transients. In: IEEE Power and Energy Society General Meeting-Conversion and Delivery of Electrical Energy in the 21st Century. IEEE/Institute of Electrical and Electronics Engineers Incorporated, pp. 1–7 (2008)
15. Rodrigues, A.P., Mello, G.D., Pai, P.Srinivasa: Selection of mother wavelet for wavelet analysis of vibration signals in machining. J. Mech. Eng. Autom. **6**(5A), 81–85 (2016)
16. Acharya, U.R., Fujita, H., Sudarshan, V.K., Bhat, S., Koh, J.E.W.: Application of entropies for automated diagnosis of epilepsy using EEG signals: A review. Knowl.-Based Syst. **88**, 85–96 (2015)

17. Noorizadeh, S., Shakerzadeh, E.: Shannon entropy as a new measure of aromaticity, Shannon aromaticity. Phys. Chem. Chem. Phys. **12**(18), 4742–4749 (2010)
18. Jain, S., Shukla, S., Wadhvani, R.: Dynamic selection of normalization techniques using data complexity measures. Expert Syst. Appl. **106**, 252–262 (2018)
19. Riaz, F., Hassan, A., Rehman, S., Niazi, I.K., Dremstrup, K.: EMD-based temporal and spectral features for the classification of EEG signals using supervised learning. IEEE Trans. Neural Syst. Rehabil. Eng. **24**(1), 28–35 (2016)
20. Kalbkhani, H., Shayesteh, M.G.: Stockwell transform for epileptic seizure detection from EEG signals. Biomed. Sig. Process. Control **38**, 108–118 (2017)
21. Andrzejak, R.G., Lehnertz, K., Mormann, F., Rieke, C., David, P., Elger, C.E.: Indications of nonlinear deterministic and finite-dimensional structures in time series of brain electrical activity: dependence on recording region and brain state. Phys. Rev. E **64**(6), 1–8 (2001)

Neuro-Fuzzy Control Algorithm for Harmonic Compensation of Quality Improvement for Grid Interconnected Photovoltaic System

B. Pragathi, Deepak Kumar Nayak and Ramesh Chandra Poonia

Abstract Current quality compensation is the major task in solar photovoltaic system. Several control algorithms have been discussed in the literature survey for reducing the power quality compensation. In the proposed system, neuro-fuzzy algorithm is implemented for reactive power compensation. The incremental conductances technique is used for extracting maximum power from the PV system by adjusting the duty cycle of the IGBT. The DC-DC boost converter is used for increasing the extracted power from PV system. The DC bus capacitor is used for maintaining constant PV voltage in the system. The voltage source converter is used for DC to AC conversion. The IGBT section of the VSC is controlled by the neuro-fuzzy controller. The neural network control algorithm is used for extracting reference currents for ZVR operation.

Keywords PV panels · Artificial neural-fuzzy control algorithm · Incremental conductance · MPPT

1 Introduction

Renewable energy resources like solar, hydro, thermal, nuclear, etc. are used for power generation in various forms. The main drawback of some renewable energy resources is effecting the global warming by inserting various gases into the atmosphere. The solar renewable resources convert light energy into electrical energy by photovoltaic effect [1–4].

B. Pragathi (✉) · D. K. Nayak
Department of ECE, Koneru Lakshmaiah Education Foundation, Vaddeswaram, Guntur, India
e-mail: pragathibellamkonda@gmail.com

D. K. Nayak
e-mail: nayak@kluniversity.in

R. C. Poonia
Amity Institute of Information Technology Amity University Rajasthan, Jaipur, India
e-mail: rameshcpoonia@gmail.com

© Springer Nature Singapore Pte Ltd. 2020
A. K. Luhach et al. (eds.), *First International Conference on Sustainable Technologies for Computational Intelligence*, Advances in Intelligent Systems and Computing 1045, https://doi.org/10.1007/978-981-15-0029-9_49

Nowadays, power quality issues like reactive power, unbalance load, harmonics in current, voltage saw swell and harmonic in source are the critical problems in distribution systems. The main accept in electric system is to provide constant voltage, frequency and current inspite of sensitive load variations. Custom power devices, UPQC, are used to mitigate several quality issues in AC [5]. Variations in signal cause unwanted effect in load voltage which can be compensated using state-space modelling of UPQC by reducing the DC link voltage [6]. The source voltage quality is improved by open control scheme of unified power quality conditioner [7]. Various current compensation techniques are used to eliminate harmonics in currents [8]. The fundamental active and reactive components are extracted using synchronous reference frame (SRF) theory algorithm [9]. The shunt compensator is used for elimination of harmonics by extracting unit templates using references current by least mean square error algorithm trained by ADALINE [10, 11]. Negative sequences and reactive power are eliminated by thyristor controlled capacitor bank and testing with and without filter [12]. Fuzzy-based control algorithm is used for voltage compensation, thereby reducing voltage swag, swell, flicker and harmonics in load voltage [13]. The current and voltages are simultaneously compensated using series-DSTATCOM and shunt-DVR compensator generating unit vector signals for parks transformation by PLL technique [14–16]. Synchronous reference frame theory [17], fuzzy control algorithm [18]. Along with several power quality issues, compensation neutral current generation is the major task, which is accomplished by three-phase four-wire system [19, 20] and the control algorithm is discussed in paper [21].

2 Proposed System

The proposed system consists of solar photovoltaic panel, MPPT connected to DC-DC converter to maintain constant DC voltage, interfacing inductors, VSC sections, isolation transformer and ripple filter. The main objective of the proposed system is to compensate current quality problems. The DSTATCOM (shunt compensator) is used to mitigate reactive power compensation, neutral current, load balancing, harmonic elimination and is connected parallel to the load controlled by PWM current controller for driving the VSC section of DSTATCOM. There are two control algorithms used in the proposed system.

(1) Incremental conductances MPPT algorithm for extracting maximum power from solar PV system.
(2) Control of DSTATCOM using neural network control algorithm for generating reference grid currents.

The proposed system consists of history and background work in Sect. 2, grid-connected PV system configuration, incremental conductance-based MPPT algorithm in Sect. 3, proposed artificial neural-fuzzy logic controller in Sect. 4 and discussed MATLAB simulation results in Sect. 5.

3 Grid-Connected PV System

Grid-connected solar system is shown in Figs. 1 and 2.

3.1 Modeling Solar Cell

Solar cell is used to convert light energy into electrical energy by photovoltaic effect and is equated as:

$$I = I_{ph} - I_s\left(\exp q\,\frac{(v + R_s I)}{NKT}\right) - 1 - \left(\frac{v + R_s I}{R_{sh}}\right) \tag{1}$$

where

I_p	photovoltaic current,
I_{rs}	reverse saturation current,
q	charge,
v	diode voltage,
K	Boltzmann's constant,
T	junction temperature,
N	ideality factor,

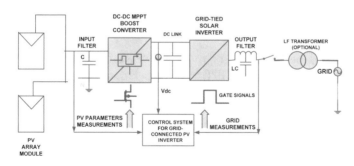

Fig. 1 Grid-connected PV system

Fig. 2 Equivalent circuit of PV cell

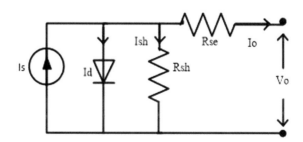

Fig. 3 DC-DC boost
converter circuit

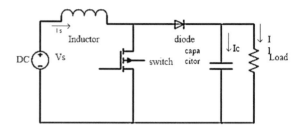

R_s and R_h series and shunt resistors.

3.2 DC-DC Boost Converter

DC-DC converters are used to extract maximum power from solar PV system using MPPT algorithm. The perturb and observe MPPT algorithm is used to maintain constant solar power and apply gating pulses for DC-DC converter. Maximum power is extracted by varying the duty cycle in the DC-DC converter. The basic circuit of DC-DC boost converter consists of inductor, MOSFET as switch, unidirectional diode, capacitor and output equation is given as $V_{out} > V_{in}$ (Fig. 3).

3.3 Maximum Power Point Tracking (MPPT)

The maximum power is extracted from the solar PV system using MPPT technique by adjusting the duty cycle of the IGBT. In the proposed method, incremental conductances algorithm is used to increase the efficiency of the PV cell. The power-voltage and current-voltage characteristics of PV cell are shown in Fig. 4 (Fig. 5).

Fig. 4 P-V, I-V
characteristics of the PV cell

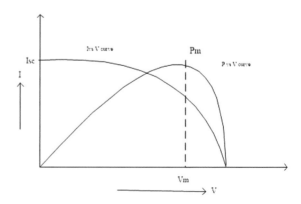

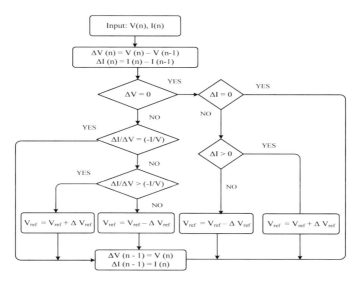

Fig. 5 Flow chart diagram for incremental conductance method

(1) *Neural Network Control Algorithm for DSTATCOM*

The neural network control algorithm is used for generation of reference solar grid currents for controlling of DSTATCOM shown in figure.

Direct-axis Components and quadrature-axis components for references grid currents:

(i) The direct-axis components (u_{rd}, u_{bd}, u_{yd}) for reference grid currents are estimated by PV grid source currents (i_{sr}, i_{sb}, i_{sy}) and unit vector components in direct axis are given in Eq. (2) as

$$u_{rd} = v_{sr}/v_t; \quad u_{bd} = v_{sb}/v_t; \quad u_{yd} = v_{sy}/v_t; \tag{2}$$

where v_t is the magnitude of reference solar PV system at the PCC is estimated as

$$v_t = \sqrt{\frac{2}{3}} * \left(v_{sr}^2 + v_{sb}^2 + v_{sy}^2\right)^{1/2} \tag{3}$$

$$v_{er}(k) = v_{ref}^*(k) - v_{dc}(k) \tag{4}$$

where $v_{er}(k)$ is the solar photovoltaic DC bus voltage at kth sample instant. $v_{ref}^*(k)$ is solar panel DC bus reference voltage, $v_{dc}(k)$ is sensed DC bus voltage of solar panel at kth sample. The $v_{er}(k)$ is given as the input to PI controller and the output is taken as the weight loss error as

$$w_{erd}(k) = w_{erd}(k-1) + k_{pd}\{v_{er}(k) - v_{er}(k-1)\} + k_{id}v_{er}(k) \qquad (5)$$

where k_{pd} and k_{id} are the DC bus proportional and integral gains. The direct-axis components of the reference source current are estimated as weight components at kth sample instant as

$$w_{ld}(k) = \{w_{erd}(k) + w_{rd}(k) + w_{bd}(k) + w_{yd}(k)\}/3 \qquad (6)$$

Least mean square error trained by ADALINE is used to extract fundamental direct and quadrature-axis weight components. Extracted direct-axis weight components are:

$$w_{rd}(k+1) = w_{rd}(k) + \mu e_r(k)u_{rd}(k) \qquad (7)$$

$$w_{bd}(k+1) = w_{bd}(k) + \mu e_b(k)u_{bd}(k) \qquad (8)$$

$$w_{yd}(k+1) = w_{yd}(k) + \mu e_y(k)u_{yd}(k) \qquad (9)$$

$$e_r(k) = \{i_{rl}(k) - i_{rd}^*\} \qquad (10)$$

$$e_b(k) = \{i_{bl}(k) - i_{bd}^*\} \qquad (11)$$

$$e_y(k) = \{i_{yl}(k) - i_{yd}^*\} \qquad (12)$$

where $e_r(k)$, $e_y(k)$, $e_b(k)$ are the error outputs between the load and fundamental reference currents of phase r, b and y. The accuracy of the system depends on the convergence factor (μ) which varying between 0.01 and 1.0 for the proposed system, convergence factor (μ) is taken as 0.2. The fundamental reference currents of direct axis are estimated as

$$i_{rd}^* = w_{rd}(k)u_{rd}(k); \quad i_{bd}^* = w_{bd}(k)u_{bd}(k);$$
$$i_{yd}^* = w_{yd}(k)u_{yd}(k) \qquad (13)$$

(ii) The quadrature-axis components (u_{rq}, u_{bq}, u_{yq}) for reference grid currents are estimated by PV grid source currents (i_{sr}, i_{sb}, i_{sy}) and unit vector components in direct axis are given in Eqs. (13), (14) and (15) as

$$u_{rq} = -u_{bd}/\sqrt{3} + u_{yd}/\sqrt{3} \qquad (14)$$

$$u_{bq} = \sqrt{3}u_{rd}/2 + (u_{bd} - u_{yd})/2\sqrt{3} \qquad (15)$$

$$u_{yq} = \sqrt{3}u_{rd}/2 + (u_{bd} - u_{yd})/2\sqrt{3} \tag{16}$$

The $v_{tref}(k)$ is given as the input to AC-PI controller and the output is taken as the weight loss error as

$$w_{erq}(k) = w_{erq}(k-1) + k_{pq}\{v_{tref}(k) - v_{tref}(k-1)\} + k_{iq}v_{tref}(k) \tag{17}$$

where k_{pq} and k_{iq} are the AC bus proportional and integral gains. The quadrature-axis components of the reference source current are estimated as weight components at kth sample instant as

$$w_{lq}(k) = \{w_{erq}(k) + w_{rq}(k) + w_{bq}(k)w_{yq}(k)\}/3 \tag{18}$$

Least mean square error trained by ADALINE is used to extract fundamental direct and quadrature-axis weight components.

The direct-axis weight components are:

$$w_{rq}(k+1) = w_{rq}(k) + \mu e_r(k)u_{rq} \tag{19}$$

$$w_{bq}(k+1) = w_{bq}(k) + \mu e_b(k)u_{bq}(k) \tag{20}$$

$$w_{yq}(k+1)w_{yq}(k) + \mu e_y(k)u_{yq}(k) \tag{21}$$

$$e_r(k) = \{i_{rl}(k) - i_{rd}^*\} \tag{22}$$

$$e_b(k) = \{i_{bl}(k) - i_{bd}^*\} \tag{23}$$

$$e_y(k) = \{i_{yl}(k) - i_{yd}^*\} \tag{24}$$

where $e_r(k)$, $e_y(k)$, $e_b(k)$ are the error output between the load and fundamental reference currents of phase r, b and y is the convergence factor. The accuracy of the system depends on the convergence factor (μ) which varying between 0.01 and 1.0 for the proposed system convergence factor (μ) is taken as 0.2. The fundamental reference currents of direct axis are estimated as

$$i_{rq}^* = w_{rq}(k)u_{rq}(k); \quad i_{bq}^* = w_{bq}(k)u_{bd}(k);$$
$$i_{yq}^* = w_{yd}(k)u_{yd}(k) \tag{25}$$

The references source currents of grid are estimated as sum of fundamental reference currents of direct and quadrature axis.

$$i_{er}^* = i_{rd}^* - i_{rq}^*; \quad i_{eb}^* = i_{bd}^* - i_{bq}^*;$$

$$i_{ey}^* = i_{yd}^* - i_{yq}^* \tag{26}$$

The switching pulses for DSTATCOM are generated by comparing the source error currents with triangular wave of 10 kHz frequency and the source error currents are estimated as

$$i_{esr}^* = i_{er}^* - i_{sr}, \ i_{esb}^* = i_{eb}^* - i_{sb};$$
$$i_{esy}^* = i_{ey}^* - i_{sy} \tag{27}$$

4 Neural Network

The DC bus capacitor is used to maintain constant MPPT voltage, the DC voltage and references DC voltage are compared and given as input to PI controller. The output of PI controller produces the proportional and integral gain for the system which is fed as input to neural network. The network consists of input layer, hidden layer and output layer. In the proposed method, x(1) is the input layer, layer 1, 2, 3 are the hidden layers and y(1) is the output layer as shown in Fig. 6.
Processing of layer 1:

Layer 1—x(1) is given as input to layer 1 and a(1) is the output of layer 1. Internal processing of layer 1 is given as p(1) given to weight which is summed up with bias and processed using tansig transfer function producing the a(1) output as shown in Fig. 7a and 7b.

The weighting function consists of two weights multiplied with the input p(1) and multiplexed as a single output.

Fig. 6 Neural network

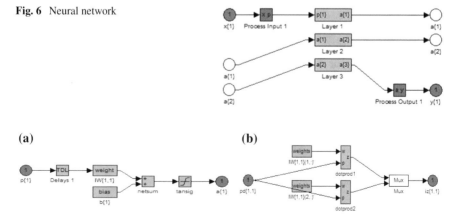

(a)   (b)

Fig. 7 a, Processing of layer 1 and **b** weights of layer 1

(a) (b)

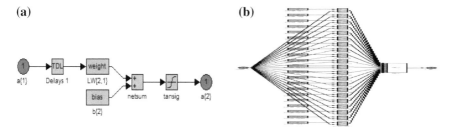

Fig. 8 **a**, **b** Processing of layer 2

(a) (b)

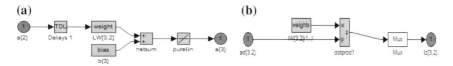

Fig. 9 **a**, **b** Processing of layer 3

Processing of layer 2:

Layer 2—a(1) is given as input to layer 2 and a(2) is the output of layer 2. Internal processing of layer one is given as a(1) given to weight which is summed up with bias and processed using tansig transfer function producing the a(2) output as shown in Fig. 8a and 8b.

The weighting function consists of 21 weights multiplied with the input a(1) and multiplexed as a single output.
Processing of layer 3:

Layer 3—a(2) is given as input to layer 2 and a(3) is the output of layer 3, processed as final output y(1). Internal processing of layer 1 is given as a(2) given to weight which is summed up with bias and processed using tansig transfer function producing the y(1) output as shown in Fig. 9a and 9b.

The weighting function consists of one weight multiplied with the input a(3) and multiplexed as a single output

The output of neural network is given as input to the fuzzy controller for rectifying the errors. The fuzzy system consists of membership function for load voltage, load current and firing angle as shown in Fig. 10a, 10b and 10c given with 25 rules. The ouput of fuzzy is used to drive the inverter section of the proposed system.

5 Results

The proposed system is designed in MATLAB/Simulink and is shown in Figs. 11 and 12. The output waveform for phase current, phase voltage and PV power and

(a)   **(b)**   **(c)**

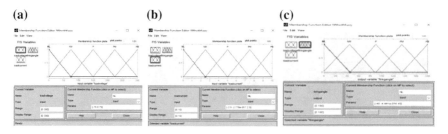

Fig. 10 **a** MF of load voltage, **b** MF of load current and **c** MF of firing angle

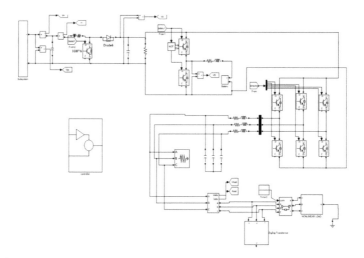

Fig. 11 Simulink model

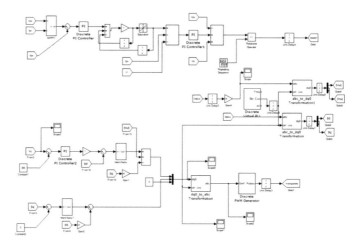

Fig. 12 Proposed control

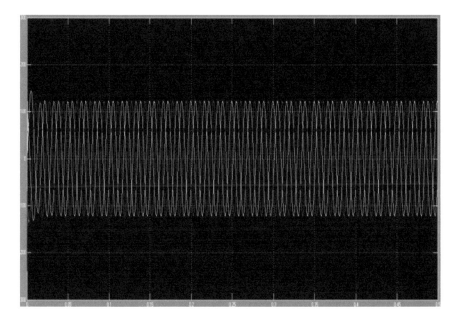

Fig. 13 Phase current

irradiances is shown in Figs. 13, 14 and 15. The THD calculation for load current and source current is shown in Figs. 16 and 17.

Conclusion

The reactive compensation with ZVR is proposed in the given method, thereby reducing the harmonics. The neuro-fuzzy control algorithm is used for VSC section controlling. The fuzzy is trained by the membership function and rules defined for the algorithm. Steady-state DC is analysed under linear and non-linear loads. The incremental conductances (IC) MPPT method is used for maintaining constant voltage.

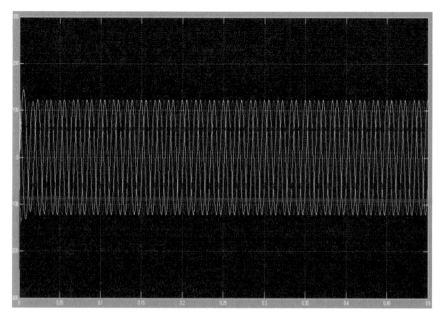

Fig. 14 Phase voltage

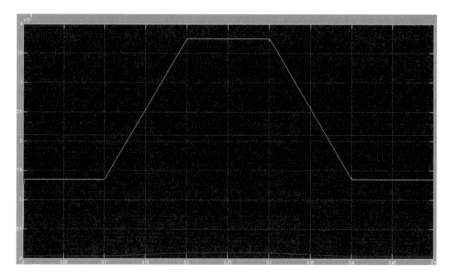

Fig. 15 PV power and irradiance

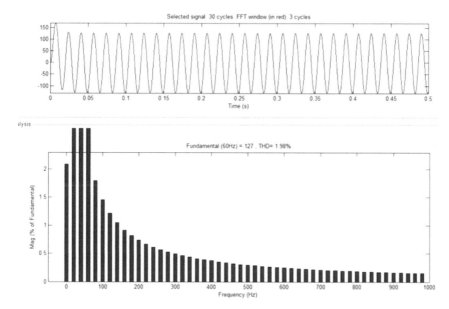

Fig. 16 THD level of load current

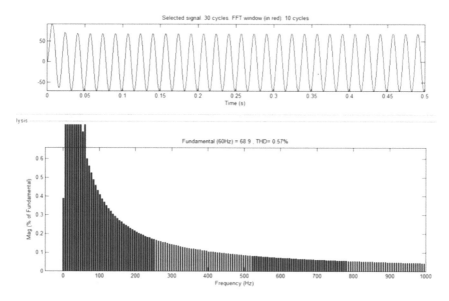

Fig. 17 THD level of source current

References

1. Alawadhi, N., Elnady, A.: Mitigation of power quality problems using unified power quality conditioner by an improved disturbance extraction technique. In: International Conference on Electrical and Computing Technologies and Applications (ICECTA) (2017)
2. Sukumaran, J., Thomas, A., Bhattacharya, A.: A reduced voltage rated unified power quality conditioner for harmonic compensations. In: IEEE 7th Power India International Conference (PIICON) (2016)
3. Wei, T., Jia, D.: A new topology of OPEN UPQC. In: 9th IEEE Conference on Industrial Electronics and Applications. IEEE (2014)
4. Hembram, M., Tudu, A.K.: Mitigation of power quality problems using unified power quality conditioner (UPQC). In: Proceedings of Third International Conference on Computer, Communication, Control and Information Technology (C3IT) (2015)
5. Subudhi, B., Pradhan, R.: A comparative study on maximum power point tracking techniques for photovoltaic systems. IEEE Trans. Sustain. Energy **4**(1), 89–98 (2013)
6. Yu, H., Pan, J., Xiang, A.: A multifunction grid connected PV system with reactive power compensation for the grid. Sci. Direct-Solar Energy **79**, 101–106 (2005)
7. Kelesidas, K., Adamidis, G., Tsengenes, G.: Investigation of a Control Scheme Based on Modified p-q Theory for Single Phase Single Stage Grid Connected PV, pp. 535–540
8. Liu, J., Yang, J., Wong, Z.: A New Approach for Single Phase Grid Connected Harmonic Current Detection and its Application in Hybrid Active Power Filter
9. Gevorkian, P.: Large Scale Power System Design. USA (2011)
10. Prabha, K., Malathi, T., Muruganandam, M.: Power Quality Improvement in Single Phase Grid Connected Nonlinear Loads, vol.4, no. 3, pp. 1269–1276 (2015)
11. Hongpeng, L., Shingong, J., Wei, W.: Maximum Power Point Tracking Based on Double Index Model of PV Cell, pp. 2113–2116 (2009)
12. Krishna, R., Sood, Y.R., Kumar, U.: Simulation and Design for Analysis of Photovoltaic System Based on Mat Lab, pp. 647–651 (2011)
13. Zhou, X., Song, D., Ma, Y., Cheng, D.: Simulation and design based for MPPT of PV system based on incremental conductance method. In: International Conference on Information Engineering, pp. 314–317 (2010)
14. Charles, S., Bhuvaneswari, G.: Comparison of three phase shunt active power filter algorithms. Int. J. Comput. Electr. Eng. **2**(1), 175–180 (2010)
15. Sahoo, N.C., Elamvazuthi, I., Nor, N.M., Sebastian, P., Lim, B.P.: PV Panel Modelling using Simscape, pp. 10–14 (2011)
16. Femia, N., Petrone, G., Spagnuolo, V., Vitelli, M.: Optimization of perturb and observe maximum power point tracking method. IEEE Trans. Power Electron. **20**(4), 963–973 (2005)
17. Zhihao, Y., Xiaobo, W.: Compensation loop design of a photovoltaic system based on constant voltage MPPT. In: Power and Energy Engineering Conference, APPEEC 2009, Asia-Pacific, pp. 1– 4, Mar 2009
18. Rezvani, A., Gandomkar, M., Izadbakhsh, M., Vafaei, S.: Optimal power tracker for photovoltaic system using ANN-GA. Int. J. Curr. Life Sci. **4**(9), 107–111 (2014)
19. Vincheh, M.R., Kargar, A., Markadeh, G.A.: A hybrid control method for maximum power point tracking (MPPT) in photovoltaic systems. Arabian J. Sci. Eng. **39**(6), 4715–4725 (2014)
20. Ramaprabha, R., Gothandaraman, V., Kanimozhi, K., Divya, R., Mathur, B.L.: Maximum power point tracking using GA-optimized artificial neural network for Solar PV system. In:Electrical Energy Systems (ICEES), pp. 264–268 (2011)
21. Yang, J., Donavan, V.: Feature subset selection using a genetic algorithm. IEEE Intell. Syst. **13**(2), 44–49 (1998)

Network Intrusion Detection System Using Random Forest and Decision Tree Machine Learning Techniques

T. Tulasi Bhavani, M. Kameswara Rao and A. Manohar Reddy

Abstract In the network communications, network interruption is the most vital concern these days. The expanding event of the system assaults is a staggering issue for system administrations. Different research works are now directed to locate a successful and productive answer for forestall interruption in the system so as to guarantee to arrange security and protection. Machine learning is a successful investigation device to identify any irregular occasions happened in the system traffic stream. In this paper, a mix of the decision tree and random forest algorithms is proposed to order any strange conduct in the system traffic.

Keywords Machine learning · Random forest · Decision tree

1 Introduction

With the movement in the development, countless are as of now connected with each other through one or other kind of framework where they share packs of crucial data. In this manner, the need of security to ensure data reliability and mystery is extended rapidly. Regardless of the way that efforts have been made to grapple data transmission, strike procedure for breaking the framework continued creating. Subsequently, it prompts the need of such a system which can modify with these reliably changing strike methodologies. In this paper, we have utilized irregular woodland and choice tree machine learning calculations. Machine learning gives adaptability to the framework which is utilized to foresee the assaults that happen in future.

T. T. Bhavani (✉) · M. K. Rao · A. M. Reddy
KL Deemed to be University, Green Fields, Vaddeswaram, Guntur Dist,
Andhra Pradesh 522502, India
e-mail: tulasi.bhavani1998@gmail.com

M. K. Rao
e-mail: kamesh.manchiraju@kluniversity.in

A. M. Reddy
e-mail: manohar.oct8@gmail.com

© Springer Nature Singapore Pte Ltd. 2020   637
A. K. Luhach et al. (eds.), *First International Conference on Sustainable Technologies for Computational Intelligence*, Advances in Intelligent Systems and Computing 1045,
https://doi.org/10.1007/978-981-15-0029-9_50

2 Related Works

In [1], Rafath and Vasumathi grouped the interruption identification framework into two sorts to be specific network-based IDS and host IDS. The last screens every one of the exercises of investigated bundles and assets that are being used by the projects. If there should be an occurrence of any modification in systems, the client gets a system alert. HIDS is fused into the PC system to distinguish the irregularities and shield the data from the gate crasher. Then again, NIDS is the trait capacity of target framework. It utilizes hostile to string programming to control approaching and active strings. It comprises mark-based order, which helps in distinguishing the variations from the norm by contrasting it and log records and past mark.

In [2], the IDS depends on inconsistency recognition strategy. In such procedure, a framework endeavors to evaluate the 'ordinary' condition of the system and produces a ready when any exercises veer off from this 'typical' state. The principle advantage of irregularity-based framework is that it can identify beforehand inconspicuous interruption occasions. They have grouped identification methods into three classifications: measurable-based, information-based, and machine learning-based. In factual-based method, an irregular perspective is utilized to speak to the conduct of the framework. While learning-based procedure, use the accessible framework information to catch the conduct of framework. At last, the machine learning-based procedure utilizes an unequivocal or understood model to empower order of the dissected example.

In [3], different machine learning methods can result in higher location rates, bring down false caution rates, sensible calculation, and correspondence costs in interruption discovery. In this paper, Mahdi Zamani and Mahnush Movahedi considered a few such methods and plans to look at all their execution. They isolate the plans into strategies dependent on traditional computational insight (CI) and man-made brainpower (AI). They clarify how a few highlights of CI systems can be utilized to assemble current and effective IDS.

3 Model Classification

3.1 Random Forest

Random forest algorithm is a managed request calculation. We can understand it from its name, which is to create woods in a way and make it unpredictable. There is a prompt association between the number of trees that results it can get: The greater the number of trees, the more claims the result. Regardless, the creation the forest is not equivalent to developing the choice with data gain or gain list approach. Random forest, the procedure which is of searching the root hub and part the component hubs, will run in a manner. There are two phases in the random forest algorithm, that is irregular random forest creation, and making a prediction from the random forest classifier made in the primary stage.

Creation:

(1) Select K from where $k \ll m$.
(2) Among the K, hub d is using the better part of point.
(3) Divide the hub into useful division.
(4) Repeating the process a to c until L number of hubs to come.
(5) Repeat steps from a to d.

Prediction:

(1) Steps through the examination highlights and utilize the standards of each haphazardly settled on decided tree to anticipate the result and saves the predicted result as the target.
(2) Count the votes for each predicted result.
(3) Take into consideration of high threw a poll assumed center as the last estimate from the sporadic boondocks figuring.

3.2 Decision Tree

Decision tree is a sort of supervised machine learning where the data is always part of a particular parameter. The tree can be elucidated by two components, to be explicit decision leaves and nodes. The leaves are the definitive or decision outcomes, whereas their nodes are the place where the data is a part.

Types

(1) Classification trees
(2) Regression trees.

Entropy:

Entropy, likewise known as Shannon entropy, is indicated by $H(S)$ for a limited set S and is the proportion of the measure of vulnerability or arbitrariness in information.

$$H(S) = \sum_{x \in X} p(x) \log 2 \frac{1}{p(x)}$$

Instinctively, it enlightens us with respect to the consistency of a specific occasion. Here the entropy is the most stunningly conceivable, since there is no plausibility to get of comprehending what the result may be. On the other hand, by the day's end, this event has no anomaly; thus, its entropy is zero. In particular, cut-down characteristics induce less weakness while higher characteristics propose high defenselessness.

Information Gain:

Data gain is distinction demonstrated by IG(S, A) for a set s is the suitable for change in entropy in the wake of settling on a particular property a. It evaluates the relative change in the entropy with respect to the free factors.

$$IG(S, A) = H(S) - \sum_{i=0}^{n} P(x) * H(x)$$

Alternatively,

$$IG(S, A) = H(S) - \sum_{i=0}^{n} P(x) * H(x)$$

Here IG applying feature a. $H(S)$ is the entropy of the set, while the another term finds out the entropy by applying for the segment, a, $P(x) =$ probability of event there are various computations out there which fabricate Decision Trees, anyway a champion among the Frequent used algorithm is called as ID3 (Iterative Dichotomies 3) Algorithm.

Iteration performed by ID3 is:

(1) For the tree, center point is created
(2) Arrival leaf center will be positive if all points of reference are certain.
(3) Else if all points of reference are not positive, return leaf center negative.
(4) At the state of $H(S)$, calculate the entropy.
(5) The entropy should me marked as x for every property that implied by $H(S, x)$.
(6) Select the property that has most estimation of IG(S, x).
(7) Eliminate the property from the most shocking IG from the course of action of characteristics.
(8) Repeat until the moment that we miss the mark on all characteristics, or the decision tree has all leaf center points.

The underlying advance is to compute $H(S)$, the entropy of the present state.

$$\text{Entropy}(S) = \sum_{x \in X} \log^2 \frac{1}{p(x)}$$

Entropy is 0 if all individuals have a place with a similar class, and 1 when half of them have a place with one class, and other half have a place with different class that is impeccable haphazardness.

Directly, the accompanying stage is to select the qualities that give us most possible information gain which is selected as the root center point.

$$IG(S, \text{Wind}) = H(S) - \sum_{i=0}^{n} P(x) * H(x)$$

where x is the possible values for an attribute. Here, attribute 'wind' takes two possible values in the sample data and $x = $ weak, strong.

Calculate:

(1) $H(S_{\text{Weak}})$
(2) $H(S_{\text{Strong}})$
(3) $P(S_{\text{Strong}})$
(4) $P(S_{\text{Weak}})$
(5) $H(S)$

$$P(S_{\text{Weak}}) = \frac{\text{Number of weak}}{\text{Total}}$$

$$P(S_{\text{Strong}}) = \frac{\text{Number of Strong}}{\text{Total}}$$

$$\text{IG}(S, \text{wind}) = H(S) - \sum_{i=0}^{n} (P(x) * H(x))$$

This discloses to the information gain by considering wind as the element. Entropy to quantify prejudicial intensity of a characteristic for arrangement errand. It describes the proportion of haphazardness attribute for request task. Entropy is insignificant strategies that credit appears to be close to one class and has a tolerable biased power for grouping.

4 Simulation Result

See Figs. 1, 2, 3 and 4.

5 Conclusion

In this paper, a combination of random forest and decision tree machine learning techniques is utilized for system interruption discovery. Random forest calculation will stay away from the over-fitting issue. It is a powerful supervised machine learning algorithm that is capable of performing regression and classification tasks. Simplicity of coding, tending to nonlinearity, and construing connection terms should be possible utilizing decision tree calculation.

frequency_percent_train
frequency_percent_test
50
40
30
20
10
0
DoS
Normal
Probe
R2L
U2R

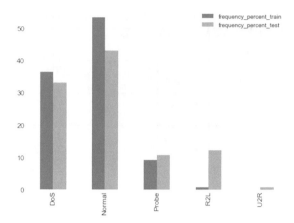

Fig. 1 Attack class distribution

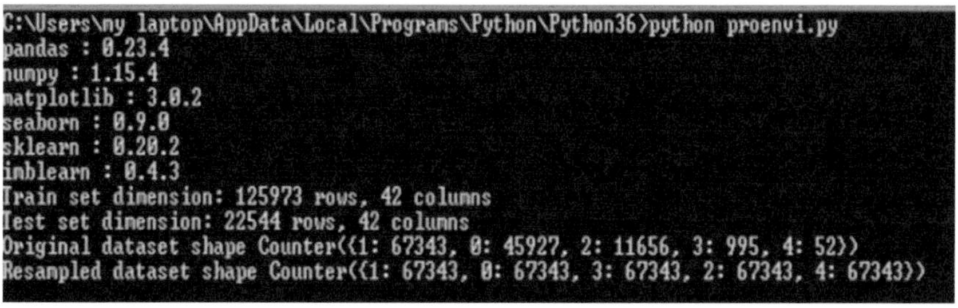

C:\Users\my laptop\AppData\Local\Programs\Python\Python36>python proenvi.py
pandas : 0.23.4
numpy : 1.15.4
matplotlib : 3.0.2
seaborn : 0.9.0
sklearn : 0.20.2
imblearn : 0.4.3
Train set dimension: 125973 rows, 42 columns
Test set dimension: 22544 rows, 42 columns
Original dataset shape Counter({1: 67343, 0: 45927, 2: 11656, 3: 995, 4: 52})
Resampled dataset shape Counter({1: 67343, 0: 67343, 3: 67343, 2: 67343, 4: 67343})

Fig. 2 Loading data

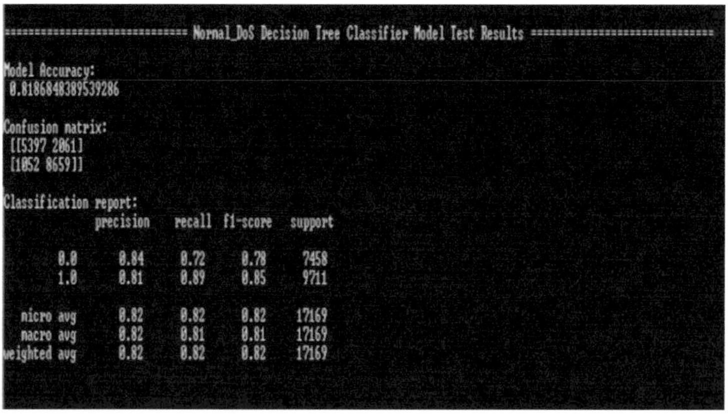

============================== Normal_DoS Decision Tree Classifier Model Test Results ==============================
Model Accuracy:
0.8186848389539286
Confusion matrix:
[[5397 2061]
[1052 8659]]
Classification report:
precision recall f1-score support
0.0 0.84 0.72 0.78 7458
1.0 0.81 0.89 0.85 9711
micro avg 0.82 0.82 0.82 17169
macro avg 0.82 0.81 0.81 17169
weighted avg 0.82 0.82 0.82 17169

Fig. 3 Decision tree algorithm model results

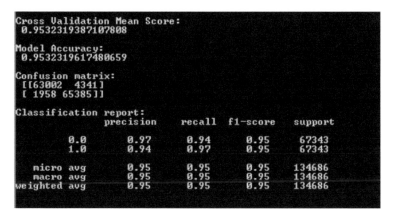

Fig. 4 Random forest algorithm model results

References

1. Samrin, S., Vasumathi, D.: Review on Anomaly based Network Intrusion Detection System. Computer Science and Engineering ISL Engineering College Hyderabad, India Computer Science and Engineering JNTUH Hyderabad, India, 978-1-5386-2362-6 (2017)
2. Garcia-Teodoroa, P., Diaz-Verdejoa, J., Macia- Fernandeza, G., Vazquezb, E.: Anomaly-Based Network Intrusion Detection: Techniques, Systems and Challenges. Department of Telematic Engineering, Universidad Politécnica de Madrid, Madrid, Spain (2009)
3. Zamani, M., Movahedi, M.: Machine Learning Techniques for Intrusion Detection. Cornell University, 2 (2015)
4. Tiwari, M., Kumar, R., Bharti, A., Kisan, J.: Intrusion detection system. Int. J. Tech. Res. Appl. **5**, 38–44 (2017)

EHA-RPL: A Composite Routing Technique in IoT Application Networks

Soumya Nandan Mishra, Manu Elappila and Suchismita Chinara

Abstract IoT solutions are based on low power and lossy networks (LLNs). LLN network consists of nodes having limited processing power, memory and battery capacity. The Routing Over Low Power And Lossy Networks (ROLL) group have developed an RPL protocol for LLN. To fulfil the QoS requirements for different applications, the RPL needs to be optimized and the resources have to be utilized efficiently. The RPL protocol uses an objective function that helps to calculate the rank of a node and select an optimized parent through which data is forwarded. The single routing metric used by RPL does not satisfy all the QoS requirements. So, we have proposed a new composite routing technique called EHA-RPL which combines three routing metrics, i.e. Expected Transmission Count (ETX), hop count and available energy using two approaches, namely lexicographic and additive approach. It is observed that our approach shows better performance in terms of energy consumption, network latency and packet delivery ratio as compared to the traditional approaches.

Keywords IoT · RPL · Composite · Lexicographic · Additive

1 Introduction

To make human life easier and comfortable, technologies are being made and developed. But to get benefit from these technologies, the devices must communicate with each other in real time. Internet of things (IoT) provides a glimpse on how different physical devices communicate among themselves and connect to the Internet so that

S. N. Mishra (✉) · M. Elappila · S. Chinara
National Institute of Technology, Rourkela 769008, India
e-mail: 617cs3001@nitrkl.ac.in

M. Elappila
e-mail: 514cs1009@nitrkl.ac.in

S. Chinara
e-mail: suchismita@nitrkl.ac.in

© Springer Nature Singapore Pte Ltd. 2020
A. K. Luhach et al. (eds.), *First International Conference on Sustainable Technologies for Computational Intelligence*, Advances in Intelligent Systems and Computing 1045, https://doi.org/10.1007/978-981-15-0029-9_51

they can be used in a wide range of applications like healthcare, agriculture sectors, home automation for management of daily tasks, etc. The low power and lossy network (LLN) formed by these physical devices and objects have the following desired characteristics [1]:

– Automation: devices should be able to automate their tasks like collecting data, processing the gathered data, exchange it with other IoT devices, make decisions based on the gathered data, etc.
– Intelligence: objects should be smart enough to act according to the real-time circumstances without any human intervention.
– Dynamicity: dynamicity ensures that an object is free to move from one place to another place and the IoT system should be able to recognize these dynamic scenarios inside the network.
– Zero configurations: users may face difficulty in configuring some devices. So plug-and-play feature should be available for such devices.

It becomes a challenge to control these devices when hundreds and thousands of them try to communicate together in a real-world situation, and the device may get added or removed according to the application requirements. Then, it is important to have scalable and energy-efficient routing protocols to meet the challenge. Some of the routing protocol requirements for LLN applications are listed below [2]:

– Scalability: the routing protocol must be designed in such a way that it performs as desired even with an increasing number of nodes in a particular region.
– Parameter-constrained routing: parameters such as remaining battery power, memory size and CPU should be advertised frequently by the protocol to make routing decisions. Depending on the nature of the traffic, the nodes deployed on the fields should dynamically compute, install and select different paths towards the destination.
– Support for highly directed information flows: nodes near the border router become victims of highly concentrated traffic, due to which load imbalance occurs in some of them. The routing protocol must utilize the highly directed information flows and select multiple paths towards the same destination.
– Latency: the protocol must have the ability to select a route and forward the packet based on some latency metric for some delay constrained applications.

The wireless sensor nodes forming an LLN network have limited processing power, memory and battery capacity. Traditional routing protocols like Ad-Hoc On-Demand Vector (AODV) and Optimized Link State Routing (OLSR) cannot be well fitted in such networks. There are several routing protocols developed for IoT-based scenario that work on reliable data transmission from source to destination. RPL [3] is one of the IoT-based routing protocols which is designed for low power and lossy networks. The routing process in RPL starts by forming a Destination-Oriented Directed Acyclic Graph (DODAG) topology containing a single root known as DODAG root. Networks can have more than one DODAG's, each identified by a unique DODAG ID. The root multicasts a DODAG Information Object (DIO) message in the network. When a node receives the DIO message, it calculates its rank based on the objective function used in the protocol.

Fig. 1 Calculation of node's rank

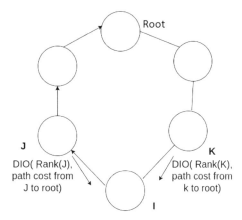

The calculation of rank can be explained with a diagram as shown in Fig. 1. In the figure, $R(I) = Rank(J) + c(I,J)$ and $R'(I) = Rank(K) + c(I,K)$, where $R(I)$, $R'(I)$ are rank of I through J and K, respectively, and $c(I,J)$, $c(I,K)$ are functions of routing metrics used by the objective function for calculating the link cost from node I to node J and from node I to node K, respectively. Suppose, $R(I) < R'(I)$ then node I will choose node J as its parent, and $Rank(I) = min\{R(I), R'(I)\}$.

RPL nodes choose its parent based on the rank value calculated from the objective function and the routing information provided in the DIO message of the sender. Nodes use DODAG Destination Advertisement Object (DAO) control message to send the destination information upward to the root. The main objective of RPL is to reduce power consumption and decrease the network set-up time. RPL uses the Trickle algorithm to achieve the same, by reducing the number of control messages transmitted while constructing the DODAG topology.

The rest of the paper is organized as follows: In Sect. 2 all the related works are mentioned. In Sect. 3, the proposed work is described in detail with example diagrams. In Sect. 4, the performance of the proposed approach is compared with the existing approaches. Finally, the paper is concluded in Sect. 5.

2 Related Work

The existing RPL considers only one single routing metric, i.e. either hop count [4] or ETX [5] to select a path for data transmission. However, taking only a single metric into account does not satisfy all the QoS requirements for different applications. There are several works in the literature where authors have proposed a combination of more than one metrics to satisfy QoS requirements for LLNs.

Mohamed et al. [6] proposed an objective function which chooses a path having a high transition probability. The transition probability is calculated by taking two metrics into account, i.e. transmission delay and residual energy. However, it did not consider the ETX metric for detecting lossy links. So, it may result in choosing inefficient routes.

The authors in [7] have combined four routing metrics, namely ETX of the link, number of packets received by a node n (REC_n), rank of a node and minimized delay metric, to select the most optimal path for data transmission. However, the energy consumption of the node has not been taken into account for studying the network lifetime of nodes.

Iova et al. have proposed an Expected Lifetime metric [8] which evaluates the residual time of each node, i.e. the time before which the first node runs out of energy. It aims to maximize the lifetime of each node. The authors have compared their proposed approach with several other routing metrics and found their method to be better in terms of longevity of the network. However, the performance metrics like packet delivery ratio and end-to-end delay do not offer good performance.

Kamgueu et al. [9] have proposed an energy-based routing metric to be used by the objective function in RPL. Although the protocol performs better in terms of energy consumption, there is no consideration of link quality metrics for path selection, which may result in choosing lossy and inefficient routes.

Sanmartin et al. [10] have proposed a sigma-ETX metric to solve long-hop problem. The standard deviation of ETX value for each path is calculated, and the path having the lowest standard deviation is selected as the best path. However, energy metric is not taken into consideration for selecting the path, which can result in faster energy depletion of some nodes.

Combining routing metrics of different types may lead to routing loops or selection of non-optimal paths. So the objective function must be carefully designed to satisfy the monotonic and isotonic properties of the routing metrics. Monotonic property ensures that the topology is loop-free. Isotonic property ensures that the optimal path is chosen by the source for transmitting data to the destination.

Karkazis et al. [11] have proposed several ways to combine the routing metrics and used them in a lexical or additive manner.

Velivasaki et al. [12] have proposed a paper in which two metrics, packet forwarding indication (PFI) and ETX, have been combined lexically and additively and their performance has been analysed. However, the remaining energy metric has not been taken into account, which is the main constraint in WSN deployments.

In this paper, we analyse the performance of proposed EHA-RPL technique which combines three routing metrics (ETX, hop count and available energy) into one composite routing metric using two approaches, i.e. lexicographic and additive approach. The results show that the proposed technique outperforms the single routing metric techniques, and also the additive approach combination shows better performance than the lexicographic approach combination in terms of packet delivery ratio, average latency and energy consumption.

3   Proposed Work

The RPL protocol, by default, uses a single metric (ETX or hop count) in the formation of the objective function to calculate the rank of a node. It chooses the optimized parent using this objective function to forward the data packets to the destination. The problem of load balancing and network bottleneck occur when only the hop count is used as the metric for rank calculation. And if only ETX metric is used, then the packets may have to travel longer hops and hence longer delay in the packet delivery. So by considering only a single metric, we cannot satisfy all the QoS requirements needed by different application domains.

This provided the motivation to work on the designing of a composite routing metric in which, more than one basic routing metrics can be taken into account for creating an objective function. In the work proposed, three routing metrics, namely ETX, hop count and available energy, are considered. The performance of RPL is analysed based on the combination of these three to a single composite metric. We use three basic routing metrics and different types of combination approaches. Based on the QoS requirement of the application, appropriate routing metrics can be used with the approaches explained in algorithm 1 and algorithm 2.

There are two approaches to combine these metrics, they are:

– Lexicographic approach.
– Additive approach.

In the lexicographic approach, a composite metric is composed with an order of preferences among the component metrics. The first component metric is considered and the path having the highest/lowest value for that metric is preferred for data transmission. If it is found identical, the second metric is considered and so on.

In the additive approach, weight factors are concatenated with each of the primary routing metrics. The weights are adjusted according to the user requirements based on the application being used.

Algorithm 1 Algorithm for the Lexicographic approach for parent selection

---

1: Let N1 and N2 be two neighbours of a node, and ETX_i and ETX_j be the ETX value for each link in the path containing N1 and N2 as its nodes respectively.
2: **if** Maximum (ETX_i) and Maximum (ETX_j) are equal or less than a particular threshold **then**
3:    **if** Hop count for paths through N1, N2 are equal or exceeds the maximum limit for hop count **then**
4:      Select parent having maximum available energy (third criteria)
5:    **else if** Hop count for N1, N2 are not equal **then**
6:      Select parent with minimum hop count (second criteria)
7:    **else**
8:      Set preferred parent to NULL
9:    **end if**
10: **else if** Maximum (ETX_i) and Maximum (ETX_j) are not equal **then**
11:    Select preferred parent with [Minimum (Maximum (ETX_i,ETX_j)] (first criteria)
12: **else**
13:    Set preferred parent to NULL
14: **end if**
15: RETURN Preferred parent

In the algorithm 1, the first criteria for selecting the parent are chosen by using the ETX metric. If the ETX value for both the neighbours is equal or below a particular threshold, then it checks the hop count metric. If the hop count for both neighbours is equal or exceeds a maximum limit, then it selects the parent on the basis of available energy metric. Here, while choosing parent based on the ETX metric, the path having the minimum of the maximum ETX values for both the paths are chosen as the preferred path for data transmission. The ETX value for each link is calculated using the following formula:

$$ETX_i = \frac{s+f}{s} \tag{1}$$

where s denotes the number of packets successfully delivered to the neighbour node and f denotes the number of packets failed to be delivered.

The lexicographic algorithm can be explained better with the help of Figs. 2, 3 and 4. In Fig. 2, node 7 is the sink. The sink node multicast DIO messages to node 4 and 6. Node 4 and 6 choose node 7 as the common parent by sending DAO message to the sink node. Similarly, node 2, 3 and 5 choose node 4, 5 and 6 as their respective parent to forward the data to the destination. But node 1 can either choose node 2 or node 3 as its next-hop parent. According to the lexicographic algorithm, the first criteria for node 1 to choose its parent would be the value of the ETX metric. It must choose the node through which the ETX path value would be minimum among all the possible paths.

If node 1 chooses node 2 as its parent, the link having the maximum ETX value in that path would be 4→7 having a value of 4.7, and if it chooses node 3 as the parent, the link having the maximum ETX value in that path would be 5→6 having a value of 3.3. Since the minimum among the ETX values is to be considered, node 1 choose node 3 as its preferred parent for data transmission. In the algorithm 1, the ETX value is compared with the maximum threshold value of ETX. It might happen that the ETX value of paths through node 2 and node 3 never be the same, so second and third criteria, i.e. hop count and available energy, will never be considered as a metric for

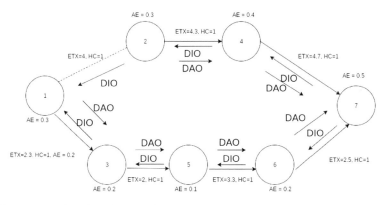

Fig. 2 Path gets selected on the basis of ETX value

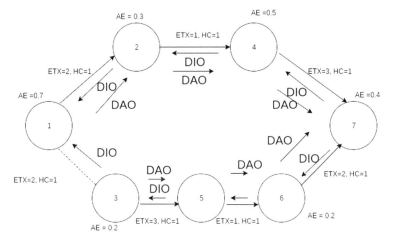

Fig. 3 Path gets selected on the basis of hop count

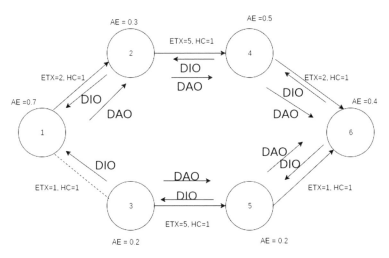

Fig. 4 Path gets selected on the basis of available energy

path selection. Similarly, in the case of hop count metric, we have considered that if the number of hops for both paths exceeds the maximum hop limit, then the path is selected by taking the available energy of nodes as the last criteria.

In Fig. 3, node 1 chooses node 2 as its parent for data forwarding. Here, it is observed that if node 1 chooses node 2 as the parent, the maximum ETX value in the path is 3, and if it chooses node 3 as its parent, then the value would be also 3. As the algorithm says that if the ETX value for both paths is equal, the second criteria (hop count) are used as the metric for selecting parent. In this case, node 1 chooses node 2 as the preferred parent because the number of hops to be traversed by node 1 is 3 which is less than the number of hops it needs to traverse while choosing node 3 as the parent.

In Fig. 4, it is observed that node 1 chooses available energy as the metric for path selection. In this case, the maximum ETX value for two possible paths from node 1 is same, i.e. 5. So, it goes for checking the second criteria (hop count), and it is found that for both paths, the number of hops it needs to traverse to reach the destination 7 is equal to 3. So, according to the algorithm the path needs to be selected on the basis of the available energy of the nodes. The path having the maximum available energy is selected as the optimal path.

Here, node 1 chooses node 2 as the parent because it finds that by summing all the available energy of the nodes while going through that path would be (0.3 + 0.5 + 0.4)=1.2, whereas if it chooses node 3, then the total available energy on that path would be (0.2 + 0.2 + 0.4)=0.8.

In additive approach, the primary routing metrics, i.e ETX, hop count and available energy, must be combined in such a way that each metric satisfies the same order relation as mentioned in the design guidelines for routing metrics composition. Unlike the lexicographic approach, metrics cannot be taken as a combination of different order relation. So, the primary routing metrics need to be either maximizable or minimizable in order to be used as an additive manner.

In the proposed approach, ETX and hop count are minimizable metrics, whereas available energy is maximizable. Therefore, the metric is transformed into a derived metric using the following formula:

$$AE(p) = \frac{\sum_{i=1}^{n} \frac{1}{AE_i}}{n} \tag{2}$$

In the equation, AE(p) denotes the available energy for a particular path p which is a ratio of two quantities, namely available energy of a node (AE_i) and the total number of hops (n) in that path.

Algorithm 2 depicts the parent selection by a node using an additive approach. The algorithm has already been described in [13]. In the algorithm, c(i,j) is a function of link metric between child(i) and parent(j) and node metrics of parent(j). It is calculated by adding the path cost of ETX ($c_{ETX}(p)$), hop count ($c_{HC}(p)$) and available energy ($c_{AE}(p)$). Here, α_1, α_2 and α_3 are weight factors which can be adjusted according to the requirements of LLN's applications. The algorithm chooses path through which minimum cost is achieved for data transmission.

Algorithm 2 Algorithm for parent selection in additive approach

---

Require: Node ID and rank of parents
Ensure: Select the parent through which path cost is minimum
1: Let $P_1, P_2,..., P_n$ be the parent list for a node X.
2: $Preferred\_Node\_Rank$ = INFINITY
3: **for** $Parent(P) \in P_1, P_2, P_3,..., P_n$ **do**
4:    $Rank(Child) = Rank(P) + c(i, j)$
5:    $c(i, j) = \alpha_1 * c_{ETX}(p) + \alpha_2 * c_{HC}(p) + \alpha_3 * c_{AE}(p)$
6:    **if** $Preferred\_Node\_Rank > Rank(Child)$ **then**
7:       $Preferred\_Node\_Rank = Rank(Child)$
8:       $Select\_Parent = Preferred\_Parent\_Id$
9:    **end if**
10: **end for**
11: RETURN $Select\_Parent$

---

4 Performance Evaluations

The proposed work is simulated in Cooja simulator by using Contiki OS. In Fig. 5, there are 75 client nodes coloured yellow and one server node coloured green representing the root of the DODAG, arranged in a random fashion. The green region represents the transmission range of the server node, and the grey region represents its interference range. The values of the simulation parameter are shown in Table 1.

In Fig. 5, the values in % beside each of the client nodes show the success probability of packet reception at that node. As we can see that the client nodes near the server node 1 have a higher probability of success to receive the packets than the client nodes far away. Client nodes at the edge of the transmission region have a probability equal to the packet reception or RX ratio of the network. For the given figure, RX ratio is set to 80%, that means the nodes lying at the edge of the transmission

Fig. 5 Topology

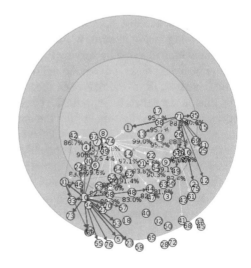

Table 1 Simulation parameters

Parameters	Value
Routing metric	ETX, HC, AE
Radio medium	UDGM
DIO min	12
DIO doublings	8
RDC channel check rate	16 Hz
RX ratio	10–100%
TX ratio	100%
TX range	50 m
Interference range	80 m

Fig. 6 Energy consumption

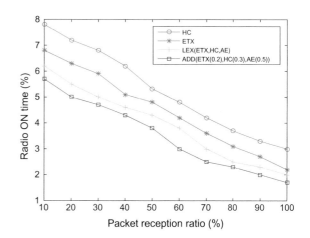

range have 80% probability to successfully receive the packets. The proposed lexicographic and additive combination of three metrics is compared with the existing single routing metric in terms of three performance parameters as shown in figures. The simulation is tested by varying the packet reception ratio from 10 to 100%.

In Fig. 6, it is observed that the radio ON time decreases as the packet reception ratio increases. Here, the radio ON time is taken as the average of all node's radio ON time in the network. Suppose the packet reception ratio is 10%, that means, the nodes positioned at the edge of the transmission region have only 10% probability to successfully receive link layer acknowledgements from the server and the nodes closer to the server have a high probability of receiving packets successfully. So the nodes far away have less chance to receive acknowledgement packets and remain awake by keeping their radio ON until they receive the acknowledgements. Now if we set the packet reception ratio to 100%, then all the nodes have same packet receiving capacity as 100%, and they will receive much faster acknowledgements and go to sleep state early, thereby decreasing the radio ON time. It is observed that the combination of three metrics in two approaches outperforms the single routing metric in terms of energy consumption.

When only hop count is taken as the routing metric, the energy consumption is higher because it tries to forward data to the node that is closer to the root and does not consider any energy constraint. So there is a possibility that nodes closer to the sink go out of energy much faster than the nodes far away from the sink. The ETX metric performs better than hop count, as the ETX metrics choose links having less number of transmissions, thereby building a load-balanced network and indirectly decreasing the energy consumption of the network. But still, only a single metric cannot satisfy all the QoS requirements.

The additive approach performs better than the lexicographic approach because more emphasis is given to the available energy metric of the node by taking the weight factor as 0.5, whereas in lexicographic approach, more preference is given to the ETX metric.

In Fig. 7, the network latency can be defined as the difference between the time that the packet was sent and the time when the data was received by the sink. The sink sends the acknowledgement if it successfully received the packet. Here, the average of all latencies of packets generated by all the nodes in the network is taken. Here, it is seen that the network latency decreases with the increase in packet reception ratio. The reason for this is that as we increase the packet reception ratio, more nodes are capable of successfully receiving the packets in a small amount of time. Thus, more number of packets are received by the sink within a small period of time, thereby decreasing the overall latency of the network. The additive approach performs better than all other approaches when we give more weight factor to the hop count. This is because it prefers to send data through a lesser number of hops, thereby reducing the latency time of the packet.

In Fig. 8, the packet delivery ratio is calculated as the ratio between the number of packets received by the sink to the number of packets sent. When the packet reception ratio is increased, it is observed that the packet delivery ratio is also increased. The reason for this is that the packet success ratio of nodes increases due to the increase in packet reception ratio. Hence, more packets are successfully received at the sink.

Fig. 7 Network latency

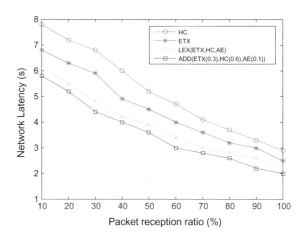

Fig. 8 Packet delivery ratio

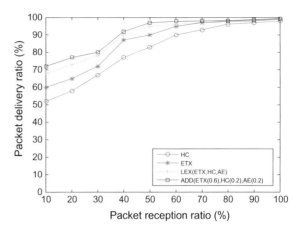

Here, we can also see that the additive approach performs better than others when we put more weight on the ETX metric. So, it is concluded that due to flexible nature of the additive approach, which comes from the concatenation of variable weight factor values, the approach is better suited to QoS requirements of LLN's applications than the other approach.

5 Conclusion

Routing data over a network comprising of a large number of energy-constrained nodes is a major challenge faced by IoT routing protocols. To address this challenge, various parameters like the residual energy of the node and link quality need to be considered while making routing decisions. RPL protocol uses either OF0 [4] or MRHOF [5] as an objective function to calculate the node's rank and to select the preferred parent. By taking only a single metric, RPL cannot meet the requirements of all the LLN-based applications. In this paper, the proposed EHA metric helps to address this issue, by combining the three metrics lexically and additively. The combination of these metrics has been shown, and the result shows that by combining these metrics, better performance is achieved than using a single metric.

References

1. Lee, G.M., Kim, J.Y.: The internet of things-a problem statement. In: 2010 International Conference on Information and Communication Technology Convergence (ICTC), IEEE, pp. 517–518 (2010)
2. Dohler, M., Watteyne, T., Winter, T., Barthel, D.: Routing Requirements for Urban Low-Power and Lossy Networks. Technical report (2009)

3. Winter, T., Thubert, P., Brandt, A., Hui, J., Kelsey, R., Levis, P., Pister, K., Struik, R., Vasseur, J.P., Alexander, R.: Rpl: Ipv6 Routing Protocol for Low-Power and Lossy Networks. Technical report (2012)
4. Thubert, P.: Objective Function Zero for the Routing Protocol for Low-Power and Lossy Networks (rpl). Technical report (2012)
5. Gnawali, O., Levis, P.: The Minimum Rank with Hysteresis Objective Function. Technical report (2012)
6. Mohamed, B., Mohamed, F.: Qos routing rpl for low power and lossy networks. Int. J. Distrib. Sens. Netw. **11**(11), 971545 (2015)
7. Tang, W., Ma, X., Huang, J., Wei, J.: Toward improved rpl: A congestion avoidance multipath routing protocol with time factor for wireless sensor networks. J. Sens. **2016** (2016)
8. Iova, O., Theoleyre, F., Noel, T.: Using multiparent routing in rpl to increase the stability and the lifetime of the network. Ad Hoc Netw. **29**, 45–62 (2015)
9. Kamgueu, P.O., Nataf, E., Ndié, T.D., Festor, O.: Energy-Based Routing Metric for RPL. Ph.D. thesis, INRIA (2013)
10. Sanmartin, P., Rojas, A., Fernandez, L., Avila, K., Jabba, D., Valle, S.: Sigma routing metric for rpl protocol. Sensors **18**(4), 1277 (2018)
11. Karkazis, P., Leligou, H.C., Sarakis, L., Zahariadis, T., Trakadas, P., Velivassaki, T.H., Capsalis, C.: Design of primary and composite routing metrics for rpl-compliant wireless sensor networks. In: 2012 International Conference on Telecommunications and Multimedia (TEMU), IEEE, pp. 13–18 (2012)
12. Velivasaki, T.H.N., Karkazis, P., Zahariadis, T.V., Trakadas, P.T., Capsalis, C.N.: Trust-aware and link-reliable routing metric composition for wireless sensor networks. Trans. Emerg. Telecommun. Technol. **25**(5), 539–554 (2014)
13. Mishra, S.N., Chinara, S.: Ca-rpl: A clustered additive approach in rpl for iot based scalable networks. In: International Conference on Ubiquitous Communications and Network Computing, Springer, pp. 103–114 (2019)

Sustainable IPR Practices to Address Risk Capital Finance in Software Industries in Developing Countries

Feroz Eftakhar Nowaz, Mahady Hasan, Nuzhat Nahar and M. Rokonuzzaman

Abstract 'Innovation' is the most common catchphrase around the globe for a great reason; likewise, it is one of the major approaches to power an economy and employment. The innovation process requires adequate fund, resources and a law for sufficient protection of the intellectual properties. Venture capitals (VCs) from risk capital finances (RCFs) are one of the best sources of gathering private equity for small companies. RCFs back those firms which are more innovative and produce more valuable patents which are protected by intellectual property rights (IPR). However, sheer negligence and inadequately unbalanced IPR culture within the software industries fail to communicate the commercial value of those assets to potential investors and business partners to grab the finances. As a result, the software companies fall short of turning their ideas into commercially viable product. Yet, a balanced IPR culture must take into account the cases of IPR abuse and secure the risk capitals. This paper researches the need of balanced IPR practices within the local small industry, predominantly software firms, which can be benefitted from it by securing the risk capital finance for their protected ideas and join the innovation race to maximise their wealth.

Keywords Intellectual property · Risk capital finance · Venture capital finance · IPR infringement · Trademarks · Patents · Copyright · Balanced IPR · Wealth creation · Innovation race

F. E. Nowaz · M. Hasan (✉) · N. Nahar
Department of Computer Science and Engineering,
Independent University Bangladesh, Dhaka, Bangladesh
e-mail: mahady@iub.edu.bd

M. Rokonuzzaman
Department of Electrical & Computer Engineering,
North South University, Dhaka, Bangladesh

© Springer Nature Singapore Pte Ltd. 2020
A. K. Luhach et al. (eds.), *First International Conference on Sustainable Technologies for Computational Intelligence*, Advances in Intelligent Systems and Computing 1045,
https://doi.org/10.1007/978-981-15-0029-9_52

1 Introduction

A company should improve the quality of their products and reduce the cost of production to expand economic output. Intellectual property lies at the focal point of the modern company's financial success or failure [1]. However, to join the innovation race, companies need to develop good business models to support this process which requires a huge investment in the innovation and R&D sectors. Due to the lack of R&D investment in software industries in the developing countries, fewer intellectual properties are produced which appears to have a low association with the economic growth of the nation [2].

Venture capitals from risk capital finances (RCFs) are professionally managed funds that put resources into companies that have tremendous potential. VCs carefully screen firms, structure contracts to strengthen incentives, and monitor firms. The essential precondition from venture capital firms (VCF) is that a product item ought to have an intellectual property right (IPR) which facilitates them to estimate the IP's potentiality [3]. But the continuous growth of IP infringement in software products is a typical concern among the RCFs to invest their resources in this sector. There are a number of issues which lie in the existing IPR culture which discourage the software industry from filing for the IPR protection. Firstly, intellectual property is protected for a limited period of time, during which the holder of such rights may solely misuse them. Secondly, protecting the rights can be extremely expensive. In addition to these, governments of developing countries often find it difficult to allocate resources to strengthening the IPR regime [4]. In the software industry, copyright, piracy, and trademark counterfeiting are allowed to reproduce the product. These are minor offenses against wealthy multinational concerns that can easily afford the loss. As a result, most of the software firms are not interested in filing for IPR protection [5].

IPR encourages investors to put their resources at risk and thereby enables value creation in the rapidly growing technology-based companies. IPRs encourage the right holder to develop and propel innovation and enable the originator to reap the benefits. It turns out to be exceptionally troublesome for the risk capital finance to make valuation of the IP product without IPR protection. It can only be possible if there is a balanced IPR practice that will encourage software firms to follow to protect their products and secure the risk capital finance. An improved balanced IPR culture among all stakeholders of the software industry will address the information asymmetry, which is critically needed for market expansion and investment mobilization to join the innovation race to maximize the wealth creation for them as well as for the nation [6].

2 Literature Review

Local technology and innovation capability are vital to attain the highest level of success. But it can only be possible when the technology-based company can invest more money in their innovation and R&D sector. VC money is one of the best sources of funds to invest in this sector. Venture capitalists usually invest in a business against equity and exit when there is an initial public offering (IPO) or an acquisition [7]. VCs also provide expertise and mentorship and acts as a litmus test of where the organization is going, evaluating the business from the sustainability and scalability point of view [1]. VCF are only attracted by those companies' products that have a tremendous potentiality and the right IPR protection.

According to ICC's (International Chamber of Commerce) intellectual property policy, intellectual property protection encourages innovation and the development of knowledge-based industries, inspires international trade, and creates a favorable climate for foreign direct investment and technology transfer [3].

Thus, the relationship between these two parties is vital. However, there are three major sectors in which IPR are a big concern in developing countries like Bangladesh.

2.1 Lack of Proper IPR Enforcement

Lack of effective implementation of strong IPR system in Bangladesh gives least priority to settlement of IP matters. Intellectual property rights apply on the Internet but the main issue is to make them enforceable. The TRIPs agreement was implemented in 1993 among WTO nations, but this agreement is hardly followed in developing countries [8].

The technological activities in Bangladesh are insignificant, and a strong backdrop is needed to implement a strong IPR culture. Since there is a solid opinion that if a country has no important innovative movement, then for that nation intellectual property rights are irrelevant and spending money on IPR culture will create an economic burden on the nation. Still the current IP laws in Bangladesh are exceptionally small in number and untimely in shape. As a result, vast areas of IP rights cannot be protected, especially in software products [9].

2.2 Expensive to Protect IP

It has been reported by the daily newspaper, the Telegraph (Tuesday, December 22, 2015), that securing ideas and innovation 'too expensive' and complex for small firms [6]. IPR legal counselors fees, court costs, settlement charges, filing fees and various different costs can mount very quickly, making protection of intellectual property rights expensive for even very large companies. According to Krummenacker, "the only party in this copyright and patent game, which consistently benefits from the current situation, are the Intellectual Property lawyers" [10].

2.3 Threats of Competitors

According to an article published in Forbes, for most of the small firms IPR works like a sword, not like a shield. It does not give them ample right to use technology with peace of mind. There is always a threat of infringement on their protected products from competitors specially form big companies. To protect their innovations requires huge amounts of money for legal fees, and there is a loss of revenue. Inevitably, large firms win over small firms, even if they lose in the court. Particularly in the USA, for a start-up, it is a nightmare to file a law suit against IPR infringement. It takes 3–5 years and needs $3–$5 million to get a judgment from the court. As a result, more and more small firms are not interested in applying for IPR protection for their invention [6].

For rapid technology changes and for uncertainty in the futuristic evaluations have a great blow in the growth of telecommunication sector. The article proposes a real option model which would help delay mobile telecommunication network that invests decisions. It also proposes a method for calculating the real option impacts on mobile service costs [11].

Moreover, body of knowledge in software engineering and project management incorporating additional knowledge area and for IPR we may need to incorporate similar area so that IP of the software can be protected and the company can grow [12–14].

3 Problem Statement

Bangladesh is a strong emerging economy and a culturally enriched nation. Absence and sheer negligence to IPRs culture has created enormous difficulty and potential risk on venture capital investors to make valuation of primary assets of software firms and make investment to take ownership.

In the absence of collateral and legally defendable IPR, this fails to raise financing both from conventional debt financing institutions and risk financing sources. The EEF (Entrepreneur Equity Fund) experience indicates that lack of IPR culture makes it extremely difficult for the software company to turn ideas to commercial products from most of the intellectual property rights of their developed assets or the assets they want to develop.

Entrepreneurs, software firms, and early participation of local angel investors do not scale up to institutional risk capital financing for the software sector. As a result, the society itself lags behind the innovation race to maximize the wealth creation process.

Researchers believe that Strong IPR laws are essential—however aimless without effective enforcement. Strengthening IPRs may enhance economic growth prospects under a couple of conditions yet offer no change or even discourage the development in some conditions. Developing better tools for understanding the current IPR systems and gaining an understanding of its limitations should be a central focus in IPR research. The following questions ought to be explored more which are important to address in this respect:

- How does the current IPR system relate to the social and economic impacts that obstruct to secure the risk capital finance and economic scale up for the innovators?
- How a strong IPR likewise discourages the developing nation's innovation interest?
- What is the right balance between the 'public versus private interests' in the innovation process?

The violation of intellectual property rights are basically ignored in Bangladesh. The historical context of Bangladesh reveals that the IP protection laws have a long history and it goes back to the century of British India. Bangladesh has a few IP Laws which have been acquired from the British period. Nevertheless, one of the weaknesses of these laws is that the Patent Law doesn't particularly cover the area of intellectual property [15].

New development of technology have made new potential types of intellectual property rights and made old rights unenforceable. We have to reexamine in a general sense what ought to and ought not to be applicable for intellectual property. In the meantime, we have to produce new thoughts and advances to offer successful insurance of intellectual property rights.

3.1 Powerless IPR Is an Obstruction to Risk Capital Financing

While conducting a study, a few obstacles have been found due to the week IPR practices:

Absence of a strong IPR culture, potential business estimation of software assets which are being produced by software firms to create future income is hard to evaluate. Without legitimate protection to prevent unauthorized copying, the obvious estimation of software assets for commercialisation appears to be unfeasible. In a situation like this, it is practically impossible for individuals to mobilise risk capital finance, whether from individual, or VCFs [16].

3.2 Weak IPR Is an Obstruction to Economic Scale Up

IPR practices are important for empowering firms to put resources into the innovation. Due to zero cost of copying software products without the protection from IPRs, the risk of covering the investment in developing software products is extremely high. As a result, risk capital finance for developing such products simply becomes impossible to mobilize [16].

Without the presence of adequate IPRs in the local environment, foreign software firms also find it very risky to set up development centres in Bangladesh.

3.3   Extreme IPR Culture Limits Innovation Interest

Extreme measurements of IPRs do not appear to be free from negative implications. In the event that there is no restriction of imitation, risk of investment in innovation is extremely high. As a result, the risk capital availability for new wealth creation through innovation will simply go away.

It is also argued that extreme enforcement of IPR could lead to a monopoly which can provide incentives for the innovators but it can also push the former innovators to gain high return on their existing products rather than pushing them towards further new innovation. This situation eventually weakens the advancement of technological innovation which is the barrier for wealth creation for the individual, firms and the country [16].

Moreover, a strict IPR system also allows the big firms to set up barriers for their rivals to enter the market. In the end, society will lose the chance to profit by rivalry to enhance those thoughts and locate their numerous uses for new wealth creation.

4   Proposed Framework

In light of the above discussion, a balanced IPR culture is vital for a developing nation for their software industries progression. It could bring the following opportunities to the software industries in developing countries like Bangladesh.

Developing countries like Bangladesh should immediately amend the laws on intellectual property to develop a harmonious system of IP rights protection.

Most of the developed nation has formalised their IPR practices in such a way which restrict the other nations to enter their innovation race. They made their rules as an 'international rule' which forces other nations to follow them. In the end, they help the developing nation. The Western nations spent a huge amount in their research and development sector to enable their firms to come up with new ideas and monopolize the innovation market. Bangladesh should adapt the same strategy to grasp the opportunities.

Copyright laws ought to give the fundamental protection on software IPs because a week and unbalanced IPR culture is responsible for poor flow of foreign clients and a low rate of export of software.

Studies have shown that patent ownership in early-stage high-innovative companies positively affects the planning and estimation of venture financing received, and on the probability of pulling in a prominent capital investor.

In Patent Signaling, Entrepreneurial Performance, and Venture Capital Financing research work analyse more than 10,000 VC—financed US start-up companies in various sectors from 1976 through 2005, found that new businesses which effectively file patents before accepting investment from VCs receive substantially more VC funding, which attracts a larger number of more prominent VCs, and are more likely to complete initial public offerings (IPOs). The review recommended that licenses give a vital flagging capacity as to an enterprise's performance [17].

Bangladesh ought to begin an investigation to come up with a fine balanced IPR so that the local industry, mainly software firms, can be benefited from it. The possible strategy should be neither dismissal nor acceptance; it should rather be ideal adjustment to worldwide practices of IPR that is most appropriate to local conditions to make the most out of the wealth creation from innovations.

5 Conclusion

It is assumed that there is a positive correlation between innovation and economic growth. Bangladesh must grasp the excellent opportunity to create a dynamic hybrid, versatile system, leveraging and learning from mistakes and successes of other relevant strong IPR regimes like India and the USA. Bangladesh should focus on the legal aspect of IPRs, as well as on creating the awareness and capacity of growing legitimately protectable intellectual/software assets and preventing their commercial misuse. Such capacity of the ecosystem is basically essential to support the risk capital finance to invest on the creation of software assets for commercial benefits.

The study reveals that the absence of risk capital finance is slowing down the advancement of the software industry of Bangladesh. The company struggles to find the risk capital finance to invest on the R&D sector which prevents them from getting into the innovation process.

To finish up, we would like to cite Joff Wild, who splendidly perceived that "a solid intellectual property position may not ensure start-up software firms will be successful, however it will be discovered that it is a whole lot harder to succeed if it doesn't have one. Furthermore, critically, it is not quite the ownership of intellectual property that is imperative; it is the understanding that intellectual property is a key" [18]. It is understandable that intellectual property plays a significant role in the domain of venture capital transactions.

References

1. Thurow, L.C.: Needed: a new system of intellectual property rights. Harvard Bus. Rev. September–October (1997)
2. Bangladesh's next phase of growth: Build Intellectual Asset Stockpile?, http://opedsarchive.blogspot.com/
3. Dixon, A.N.: Intellectual Property: Powerhouse for Innovation and Economic Growth. IIPTC—Intellectual Property and technology consulting, International Chamber of Commerce, The World Business Organisation, London
4. Biancini, S., Paillacar, R., Auriol, E.: Intellectual Property Rights Protection in Developing Countries (2012)
5. Intellectual Property and Developing Countries: An Overview, Brief paper. USAID, Washington, Dec 2003
6. SME Foundation: Study on Software Development Sector of Bangladesh (2015)
7. Zider, B.: How venture capital works. Harvard Bus. Rev. November–December (1998)
8. The Enforcement of Intellectual Property Rights—ACAse Book, 3rd edn. WIPO (2012)
9. Naznin, S.M.A.: Protecting intellectual property rights in Bangladesh: an overview. Bangladesh Res. Publ. J. **6**(1), 12–21 (2011). ISSN: 1998–2003
10. Krummenacker, M.: Are "Intellectual Property Rights" Justified? (1995)
11. Tahon, M., Verbrugge, S., Willis, P. J., Botham, P., Colle, D., Pickavet, M., & Demeester, P.: Real options in telecom infrastructure projects—A tutorial. IEEE Communications Surveys & Tutorials (Jan, 2014)
12. Morshed, M.M., Hasan, M., & Rokonuzzaman, M.: Software architecture decision-making practices and recommendations. In: Advances in Computer Communication and Computational Sciences, pp. 3–9. Springer, Singapore (2019)
13. Tahsin, S., et al.: Market Analysis as a possible activity of Software Project Management. In: 2018 IEEE 16th international conference on Software Engineering Research, Management and Applications (SERA). IEEE (2018)
14. Raha, L.N., et al.: A guide for building the knowledgebase for software entrepreneurs, firms, and professional students. In: 2018 IEEE 16th international conference on Software Engineering Research, Management and Applications (SERA). IEEE (2018)
15. Rahman, M.M.: Intellectual Property Protection in Bangladesh: An Overview. Department of Patents, Designs and Trademarks, Government of Bangladesh, p. 2
16. Need for balanced IPR practices http://opedsarchive.blogspot.com/
17. Cao, J.X., Hsu, P.: Patent Signaling, Entrepreneurial Performance, and Venture Capital Financing. SSRN's eLibrary, 17 Sept 2010
18. Wild, J.: Why VCs and Start-ups Should Love IP. Intellectual Asset Management Magazine. [Online]. http://www.iammagazine.com/blog/detail.aspx?g=ae1d78db-9c5e-41aa-a4f9-161c7de6c9fb (2010)

ConFuzz—A Concurrency Fuzzer

Nischai Vinesh and M. Sethumadhavan

Abstract Concurrency bugs are as equally vulnerable as the bugs found in the single-threaded programs and these bugs can be exploited using concurrency attacks. Unfortunately, there is not much literature available in detecting various kinds of concurrency issues in a multi-threaded program due to its complexity and uncertainty. In this paper, we aim at detecting concurrency bugs by using directed evolutionary fuzzing with the help of static analysis of the source code. Concurrency bug detection involves two main entities: an input and a particular thread execution order. The evolutionary part of fuzzing will prefer inputs that involve memory access patterns across threads (data flow interleaving) and thread ordering that disturb the data dependence more and direct them to trigger concurrency bugs. This paper suggests the idea of a concurrency fuzzer, which is first of its kind. We use a combination of LLVM, Thread Sanitizer and fuzzing techniques to detect various concurrency issues in an application. The source code of the application is statically analyzed for various paths, from the different thread related function calls to the main function. Every basic block in these paths are assigned a unique ID and a weight based on the distance of the basic block from the thread function calls. These basic blocks are instrumented to print their ID and weight upon execution. The knowledge about the basic blocks in the sliced paths are used to generate new sets of inputs from the old ones, thus covering even more basic blocks in the path and thereby increasing the chances of hitting a concurrency warning. We use Thread Sanitizer present in the LLVM compiler infrastructure to detect the concurrency bug warnings while executing each input. The inputs are directed to discover even new address locations with possible concurrency issues. The system was tested on three simple multi-threaded applications pigz, pbzip2, and pixz. The results show a quicker detection of unique addresses in the application with possible concurrency issues.

The original version of this chapter was revised: Updated with the author group "Nischai Vinesh and M. Sethumadhavan". The correction to this chapter is available at https://doi.org/10.1007/978-981-15-0029-9_66

N. Vinesh (✉) · M. Sethumadhavan
TIFAC-CORE in Cyber Security, Amrita School of Engineering, Amrita Vishwa Vidyapeetham, Coimbatore, India
e-mail: nischai.vinesh@gmail.com

© Springer Nature Singapore Pte Ltd. 2020, corrected publication 2020
A. K. Luhach et al. (eds.), *First International Conference on Sustainable Technologies for Computational Intelligence*, Advances in Intelligent Systems and Computing 1045, https://doi.org/10.1007/978-981-15-0029-9_53

Keywords Concurrency fuzzing · Concurrency bugs · LLVM · Fuzzing · Static analysis · Source code analysis

1 Introduction

The testing phase of any software development life cycle typically deals with checking the different functionalities of the software or the application. It is done by either manually testing the software or automated unit testing. Also, most of the software takes input from outside the system. Even though basic sanity checks are made to make sure proper inputs are passed to the application, an attacker can still craft inputs in such a way that it can crash the system or can be used to exploit the system. The external inputs to a system have always been an important attack vector. This is where the significance of fuzzing comes into play. Fuzzing is a software testing technique, initially developed to find implementation bugs using malformed or semi-malformed inputs given to the software/application in an automated way. It was further developed to find bugs at an earlier stage before they turn into vulnerabilities.

There were many researches on finding and eliminating bugs in single-threaded applications and there were variety of fuzzing techniques suggested for the same. As the bugs in the single-threaded applications can be exploited, bugs in the concurrency applications can also lead to concurrency attacks. With the increase in complexity and size of programs, the increase in multi-threaded programs in the recent past has been enormous. At the same time, there are not many studies on detecting concurrency bugs in these multi-threaded applications.

With the increase in size and the usage of processing power by different applications and the implementation of multithreading for more efficiency, the concurrency issues have become more prominent nowadays. Unlike generic memory corruption bugs, concurrency bugs are hard to find mainly due to non-determinism of thread scheduling. To add to the difficulty, the execution of the (multi-threaded) application on a given input should also involve patterns that may lead to an unintended situation, like use-after-free bug. Hence, the input (which triggers the concurrency bug) and the particular thread execution order are very important in order to detect these kinds of bugs. Though not many researches have been done to detect concurrency bugs, most of those done have a focus on manually controlling the thread scheduling to trigger a specific set of thread interleavings so that they will result in a concurrency issue. But this approach has its own pros and cons. In the real-time application, this particular set of thread interleavings may not occur in the specific order that it was assumed and the interleavings happen in a non-deterministic way. So instead of relying on controlling the thread interleavings, finding inputs that can trigger the same concurrency issues are more significant in solving the concurrency bugs in a multi-threaded application.

Finding such inputs, which takes a particular execution order, is not possible manually and hence fuzzing is an obvious direction to look forward to. The fuzzing of single-threaded applications to find bugs was a huge success and much critical vulnerability were discovered with this technique. But fuzzing a multi-threaded application blindly with random inputs and expecting it to hit a specific address with possible concurrency issue is not a practical solution. In case of concurrency bugs, the fuzzing technique has to be refined even more for very specific conditions. Instead of fuzzing randomly, specific addresses in the application with possible concurrency issues have to be detected first and then modify the inputs in such a way that they hit these initially detected memory addresses. This way of fuzzing is known as directed fuzzing. Source code analysis is the best and easy way of detecting all these significant memory addresses. Analyzing the application at binary level for this will not be feasible for very large applications. Also, source code can give even more information like the different paths from main function to these memory addresses, distances of these paths, etc.

LLVM [1] is a popular compiler infrastructure that provides various front end and back end tools for analysis and optimizations. There are different sanitizer tools, like Address sanitizer and Memory sanitizer in LLVM that performs analysis to find vulnerabilities in the code. Out of all these sanitizers, the Thread Sanitizer [2] looks apt for our purpose. TSAN gives runtime warnings about possible concurrency bugs including data races, deadlocks, thread leaks, etc., for a specific input execution.

Combining these possibilities, we arrived at a research question, whether the Thread Sanitizer in LLVM can be used to detect inputs that can trigger concurrency bugs in an application and LLVM infrastructure to instrument source code of the application to analyze the significant addresses and the input execution. LLVM can also be used to help generate more inputs that execute the paths to the initially detected significant memory addresses and trigger a concurrency bug.

2 Overview

This paper is divided into two parts—the instrumentation phase and the fuzzing phase. The instrumentation phase consists of instrumenting the source code with a custom LLVM IR pass and Thread Sanitizer (TSAN) of LLVM. The source code of the program is compiled into LLVM bitcode format and passed onto the custom pass. The custom LLVM pass inserts instructions at IR level. This modified bitcode file with instrumentations is recompiled along with thread sanitizer instrumentations into an executable. Then the final instrumented executable and a set of seed input files are passed to the fuzzing phase (Fig. 1).

In the fuzzing phase, the instrumented executable is run with the provided seed inputs. The fuzzer gets information about the execution path during each input execution and generates a report based on that. The thread sanitizer also generates runtime warning messages if any concurrency issues are detected. Based on these reports, next generations of the inputs are created.

We propose a fuzzing technique that is both evolutionary and application-aware. The source code analysis and instrumentation part of the system helps fuzzer in

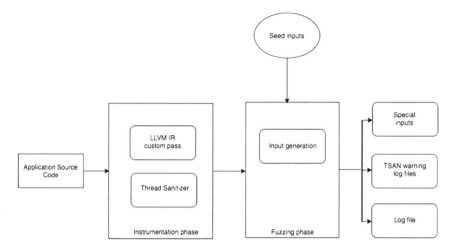

Fig. 1 Block diagram of overall system

understanding the information regarding the execution paths that lead to concurrency issues. It records all thread related function calls in the program and does a backward slicing to record the various paths leading to the call sites. The instrumentation pass takes care of loops and duplicate paths. The instrumentation also gives out important information to improve the next generation of inputs at runtime. The fuzzing component uses this information from the instrumentation phase to detect concurrency bugs at runtime for each input execution and saves the inputs that trigger it. The Fuzzer also generates specific number of new generation inputs using this information and passes it to the instrumented executable. This is repeated for a specific number of times which is configurable in the Fuzzer's config file. The Fuzzer makes sure it reports only unique bugs.

3   Concurrency and Multithreading

Concurrency is the ability of a program to run more than one task at a time. This allows for parallel execution of different tasks out of order, and the result would still be the same as that of those executed in order. The concept of concurrency and parallel processing in computing have become pervasive and critical because of the enormous usage of multicore processors and deployment of large scale distributed systems. Concurrent programs are written using threads and multiple threads are defined to carry out various tasks in parallel. Writing, testing and debugging the multi-threaded programs have continued to remain difficult since a long time and this impediment has led to many vulnerabilities and bugs in the concurrent programs. Addressing these challenges require advancements in many directions like concurrent program testing, concurrent program model design, concurrent bug detection, etc. Since concurrent programs exhibit more non-deterministic behavior, so do the bugs found in the concurrent programs and non-deterministic bugs are much more challenging when compared to normal bugs. Writing a bug-free concurrent program requires the developers to keep track of all the patterns in the interleavings between different threads that carry out concurrent tasks and the concurrently overlapping executions of different execution threads that use the shared memory. Creating test cases for all these scenarios and debugging a multi-threaded program execution makes it a notoriously difficult task.

3.1 Concurrency Bugs and Attacks

Testing is a common practice to discover bugs in a software but the existing testing techniques mainly focus on the sequential aspects of a program such as statement, branches, etc. and cannot address concurrency aspects like multi-thread interleavings within a concurrent program [3]. The exponential interleaving space of concurrent programs makes it even more challenging and exposing these concurrency bugs require a bug-exposing input along with the bug-triggering execution interleaving. Just as the errors in a sequential program can be exploited, concurrency errors can also lead to a security exploit and thereby the possibility of using it to carry out a concurrency attack, which can violate confidentiality, integrity, and availability of the system. Deadlocks in a concurrent program are a major issue, but most of the deadlocks can be avoided by the use of appropriate locks in the right places. Deadlock occurs when two or more operations circularly wait for each other to release the acquired resource and Datarace occurs when two conflicting accesses to one shared variable are executed without proper synchronization. Almost all the non-deadlock concurrency bugs can be classified into three main categories Atomicity violation, Order violation, and others. A study on Concurrency Bugs [4] describes that out of these three categories, atomicity and order violation constitutes the majority of the existing concurrency issues. Atomicity violations are caused by concurrent execution unexpectedly violating the atomicity of a certain code region. The order violations are caused when the program is executed in an order that does not follow the programmers intended order. The programmers atomicity and order assumption while writing the program could lead to most of these violations as the multi-threaded applications' thread interleavings are completely non-deterministic and it is very difficult to capture all these combinations. It is very common among the programmers to assume an order but enforcing that order is difficult, sometimes they may even forget to do it where it was possible. Most of these concurrency issues manifest due to the involvement of very small number of threads, in most cases there are just two threads.

The main reason behind this finding is that even though there are hundreds of threads used in a program, most threads do not closely interact with each other and most of the thread interactions happen between two or very small number of threads. Hence the vector space of concurrency issues can be reduced to a large extent by considering those threads that interact or collaborate. It is very much evident from the CVE database [5] that the concurrency bugs can be used to carry out a viable concurrency attack. A recent study [6] discusses and shows how the existing concurrency errors can be exploited to carry out an attack. To carry out such attacks, understanding and reproducing the concurrency errors produces a challenge due to its non-deterministic nature. The ability to exploit a concurrency bug depends on the vulnerability window. The Vulnerability Window is the size of the timing window within which the error may occur. The vulnerability window can be measured in terms of human time, disk access time and memory access time. An attacker using carefully crafted inputs can extend this vulnerability window to effectively carry out the concurrency attack. In some cases, these can be done using a sequence of UI events and in some other, the attacker can re-run a command few times, possible using a shell script. Exploiting the errors during the memory access time is more difficult compared to the other two as the offending events should occur within small timing windows along with the hardware cache leases or CPU time slices being larger than the timing windows masks the errors.

The current concurrency bug detectors are not mature enough as that of sequential tools and most of the bug detectors and the defense techniques are created to serve the sequential programs. These commonly used sequential defenses become unsafe when applied to concurrent programs. Through ConFuzz, we are trying for the first time to build a fuzzer that can detect concurrency bugs.

3.2 Fuzzing

Fuzzing is an automated way of software testing that involves providing invalid, unexpected or random inputs to the program and trying to discover various issues with the combinations of code and data along with the knowledge of protocols and heuristics. It is used to detect exploitable bugs in the code and check for corner cases by providing random inputs to check for crashes. The most significant part of a fuzzer is the ability to generate inputs that can trigger a bug or crash the application. The first form of fuzzing was carried out by Barton Miller from University of Wisconsin, Madison [7] by supplying random inputs to Unix utilities and then observing for crashes. This resulted in the discovery of various exploitable vulnerabilities. Since then the complexity of code has increased over the years and a set of random inputs may not be enough to cover the application well enough to reach the deeply buried bugs. Even though there have been lots of improvements in the field of testing, we need more heuristics about the application for a fuzzer to be efficient.

A Fuzzing system mainly has three components—a generator, a delivery mechanism, and a monitor. These components carry out the three major tasks in the process of fuzzing. The generator is responsible for generating inputs, random, malformed or crafted, which have to be passed on to the system under test. There are various methods to generate inputs depending on the bugs that the fuzzer is trying to detect. The passing of inputs to the system is handled by the delivery mechanism. The delivery mechanisms may vary depending on the target applications way of receiving the inputs. Once the application to be tested is run with the generated inputs, the monitor will identify errors, bugs, and crashes. It also keeps track of the inputs that triggered the bugs, type of inputs that can trigger the errors, code coverage and other general statistics required for analysis and improvements.

Code coverage is one of the most important factors among the metrics that can be used to evaluate a fuzzer. It specifies the amount of code within the application that is tested. The higher the code coverage, the better the chances of finding vulnerabilities. Input space is another factor to measure the capability of a fuzzer. It consists of various combinations of inputs that the application accepts or that can be passed to the testing application. The possibilities of such an input space is large and hence different heuristics are used to constraint the space.

Many attack heuristics like buffer overflows, format string attacks, parse errors, etc. can be used to find bugs easily. Since the various inputs within the generated input space can be similar to each other, there is always a chance of the application taking the same execution path that leads to a crash. Hence only unique crashes need to be logged and the inputs that trigger duplicate crashes can be discarded. The number of unique crashes found by a fuzzer can also be used as a factor to evaluate fuzzer.

4   Design

The design of the concurrency fuzzer is divided into two parts, instrumenting the application for coverage calculation and concurrency detection, and generating inputs and fuzzing the application. The first part of the system is instrumenting the application. There are two types of instrumentations needed for the concurrency fuzzer. First, the source code of application has to be converted into bitcode format so that LLVM can apply the custom pass instrumentations at the IR level. Hence, all the files are to be converted and combined into one single bitcode file using LLVM tools. The custom LLVM pass, at the IR level, will find all paths from significant thread related function call sites to the main function using backward slicing and instrument each basic block within these paths. The thread related functions, which need to be considered, could be provided through a config file. A rank is calculated for each input based on the execution path that it takes. The application code also needs to be instrumented with LLVM Thread Sanitizer for detecting concurrency issues at runtime. We use TSAN warnings to detect and analyze concurrency issues at runtime for each input. After both the instrumentations are complete, the final bitcode file with instrumentations is to be converted into an executable. The fuzzing part of the system does the major part in detection of possible concurrency issues. It runs the instrumented application and reads the information output from the custom pass instrumented codes. This information is used in ranking each input. The parents for the next generation of inputs are considered based on this rank. The fuzzer also reads and analyzes the warning messages thrown by the TSAN during each run.

The input generation, a part of the fuzzing component, is mutation-based that requires minimum number of seed inputs to start with. The seed inputs are initially run and ranked to consider for creating the next generation of inputs. We use a similar strategy as that of VUzzer [8] for randomly selecting bytes. Then the seeds are passed to crossover function to generate new inputs and mutate them. The fuzzer monitors each input execution for detection of concurrency issues and discovery of new basic blocks within the analyzed paths to thread function call sites. These information are used to optimize the generation of next set of inputs.

4.1   LLVM Custom Pass

The custom LLVM pass is a transformation module pass, which inserts a function call at the specified target within a basic block. A separate C file with a custom print function is compiled, converted to bitcode file and then combined with the application bitcode file before any instrumentation is done. The print function accepts two parameters and prints them into a file, which is later used to calculate the rank of the input. The first parameter in the print function is an integer value, which will be the unique ID assigned to the particular basic block, and the second is a float value, which is the calculated weight of the basic block. The custom pass loops through each instructions and looks for critical sections on the code, which are the specific thread function calls (mutex lock/unlock, pthread lock/unlock, etc.) mentioned in the ConFuzz's config file, and creates a multi-linked-list of call traces from the function call sites to the main function using backward slicing. Once all the different paths are found, the basic blocks from the overlapping or repeated paths are avoided and only the unique basic blocks are kept. Each unique basic block is assigned with an ID and a Weight is calculated for it. The Weight of a basic block is based on how far it is from the thread function call site. The farther the basic block, the lesser it's weight. Whenever an instrumented basic block is executed, the custom print function is called and the respective ID and Weight are printed in file in the disk. After every input execution, this file is read by the fuzzer to calculate the rank by summing up the weights of basic blocks executed. The fuzzer also detects whether any new basic block is executed from the previous executions and keeps track of it. A high-level logic is explained in Fig. 2.

The custom pass takes care of loops of basic blocks within a function and loops of functions within the module. Also, the pass is designed to print only unique basic blocks to avoid overhead. The pass also skips TSAN related functions and function calls as they may trigger false datarace warnings.

Initialize TARGETS (Set of critical section/thread function call sites);
Initialize PATH, A multi linked list where all the paths from critical sections to main
  function are stored;
Initialize FUNCTION_NAMES, a set of thread function names specified in the config
  file;
while *within the Module - Loop though Functions in the module* **do**
    while *within a function - Loop through Basic blocks of the function* **do**
        while *within a Basic Block(BB) Loop through instructions of the BB* **do**
            if *instruction EQUALS function call of fn, fn ∈ FUNCTION_NAMES AND
            current BB not in TARGETS* **then**
                Add the BB to TARGETS;
                Add BB as the head node of a new list in PATHS;
                GetPredecessors(BB) and add them to PATHS under the list;
                while *While current function name NOT EQUALS main* **do**
                    Find the parent function(s) and the call site of the current function;
                    Repeat the steps (10-13) for the parent function(s);
                end
            end
        end
    end
end
while *Loop through the list of recorded BBs in the PATH* **do**
    Calculate and Assign weight to BB;
    Instrument BB insert instruction to call the custom print function;
end

Fig. 2 Algorithm of static analysis LLVM pass

4.2 LLVM Thread Sanitizer

As mentioned earlier, Thread Sanitizer is a LLVM tool to detect possible dataraces during runtime. When application code is instrumented with TSAN, it adds many TSAN related function calls into the source code to check for possible concurrency issues upon the execution of the application. During the execution of the instrumented code, TSAN raises warning messages when it detects a possible datarace or any other type of concurrency bugs. TSAN has the capability to detect many concurrency related issues including dataraces, thread leaks, deadlocks and many more by detecting shared memory accesses by two or more threads. It records information about each memory accesses and checks whether the memory access participates in a race. Each thread is instrumented to store its own timestamp and the timestamps for other threads for synchronization purpose. Timestamps are incremented every time a shared memory is accessed. By checking for the consistency of memory accesses across threads, data races can be detected independent of the actual timing of the access. Therefore, the Thread Sanitizer can detect races even if they didn't manifest during a particular run [9].

    The concurrency fuzzer makes use of the TSAN warning messages to detect if the input generated can trigger a datarace or not. It also gives more priority for such inputs in the process of creating next-generation of inputs. There is a known slowdown of $5\times-15\times$ and a memory overhead of $5\times-10\times$ when running any application with TSAN instrumentation.

4.3 Input Generation

Fuzzing a complex application and finding bugs, which can be found in the deeper execution paths, is extremely difficult. It is even more difficult when it comes to detecting concurrency bugs. To find such bugs, prior information of the program structure is required and the inputs have to be crafted in such a way that their execution will reach the specific locations or follow specific input execution paths. There are different ways of achieving this. Fuzzers like AFL keep track of basic blocks executed and looks for inputs that explore new basic blocks with the help of a feedback loop. It then mutates those bytes in the input that do not help in detecting new basic block for better coverage. Symbolic/concolic execution is more efficient in finding those offsets in the input that are to be mutated and with what value, but the overhead is too high when it comes to large application. In case of fuzzing concurrency applications, the shared memory address along with the execution path need to be known and the inputs have to be directed towards it to have a chance of triggering the bug. Practically, this is extremely difficult without a proper static analysis of the application's source code.

We propose a fuzzer that is designed to be application-aware. It uses an evolutionary fuzzing strategy based on static analysis of the source code. The strategy is mutation-based which makes use of valid seed inputs provided to mutate and generate a new set of inputs. It makes use of information about the memory locations in the application that is shared by more than one thread and the different execution paths towards them. This is done by looking for *pthread* related function calls in the application and marking the basic blocks that contain the *pthread* related function calls as important targets to fuzz. The focus is also given to inputs that explore new paths within the previously gained set of execution paths.

The evolutionary fuzzing strategy starts with valid seed inputs that are mutated or crossover multiple times. The magic bytes of the seed inputs are maintained as such using the ConFuzzs config file so that the execution of input with the application does not crash due to invalid format. The crossover technique selects two-parent inputs and produces two children, while the mutation technique transforms one parent into a child. A combination of these two techniques, along with single or double crossover, is used to improvise the input generation. Each input that is generated is passed to the application to execute and are ranked based on its coverage of the recorded execution paths towards the shared memory locations. These ranks are used when considering parent inputs for the next generation of inputs. This process of execution, ranking, input generation, and monitoring continues until a concurrency issue is found or till a specified number of generations is over. This strategy is very similar to that of

VUzzer [8], which also weighs basic blocks and use that information to better the input generation.

5   Implementation

The implementation of the concurrency fuzzer and its different components are explained in this section. We will also discuss the approaches that we considered in implementing the fuzzer.

5.1   Instrumentation—Source Code Analysis

The implementation of concurrency fuzzer is divided into 2 parts. The first part is to analyze the source code of the application to be fuzzed. Depending on each application's build methods and the structure, all files of the application are converted into bitcode format using clang [10] emit-llvm and combine all bitcode files into one single file using llvm-link. This final bitcode file will also contain the defined print function that needs to be instrumented into the source code of the application. This way of conversion is applicable only to those applications that have a final single executable generated once it is built. In case of applications that generate multiple executables or use other ways like shared objects, we have to tweak the process to suit it or make use of the flto and llvm-gold plugin in the build scripts of the applications to generate bitcode files automatically. This method will generate single/multiple (according to the structure of the program) pre-codegen bitcode files before it creates the final executable(s). Hence, we just have to use these pre-codegen bitcode files and combine them together to make it a single final bitcode file to be instrumented.

5.2   LLVM IR Pass

Once the final bitcode file is ready, this file needs to be analyzed. Since the bitcode file is in LLVM IR level, it is easy to analyze with the help of llvm defined functions and properties. We wrote a custom llvm transformation module pass to analyze the generated final bitcode file.

The pass will loop through all the functions and basic blocks of the module and will look for specific function calls in the instructions list. This function calls can be configured using a PassConfig.txt text file. The names of the functions can be mentioned in this text file with a newline. If the file is empty or there is no file found with this name, the pass will consider the default values mentioned in the code. In our case, the function names will be thread related calls like pthread create(), pthread join(), pthread mutext lock(), pthread mutex unlock(), etc. Once a particular function call site is found, the pass will do a backward slice of the program from the function call site to the main function of the program. It records different paths separately in a connected linked-list. Since the paths are dynamically added during the analysis time, linked-list is preferred as each node of it can store the details of basic block like the address, the parent function and unique id assigned to every unique basic block. The pass looks out for loops at the instruction level, the basic block level, and the function level and bypasses it when recording the sliced path. Once all the sliced paths are found for every function mentioned in the PassConfig.txt, the unique basic blocks in these paths are instrumented with a function call instruction to the PrintID function. These basic blocks are also assigned with a weight depending on its distance from the thread function call. The nearest basic block will have a weight of one, the next nearest will have 1/2, the next will have 1/3 and so on. These weights are calculated in decimal values. The PrintID function takes two arguments, an integer value, and a float value. The first integer value is the basic block ID, which is unique for each basic block in the recorded sliced paths. The second argument is the weight of the particular basic block. The PrintID will print these two values with a space in between into a file, WeightID.txt, in the current path at runtime, whenever the instrumented basic block is executed.

5.3 Fuzzing Technique

The second part of the implementation is the fuzzer. We used Python 2.7 to implement the fuzzer. The fuzzer depends on two files, a config file, and the weightID.txt, to fuzz the application. The values of the configurable variables are taken from the config files by the fuzzer. The fuzzing technique can be divided into different steps as follows:

Configuration: The config file needs to be updated with proper values for the variables defined in it, before the fuzzing process starts. The config file contains the variables for the seed inputs folder path, generated inputs folder path, fuzzer log path, folder path for TSAN warnings, number of generations to be created, total population size to be maintained, percentage of best parents to be considered for creation of each subsequent generation. The magic bytes that need to be skipped while randomizing the input bytes can also be configured in the config file depending on the application and the input file format the application supports.

Initialization: The fuzzer starts with verifying whether the configuration variables are set properly. It also creates separate folders for keeping the inputs that explore new paths and for those that triggers TSAN warning(s). The command-line arguments, if any, for running the application needs to be defined in the fuzzer script itself.

Fuzzing: The fuzzer runs the instrumented application executable passed to it as runtime argument with the input files in the seed input folder. This dry run is to check if the application binary is working fine or not. Once the application is run with all the seed inputs, the fuzzer ranks each input using the weightID.txt file. The rank of an input file is calculated as the total sum of the weights of the unique basic blocks executed. This information is gathered from the weightID.txt, as the weights of all basic blocks executed will be printed into this file during runtime. It also records how many unique basic blocks from within the sliced path are executed by reading the IDs of the executed basic blocks from the weightID.txt file. Once all the seed inputs are successfully executed with the application and the rankings of the inputs done, the fuzzer starts creating the next generation of inputs from the seeds based on their ranks. After the next generation is over, the fuzzer repeats the execution process until the number of generations mentioned in the config file.

Check for TSAN warnings: Upon every execution of the instrumented application binary, the fuzzer checks for any newly generated TSAN warning log files in the TSAN log folder mentioned in the config file. Whenever there is a concurrency issue, say a datarace, the thread sanitizer instrumented binary will generate a log file at the specified path with the details of the concurrency bug. Hence the fuzzer checks for such log files after the execution of each input file. The fuzzer reports the concurrency bug to the console only if it is a unique one. There is a chance of thread sanitizer warning about the same concurrency issue upon execution of application binary with different inputs. Hence the fuzzer has to check for unique TSAN warnings by reading the TSAN log files for the addresses of significance and record the same for comparisons in future.

Input Generation: The fuzzer uses a mutation-based evolutionary way of input generation. The approach is similar to the VUzzer's [8] logic of input generation. A set of parents is chosen based on elitism approach from the pool of inputs to create the next generation of inputs. The parents are chosen based on their ranks, calculated after the execution. There are different mutation and crossover operations defined and the parents undergo mutation and crossover operations randomly chosen by the mutation and crossover functions for creating new child input(s). The number of magic bytes defined in the config file is maintained as such for every parent inputs to avoid crash of the application due to invalid format. If an input triggers a TSAN warning or explores a new basic block from the sliced paths, the input will be copied to special folders and will be considered for each of the subsequent generations of inputs.

Statistics collection: The fuzzer maintains a log file for each fuzzing run. It logs information about the duration of fuzzing, number of files created for each generation and time taken for creating them, time of execution of inputs, number of TSAN warnings generated and the possible addresses which are vulnerable to concurrency bug and time taken to trigger these warnings. The fuzzer also records the number of basic blocks executed from the sliced path and logs if any input explores a new basic block out of this recorded list. The special folders contain those inputs that triggered the TSAN warnings and explored new basic blocks, which can be used to reproduce the concurrency issue and produce a TSAN warning.

6   Evaluation and Results

The evaluation of the concurrency fuzzing system done on three different applications is discussed below. The applications are as follows:

1. pbzip2 [11] parallel bzip
2. pigz [12] parallel gzip
3. pixz [13] parallel xz.

The evaluation was done by fuzzing these applications for a long time and for large number of generations and then observing the logs. The selection of applications had to be limited to open source (so that we could do the source code analysis), comparatively small (so that it is easy to instrument) and multi-threaded (which uses thread related functions) which takes a file as input. The application with known concurrency issues like datarace would be even better for us to verify and discover more bugs at the same time. The input file size also had to be kept as small as possible to keep the fuzzing process effective, at the same time maintaining the minimum input size to trigger the parallel processing.

6.1   pbzip2-1.1.13

bzip/bzip2 [14] is an extensively used open-source compressing utility, which has many advantages over other compressing applications. pbzip2 is the parallel version of bzip2 that uses pthreads to parallelize the compressing process. It takes file(s) as input and outputs compressed files compatible with bzip2. It uses thread related function calls like pthread create, pthread mutex lock, pthread mutex unlock, pthread cond wait and many more. This makes it a right candidate for evaluating the ConFuzz as it satisfies the conditions.

Experimentation

The application source code consists of mainly three files and it uses the bzlib library to do the compression. Since bzip does not use threads, we need not have to instrument bzlib library. So the three main source code files are converted to bitcode files, combined along with the bitcode file of printID into a single final bitcode file. All these files were instrumented with thread sanitizer also. The final bitcode file is passed to the custom IR pass for instrumentation. The instrumented final bitcode file is converted into an executable.

We considered seven files of varied sizes, from 2 to 500 MB, as seed inputs, as it was noticed that the dataraces were not triggered with smaller sized inputs. Hence we wanted to use a combination of files with different sizes and different file types like pdf, mp4, and text les. We have set the number of generations flag in ConFuzz to 1000 and the population size flag per generation to 100. The other variables in the ConFuzz's config file for different folder paths were also set like log path for Thread sanitizer warnings, input folder path, folder path for inputs that trigger concurrency issues and for inputs that explore new basic blocks during fuzzing. We have also passed the following parameters to the binary so that it is optimized for multithreading:

- b5: block size in 500k steps
- r: read complete file into RAM and split between the processors
- m500: Max memory usage of 500 mb

The output of each execution, which is a. bz2 compressed file, is removed before the next execution, to save the disk space, as we are dealing with large-sized files for fuzzing.

Results

The ConFuzz took more than 12 h to run 1000 generations. The tsan log files in the log folder gives the details of all the concurrency issues found per input execution and the ConFuzz log file gives information about the number of unique concurrency issues and the time taken to find it. The total summary of the run is also provided at the end of execution in the log file. There were 55 basic blocks instrumented in the various sliced paths from different thread function call sites to the main function and a total of 2197 basic blocks instrumented outside these sliced paths. Upon fuzzing, 56% of the basic blocks were executed from the sliced paths and around 23% of basic blocks outside the sliced paths were executed on an average. ConFuzz detected 15 unique dataraces during the run.

We compared the results of ConFuzz with our own implementation of a dumb fuzzer, which does not have any intelligence and just randomly changes the inputs and fuzzes the given application. Even though the source code of this dumb fuzzer is derived from ConFuzz, the logic of directed fuzzing by considering the executed basic blocks from the sliced path and the rank calculation is removed and all the basic blocks are instrumented evenly. The same inputs used in ConFuzz are also used for dumb fuzzer with similar configurations. It was noticed that the dumb fuzzer also found 13 unique dataraces in 6 min from the start of fuzzing, ConFuzz found 15 dataraces in 2 min and 10 s, which proves the efficiency of ConFuzz over the dumb fuzzer. We have also noticed that the maximum number of basic blocks from the sliced path is covered during the first few generations and they tend to remain constant for the rest of the run.

6.2 pigz-2.3.4

gzip is another popular data compression program with a superior compression ratio. pigz is the parallel implementation of gzip that makes use of multiple processors and multiple cores when compressing data. pigz also takes file(s) as input and outputs a .gz compressed file and uses pthreads for multithreading, thus satisfying the conditions of a right candidate to evaluate ConFuzz.

Experimentation

Pigz has twelve files that have to be converted into bitcode files and combine them all together with bitcode file of printID to form a single bitcode file. This final bitcode file is passed to our custom IR pass for instrumentation and the output bitcode file is converted to an executable. There were 466 basic blocks instrumented within the various sliced paths from the pthread function call sites to the main function and 2731 basic blocks outside the sliced paths.

In this case also, we have considered seed inputs of comparatively large size, but didn't require as large as the inputs we used for pbzip2. We started with 5 inputs of sizes from 1 mb and varied up to 5 mb consisting of image file, pdf, text file, and an avi file. We have kept the ConFuzz configuration constant at 1000 generation with 100 population size. We did not pass any parameters while executing the binary, as pigz will make use of the number of CPUs available by default.

Results

ConFuzz ran for more than 12 h. The ConFuzz log file reports a coverage of 21% of the instrumented basic blocks within the sliced path and around 9% of basic blocks outside the sliced path. ConFuzz could detect 44 unique dataraces and generated more than 5000 TSAN warning files.

When we compared the results of ConFuzz with our own dumb fuzzer using the same inputs, it was found that the dumb fuzzer also could detect 42 unique dataraces. The number of basic blocks covered by the dumb fuzzer is also similar to the ConFuzz. But the difference between the ConFuzz and the dumb fuzzer can be found in the time taken to find these dataraces. ConFuzz's efficiency was in finding the first datarace on the 19th second from the start of the fuzzing and the second datarace on the 20th second, while the dumb fuzzer could detect the first datarace only on the 43rd second. The ConFuzz log shows more information about the time taken to detect the concurrency issues by ConFuzz as well as the dumb fuzzer. Since the logic of the dumb fuzzer is similar to ConFuzz, except for considering the ranking of the inputs used for execution, we assume that this can be the reason for the similarities in the number of concurrency issues detected and the basic blocks coverage over the time.

6.3 pixz-1.0.6

pixz is the parallel, indexing version of xz, which is another lossless compression program. pixz automatically indexes tarballs during compression and uses all the available CPU cores by default. Like the above candidates, the basic operation of pixz also takes an input file and outputs a compressed file in .xz format. It also uses pthread for multithreading, hence making it a right candidate for evaluating ConFuzz.

Experimentation

Since the source code of pixz is a little more complicated than the previous two utilities, we have to pass the parameters to the configure script and use the—flto

Table 1 Result summary

Application	Fuzzed time	ConFuzz	DumbFuzz
pbzip2	13 h	15 in 2.10 min	13 in 6 min
pigz	12+ hours	44 (1st in 20 s)	42 (1st in 43 s)
pixz	12+ hours	0	0

option along with llvm-gold plugin. We use 'plugin-opt save-temps' to create bitcode of various module partitions. The generated bitcode files are used, after the 'make' command is called, to combine it with the bitcode file of printID to form the final single bitcode file. This final bitcode file is then passed to the custom IR pass for instrumentation and the output is converted to an executable. In the source code, 466 basic blocks were instrumented within the sliced path between different *pthread* function call sites to the main function and 2731 basic blocks outside the sliced paths. We have considered seed inputs of various sizes from 200 kb to 500 mb and same ConFuzz configurations as the previous experiments of 1000 generations with 100 population size. pixz uses the available CPU cores by default and hence it does not have to pass as parameter. We can set the number of blocks to be allocated for the compression queue by setting the −q size parameter and change the size of each compression block by setting the parameter '−f fraction' [15]. We have set the default values (−q 1.3*cores + 2, rounded up and −f 2.0)[1].

Results

We fuzzed pixz with ConFuzz for more than 12 h and the ConFuzz log reported coverage of 41% of basic blocks from the sliced paths and 21.7% of basic blocks from the non-sliced paths. But, unfortunately, ConFuzz could not trigger any TSAN warnings for concurrency issues. We compared the results with our own-implemented dumb fuzzer and as expected, dumb fuzzer also did not trigger any TSAN warnings.

Table 1 summarizes the evaluation results of the ConFuzz fuzzing on the three applications. It showcases the efficiency of ConFuzz on DumbFuzz with respect to the time taken to detect concurrency issues in each application. The overall results also point to the fact that after some time of fuzzing, it stops detecting unique concurrency issues.

7 Discussion

We considered many multi-threaded applications including libav [16], vlc player [17], ffmpeg [18], and ImageMagick [19] and all these applications including the ones we considered for evaluation, threw TSAN warnings for concurrency issues even upon execution with thread sanitizer instrumented. Hence fuzzing was not needed for these to find concurrency issues. That is when we decided to fuzz these applications to find even deeper concurrency issues, which would not be detected in a simple run. But due to the design constraints of ConFuzz, many of these applications could not be instrumented with our custom IR pass, only the above mentioned three applications could be done. It is easier to generate bitcode files of the applications for instrumentation, which do not have any dependencies and have a straightforward way of building the source code. These methods vary from application to application.

One of the challenges we faced while searching for an appropriate application for the evaluation of ConFuzz was that almost all the applications already had concurrency issues and the thread sanitizer threw TSAN warnings due to concurrency issues. So there was no requirement to fuzz the application to find out a new concurrency issue.

The major challenge we encountered in finding a suitable application was to convert all the source code files into respective bitcode formats and then combine all the bitcode files into a single bitcode file, which can then be passed to the custom IR pass for instrumentation. Every application has its own method of building the source code and creating the executable. Also the applications will have dependencies on other libraries, statically and dynamically, which have to be considered while generating bitcode files of the source code. Again, it should also be verified that all the bitcode files generated from the source code files have to be combined into one single bitcode file and then be able to generate the executable from that single bitcode file by linking the respective binaries. We have to consider only those applications which have one single executable generated at the end of building the source code. For example, ImageMagick version 6 generated multiple utility executables and each executable has some common and some specific object files to be combined while generating the executable. This was really difficult to figure out from the source code Makefile and such applications are not suitable for the current design of ConFuzz. Similarly, ImageMagick links with some object files dynamically, which also is a scenario that cannot be accommodated with ConFuzz's current design.

Another roadblock we faced during evaluating the ConFuzz was finding a fuzzer with which we could compare our results. Initially, we considered AFL-fuzz to compare the crashes it could find against the unique concurrency issues ConFuzz could find. But there were many factors, which made that infeasible. First of all AFL-fuzz is more of a general fuzzer and ConFuzz is a specific purpose fuzzer. The crashes that AFL reports may not be related to concurrency issues and requires more analysis of the detailed report to figure out. Also, AFL can throw multiple TSAN warnings from the same address and report as unique crashes, which complicates the comparison. As of now, as per the AFL team, AFL does not support thread sanitizer, even though it supports address sanitizer (ASAN) and memory sanitizer (MSAN). But the AFL source code can be tweaked a little bit to add support for thread sanitizer (TSAN) as well. Even though we made these changes, it was not reliable, as AFL does not report TSAN warnings all the time. Also, AFL is at its best when the seed input size is less than 1 KB and it does not run if we add seed inputs of bigger size. This can also be tweaked in the AFL's config file to support bigger sized inputs, but the AFL's efficiency could be compromised. Sometimes AFL does not run with bigger seed inputs and simply crashes. Hence we decided to simulate the basic fuzzing technique of AFL in our own implementation of the dumb fuzzer and compare the results with those of the ConFuzz.

8 Related Works

We reviewed some of the existing works done and its advantages and disadvantages. We will discuss these and how our design and implementation differ from them.

8.1 Concurrency Bug Detection

With the increase in demand for speed and efficiency of the computing systems in this fast-growing world, parallel programming and multithreading have become prevalent. With these arise the problem of concurrency (shared memory accesses without proper synchronization among threads) and such programs can lead to very severe consequences like memory corruption, wrong outputs and program crashes [4]. Worse, a prior study [6] showed that many real-world concurrency bugs can lead to concurrency attacks: once a concurrency bug is triggered, attackers can leverage the memory corrupted by this bug to construct various violations, including privilege escalations [20, 21], malicious code injections [22], and bypassing security authentications [23–25].

The detection of these bugs in concurrency programs is very difficult as they are non-deterministic in nature. There were many approaches in the recent past towards this direction of detecting concurrency bugs [26–35], their diagnosis [36–40] and correction [41–44]. Some of these approaches are targeted towards a specific type of application like .Net based [27] or JAVA based [45] and approaches like that of Eraser [26], RacerX [28], RaceMob [34] and even Helgrind [46], the race detector of Valgrind, uses lockset algorithm to record locks and analyze the programs based on these locksets. ConMem [32] targets only those bugs which crashes the program. Unlike these, there are other approaches that make use of hardware to detect concurrency bugs [47, 48]. Most of the other approaches tackle the problem of non-determinism of concurrency bugs by changing the thread schedulers (Eg. NeedlePoint [35] and Concuerror [49]). Since the concurrency bugs are caused due to inappropriate thread interleavings, manually controlling the schedulers was obviously a solution to reproduce them. The backward approach of ConSeq [31] in the bugs life cycle is similar to that of ConFuzzs except that ConSeq deals with binaries and not source code. ConSeq finds the error sites, like an assert call, within the binary and tries to connect critical memory read instructions to these error sites. But in the end, ConSeq detects alternative interleavings from a correct run of the program and then tries out these suspicious interleavings by perturbing the programs re-execution. But these approaches ignore the possibility of an external input triggering a concurrency bug.

8.2   Fuzzing

Fuzzing has been used to test softwares from 1990 [50] and since then there were many methods discovered to fuzz applications. Even though it started as a vulnerability-testing tool, it has become a common method for testing all kinds of bugs in various applications. It started with randomly generating inputs and mutating valid inputs to create new ones. This is known as Traditional fuzzing. Traditional fuzzing is very simple and easy to implement but may generate more number of invalid inputs than the number of inputs of interest and the huge set of inputs generated may trigger the same bug. Blackbox fuzzers like Spike [51] and FileFuzz [52] uses such methods and is good only to find shallow bugs and have less probability of penetrating deep into the application. Whitebox fuzzers like BuzzFuzz [53] tries to overcome the drawbacks of blackbox fuzzers by accessing the internal working and design of the applications. Methods like dynamic taint analysis at the source code level are used for this purpose. Greybox fuzzing is another classification of fuzzers that uses different dynamic and static analysis with no access to source code to understand the working and design of the application to be tested. Fuzzers like Mayhem [54] employs techniques like symbolic/concolic execution to analyze the application. These techniques help in generating valid inputs that can penetrate deeper into the applications' control flow to trigger bugs. Apart from these general classifications of fuzzers, there is another way of fuzzing application known as Directed Fuzzing. Dowser [55] uses this approach with the help of static analysis to detect the regions with probable buffer overflows and the inputs are tainted and traced to find regions that process the inputs, which are considered as probable regions that can lead to a crash. AFL-Fuzz [56] is also instrumentation guided fuzzer that uses the compiler to instrument the code being fuzzed and mutate the inputs to explore all the code paths. It makes use of a feedback loop to retain inputs that contribute to a new execution path and mutate other inputs for better coverage. Symbolic execution is used to get information on what offsets to mutate and with what value [57]. But it does not scale well with large projects. VUzzer [8] overcomes these pitfalls by using the control flow features to prioritize interesting execution paths and the dataflow features to decide which part of inputs to be mutated to explore new execution paths. This makes VUzzer fast, scalable and can penetrate deep into the program code to detect bugs.

8.3   Other Techniques

There were enormous researches on defense techniques for sequential programs like taint tracking, anomaly detection and static analysis. But most of these techniques or tools can be weakened or even bypassed [6] with the implementation of multi-threading. There are already many static vulnerability detection techniques [58–63].

Some of them target general programs and some others focus on specific behaviors in specific programs for scalability and accuracy. The vulnerability detection approaches for general programs are not sufficient to cope with concurrency bugs and the concurrency attacks that follow [64]. Similarly, the approaches that focus on specific behaviors in specific programs check web application logics, android applications, cross-checking security APIs, verifying the Linux security module, etc. These approaches are complimentary to ConFuzz and can be further used in improving the efficiency of detecting concurrency bugs.

9 Conclusion

In this paper, we try to explore the possibilities of detecting concurrency issues by fuzzing the application. We use the static analysis of the source code of the application along with the Thread sanitizer present in the LLVM compiler framework. The paper looks at a design to use fuzzing techniques in detecting the vulnerable memory locations with the possibility of concurrency issues. The design aims to leverage the information from the source code analysis about the executions paths and the thread related function call sites to rank the inputs and use the high ranked inputs to trigger thread sanitizer warnings.

We gained insights into the design of thread sanitizer and its working during this research. We designed a custom LLVM pass to find all the execution paths from various thread function call sites to the main function and rank the basic blocks between them. This pass also helped in calculating the total rank of input after its execution. This information is used in fuzzing to prioritize inputs for the next generations. The unique TSAN warnings are calculated by reading the memory locations in the TSAN warning messages.

There were numerous implementation difficulties during the research into the approach presented, with regards to source-level instrumentation. It was not easy to convert all the source code files into bitcodes especially when the application was really huge and there were dependencies with static and dynamic binaries. It was also difficult to find suitable applications to fuzz using this approach, as the application should be multi-threaded and small enough as to easily instrument. The input sizes were also a constraint as most of the parallel processing in these multi-threaded applications are triggered only when the input size is above a threshold and it is not suitable to fuzz with large sized inputs.

We evaluated ConFuzz on three different open source applications, namely pbzip2, pigz, and pixz. All three applications were fuzzed with around 1000 generation for more than 12 h. The output showed that the fuzzer reached its maximum coverage after numerous generations and then the coverage of the basic blocks within the sliced paths remained almost constant. It recorded many TSAN warnings during these runs and unique TSAN warnings were identified. The inputs that triggered unique warnings were separately stored in a different folder, which can be used for further analysis. The results from the evaluation show a potential in the approach if we can fine-tune it a little more and the discussed implementation constraints worked out.

10 Future Works

ConFuzz fuzzes the multi-threaded applications to detect possible memory locations where concurrency bugs can happen. Even though it operates under a number of constraints, there holds tremendous scope for improvement at the design as well as implementation levels. We can use DFSan [65] available in the LLVM framework to compute the taint dependency of the memory accesses made by the target calls, detected during the static analysis of the source code, and mutate those bytes to check if they trigger TSAN warnings. As per our current design, we are monitoring only a few target calls (thread function call sites). Instead, we can design the current system to monitor the memory accesses made between a pair of calls, say lock/unlock function calls. A static analysis can also be performed to find other basic blocks that access these memory locations. From the newfound set of basic blocks, we can do similar slicing and weigh the sliced paths, so that we can use them in the fuzzing strategy to rank the inputs. We can also use DFSan to compute the taint dependency of the calculated slices and mutate the inputs at those offsets to drive the input generation. We can also make use of the OWL [64] application to verify if we can actually carry out a concurrency attack on the detected vulnerable memory locations.

The access to source code provides us with a number of possibilities to improve on and the above-mentioned ones are just a few of them. The many existing as well as developing tools within the LLVM framework can also be put to use to further explore the possibilities of improving ConFuzz.

References

1. Llvm. https://llvm.org/
2. Thread sanitizer. url. http://clang.llvm.org/docs/threadsanitizer.html
3. Musuvathi, M., Qadeer, S.: Iterative context bounding for systematic testing of multithreaded programs. In: pldi (2007)
4. Seo, E., Zhou, Y., Lu, S., Park, S.: Learning from mistakes a comprehensive study on real world concurrency bug characteristics. ACM Trans. Comput. Syst. 2(4), 277–288 (2008). ISSN 0734-2071
5. Common vulnerabilities and exposures database. http://cvedetails.com
6. Stolfo, S., Sethumadhavan, S., Yang, J., Cui, A.: Concurrency attacks. In: Fourth USENIX Workshop on Hot Topics in Parallelism (HOTPAR 12) (2012)
7. Fredriksen, L., Miller, B.P., So, B.: An empirical study of the reliability of unix utilities. Commun. ACM 33(12), 3244 (1990)
8. Kumar, A., Cojocar, L., Giuffrida, C., Rawat, S., Jain, V., Bos, H.: Vuzzer: application-aware evolutionary fuzzing. In: Proceedings of the Network and Distributed System Security Symposium (NDSS) (2017)

9. Apple developer page for llvm thread sanitizer
10. Clang. http://clang.llvm.org/index.html
11. Pbzip2. http://compression.ca/pbzip2/
12. Pigz. https://zlib.net/pigz/
13. Pixz. https://github.com/vasi/pixz
14. Bzip. http://www.bzip.org/
15. Pixz man page. https://www.mankier.com/1/pixz
16. Libavi. https://libav.org/
17. Vlc. https://www.videolan.org/vlc/download-sources.html
18. Ffmpeg. https://www.ffmpeg.org/download.html
19. Imagemagick. https://www.imagemagick.org/script/download.php
20. Linux kernel bug on uselib(). http://osvdb.org/show/osvdb/12791
21. Mysql bug 24988. https://bugs.mysql.com/bug.php?id=24988
22. Msie javaprxy.dll com object exploit. http:// www.exploit-db.com/exploits/1079/
23. Cve-2010-0923. http://www.cvedetails.com/cve/cve-2010-0923
24. Cve-2008-0034. http://www.cvedetails.com/cve/cve-2008-0034/
25. Cve-2010-1754. http://www.cvedetails.com/cve/cve-2010-1754/
26. Nelson, G., Sobalvarro, P., Anderson, T., Savage, S., Burrows, M.: Eraser: a dynamic data race detector for multithreaded programs. ACM Trans. Comput. Syst. **15**(4), 391–411 (1997)
27. Chen, W., Yu, Y., Rodeheffer, T.: Racetrack: efficient detection of data race conditions via adaptive tracking. In: Proceedings of the 20th ACM Symposium on Operating Systems Principles (SOSP 05), pp. 221–234 (2005)
28. Ashcraft, K., Engler, D.: Racerx: effective, static detection of race conditions and deadlocks. In: Proceedings of the 19th ACM Symposium on Operating Systems Principles (SOSP 03), pp. 237–252 (2003)
29. Hu, C., Ma, X., Jiang, W., Li, Z., Popa, R.A., Lu, S., Park, S., Zhou, Y.: Muvi: automatically inferring multivariable access correlations and detecting related semantic and concurrency bugs. In: Proceedings of the 21st ACM Symposium on Operating Systems Principles (SOSP 07), pp. 103–116 (2007)
30. Qin, F., Lu, S., Tucek, J., Zhou, Y.: Avio: detecting atomicity violations via access interleaving invariants. In: Twelfth International Conference on Architecture Support for Programming Languages and Operating Systems (ASPLOS 06), pp. 37–48 (2006)
31. Olichandran, R., Scherpelz, J., Jin, G., Lu, S., Zhang, W., Lim, J., Reps, T.: Conseq: detecting concurrency bugs through sequential errors. In: Sixteenth International Conference on Architecture Support for Program- ming Languages and Operating Systems (ASPLOS 11), pp. 251–264 (2011)
32. Sun, C., Zhang, W., Lu, S.: Conmem: detecting severe concurrency bugs through an effect-oriented approach. In: Fifteenth International Conference on Architecture Support for Programming Languages and Operating Systems (ASPLOS 10), pp. 179–192 (2010)
33. Chen, P.M., Flinn, J., Wester, B., Devecsery, D., Narayanasamy, S.: Parallelizing data race detection. In: Eighteenth International Conference on Architecture Support for Programming Languages and Operating Systems (ASPLOS 13), pp. 27–38 (2013)
34. Zamfir, C., Kasikci. B., Candea, G.: Racemob: crowdsourced data race detection. In: Proceedings of the 24th ACM Symposium on Operating Systems Principles (SOSP 13) (2013)
35. Martin, M.M.K., Nagarakatte, S., Burckhardt, S., Musuvathi, M.: Multicore acceleration of priority-based schedulers for concurrency bug detection. In: Proceedings of the 33rd ACM SIGPLAN Conference on Programming Language Design and Implementation(PLDI '12) (2012)
36. Lu, S., Park, S., Zhou, Y.: Ctrigger: exposing atomicity violation bugs from their hiding places. In: Fourteenth International Conference on Architecture Support for Programming Languages and Operating Systems (ASPLOS 09), pp. 25–36 (2009)
37. Park, C.-S., Sen K.: Randomized active atomicity violation detection in concurrent programs. In: Proceedings of the 16th ACM SIGSOFT International Symposium on Foundations of Software Engineering (SIG- SOFT 08/FSE-16), pp. 135–145 (2008)

38. Sen, K.: Race directed random testing of concurrent programs. In: Proceedings of the ACM SIGPLAN 2008 Conference on Programming Language Design and Implementation (PLDI 08), pp. 11–21 (2008)
39. Pereira, C., Pokam, G., Kasikci, B., Schubert, B., Candea, G.: Failure sketching: a technique for automated root cause diagnosis of inproduction failures. In: Proceedings of the 25th ACM Symposium on Operating Systems Principles (SOSP 15) (2015)
40. Chow, M., Attariyan, M., Flinn, J.: X-ray: automat- ing root-cause diagnosis of performance anomalies in production software. In: OSDI (2012)
41. Deng, D., Liblit, B., Jin, G., Zhang, W., Lu, S.: Automated concurrency bug fixing. In: Proceedings of the Tenth Symposium on Operating Systems Design and Implementation (OSDI 12), pp. 221–236 (2012)
42. Cristian, Z., Jula, H., Tralamazza, D., George, C.: Deadlock immunity: enabling systems to defend against deadlocks. In: Proceedings of the Eighth Symposium on Operating Systems Design and Implementation (OSDI 08), pp. 295–308 (2008)
43. Kudlur, M., Lafortune, S., Wang, Y., Kelly, T., Mahlke, S.: Gadara: dynamic deadlock avoidance for multithreaded programs. In: Proceedings of the Eighth Symposium on Operating Systems Design and Implementation (OSDI 08), pp. 281–294 (2008)
44. Cui, H., Wu, J., Yang, J.: Bypassing races in live applications with execution filters. In: Proceedings of the Ninth Symposium on Operating Systems Design and Implementation (OSDI 10) (2010)
45. Whaley, J., Naik, M., Aiken, A.: Effective static race detection for java. In: Proceedings of the 27th ACM SIGPLAN Conference on Programming Language Design and Implementation (PLDI 06), pp. 308–319 (2006)
46. Valgrind. http://valgrind.org/docs/manual/hg-manual.html
47. Zhang, Weihua, Yu, Shiqiang, Wang, Haojun, Dai, Zhuofang, Chen, Haibo: Hardware support for concurrent detection of multiple concurrency bugs on fused cpu-gpu architectures. IEEE Trans. Comput. **65**, 3083–3095 (2016)
48. Alam, M.U., Begam, R., Rahman, S., Muzahid, A.: Concurrency bug detection and avoidance through continuous learning of invariants using neural networks in hardware (2013)
49. Gotovos, A., Christakis, M., Sagonas, K.: Systematic testing for detecting concurrency errors in erlang programs. In: Proceedings of the 2013 IEEE Sixth International Conference on Software Testing, Verification and Validation (ICST '13), pp. 154–163 (2013)
50. Fredriksen, L., Miller, B.P., So, B.: An empirical study of the reliability of unix utilities. Commun. ACM **33**(12), 32–44 (1990)
51. Aitel, D.: An introduction to spike, the fuzzer creation kit. (presentation slides) (2002)
52. Sutton, M., Greene, A.: The art of file format fuzzing. In: Blackhat USA Conference (2005)
53. Leek, T., Ganesh, V., Rinard, M.: Taint-based directed whitebox fuzzing. In: Proceedings of the 31st International Conference on Software Engineering, pp. 474–484. IEEE Computer Society (2009)
54. Rebert, A., Cha, S.K., Avgerinos, T., Brumley, D.: Unleashing mayhem on binary code. In: 2012 IEEE Symposium on Security and Privacy (SP), pp. 380–394 (2012)
55. Neugschwandtner, M., Haller, I., Slowinska, A., Bos, H.: Dowsing for overflows: a guided fuzzer to find buffer boundary violations. In: USENIX Security Symposium, pp. 49–64 (2013)
56. American fuzzy loop (afl-fuzz). https://github.com/rc0r/afl-fuzz
57. Salls, C., Dutcher, A., Wang, R., Corbetta, J., Shoshitaishvili, Y., Kruegel, C., Stephens, N., Grosen, J., Vigna, G.: Driller: augmenting fuzzing through selective symbolic execution. In: NDSS, vol. 16, pp. 1–16 (2016)
58. Livshits, V.B., Lam, M.S.: Finding security errors in java programs with static analysis. In: Proceedings of the 14th Usenix Security Symposium, pp. 271–286 (2005)
59. Arp, D., Yamaguchi, F., Golde, N., Rieck, K.: Modeling and discovering vulnerabilities with code property graphs. In: Proceedings of the 2014 IEEE Symposium on Security and Privacy (SP 14), pp. 590–604 (2014)
60. Kruegel, C., Felmetsger, V., Cavedon, L., Vigna, G.: Toward automated detection of logic vulnerabilities in web applications. In: Proceedings of the 19th USENIX Conference on Security (USENIX Security 10), pp. 1010 (2010)

61. Fritz, C., Bodden, E., Bartel, A., Klein, J., Le Traon, Y., Octeau, D., Arzt, S., Rasthofer, S., McDaniel, P.: Flowdroid: precise context, flow, field, object-sensitive and lifecycle-aware taint analysis for android apps. In: Proceedings of the 35th ACM SIGPLAN Conference on Programming Language Design and Implementation (PLDI14), pp. 259–269 (2014)
62. McKinley, K.S., Srivastava, V., Bond, M.D., Shmatikov, V.: A security policy oracle: detecting security holes using multiple api implementations. In: Proceedings of the 32nd ACM SIGPLAN Conference on Programming Language Design and Implementation (PLDI 11), pp. 343–354 (2011)
63. Edwards, A., Zhang, X., Jaeger, T.: Using cqual for static analysis of authorization hook placement. In: Proceedings of the 11th USENIX Security Symposium, page p. 33–48 (2002)
64. Zhao, J., Ning, Y., Cui, H., Yang, J., Gu, R., Gan, B.: Understanding and Detecting Concurrency Attacks
65. Data flow sanitizer. http://clang.llvm.org/docs/dataflowsanitizer.html

Sifar: An Attempt to Develop Interactive Machine Translation System for English to Hindi

Meenal Jain, Mehvish Syed, Nidhi Sharma, Shambhavi Seth
and Nisheeth Joshi

Abstract The presently available machine translation systems are still far from being perfect, and to improve their performance the concept of interactive machine translation (IMT) was introduced. This paper proposes Sifar, an IMT system, which uses statistical machine translation and a bilingual corpus on which several algorithms (Word error rate, Position Independent Error Rate, Translation Error Rate, n-grams) are implemented to translate text from English to Hindi. The proposed system improves both the speed and productivity of the human translators as found through experiments.

Keywords Machine translation · Statistical machine translation · Computer aided translation · Interactive machine translation · Position independent error rate · Word error rate · Translation error rate

1 Introduction

Since the very advent of languages, translations emerged as a very important aspect to minimize the communication gap. With time and computational capacity, a vast number of natural languages are being digitized. In India alone, we have such large

M. Jain · M. Syed · N. Sharma · S. Seth (✉) · N. Joshi
Department of Computer Science, Banasthali Vidyapith, Vanasthali, Rajasthan, India
e-mail: shambhavi22seth@gmail.com

M. Jain
e-mail: mpbjain@gmail.com

M. Syed
e-mail: mehvishsyed97@gmail.com

N. Sharma
e-mail: vats.nidhi10@gmail.com

N. Joshi
e-mail: jnisheeth@banasthali.in

© Springer Nature Singapore Pte Ltd. 2020
A. K. Luhach et al. (eds.), *First International Conference on Sustainable Technologies for Computational Intelligence*, Advances in Intelligent Systems and Computing 1045, https://doi.org/10.1007/978-981-15-0029-9_54

multilingual diversity. It becomes a necessity to have translators for efficient communication and having human translators was tedious and both resource and time-consuming. This gave birth to machine translation systems. In such a fast-paced world where time plays a significant role, MT requires certain enhancements which were introduced as CAT (computer-aided translation) tools. The current state-of-the-art MT systems do not produce ready-to-use translations; thus, human post-editing is required to achieve high-quality translations. To address this issue, the MT systems are augmented with human translators to give birth to a new technology, namely interactive machine translation which provides a framework for adaptive learning to help the user with sentence and phrase completion suggestions during the translation process. Many IMT systems have so far been developed for foreign languages, but negligible efforts have been made for Indian languages. So, we thought of developing a tool that could allow an English speaker to translate in Hindi and developed translation systems lack good quality in open domain automatic translation. Since they do not have an open domain ability to translate text, we have to move towards a translation technique where a first-hand translation can be co-headed by machine and then can be perfected by a human. Hence, Sifar is an attempt to develop a system for English to languages translation, and thus it is an IMT so that user can get the translation with multiple suggestions. Sifar aims at assisting human translators to accelerate the overall translation process with the help of an example database. As the user translates the text, the translations are added to the example database, and when a similar sentence reoccurs, the previous translation is inserted into the translated document, in turn saving the user's effort of re-translating that sentence, and is particularly useful when translating a revision of a previously translated text.

Section 2 describes the related work done in this area. Section 3 discusses our methodology where we have shown the development process of Sifar. Section 4 evaluates the system, and Sect. 5 concludes the same.

2 Related Work

Nagao [10] proposed the idea of machine translation based on the examples, which were made available to the system prior to the translation process. A bilingual corpus was used which was comprised of source language text and their corresponding target language text. This approach was based on how humans process a sentence in source language and translate it to target language. Machine translation system use natural languages, which are highly complex (use of homonyms, different grammatical rules), to make decisions. In RBMT, defining rules is a difficult task and cost of building dictionaries and making changes is high. To overcome the problem of knowledge acquisition, Corpus-based MT comprising of SMT and EBMT was used. Languages with varying word orders proved to be a challenge for SMT. However, for the production of dependency trees for sentence analysis and example database, EBMT requires analysis and generation modules. Hybrid approach minimizes the challenges of other approaches [11]. Block transpositions [9] are used to extend edit distance to define inversion edit distance as a metric of the cost of parsing for a sentence pair within an inversion grammar. Experiments showed that at system level, correlation of automatic evaluation with human judgment is appropriate. DerivTool [2], an interactive translation visualization tool provides with a myriad number of options for the user to choose from and allows her to look into the decoding process where syntax-based framework for translation has been used. Moses [5], an open source toolkit for SMT, featured the phrase-based translation with factors and confusion network decoding, which allowed translation of ambiguous input, along with the capabilities of Pharaoh decoder [6]. Caitra, developed by Koehn et al. [7] based on the TransType Project [8], a web-based IMT tool, with the help of Moses decoder, featured making suggestions for sentence completion, alternative word provision and phrase translation, and giving post editing options with key stroke logging for detailed analysis. Cunei [12] proposed a hybrid system comprising of features of both EBMT and SMT, by using the example base to model each phrase pair at run time and perform recombination to get the translations. The system was tested using three language pairs Finnish to English, French to English, and German to English against the Moses model, where Cunei displayed unexpectedly better results for German to English. Due to German compounding, it becomes difficult for the one to many alignment and phrase extraction, but Cunei had advantage due to its runtime modeling as it adapts itself over time by updating weights. The complex lexicalized distortion model gave nearly the same result for the reordering model which was informed only about how far and frequently the phrases were moved during decoding. CASMACAT [1], a modular, web-based translation workbench with advanced functionality and editing features provided us with TM for raw match and automatic translations from MT server for post editing, consisting of a GUI, backend, an MT server, and a CAT server. The main features included interactive translation prediction (ITP), confidence measures, prediction length control, search and replace, word alignment information, one clicks rejection, replay mode and logging function, and e-pen for handwritten interaction. The system was evaluated at Celer Soluciones, where 9 professors participated to carry out post-editing for English–French translations. Three workbench features were tested: post-editing, intelligent autocompletion (IA), and ITP where ITP had a slower rate due to the unfamiliarity with the IA system. Also due to the lack of visual aid to control IA, double checking had to be done even for small post-editing tasks. Dungarwal et al. [3] proposed a Hindi–English Translation consisting of phrase based and factored statistical machine translation (SMT) systems that considered number, case, and tree adjusting grammar information as factors. Translation of out of vocabulary and transliteration of named entity are done at the preprocessing stage. Merging the chunks, preposition chunk reordering, verb chunk reordering was used as the rules of development. Wang et al. [14] proposed CytonMT, an open source toolkit which emphasize on neural machine translation. The system use attention-based RNN encoder-decoder and is developed on C++. It uses NVIDIA GPU accelerated libraries. The proposed system proved to be efficient on BLUE score and provides a considerable speed-up in training phase of the system. Helcl et al. [4] presented a CUNI system which supports multimodal machine translation (MMT) task and incorporates self-attentive neural network instead of a recurrent neural network which proves to be an effective measure. The proposed system works on doubly attention transformer architecture and imagination concept, formerly introduced for sequence to sequence RNN model.

3 Methodology

To implement Sifar, we integrated out translator with an existing machine translation system, which has been trained statistically and is based on Moses machine translation toolkit. In order to train this statistical system, we used a mix-domain corpus which comprises of translations from tourism, health, agriculture and administrative domains. We preferred mix-domain training to achieve open domain translations. Another important factor we incorporated in our system is a bilingual corpus on which our translator is trained. The chunks in the corpus are taken from the tourism domain. When source text is provided to the system, it first looks for the sentences in the corpus, evaluates the sentences one at time, to check for their similarity with the input text provided. In case, the similarity score is greater than the set threshold value, the target text corresponding to that source text as well as the output from SMT system is joined and the first-order output is proposed to the user. As we claim our system to be interactive, we allow the user to make changes as required in the target text of most suitable candidate target text and further, add this new source–target pair as a new tuple in the corpus so that the forthcoming translations can make a use of it. Sifar calculates the similarity score of hypotheses (translation produced by system) and reference sentence (example translation available in knowledge base) using various algorithms. These algorithms are named as Word Error Rate, Position Independent Error Rate, Translation Error Rate, n-grams similarity, and all of these are discussed further.

3.1 Word Error Rate (WER)

Word error rate [13] is defined as the minimum number of edits (insertion, deletion, and substitution) required to transform the hypothesis string into reference string. It is calculated as the Levenshtein's Edit distance between the hypothesis string and the reference string, normalized by the length of reference string.

Algorithm-1 to calculate WER

Input: REFERENCES R, HYPOTHESIS h.
for all *r* in *R* **do**
   h' h
   e' 0
   e' min_edit_distance(h'; r)/length(r)
end for
return *e*

Example
Hypothesis: The best time to visit Jaipur city is around October to March.
Result: जयपुर शहर घूमने के लिए अक्टूबर से मार्च तक तक का समय उत्तम है। 50%.
जयपुर संगमरमर की मूर्तियों, नीले मिट्टी के बर्तन और राजस्थानी जूतियों के लिए भी
प्रसिद्ध है। 32%.

3.2 Position Independent Error Rate (PER)

Position independent error rate [11] is similar to WER except the fact that PER
does not take word order into consideration. PER counts the number of non-similar
words between the hypothesis and the reference string as the number of substitutions
required. The number of insertions or deletions required depends upon the difference
between the length of the reference and hypothesis string. The total error rate is
the sum of insertions, deletions, and substitutions normalized by the length of the
reference string.

Algorithm-2 to calculate PER

Input: REFERENCES R, HYPOTHESIS h
for all *r* in *R* **do**
   h' h
   e 0
   for all *words;words_hypothesis*in *h* **do**
      e 0
      if *word - hypothesis* is present in *r* **then**
         e e+ 1
      end if
   end for
end for
max_length max
e<-(max_length□ e)/ length of reference string
return *e*

Example

Hypothesis: The best time to visit Jaipur city is around October to March.

Result: जयपुर का बाज़ार चमकीला; और दुकानें रंग बिरंगी वस्तुओं, जिनमें हस्तकला की वस्तुएँ, बहुमूल्य पत्थरों मीनाकारी की वस्तुओं गहनों व राजस्थानी चित्रकला आदि से भरे हैं। 78.788 %

नापो, गौलेरास, कुटुक और कॉडोर की शृंखलाएँ यहाँ स्थित हैं। 42.851%.

3.3 *Translation Error Rate (TER)*

Translation error rate is calculated as the minimum number of edits (insertion, deletion, substitution) required on individual words as well as sequence of words, normalized by the length of reference string, to transform hypothesis string into reference string.

Algorithm-3 to calculate TER

Input: REFERENCES R, HYPOTHESIS h
for all r in R **do**
  h' h
  e 0
  repeat
    if s reduces edit distance **then**
      h apply s to h'
      e e+ 1
    end if
  until No shifts that reduce edit distance
  e e+ *min_edit_distance*(h'; r)
end for
return e

Example

Hypothesis: The best time to visit Jaipur city is around October to March.

Result: जयपुर शहर घूमने के लिए ओक्टोबर से मार्च तक का समय उत्तम है। 75%. अक्तूबर से फरवरी जैसलमर भ्रमण का श्रेष्ठ समय माना जाता है। 50%.

3.4 *N-Grams*

The n-gram algorithm divides the reference and the hypothesis string into chunks of n words each and then compares the chunks of the hypothesis string with those of the reference string. Sifar uses n-grams algorithm as 3-grams and 4-grams.

Algorithm-4 for N-Grams

Input: REFERENCES R, HYPOTHESIS h
h' h
hyp[] divide h' into substrings of length n each
for all *r* in *R* **do**
    ref[] divide r into substrings of length n each
    for all *t* in *hyp*[] **do**
      q <-0
      repeat
        if *t* **equals** ref[q] **then**
          cost+ 1
          q+ 1
        end if
      until *q* **equals** length(hyp[ ])
      result cost / max(length(hyp[]), length(ref[]))
    end for
end for
return *result*

Example
Hypothesis: The best time to visit Jaipur city is around October to March.
Result: जयपुर शहर घूमने के लिए अक्तूबर से मार्च तक का समय उत्तम है। 69.230%.

Our IMT system (Sifar) made use of these algorithms to get the example translations and ranked them according to similarity. It also took the translation of an MT system and adds to the list of possible translations. These all are provided to the human translation who can select the one which the human translator feels most appropriate and perform post editing to generate a final high-quality translation. The working of this entire system is shown in Fig. 1 (Fig. 2).

4 Evaluation

We performed our evaluation on 5 human translators who were asked to translate 1000 sentences using Sifar. For the first 500 sentences, we did not provide them with any suggestive translations and where are to do translations on their own. For the rest 500 sentences, we provided the translators with suggestions. They analyzed the results based on speed and productivity. The results are as follows.

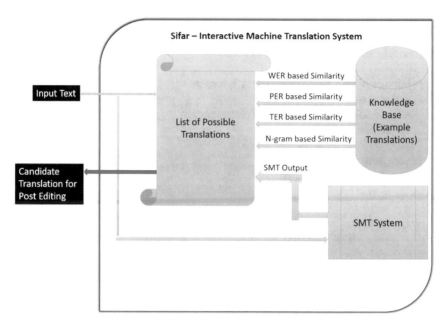

Fig. 1 Working of Sifar-interactive machine translation system

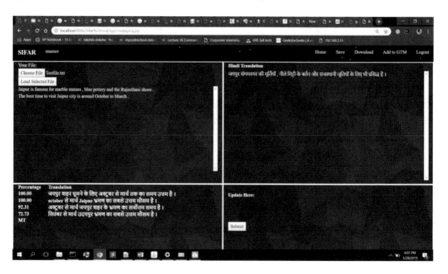

Fig. 2 Screenshot: Sifar-interative machine translation system

Table 1 Time taken to complete the translations

Human translations	Without suggestive translators (s/word)	With suggestive translations (s/word)	Difference (s/word)
H1	6.3	3.2	3.1
H2	5.3	3.4	1.9
H3	3.3	2.3	1
H4	6.5	2.6	3.9
H5	6.1	3.5	2.6

4.1 Speed

All the five translators performed very well, when they were provided with the suggestive list of possible translations. This improved their speed to doing translations. Table 1 shows the results of this experiment. In this, we calculated the speed as well as the total time taken to complete the task divided by total no. of words in the documents. This is shown in Eq. (1). Here the calculated speed of doing translations when no suggestions were provided and when the suggestions were provided.

$$\text{Speed} = \frac{\text{Total Time Taken}}{\text{Total No. of Words}} \tag{1}$$

4.2 Productivity

In order to ascertain productivity, we calculated the score of HTER (human translation edit rate). In this, we calculated the total number of edits (inserts, deletes, substitutes, sifts) performed by human translator to correct the translation which was selected from the list of suggestive translations. This was done on only last 500 sentences, as these were the sentences on which suggestive translations were provided. In all the cases, the edit rate of the translation was very less. This confirms that the use of Sifar improves the productivity of the human translators. Table 2 shows the results of this study.

Table 2 HTER score

Human translators	HTER score
H1	0.3452
H2	0.1542
H3	0.1453
H4	0.3421
H5	0.2568

5 Conclusion and Future Work

In this paper, we have shown the design of an interactive machine translation system which combines the strengths of both human and machine and which provides a high-quality machine assisted translation. Through experiments, we have verified our claim, as with the help of our system the speed of the human translators also improved. It also improved the productivity of the translators. As an extension to this study, we wish to improve the user friendliness of the system by providing several features like providing a human translator an option to add their example translations in the knowledge base, before doing actual translations. We also wish to perform a more thorough evaluation to understand the expectations of the human translators from this system. One of the possible expectations is to reduce the key in effort. We shall perform the usability evaluation of the system to analyze such expectations and incorporate the same in subsequent versions of Sifar.

We plan to improve our translation system by implementing certain modules. The first module would include expanding the translation system for other Indian languages as our current system works only on English to Hindi translations. This would increase the scope and utility of our system. The second module would incorporate the concept of neural networks to the present system. Neural machine translation systems produce better outputs as they understand and establish the similarities between the words and produce more fluent outputs. Hence, if used with our system, it would increase our system's efficiency.

References

1. Alabau, V., Buck, C., Carl, M., Casacuberta, F., García-Martínez, M., Germann, U., Mesa-Lao, B.: Casmacat: a computer-assisted translation workbench. In: Proceedings of the Demonstrations at the 14th Conference of the European Chapter of the Association for Computational Linguistics, pp. 25–28 (2014)
2. DeNeefe, S., Knight, K., Chan, H. H.: Interactively exploring a machine translation model. In: Proceedings of the ACL 2005 on Interactive poster and demonstration sessions. Association for Computational Linguistics, pp. 97–100 (2005)
3. Dungarwal, P., Chatterjee, R., Mishra, A., Kunchukuttan, A., Shah, R., Bhattacharyya, P.: The iitbombayhindi-english translation system at wmt 2014. In: Proceedings of the Ninth Workshop on Statistical Machine Translation, pp. 90–96 (2014)
4. Helcl, J., Libovický, J., Variš, D.: CUNI System for the WMT18 Multimodal Translation Task. arXiv preprint arXiv:1811.04697 (2018)
5. Koehn, P., Hoang, H., Birch, A., Callison-Burch, C., Federico, M., Bertoldi, N., Cowan, B., Shen, W., Moran, C., Zens, R., Dyer, C.: Moses: open source toolkit for statistical machine translation. In: Proceedings of the 45th Annual Meeting of the ACL on Interactive Poster and Demonstration Sessions. Association for Computational Linguistics, pp. 177–180 (2007)
6. Koehn, P.: Pharaoh: a beam search decoder for phrase-based statistical machine translation models. In: Conference of the Association for Machine Translation in the Americas, pp. 115–124. Springer, Berlin, Heidelberg (2004)
7. Koehn, P.: A web-based interactive computer aided translation tool. In: Proceedings of the ACL-IJCNLP 2009 Software Demonstrations, pp. 17–20 (2009)

8. Langlais, P., Foster, G., Lapalme, G.: TransType: a computer-aided translation typing system. In: Proceedings of the 2000 NAACL-ANLP Workshop on Embedded machine translation systems, vol. 5. Association for Computational Linguistics, pp. 46–51 (2000)
9. Leusch, G., Ueffing, N., Ney, H.: A novel string-to-string distance measure with applications to machine translation evaluation. In: Proceedings of Mt Summit IX, pp. 240–247 (2003)
10. Nagao, M.: A framework of a mechanical translation between Japanese and English by analogy principle. In: Artificial and Human Intelligence, pp. 351–354 (1984)
11. Okpor, M.D.: Machine translation approaches: issues and challenges. Int. J. Comput. Sci. Issues (IJCSI) **11**(5), 159 (2014)
12. Phillips, A.B., Brown, R.D.: Cunei machine translation platform: system description. In: 3rd Workshop on Example-Based Machine Translation (2009)
13. Tillmann, C., Vogel, S., Ney, H., Zubiaga, A., Sawaf, H.: Accelerated DP based search for statistical translation. In: Fifth European Conference on Speech Communication and Technology, pp. 2667–2670 (1997)
14. Wang, X., Utiyama, M., Sumita, E.: CytonMT: An Efficient Neural Machine Translation Open-source Toolkit Implemented in C++. arXiv preprint arXiv:1802.07170 (2018)

MinMaxScaler Binary PSO for Feature Selection

Hera Shaheen, Shikha Agarwal and Prabhat Ranjan

Abstract Particle Swarm Optimization (PSO) is a widely accepted optimization method which has shown merits in many fields. Particle Swarm Optimization has been modified several times to get the desired result in different problems. This paper presents a survey of developments and modifications of PSO, and from the survey, it is found that the major drawback of PSO is the trapping in local minima, due to which premature convergence takes place. Its performance is poor in high-dimensional data and in complex problems. That's why PSO for feature selection problem has been surveyed further and a new MinMaxScaler Binary PSO is proposed for feature selection.

Keywords Particle Swarm Optimization · Feature selection · Machine learning · Soft computing · Bio-inspired computing

1 Introduction

Particle Swarm Optimization (PSO) was developed by Kennedy and Eberhart. Each single solution in PSO is considered as a particle. Every particle uses its own information and the information gained by the whole swarm. Every particle adjusts or updates its position by changing its velocity. Particles move through the problem space by following current optimum particle. This process is iteratively done some fixed number of times until some predefined minimum error [1, 2]. PSO was initially introduced for optimization of real number spaces, and it has been used in various

H. Shaheen · S. Agarwal (✉) · P. Ranjan
Department of Computer Science, Central University of South Bihar, Gaya, Bihar, India
e-mail: shikhaagarwal@cub.ac.in

H. Shaheen
e-mail: shaheen.hera.netcs@gamil.com

P. Ranjan
e-mail: prabhatranjan@cub.ac.in

© Springer Nature Singapore Pte Ltd. 2020
A. K. Luhach et al. (eds.), *First International Conference on Sustainable Technologies for Computational Intelligence*, Advances in Intelligent Systems and Computing 1045, https://doi.org/10.1007/978-981-15-0029-9_55

areas like optimization of function, training artificial neural network, fuzzy system control, image recognition system, speech recognition system.

In this paper, a survey of developments, modifications of basic PSO, and hybrids of PSO with different methods have been presented. Then the application areas of PSO have been surveyed which is based on modifications of PSO and hybrids of PSO only. From the survey, it is found that the major drawback of PSO is the trapping in local minima, due to which premature convergence takes place. Its performance is poor in high-dimensional data and in complex problems. In survey of modifications of PSO, the change in velocity equation of PSO [3, 4], the change in position equation of PSO [5], change in both velocity and position equation, change in particle structure of PSO, quantum-behaved PSO, topology PSO, and multi-objective PSO [6, 12] literature have been surveyed. The problem encountered here is trapping in local optima. From the survey, it is found that most of the time authors have given change in inertia weight as a solution. And tried to set its value using various methods. But the solution still lacks the optimal value of parameters because they are just updating the inertia weight and not using learning induced methods.

In hybrids of PSO [13, 14], PSO with soft computing, simulated annealing, machine learning, bio-inspired computing has been surveyed. Hybrid methods improve solution to some complex problems. It helps in large-scale optimization and parallel implementation. Then the applications of PSO using the modified and hybrid PSO have been surveyed and discussed. PSO application in software engineering, networking, speech recognition, image processing, data mining, scheduling, stock market, biological and medical application, and in satellite data. It is found that PSO is most popular in biological and medical data and is least popular in the area of stock market. High-dimensional data is a major problem in data mining techniques which may gives wrong result or prediction. High-dimensional data means that the number of dimensions are so large that predictions become extremely difficult. Searching for optimal feature subset from a high-dimensional data is a NP-complete problem. Then another most challenging task is to optimize parameters of basic PSO. The parameters of PSO are inertia weight, cognitive, and social parameters and to find the optimal values of these parameters are itself a challenging task. And hybrid with machine learning methodologies has improved the performance of PSO in many areas. So, hybrid with machine learning methodologies may have some path to optimal parameter selection and will finally give optimal solution and may improve classification accuracy.

2 Particle Swarm Optimization

Particle of PSO is defined by position and velocity. Position and velocity vector of ith particle are defined as $X_i = (x_{i1}, x_{i2}, \ldots, x_{id})$ and $V_i = (v_{i1}, v_{i2}, \ldots, v_{id})$ where d is the dimension. Pbest(pb_i) and Gbest(gb_i) are the personal best and global best obtained by any particle.

The equations to update position and velocity of each particle which are given as Eqs. 1 and 2

$$v_{id}^{new} = v_{id}^{old} + c_1 * r_1 * (pb_{id} - x_{id}^{old}) + c_2 * r_2 * (gb_d - x_{id}^{old}) \qquad (1)$$

$$x_{id}^{new} = x_{id}^{old} + v_{id}^{new} \qquad (2)$$

where r_1, r_2 are random values in the range of (0,1). c_1 and c_2 are cognitive and social parameters. Survey has been done in these respects: (i) modifications of PSO, (ii) hybridizations of PSO, and (iii) applications of PSO.

3 Modifications of PSO

In modifications of PSO, we have surveyed the developments and modifications in basic PSO. Here modifications have been performed in the standard equations of velocity and position update of basic PSO. Change in velocity equation deals with the studies of PSO in which authors have tried to improve PSO by making some changes in Eq. (1) of velocity. It improves convergence and gives better optimal solution [3, 4]. Larger inertia weight helps in global exploration, while smaller helps in local exploitation. It is good for NP-hard problems. Here inertia weight could be improved more by learning induced methods. Change in position equation deals with the studies of PSO in which authors have tried to improve PSO by making some changes in Eq. (2) of position. It enhances global search ability, and diversity of swarm also improves [5]. It helps in fine tuning and good in knapsack problem. Change in velocity and position equation both gives guaranteed convergence, good for niching algorithms, enhances optimization precision and success rate, handles high-dimensional cases [6]. Weaknesses of these variants are that parameter optimization is not done most of the time.

Change in particle structure has been studied where change has been done in the structure of particle. It helps in finding global optimal solution, adaptively adapt inertia weight [7].

Quantum particle swarm optimization is inspired by the theory of quantum mechanics which is probabilistic optimization method. It helps in optimal parameter selection, provides better position in searching space, balances global and local search, and overcomes premature convergence [8, 9] and search strategy (Topology PSO); population topology tells particles about its neighborhood [10, 11]. From their neighborhood, particles learn and share information.

In multi-objective PSO, when we have a problem with two or more objectives that should be optimized simultaneously,and these objectives are generally conflicting in nature, and then this is called multi-objective optimization problem. It is used mostly in feature selection problem [12]. It gives optimal solution, handles difficult cases of Pareto front, and has remarkable performance.

4 Hybrid of PSO

It deals with the studies of PSO, where PSO has been used along with the other popular techniques.

In hybridization with soft computing, PSO has been hybridized with soft computing methods like fuzzy logic and genetic algorithm. It improves performance of PSO [13, 14], and it is mostly used in traveling salesman problem. It is efficient and easy to implement. Here MapReduce with fuzzy PSO helps in parallel implementation, and GA mutation operator helps in preventing premature convergence and good in cancer dataset and large-scale optimization performance.

In hybridization with simulated annealing, PSO has been combined with simulated annealing techniques to enhance its performance. It helps in better convergence, and information exchange is also improved [15, 16]. It improves convergence ability, global search ability, and convergence precision.

Hybridization with machine learning deals with the studies of PSO where PSO has been combined with machine learning techniques to improve its performance [17]. Here PSO's performance is enhanced, helps in improvement of generalization power, and gives better performance. The classification ability is high.

Hybridization with bio-inspired computing helps in improving quality of result. For the detection of bundle branch block, abnormal cardiac beat identification and many other problems have been solved with this. It is widely used for NP-hard problems [18].

5 Applications of PSO

Applications of PSO are based on the two concepts discussed above only, i.e., modifications of PSO and hybrids of PSO. PSO has a wide range of applications [19, 20] in almost all fields of engineering and medical applications.

All the applications of PSO discussed here are either using modifications of PSO or hybrids of PSO. Some of the areas of applications of PSO are given in Table 1.

6 Particle Swarm Optimization for Feature Selection

Irrelevant features are opposite of relevant features. These features do not give any contribution to result. That is why it may produce incorrect result. Therefore, feature selection is performed. Feature selection, also called variable selection or variable subset selection, is the process of selecting a subset of relevant features (variables, predictors) for use in model construction.

Feature selection methods are mainly divided into two categories: filter approach and wrapper approach. Wrapper approach uses a classification algorithm as a part

Table 1 Summary of studies of applications of PSO

Applications of PSO	Conclusion
Software engineering [19, 20]	Gives objective function for partitioning. It finds target very fast. Classification performance enhanced. It helps in producing software on time and manage staffing. it has remarkable performance and improves accuracy and prediction
Networking [21]	Effective and promising, clustering performance enhanced, good for network community detection problem
Speech recognition [22]	Gives global optimal solution using hybrid of SVM with PSO is good here to train the model and get optimal result
Image processing [23]	Used in clustering; it minimizes quantization error, and intracluster distances easy to converge to the optimal solution
Data mining [24]	Helps in medical decision support system. SVM classifier is good here. Effective in clustering of gene expression data
Scheduling [25]	It is effective considering computation cost and time, task of machine assignment and operation sequencing problem
Stock market [26, 27]	Hybrid of genetic algorithm and PSO, gives better performance, works well for short term but accuracy reduces for long term
Biological applications [28]	Hybrid of PSO with neural network, experiment performed on three multiple classes and cancer dataset shows better result, finds global optimal solution, helps in taking quick decision and less cost
Satellite [29]	Reduces complexity, uses dynamic depth first search. Parallel PSO is used and improves convergence; on real-time satellite tracking, self-adapting strategy in parameter selection, better prediction ability

of the evaluation function to determine the goodness of the selected feature subsets. Filter approach uses statistical characteristics of the data for evaluation, and the feature selection is independent of any classification algorithm.

PSO has been used widely for feature selection. Some of the studies is shown in Table 2. Feature selection is a broad area which done successfully, enhances classification accuracy in less time. It makes task easier.

Table 2 Summary of survey of particle swarm optimization for feature selection

Author	Study
Xue et al. [30]	Multi-objective particle swarm optimization for feature Selection, CMDPSOFS, which is based on the ideas of crowding, mutation, and dominance method designed
Ghamisi and Benediktsson [31]	Hybrid of PSO with GA used for feature selection
Mistry et al. [32]	Facial expression recognition system developed using micro-GA with PSO
Kumar and Inbarani [33]	Brain–computer interface developed for multiclass motor imagery task, here PSO and neighborhood rough set are used for classification
Agarwal et al. [13]	Fuzzy rule-based binary PSO (FRBPSO) Fuzzy Logic used for decision-making
Agarwal and Ranjan [14]	MapReduce-based ternary PSO for better exploration and fast processing
Li et al. [34]	Hybrid of PSO with random forest in intrusion detection system
Liu et al. [35]	Gene selection using hybrid of PSO with Relief
Ahila et al. [36]	Feature selection and parameter optimization for power system disturbance Classification PSO with probabilistic neural network is used
Hamed et al. [37]	Feature selection and parameter optimization
Qasim and Algamal [17]	Feature selection method is proposed using particle swarm optimization and logistic regression

7   Discussion on Survey

PSO has poor performance most of the time when complex problems are encountered and in large-scale optimization. And also because parameters are not tuned. If tuned, they are just updated, learning update might be explored. Many powerful machine learning optimizations have open path and that to be explored. From the literature survey, it is clear that PSO for feature selection has great performance with hybrid methods. But as far as my knowledge from the survey, no one has done parameter optimization of PSO hybrid with machine learning in which learning-induced methods are utilized. Some author has tried to use, but not done well. So, there is a need of better feature subset and classification accuracy by exploring and experimenting machine learning methodologies.

8 Proposed Modification in PSO

In general, feature selection using PSO [38] uses sigmoid function given by:

$$Sig(v) = \frac{1}{1 + e^{(-v)}}$$

Here, velocity is passed as an input. Then decision is done for feature selection using position equation on the basis of some threshold value. Sigmoid function is used as an activation function having characteristics of S-shaped curve called sigmoid function. It transfers the output in the range of 0 and 1. Then by setting a threshold value, decision is taken. In the proposed method, we are replacing sigmoid function with MinMaxScaler function which is discussed in next section.

8.1 MinMaxScaler PSO

The MinMaxScaler is the scaling function, which has following formula for each feature:

$$MinMaxScaler(v) = \frac{v_{id} - min(v)}{max(v) - min(v)} \tag{3}$$

where min(v) is minimum velocity and max(v) is the maximum velocity and v_{id}^{new} is defined by

$$v_{id}^{new} = v_{id}^{old} + c_1 * r_1 * (pb_{id} - x_{id}^{old}) + c_2 * r_2 * (gb_d - x_{id}^{old})$$

Then, for feature selection based on the following rule is used.

$$x_{id} = \begin{cases} 1 & if \quad MinMaxScaler(v) >= 0.5 \\ 0 & otherwise \end{cases}$$

When $x_{id} = 1$ means feature is selected, and $x_{id} = 0$ means feature is not selected. So, here in place of sigmoid function, we will use MinMaxScaler function. Then, position is evaluated and found that whether it is selected or rejected.

9 Algorithm

Algorithm 1: Proposed PSO for feature selection

Input: High Dimensional Data=$(S_1, S_2, ..., S_m)$,$(S_i \epsilon R^d)$ and associate class is
$\quad\quad C = (C_1, C_2, .., C_m)$
Output: Fitness of gbest fit_{gb}, position of gbest (gb)
1 Initialize the position of particles (x), velocity (v), pbest (pb), gbest (gb), fitness of pb (fit_{pb}), fitness of $gb(fit_{gb})$, vmin and vmax
2 Initialize the inertia weight (w) and number of particles(p)
3 **begin**
4 $\quad$ **while** *(Iter≤MaxIter)* **do**
5 $\quad\quad$ **for** *(j=1 to p)* **do**
6 $\quad\quad\quad$ calculate fitness of particles using LOOCV-KNN
7 $\quad\quad$ **end**
8 $\quad\quad$ **for** *(i=1 to p)* **do**
9 $\quad\quad\quad$ **if** $(fit_i \geq fit_{pb_i})$ **then**
10 $\quad\quad\quad\quad$ pb_i=x_i
11 $\quad\quad\quad$ **else**
12 $\quad\quad\quad\quad$ No change in pb
13 $\quad\quad\quad$ **end**
14 $\quad\quad\quad$ **if** $(fit_i \geq fit_{gb})$ **then**
15 $\quad\quad\quad\quad$ $gb = pb_i$
16 $\quad\quad\quad$ **else**
17 $\quad\quad\quad\quad$ No change in gb
18 $\quad\quad\quad$ **end**
19 $\quad\quad\quad$ **for** *(d = 1 to number of features)* **do**
20 $\quad\quad\quad\quad$ /*Update the particle */
21 $\quad\quad\quad\quad$ $v_{id}^{new} = \omega v_{id}^{old} + c_1 * r_1 * (pb_{id} - x_{id}^{old}) + c_2 * r_2 * (gb_d - x_{id}^{old})$
22 $\quad\quad\quad\quad$ /*Standardize the algorithm using MinMaxScaler PSO as given by equation*/
23 $\quad\quad\quad\quad$ $v_{id}^{scaled} = \dfrac{v_{id}^{new} - vmin}{vmax - vmin}$
24 $\quad\quad\quad\quad$ $x_{id}(t + 1) = \begin{cases} 1, & \text{if } v_{id}^{scaled} \geq 0.5 \\ 0, & \text{if } v_{id}^{scaled} < 0.5 \end{cases}$
25 $\quad\quad\quad$ **end**
26 $\quad\quad$ **end**
27 $\quad$ **end**
28 **end**

---

10 Experimental Design

To show the curse of dimensionality, our proposed method is applied on gene expression profile dataset (SRBCT dataset) taken from gene expression model selector(GEMS) [40]. The **description of dataset SRBCT** is that it has 83 number of samples, 2308 features, 4 number of classes. Fitness of each particle is obtained using K-nearest neighbor classifier which leaves out one cross-validation. Value of k is taken as 1. The performance of algorithm is justified according to the classification accuracy and selected feature metrics.

The simulations performed on MATLAB 7.11.1.866 R2010b, License No. 691568. Cognitive and social parameters c_1 and c_2 are chosen as the value of 2. Inertia weight is assumed as $\omega = 0.5$. The parameter chosen from [39]. The value of vmin and vmax chosen as $[vmin, vmax] = [-6, 6]$ and criteria to stop are chosen as maximum iteration to be 150 or maximum accuracy to be 100. Number of particles are chosen as 40.

The data (IBPSO, EVPSO,Genetic Algorithm, KNN) in the comparision table has been taken from [41].

11 Result

Experiment was performed on SRBCT dataset having **2308** features. Total four runs were performed. We have found classification accuracy and number of features selected in each run. In the first run, we got **97.59 %** accuracy and number of features selected is **1130**. In the second run, accuracy obtained is **98.8 %** and number of features selected is **1134**. In the third run, accuracy obtained is **97.59 %** and number of features selected is **1138**. In the fourth run, accuracy obtained is **97.59 %** and number of features selected is **1148**. Then we found the average of all the runs and found that the average of accuracy is **97.89 %** and the average of number of features selected is **1137**. It is shown in Table 3. Therefore, we can say that there is consistency in all four runs. There is very less deviation in result. Average classification accuracy is good. There is good amount of feature reduction, i.e., **50.7 %**. From the result of simulation Table 3 and comparision Table 4, it is clear that this MinMaxScaler Binary PSO is performing better among them. We have tried to replace sigmoid function with MinMaxScaler Binary PSO since this function also maps the output in the range of 0 and 1. That is why, it has been explored further in the hope of getting better result.

Table 3 Summary of accuracy and number of genes selected in MinMaxScaler PSO

Dataset SRBCT	Run1	Run2	Run3	Run4	Average
Acc	97.59	98.8	97.59	97.59	97.89
Selected features	1130	1134	1138	1148	1137

Table 4 Comparision table

Dataset SRBCT	IBPSO	EVPSO	Genetic algorithm	KNN	MinMaxScaler PSO
Acc	97.59	95.66	81.9277	91.57	97.89
Selected genes	1124	1050	46	0	1137

And, it is showing some improvement. Its performnance would have improved more if vmax and vmin values have not been taken as fixed value. In future, we will try to work on this more to enhance its capability. So, this may be beneficial in some cases (e.g., here we have worked on cancer data) and may benefit the society by predicting more accurate result.

12 Conclusion

Irrelevent features sometimes reduce classification accuracy. That is why feature selection is used. Feature selection is a preprocessing task whose objective is to select minimum relevant features and gives maximum classification accuracy. This paper presents a small survey of various developments and modifications of particle swarm optimization and proposed a new MinMaxScaler binary particle swarm optimization assuming that the function of MinMaxScaler maps output in the range of 0 and 1 and might improve accuracy. Here the change is done only after calculating updated velocity. The proposed method is compared with other PSO variants and some feature selection methods. Experimental result shows comparable classification accuracy to other methods. If this method is incorporated with some other improved techniques, then its performance in future may improve.

References

1. Eberhart, R.C., Shi, Y., Kennedy, J.: Swarm Intelligence. Elsevier (2001)
2. Chuang, L., Chang, H., Tu, C., Yang, C.: Improved binary PSO for feature selection using gene expression data. Comput. Biol. Chem. **32**, 29–38 (2008)
3. Agarwal, S., Rajesh, R., Ranjan, P.: Enhanced velocity BPSO and convergence analysis on dimensionality reduction. In: Recent Advances in Mathematics, Statistics and Computer Science, pp. 413–421 (2016)
4. Norouzi, N., Sadegh-Amalnick, M., Tavakkoli-Moghaddam, R.: Modified particle swarm optimization in a time-dependent vehicle routing problem: minimizing fuel consumption. Optim. Lett. **11**, 121–134 (2016)
5. Bansal, J., Deep, K.: A modified binary particle swarm optimization for knapsack problems. Appl. Math. Comput. **218**, 11042–11061 (2012)
6. Han, H., Lu, W., Qiao, J.: An adaptive multiobjective particle swarm optimization based on multiple adaptive methods. IEEE Trans. Cybern. **47**, 2754–2767 (2017)
7. Agarwal, S., Ranjan, P.: Optimum feature selection using new ternary particle swarm optimization in two phases. J. Intel. Fuzzy Syst. **33**, 2095–2107 (2017)
8. Chiang, C.: Quantum-behaved particle swarm optimization for economic/emission dispatch problem of power system. DEStech Trans. Comput. Sci. Eng. (2018)
9. Wu, A., Yang, Z.: An Elitist Transposon quantum-based particle swarm optimization algorithm for economic dispatch problems. Complexity **2018**, 1–15 (2018)
10. Lin, A., Sun, W., Yu, H., Wu, G., Tang, H.: Global genetic learning particle swarm optimization with diversity enhancement by ring topology. Swarm Evol. Comput. **44**, 571–583 (2019)
11. Xia, X., Gui, L., Zhan, Z.: A multi-swarm particle swarm optimization algorithm based on dynamical topology and purposeful detecting. Appl. Soft Comput. **67**, 126–140 (2018)

12. Zhang, Y., Gong, D., Cheng, J.: Multi-objective particle swarm optimization approach for cost-based feature selection in classification. IEEE/ACM Trans. Comput. Biol. Bioinf. **14**, 64–75 (2017)
13. Agarwal, S., Rajesh, R., Ranjan, P.: FRBPSO: a fuzzy rule based binary PSO for feature selection. Proc. Natl. Acad. Sci. India A. Phys. Sci. **87**, 221–233 (2017)
14. Agarwal, S., Ranjan, P.: Map reduce fuzzy ternary particle swarm optimization for feature selection. J. Stat. Manage. Syst. **20**, 601–609 (2017)
15. Shu-ting, L., Xian-wen, G.: Adaptive simulated annealing particle swarm optimization for catalyst protected region parameter identification. In: 29th IEEE Chinese Control and Decision Conference (CCDC), pp. 1580–1585 (2017)
16. Wu, Z., Zhang, S., Wang, T.: A cooperative particle swarm optimization with constriction factor based on simulated annealing. Computing **100**(8), 861–880 (2018)
17. Qasim, O., Algamal, Z.: Feature selection using particle swarm optimization-based logistic regression model. Chemometr. Intel. Lab. Syst. **182**, 41–46 (2018)
18. Khan, I., Maiti, M. K., Maiti, M.: Coordinating particle swarm optimization, ant colony optimization and K-Opt algorithm for traveling salesman problem. In: International Conference on Mathematics and Computing, pp. 103–119. Springer, Singapore (2017)
19. Tong, Q., Zou, X., Zhang, Q., Gao, F., Tong, H.: The hardware/software partitioning in embedded system by improved particle swarm optimization algorithm. In: Fifth IEEE International Symposium on Embedded Computing, SEC'08, pp. 43–46 (2008)
20. Pandey, S., Wu, L., Guru, S. M., Buyya, R.: A particle swarm optimization-based heuristic for scheduling workflow applications in cloud computing environments. In: 24th IEEE International Conference on Advanced Information Networking and Applications (AINA), pp. 400–407 (2010)
21. Li, Z., He, L., Li, Y.: A novel multiobjective particle swarm optimization algorithm for signed network community detection. Appl. Intell. **44**(3), 621–633 (2016)
22. Batista, G.C., Silva, W.L.S., Menezes, A.G.: Automatic speech recognition using support vector machine and particle swarm optimization. In: IEEE Symposium Series on Computational Intelligence (SSCI), pp. 1–6 (2016)
23. Wang, R.: Research on image processing based on improved particle swarm optimization. In: 10th IEEE International Conference on Measuring Technology and Mechatronics Automation (ICMTMA), pp. 538–540 (2018)
24. Chen, W., Panahi, M., Pourghasemi, H.R.: Performance evaluation of GIS-based new ensemble data mining techniques of adaptive neuro-fuzzy inference system (ANFIS) with genetic algorithm (GA), differential evolution (DE), and particle swarm optimization (PSO) for landslide spatial modelling. Catena **157**, 310–324 (2017)
25. Nouiri, M., Bekrar, A., Jemai, A., Niar, S., Ammari, A.C.: An effective and distributed particle swarm optimization algorithm for flexible job-shop scheduling problem. J. Intel. Manuf. **29**(3), 603–615 (2018)
26. Aboueldahab, T., Fakhreldin, M.: Prediction of stock market indices using hybrid genetic algorithm/particle swarm optimization with perturbation term. In: International Conference on Swarm Intelligence, ICSI (2011)
27. Bin Shalan, S., Ykhlef, M.: Solving multi-objective portfolio optimization problem for Saudi Arabia stock market using hybrid clonal selection and particle swarm optimization. Arab. J. Sci. Eng. **40**, 2407–2421 (2015)
28. Lin, K., Hsieh, Y.: Classification of medical datasets using SVMs with hybrid evolutionary algorithms based on endocrine-based particle swarm optimization and artificial bee colony algorithms. J. Med. Syst. **39**(10), 119 (2015)
29. Alizadeh Naeini, A., Babadi, M., Mirzadeh, S., Amini, S.: Particle swarm optimization for object-based feature selection of VHSR satellite images. IEEE Geosci. Remote Sens. Lett. **15**, 379–383 (2018)
30. Xue, B., Zhang, M., Browne, W.: Particle swarm optimization for feature selection in classification: a multi-objective approach. IEEE Trans. Cybern. **43**, 1656–1671 (2013)

31. Ghamisi, P., Benediktsson, J.: Feature selection based on hybridization of genetic algorithm and particle swarm optimization. IEEE Geosci. Remote Sens. Lett. **12**, 309–313 (2015)
32. Mistry, K., Zhang, L., Neoh, S., Lim, C., Fielding, B.: A Micro-GA embedded PSO feature selection approach to intelligent facial emotion recognition. IEEE Trans. Cybern. **47**, 1496–1509 (2017)
33. Udhaya Kumar, S., Hannah Inbarani, H.: PSO-based feature selection and neighborhood rough set-based classification for BCI multiclass motor imagery task. Neural Comput. Appl. **28**, 3239–3258 (2017)
34. Li, H., Guo, W., Wu, G., Li, Y.: A RF-PSO based hybrid feature selection model in intrusion detection system. In: 2018 IEEE Third International Conference on Data Science in Cyberspace (DSC), pp. 795–802 (2018)
35. Liu, M., Xu, L., Yi, J., Huang, J.: A feature gene selection method based on ReliefF and PSO. In: IEEE 2018 10th International Conference on Measuring Technology and Mechatronics Automation (ICMTMA), pp. 298–301 (2018)
36. Ahila, R., Sadasivam, V., Manimala, K.: Particle swarm optimization-based feature selection and parameter optimization for power system disturbances classification. Appl. Artif. Intell. **26**, 832–861 (2012)
37. Hamed, H.N.A., Kasabov, N., Shamsuddin, S.M.: Integrated feature selection and parameter optimization for evolving spiking neural networks using quantum inspired particle swarm optimization. In: IEEE International Conference on Soft Computing and Pattern Recognition, SOCPAR'09, pp. 695–698 (2009)
38. Siqueira, H., Figueiredo, E., Macedo, M., Santana, C.J., Santos, P., Bastos-Filho, C.J., Gokhale, A.A.: Double-swarm binary particle swarm optimization. In: 2018 IEEE Congress on Evolutionary Computation (CEC), pp. 1–8 (2018)
39. Shi, Y., Eberhart, R.: A modified particle swarm optimizer. In: 1998 IEEE International Conference on Evolutionary Computation Proceedings. IEEE World Congress on Computational Intelligence (Cat. No. 98TH8360), pp. 69–73 (1998)
40. Statnikov, A.: Gene Expression Model Selector. www.gems-system.org (2005)
41. Agarwal, S., Ranjan, P.: Optimum feature selection using new ternary particle swarm optimization in two phases. J. Intell. Fuzzy Syst. **33**(4), 2095–2107 (2017)

Predicting the Stock Market Behavior Using Historic Data Analysis and News Sentiment Analysis in R

A. C. Jishag, A. P. Athira, Muchintala Shailaja and S. Thara

Abstract Predicting the stock market has always been an attractive topic, mainly due to its vitality in the economic and financial sectors. Yet, predictions of the stock market pose a challenging exercise, even to the brightest and sharpest minds in the business. Prediction of stock market is never an easy task, due to the complexity and dynamic characteristics of the data it deals with. Bulk amount of the data output generated by the stock market is considered to be a treasure house of knowledge for investors; several studies have been conducted in an attempt to predict the stock market trends. Hence, it is imminent to uncover the behavior of the stock market data in order to avoid future investment risks for the investors. Here we tried a different approach for solving this problem by combining two different components: sentiment analysis on stock-related news reports and historic data analysis. The primary aim of this study was to construct an efficient model to predict trends in the stock market, with minimum error ratio and with maximum accuracy possible for the prediction. This model achieved notably better accuracy as compared to the models created in the previous studies. Two datasets were used in this study. A historical dataset containing the stock values of over ten 11, xxxx companies in the previous years, and a sentiment dataset containing the stock market news reports from social media and other online sources. The first step was to analyze the stock reports and classify them either as a positive or a negative sentiment. Lexicon method of text sentiment classification was used for this purpose. Predictions at this stage achieved an accuracy of 67.14%. The second step of this study used ts and ARIMA functions to predict stock trend, using the historical dataset. In the final step, results from both the components were combined together, to predict stock prices in future. This improved the prediction accuracy up to 89.80%.

Keywords Historic data analysis · Lexicon method · Sentiment analysis · Stock market prediction

A. C. Jishag (✉) · A. P. Athira · M. Shailaja · S. Thara
Amrita Vishwa Vidyapeetham, Amritapuri, India
e-mail: jishagac@gmail.com

S. Thara
e-mail: thara.amrita06@gmail.com

© Springer Nature Singapore Pte Ltd. 2020   717
A. K. Luhach et al. (eds.), *First International Conference on Sustainable Technologies for Computational Intelligence*, Advances in Intelligent Systems and Computing 1045,
https://doi.org/10.1007/978-981-15-0029-9_56

1 Introduction

Stock market also known as a share market or an equity market is a digital marketplace, where the sellers and buyers, from various parts of the globe, meet for economic transactions. These transactions are carried out with a discrete entity of stocks or shares. These entities represent ownership claims on various business organizations. It might also represent securities listed on the public as well as privately traded stock exchanges. Examples of privately traded stock exchanges include shares of various private companies. These shares are sold to the investors or buyers through various equity crowd funding platforms.

The complex behavior and volatility in the data make the decision-making in stock market a tedious work for investors. This points to an important need to explore and uncover the bulk amount of valuable data generated everyday by the stock market. Investors tend to look for ways of predicting the future stock prices. Such predictions help them determine the best time to carry out a transaction. A transaction carried out at an opportune time can lead to the best results on the investments. Since the later part of the last century, stock trading can be carried out either digitally or physically. Stock market prediction provides strong insights into the future stock behavior to investors, helping them in their investment decisions.

Stock market prediction has always been of immense interest to investors. Stock market windows produce bulk amount of data, and these are considered as a treasure of knowledge for investors. This is the major risk in the prediction of the stock market. The bulkiness of data makes the stock values vulnerable to any incident happening in the economic market. Prediction gets more complicated when all of these factors affecting the stock prices have to be considered for prediction. In this study, we aim to construct an effective and efficient model that can predict the future trends in the stock market with the minimum error ratio and maximum possible accuracy for the prediction. The prediction model is primarily based on two components: sentiment analysis and historical data analysis. Sentiment analysis component would find the polarity of stock market-related news articles from various sources. Historical data analysis component would predict a function based on the stock prices from the previous years.

2 Literature Survey

Stock market prediction is a method of understanding the upcoming fluctuations in the stock prices of a company. Several researches have gone into this field. Some of them focused on methods to improve the accuracy of the prediction based on sentiment analysis of news related to stock trends, while others focused on the prediction of price differentials with different phases. Studies have proved the presence of a strong correlation between stock-related news and fluctuation in stock prices.

2.1 Studies that Rely on the Online Media Stock Information

Bing in [1] proposed an algorithm for an accurate identification of stock prices by studying social media data. In this study, Bing used several NLP routines, concurrent with data mining techniques. He successfully deduced a relationship between public sentimental values and numeric stock prices, in multilayer hierarchical structures. He also found a relationship within the lower layers and top layer of unstructured data.

Cara in [2] proposed a model to predict the Indonesian stock market based on online articles using sentimental analysis. This study had three main objectives— prediction of stock? price fluctuation, computations on margin percentage, and predicting the future stock prices. He made use of five unsupervised algorithms: support vector machine (SVM), decision tree, random forest, naive Bayes, and neutral networks. Cara observed that the random forest and naive Bayes algorithms performed much better compared to their counterparts. But a conspicuous limitation of this study was that the derived prediction model factored in only the stock prices of the five records that existed in the past.

Hana Hasans approach in [3] predicted stock trends, of a rise or a fall, combining data hourly reports from online Web sites, breaking news on financial channels, along with one-hour charts of stock prices, in the research demonstrated that the use of logistic regression along with a keyword performed well in the prediction.

2.2 Studies Related to News Analysis

Patrick et al. in [4] deployed several text mining techniques in performing sentiment analysis in the financial domain by integrating the word and the resources to understand and study the stock market reports. Their study was focused on German language, which was used as a tool for sentiment analysis, at various levels. Stock values were compared to the sentiment measured models, in identifying the person who invested. This was later used in publishing investment recommendations, to avoid the future investment risks.

Shynkevich in [5] made use of the multiple kernel language technique in investigating the usage of two different classes of articles related to stock prices that focused on targets. Shynkevich observed that these two classes could enhance the accuracy in prediction of stock trends, by using data from news reports and historical stock prices. He made use of open and close attributes. He also showed that usage of SVM AND K-NN algorithms? lead to share drops in the prediction accuracy. Rule mining was also used to uncover the stock market and generate the regulations in the prediction of stock price. Naive Bayes algorithm was used in this study to predict the class labels.

Hoang and Phayung proposed a model in [6] to identify upcoming stock trends on stock exchanges in Vietnam, using news articles from various publications. The author's combined algorithms from general SVM with linear SVM in their predictions, based on the closing stock prices.

Jageshwar and Shagufa in [7] superimposed financial news reports on the observed daily changes in stock prices. The main intent of this scheme was to increase the prediction accuracy, by combining elements of technical analytics with rule-based classifiers.

In [8], Ruchi and Gandhi conceived a model to predict stock trends observed in the stock market, by employing non-quantifiable information from articles. The authors developed a behavioral model based on statistical parameters, which helped increase the prediction accuracy.

Saadi in [9] observed a relationship in economic news reports, using time series methods. He made use of closing stock market values with ten methods, for time serializing the data, followed by integrating them with machine learning algorithms like SVM and K-NN.

In [10], Kim studied the stock market values. His research was focused on mining opinion analytics. The study proved a strong correlation between news reports and the stock prices.

Abdulla [11] scrutinized the Bangladesh stock market using NLP and text mining techniques derived the information for his research. This study made use of an information parser algorithm along with Open NLP, to familiarize the investor with risks in selling or buying a particular stock.

Thara and Krishna in [12] used random Fourier features in the detection of sentiment analysis. This study made a selective choice from various relevant features and was successful in showing a considerable improvement in the accuracy of the detection of aspect-based polarities. This model was able to yield an accuracy of 90%.

In [13], Thara and Sidharth focused on the classification of data into different categories of polarity. They were able to present a comparative result on the classification. This study concluded that the SVM, algorithm along with polynomial and RBF outperformed all other algorithms.

Veena et al. in [14] identified and characterized cells as malignant and not-malignant in association with mini-chromosomes. This study assessed the potential of this association as the biomarkers of malignant and benign tissues of lungs. They used lexicon method as a factor of classification.

Priyanka et al. in [15] conducted a study on methanolic extracts of turmeric cultivars. This investigation pointed to a statistically significant radical scavenging activity. The authors were able to establish Prathibha, a variety, displayed a very significant curing method. This study also depended upon the algorithms of SVM and random forest routines for the classification.

3 Proposed Method

Stock market is an ever-vibrant section of financial management. There are tons of Web sites and agencies, which promise stock predictions to a great level of accuracy to attract customers. Prediction of stock market performance is an impossible task, theoretically. Voluminous stock transactions and the spontaneity of the factors affecting the data bear testimony to the theory. Spanning from the wide range of the unexpected changes in the business climate to the mood outlandish swings of capricious CEOs, anything and everything can affect stock prices.

Several attempts have been made in the past to enhance accuracy in the prediction of stock prices as well as overall performance of stock exchanges. Most of this prediction was models aimed to predict future performance of stocks, on from the analysis of historic data. None of these models succeeded in their prediction accuracy of at least 70–75%. To minimize the errors of prediction, researchers resorted to methods of machine learning, and sentiment analysis of stock reports on the social media. Such measures scarcely achieved their objectives due to uncontrollable trolls of unauthenticated reports on social media and polarized viewpoints of bloggers.

This paper describes a model combining these two ideologies: sentiment analysis of social media reports and analytics of historic stock prices. Figure 1 shows a block diagram of the proposed algorithm, in a step-by-step approach.

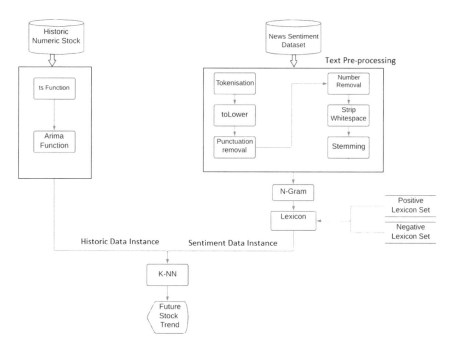

Fig. 1 Data flow diagram—proposed algorithm

3.1 Data Description

Data of over ten companies traded on the NASDAQ stock exchange in New York, USA, was used for this study. NASDAQ is one of the global electronic marketplaces for selling and buying of stocks and securities. It is also considered to be the benchmark index for the US technology stock exchanges. Two types of datasets were collected for this study—news sentiment dataset and historic numeric dataset. Historic data is constituted of stock-related numeric data from NASDAQ. Sentiment dataset was collected from various online sources such as nasdaq.com, the Wall Street Journal, and Yahoo Finance. These sources post each days news about various stock exchanges, concerning several companies. The data considered from these sources was articles, about stocks dividends, company splitting, mergers and acquisitions, and reports from financial experts The news articles culled for the present research were specific to the companies selected for this study.

3.2 Proposed Model Description

This section describes details of each component of the model proposed in this paper. This model is comprised of two components, namely sentiment analysis and historic data analysis.

(1) *Sentiment Analysis*: Sentiment analysis component explains the analysis of stock-related news. The primary objective is to classify the reports to be either positive or negative. Several data preprocessing routines are performed on the stock reports. This action is followed by the classification of the news data using the lexical approach. The following section describes the preprocessing steps in detail:

- Text preprocessing
  Tokenization: It is the process of splitting textual contents of a document into atomic components, tokens. In the context of this probe, tokenization was used to split the stock reports into meaningful words.
  Data Standardization: It is the process used in data analytics to bring all the data within a dataset to a unique common format. For this study, Data standardization was used to transform all the words in the stock report into lower case.
  Stop-word-removal: Certain words in stock reports may not convey significant meaning in the context of the stock prices, such as the, a, of etc. Stop-word-removal is the process of removing these words from the stock report or document.
  Stemming: It is the process of removing suffixes, such as -ed, -ing, -ion, from a root word. This process helps reduce the complexity of the document and minimize the processing time, which improves the model's performance.
- Lexicon Method There are primarily two approaches for creation of a sentiment analysis model—machine learning and lexical approach. In the machine learning

approach, the older posts and tweets, from online platforms, are used as a training set to train a model. This model is used to predict the upcoming stock trend, by passing the latest tweets through this model. On the other hand, in the lexicon method, a pre-defined library containing the set of English words is used, categorized as either positive or negative. A weight is assigned to each of these words with respect to its positivity or negativity, in the context of stock market. Each word, from the online posts and tweets, is compared with this positive and negative list, and the overall sentiment of the post is computed. Lexicon method is far simpler and takes lesser time to evaluate as compared to the machine learning approach, but the latter provides a higher degree of accuracy. The algorithm discussed in this paper preferred the simpler lexicon method.

Data obtained from the online sources was first preprocessed, prior to being subjected to sentiment analysis. The raw data is first made into tokens or words by the process of tokenization. The resultant tokens are recast into a consistent format by conversion of all the upper case letters to the lower case. Now the use of punctuations is taken care of and the numbers are removed. Any white spaces are then stripped off, before the data is ready for sentiment analysis.

- N-Gram: This algorithm is applied upon the preprocessed data, just before sentiment analysis, to compensate for the low accuracy of the lexicon method. N-gram finds all combinations of adjacent words or letters of length n that constitute the dataset, thereby enhancing the accuracy.

Upon the successful implementation of sentiment analysis routines, a graph is plotted, stating the positive and negative sentiment values, as shown in Fig. 2.

Fig. 2 Final result from sentiment analysis

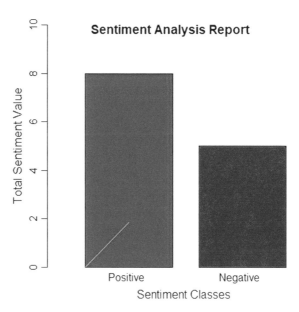

(2) *Historic Data Analysis*: Historic data analysis is an inevitable part of stock market prediction. It processes the part stock trend and models a graph predicting the upcoming market trend. The algorithm discussed in this paper used autoregressive integrated moving average (ARIMA) and time series (ts) functions in R Studio for this purpose.

- Time Series Function Time series function or ts function is used to create time series objects. These time series objects are matrices or vector with class of ts, which represent the data, sampled at periodic points in time. In the context of this paper, ts function intakes the historic dataset and serializes the data according to time. This helps in the creation of a function with respect to the existing stock trend. The resultant function is provided to ARIMA function.
- ARIMA Function ARIMA stands for autoregressive integrated moving average. It is a model class that captures a suite of different standard temporal structures in the time series data. There are different definitions available for ARIMA as it is a combination of three different entities or classes: auto regression, integrated, and the moving average. Also, each definition of the ARMA models has different signs for the AR and/or MA coefficients.

$$X_t = a_1 X_{t-1} + \cdots + a_p X_{t-p} + e_t + b_1 e_{t-1} + \cdots + b_q e_{t-1} \qquad (1)$$

The differenced series follows a zero-mean ARMA model for the ARIMA models with differencing. In the aspects of our study, this function helps us in identifying a function to be plotted as a graph for the historic trend of the past historic data as in Fig. 3.

Fig. 3 Forcasted output from ARIMA function

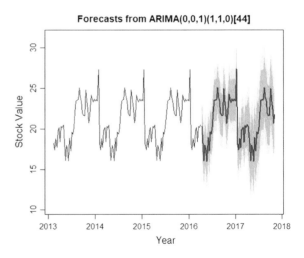

Finally, the results from both sentiment data analysis and historic data analysis are combined showcased in the shiny application. In the users perspective, this gives them an easier way of understanding the both historic and sentiment aspects of the company at a single stretch.

4 Experimentation and Result

This section describes the results of the various experiments, performed in this study, to predict trends in the stock market, using sentiment analysis and historic data analysis. The trials can broadly be categorized into two phases—the initial phase covering results from the sentiment analysis component that classified the stock reports and the subsequent phase describing the historic data analysis component that processed the historic stock values and plotted a function for the same. Stock values of ten companies were taken into consideration for either of the phases.

4.1 Results of Sentiment Analysis

Table 1 shows the results of the experiments on the sentiment analysis component. The result shown is a cumulative accuracy the prediction of the sentiment analysis component corresponding to each company.

The results from these experiments stated that the model proposed in this paper is able to achieve a better accuracy when compared with the models from previous studies in the field of stock market. Also, it was observed that the execution time was considerably low, compared to the existing models. Never did we have a system for sentiment analysis in the market, which could surpass an accuracy of 80%. Almost all of the results for stock market news sentiment analysis models gathered an accuracy in the range of 71–75%. We were able to obtain accuracy ranging from 71.95 to 86.21%.

Three algorithms were compared upon the datasets of the selected companies so as to ensure a high accuracy in the results. The algorithms selected for this purpose were naive Bayes classifier, K-nearest neighbors (K-NN), and support vector machine (SVM).

The results from this comparison are summarized in Table 2. The table contains the accuracy obtained from each of the classifier algorithms ran on each company's

Table 1 Sentiment analysis result based on each company

Company	Yahoo	Msft	Fb	AAL	Apple	Adobe
Accuracy	86.21	72.73	82.76	80.52	76.32	81.22

Table 2 Accuracy of each classifier upon each dataset

	K-NN (%)	SVM (%)	Naive Bayes (%)
Yahoo	78.75	58.62	86.17
Msft	69.12	54.48	76.18
Fb	75.15	70.14	79.00
AAL	80.73	62.49	81.58
Apple	71.88	49.76	84.56
Adobe	70.01	60.09	76.11

dataset. The results clearly exhibit that the naive Bayes algorithm outperformed both the SVM classifier and the K-NN algorithm with respect to textual data. It was also observed that the SVM classifier registered the least accuracy when dealing with textual data in comparison with K-NN and naive Bayes for our experimental dataset.

4.2 Results of Historic Data Analysis Component

Table 3 shows the prediction accuracy with and without the use of historic data component. It is evident from the results that the historic data analysis gives a big boost in the accuracy of the prediction model. Hence, it can be validated that there is a strong relation between the history of the stock prices and the stock news reports for these stock prices. Concerning prediction of the stock market, hardly any of the researchers have ever combined these two components together for the prediction of stock prices. And hence most of these researchers achieved an accuracy ranging from 74 to 81%, for the prediction on stock performance trends.

This study clearly showcases that the sentiment analysis component alone cannot identify the future scope of the stock market. While sentiment analysis component alone achieves an accuracy of 59.18%, sentiment analysis component, when combined with the analysis of the past stock data achieved accuracy up to 89.79%. The prediction model mainly depends on the K-NN algorithm, which earlier proved to be the most efficient when dealing with textual and numeric data.

Thus, the results so far clearly depict that the model described in this paper is an effective and efficient way to predict the future stock trends with a relatively higher accuracy in comparison with the existing models. To verify this assumption,

Table 3 Accuracy with and without the historic component for the proposed model

Measurement	Sentiment component alone	Sentiment and historic component
Accuracy (%)	67.14	89.80

Table 4 Comparison of the proposed model with the existing models

Previous studies	Obtained accuracy (%)
Model by Bing [1]	76.12
Cara's model [2]	60.39–67.73
Shynkevichl Model [5]	79.59
Model proposed by Phyng [11]	75
Kim's model [10]	60–65
Model proposed in this paper	89.80

the proposed model is compared with the accuracy obtained by the existing models and the models from the previous researches. Results from Table 4 demonstrate that our proposed model outperforms all the other previous studies and models existing currently in the market.

5   Conclusion

The proposed model conducted an investigation upon the effects of combining two different types of analyses for predicting the future stock trend, namely news sentiment analysis and historical stock data analysis.

This paper has proposed model that showed a considerable increase in the accuracy of the prediction when the historic analysis was combined with the sentiment stock report component. Sentiment analysis component consisted of news reports collected about various companies over the social media and other online platforms. Historical stock data component was comprised of the historical data trend of each company for the previous years. The first stage of the proposed model was to polarize the news reports to either positive or negative using lexicon method. The second stage was to predict a future stock trend using ARIMA and ts functions, upon the historic stock dataset, for each company, obtained from NASDAQ. This was followed by incorporation of the outputs from the first and second stages as the input in the prediction of the future stock market using K-NN algorithm. The results of the proposed model achieved an accuracy of 89.80%. The results obtained by this study also pointed toward a strong relation between the sentiment analysis report of the stock reports and the historic stock market trend. This model can be further refined by the inclusion of machine learning or neural network algorithms in order to recognize the underlying emotions in the classification of the news as positive or negative.

References

1. Bing, L.I., Ou, C.: Public sentiment analysis in Twitter data for prediction of a company s stock price movements. In: IEEE 11th International Public, Conference E-bus. Eng (2014)
2. Yahya Eru Cara, B.D.T.: Stock price prediction using linear regression based on sentiment analysis. In: International Conference Advance Computer Science Information System, pp. 147154 (2015)
3. Hana Alostad, H.D.: Directional prediction of stock prices using breaking news on Twitter. In: IEEE/WIC/ACM International Conference on Web Intelligence Intelligent Agent Technology, pp. 07 (2015)
4. Patrick Uhr, M.F., Zenkert, J.: Sentiment analysis in financial market. In: IEEE International Conference System Man, Cybernetics, pp. 912917 (2014)
5. Shynkevichl, Y., Mcginnityl, T.M., Colemanl, S., Belatrechel, A.: Stock Price Prediction based on StockSpecific and Sub-Industry-Specific News Articles (2015)
6. Abdullah, S.S., Rahaman, M.S., Rahman, M.S.: Analysis of stock market using text mining and natural language processing. In: 2013 International Conference Informatics, Electronics Vis, pp. 16 (2013)
7. Price, S.M., Shriwas, J., Farzana, S.: Using text mining and rule based technique for prediction. Int. J. Emerg. Technol. Adv. Eng. **4**(1) (2014)
8. Desai, R.: Stock market prediction using data mining 1, **2**(2), 27802784 (2014)
9. Journal, I., Social, O.F., Studies, H.: TIME SERIES ANALYSIS ON STOCK MARKET FOR TEXT MINING **6**(1), 69–91 (2014)
10. Kim, Y., Jeong, S.R., Ghani, I.: Text Opinion Mining to Analyze News for Stock Market Prediction Int. J. Adv. Soft Comput. Its Appl. **6**(1), 113 (2014)
11. Thanh, H.T.P., Meesad, P.: Stock market trend prediction based on text mining of corporate web and time series data. J. Adv. Comput. Intell. Intell. Informatics **18**(1) (2014)
12. Thara, S., Athul Krishna, N.S.: Aspect sentiment identification using random fourier features. Int. J. Intell. Syst. Appl. **10**(9), 32–39
13. Thara, S. Sidharth, S.: Aspect based sentiment classication: SVD features. In: 2017 International Conference on Advances in Computing, Communications and Informatics, ICACCI 2017, pp. 2370–2374 (2017)
14. Rafeek, R., Remya, R., Detecting contextual word polarity using aspect-based sentiment analysis and logistic regression. In: ICSTM 2017-Proceedings, : IEEE International Conference on Smart Technologies and Management for Computing. Communication, Controls, Energy and Materials (2017)
15. Vijayan, V.K, Bindu, K.R., Parameswaran, L.A.: Comprehensive study of text classication algorithms. In: 2017 International Conference on Computing and Network Communications, CoCoNet 2015 (2015)

DDoS Attacks Detection and Mitigation Using Economic Incentive-Based Solution

Amrita Dahiya and B. B. Gupta

Abstract DDoS attack is posing an immense threat to Internet and online businesses. DDoS attack makes an online service unavailable to legitimate users by sending voluminous number of dummy requests or by exploiting some vulnerability or security flaws present in current Internet infrastructure. Information present on Internet needs to be defended against DDoS attack by not only technological means but also with the help of economic means using incentives as a tool. In this paper, we have proposed a model using a network management strategy which involves policies to maintain quality of service (QoS) during a DDoS attack. User must have sending rights in the form of contract to access server. A XML form having different fields on QoS parameters is provided by the service provider which a user has to fill with the help of policies present in policy pool. This XML form is turned into contract after evaluation by the control broker. Policy-based network management (PBNM) is used to execute proposed approach which enables user to have dynamic negotiation with the service provider on the cost and types of resources. Proposed model has been implemented on NS2. Results obtained from simulation show the supremacy of our proposed model.

Keywords DDoS attack · Economic incentives · PBNM · QoS · Risk transfer

1 Introduction

Internet is traditionally designed for functionality not for security purposes resulting in the presence of many security flaws, and DDoS attack is direct implication of this fact. Cumulative efforts to send traffic by multiple bots (compromised systems) distributed all over the Internet targeting a specific node, and service or server is called as DDoS attack. DDoS attack makes specified target paralyzed or completely down for legitimate users. Traffic from one compromised system is not huge enough to be detected by local ISPs and attacker takes advantage of this fact. DDoS is a

A. Dahiya · B. B. Gupta (✉)
National Institute of Technology, Kurukshetra, Kurukshetra, Haryana, India
e-mail: gupta.brij@gmail.com

© Springer Nature Singapore Pte Ltd. 2020
A. K. Luhach et al. (eds.), *First International Conference on Sustainable Technologies for Computational Intelligence*, Advances in Intelligent Systems and Computing 1045,
https://doi.org/10.1007/978-981-15-0029-9_57

much generalized concept, and it can be carried out at any point of network, i.e., on a bandwidth link, router or ISPs. DDoS attack is carried out in two ways: either recruiting zombies or flooding the victim with dummy requests [1]. In the process of making systems compromised, security loopholes of a network or a device are exploited. While process of flooding, the victim with dummy requests is carried out directly on victim with the help of IP spoofing. There are many evidences present in history that depicts the level of deterioration caused by DDoS attack. For example, many big e-commerce firms like Amazon, Yahoo!, E-bay, CNN, and Netflix have been attacked with DDoS attack at least once causing not only reputational loss but also loss of users or clients. There are many technical solutions proposed by researchers in the last decade [2–7], but only a few solutions present in economic domain [8, 9]. We do have many good cooperative technical solutions to combat DDoS attacks like cooperative filtering and cooperative caching. These solutions are quiet effective, but organizations lack incentives to implement them. Internet world is a collaboration of multiple entities that require strong motivation or incentives to cooperate with each other in order to provide quality service to end user. We need to move from only technological solutions to a combination of both technological and economical solutions based on properly aligned incentives given to right entity in a network.

Conventional approaches involve directly implementing packet filtering which identifies and removes malicious packets but loses QoS as a tradeoff. However, in this paper, we have proposed a PBNM [10]-based model for defending DDoS attacks which maintains QoS for legitimate users. A user is allowed to send a service request to a server only if he has sending rights. Sending rights can be obtained by having contract with server. XML form facilitates users with ability to create proposals. Resources are classified into classes, i.e., high, medium, and low priority classes. User has to specify his service class in XML form according to his budget and QoS constraints. Appropriate defending actions against DDoS attack will be taken place according to different traffic load levels.

The remaining sections of the paper are arranged as follows. Section 2 presents the literature survey analyzing the solutions against DDoS solutions in economic domain, followed by Sect. 3 presenting proposed work. Section 4 explains experimentation done on NS2, and results are obtained from simulation. Section 5 concludes the paper with future work and directions.

2 Literature Survey

As already mentioned, DDoS attacks can be tackled by not only technological solutions but with the proper blend of cooperative technological and economic solutions. More consideration must be given to economic solutions against DDoS attacks. Although very less research is present in this domain but in this section, we will try to analyze some important work that has previously been done.

Geng and Whinston in [8] presented a concept of e-postal service that inherited the incentives available in usage-based scheme. This e-postal service basically generates incentives for users so that users can cooperate in managing traffic load levels in a network. In e-postal framework, ISPs are made to cooperate with each other in order to reduce traffic cost and lessen the risk of being compromised. Computers need to buy e-stamps and store them at closest edge router. Edge router charges one e-stamp to make one traffic packet from a specific computer enter into e-postal framework. This whole process puts responsibility on the user to use e-stamps wisely. Further, authors have laid the importance of creating incentives at right party in a network. This model might prove to be an effective one but convincing users to switch from monthly or flat-based fee structure to a pricing mechanism based on Internet surfing is a difficult task.

In [11], Huang et al. have laid down very important fact that there exists good cooperative technological solution like cooperative filtering and cooperating caching, but organizations do not have incentives to execute them. Authors have analyzed the broken incentive chain in these technical solutions. Usage-based pricing and capacity provision networking have been proposed as solutions to fix the incentive chain in effective technological solutions.

Geng et al. in [9] have proposed a framework against DDoS attack in wireless environment. They have incorporated cooperative technological solutions with economical solution based on usage-based fee. A two tier solution is proposed where first tier deals with coordinated technical solutions, whereas second tier deals with economic solutions in the form of incentives that must be provided to end users. Coordinated technological solution involves device security, user-based traffic management, coordinated filters, and tracing back mechanisms. Usage-based fee is implemented as an incentive in this approach. Cost-effectiveness of this approach is found through implementation through policy-based networking (PBN).

In [12], Gupta et al. have stated that pricing Internet traffic is very important not only for defending DDoS attacks and providing incentives to users but also to provide QoS to users at the time of congestion. Authors had proposed decentralized computation of prices for online services and commodities because decentralized computation of prices facilitates minimal disruption in services during congestion period. Thus, QoS requirement is different for users and applications; therefore, pricing strategy facilitates multi-level QoS to every type of users and applications. Hence, these are some significant papers that drive us in this direction.

3 Proposed Work

DDoS attack is not only increasing in quantity but also in quality. DDoS attack tends to be more sophisticated and complex. Hiring a person for performing a cyber-attack becomes as easy as searching on Google, and it incurs negligible cost on attackers for rendering victim's services inoperable from few hours to several days. There exists a new trend too, where artificial intelligence is collaborated with IoT-based botnets for carrying out DDoS attacks. This could be very devastating as AI has power and severity to change the map of any network. Earlier there were codes that were forced to run in a loop to make a machine paralyze, and now methods of performing DDoS attacks have completely changed. So, we need to change the defending techniques to cope up with the evolution in attacking methods. Leaving whole responsibility of combating DDoS attacks with the target is the main drawback of defensive solutions proposed so far. Every device in a network must be programmed in such a way that it performs with the best of its capabilities and cooperates with other devices to compensate for the weaknesses.

Our model is a capability-based defensive approach which transfers residual risk of being attacked to third party. Residual risk is the risk left after deploying risk avoiding and risk mitigating methods. In our proposed model, broker network sits between client and server. It acts as shield for target server so that no service request from user can directly reach to server. Broker network between client and server holds accountability of contract for both parties. In order to send a service request, a user must have sending rights which he can obtain by filling a XML form provided by the service provider. After submitting XML form, he can negotiate with the service provider on the cost and types of resources. There are three different traffic load levels, i.e., normal, cautious and alert levels, which is discussed as follows:

(a) **Normal traffic level**: Traffic under normal level can be easily handled by broker network. User need not to have contract for sending service request to server. Users are encouraged to have contract at normal level by giving them incentives of high subscription at low price.

(b) **Cautious traffic level**: Traffic under cautious level has already crossed normal level. Here, packets from users having contract will only reach to server and packets from users that are deprived of contract will get dropped. We will classify incoming traffic according to the type of service class user has specified in XML form and type of service request. Requests from users having contract with server are queued at priority level. While users not having contract are attended by broker network with a message of "set up a contract first".

(c) **Alert traffic level**: Traffic under alert level has already crossed cautious level. Here, broker network will try to find the cooperating ISPs and transfer the attack traffic to the most cost-efficient set of ISPs in return for economic incentives. At alert level, there is full flexibility to front-end broker to shape incoming traffic and drop-free riders packets and maintain queues only for the users with high or low service class or on the basis of factors specific to an organization.

Every user's knowledge about Internet and its security is different. Some users use Internet just for surfing or for basic tasks while some users use it for developing complex applications. So, for the sake of simplicity for users, server will create a form in XML having different fields which a user has to fill according to QoS constraints and send it to CB. We take certain assumptions here that every user and service provider has policy pool, i.e., policy repository and every user have a utility function. User's and server's policy pool contains various policies like sending policies (for user), receiving policies (for server), preference policies, and security policies. Policies present in policy pool help users to fill the XML form to be sent to CB in the form of proposal. Policy pool of server helps server to create its proposal which will be a standard proposal on the basis of which user's proposal is evaluated by CB. Utility function of a user depicts his pattern of sending service requests. It is the function of QoS parameters at minimum accepted rate. Summation of utility functions of all users will serve as demand function for service provider which help him to rate the services and to modify form's fields accordingly.

Form has fields like name, service class type, i.e., high, medium, or low priority classes, and type of user, i.e., home user or organization (small sized firm or big enterprise). Organization has to specify number of users that are going to use services of a service provider. According to the type of user and service class type, more different fields will get open up for getting more information about user like how much a user is willing to pay for one service request with or without congestion or what type of content a user wants to watch or wants to block. A service provider can create XML form according to profit, its type, available resources, number of users and market's observables. We are giving here just the basic idea of having forms and fields created by service provider. User can access this form through secure channels. Inherited feature of TLS/SSL protocol called mutual authentication and X.509 public key certifications are used to authenticate user as well as server.

Proposed approach can be effectively executed using policy-based network management (PBNM) in real time. PBNM leads to efficient use of resources as dynamic negotiation and regular updates from users and service providers lead to change in resource's pricing periodically. It provides controlled access to resources. PBNM enables an organization to react rapidly to security related incidents identified through a network assessment procedure.

In our proposed model, CB acts as policy decision point (PDP) while front-end broker acts as policy enforcement point (PEP). Policies help in deciding which packets to drop and which to reroute through the front-end broker. Policy repository at control broker will store all the proposals of authorized users. Figure 1 shows the flowchart of proposed model. Simulation has been carried out in NS2. Results show the supremacy of the proposed model.

4 Experimentation and Results

Proposed model is implemented on NS2. Though there are not standard parameters used by research community for evaluating a model used for defending volumetric DDoS attacks, parameters are divided according to traffic load levels in evaluating the proposed model, i.e., when traffic load level is low and when traffic load level is high. So, we have taken following parameters for the evaluation of our model.

(a) **Malicious packet drop rate**: This parameter evaluates the defensive mechanism on the basis of number of malicious packet dropped from total traffic destined

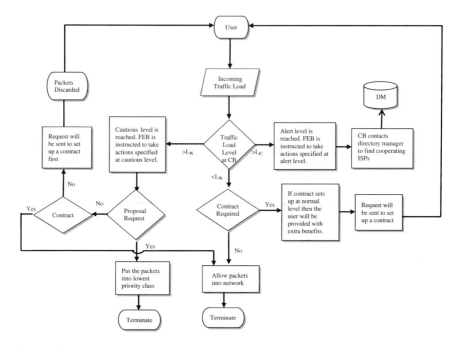

Fig. 1 Flowchart of proposed model

to the specific node. It checks the ability of defensive mechanism to control the flood traffic.

(b) **Legitimate packet drop rate**: As, we have already mentioned the fact that a defensive mechanism should not only provide protection from DDoS attack but also provide QoS to users during attack time. This parameter evaluates the defensive mechanism on the basis of number of legitimate packets reached to server (target node) to the total number of packets destined to server.

Now, we will discuss results obtained by NS2 implementation of the proposed approach. We simulate DDoS attack on a node under cautious level where broker node has to allocate sending rights to clients. Figure 2 shows the screenshot of assignment of sending rights to clients reaching cautious level. Figure 3 shows the transfer of residual risk to incentivized ISP nodes after reaching attack level. When attack level is reached, broker node finds the set of most cost-efficient ISPs to transfer the attack traffic. Many optimization algorithms have been proposed by research community but here we use whale optimization algorithm [13].

Figure 4 shows timeline and variations in legitimate packets delivered to the service provider against malicious intervention at cautious level. It can be clearly shown in this figure that from 30 to 45 s, DDoS attack is at its peak, and there is a significant drop in legitimate packets. Initial delay was incorporated to show the effectiveness of our proposed model. After 45 s, CB and FEB intervene and start working and there is a considerable rise in legitimate packets delivery. Figure 5

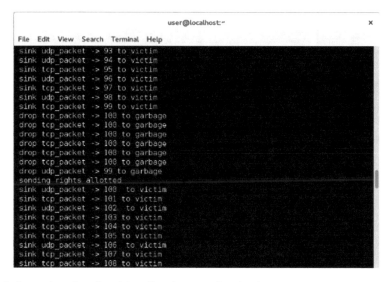

Fig. 2 Screenshot of sending rights allocation at cautious level

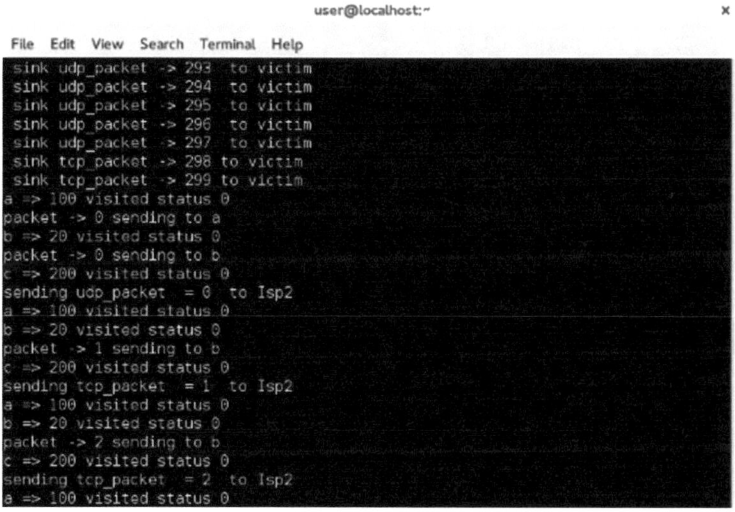

Fig. 3 Screenshot of transferring packets to ISPs at attack level

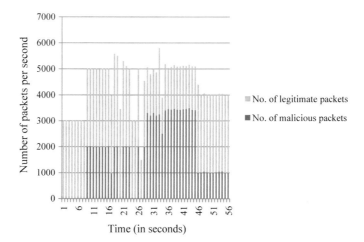

Fig. 4 Simulation of cautious level

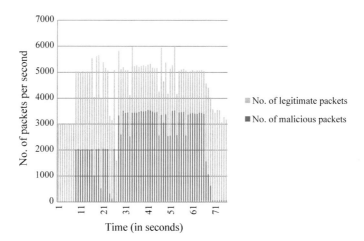

Fig. 5 Simulation of alert level

shows timeline and variations in legitimate packets delivered to the service provider at attack level. Since attack level requires transfer of attack traffic to cooperating ISPs. From 30 to 64 s, there is legitimate packet loss. At instant 64, sudden increase in legitimate packet delivery can be seen after deployment of proposed approach.

Table 1 Comparative analysis of proposed approach with existing solutions

	Mirkovic et al. [14]	DefCom [15]	DIFF [16]	Proposed model
Deployment	At source	Overlay solution	All over the network	Overlay solution
Mode of design	Centralized	Distributed	Distributed	Distributed
Objective	Prevention	Prevention through rate limiting technique	Mitigation through dynamic routing mechanism	Mitigation
Robustness	Weak	Strong	Robust against bandwidth depletion attacks	…
Remarks	Lack of incentives as it provides protection to other users not to the node that deploys it	DefCom can cause DoS attack on legitimate clients from the networks not having DefCom as a defending solution	Mitigates DoS attacks by link utilization of SDN switches	It is a capability based system which mitigates DDoS attack by having dynamic negotiation on cost and types of resources

5 Comparative Analysis

This section outlined comparative analysis of proposed model from existing solutions. We will distinguish our approach from existing standard solutions on the basis of following features (Table 1).

6 Conclusion

There is no denying the fact that DDoS attack can be combated effectively with the help of proper generation of incentives. Further, defenders have fewer incentives than attackers. Only technological solution is not enough for defending DDoS attacks. Therefore, in this paper we have proposed an approach that has the blend of technical and economical aspects. A risk transfer-based approach has been proposed where users having contract are only allowed to access server's resources. Moreover, network management through policies is effectively used for dynamic negotiation and to defend DDoS attacks. Proposed model intends to not only protect server from DDoS attack but also provide QoS to the legitimate users during attack period. Interplay between cooperating ISP's and control broker ensures that legitimate packets are delivered. Results from implementation show the effectiveness of proposed model.

Acknowledgements This research work is being supported by sponsored project grant (SB/FTP/ETA-131/2014) from SERB, DST, Government of India.

References

1. Chang, R.K.: Defending against flooding-based distributed denial-of-service attacks: a tutorial. IEEE Commun. Mag. **40**(10), 42–51 (2002)
2. Wang, X., Reiter, M. K.: Mitigating bandwidth-exhaustion attacks using congestion puzzles. In: Proceedings of the 11th ACM Conference on Computer and Communications Security, pp. 257–267. ACM (2004)
3. Badishi, G., Keidar, I., Sasson, A.: Exposing and eliminating vulnerabilities to denial of service attacks in secure gossip-based multicast. IEEE Trans. Dependable Secure Comput. **3**(1), 45–61 (2006)
4. Simmons, G. J.: Cryptanalysis and protocol failures. In: Proceedings of the 1st ACM Conference on Computer and Communications Security, pp. 213–214. ACM (1993)
5. Sandhu, R.: Role-based access control, advanced in computers. Acad. Press **46** (1998)
6. Vigna, G., Kemmerer, R.A.: NetSTAT: a network-based intrusion detection system. J. Comput. Secur. **7**(1), 37–71 (1999)
7. Ferguson, P.: Network ingress filtering: defeating denial of service attacks which employ IP source address spoofing (2000)
8. Geng, X., Whinston, A.B.: Defeating distributed denial of service attacks. IT Prof **2**(4), 36–42 (2000)
9. Geng, X., Huang, Y., Whinston, A.B.: Defending wireless infrastructure against the challenge of DDoS attacks. Mobile Netw. Appl. **7**(3), 213–223 (2002)
10. Pérez, G.M., Skarmeta, A.F.G., Zeber, S., Spagnolo, J., Symchych, T.: Dynamic policy-based network management for a secure coalition environment. IEEE Commun. Mag. **44**(11), 58–64 (2006)
11. Huang, Y., Geng, X., Whinston, A.B.: Defeating DDoS attacks by fixing the incentive chain. ACM Trans. Internet Technol. (TOIT) **7**(1), 5 (2007)
12. Gupta, A., Stahl, D.O., Whinston, A.B.: The economics of network management. Commun. ACM **42**(9), 57–63 (1999)
13. Mirjalili, S., Lewis, A.: The whale optimization algorithm. Adv. Eng. Softw. **95**, 51–67 (2016)
14. Mirković, J., Prier, G., Reiher, P.: Attacking DDoS at the source. In: Null, p. 312. IEEE (2002)
15. Robinson, M., Mirkovic, J., Michel, S., Schnaider, M., Reiher, P.: DefCOM: defensive cooperative overlay mesh. In: Proceedings DARPA Information Survivability Conference and Exposition, vol. 2, pp. 101–102. IEEE (2003)
16. Guo, Z., Xu, Y., Liu, R., Gushchin, A., Chen, K.Y., Walid, A., Chao, H.J.: Balancing flow table occupancy and link utilization in software-defined networks. Future Gener. Comput. Syst. **89**, 213–223 (2018)

Image Enhancement Using Local and Global Histogram Equalization Technique and Their Comparison

Simran Somal

Abstract This paper reviews about the different algorithms used for image enhancement. Image processing is a process of matching and overlay more than one image of any scene. This is a method of converting any image into its digital form and performs useful operations by applying some function on it. The purpose for image processing is to observe the objects, for sharpening and restore distorted image. These algorithms are improving the attribute of image and reduce the noise in the image. It has been observed that image enhancement is done before segmentation, detection of object and recognition in vision applications. There are various applications of image processing like remote sensing, tracking of moving objects, in defense field, in biomedical techniques. Current researches under image processing are like Cancer imaging, focusing on Brain imaging, different imaging technologies. This paper also provides a work for understanding that how different type of noises can remove by enhancing the image. Image enhancement basically used in Medical Images.

Keywords Digital image processing · Medical imaging · Algorithms · Enhancement

1 Introduction

A white and black sustained picture or a black bar in drawing is referred to a communication channel that is it is a carrier of information. Before we come to modern electrical, communication approach, the most regular means of transmitting message over greater distances. There are two main operators are explore that are Contour enhancement and Contour outlining [1]. When we are having images we need to process those images for different purposes. Analog and digital image processing are two types of image processing. In analog image processing all the images are analog in nature. That is they are represented in Sinusoidal waveform. In digital image processing, images are digital in nature. They are represented in discrete nature. Images

S. Somal (✉)
Chandigarh University, Mohali, India
e-mail: Simrand122@gmail.com

© Springer Nature Singapore Pte Ltd. 2020
A. K. Luhach et al. (eds.), *First International Conference on Sustainable Technologies for Computational Intelligence*, Advances in Intelligent Systems and Computing 1045,
https://doi.org/10.1007/978-981-15-0029-9_58

are used everywhere. For showing the heating effects, images are enhanced at that level. In images, we see how temperature changes by high electric fields [2]. From the various technologies, the one of the most using technology is underwater range-gated technology. By developing the digital image organization, adding different module in every imaging system, it becomes very important to us [3].

Medical imaging is one of the different techniques in digital image processing. Magnetic Resonance Imaging (MRI) is a radiology technique which is used in analyzing the functions of both strength and ailment. In this way, they are using two different modulation techniques, i.e., Spatial-Temporal Psycho visual Modulation (STVPM) and Temporal Pshychovisual Modulation (TVPM) [4]. Now the thing is where all images are processed. All images are processed on different Software. Matlab is the starting tool. After that various software are used like OpenCL, OpenCV, CUDA, LabView, etc. are used for processing images [5–7].

For processing an image, its registration is very important. One important field in image processing is image completion which addresses the problems of filling missing parts in an image [8]. Advances in image processing technique contribute to better observation of retinal microstructures and therefore more accurate detection. More specifically, in Medical image processing where the amount of data is enormous, the necessity of data reduction process that manages to maintain the relevant information content as much as possible of major importance [9–11]. Recently, a promising approach Adaptive optics was suggested to restore the 3D structure of human funds [12, 13].

A high-value optical cameras are become the usual hardware of many devices because of its prices and high resolution [14]. The image colorization technique provides a tincture image from fluorescence image and several color components referred as prototypical pixels. In encoding phase, appropriate representative pixels are stored and the fluorescence image is flattening by excellence coding performance. In decoding phase the colorization algorism recovers a tincture image for fluorescence and representative pixel [15]. The plenoptical candid camera is currently receiving a large attentiveness due to the diminished cost of hardware and the approach in microlens muster fabrication which extends the province of applications [16]. Documents are having different images. These images are used to retrieve the data and to recover the documents [17, 18]. If we need to differentiate the properties of images by different manner of same locality, perfect image certification is required and there are various algorithms to process all these images [19, 20].

In this paper, we review the various image enhancement techniques and algorithms. The basic algorithms which are used for image enhancement is histogram equalization. Histogram equalization further of two types: Local and Global which are discussed further. In Sect. 2, we will give steps of image processing because image enhancement is a part of image processing. Section 3 gives various algorithms and Sect. 4 includes Results and Conclusion about the review paper.

Digital image A digital image is a commutative presentation of a two-dimensional image. Two types of digital images are there: Raster type and Vector type. Raster type images are those which are having fixed value of pixels. Vector images are having one point in both directions. There are two main parts of an image pixel and resolution.

(a) **Pixel**: A pixel is the smallest unit of a digital image. Pixel can be of any shape i.e. triangle, dot or square.
(b) **Resolution**: resolution is density of pixel in an image. For example, if the resolution of an image is 256 * 256, that means it has 256 pixels in row and 256 pixels in column.

2 Steps of Image Processing

For processing an image various operations are applied on it. Figure 1 shows the block diagram of method of image processing:

(a) **Problem Domain**: Problem domain is an image which is to be processed. Image can be of any type. It can be analog or digital in nature basis on the requirement.
(b) **Image Acquisition**: image acquisition is referred to as transform of an optical image into the array of numerical data. That data is further sent for manipulation in the computer. Image acquisition can be achieved by suitable cameras. The cameras which are sensitive to visual spectrum are used for normal images. This is the first step for any image processing technique. Different kind of filters is used for image acquisition.

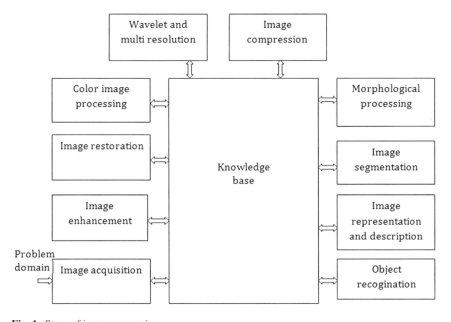

Fig. 1 Steps of image processing

(c) **Image Enhancement**: this is the method of adjusting images for suitable results of display or for image analysis.

Image enhancement is basically of two types:

(i) Geographical Domain Method
(ii) Frequency Domain Method

- **Spatial Domain Method**: In this, we are directly dealing with the pixels of an image. For desired results, pixel values are manipulated.
- **Frequency Domain Method**: Frequency domain method is very simple for image enhancement. In this method, we simply take the Fourier transform of an image and calculate the results.

In Fig. 2, we are showing the process of how Fourier transform takes place in the process of image enhancement.

(d) **Image Restoration**: This is the process of proceeds an image which is filling by noise and changes it into aboriginal better quality image. Noise can be of any type like due to cameras, blurry effects, due to environment etc.

In Fig. 3, we are representing four images, which show the process of image restoration. Image restoration can be done by different type of filters. The various kinds of filters are there:

- **Median Filter**: This method is implied by name in this we replace the values of pixels by median of the pixels which is in neighbor. This method is used for removing the noise like salt and pepper noise. By using this filter images cannot get blurred.

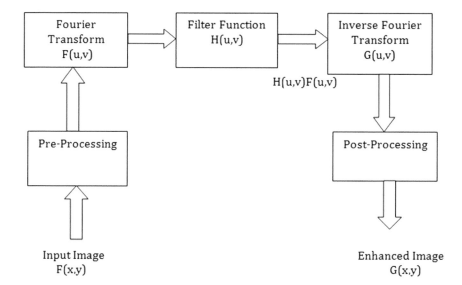

Fig. 2 Process of Fourier transform

(a) (b)

(c) (d)

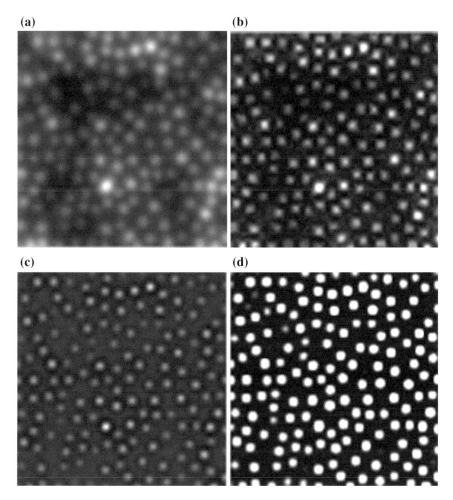

Fig. 3 Process of image restoration [9]

- **Adaptive Filter**: in this type of filter transfer function is controlled by the variable parameter. These types of filters are used for removing the impulse noise. It gives better results for suppression of noise.

(e) **Color Image processing**: it is the type of technique also. If an image does not have colors, from that image we are not able to extract the useful information.

In the Fig. 4, we are having two images, in (a) this is the original image which is to be colored and (b) is sharpened image after applying processes on it.

For this there are two problem domains:

- Color Detection
- Edge Detection.

(a)   **(b)**

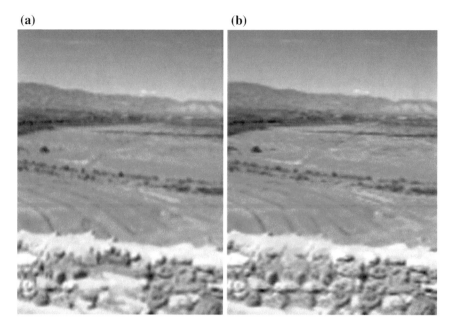

Fig. 4 **a** Original image. **b** Color added to image [8]

In edge detection, object segmentation is done and features of the image are converted. Therefore in edge detection algorithms some of the data get reduced and we are left with some small amount of data which is to be processed.

In color detection, we apply some algorithms to find out which colors are light, from which information is not visible. And we sharp those colors.

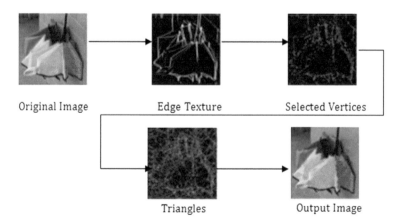

Fig. 5 Framework for edge detection, vertex selecting, triangulation, coloring [6]

In the given Fig. 5, vertex selecting is used for edge detection. The pixel is selected with the high probability of edges compose triangle. In triangulation, Delaunay triangulation is used to select vertices from triangles. It maximizes the minimum angles of triangles. After that, each triangle is fixed with some colors [6].

(f) **Wavelet and Multiresolution**: When we require resolution in frequency domain, then wavelet transformations are used. It is having both frequency and location information of image.

(g) **Image Compression**: Nowadays, memory is the main factor that is how much storage capacity a device has? Or any image or any document is carried how much space in the device. Therefore, we require compression that is to compress the storage.

In image compression, we compress the image so that it is required mall space in the memory. But when we compress the image, we have to see toward some factors of image also like its resolution, pixel, and image quality should remain like real or original image.

In [15], we compress the pixels with the algorithm named as Graph Fourier Transform. In this, a tincture image is diminished by encoding its fluorescence image by excellence coding technique like jpeg format and by reserving some components called representative pixels. In apprehend phase, tincture image is reserved from fluorescence image by image colorization technique.

In Fig. 6, we show the different images which are compressed successfully. And in Fig. 7, they show the graph of the Fourier transform of compressing image. In apprehend phase, we get similar graph that of conceal phase.

(h) **Morphological processing**: this kind of processing is depends upon pixel values and not on numerical values of pixels. This type of processing also applied on grayscale images.

This is the type of structure and formation of the element. This is basically used for recognition of objects. As like for matching fingerprints, for matching or detecting the faces.

In [12], we are using Morphological operations for fabric defect detection. We know that always there are some small defects in fabric, which are not able to detect by human eye. Therefore, we apply Morphological operations on it. Morphological

(a)　　　　　**(b)**　　　　　**(c)**　　　　　**(d)**

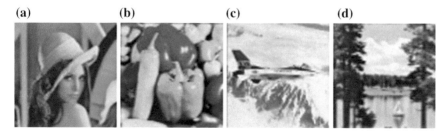

Fig. 6 Test images: **a** Lenna. **b** Peeper. **c** Airplane. **d** Sailboat [15]

Fig. 7 Graph for
compression [15]

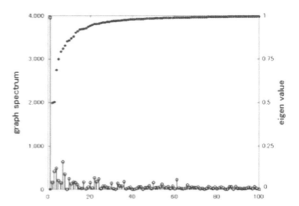

Image processing depends upon some mathematical operations. This is depending on the Local correlation process.

In first, local correlation of each component is calculated to construct a different correlation image i.e. H-Image. Then Histogram calculated for H-image to select a perfect thresholding value to make a similar Binate image, which sends to extricate perfect size and configuration of strutting portion for mathematical Morphological process.

Morphological operations having a goal to remove imperfections by accounting of the form and structure of the image.

(i) **Image Segmentation**: It's a method of segmenting the image into different parts. For doing the Morphological process we need to divide the image into segments so that it becomes a meaningful image.

In [16], they are using a new algorism for image segmentation. It generates a preliminary method of segmentation, either by escalade the potency of an image or by using a new different quantum cut based algorithm.

In Fig. 8, it shows the segmentation of images at different scaling factor (α). In this, they are scaling the intensity of image.

(j) **Image Representation and Description**: In this, we are representing the features of an image. The two terms are including here that is boundary and region.

(a)   **(b)**   **(c)**

Fig. 8 a Binate edge matrix. **b** Segmentation with $\alpha_1 = 10^2$. **c** Segmentation with $\alpha_2 = 10^{-3}$ [16]

Boundary is nothing but the shape of the object. Region is about the particular region for at which we are applying operations.

(k) **Object recognition**: In this, we recognize about the object. By looking at its properties. As in [18], they are using a logo for recognition of document.

In all these steps, from 1 to 6 input and output is an image. This level is also called low-level processing. From 7 to 10, input is an image but output is attributes of image. This is called Medium level of processing. In 11, we are sensing the image. Therefore, this is called high level of Processing.

3 Algorithms of Image Enhancement

Different algorithms are used in every different paper. The most popular used algorithms are Histogram Equalization. Histogram Equalization is the method of image enhancement for adjusting contrast using the histogram of image. By doing these adjustment intensities of an image can be distributed in equal way. Histogram Equalization is of following types:

- Global Histogram Equalization
- Local Histogram Equalization
- Adaptive Histogram Equalization
- Brightness Preserving Dynamic Histogram Equalization [21]

Form all these types the first three are used.

(a) **Global Histogram Equalization**:

Global Histogram Equalization enhances the contrast of whole image [22]. In this method, image is given as input and it enhances the image globally and at the output we get both initial and final images [21]. The formula for this can be given as ratio of brightest and darkest pixel intensities.

The transformation $c\,(r_k)$ is given as [21],

$$S_k = c(r_k) = \sum_{i=0}^{k} p(ri) = \sum_{i=0}^{n} \frac{ni}{n} \tag{1}$$

where $0 \le S_k \le 1$ and $k =$ any positive number up to $L - 1$.

In this each pixel is adjusted to improve its quality. The probability of occurrence of pixel intensities can be given as [23],

$$P(r_k) = \frac{n_k}{N * M} \tag{2}$$

where $N * M$ is number of pixels.

In the given Fig. 9, shows the image of a brain which is enhanced by local histogram equalization algorithm and Fig. 10 shows the histogram of given image.

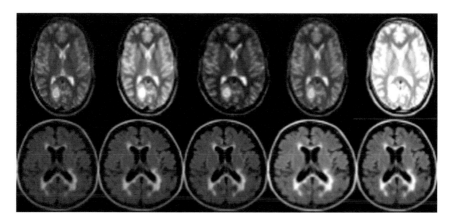

Fig. 9 Global histogram equalization enhances the contrast of brain image [21]

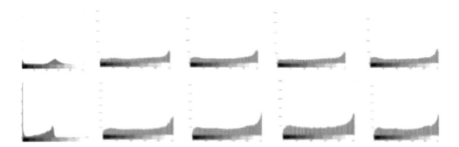

Fig. 10 Histogram of global equalization [21]

(b) **Local Histogram Equalization**:

This Histogram equalization is used for getting all small details of an image. By this histogram, useful information can be analyzed easily. In this image, original image is subtracted from blurred image to get more enhanced image.

If blurred image is $b(i, j)$ and original image is $p(i, j)$, then mask image $m(i, j)$ can be given as [22],

$$M(I, j) = p(I, j) - b(I, j) \tag{3}$$

after that sharpened image is given as [22],

$$S(I, j) = p(I, j) + w^*m(I, j) \tag{4}$$

where 'w' is the weight of image which is always greater than zero. When $w = 1$, is called high boosting filtering [28] (Figs. 11 and 12).

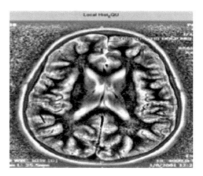

Fig. 11 Local histogram equalization enhances the contrast of brain image [24]

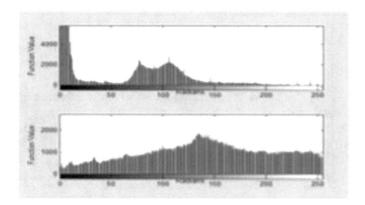

Fig. 12 Histogram of local equalization [24]

(c) **Adaptive Histogram Equalization**:

This method is useful for both medical and natural images. In adaptive Histogram Equalization, according to the neighboring region, pixels are customized [21] (Fig. 13).

Fig. 13 Adaptive histogram equalization enhances the MRI brain image [24]

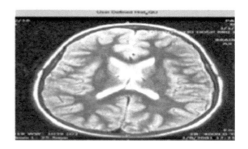

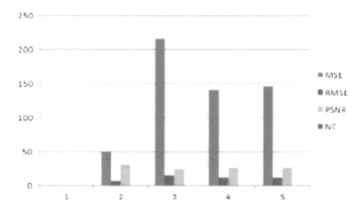

Fig. 14 Graph of average filter [23]

$$a = \frac{y - y_-}{y_+ + y_-} \quad b = \frac{x - x_-}{x_+ + x_-} \tag{5}$$

The given Fig. 14, shows the enhanced image which is enhanced by adaptive Histogram equalization.

Adaptive Speckle Filter is used for Adaptive Histogram equalization. Mapping function is used,

$$Z_{\text{out}} = 255[(Z_{\text{in}} - L)/(H - L)]^{\gamma}$$

where

Z_{in} Insert value
Z_{out} Created value
γ Positive real value parameter.

4 Results and Calculations

In every different paper, there are number of algorithms are used for enhancing the image. For enhancing the image it is necessary the there should be less noise in an image. By removing the noise we get sharper and enhanced image.

In paper [23], they are working on four main parameters which are used for removing the noise. The parameters are PSNR (Peak Signal to Noise ratio), MSE (Mean Square Error), RMSE (Root Mean Square Error) and NC (Normalized Coefficient) (Table 1).

Where NO is Neighborhood Operation, AF is Adaptive Filter IMA is Imadjust, BRX is Bilateral ratinex and SF is Sigmoid Function.

Table 1 Comparison of average filter with other used methods [23]

Parameter	NO & AF	NO & IMA	NO & BRX	NO & SF
MSE	52.1431	223.839	197.6512	0.0011
RMSE	7.2515	14.8312	14.0597	0.0321
PSNR	31.1430	24.6499	26.1718	77.8391
NC	1.0509	0.4109	0.9376	164.66

If value of MSE is low means error is less and if PSNR is high that means noise is less. If RMSE is low that means image has good contrast (Fig. 15; Table 2).

By seeing these resultant graphs we conclude that for pharmaceutical images, both Sigmoid function and Neighborhood operation are best Techniques.

In paper [23], four quality factors are used named as Weber Contrast, Michelson contrast, AMBE, and Contrast. Different Histogram equalization values are taken by different equalization methods (Table 3).

Fig. 15 Graph of neighborhood operation [23]

Table 2 Comparison of neighborhood operation with other technique [23]

Parameter	NO & AF	NO & IMA	NO & BRX	NO & SF
MSE	52.1431	223.839	195.6512	0.0012
RMSE	8.2515	15.8321	15.0579	0.0321
PSNR	32.1430	23.6499	26.1715	78.8391
NC	1.0519	0.4109	0.9378	165.66

Table 3 Comparison of neighborhood operation with other technique [23]

Method	Weber contrast	Michelson contrast	AMBE	Contrast
GHE	0.6232	0.3542	81.351	24.526
LHE	0.7534	0.5213	84.524	23.845
AHE	0.5331	1	0.4831	31.254
BPDHE	0.7245	0.7257	27.641	34.456

5 Conclusion

This review paper is based on different image enhancement techniques. The algorithms which are used for image enhancement give us long-range of approach to get satisfactory visual and viewing conditions. We conclude that Image Enhancement techniques are very useful in medical operations. In future various kinds of algorithms are used for medical imaging. For removing of noise we need our PSNR should be high and MSE should be low for removing of error. Almost all operations are done in the software MATLAB. It is also concluded that global equalization technique is better among all.

References

1. Daniell, G., Gull, S.: Maximum entropy algorithm applied to image enhancement. In: IEE Proceedings E-Computers and Digital Techniques, vol. 127, pp. 170–172 (1980)
2. Hanami, N., Katsumata, T., Aizawa, H., Honda, M., Shibasaki, M., Otsubo, K., et al.: Fluorescence thermometer based on luminescence imaging of garnet sensor. In: 2008 International Conference on Control, Automation and Systems, 2008, pp. 992–995 (2008)
3. Yu, N., Li, L., Su, Q., Weiqi, J., Kang, H., Yan, S., et al.: Underwater range-gated laser imaging system design with video enhancement processing. In: 2013 2nd International Symposium on Instrumentation and Measurement, Sensor Network and Automation (IMSNA), 2013, pp. 760–763 (2013)
4. Gao, Z., Zhai, G., Hu, C., Min, X.: Dual-view medical image visualization based on spatial-temporal psychovisual modulation. In: 2014 IEEE International Conference on Image Processing (ICIP), 2014, pp. 2168–2170 (2014)
5. Dantas, D.O., Leal, H.D.P., Sousa, D.O.B.: Fast multidimensional image processing with OpenCL. In: 2016 IEEE International Conference on Image Processing (ICIP), 2016, pp. 1779–1783 (2016)
6. Zhang, W., Xiao, S., Shi, X.: Low-poly style image and video processing. In: 2015 International Conference on Systems, Signals and Image Processing (IWSSIP), 2015, pp. 97–100 (2015)
7. Nasonova, A., Krylov, A.: Deblurred images post-processing by Poisson warping. IEEE Sig. Process. Lett. **22**, 417–420 (2015)
8. Köppel, M., Makhlouf, M.B., Müller, K., Wiegand, T.: Fast image completion method using patch offset statistics. In: 2015 IEEE International Conference on Image Processing (ICIP), 2015, pp. 1795–1799 (2015)
9. Lazareva, A., Liatsis, P., Rauscher, F.G.: An automated image processing system for the detection of photoreceptor cells in adaptive optics retinal images. In: 2015 International Conference on Systems, Signals and Image Processing (IWSSIP), 2015, pp. 196–199 (2015)

10. Sun, J., Lv, Q., Tan, Z., Liu, Y.: An image sharpening strategy based on multiframe super resolution for multispectral data. In: 2016 8th Workshop on Hyperspectral Image and Signal Processing: Evolution in Remote Sensing (WHISPERS), 2016, pp. 1–5 (2016)
11. Sotiropoulou, C.-L., Luciano, P., Gkaittzis, S., Viti, M., Giuliano, A., Citraro, S., et al.: Medical image processing using brain emulation. In: 2016 IEEE Nuclear Science Symposium, Medical Imaging Conference and Room-Temperature Semiconductor Detector Workshop (NSS/MIC/RTSD), 2016, pp. 1–3 (2016)
12. Rebhi, A., Abid, S., Fnaiech, F.: Fabric defect detection using local homogeneity and morphological image processing. In: 2016 International Image Processing, Applications and Systems (IPAS), 2016, pp. 1–5 (2016)
13. Krylov, A., Nasonov, A., Razgulin, A., Romanenko, T.: A post-processing method for 3D fundus image enhancement. In: 2016 IEEE 13th International Conference on Signal Processing (ICSP), 2016, pp. 49–52 (2016)
14. Chen, X., Zhai, G., Wang, J., Hu, C., Chen, Y.: Color guided thermal image super resolution. In: 2016 Visual Communications and Image Processing (VCIP), 2016, pp. 1–4 (2016)
15. Uruma, K., Saito, K., Takahashi, T., Konishi, K., Furukawa, T.: Representative pixels compression algorithm using graph signal processing for colorization-based image coding. In: 2017 IEEE International Conference on Image Processing (ICIP), 2017, pp. 3255–3259 (2017)
16. Schiopu, I., Gabbouj, M., Iosifidis, A., Zeng, B., Liu, S.: Subaperture image segmentation for lossless compression. In: 2017 Seventh International Conference on Image Processing Theory, Tools and Applications (IPTA), 2017, pp. 1–6 (2017)
17. Jin, K.-C., Lee, K.-S., Kim, G.-H.: High-speed FPGA-GPU processing for 3D-OCT imaging. In: 2017 3rd IEEE International Conference on Computer and Communications (ICCC), 2017, pp. 2085–2088 (2017)
18. Ketwong, P., Hongsa-arparsat, P., Srilaphat, E., Kaprasit, W.: The simple image processing scheme for document retrieval using date of issue as query. In: 2017 IEEE 2nd International Conference on Signal and Image Processing (ICSIP), 2017, pp. 288–291 (2017)
19. Ohnishi, T., Kashio, S., Ito, K., Makhanov, S.S., Yamaguchi, T., Iwadate, Y., et al.: Image feature conversion of pathological image for registration with ultrasonic image. In: 2018 International Workshop on Advanced Image Technology (IWAIT), 2018, pp. 1–2 (2018)
20. Demirović, D., Skejić, E., Šerifović–Trbalić, A.: Performance of some image processing algorithms in tensor flow. In: 2018 25th International Conference on Systems, Signals and Image Processing (IWSSIP), 2018, pp. 1–4 (2018)
21. Senthilkumaran, N., Thimmiaraja, J.: Histogram equalization for image enhancement using MRI brain images. In: 2014 World Congress on Computing and Communication Technologies, 2014, pp. 80–83 (2014)
22. Singh, K.B., Mahendra, T.V., Kurmvanshi, R.S., Rao, C.V.R.: Image enhancement with the application of local and global enhancement methods for dark images. In: 2017 International Conference on Innovations in Electronics, Signal Processing and Communication (IESC), 2017, pp. 199–202 (2017)
23. Kaur, R., Kaur, S.: Comparison of contrast enhancement techniques for medical image. In: 2016 Conference on Emerging Devices and Smart Systems (ICEDSS), 2016, pp. 155–159 (2016)
24. Kaur, H., Rani,: MRI brain image enhancement using Histogram equalization Techniques. In: 2016 International Conference on Wireless Communications, Signal Processing and Networking (WiSPNET), 2016, pp. 770–773 (2016)

Detection and Analysis of Congestion of Nodes in Many-Core Processor

Nishin Jude C. Abraham and D. Radha

Abstract The recent trend in developing embedded systems tends towards the many-core processors. The essence of the many-core can be experienced only when the communication and the allocation of the packets to different cores are considered and optimised. The increase in the number of cores within a semiconductor chip increases the complexity of the communication between the cores. The cores need to communicate to share the data in their cache instead of accessing the data from main memory. The major hindrance in the development of an effective communication system for inter-core communication is the phenomenon of congestion. The proposed work is to detect the congestion before a router causes further bottleneck for communication, rather than managing the congestion. The delay incurred by a router for processing a flit can be found by counting the number of clock cycles elapsed during the flits lifecycle on that router. A threshold on the delay may imply the upcoming congestion on the router. This delay can be useful for finding the idleness and activeness of routers. Identifying the router congestion in a many-core processor can later be useful for effective routing by managing the congestion. These results can be extended further for analysing the effects of congestion on various applications and also for developing routing techniques that can improve the multicore performance. The work uses Booksim2 simulator to analyse the system characteristics.

Keywords Congestion · Delay · Many-core processors · Packet injection · Routing

1 Introduction

With the evolution of the CMOS technology in the VLSI industry, the constraints related to the difficulty in embedding, millions of transistors in a single silicon chip were made practically possible. This advancement in the Silicon industry has made it possible to embed more and more circuits into a single chip which in early stages

N. J. C. Abraham (✉) · D. Radha
Department of Computer Science & Engineering,
Amrita School of Engineering, Amrita Vishwa Vidyapeetham, Bengaluru, India
e-mail: nishinat1994@gmail.com

© Springer Nature Singapore Pte Ltd. 2020
A. K. Luhach et al. (eds.), *First International Conference on Sustainable Technologies for Computational Intelligence*, Advances in Intelligent Systems and Computing 1045,
https://doi.org/10.1007/978-981-15-0029-9_59

755

was though practically impossible. This has enabled to build more complex circuitry inside a tiny Silicon chip. This was considered to be the ultimate path for the researchers. But as the years and decades passed by, many new technologies innovations came into like IoT, 4K resolution, demanding for Terra bytes of data storage for high-performance computing, and there, the need of high-performance computing started developing. For ensuring reliable and error-free data [1], at a faster rate multiple cores processing chunks of data in parallel, came into the market. Every new CMOS technology generation permits the look of larger and additional advanced systems on one integrated circuit. At similar circumstances, the expected performance enhancements of single-core general-purpose CPUs have slowed. These 2 factors, besides the theory of Dennard scaling, resulted in a recent paradigm shift by the semiconductor business and analysis community towards the look of multi-core processors.

By putting many easier processing cores on a similar die, architects are ready to extract additional performance by hard real-time tasks in parallel rather than serially. Today's multi-core processors habitually have four to eight general-purpose cores and recently styles with seventy-two cores, like the Intel Xeon. Even mobile phones and arrives with dual-core, quad-core and octa-core processors these days. With ever-shrinking semiconductor feature sizes and increasing reckon demands, future processors are expected to manoeuvre towards many-core styles with tons or thousands of cores on a single chip.

However, practically the chips were too complicated to design than in theory. Because as the count of cores inside a single chip increases, the need for an effective communication platform between cores also increases. In other words, reliable intra-core communication infrastructure has to be built up.

Thus, the increasing communication demands have incurred a great bottleneck in multicore designing which is an efficient communication infrastructure. Thus, if CMOS technology came up as the ultimate solution for the design of larger and complex systems on a single chip, the Network on Chip (NoC) arose as the solution for providing increasing communication demands between the growing numbers of cores on a single chip. An important point to be noted is that the focus has been shifted for improving the communication quality instead of increasing the core count. It is obvious that there will be performance improvement as the number of cores increased, however, this incurs a huge hardware implementation cost. This is why the focus has been shifted to intra-core communication rather than increasing the core count.

The most commonly used communication solutions for computer networks are busses and crossbar interconnects. But this is not feasible for multicore architecture since these solutions are not scalable to a larger number of cores and also due to the latency and the bandwidth limitations.

On the early stages of the SoC's the designers used direct point to point wiring as the communication means or the system resources (cores or routers) communicated directly through wires. This system does have any arbitration unit for assigning priority for the incoming traffic or packets. But the physical wiring is not at all a feasible solution for larger integrated circuits or larger core counts, as the increase in core counts will need large length wires that too very narrow wires, so that two wires will not cross over each other. This leads to large delays for routing and heating issues. The physical point to point wiring is expensive and complicated.

Due to the limitation of the physical wiring method, the SoC' s moved to shared bus as a means of the inter-core communication. In shared bus, all the cores or routers are connected to each other using a shared bus. Since the bus is shared the chance for contention is high. Thus, a bus arbitration system is provided which manages the contention or hotspot formation. Even if the shared bus has many advantages over the computer networks, it is still not a fair choice for the NoC architecture, as the rate of packet transfer becomes slow at times during arbitration. Also the efficiency of the shared bus will reduce, scaling beyond a certain limit. Thus a reliable infrastructure that is scalable and allows packet transfer at a low latency was required.

NoC emerged as the ultimate solution which satisfies latency, bandwidth, and the scalability constraints. It consists of three main components cores, routers or switches, and network interface. The cores are connected to each other using routers via network interface. The packets from one core (source code) are propagated to the other core (destined core), by traversing through one or more switches or routers.

The cores are interconnected in such a fashion that the communication between the individual cores in a many-core processor is made possible. The communication occurs through links or channels, routers, and switch allocators. Inter-core communication is required for handling the cache miss encountered by cores. The cache miss packets are the essence of the inter-core communication. A core generates a packet when it encounters a cache miss. This core generates a request for the required destination address in the form of packets. The basic unit of inter-core communication is flits. A packet is collection of flits. A primary flit of a packet or the first flit generated by a core for a packet is called the head flit. The other flits are called body flits. Depending on the behaviour of the application currently handled by the cores, the cores start injecting a core. It is practically infeasible for testing these communication scenarios in hardware model due to the cost and complexity constraints. BookSim2 simulator is an open-source software which can be used to study the NoC inter-core communication. This simulator supports the use of synthetic traffic patterns which mimics the real application scenarios in a real many-core processor. Many research models have been proposed for congestion management. But there is limited resource available for congestion detection. The proposed work is to identify the router congestion is a Many-core processor which can later be useful for effective routing by managing the congestion.

2   Literature Survey

Many advancements and techniques have been put forward for routing technique which will provide guaranteed delivery of packets, a very less initiative have been taken for knowing the current status of the nodes, or contention. The feasible solutions for managing the contention are reducing the packet size, increasing large-sized buffers, using less number of packets per message, scheduling the network traffic according to the traffic condition.

In [2], M. Daneshtalab, A Sobhani, and others proposed a methodology cited to minimise hotspot formation using AntNet Dynamic Routing Algorithm. Here the routing algorithm used is Antnet routing algorithm which is based on Ant colony. This algorithm when compared to the traditional XY, Odd Even and Dy-AD routing models has improved average latency and peak power and maximum temperature. AntNet is based on adaptive Distributed algorithm. In Adaptive Routing Path between the source and the destination is determined node by node depending on the current network traffic. In Distributed-The present status of the network (if congested or not) is determined among the nodes which exchange the information. So AntNet combines these two routing features. AntNet uses these two routing over the centralised and static routing because of its advantages over the latter. In Centralised routing, a central controller is responsible for the routing table of each node and in Static Routing, the path between the source and destination is determined by the source and the destination itself ex: XY Routing algorithm. Router used in this paper implements a short path adaptive router. This router selects the route path which has minimum number of hops, from the set of the minimum paths, among the routing table and selects the set with minimum no of hot spots. The AntNet algorithm implements Wormhole routing. Routing block consists of 2 data packet routing unit which helps in forwarding the data packets and control unit which determined the direction of flow. Data packet routing unit determines the output port of the data packet based on its destination using a routing table. Forward packet routing determines the output port of the forward Ant. Here popularity based is used as priority, i.e., the most popular node or visited node of the source node is used.

In [3], R. S. Reshma Raj, Abhijit Das, and John Jos proposed a model to implement and study the hotspot formation in mesh NoC's by deflection routing method. Main aim of this paper is to handle the congestion. Hotspots are non-uniform traffic formation near cores. Some cores need to handle relatively highest traffic compared to other cores. This work identifies the destination hotspots and upon identification, packets are de-routed away from these hotspot cores using some routing techniques, and thus decreasing the average latency of the packets. A novel deflection routing technique is proposed to mitigate the effects of the destination hotspots. The work is done on Mesh-based NoC. First the network traffic is dynamically analysed to identify the possible destination hotspots. A novel congestion management scheme with minimal deflection routing(less hops) is applied to de-route the packets. Here 2D mesh topology is used here because it is very simple and has short inter-core links. L1 cache miss generates a packet. A packet contains header and a payload .header contains control information and the packet payload contains the required data which provides the address which is missing in L1 cache.

In [4], Thomas Moscibroda and Onur Mutlu proposed an article which is case study for routing in buffer-less channels. Here a router architecture is introduced to prevent the live-lock problem in the deflection based NoC architecture. The router architecture uses wormhole routing. The network is prevented from the live-lock by worm truncation. When a live-lock occurs, the lower prioritised worm is truncated and the higher prioritised worm is allowed to follow to its desired route to the destination. But since the header contains only routing information, and the body flits does not contain any routing information, the flit that is following the head flit of the truncated node, is made as an head flit, so that every router will have the routing information of all the packets thus it is able to map the packets towards the output port. Here additional wires are used for transmitting the head flits. As the truncation rate increases, the performance of the system degrades. But the implementation cost of the additional wires is less than the buffered approach.

In [5], Yu-Hsin Kuo, Po-An Tsai, and others proposed an adaptive routing technique which is aware of the different possible paths that a packet can follow has been proposed. Using adaptive routing in NoC, may have imbalanced path diversity in many directions. Even if this causes difficulty in traffic balancing, it gives extra information about the network. Here adaptive routing scenario with Path diversity aware and augmented path diversity aware, are used in-order to achieve load balancing. They proposed some selection functions that has better throughput than the standard selection functions from the routing algorithms with less turn model restriction will have higher Path diversity.

In [6], at the International Workshop on Advanced Interconnect Solutions and Technologies for Emerging Computing Systems, Monobrata Debinath, Dimitris Konstantinou, and associates published a paper for congestion management, using edge and in network throttling. Traditional adaptive routing algorithms manage the congestion within the NoC by balancing the traffic load across the network links, i.e., by rerouting the packets through less congested nodes so as to evenly distribute the traffic across the cores. Such routing techniques have a disadvantage as it needs to collect the congestion information and spread this information throughout the network. And the next reason is that it requires virtual channel buffers [7] to prevent deadlock. This provides an easy way for load balancing by edge and in network traffic throttling. The proposed solution provides control to the injection rate as well as improves the performance of the system in terms of throughput and latency and requires only minimum logic to enable the traffic throttling within the routers, at low cost. The traffic throttling is done by including an additional hardware module inside every router. This module is called Throttling signal generation module. The traffic throttling signal decision is made based on the current buffer utilisation which is indicated by the credit counters within each router. The throttling of the packet is done by placing a minimal masking logic in-front of the Arbiter (VA).When congestion is present or if throttling signal T is asserted or the requesting packet is headed towards a congested direction, the masking logic masks or delays a request from a Virtual channel. But a packet is not throttled more than once to prevent live-lock or starvation. One flag bit is also used by each Virtual channel buffer, so that each virtual channel ensures reminding the throttling device not to get throttled more than once. The output port/ports that are to be throttled are determined based on the routers location. For this the routers are grouped and numbered according to the severity of the congestion. Those ports of the routers whose output are connected to the ones with higher severity levels are throttled down. The drawback of this paper is that it determined the congestion zones statically, but on real-time, grouping into congestion zones is a challenge. This work was able to provide better performance at low cost.

In [8], Blagodurov, S. Zhuravlev, and Fedorova at the ACM Transactions on Computer Systems proposed a scheduling process which is aware of contention in multicore system. The prediction for the resource contention is done as follows. When the cache lines are full and a thread requests for a cache line again, so that some cache lines should be replaced by the requested cache lines. But these cache lines may belong to that of other threads so this may cause resource contention. This occurs when 2 or more cores have a shared memory domain. The contention of the local-level cache is found in two ways. The first way is finding the LLC miss rate of the threads that share the same LLC or miss rate implies the cache line belonging to a thread should be replaced by the cache lines for a new thread or the cache resource is competing for the 2 thread. Hence contention occurs. The thread that needs more cache memory, will experience more contention, as that threads cache line will probably become a victim of other threads cache miss. They introduced another method to model cache contention. This method used memory reuse pattern of the thread because a thread hardly ever reuses its cached data, so no problem in the removal of the cached data for other thread ex: video streamlining application needs only less space to keep the data in the cache. Memory reuse profile can be found by stack distance or memory reuse profiles.

The work [9], introduced a flexible and cycle-accurate network on chip simulator called BookSim simulator. The BookSim2 simulator is a very popular Network on Chip simulator for creating a Many-core environment by accurately modelling the interconnection network for a wide range of systems. The prior version the BookSim1 simulator was not specific in case of the on-chip environment and was used for creating the performance evaluation graphs. BookSim is the most widely used NoC simulator and it is used to analyse the different topology, routing algorithms, flow control, network design and the micro router architecture. The simulator features more advance level or router architecture modelling, which takes into account all the delays such as inter- and intra-router channel delay and generation of traffic that is having mere resemblance with the real-time traffic.

3   Software Requirements and Details

The simulation stage will be covering the need for getting the communication pattern that is suited for a specific application. The stage of the many-core architecture selects the parameters such that it increases the efficiency and throughput with minimal latency. The simpler the architecture would aid in the improvement in area and the power consumption. The simulator should be able to match with the performance observations from a real-time many-core system, hence yielding a comparable performance.

Once the specification is prepared, the next procedure is to move towards the design process. Thus the simulator will get the input parameters from the design space, runs the simulation over those arguments and the next final step is to get the output in terms of efficiency throughput and latency. Thus the simulator can be considered as a black box that accepts some input, parses those inputs and produces the output. Here a modular design approach is the feasible solution since only modular designs are flexible.

Considering the huge architectural choices in the design space, it is always a dream concept to design a system that will be able to cover all those choices. Hence the ultimate design strategy will confine to designing a base system, which enables to add modules that will define each Many-core architectural behaviour. Hence the base system that has to be developed would cover majority part of the design space. For making it more simple, and easy to understand for a developer or a researcher, it is advisable to implement each module for each characteristic or property. Each property or characteristic should be able to identify the dependency between different implementation as well as adding new definitions or descriptions can be done with ease. All these requirements will conclude to designing a modular system, where each module will define each characteristic in the Many-core characteristics design space. Hence the simulator should be highly modular that is based on derived classes.

BookSim2 simulator:

The BookSim2 simulator is a very popular Network on Chip simulator for accurately modelling the interconnection network for a wide range of systems. The prior version of the simulator was BookSim1. The BookSim1 simulator was not specific in case of the on-chip environment and was used for creating the performance evaluation graphs. BookSim is the most widely used NoC simulator and it is used to analyse the different topology, routing algorithms, flow control, network design and the micro router architecture. The BookSim2 simulator features more advance level or router architecture modelling, which takes into account all the delays such as inter-and intra-router channel delay and generation of traffic that is having mere resemblance with the real-time traffic.

For creating a simulation environment with any interconnection network simulator, it requires a few configuration parameters such as;

- Topology: How the cores/switches are placed geographically.
- Routing: The routing function decides the algorithm or path to be taken for the packets to flow from the source to destination specific to the topology taken. In other words, the routing will deal with the entire whole message or packet
- Flow Control: This determines how the resources such as the buffer and the link bandwidth are shared among different cores. An example may be adjusting the injection rate.
- Router micro-architecture: Deals with the internal architecture of the routers such as crossbar allocators, buffers, etc.
- Traffic pattern: This is how the simulator is able to mimic the real-life traffic scenarios encountered within the real chip. This will decide the destination of a

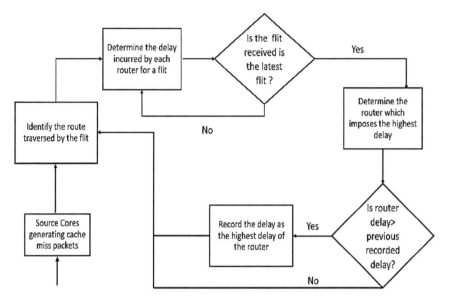

Fig. 1 The congestion detection model

packet, when a core generates a packet. In other words, the traffic pattern deals with a "packet' s" source and the destination.

The Topology will impose an initial latency and throughput constraints to the system. The Routing algorithm will affect the start-up system performance such as latency and throughput at no or zero load. Flow control and the router microarchitecture will affect the router efficiency.

The network performance is evaluated using the following parameters. Figure 1 discusses various parameters.

- Saturation Throughput: The network throughput at which the channel first go to saturation or the latency overshoots
- Zero Load Latency: It is the mean packet latency at a very negligible injection rate or load
- Saturation Load: It is the injection rate or the load, when the latency overshoots or become $5x$ the zero load latency.

BookSim2 simulates the network based on flits and clock cycles, i.e., during a clock cycle, a network channel can read single flit from its input associated router. Whenever the router gets congested, the cores associated with that router will not be able to accept or sent packets from/to that router. Every flits associated with a packet will be processed by the routers in its path. Each router will impose some delay in processing those flit before forwarding to the next router.

4   Congestion Detection Model

Figure 1 shows the model for the congestion identification which is used here. The implementation phase of the network on chip many-core architecture will finally imply the steps for finding an answer to get rid of the bottleneck in many-core traffic. The steps are as follows;

A. Finalising the parameters for the network configuration.
B. Simulating the network with the above parameters as input.
C. Analysing the path flow, each flit is taking and then identifying the delays experienced by each router for the entire lifetime of a flit.
D. The router delays are found separately for every flit passing through each router and those delay values are monitored.
E. The router delay is overwritten with the last known highest delay value of that router. And this gives an exact amount of how much time a flit will be delayed latest by that router.
F. A count of the number of times a specific router is causing highest delay to a flit is also taken into account so as to get an overall accurate statistics of the delay.
G. The delay values and the count values of each router will give a measure of the traffic congestion in a network; hence the packets or flits can be rerouted or throttled so as to maximise the performance.

Once the delay of each router is identified, the router can have certain mechanisms to route the packets towards its destination such that it will certainly improve the overall performance of that packet in-terms of latency.

5   Implementation

BookSim2 simulator is the simulator used for simulating a many-core environment. Every Core is associated with one associated router. Whenever the router gets congested, the core or cores associated with that router will not be able to accept or sent packets from/to that router. Every flits associated with a packet will be processed by the routers in its path. Each router will impose some delay in processing those flit before forwarding to the next router.

The delay incurred by each flit on the routers on its route is found. The delay incurred by a router for processing a flit can be found by counting the number of clock cycles elapsed during the flits lifecycle on that router. The processing delay is found by the time taken by the router to send the flit to the output channel after receiving the flit in one of its input channels

The router that is having the highest delay compared to the other routers is said to be congested with respect to the other routers and the router having the least delay is said to be congestion-free.

The simulation is carried out for various number of routers per dimension, various traffic models like uniform, transpose, etc. and routing functions as DOR, Adaptive XY-YX and XY-YX and in various injection rates.

6 Results

The router delay for processing each flit has been found. Each router is analysed for how much time, the flit is being held for each router, to get transferred to the next. Each individual flits will have to follow a definite path decided by the routing function used. So each flit will have to pass through a number of routers to finally reach its destination once it is being generated from the source core. During this traversing, each router will be incurring different wait times for the flit depending on many factors like buffer credit, incoming traffic and outgoing traffic.

Table 1 discusses the delay contribution of the routers for Adaptive XY-YX routing function, XY-YX routing function and normal deterministic XY routing. A graph has been plotted for analysing the results.

Figure 2 is the graph plotted from the values obtained for a Uniform mesh NoC with uniform traffic for varying injection rates. The figure describes the results analysed while testing with different values of injection rates. Here each router is causing different delays for the flits passing through them. As the injection rate increases the router delay also increases. The routers with the highest delay values and the lowest delay value has been found. This is observable from Fig. 2. Another noteworthy observation is that the routers that are placed at the boundary of the network, seemingly tends to have the lowest delays irrespective of the traffic patterns and the injection rates.

Figure 3 shows the comparison for the Router ID-Delay characteristics of Adaptive XY-YX and the deterministic XY routing algorithms, for injection rate = 0.008, among the routers occupied in the first row, namely router ID's 0–7, respectively, most of the high delays are experienced at ID's 2, 3 and for the 2nd row, the highest delays are experienced at ID's 10, 13, 14 and for the 3rd row, the ID's 19, 20. These results directly imply that compared to the intermediate routers the edge routers are least congested. The normal deterministic routing shows higher values of delays compared to the XY-YX routing and the adaptive XY routing. For adaptive routing, the latencies are almost uniformly distributed compared among the nodes and the packets experience lower delays compared to the deterministic and XY-YX routing.

7 Observation

The maximum router delay experienced by a flit has been tabulated and the tables are analysed thoroughly. The major axis has been divided in terms of the multiple of 8 that is, the radix number. The lowest values are observed at the point majorly at the

Table 1 Router ID—delay characteristics for adaptive XY-YX routing, XY-YX routing and normal XY routing

Router ID	AD_XY	XY_YX	XY	Router ID	AD_XY	XY_YX	XY
0	15	21	6	32	19	20	10
1	22	10	26	33	21	26	10
2	25	19	34	34	14	19	14
3	24	19	34	35	21	15	14
4	15	19	32	36	29	18	6
5	15	22	27	37	23	25	11
6	18	12	18	38	26	25	11
7	16	18	8	39	19	26	8
8	22	13	7	40	18	23	9
9	19	25	6	41	21	12	17
10	22	25	26	42	21	14	24
11	24	14	23	43	27	23	15
12	18	11	21	44	26	26	20
13	14	22	22	45	24	27	6
14	18	12	25	46	11	26	8
15	27	15	10	47	20	26	9
16	16	11	9	48	21	8	8
17	22	25	7	49	10	10	14
18	25	26	7	50	18	17	40
19	24	18	30	51	11	20	56
20	20	21	30	52	9	24	53
21	14	22	33	53	22	13	29
22	15	26	24	54	21	21	6
23	16	15	9	55	23	9	6
24	10	20	9	56	12	10	43
25	16	23	13	57	11	10	42
26	22	20	12	58	20	11	42
27	19	17	6	59	21	14	30
28	21	18	14	60	21	21	30
29	17	22	12	61	19	13	35
30	19	23	11	62	26	13	28
31	16	15	9	63	21	8	6

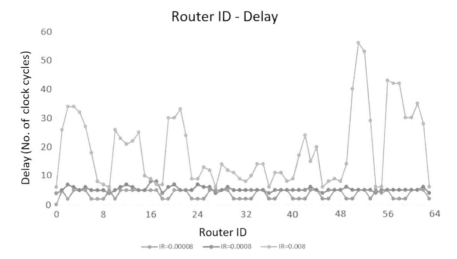

Fig. 2 Graph obtained for router ID—delay characteristics for a mesh NoC with uniform traffic for different injection rates

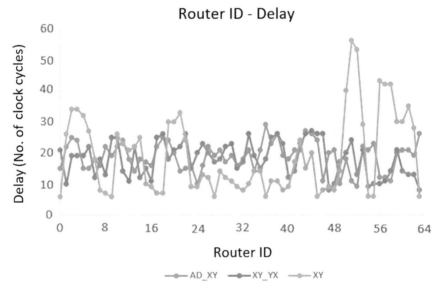

Fig. 3 Graph obtained for router ID—delay characteristics for adaptive XY, XY-YX and normal deterministic XY routing for injection rate of 0.008

router ID's which are near to multiples of 8 or the radix points. Hence the minimal delays are observed at the boundary location of the routers. From Fig. 2, it can be asserted that the delay is least for the boundary neighbourhood routers compared to all other nodes. So the boundary nodes are the least congested nodes in the network.

Figure 3 discusses various routing functions and its effects on the routing performance. The normal deterministic routing shows higher values of delays compared to the XY-YX routing and the Adaptive XY routing. For adaptive routing, the latencies are almost uniformly distributed compared among the nodes and the packets experience lower delays compared to the deterministic and XY-YX routing. As expected, the adaptive routing function performs better compared to the deterministic routing. So the uniformity for the graph is better for the adaptive routing function compared to the others. Compared to the other routing techniques, for the adaptive routing technique, the processing delay of the flits falls within a specific range of value whose threshold is found to be less than that of the delays experience due to other routing techniques provided same injection rate and same application scenarios.

8 Conclusion

An effective way to evaluate the congestion in many-core processors has been found and implemented. The router delays are associated with the time it takes to forward the packet to the next router. The delay incurred by a router for processing a flit can be found by counting the number of clock cycles elapsed during the flits lifecycle on that router. The maximum delay implies congestion in the router. For higher injection rates, the intermediate routers will become more and more congested. Generally, the least congested nodes are found to be present at the boundaries of the network. Minimal or non-minimal routing can be made adaptive with the knowledge of delay in each of the routers on the way or the packets/flits can be throttled with the knowledge of delay in the routers. The boundary to boundary communication can be done through the edges of the mesh network with the information collected about the delay of the routers in the network. Depending on the delay that is on the routers, balancing of load on the cores can be performed to avoid power-related issues. Hence, delay is the most important factor to be found in various solutions for efficient communication.

References

1. Teja, S.T., Narayana, T.V.V.S., Vinodhini, M., Murty, N.S.: Joint crosstalk avoidance with multiple bit error correction coding technique for NoC interconnect. In: 7th IEEE International conference on Advances in Computing, Communications and Informatics (ICACCI), PES Institute of Technology, Bengaluru, South campus, India (2018)
2. Daneshtalab, M., Sobhani, A., Afzali-Kusha, A., Fatemi, O., NavabiM, Z.: NoC Hot spot minimization using AntNet dynamic routing algorithm. In: IEEE 17th International Conference on

Application-Specific Systems, Architectures and Processors (ASAP'06), 11–13 Sept. 2006
3. Raj, R.S.R., Das, A., Jos, J.: Implementation and analysis of hotspot mitigation in mesh NoC's by cost-effective deflection routing technique. In: 2017 IFIP/IEEE International Conference on Very Large Scale Integration (VLSI-SoC), 23–25 Oct. 2017
4. Moscibroda, T., Mutlu, O.: A case for bufferless routing in on-chip networks. In: Proceedings of the 36th Annual International Symposium on Computer Architecture, 20–24 June 2009
5. Kuo, Y.-H., Tsai, P.-A., Ho, H.-P., Chang, E.-J., Hsin, H.-K., Wu, A.-Y.: Path diversity-aware adaptive routing in network-on-chip systems. In: 2012 IEEE 6th International Symposium on Embedded Multicore SoC, 20–22 Sept. 2012
6. Debinath, M., Konstantinou, D.: Low cost congestion management in networks-on-chip using edge and in-network traffic throttling. In: Proceedings of the 2nd International Workshop on Advanced Interconnect Solutions and Technologies for Emerging Computing Systems, 20–25 Jan. 2017
7. Avani, P., Agrawal, S.: Efficient dynamic virtual channel architecture for NoC. In: Symposium on VLSI Design and Embedded Computing (VDEC'18), co-affiliated with Seventh International Conference on Advances in Computing, Communications and Informatics (ICACCI-2018), PES Institute of Technology, Bengaluru, South Campus, India, 2018
8. Blagodurov, S Zhuravlev, & Fedorova, A.: Contention-aware scheduling on multicore systems. ACM Trans. Comput. Syst. 28(4) 4 Dec. 2010
9. Jiang, N., Becker, D.U., Michelogiannakis, G., Balfour, J., Towles, B., Shaw, D.E., Kim, J., Dally, W.J.: A detailed and flexible cycle-accurate network-on-Chip simulator. In: IEEE International Symposium on Performance Analysis of Systems and Software (ISPASS), 15 Jul. 2013

Review of Register Transfer Language and Micro-operations for Digital Systems

Vijander Singh, Ramesh Chandra Poonia, Linesh Raja, Isha Choudhary and Sai Satya Jain

Abstract Several representations have been proposed from last many years to represent Register Transfer Language (RTL). All description has their different level of success, views, design, and complexity. The primary objective of this paper is to provide a common platform or kernel to discuss all notation of register transfer language that can be described by particular representation. Register Transfer language is an abstraction of circuit switching and able to specify a hardware system from a very low level. The paper also presents the design and control of activities related to RTL. The RTL can be procedural and non-procedural in nature. The behavior of asynchronous control modules architecture has been discussed in further part of the paper. The next section deals with the application to design the complex systems using the register transfer language, instruction set architecture, and organization of the system. Remarks related to simulations, implementation automation capability of the system are given. Finally, the RTLs are analyzed concerning requirements, writing complexity, learning complexity and structural description of a hardware system.

Keywords RTL · Register · Micro-operation · PC · IR

V. Singh (✉) · L. Raja
Manipal University Jaipur, Jaipur, Rajasthan, India
e-mail: vijan2005@gmail.com

L. Raja
e-mail: lineshraja@gmail.com

R. C. Poonia · I. Choudhary · S. S. Jain
Amity University Rajasthan, Jaipur, Rajasthan, India
e-mail: rameshcpoonia@gmail.com

I. Choudhary
e-mail: ishachoudhary68@gmail.com

S. S. Jain
e-mail: sathyajain9@gmail.com

R. C. Poonia
Norwegian University of Science and Technology (NTNU), Alesund, Norway

© Springer Nature Singapore Pte Ltd. 2020
A. K. Luhach et al. (eds.), *First International Conference on Sustainable Technologies for Computational Intelligence*, Advances in Intelligent Systems and Computing 1045,
https://doi.org/10.1007/978-981-15-0029-9_60

1   Introduction

Similar to programming languages register transfer languages used to assign values to registers. Being hardware friendly RTLs are a bit complex concerning high-level programming languages. Some low-level languages like FORTRAN are used to represent notation of RTL. The algorithms are described as a language which controls the complexity and clarity of programming [1].

The essential attributes of hardware systems and an RT language are complementary to each other. Concurrency, parallelism, recursiveness, control flow, and timing issues are achieved or solved using the best combination of hardware and register transfer language. Digital system assembled using multiple functional units which requires interaction, communication, synchronization among all units as co-routines and sub-routines or symmetric relationships or hierarchical relationships. There must be a sequence of actions or activities which describe the behavior of the system and also requires a mechanism to guarantee synchronized cooperation among businesses.

Control operations of the RTL provide support to the system for synchronized functioning. The efficiency of and RTL depends on how generalized the control operations are, available primitive operations in RTL, etc. [2–5].

2   Design Hierarchy

There are five levels exist to denote and define the digital system hierarchy [3–7]:

- System Level: It describes the hardware components of the system. Hardware is the top level of the system hierarchy. Different design components of system-level hierarchy are memories, processors, peripheral, switches, devices, etc. and evaluation criteria are memory capacity, speed, access time, bandwidth, flow rate, and power consumption, etc.
- Programming Level: This level deals with the behavior of a system with the help of programming language and algorithm. The system can be simulated after defining its behavior. The algorithm level consists of machine instruction, instruction cycle and basic operations described in the register transfer language. The tools used for modeling are native C and C++ supported by hardware class library.
- Register Transfer Level (Functional Level): It defines microarchitecture by separating control and data path, which is enrooted through functional units and their interconnections. RTL aims to explain data transfer and their synchronization among functional units ALU, register and files, etc.
- Logical Level (Sequential and Combinational Sublevels): Interconnected Logic gates represent all functional units and data storage. The collection of gates; flip-flops and Boolean equation defines the system structure. Other responsibilities of a logical level of a system include handling hazards, logic realization, and sequential/combinational level optimization of functional units.

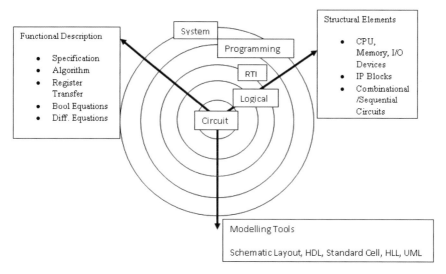

Fig. 1 Abstraction levels

- Circuit Level: Interconnected transistors implement the functionality of the logical level and the model is called the standard cell. The interconnection of a diode; transistors; registers, etc. forms the logic gates as per law of electricity (Fig. 1).

3 Register Transfer Language

The Register Transfer Language (RTL) describes the micro-operations transferred among registers. The accessibility of hardware logic circuits that execute micro-operations and transfer the result of the operations to the same or other register is termed as register transfer. However, the term 'language' is taken from the programming language. It is the procedure for writing symbols to specify a given computational process [3–9].

4 Types of Registers

The various types of the register are discussed in below section [3–5]:

- Accumulator: It is a type of common register which store the data fetched from memory.
- General Purpose Registers: These registers are used to store intermediary data and their results, while the program is in the execution state. It can be done through the help of assembly programming.

Table 1 Micro-operations and their utility

S. No.	Name of the micro-operation	Utility
1	Register transfer micro-operations	Relocate binary information from one register to another register
2	Arithmetic micro-operations	Performs arithmetic micro-operations on numerical data stored in the registers
3	Logic micro-operations	Performs bit manipulation operation on non-numerical data stored in the registers
4	Shift micro-operations	Performs shift micro-operations on the data

- Special Purpose Register: These registers are used by the computer system, for example:

  - Program Counter (PC) "points to the next instruction to be executed".
  - Instruction Register (IR) "holds the instruction to be next executed".

5 Micro-operations

The operations executed on data stored in registers are termed as micro-operations. These are the primary operation performed on the information stored in one or more registers. The examples include shift; count; clear and load [3–9].

The operations in digital computers are of four types. The details provided in Table 1:

6 History of Micro-operations

The various important micro-operations are discussed in the below section [3–11]:

- Intel iAPX432

  - The most desirable for seven years; began in 1975 and was released in 1981
  - It is a built-in building with the 32-bit capacity
  - High performance, complicated with working problems (Fig. 2)

- Motorola 68000 (1979, 8 MHz, 68,000 transistors)

  - Heavy Micro- and Nanocoded, respectively
  - 32-bit with 24 address pins general-purpose register architecture
  - Includes eight address and data registers, respectively

- Intel 8086 (1978, 8 MHz, 29,000 transistors)

  - 16-bit processor

Fig. 2 Intel iAPX432 [14]

– Includes extended accumulator architecture and assembly-compatibility with 8080
– Segmented addressing scheme with 20-bit addressing

• IBM PC, 198

– Prototypes developed in 1979
– Based on "stopgap" prototypes using 8088 boards from display writer word processor
– 8088 is "8-bit bus version of 8086" that allows the cheaper system.

7 Application of Micro-operation

The reason why there is a need of micro-operation is that in computer central processing units, micro-operations (also known as micro-ops) are the atomic structure on which the whole system works, in short, the operations of a processor. These are levels of instructions that are there: low-level instructions used in some designs to implement complex machine instructions. As the basic unit of the computer is the operations performed, hence there is a need for an instrument or a technique to do the job; therefore we need micro-operations in computer architecture [3–15].

7.1 Register Transfer

This register transfer micro-operation transfers from one register to another. It uses a substitution operator.

$$R2 \leftarrow R1$$

It above statement represents the transfer of the data from register $R1$ into $R2$.

7.2 Arithmetic Micro-operations

The fundamental Arithmetic micro-operations include the following:

- Addition; Subtraction and Increment and Decrement.

Add Micro-operation

$$R3 \rightarrow R1 + R2$$

It is simply performing addition. $R1$ to be added to data or content of register $R2$. Sum should be transferred into register $R3$.

Subtract Micro-operation

$$R3 \rightarrow R1 + R2' + 1$$

For this operation, you need a 1's complement instead of a subtraction symbol. Instead of using minus operator this operation takes 1's complement. It also adds 1 to the register which gets subtracted. Here $R1 - R2$ is equivalent to $R3 \rightarrow R1 + R2' + 1$. To put it in simple terms, the content of $R2$ is being subtracted using the compliment from $R1$ and then transferred to $R3$.

Increment/Decrement Micro-operation
This operation is all about either increasing the value by 1 unit or decreasing it. This operation is performed by adding and subtracting 1 to and from the register, respectively.

In the screenshot, $R1$ is incremented by adding 1 and then the incremented value is transferred to $R1$ itself. The second time it is decremented, by subtracting 1 from it and then transferred to $R1$, respectively.

Take a careful look at given Table 2:

Table 2 Symbolic notation for register transfer [4 & 5]

Symbolic designation	Description
"$R3 \leftarrow R1 + R2$"	"Contents of $R1 + R2$ transferred to $R3$"
"$R3 \leftarrow R1 - R2$"	"Contents of $R1 - R2$ transferred to $R3$"
"$R2 \leftarrow (R2)'$"	"Compliment the contents of $R2$"
"$R2 \leftarrow (R2)' + 1$"	"2's compliment the contents of $R2$"
"$R3 \leftarrow R1 + (R2)' + 1$"	"$R1$ + the 2's compliment of $R2$ (subtraction)"
"$R1 \leftarrow R1 + 1$"	"Increment the contents of $R1$ by 1"
"$R1 \leftarrow R1 - 1$"	"Decrement the contents of $R1$ by 1"

7.3 Logic Micro-operations

Points to keep in mind while the logic micro-operation is being performed:

- The operation is performed on the bits stored in the registers.
- The operation takes each bit separately into consideration.

X-oR and has $R1$ and $R2$.

$$P:R1 \leftarrow R1\ X-oR\ R2$$

In the above statement, the author used a Control Function (:)

Let's assume $R1 = 010$ and $R2 = 100$, and let us perform the XoR function. Something likes this:

$$010 \rightarrow R1$$
$$100 \rightarrow R2$$

$$\overline{110} \rightarrow R1 \text{ after } P = 1$$

7.4 Shift Micro-operations

We use shift micro-operation for the serial transfer of data. That means shifting the contents of the register to eithersides, i.e., right or left.

We have three types of shifting as mentioned below:

- **Logical Shift**

It transfers o through the serial input.

For shifting to left—The symbol "shl" is used and for shifting to right—the symbol "shr" is used.

$$R1 \leftarrow \text{she } R1$$

$$R1 \leftarrow \text{she } R1$$

Just take care of the fact that the register symbol must be the same on both sides of arrows.

- **Circular Shift**

In this, the rotation is done without any loss of data or contents on the bits of register around the two ends i.e., right and left. The serial output of the shift register is connected with its serial input. "cil" is used for left shifting and "cir" is used for right shifting, respectively.

- **Arithmetic Shift**

Arithmetic shift, shifts the binary number to either left or right. This micro-operation leaves the sign bit unchanged because the signed number remains the same when it is multiplied or divided by 2.

8 Basic Hardware Implementation of Micro-operations

8.1 Logic Micro-operations

See Figs. 3 and 4.

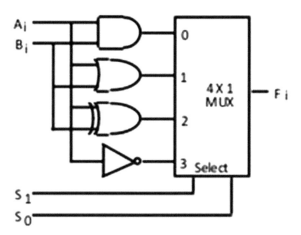

Fig. 3 Hardware implementation of logic micro-operation [4]

S_1 S_0	Output	μ-operation
0 0	$F = A \wedge B$	AND
0 1	$F = A \vee B$	OR
1 0	$F = A \oplus B$	XOR
1 1	$F = A'$	Complement

Fig. 4 Function table [5]

8.2 Arithmetic Micro-operations

See Fig. 5.

8.3 Shift Micro-operations

See Fig. 6.

9 Conclusion

The paper provides the study of all notation of register transfer language that can be described by particular representation. Register Transfer language is an abstraction of circuit switching and able to specify a hardware system from a very low level. The paper also concludes the design and control of activities related to RTL. The RTL can be procedural and non-procedural in nature. Remarks related to simulations, implementation automation capability of the system are given. Finally, the RTLs are analyzed concerning requirements, writing complexity, learning complexity and structural description of a hardware system. The operational frequency increase's slowly. In order to achieve performance and energy efficiency, the designers use the large scale and heterogeneous cores.

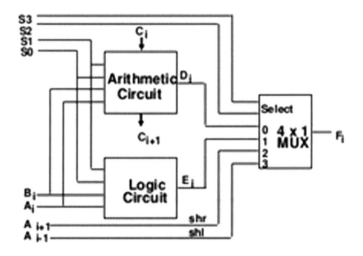

S3	S2	S1	S0	Cin	Operation	Function
0	0	0	0	0	F = A	Transfer A
0	0	0	0	1	F = A + 1	Increment A
0	0	0	1	0	F = A + B	Addition
0	0	0	1	1	F = A + B + 1	Add with carry
0	0	1	0	0	F = A + B'	Subtract with borrow
0	0	1	0	1	F = A + B'+ 1	Subtraction
0	0	1	1	0	F = A - 1	Decrement A
0	0	1	1	1	F = A	Transfer A
0	1	0	0	X	F = A ∧ B	AND
0	1	0	1	X	F = A ∨ B	OR
0	1	1	0	X	F = A ⊕ B	XOR
0	1	1	1	X	F = A'	Complement A
1	0	X	X	X	F = shr A	Shift right A into F
1	1	X	X	X	F = shl A	Shift left A into F

Fig. 5 Arithmetic logic shift unit [4 & 5]

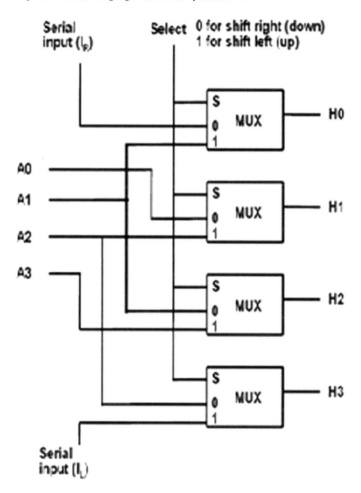

Fig. 6 Hardware implementation of shift micro-operation [4 & 5]

References

1. Barbacci, Mario, R.: A comparison of register transfer languages for describing computers and digital systems. IEEE Trans. Comput. **2**, 137–150 (1975)
2. Chu, Y.: An ALGOL-like computer design language. Commun. ACM **8**(10), 607–615 (1965)
3. Mano, M.M., Kime, C.R.: Logic and Computer Design Fundamentals. Pearson Prentice Hall, Upper Saddle River (2008)
4. Mano, M.M.: Digital Logic and Computer Design. Pearson Education India, Bengaluru (2017)
5. Mano, M.M.: Computer System Architecture. Prentice-Hall of India (2003)
6. Anceau, F., et al.: CASSANDRE: a language to describe digital systems, application to logic design. In: Software Engineering: Proceedings of the Third Symposium on Computer and Information Sciences held in Miami Beach, Florida, December 1969. Elsevier (2012)
7. Baray, M.B., Su, S.Y.H.: A digital system modeling philosophy and design language. In: Proceedings of the 8th Design Automation Workshop. ACM (1971)
8. Barbacci, M., Bell, C.G., Newell, A.: ISP: A Language to Describe Instruction sets and Other Register Transfer Systems. Carnegie-Mellon University, Department of Computer Science (1972)
9. Bell, C.G., Newell, A.: The PMS and ISP descriptive systems for computer structures. In: Proceedings of the Spring Joint Computer Conference 5–7 May 1970. ACM (1970)
10. Chu, Y.: Introducing the computer design language. In: Proceedings of the IEEE Computer Conference, COMPCON, vol. 72 (1972)
11. Ashenden, P.J., Peterson, G.D., Teegarden, D.A.: The System Designer's Guide to VHDL-AMS: Analog, Mixed-Signal, and Mixed-Technology Modeling. Elsevier (2002)
12. Radivojevic, Z., Stanisavljevic, Z., Punt, M.: Configurable simulator for computer architecture and organization. Comput. Appl. Eng. Educ. **26**(5), 1711–1724 (2018)
13. Costanzo, F., et al.: Flexible simulations of complex networks in OpenStack clouds. Int. J. Grid Util. Comput. **8**(2), 133–141 (2017)
14. Eswaraprasad, R., Raja, L.: A review of virtual machine (VM) resource scheduling algorithms in cloud computing environment. J. Stat. Manage. Syst. **20**(4), 703–711 (2017)
15. Andrews, L.J.B., Raja, L.: A study on m-health inline with the sensors applying for a real time environment. J. Stat. Manage. Syst. **20**(4), 659–667 (2017)

An Energy Efficient and QoS Achieved Through MapReduce in Cloud Environment

Sandeep Rai, Aishwarya Namdev, Praneet Saurabh and Rajesh Boghey

Abstract This paper presents an energy-efficient and quality of service achieved through MapReduce in cloud environment. Although decrease in operating costs residue to be a key desire for movement to Cloud environments, Power consumption is a big thing for data centers and cloud service providers. Many big data applications execute on Hadoop MapReduce framework for handling large workloads. Cloud computing, Virtual machine allocation policy is an important aspect while dealing with multiple component architectures. Data allocation, privacy allocation, resource optimization are challenging task in any architecture. Traditional algorithm faces either a long computation time or high cost over the computation. In this approach which is Hadoop driven matrix based architecture is used for data processing. A hybrid combination of both the approach makes user able to process data effectively with time and cost. The approach is performed on Ubuntu 16.04, HDFS and Java API and maximum power consumption of 112 W as a result.

Keywords Power consumption · Quality of service · Twitter analysis · VM allocation and VM scheduling

S. Rai (✉) · A. Namdev · R. Boghey
Department of Computer Science & Engineering, Technocrats Institute of Technology (Excellence), Bhopal, India
e-mail: sandtec@gmail.com

A. Namdev
e-mail: a.namdev2802@gmail.com

R. Boghey
e-mail: rajeshboghey@gmail.com

P. Saurabh
Department of Computer Science & Engineering, TIT, Bhopal, India
e-mail: praneetsaurabh@gmail.com

© Springer Nature Singapore Pte Ltd. 2020
A. K. Luhach et al. (eds.), *First International Conference on Sustainable Technologies for Computational Intelligence*, Advances in Intelligent Systems and Computing 1045, https://doi.org/10.1007/978-981-15-0029-9_61

1   Introduction

Big data [1] is the term for an assortment of data sets with sizes beyond the ability of commonly used software tools to capture, accurate, control and process the data in an allowable elapsed time. Big data sizes are ranging from a few dozen terabytes of many petabytes of data in a single data set. Big data is data that is too big, too hard or too fast for existing algorithms and systems to handle. Properly managed Big Data are reliable, secure, accessible and manageable. Hence, Big Data applications can be enforced in various complex scientific restraint including atmospheric science, medicine astronomy, biology, bio geochemistry, genomics. Big data technology improves performance, provides decision making support and promotes modernization in the product services of business models. Big data technology aims to decrease hardware and processing costs and to check the value of Big Data before committing significant company resources.

Even though the cloud has greatly clarified the capacity provisioning process, it poses several novel disputes in the field of Quality-of-Service (QoS) management. Quality-of-Service (QoS) [2, 3] stands for the levels of performance, reliability, and availability provided by an operation and by the platform or infrastructure that hosts it. QoS is vital for cloud users, who anticipate jobholder to convey the exhibited quality characteristics, and for cloud benefactor, who need to find the appropriate trades offs between QoS levels and working costs. Nonetheless, finding optimal tradeoff is a tough result problem, often aggravated by the existence of service level agreements (SLAs) establishing QoS targets and economical dues correlated to SLA violations [4]. While QoS properties have accepted constant consideration well before the appearance of cloud computing, performance diversity and resource isolation mechanisms of cloud platforms have significantly intricate QoS analysis, prophecy, and support. This is causing several researchers to probe automated QoS management methods that can support the high programmability of hardware and software resources in the cloud [5]. This paper gives the result based on performance parameters which shows the maximum power consumption of 112 W and enhances the quality of service.

Figure 1 shows Quality-of-Service (QoS) management for the levels of performance, reliability, and availability provided by an application.

The rest of this article is organized as follows. Section 2 presents literature review. In Sect. 3, the MaHaVM methodology Matrix HDFS virtual machine distribution (MaHaVM) is explained. The experiments and result evaluation is described in Sect. 4. The conclusion and future work is subsequently outlined in Sect. 5.

2   Literature Review

The key factors of data center are power and energy consumption. These data centers abode thousand of server and maintain infrastructures for reducing temperature.

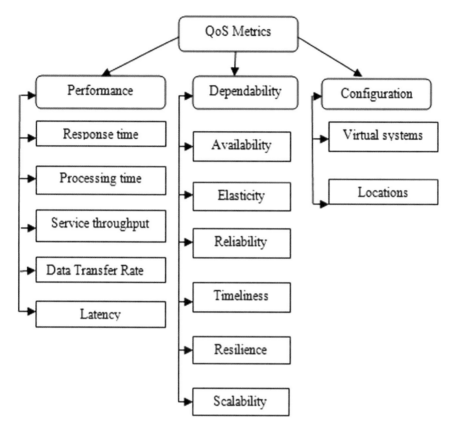

Fig. 1 Quality of service metrics

Researchers composing aim to save energy in servers. They have been given these benefits, by calculating the maximum power utilizing HP's power calculator the power consumption for each server can be found. Then we can pursue the convocation which averages power usage [6] either for midrange or for high-end servers which is approximately 66% of the greatest strength. Hard disk arrays comprise stimulating the operations like cache recall, disk array controllers, disk enclosures and unnecessary power supplies. When we articulate about [2] cloud computing data centers the storage [4] spaces which have in the data center is developed and hard disk usage is centrally coordinate. Multiple numbers of users can share a single server through server virtualization, which finally increases resource utilization and in turn decreases the total number of server's determine.

Users do not need to awake the operations being achieved by other users and can utilize the server celebrating themselves to be the only deploy on that server. Where in some servers come into a sleep mode, when they are not in request, which finally reduces energy consumption. There has remained a big quantity of previous work in energy-efficient computing systems. Comprehensively reviewing the prevailing literature closely associated with our work. A new architecture referred as Advanced Control Distributed Process Architecture (ACDPA) by relating the abstraction property of Software Defined Networking (SDN) and also the distributed process power Hadoop. Software-Defined Networking (SDN) [7] is access where abstraction is used to simplify the network in 2 layers: (1) Used for controlling the traffic. (2) Used for forwarding the traffic.

A Big Data Analytics: challenge and solution [8] using Hadoop, MapReduce [9] have focused on various methods predict to the problems in hand through MapReduce framework over Hadoop distributed file system (HDFS). MapReduce is detraction procedure makes use of file indexing with mapping, sorting, snuffing and finally detract MapReduce procedure is executed for big data analysis using HDFS author also target on big data opportunity and challenges, etc.

Big Data and Hadoop have targeted on the concepts of big data as well as 3v's., volume, velocity, and variety of big data and target on big data processing dilemma and also MapReduce HDFS architecture. An efficient MapReduce scheduling algorithm in Hadoop has MaHaVM a method to improve efficiency of MapReduce scheduling Algorithms and now it works improved than Sentiment analysis. MapReduce Scheduling Algorithms by taking less calculation and gives high accuracy.

In order to work toward QoS [2] in data processing literature is investigated and thus problem formulation is derived. The following are the observed points which described as a problem and further evaluated and performed further with improvements.

In previous techniques, all component of HDFS or other VM is participating at a time, even when there is no such requirement. Thus an optimized system is required to process it. Cloud component, virtual machine, data center, and other related component required software to understand the burden or working in each component and their current usage. It helps in taking decision for allocation policy, but it takes long bandwidth and computational cost. Different power consumption level and derivations are defined in previous approaches, which give the proper recommendation to proceed with efficient resources. This mechanism is lacking in previous available approach.

Given the previous technique having two assumptions in running the unit, either best fit or worst fit. This is again the human skill part is required to decide which step need to be taken and to observe the output. In previous techniques [7] the response time, balance utilization of resources of a host and load distribution on all the hosts is only based on the VM scheduling techniques, which further refined to upgraded technique with large data processing.

Thus in order to propose a better prediction model using Matrix approach using HDFS Hadoop platform and further combine approaches requirement is to further acquire a scheme which contributes on getting better outcome and system, here our MaHaVM methodology heuristic is utilized to scheme in place of traditional scheduling approach.

3 Proposed Methodology

This section introduces the MaHaVM methodology matrix HDFS virtual machine (MaHaVm) to measure the work of any algorithm in the virtual environment some parameters like Power consumption, response time and CPU usages are considered.

3.1 Flowchart Diagram of MaHaVM Method

Figure 2 shows the flowchart diagram of MaHaVM methodology which is used in work.

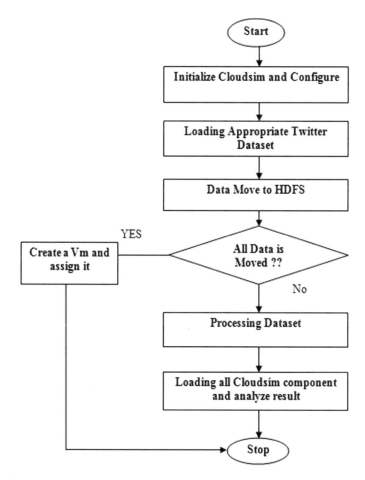

Fig. 2 Flowchart diagram of MaHaVM method

3.2 Matrix HDFS Virtual Machine Distribution (MaHaVM)

MaHaVM is a non-pre-emptive direction, in which clustering [9] of cloudlets and virtual machines is done on some benchmark like cloudlets size and accessible resources of virtual machine. And then cloudlets are mapped to relevant virtual machine for their execution which debars distracting of cloud load and cloudlet lost. This algorithm is chiefly fascinated on the proper distribution of cloudlets among the accessible virtual machines in such a way maximal achievement can be attained. This algorithm basically works in four steps. In the first step, the formation of virtual machine and cloudlets are accomplished and second it assets the virtual machine for the arrangement. In third step cloudlets circulation complete in an expected manner. And in case of uncertainty of virtual machine when all having same number of cloudlets aimlessly allow one of them for Cloudlets if the storage capacity is obtainable otherwise create new virtual machine from the storage. Eventually the complete data processing is performed using the HDFS platform of Hadoop with the stated twitter dataset. This algorithm uses the statistics promoted through matrix computation and derives the allocation policy through it. It gains the comprehensive working of the proposed algorithm is shown below.

3.3 Working of MaHaVM for Cloud Computing

Step 1: [Initializing cloudsim API]
Configure various configuration entities, number of Virtual machine, cloud data center, and other required communication components.
Step 2: [Start simulation and loading complete dataset]
Next Available data which is from twitter having tweets more than 1 crore is loaded and to process over the Hadoop component.
Step 3: [HDFS processing Dataset CSV file]
The complete csv file of twitter dataset is moved to the Hadoop sqoob environment which keeps a copy of complete dataset to proceed via Hadoop component.
Step 4: [Hadoop processing of Dataset to determine word count job]
All the simulated loaded data is then processed by the Hadoop MapReduce component to process the job in between virtual machine given by cloudsim environment.
Step 5: [Finish simulation and generating stats, results]
On processing the complete data, a processing statistics are generated and computed results were plotted over the Jfree java graphical API.

3.4 MaHaVM Algorithm

Involve four phases for Virtual Machine Management:

1. Process Evolutions
2. VM Evolutions
3. Scheduler for VMs
4. Deallocation of VMs

Input: Data load input, timeout t, current process status, current VM status, free VMs.

Output: linear proportion of VMs related to processes, higher resource values, less power usage at peak, less sudden drops.

Steps
Begin:

1. *Load All Ds*
   For(Dsi − n)
2. *Start Mapping(Ds);*
   Initialize Mt
   Mt(ki j) = Matrix Computation Mt(ki, Mtij) //(EXECUTION TIME = Final time − initial time)
   Mt(ki j) = t //time out updation.
3. *Matrix Updation Mt(ij)*
   Return VM Scheduler
4. *If input process ipn is greater (ipn > 0) //Phase 1:Process Evolution*
   VM Scheduler(ipn) //Phase 3: Scheduler for VM
5. *If(nVM ==ipn) then*
   Assign: foreach(ipi ⟶ VMi)
6. *Current Vm will be counted //Phase 2: VM Evolution*
   else if(nVM < ipn)
   calculate required VM
7. *Create new Vm and repeat step 5*
   else if(nVM > ipn)
   calculate free VM //Phase 4:deallocation of VM
   destroy VM
   powerup resources (Energy = power × time)
8. *Free resources will help current OS and current VMs which requires more power with CPU and RAM availability*
9. *If(Success)*
   Return Ci
   Return Throughput th (THROUGHPUT = Amount of work /Time period) [BEST CASE,AVERAGE CASE,WORST CASE]
   Update Matrix
   SLA
10. *End:*

In order to perform this approach multiple processes to generate VMs in a proper manner.

1. A dynamic matrix of mapping sources will require.

2. This matrix will go through scheduler and will start the estimation for first batch required VMs.
3. Current VMs will be counted and if both are same then each VM will get assigned a process.
4. Else VM scheduler will create new VMs in required proportion and step 3 will be followed.
5. As the first batch of matrix get finished processing or after a desired time, any VM get free, VM Scheduler will map another available process to it.
6. If no process is available with current proportion of VMs then VM Scheduler will proceed to destroy the VM and free up that resources.
7. Free resources will help current OS and current VMs which require more power with CPU and RAM availability.

4 Experiment and Result Evaluation

4.1 Dataset

A large dataset of 1 crore tweets is extracted from the dataset given in web resource of a social media platform Twitter. Dataset contains different attribute related to user id, user tweets, their hashtag and other multiple attributes on which a processing is performed over first 5 attribute.

4.2 Comparison Analysis

An experiment performed with Ubuntu 16.x, 4 GB RAM, 1 TB HDD with cloudsim API. Four different twitter data size is created. These comparisons are explained in tabular form.

4.3 Parameters for Result Checking

Different parameters Cloud component, virtual machine, data center, and Hadoop is used to determine the result.

4.4   Calculation Significance

The calculation is based on the Matrix HDFS virtual machine (MaHaVm) used to calculate the throughput to improve quality of service and to reduce power consumption.

Table 1 shows the comparison among different considered algorithms with respect to completion time. Here results display that MaHaVM algorithm is taking less time as comparison to different number of tweets up to 1 Crore in Cloud Analyst simulation environment.

Figure 3 demonstrates time comparison among algorithm. This time comparison graph represents graph between time (milliseconds) on Y axis and no of tweets (in lacs) on X axis.

Table 2 shows comparison among different measured algorithm with respect to Analysis of Power Consumption. This result illustrates that proposed algorithm takes less power consumption with respect to existing Virtual machine distribution with varying no of tweets (in Lac) in Cloudsim Analyst simulation with HDFS processing.

Table 1 Comparison among different considered algorithms	Number of tweets	Sentiment analysis result (ms)	MaHaVm result (ms)	Variability
	T set 1	1013	1692	679
	T set 2	1017	1649	632
	T set 3	1025	1662	637
	T set 4	971	1660	689

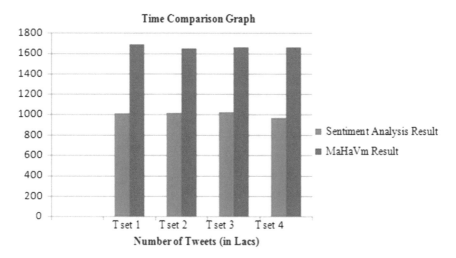

Fig. 3   Time comparison among algorithm

Table 2 Analysis of power consumption among algorithm

Number of tweets (in lakh)	Number of VM	Min power	Max power	Average	Variability	Variability (%)
T set 1	1	117.87	118.33	118.33	±0.92	±0.78
T set 1	2	112.95	117.85	115.4	±2.45	±2.13
T set 1	4	118.89	122.79	120.84	±1.95	±1.62
T set 1	8	113.97	117.71	115.84	±1.87	±1.61
T set 2	1	112.42	113.44	112.93	±0.51	±0.46
T set 2	2	109.57	112.12	110.84	±1.28	±1.16
T set 2	4	117.85	118.79	118.32	±0.47	±0.39
T set 2	8	117.71	118.89	118.30	±0.59	±0.50
T set 3	1	118.69	122.17	120.43	±1.74	±1.45
T set 3	2	115.87	120.84	118.36	±2.48	±2.09
T set 3	4	113.97	115.54	114.76	±0.78	±0.68
T set 3	8	112.86	113.81	113.34	±0.46	±0.41
T set 4	1	114.21	118.00	116.11	±1.87	±1.62
T set 4	2	117.18	118.30	117.74	±0.56	±0.48
T set 4	4	112.13	115.42	113.78	±1.64	±1.45
T set 4	8	109.66	110.25	109.96	±0.3	±0.28

Figure 4 demonstrates the graph between number of virtual machine on X axis and power consumption in watts on Y axis. The graph represents comparison between existing result and proposed result.

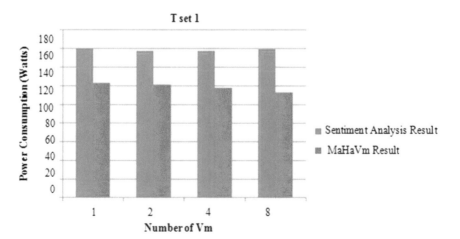

Fig. 4 Power consumption measure on T set 1

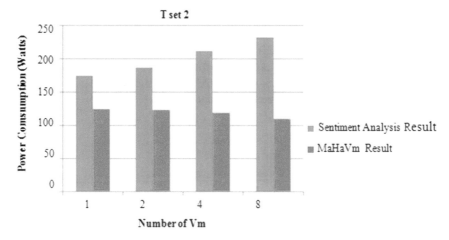

Fig. 5 Power consumption measure on T set 2

Figure 5 demonstrates the graph between algorithm with respect to Power consumption for several no. of Tweets (in Lakhs) with several no. of Virtual machine (1, 2, 4, 8). In this graph X axis serve number of Virtual Machine use at any time instance and Y axis represents the Power Consumption in watts.

Thus the overall comparison between proposed and existing two algorithms in cloud VM Hadoop environment and it also represent the graph of proposed algorithm and existing approach with respect to time and cost.

5 Conclusion and Future Work

Data processing with the multiple environment and component is derived by various units. Large data processing is always a challenging issue in working strategy processing. In this dissertation work large data processing over the multiple virtual machine environments is proposed. An algorithm named MaHaVm is proposed which takes advantage of matrix computation before processing of any number of input data processing. Matrix keeps the information regarding the current utilization statistics. Matrix keeps a row of understanding of VM, DC load availability. HDFS processing of 10 lakh to 1 crore tweets is taken into consideration which is processing for sentiment analysis of twitter data with the help of MapReduce word count process and through which an efficient outcome is monitored. Our proposed approach is setup is Hadoop setup, Netbeans tool set up with JDK 8 on Ubuntu x64 Machine with 8 GB RAM and 1 TB HDD. A computation time, computation cost and throughput is monitored for system comparison and performance. A work is performed with the cloudsim using the HDFS Hadoop platform for work processing over the multiple virtual machine environment, which is a platform to test efficiency of our algorithm.

A further extension is also left to perform with multiple environments and working on cloud availability and their different data size and distance measure concept for communication and monitor to efficiency algorithm.

Further work can be done to make our approach more realistic on applying such multi-tenant, distributed and multi-authority architecture such that it can be processed with real-time data and Virtual machine vendor.

References

1. Rathi, R., Lohiya, S.: Big data and hadoop. Int. J. Adv. Res. Comput. Sci. Technol. (IJARCST 2014) **2**, 214–217 (2014)
2. Patokar, A.A., Patil, V.M.: Efficient analysis of big data by using Hadoop in cloud computing by map reducing. In: National Conference on Innovative Trends in Science and Engineering (NC-ITSE), vol. 4(7), pp. 378–381 (2016)
3. Osman, R., Pérez, J. F., Casale, G.: Quantifying the impact of replication on the quality-of-service in cloud databases. In: 2016 IEEE International Conference on Software Quality, Reliability and Security (QRS), Vienna, pp. 286–297 (2017)
4. Energy star enterprise storage specification. United States Environmental Protection Agency, http://www.energystar.gov/index.cfm?c=new specs. Enterprise storage, under development, Apr. 2009
5. Green Grid data center efficiency metrics: PUE and DCIE. White Paper, The Green Grid, December (2008)
6. Fan, X., Weber, W.D., Barroso, L.A.: Power provisioning for a warehouse-sized computer. In: ISCA '07: Proceedings of the 34th Annual International Symposium on Computer Architecture, pp. 13–23. ACM, New York, USA (2007)
7. Su, C.-S., Yen, T.-C.: An SDN based cloud computing architecture and its mathematical model. Inf. Sci. Electron. Electr. Eng. (ISEEE), 1728–1731 (2014)
8. Dhavapriya, M., Yasodha, N.: Big data analytics: challenges and solutions using Hadoop, MapReduce and big table. Int. J. Comput. Sci. Trends Technol. (IJCST) **4**(1), 5–14 (2016)
9. Ghemawat, S., Dean, J.: MapReduce: simplified data processing on large clusters. Commun. ACM **51**, 107–113 (2008)

Static and Dynamic Malware Analysis Using Machine Learning

Chandni Raghuraman, Sandhya Suresh, Suraj Shivshankar and Radhika Chapaneri

Abstract Malware is a section of code written with the intention of harming a device. Attacks on the Android operating system have been on the rise of late as there are plenty of applications on the Internet that possess malware. To analyze these attacks, machine learning can be used to make the process more efficient. This paper demonstrates static and dynamic analysis of Android malware. By identifying patterns from datasets created and using a myriad of classifiers, the results have been compared to infer the most optimal method of malware analysis. Various machine learning classifier algorithms are implemented, with Random Forest and Decision Tree giving the best accuracy and F1-Score of 94% in static analysis. Support Vector Machine and Neural Network have given the highest accuracies of about 99% after implementing Principal Component Analysis in dynamic analysis.

Keywords Android malware detection · Static analysis · Dynamic analysis · Machine learning · Principal component analysis

1 Introduction

The value of mobile devices in today's technologically savvy world is indispensable. Functionalities of mobile applications have made even remote bank transactions possible. With the increase in the number of smart devices in recent times, the amount of malware affecting these devices has also increased [1]. Many hackers can extract

C. Raghuraman (✉) · S. Suresh · S. Shivshankar · R. Chapaneri
Mukesh Patel School of Technology, Management and Engineering, NMIMS University, Mumbai, India
e-mail: chand1003@gmail.com

S. Suresh
e-mail: sandhyasuresh2525@gmail.com

S. Shivshankar
e-mail: surajshivshankar97@gmail.com

R. Chapaneri
e-mail: Radhika.Chapaneri@nmims.edu

© Springer Nature Singapore Pte Ltd. 2020
A. K. Luhach et al. (eds.), *First International Conference on Sustainable Technologies for Computational Intelligence*, Advances in Intelligent Systems and Computing 1045, https://doi.org/10.1007/978-981-15-0029-9_62

relevant data from either an individual or a corporation, such as bank details. It is so rampant that as of 2018, the average cost of a malware attack on a company is $2.4 million [2]. While there are plenty of antiviruses available in the market, 4 out of 5 people believe that only their antivirus cannot protect their devices from malware attacks. Therefore, a better way of cybersecurity needs to be developed to deal with the threats arising from malware.

While smartphones today are manufactured using various operating systems, research indicates that Android is the most common operating system in the world owing to its ease of use and in general affordability. This increased popularity of Android also means an increase in the amount of malware attacks that have occurred in the past few years [3]. As a result, this paper focuses exclusively on malware analysis for the Android operating system.

There are different types of attacks that can affect an Android device. The attacks may be triggered by injecting malware into it. Nature of malware attacks may vary based on how the attacker wants to affect the system and cause damage. The different types of malware that can affect a system are listed in Table 1.

This paper aims to implement various machine learning algorithms and compare their results. The performance metrics determine the probability of the malware getting detected and hence the choice of classifiers also plays a big role in detection. This paper broadly looks at various machine learning algorithms, namely—Decision Tree, Naïve Bayes, K-Nearest Neighbour, Support Vector Machine and Random Forest.

Table 1 Types of malware attacks

Malware	Description
Virus	A virus replicates itself and spreads within the operating system. Human interaction is necessary for a virus to cause damage
Spyware	Spyware targets the user specifically and it can track everything a user types. It is silently installed in the background without the user realizing
Adware	It tracks what the users do online and shows advertisements accordingly. This kind of malware significantly slows down the system
Scareware	This malware scares the users into buying software that they do not require. It might come along with other types of malware as well
Ransomware	This malware takes a system as hostage and demands a ransom in exchange for the users' data. Since a very strong encryption is used, the users generally end up paying the amount to get their data back
Worms	Worms do not require any human intervention. It can travel from one device to another on its own and it has a very negative impact on the system once it is installed
Trojans	Trojans create a backdoor with the help of which a hacker can take control of the computer. They can steal files and the attacked device can be used as a proxy for the hacker to hide behind

Malware Analysis is done in two main forms: Static Analysis and Dynamic Analysis [4, 5]. Both analysis models have their advantages and disadvantages, due to their varying observing capacity during different stages of malware introduction and growth.

2 Related Work

Android is an open-source smartphone operating system that is being used by billions across the globe as of 2018. Malware attacks originate from the root files of Android applications that are in the format 'Android Packaging Kit' (APK) as ZIP archives. These archives contain files such as the Android Manifest file (.xml) that holds important permission calls required by the application and dex files that contain Java codes. The final format of an Android dex file is also known as Dalvik, as it is run on Dalvik Virtual Machine, a Linux based kernel [3]. Although security patches provided by Android and the strict policies of Google Play Store (the official vendor for Android applications) have strict guidelines, malicious applications can still be downloaded by users via third-party vendors or sophisticated encoding in seemingly benign applications. To understand the methods of malware attacks in Android smartphones and remedies to curb this issue, analysis can be done in two ways: Static Analysis and Dynamic Analysis [4, 5].

2.1 Static Analysis in Malware Detection

Static Analysis is performed during the compile time of an Android application. Signature-based detection and pattern recognition are some key methods used in such analysis. The advantage of static analysis is that it can be done in a short runtime and it causes minimal kernel overhead. However, it fails to accurately detect malware types that tend to change their behaviour over time. The accuracy of malware detection based on behavioural anomalies is also less in Static Analysis [4]. The malware analysis environment DREBIN [5], the Malware Genome Project [4], Droid API Miner [6] and other such initiatives have contributed extensively to static analysis research and development for malware detection. Various organizations have also released binary datasets to aid researchers in this regard—some of them are AndroZoo [7] and AndroMalShare [8].

The paper [5] was very resourceful with regard to the creation and applications of the DREBIN Dataset, which consists of a myriad of malware as well as benign data samples. The paper [9] gave an overview of the different machine learning classifiers used on malware datasets and how their results vary in terms of various precision metrics. These results inspired us to conduct our own analysis on a custom dataset. Urcuqui-López and Cadavid [10] focused on static analysis for malware detection and explained the nuances of conducting independent research in this domain. A sample of the dataset used in this paper was shared on Kaggle for experimentation purposes. Commonly used extraction tools to gather data from Android applications are illustrated in paper [6].

2.2 Dynamic Analysis in Malware Detection

Dynamic Analysis is done during the runtime of an Android application [11]. Various sophisticated malware detection projects such as Andromaly and CrowdDroid are based on dynamic analysis of Android apps [12]. This type of analysis can deal with sophisticated malware and also adapt to the creative ways with which malicious breaches are made. Although dynamic analysis provides more accurate results for the mainstream malicious applications, it causes significant overhead in the kernel and can also slow down the system during the analysis phase. As a result, it is used sparingly, where hardware specifications of the device match the requirements to contain the runtime as well as overhead [13]. Running analysis on network protocols used by APKs is a method used for dynamic analysis in malware detection.

In paper [14], a new detection and characterization system for detecting meaningful deviations in the network behaviour of a smartphone application is proposed, which is a dynamic form of malware analysis. The main goal of the proposed system is to protect mobile device users and cellular infrastructure companies from malicious applications that are downloaded on Android smartphones. Sandbox environment experiments as well as datasets used by the authors of this paper are shared available, and we have used these resources to conduct our experiment in this domain.

3 Methodology

3.1 Methodology for Static Analysis

While reading survey and research papers, we found that most authors used the following strategy to procure data samples and run analysis for malware detection [4, 5]:

I.   Analyzing the types of malware attacking Android phones via malicious application downloads
II. Creating a feature list of system calls that might indicate if the device has been attacked by the malware
III. Extracting data from potentially malicious software and analyzing it using available software tools
IV. Comparing extracted data with the feature list and discarding unimportant data

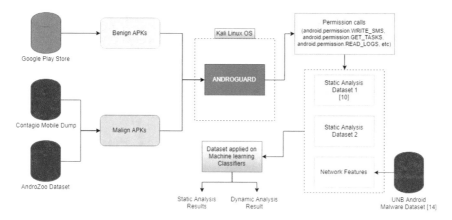

Fig. 1 Framework for static and dynamic analysis of Android malware

V. Running machine learning algorithms on the reduced data and producing the results on the User Interface

VI. Provide a simple explanation to the results obtained on the User Interface.

Data Collection

For static analysis, we prepared a dataset using permission calls extracted from Android APKs of both benign and malign applications. The benign applications were collected from Google Play Store and the malign applications were collected from various sources such as the DREBIN dataset, the Androzoo Dataset and Contagio Dump [5–7]. The permission calls are unpackaged and viewed using various extraction tools. Malware applications mainly originated from third-party vendors and Chinese markets. The dataset was prepared by entering a '1' against permission calls that were present in the application and thus, a binary dataset was created in CSV format. The dataset consisted of 257 applications of unbiased malign and benign distribution and 333 features. Figure 1 illustrates our malware analysis framework.

It was observed that certain permission calls are highly used in malicious applications. However, this method alone cannot determine if an application is malign or benign, it can be only one of the factors taken into consideration.

Figure 2 shows some of the most commonly used permissions by applications, both malign and benign. We used a dataset of around 400 applications and found out the percentage of malign applications using a particular permission with respect to benign applications using the same permission call. The X-axis represents the various permission calls accessed by an Android system and the Y-axis shows the percentage distribution of the applications. For example, out of all the applications requesting the permission call 'com.android.permission.WRITE_SMS', 99% is malign. Similar observations can be made for the other permission calls.

While creating a static analysis dataset, we must ensure that the most relevant permission calls are considered. There are many instances where a benign application makes use of permission calls that are also used by malign applications. In such a

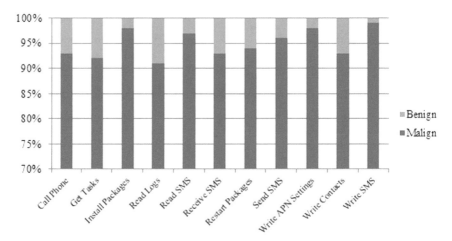

Fig. 2 Percentage ratio of malign and benign applications using permission calls

scenario, the application must not be incorrectly labelled as malicious, as it would hinder the credibility of the malware detection system. Therefore, dynamic analysis for malware detection is more reliable and adaptable [4, 5, 15].

Owing to this trend in malware analysis, we conducted dynamic analysis using network protocols observed in Android applications when they connect to the server. Based on network activity, the nature of the application could be determined.

Feature Extraction

When an Android application needs to be tested to see if it is malign or benign, the testing input is the permission calls that are required for the application to run. Feature extraction is the process where permission calls coded into an Android APK are retrieved for analysis purposes by disassembling the application. The permission calls retrieved post extractions are the features for the malware dataset, which will then be analysed by machine learning algorithms to produce the end result [13, 15, 16].

In the case of a disassembled Android application, the permissions can be found on the Android manifest file. These permissions can be used to judge the nature of an application. A malicious application might require more permissions than necessary while benign applications will only request permissions that are well and truly required for the proper functioning of the application. Some of these permissions include 'Send SMS', 'Access Network State', 'Write APN Settings', etc. There are many reverse engineering tools available on Kali Linux OS and Internet. We have used the following tools for feature extraction.

- **APKTool**

APKTool is an application on Kali Linux that can be used for reverse engineering any APK. Once an APK is uploaded, APKTool disassembles it and returns all the files that constitute it. Out of these, the Android Manifest file contains all the permission calls that the application requires to function. These permissions can be noted to see if the application is receiving permissions for something that may be used for malicious purposes [17]. We used this feature to retrieve permission calls from malign and benign applications.

- **Androguard**

  Androguard is an open-source static analysis tool used to disassemble Android applications [13]. Some features of Androguard include detecting code similarity between applications and disassembling codes to show permission calls, class activities, etc. Androguard can be run on the Windows OS as well as Kali Linux. It has various tools that can be used to unpack APKs, compare APKs, add permission calls or features to an APK and repackage it, as well as convert the internal libraries of APKs to other formats for editing and observation.

Feature Selection

After obtaining the dataset, we observed that for a few applications, the features were not making any significant contributions. There were many feature columns that had only 0 as its value. So, we wrote a Python script that removed all columns that had only 0s. This reduced the number of columns in the dataset and resulted in better performance of the classifiers.

Classification Algorithms

Machine learning classifiers have always proved to be helpful for training systems and making them intelligent machines. With a set of data served as input, the classifier is able to construct a model with the help of which it can accurately classify the new data given to it. The classifiers used are:

- **k-Nearest Neighbour (kNN)** is a lazy learning algorithm whose aim is to classify a new object based on the already existing classes in which the previous training points are classified. It classifies the new data points based on similarity measures. More the neighbours belonging to the same class, the data gets classified into that class. kNN was chosen to be applied on our datasets because it is easy to interpret, has lesser calculation time and has good predictive power.
- **Naive Bayes** is based on the Bayes Theorem which works on conditional probability. Naive Bayes classifier predicts the probabilities for each of the output class based on some predefined conditions for the input classes. The output class with the highest probability is considered as the most likely class. Naïve Bayes usually needs lesser training data and is a powerful algorithm for classification tasks.
- **Support Vector Machine (SVM)** is a classifier that builds a model by defining a hyperplane. This plane is constructed by analyzing the input data.
- **Decision Tree** is a tree-like model that is used to visually represent the decision-making process. When a set of input data is fed to the classifier, it builds a decision tree with the help of which the new data given for testing the algorithm is classified. A decision tree consists of a root node, which is the first feature based on which the rest of the tree is built. One of its main advantages is its transparent structure. It explores all possible directions and comes up with a decision.

- **Random Forest** is a type of ensemble learning in which multiple learners are combined to produce optimum results. In random forest, during the training phase, the model is constructed by building multiple decision trees. These trees classify the same data with respect to different parameters. The trees are then merged to get results that are more accurate and stable. Whenever a decision has to be made, the result predicted by majority of the trees is chosen.
- **Neural Network** is a single layer feed-forward network that doesn't have any hidden layers and it is used for binary classifications. It has a set of inputs in an input layer with all the inputs connected to the output.

3.2 Methodology for Dynamic Analysis

Dataset

We used the dataset provided by Lashkari et al. [14] and run the machine learning classifiers to obtain independent results.

Dimensionality Reduction

Having more features does not always mean that we have more information and hence more classification power. Some features can be irrelevant or redundant, especially when there are less training samples. This is referred to as the curse of dimensionality [9]. Feature selection is the process in which those features are selected that contributes most towards the output or the result. It has an impact on the model and it is used because having irrelevant features can reduce the accuracy of the system. Variance of the data in the lower dimensional space should be maximum. Some of the results of successful feature selection can be improved accuracy and reduced runtime of the model. The feature selection model applied on our dataset is elaborated below. Principal Component Analysis (PCA) is a mathematical procedure in which a set of features that are highly correlated with each other are converted to features that are uncorrelated. PCA is used to focus on only a few variables instead of increasing the running time of the model by working with all the features or variables. Suppose a vector z consists of b random variables, it is not useful to look at all the variances of b unless it is very small. Hence, an alternate approach is to look at only a few variables ($\ll b$) that preserve most information about the vector [18]. PCA provides an approximation of any dataset or data matrix $\mathbf{X}$ in terms of $\mathbf{T}$ and $\mathbf{P'}$ where $\mathbf{T}$ and $\mathbf{P'}$ capture the essential features of the data matrix $\mathbf{X}$ [19].

4 Experiment and Results

4.1 Performance Metrics

The performance metrics that are used to evaluate and compare the classifiers are accuracy, precision, F1 score and ROC curve accompanied by the ROC-AUC Score.

- **Accuracy** is the number of correct predictions made by the model. It is compared to all the predictions that have been made.

$$\text{Accuracy} = (TP + TN)/(TP + FN + FP + TN) \qquad (1)$$

- **Precision** is a measure that tells us how many of the truly predicted values actually turned out to be positive.

$$\text{Precision} = TP/(TP + TN) \qquad (2)$$

- **Recall** is a measure that tells us what proportion of values that actually predicted as positive was predicted by the algorithm as positive.

$$\text{Recall} = TP/(TP + FN) \qquad (3)$$

- **F1 Score** is the harmonic mean of precision and recall.

$$\text{F1 Score} = (2 * \text{Precision} * \text{Recall})/(\text{Precision} + \text{Recall}) \qquad (4)$$

4.2 Results and Implementation for Static Datasets

Dataset 1

The first experiment was conducted using 398 samples of Android APKs, out of which 199 were malicious, denoted by 1 and the rest were benign, denoted by 0. The experiment demonstrates that the decision tree classifier and random forest have outperformed all the other classifiers with an accuracy and F1-Score of 94%, as displayed in Table 2. It also indicates that Gaussian Naïve Bayes and the Neural Network are less effective compared to the other classifiers for this dataset. The ROC is displayed in Fig. 3.

Dataset 2

The results for the second experiment, conducted using 100 malware samples and 100 benign samples which were extracted by us, are presented in Table 3. The k-NN classifier has performed better than the rest of the classifiers with respect to accuracy

Table 2 Comparison of performance metrics for first static dataset

Classifier	Accuracy (%)	Precision (%)	Recall (%)	F1-score (%)
k-NN	93.33	94.73	91.52	93.10
Decision tree	93.33	95.31	92.42	93.84
Gaussian Naïve Bayes	92.50	96.49	88.70	92.43
Bernoulli Naïve Bayes	93.33	95.08	92.06	93.54
Neural network	90.00	81.35	97.95	88.88
SVM	94.16	93.33	94.91	94.11
Random forest	94.16	89.47	98.07	93.57

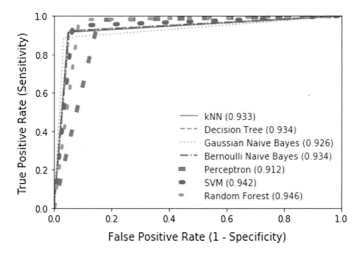

Fig. 3 ROC curve for dataset 1

and precision. The Gaussian Naïve Bayes classifier has produced least satisfactory results with respect to accuracy, recall and F1-Score. The ROC is displayed in Fig. 4.

Table 3 Comparison of performance metrics for second static dataset

Classifier	Accuracy (%)	Precision (%)	Recall (%)	F1-score (%)
k-NN	85.00	83.33	80.00	81.63
Decision tree	73.33	72.00	66.66	69.23
Gaussian Naïve Bayes	68.33	81.81	54.54	65.45
Bernoulli Naïve Bayes	70.00	75.86	66.66	70.96
Neural network	75.00	80.95	60.71	69.38
SVM	78.33	84.37	77.14	80.59
Random forest	83.33	82.75	82.75	82.75

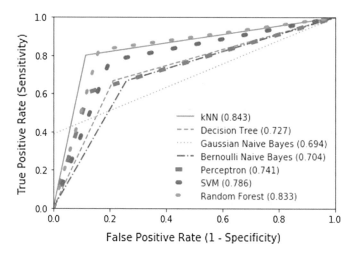

Fig. 4 ROC curve for dataset 2

4.3 Results and Implementation for Dynamic Dataset

The final experiment uses a dataset taken from the University of New Brunswick and it comprises of 4000 samples each, belonging to benign and adware malware classes. The ROC curve for this experiment is presented in Fig. 5. Based on the results presented in Table 4, SVM has once again outperformed the rest of the classifiers with respect to the F1-Score. Decision tree has the second-best accuracy along with

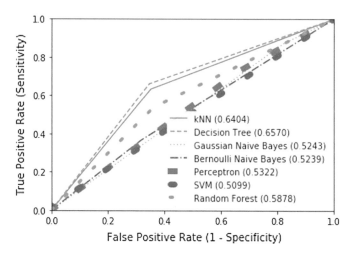

Fig. 5 ROC curves before applying PCA

Table 4 Comparison of performance metrics for dynamic dataset

Classifier	Accuracy (%)	Precision (%)	Recall (%)	F1-score (%)
k-NN	60.08	61.77	54.01	57.68
Decision Tree	61.16	62.39	61.28	61.83
Gaussian Naïve Bayes	54.04	56.05	43.00	48.67
Bernoulli Naïve Bayes	50.66	50.47	39.89	44.56
Neural Network	50.95	50.57	88.34	64.32
SVM	59.70	57.31	74.45	64.77
Random forest	53.70	55.71	45.83	50.02

Table 5 Comparison of performance metrics after applying PCA

Classifier	Accuracy (%)	Precision (%)	Recall (%)	F1-score (%)
k-NN	97.41	96.72	98.17	97.44
Decision tree	99.21	98.91	99.49	99.20
Gaussian Naïve Bayes	96.13	95.37	97.02	96.19
Bernoulli Naïve Bayes	96.20	96.17	96.09	96.13
Neural network	99.62	99.92	99.26	99.63
SVM	99.60	99.75	99.80	99.90
Random forest	97.20	95.08	99.67	97.32

the best results for Precision. The Neural Network has a recall rate of 88% but its other parameters suggest that it might not be suitable for this dataset taken by us.

After applying various classifiers on the dataset, we used the concept of Principal Component Analysis, as mentioned in Sect. 3.2, and got better results as shown in Table 5. The ROC curve for this result is shown in Fig. 6.

PCA has given us the best results for the dynamic dataset but at the same time, it did not improve the accuracy for any of the other datasets.

Figure 7 shows a comparison in the accuracies generated by the machine learning classifiers used before and after applying PCA on Dataset 2.

5 Conclusion and Future Scope

Static as well as dynamic analysis has been performed for malware detection in Android OS using machine learning algorithms. From the obtained results, the classification algorithms Random Forest, SVM and Decision Tree have performed in the most optimum manner. The use of Principal Component Analysis has improved the results substantially from 59.70 to 99.60% for the SVM classifier. Dynamic analysis using network protocols has been demonstrated in this paper. There are various other parameters for malware analysis which can be explored to get accurate results.

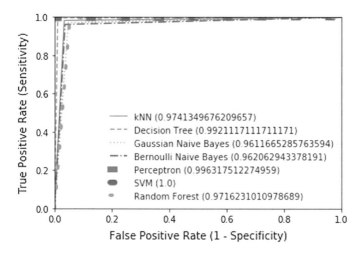

Fig. 6 ROC curves after applying PCA

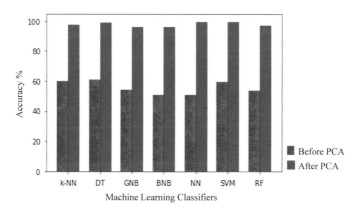

Fig. 7 Accuracies obtained from dataset 2 before and after applying PCA

Consequently, the aim is to assess malware datasets with deep learning approaches as well, since it would provide a novel and more detailed perspective to Android malware analysis. Deep learning holds the key to completely securing the Android OS from malware attacks as it can work with a vast dataset and also showcases adaptability, which is very essential due to the complex coding that is present in malware today.

References

1. Tupakula, U., Varadharajan, V., Akku, N.: Intrusion detection techniques for infrastructure as a service cloud. In: Ninth IEEE International Conference on Dependable, Autonomic and Secure Computing, pp. 744–751. IEEE (2011)
2. Mcafee.com.: McAfee Labs Threats Report (online) (2018). Available at: https://www.mcafee.com/enterprise/en-us/assets/reports/rp-quarterly-threats-dec-2018.pdf. Accessed 3 Feb. 2019
3. KOÇ, U.: Introduction to Android Malware Analysis. (Blog) Uceka (2013). Available at: https://uceka.com/2013/08/06/introduction-to-android-malware-analysis/. Accessed 18 Dec. 2018
4. Sahs, J., Khan, L.: A machine learning approach to android malware detection. In: European Intelligence and Security Informatics Conference, pp. 141–147. IEEE Computer Society (2012)
5. Arp, D., Spreitzenbarth, M., Hubner, M., Gascon, H., Rieck, K., Siemens, C.E.R.T.: Drebin: effective and explainable detection of android malware in your pocket. In: Ndss, vol. 14, pp. 23–26 (2014)
6. Aafer, Y., Du, W., Yin, H.: Droidapiminer: mining api-level features for robust malware detection in android. In: International Conference on Security and Privacy in Communication Systems, pp. 86–103. Springer, Cham (2013)
7. Allix, K., Bissyandé, T.F., Klein, J., Le Traon, Y.: Androzoo: Collecting millions of android apps for the research community. In: 2016 IEEE/ACM 13th Working Conference on Mining Software Repositories (MSR), pp. 468–471. IEEE (2016)
8. Android Malware Dataset. (n.d.). Android Malware Dataset (online). Available at: http://amd.arguslab.org. Accessed 15 Jan. 2019
9. Rana, M.S., Rahman, S.S.M.M., Sung, A.H.: Evaluation of tree based machine learning classifiers for android malware detection. In: International Conference on Computational Collective Intelligence, pp. 377–385. Springer, Cham (2018)
10. Urcuqui-López, C., Cadavid, A.N.: Framework for malware analysis in android. Sistemas Telemática **14**(37), 45–56 (2016)
11. Willems, C., Holz, T., Freiling, F.: Toward automated dynamic malware analysis using cwsandbox. IEEE Secur. Priv. **5**(2), 32–39 (2007)
12. Aung, Z., Zaw, W.: Permission-based android malware detection. Int. J. Sci. Technol. Res. **2**(3), 228–234 (2013)
13. Desnos, A.: Androguard (2019)
14. Lashkari, A.H., Kadir, A.F.A., Gonzalez, H., Mbah, K.F., Ghorbani, A.A.: Towards a network-based framework for android malware detection and characterization. In: 2017 15th Annual Conference on Privacy, Security and Trust (PST), pp. 233–23309. IEEE (2017)
15. Choi, S., Bijou, M., Sun, K., Jung, E.: API tracing tool for android-based mobile devices. Int. J. Inf. Educ. Technol. **5**(6), 460 (2015)
16. Feizollah, A., Anuar, N.B., Salleh, R., Wahab, A.W.A.: A review on feature selection in mobile malware detection. Digit. Inv. **13**, 22–37 (2015)
17. Javadecompilers.com. (n.d.). APK Decompiler (online). Available at: http://www.javadecompilers.com/apk. Accessed 5 Feb. 2019
18. Geladi, P., Isaksson, H., Lindqvist, L., Wold, S., Esbensen, K.: Principal component analysis of multivariate images. Chemometr. Intell. Lab. Syst. **5**(3), 209–220 (1989)
19. Jolliffe, I.: Principal Component Analysis, pp. 1094–1096. Springer, Berlin (2011)

Review-Based Sentiment Prediction of Rating Using Natural Language Processing Sentence-Level Sentiment Analysis with Bag-of-Words Approach

K. Venkata Raju and M. Sridhar

Abstract User's opinion plays a vital role in the global world of Internet as they are given freedom to express their feedback. A lot of hidden information and feeling of the user are expressed in the words that he/she used. Extracting this hidden information can help the service industry to better serve the need as per the user acceptance and to outfit the competition in the market. Methodology of this paper is intended to define the rating of the review given by the user. Method of bag-of-words approach taking the sentiment score and magnitude of the sentence using natural language processing is applied. The scale of one to five is considered in this experimental classification on the data set of hotel reviews. The results have shown that around 60% of the ratings can be predicted and 40% unpredicted with the given review. The Data set used in this experimental analysis is the Datafiniti's hotel reviews.

Keywords Bag-of-words · Sentimental analysis · Review · Prediction · Natural language processing · Sentiment prediction · Statistical methods · Opinion rating

1 Introduction

Decision making in day-to-day life is driven by the rating system with the rapid growth and dependency on the Web 2.0. In the world of Internet-driven system, a large scale of data has a need of summarization automatically. Customer decides his travel destination or planning of hotel booking in the trip depending on the feedback given by the users, which is playing vital role in tourism eco system. In the system of purchasing of the products, customers are showing a lot of interest on the feedback

K. Venkata Raju (✉)
Department of Computer Science and Engineering, Acharya Nagarjuna University Guntur, Guntur, India
e-mail: venkatsagar05@gmail.com

M. Sridhar
Department of Computer Applications, R.V.R & J.C College of Engineering, Guntur, India
e-mail: mandapati12@gmail.com

© Springer Nature Singapore Pte Ltd. 2020   807
A. K. Luhach et al. (eds.), *First International Conference on Sustainable Technologies for Computational Intelligence*, Advances in Intelligent Systems and Computing 1045,
https://doi.org/10.1007/978-981-15-0029-9_63

of the product before they opt for the product. An individual or the corporate system in the business is driven by the opinionated data analysis available on the Web. Opining rating prediction is a challenging task since there is no aforementioned sequence about the aspects expressed by the user and the existing relation between the reviewer's textual aspects to the overall rating. Sentimental analysis (SA) is broadly classified as lexicon-based (LB) and machine learning (ML) systems. A further classification of the machine learning system has evolved as unsupervised and supervised approaches. On the other side, the lexicon-based systems are evolved as corpus- and dictionary-based approaches. The view of sentiment classification methods is shown in Fig. 1 [8].

Analyzing the feedback and the rating is the driving factor of the present industry. Rating is one factor where the user is more attracted to gain the information at a glance. Since the user will not spare to read all the reviews as the data is in tons which will throw the user into confused mind in decision making. Human psychosomatic system is most attracted to the readymade-driven data of how the user rates the product over a scale of 1–10 or on the scale of 1–5. Considering 1 as worst or not accepted case and 10 or 5 as best, over the given scales, the rating is the parameterized system given by considering all the parameters in the system or to a particular parameter of the system. The rest of the paper is structured as follows: The details of the methods that are implemented in this area are briefly discussed in Sect. 2. Model of unigram and bigram is discussed in Sect. 3. Model of statistical approach is discussed in Sect. 4. Model using sentiment prediction of rating using natural language processing (NLP) sentence-level sentiment analysis with bag-of-words approach is discussed in Sect. 5. The results of prediction system are discussed in Sect. 6. Finally, the paper is concluded with future research directions in Sect. 7.

2 Literature Review

Han-Xiao Shi et al., Frequency and TF-IDF discussed about the unigram feature to realize the polarity classification of the Documents. Frequency and Term Frequency-Inverse Document Frequency are applied with supervised Learning method SVM (Support Vector Machine) for predicting the online Review as positive or negative. The study of unigram features is segmented Chinese words, with a corpus of 4000 reviews and 12,745 unigrams appeared at least one time. TF-IDF has shown more effective results when compared to frequency [7]. Walter Kasper and Mihaela Vela proposed a system that collects reviews and comments on hotel from Web and facilitates the information creating classified and structured overview. This work is submitted as a part of BESAHOT (BEwertung SAarlandischer HOTels), project from the Saarland ministry of Economics. The project handles acquisition of data, analysis, and storage. Data acquisition is done using web crawler. Text segment is analyzed by polarity values as positive, negative, or neutral opinion. Since a statement can have multiple polarity on different aspect, it is disregarded in global polarity assignment. Evaluation of the system is done on a corpus of 1559 reviews containing

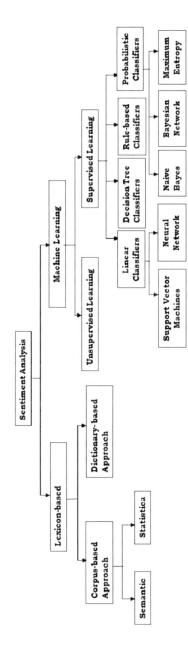

Fig. 1 Sentiment classification methods

4792 text segments [2]. Anshuman et al., Rating-based approach has discussed about the rating-based approach using sentiment analysis on food recipes. The data used for the analysis is based on the reviews written sentiment. Bag-of-words are created over all the crawled hyperlinks based on the ingredient name provided by the user. Weight for each word is calculated using the stored words and total will be evaluated. The results are presented with Foodoholic a mobile application. The output displayed is the recipes ordered list based on the core ingredient input [6]. Ging Li et al. have proposed sentimental analysis on movie reviews using clustering method. Due to the drawback of low accuracy rate by the existing system, Term Frequency-Inverse Document Frequency is used to improve the accuracy. Clusters are applied with the mechanism of voting to get stabilized. The analysis is used to judge whether the movie is good/bad, achieved stable results by overcoming the problem of low accuracy [3]. Wenjuan Luo et al. have performed predicting the rating of the unrated reviews by considering the sentimental ratable aspects and proposed the model of latent Dirichlet allocation (LDA) with supervision indirectly. In modeling, the association between the modifiers and rating is done using the quad-tuples of head, modifier, rating, and entity. The reviews that are unrated are decomposed into three steps: (1) Using the training reviews, ratable aspects are generated over sentiments; (2) rating of unrated reviews and inference of identifying aspects, and (3) predicted the ratings of the unrated ratings using the joint probabilities through generative process. The effectiveness of this method is tested on data set of TripAdvisor [5]. Yue Le et al. have discussed short comments rated aspects by which decomposes the short comments overall rating. It provides users with the information of different entities to find the target entity. The problem is decomposed into three steps: (1) Opinion of different aspects is decomposed into aspect or feature as head term and opinion toward the aspects is modifier. Identifying the aspects and clustering head terms into aspects; (2) predicting the rating of the aspect depending on the overall rating without any external knowledge; and (3) to give better understanding to the user take out some phrases for predicting the aspect rating. Experimentally studied on the eBay seller feedback, this method can be applied on any short comments to generate rated aspects summarization automatically [4]. Erfan Ahmed et. al. have discussed challenges and comparative analysis on movie reviews using machine learning and proposed methodology for predicting sentiment. Sentiment analysis is done at sentence level, for this investigation and applied three different ML algorithms: Naive Bayes, Multi Layer Perceptron (MLP), and SVM for analysis. To improve the accuracy of classification, (1) then applied SVM experimentation using two kernels of SMO Poly and RBF. (2) Naïve-based classifier is applied with multinomial updatable and Naïve Bayes net. (3) Layer- "0" MLP. Polarized the Sentences into +ve / −ve classes, which express the subjective information. The data set used is movie reviews from acllmdb. Discussed on the impact of stop words and number of attributes impacting the accuracy of the system [1]. Consolidation of the literature with methods/algorithms used with output achieved is given in Table 1.

Table 1 A consolidated literature in the area of sentimental analysis

S. No.	Author name	Method or algorithm used	Output achieved
1	Han-Xiao	Unigram features using frequency and TF-IDF polarity classification	Classification of reviews as positive or negative
2	Walter Kasper	BESAHOT project	Classification of reviews as positive, negative or neutral
3	Anshuman	Bag-of-words approach	Recipes ordered list based on the core ingredient input
4	Gang Li, Fei Liu	TF-IDF weighting method	Accuracy is improved with stability when compared to supervised learning approach to judge movie is good or bad
5	Wenjuan Luo et al.	Latent Dirichlet allocation (LDA)	Predicting the rating of the unrated reviews
6	Yue Le et al.	Decomposes the short comments overall rating	Short comments rated aspects
7	Erfan Ahmed et al.	Applied three different ML algorithms: naive Bayes, MLP, and SVM	To improve the accuracy of the classification

3 Unigram and Bigram Approach

In this approach, we have taken the corpus of unigrams and bigrams from the feedback given by the users. The formulation of term frequency and the cleaning process of the review corpus are done by using the R language text mining package (TM) which is shown in Fig. 2.

Once the reviews are preprocessed, then they are used for creating the term document matrix (TDM). The total frequency of the term is calculated as shown in Eq. 1. t_i is the unique terms in the document, where $T_i = K$ or $T_i = 0$, K is the number of times the term occurred in the document k = Number of reviews used in rating

$$\text{Term Frequency}(t_i) = \sum_{K=1}^{n} T_i \qquad (1)$$

Term corpus is prepared for all the rating categories. TDM created for rating 5 is shown in Fig. 3. Same process is applied for all the rating categories.

Consider the Feedback: (1) "The rooms were OK, but there was no insulation or soundproofing. You could hear every noise both upstairs and next door, not only that but every time a car started outside, it sounded like it was in your room!". In this, the user is not happy with the sound proofing system of the hotel stating the environment of the hotel in the negative sense, but the rating is given as "5" for the hotel that in turn specifies that the review he given in writing is no way having relation with the

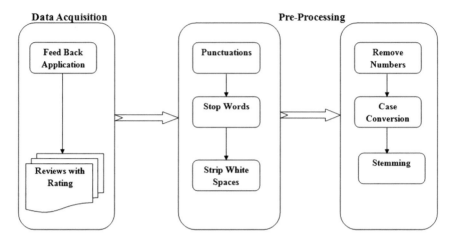

Fig. 2 Data acquisition and preprocessing process

Terms	1	2	3	4	5	6	7	8	9	10	11	12	13	14	15	16	17	18	19	20	21	22	23	24	25
ability	0	0	0	0	0	0	0	0	0	0	0	0	0	0	0	0	0	0	0	0	0	0	0	0	0
about	0	0	0	0	0	0	0	0	0	0	0	0	0	0	0	0	0	0	0	0	0	0	0	0	0
above	1	0	0	0	0	0	0	0	0	0	0	0	0	0	0	0	0	0	0	0	0	0	0	0	0
absolutely	0	0	0	0	0	0	0	0	0	0	0	0	0	0	0	0	0	0	0	0	0	0	0	0	0
accommodating	0	0	0	0	0	0	0	0	0	0	0	0	0	0	0	0	0	0	0	0	0	0	1	0	0
accommodations	0	0	0	0	0	0	0	0	0	0	0	0	0	0	0	0	0	0	0	0	0	0	0	0	0
according	0	0	0	0	0	0	0	0	0	0	0	0	0	0	0	0	0	0	0	0	0	0	0	0	0
accurate	0	0	0	0	0	0	0	0	0	0	0	0	0	0	0	0	0	0	0	0	0	0	0	0	0
across	0	0	0	0	0	0	0	0	0	0	0	0	0	0	0	0	0	0	0	0	0	0	0	0	0
actually	0	0	0	0	0	0	0	0	0	0	0	0	0	0	0	0	0	0	0	0	0	0	0	0	0
address	0	0	0	0	0	0	0	0	0	0	0	0	0	0	0	0	0	0	0	0	0	0	0	0	0
adults	0	0	0	0	0	0	0	0	0	0	0	0	0	0	0	0	0	0	0	0	0	0	0	0	0
affordable	0	0	0	0	0	0	0	0	0	0	0	0	0	0	0	0	0	0	0	0	0	0	0	0	0
after	0	0	0	0	1	0	0	0	0	0	0	0	0	0	0	0	0	0	0	0	0	0	0	0	0
again	0	0	0	0	1	0	0	0	0	0	0	0	0	0	0	0	0	0	0	0	0	0	1	1	0
agent	0	0	0	0	0	0	0	0	0	0	0	0	0	0	0	0	0	0	0	0	0	0	0	0	0
ago	0	0	0	0	0	0	0	0	0	0	0	0	0	0	0	0	0	0	0	0	0	0	0	0	0
agree	0	0	0	0	0	0	0	0	0	0	0	0	0	0	0	0	0	0	0	0	0	1	0	0	0
air	0	0	0	0	0	0	0	0	0	0	0	0	0	0	0	0	0	0	0	0	0	0	0	0	0
ajan	0	0	0	0	0	0	0	0	0	0	0	0	0	0	0	0	0	0	0	0	0	0	0	0	0
alexandria	0	0	0	0	1	2	1	1	0	0	0	0	0	0	0	1	0	0	0	0	0	1	0	0	0
all	0	0	0	0	0	0	0	0	0	0	0	0	0	0	0	1	0	0	0	2	0	0	0	0	0
alone	0	0	0	0	0	0	0	0	1	0	0	0	0	0	0	0	0	0	0	0	0	0	0	0	0
along	0	0	0	0	0	0	0	0	0	0	0	0	0	0	0	0	0	0	0	0	0	0	0	0	0
already	0	0	0	0	0	0	0	0	0	0	0	0	0	0	0	0	0	0	0	0	0	0	0	0	0
also	0	0	0	0	1	0	0	0	0	0	0	0	0	0	0	0	0	0	0	0	0	1	0	0	0
always	0	0	0	0	0	0	0	0	0	0	0	0	0	0	0	3	0	0	0	0	0	0	0	0	0
amazing	0	0	0	0	0	0	0	0	0	0	0	0	0	1	0	0	0	0	0	0	0	1	0	0	0
amenties	0	0	0	0	0	0	0	0	0	0	0	0	0	0	0	1	0	0	0	0	0	0	0	0	0
american	0	0	0	0	0	0	0	0	0	0	0	0	0	0	0	0	0	0	0	0	0	0	0	0	0
americas	0	0	0	0	0	0	0	0	0	0	0	0	0	0	0	0	0	0	0	0	0	0	0	0	0
among	0	0	0	0	0	0	0	0	0	0	0	0	0	0	0	0	0	0	0	1	0	0	0	0	0
and	1	0	0	1	5	2	1	1	4	2	0	0	0	0	0	2	1	1	2	2	3	2	1	2	0
another	0	0	0	0	0	0	0	0	0	0	0	0	0	0	0	0	0	0	0	0	0	0	0	0	0
any	0	0	0	0	0	0	0	0	0	0	0	0	0	0	0	0	0	0	0	0	0	0	0	0	0

Fig. 3 TDM for rating 5

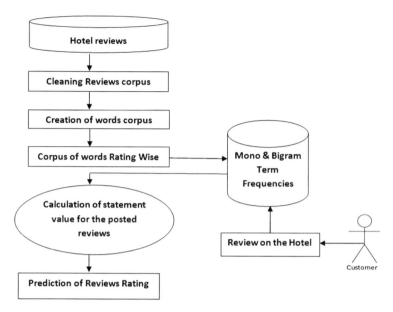

Fig. 4 Model of monogram and bigram approaches

rating that he has given. (2) "Rooms were comfortable and clean and the staff was very peasant, great stay for the price. Had dinner one night in the restaurant which wasn't very good but there are some local pubs in the area which were very good. Breakfast in the restaurant was good. Overall the stay was a good value." In this feedback, the user has given he is happy with the environment and but little unhappy with the food. The rating that he has given is very poor which is "1." The model of unigram and bigram is given in Fig. 4 [10]. The results of the system have shown a skewed results even when applied with Naïve-based approach. The analysis has given that mismatch of what the user intends his feeling in words is not matching with the review rating that he is giving in physical scale of 1–5.

The results of unigram and bigram are given in the Fig. 5 which displayed the percentage of successful prediction of all the rating categories. The data used for training and testing is considered in three batches considering the data from 35 different hotels' feedback from the data set of Datafiniti. Each of the three batches has 100 training and 25 testing cases. In Batch 1, we have considered 15 hotels, Batch 2 we have considered 2 hotels, and Batch 3, 18 hotels. The graphs have also presented a comparison with the Naïve-based approach in both monograms and bigrams. Bigrams have shown better results when compared to the monogram approach.

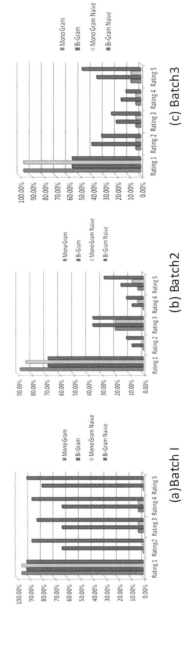

Fig. 5 Unigram and bigram outputs on three batches

4 Statistical Approaches in Prediction

To over the issues of the skewed results, we have implemented a statistical Six Gram Model which has considered the statement values in six variations of monogram, bigram, and trigram with original term frequencies and with uniform frequencies (i.e., all the terms will have the weight of 1). The model of Six Gram is shown in Fig. 6 [9].

The Review given by the user is processed into each vertical of the model, and it produces six outputs given by: $O1$—score of review given by the user when processed with unigram term frequencies. $O2$—score of review given by the user when processed with bigram term frequencies. $O3$—score of review given by the user when processed with trigram term frequencies. Similarly, the outputs of $O4$, $O5$, and $O6$ are the scores of the statement when considered with uniform term frequencies. Applied the statistical approaches on the score of the statement, they are:

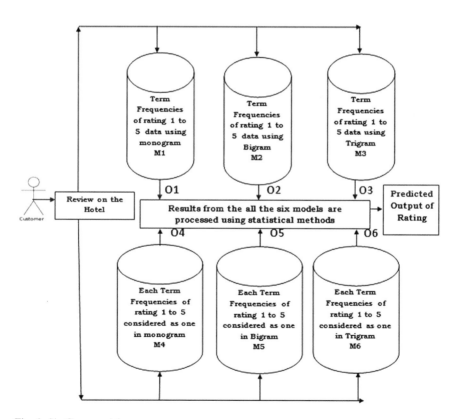

Fig. 6 Six Gram model

1. Mean
2. Median
3. Min
4. Max

The process of predicting the rating in the statistical methods is defined as follows: Step1: The given feedback is processed into every vertical corpus of monogram, bigram, and trigram with original and uniform term frequencies. Step2: The sum values of the term weights used in the feedback are considered as the output of each rating; the vertical output is applied with the statistical method. $O_i = StatiticalMethod(r1, r2, r3, r4, r5)$, where $r1, r2, r3, r4$, and $r5$ are the rating outputs from the five rating scaled corpus. Step3: Statistical methods are applied on the output of all the six verticals to predict the rating of the statement given as follows $Predictedrating = StatiticalMethod(O1, O2, O3, O4, O5, O6)$. The statistical process has recovered the skewed result feature to an extent, but the processing system needs to be fine-tuned to fit into a perfect system. In some cases, mean and median have shown better outputs and in some cases, min-max has shown better results. The output of the three batches of the data is shown in Fig. 7.

5 Sentiment Prediction of Rating

To give a perfect prediction system with the corpus, we have used Google NLP prediction API and the monogram-based score of the terms in combination to give the best prediction results. It supports the management in the following aspects:

- This system will give accurate prediction results as we have combined both NLP-and term-based system.
- This system helps the management/application user which feedbacks are considerable feedbacks.
- This helps the management/application to check who are the genuine feedback given customers with whom they can go with post review process if needed.
- This helps the management/application which feedback is to be considered and used to make the update in the database of their system.

The model of the sentiment prediction of rating is shown in Fig. 8 which is defined in two phases as follows:

- Corpus preparation
- Prediction system

The corpus preparation process takes the training data and processes it through Google NLP API to get the score and magnitude of the review statements of the statement with which the system gives the management a choice of selecting the perfect reviews for the corpus build in the system. If the system is not having any previous feedback with rating combinations, we consider in the beginning the positive

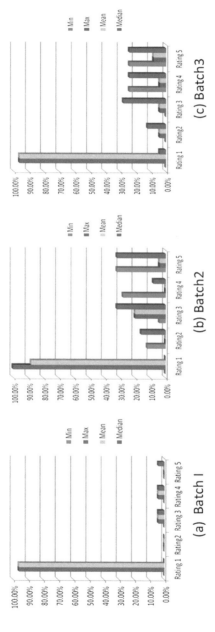

Fig. 7 Unigram and bigram outputs on three batches

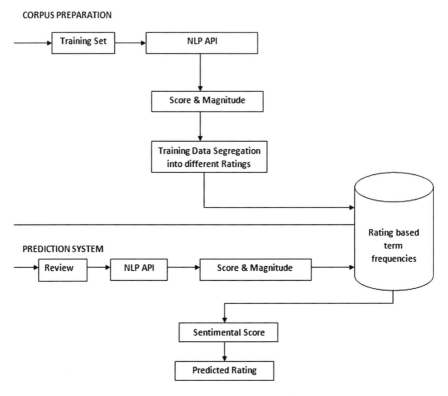

Fig. 8 Model for sentiment prediction of rating

and negative words of the universal accepted dictionary's as a monogram terms. The prediction system is used to predict the rating of the review given the user which is done as follows:

- The review is through Google NLP.
- The statement-level score of the statement will decide the internal process.
- If the score is positive, it switches to the word corpus of 4 and 5. If it is negative, it switches to 1 and 2, and if it is neutral, it takes the word corpus of 3.
- Depending on the score achieved maximum will be considered and the output of the rating is defined.

6   Experimental Results and Analysis

A comparative study is made by taking the NLP output of the Google for the given review and the proposed model. A better result has been achieved by taking the combination of NLP output of Google and monogram bag of corpus. The results of

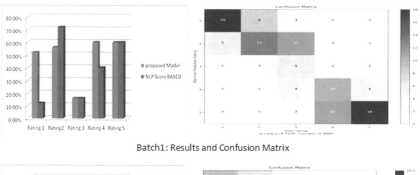

Batch1: Results and Confusion Matrix

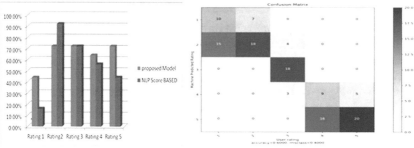

Batch2: Results and Confusion Matrix

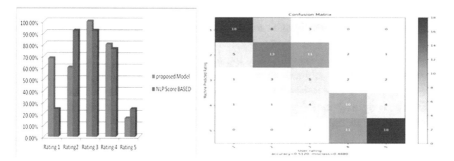

Batch3: Results and Confusion Matrix

Fig. 9 Sentiment prediction of rating for the three batches

both the models in comparison with all the three batches of data are given in Fig. 9 with the analysis of the review data with confusion matrix showing the values of precession and recall and accuracy of the data batch-wise.

As we have considered the magnitude of the statement into analysis which represents the strength of the review that the user has given around 0–8 % difference in prediction of rating in some cases of analysis has shown by the system. The values of the precession and recall for all the rating in all the three batches that we have considered are shown in Table 2.

Table 2 Precision and recall values of three batches data

Rating	Batch 1		Batch 2		Batch 3	
	Precision	Recall	Precision	Recall	Precision	Recall
Rating 1	0.72	0.62	0.4	0.58	0.68	0.58
Rating 2	0.52	0.40	0.72	0.48	0.52	0.61
Rating 3	0.2	0.38	0.72	1	1	1
Rating 4	0.4	0.5	0.36	0.52	0.36	0.47
Rating 5	0.72	0.58	0.8	0.55	0.6	0.48

If we look into the results produced by the system, the outputs of the systems are in alignment with respect to the precision values of the system. It reflects that the system has performed as per the user feedback perfectly defined the rating that he need to give as per his grief on the system that he/she experienced. The system that is designed is giving perfect solution rather than user flick rating without any relation with his words that he has given in the review.

7 Conclusion and Future Research Directions

The proposed system has given improved prediction result of the rating depending on the bag-of-words approach when compared to the NLP-based system which directly works with the score and magnitude of the statement. A further application of neural-based system can be designed which can use the score and magnitude to decide the threshold and guess the perfect rating process. An alternative to the process is directly giving the weight age to the positive and negative words of the system, and the rating can be guessed by using the neural network-based system.

References

1. Ahmed, E., Sazzad, M.A.U., Islam, M.T., Azad, M., Islam, S., Ali, M.H.: Challenges, comparative analysis and a proposed methodology to predict sentiment from movie reviews using machine learning. In: 2017 International Conference on Big Data Analytics and Computational Intelligence (ICBDAC), pp. 86–91. IEEE (2017)
2. Kasper, W., Vela, M.: Sentiment analysis for hotel reviews. In: Computational Linguistics-Applications Conference, vol. 231527, pp. 45–52 (2011)
3. Li, G., Liu, F.: A clustering-based approach on sentiment analysis. In: 2010 IEEE International Conference on Intelligent Systems and Knowledge Engineering, pp. 331–337. IEEE (2010)
4. Lu, Y., Zhai, C., Sundaresan, N.: Rated aspect summarization of short comments. In: Proceedings of the 18th International Conference on World wide web, pp. 131–140. ACM (2009)
5. Luo, W., Zhuang, F., Cheng, X., He, Q., Shi, Z.: Ratable aspects over sentiments: predicting ratings for unrated reviews. In: 2014 IEEE International Conference on Data Mining, pp. 380–389. IEEE (2014)

6. Rao, S., Kakkar, M., et al.: A rating approach based on sentiment analysis. In: 2017 7th International Conference on Cloud Computing, Data Science & Engineering-Confluence, pp. 557–562. IEEE (2017)
7. Shi, H.X., Li, X.J.: A sentiment analysis model for hotel reviews based on supervised learning. In: 2011 International Conference on Machine Learning and Cybernetics, vol. 3, pp. 950–954. IEEE (2011)
8. Venkata Raju K.D.M.: Rating based sentimental prediction: a six gram statistical mean-medial approach. J. Adv. Database Manage. Syst. **5**, 33–42 (2018)
9. Venkata Raju K.D.M.: Rating based sentimental prediction: a six gram statistical min-max approach. Int. J. Pure Appl. Math., 1456–1462 (2018)
10. Venkata Raju K.D.M.: Rating based sentimental prediction-bag of words approach using monogram and bigram. J. Adv. Res. Dyn. Control Syst., 663–671 (2018)

A Bio-immunology Inspired Industrial Control System Security Model

Mercy Chitauro, Hippolyte Muyingi, Samuel John and Shadreck Chitauro

Abstract Industrial Control System (ICS) security is inadequate to protect ICS from Advanced Persistent Threats (APT). APTs attack ICS in such a way that they are detected after a long time in the system. This paper proposes the use of the biological immune system as a foundation for developing ICS security architecture because the biological immune system is renowned for defending the body from pathogens. The paper compares how the Biological Immune System (BIS) defends the body from pathogens and how ICS are secured from invasion by any attack. By considering the similarities and differences between ICS security and the BIS operation and taking into consideration current research on ICS security a bio-immunology inspired security model to defend ICS was designed. The proposed model was designed using design science research and initial results are presented in this paper.

Keywords Industrial control system · Security · Biological immune system · Advanced persistent threats

1 Introduction

Industrial Control Systems (ICS) is a generic term for systems that manage the automation of industrial processes. ICS are vulnerable to network attacks and thus, robust mechanisms for securing ICS; ICS security standards and frameworks were drafted are used as general guidelines on the best way to secure ICS [1]. ICS security

M. Chitauro (✉) · H. Muyingi · S. John · S. Chitauro
Namibia University of Science and Technology, Windhoek, Namibia
e-mail: mchitauro@nust.na

H. Muyingi
e-mail: hmuyingi@nust.na

S. John
e-mail: sjohn@nust.na

S. Chitauro
e-mail: schitauro@nust.na

© Springer Nature Singapore Pte Ltd. 2020
A. K. Luhach et al. (eds.), *First International Conference on Sustainable Technologies for Computational Intelligence*, Advances in Intelligent Systems and Computing 1045,
https://doi.org/10.1007/978-981-15-0029-9_64

and frameworks guidelines provide a basis for securing ICS from any computing-related threats.

Although the guidelines mentioned in ICS standards and frameworks are good recommendations for securing ICS, their implementation still keeps ICS vulnerable to attacks from Advanced Persistent Threat (APT) [2]. For example, Stuxnet which was discovered in 2010 used among others zero-day exploits to sabotage Iranian Natanz Nuclear Enrichment facility operations. Ever since a plethora of other APTs were discovered, for example, Flame in 2012, Miniduke in 2013, Red October 2013, Desert Falcons in 2014, Crounching Yeti in 2015 and Poseidon in 2015 [3].

To mitigate the APT attack in ICS [4] propose the use of a Biological Immune System (BIS) security emulation to detect APT that target ICS. The BIS was proposed as a suitable emulation for security systems because it effectively detects and protects the body from pathogens that enter into the body.

This paper is organised as follows: Sect. 2 discusses the basic principles of a BIS. Section 3 describes ICS security. Section 4 gives a comparison of BIS and ICS security, which is followed by Sect. 5 that showcases current ICS security research. Section 6 is a brief discussion on ICS security and BIS that ends with the presentation of the proposed ICS security model. Section 7 is a brief discussion of the validation and initial results and finally, Sect. 7 will conclude the paper and discuss future work.

2   The Biological Immune System (BIS)

2.1   Overview of Biological Immune System

A BIS is a complex system that defends the body from pathogens through a combination of physical, chemical and cellular components [5]. The BIS is of two types namely the innate immune system and the adaptive immune system. Innate immunity also called natural or native immunity is that which is determined by the genes inherited from parents and is always present in healthy individuals [6, 7]. Adaptive immunity known as specific or acquired immunity is 'organised around an ongoing infection and adapts to the nuances of the infecting pathogen' [6, 7].

The innate immune system is responsible for physically blocking and chemically destroying pathogens. The innate immune system is also responsible for inflammation. Inflammation is a process which involves recruiting to the infected site cells and molecules of the innate immune system [8]. In addition, the innate immune system is a component of the complement system. 'Complement system is a set of plasma proteins (plasma is fluid component of blood) that act together to attack pathogens' [8].

On the other side, the adaptive immune system is responsible for humoral immunity; cell-mediated immunity and is part of the complement system [5–9]. Antibodies found in blood fluid or plasma are the main actors of humoral immunity. Humoral immunity is named as such because historically blood fluids were known as humours [8]. Cell-mediated immunity which is championed by T cells involves targeted destruction of pathogens in cells. The adaptive immune system works differently in that it does not mount the same kind of defence against all pathogens like the innate system. The adaptive immune system gets intelligence about pathogens, keeps a record of the information and finds ways to defend the body. To do this are two lymphocytes named T cells and B cells.

It is possible to derive some properties of the BIS that enable it to protect, detect and eradicate pathogens. These properties are discussed next.

2.2 Biological Immune System Properties

Properties of the BIS derived from the way it works to protect and defend the body from pathogens are it is:

- *Distributed*—Distributed control of the BIS means that there is no central control that governs how immune cells work but rather it works by local interactions of the cells and antigens [10].
- *Intelligent*—The BIS also utilises intelligence and collaboration properties. BIS players are intelligent entities which can respond to various situations and past responses and collaboratively work together to eliminate pathogens [6–8].
- *Messaging*—In can be viewed as being capable of communicating message transfer through physical and chemical entities and their mobilities as can be exemplified by interleukin, opsonisation, antigen presentation, etc. [6–8].
- *Resilient*—The BIS is a resilient system that can develop large rescue resources whilst fighting against attacks. This also means that BIS enables the body to completely recover after attack by pathogens [6–8].
- *Environmental self-awareness*—In the human body there are no body parts or tissues that get a different type of defence from what another part is getting. All get the same attention via the BIS. Consequently, this means all body components have the same kind of localised defence strategies. All body parts and tissues know how to act whenever pathogens are detected.
- *Defence-in-depth*—The BIS uses a layered defence approach.

The processes and the operations of the BIS have inspired the emergence of Artificial Immune Systems (AIS). AIS are 'computational systems inspired by the principles of the BIS' [11]. Most of the models emulating the BIS model the processes of the adaptive immune system.

2.3 Artificial Immune System (AIS)

AIS have emerged from observing and learning about the BIS. AIS are 'computational systems inspired by the principles of the natural immune system' [11]. The majority of computational models model adaptive immune system processes like immune network theory, negative selection, clonal selection theory and the danger theory. AIS principles and mechanisms are used to solve computing problems such as learning, anomaly detection and optimisation [12].

In AIS infancy mostly negative selection, clonal selection and immune network theories were being used to create computational models. The AIS community was mostly inspired by the adaptive immune system to solve computational problems. By learning about the BIS it is clear that many designs can be extracted from the BIS, but over the years success of AIS has not been obvious because: [4, 13, 14]

- Inspiration is drawn from biological principles that are not fully understood;
- Most AIS solutions solve problems that have been solved already by other paradigms;
- It is not always obvious from which BIS functionality to draw inspiration from;
- Most inspirations were drawn only from the adaptive immune system whilst ignoring the innate immune system.

These problems can be solved by conclusions from [4, 13, 14]; who conclude that for AIS to be more successful it must be fully understood. When the BIS is fully understood it can be effectively implemented so that it works as good as it does in the human body. But, how sure are we that the functionality being modelled has been fully understood? [15]. Biologists do not even agree on how the BIS functions [15]. Instead of trying to model BIS functionalities by using algorithms the researchers propose emulating BIS properties to secure systems. Ensuring that all properties exuded by BIS are imported to a system that needs protection promises to be more effective. What it means is that inspiration is drawn from BIS properties established like being distributed, intelligent, messaging and resilient but the implementation of these properties is achieved by using tried and tested mechanisms that can bring the said property to the system that needs securing.

3   Industrial Control System Security

Bere [2] mentions that ICS are secured following ICS frameworks and ICS security standards. Bere [2] categorised security mechanisms in three namely; mitigating reconnaissance attacks, access control mechanisms, and auditing and accountability.

To mitigate reconnaissance attacks [2] states that ICS users are being trained to raise awareness and to have policies and procedures that enforce security training. Access control mechanisms in ICS are such that users must have a means of being identified. Users must be given access to only what they need, access of the control system from the corporate network should be through a firewall in demilitarised zones, no email access in the control system and physical access to the control system and devices that display ICS information should be restricted [2].

To audit and monitor systems ICS security standards, regular risk vulnerability and risk management must be conducted, use intrusion prevention systems, monitor firewalls. In addition, conduct audit that verifies ICS systems are working as they should. Stouffer et al. [17] state that to achieve security in ICS then there is need to implement, network segmentation, network security controls and host security and access controls.

4 Comparison of ICS Security and BIS

Since the BIS is used to defend the body from pathogens it would be reasonable to compare it to the way ICS are secured in order to find any similarities and differences between the two.

1. The innate immune system has physical barriers such as the skin, mucus, lyso-somes in tears, saliva, etc. which prevent pathogens from entering the body by exuding unfavourable conditions for pathogens. These physical barriers make the BIS environment hostile to pathogens. Similarly, in the ICS, there are barriers that prevent physical and logical intrusion. These physical barriers in the ICS have the form of physical locks to the control rooms, firewalls, cameras and access control mechanisms.

2. In the BIS when pathogens have somehow bypassed physical and chemical barriers, phagocytic response is mounted. Phagocytes identify pathogens and work to eliminate them. Similarly, in ICS security, after intrusions have bypassed access control mechanisms they can be detected by antivirus software and intrusion prevention systems which detect intrusions and eliminate them.

3. Healthy cells exhibit MHC1 which stops them from being phagocytosed. Cells that do not exhibit MHC1 have been 'attacked' and are eliminated together with the pathogens attacking them. This works the same as signature-based intrusion detection systems and antiviruses, which look for patterns of potential attacks. In ICS, intrusion prevention systems can detect when a system is not working as it should. When cells are attacked by a virus they produce interferons which act as indicators to neighbouring cells to up their viral defences. This prevents viral replication. In the ICS, an intrusion detection system can perform the same task of sending alarms to signal the presence of an intrusion so that other entities can eliminate the intrusion.

4. Another security measure found in ICS security is that of application whitelisting. This means only preselected applications are allowed to run in the ICS while everything else is blocked. The BIS does not whitelist like in ICS security. The BIS allows certain pathogens to co-exist in the body so long as their operation does not make the body deviate from normal behaviour [15]. This behaviour though is comparable to how safety instrumented systems monitor ICS to make sure they are always in a safe state.

5. Dendritic cell phagocytes, after phagocytosing a pathogen keep records of the pathogen for the adaptive immune system to quickly recognise this type of pathogen in the future. Similarly, in the ICS intrusion prevention systems, logs

and antivirus software keep records of intrusion and attacks so that in the future they are quickly eliminated.

6. ICS prioritise availability of data over integrity and confidentiality [17]. The researchers take the position that this is also a similar stance taken by the BIS which strives to keep the body healthy and functional. Mobility of blood fluid or chemical agents ensures availability of BIS agents and body tissues, while binding and cloning resemble integrity and confidentiality.

Despite all the measures discussed in Sect. 3 to defend ICS from attackers, attackers still gain access to ICS. How then do attackers gain access? APTs target any access measure to gain entry into ICS so that they can execute their payload. In the BIS pathogens have many entry points that they can use to gain entry into the body and in most of these cases, the pathogens are detected and eradicated. If anything else other than the BIS are attacked the BIS can kick in and defend the body from invasion. This behaviour is unlike in ICS, where despite which entry point is used to gain access by the APT, the attack is almost always detected after the APT has executed its payload. We believe that this is where the BIS prevails over ICS security in that the BIS 'knows' how all the different body tissues should function as well as the extend or range of deviation from normal behaviour, anything else is treated as a harmful invasion.

What makes the body resilient to attacks? Where is the difference? What makes the BIS more robust in protecting the body from pathogens compared to how the ICS operates? The average person will live a life of 71.4 years [18]. This means the average human being will fend off attacks for much longer than when ICS are attacked by APTs. Thus, we need to ask; what kind of ICS security would be attack-type (threat) independent and resilient so that the most likely or the least likely type of attack is detected early to minimise the effects?

5 Current ICS Defence from APT Research

Current research on ICS defence from APT can be loosely divided into four categories namely: using anomaly detection, tightening security controls, using defence-in-depth strategies, increasing user awareness or protecting the user plane. Many of these research solutions actually employ a hybrid of any of these security strategies. These strategies are not only recommend for ICS but for other IT systems that are also being attacked by APT.

The first category utilises anomaly detection techniques. Anomaly detection is when the system is trained to learn normal system behaviour over time and when any behaviour or circumstance occurs that is perceived as not normal then the system raises an alarm to signal presence of an attack. This can be exemplified by [19–21]. These researchers believe to best defend ICS from APT using anomalous defence techniques would be the way forward. The difference in their research is in what is used as a baseline from which to determine abnormal conditions.

The second category involves tightening already existing security techniques. This means from the security methods that are already ensured they are used correctly and/or change security combinations in order to protect the ICS. This type of thinking is put forward by [22–24].

The third type of defence strategy is making sure the user is well trained and prepared to recognise APT attack. This strategy also makes sure that the user plane into the system is well secured. This concept is exemplified by [25–27]

The most popular defence strategy among ICS vs APT security researchers in the papers reviewed is that of defence-in-depth [28–34] are some of the researchers that are developing the concept of using defence-in-depth strategies to safeguard ICS from APT. The defence-in-depth strategy is a layered defence strategy whereby many security controls or mechanisms are used in conjunction to deter APT.

6 Discussion

Although most of the researches above report that they have been successful in what their designs were designed for, the researchers believe the defence-in-depth method is the best method to deter APT. The reason being that it enables many layers of security. Layering security methods makes it difficult for an APT to gain entry into an ICS. Even if an APT did gain entry into a system having layered security means it would also be difficult to communicate with its command and control centre. It will also be equally hard to execute its payload because it has to circumvent many security parameters in the system before it can execute its payload. The defence-in-depth method gets rid of the notion of perimeter only security which makes it possible to combine many human and technical techniques in ICS security so that ICS achieves the ability to detect, prevent, recover and deflect APT whenever required to do so.

Section 2 highlights that all body parts are able to protect, deflect, detect and recover from attacks. This implies that BIS employs a layered security mechanism. Comparison of ICS security and BIS properties yielded that ICS security lacks environmental self-awareness and resilience properties. Thus, in order for ICS security to have the same security properties as discovered in this research, there is need to add these two missing properties to ICS security.

But how can a controlled process be environmentally self-ware and resilient? As stated in Sect. 3 environmental self-awareness is when 'all body components have the same kind of localised defence strategies and when body parts and tissues know how to act whenever pathogens are detected.' To bring this idea to ICS security the researchers propose embedding a model of the controlled process in the control parameters of the controlled process. In this manner, 'knowledge about the process is added in, thus, the system will now be self-aware'. A model of the system is precompiled process behaviour that enables the system to have prior understanding of how the controlled process should perform.

Similar work is done by Lerner et al. [35]. In their design, the prediction module adds self-awareness. A model of the process found in the prediction module is used to continuously match behaviour of the actual process state and expected states. This research from [35] mainly concentrates on keeping the process stable. Follow up research to [35] by Lerner et al. [36], Franklin et al. [37] also focus on the idea that makes sure that the process remains stable even in the event of an attack. But we argue that there is need to add more security layers as recommended by defence-in-depth security strategy. Thus, the researchers propose to add additional security devices to the ones in the research by Lerner et al. [35], an intrusion detection system that adds a layer of protection to the process and prediction controllers from attacks directed to them and not the actual process. In addition, add a firewall/filter that tracks the reference input before they are accepted in the system. The result a bio-immunology inspired security model is shown Fig. 1.

The firewall should monitor reference input so that it is correct all the time. The prediction module should have a model of the process that is used to verify the process controller output. If there are problems then control should be given to the fall-back controller. An IDS is used to further monitor the controllers for any attacks that are targeting them instead of the process itself.

The bio-immunology inspired security model is a culmination of the BIS properties but to verify the model only the concept of self-awareness was tested since the usefulness of intrusion detection and firewall have been tested and proven for use. Bere [2] states that ICS security standards and frameworks provide reliable methods

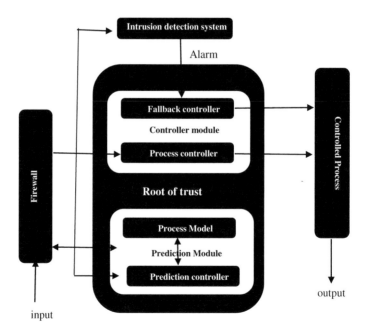

Fig. 1 Bio-immunology inspired security model

to securing systems and the standards' recommendations have been proven through research and evaluation. ICS standards [17, 38, 39] already recommend the use of firewalls and intrusion detection systems in ICS.

7 Initial Results

To evaluate the bio-immunology inspired security model the researchers used a jacketed non-adiabatic tank Continuously Stirred Tank Reactor (CSTR) model commonly used in the process industry [40]. An adiabatic process is one that occurs without transfer of heat [40]. The model was implemented in MATLAB Simulink.

Next section describes a CSTR as done by [40] used for testing.

The vessel is assumed to be perfectly mixed and single a single first-order exothermic and irreversible reaction $A \rightarrow B$ takes place where A the reactant is converted to B the product. An inlet of reactant feeds into the tank at a constant rate. The exothermic reaction takes place to produce the product stream, which exits the reactor at the same rate as the input stream.

It is also assumed that

- *perfect mixing (product stream values are the same as the bulk reactor fluid)*
- *Constant volume*
- *Constant reaction parameter values.*

We used the following two CSTR state equations; concentration and temperature.

$$\frac{dC_A}{dt} = \frac{F}{V}(C_{feed} - C_A) - r \tag{1}$$

And

$$\frac{dT}{dt} = \frac{F}{V}(T_{feed} - T) + \frac{Q_{generated}}{\rho C_p V} \tag{2}$$

where: $r = 3C_A$.

Two controllers were used for comparison. A PID controller commonly used in the process industry [42] and a model predictive controller. A model predictive controller (MPC) is the controller that was chosen to introduce the concept of environmental self-awareness. At first the process was run as normal without an attack to build baseline behaviour. The process was then run using a step disturbance as an attack. The step disturbance was introduced at the beginning of the process, i.e. at $t = 0$ s. To compare the results of the different stages the following parameters where chosen for comparison.

- Rise Time (t_r)—the time taken for the output to go from 10 to 90% of the final value.

Table 1 Initial results

		Normal controlled Process by:		APT attacked process by:	
	Normal process	PID controlled process	MPC controlled process	PID	MPC
t_r	3	0.7	0.5	1	0.5
t_s	9	3.5	1.1	3.5	4
e_{ss}	0	0	0	0	0
Overshoot	0	8%	0.1%	8%	3.2%

- Settling Time (t_s)—The time taken for the process signal to be bounded to within a tolerance of 99% of the steady-state value.
- Steady State Error (e_{ss})—The difference between the input step value and the final value. This is the difference between the desired value and the final signal value.
- Overshoot − (max value − final value)/final value × 100.

Table 1 shows the initial results.

At time t_0 for step disturbance of 25 when we compare the results without APT rise time for both controllers' changes by 0.3 for PID controlled process and by 0 for MPC controlled process. This means when APT attacks the controlled system at the beginning of the process then the rise time of the MPC controlled process is not affected.

At time t_0 for step disturbance of 25 settling time and we find that the PID controller has a superior settling time compared to that of the MPC controller. Settling time for PID controlled process was 3.5 and that of MPC control was 4 s.

At time t_0 for step disturbance of 25 for the overshoot MPC controller is better compared to PID controller. This means that when APT attacks a controlled process MPC will not allow the final values of the process to go way beyond the expected parameters. This is the more desirable condition.

In addition, at time t_0 for step disturbance of 25 for both PID and MPC controlled processes under attack a steady-state error of 0 was recorded. This means that both PID are able to return the process to the desired input values.

8   Conclusion and Future Work

It is of utmost importance that APT be detected in ICS before they cause any harm in the system. Malfunctioning ICS may have detrimental effects on the environment and to human life. Since the BIS is very good at protecting the body from pathogens for a quite a number of years until the body dies, the researchers sought to emulate properties of the BIS that make it resilient to attacks. Some of the properties that make the BIS robust are it is decentralised, it uses defence-in-depth, it is resilient, it is an intelligent and collaborative system and it is environmentally self-aware.

By being self-aware the body knows when it has been invaded, what it needs to do to successfully fight the invaders. Being self-aware is a result of intelligence, collaboration and decentralisation of BIS which can be summed up to resilience, but how can the ICS be resilient? How can ICS be able to identify attacks, eliminate attacks whilst still running as normal? ICS can be resilient by implementing 'self-aware' process models. By implementing 'self-aware' process models properties ICS will 'know' how the controlled process should behave. If there are deviations from normal behaviour, then intrusion will have been detected and proper measures can be taken.

Initial results of experiments being carried out show that a bio-immunology inspired controller (MPC) performs better than a PID controller in the case of the CSTR described in this paper. Further results are being run to evaluate how the bio-immunology inspired controller will perform when the system is being attacked. The researchers are excited by the initial results thus believe the bio-immunology inspired controller will fare well when disturbances are added to the system.

References

1. Piggin, R.S.H.: Emerging good practice for cyber security of industrial control systems and SCADA. In: 7th IET International Conference on System Safety, Incorporating the Cyber Security Conference 2012, pp. 1–6. IET (2012)
2. Bere, M.: A preliminary review of ICS security frameworks and standards versus advanced persistent threats. In: Iccws 2015—The Proceedings of the 10th International Conference on Cyber Warfare and Security: ICCWS2015 (2015)
3. Targeted cyberattacks logbook, https://apt.securelist.com/#!/threats/
4. Bere, M., Muyingi, H.: Initial investigation of industrial control system (ICS) security using artificial immune system (AIS). In: 2015 International Conference on Emerging Trends in Networks and Computer Communications (ETNCC), pp. 79–84 (2015)
5. Edgar, D.J.M.: Master Medicine: Immunology: A Core Text with Self-Assessment
6. Parham, P.: The Immune System. Garland Science, London (2009)
7. Abbas, A.K., Lichtman, A.H., Pillai, S., Baker, D.L., Baker, A.: Basic Immunology: Functions and Disorders of the Immune System. Elsevier, St. Louis, Missouri (2016)
8. Murphy, K.P., Travers, P., Walport, M., Ehrenstein, M., Janeway, C. (eds.): Janeway's Immunobiology. Garland Science, New York, NY (2008)
9. Segel, L.A., Cohen, I.R. (eds.): Design Principles for the Immune System and Other Distributed Autonomous Systems. Oxford University Press, Oxford (2001)
10. Aickelin, U., Dasgupta, D., Gu, F.: Artificial immune systems. In: Search Methodologies, pp. 187–211. Springer (2014)
11. Mohamed Elsayed, S.A., Ammar, R.A., Rajasekaran, S.: Artificial immune systems: models, applications, and challenges. In: Proceedings of the 27th Annual ACM Symposium on Applied Computing—SAC '12, p. 256. ACM Press, Trento, Italy (2012)
12. Hart, E., Timmis, J.: Application Areas of AIS: The Past, The Present and The Future, 15
13. Zheng, J., Chen, Y., Zhang, W.: A survey of artificial immune applications. Artif. Intell. Rev. **34**, 19–34 (2010)
14. Andrews, P.S., Timmis, J.: Inspiration for the next generation of artificial immune systems. In: Jacob, C., Pilat, M.L., Bentley, P.J., Timmis, J.I. (eds.) Artificial Immune Systems, pp. 126–138. Springer, Berlin Heidelberg (2005)

15. Somayaji, A., Locasto, M., Feyereisl, J.: The future of biologically-inspired security: is there anything left to learn? 6

16. Knapp, E.D., Langill, J.T.: Industrial Network Security: Securing Critical Infrastructure Networks for Smart Grid, SCADA, and Other Industrial Control Systems. Elsevier, Syngress, Amsterdam (2015)

17. Stouffer, K., Pillitteri, V., Lightman, S., Abrams, M., Hahn, A.: Guide to Industrial Control Systems (ICS) Security. National Institute of Standards and Technology (2015)

18. WHO | Life expectancy, http://www.who.int/gho/mortality_burden_disease/life_tables/situation_trends_text/en/

19. de Vries, J., Hoogstraaten, H., van den Berg, J., Daskapan, S.: Systems for Detecting Advanced Persistent Threats: A Development Roadmap Using Intelligent Data Analysis. Presented at the December (2012)

20. Averbuch, A., Siboni, G.: The classic cyber defense methods have failed-what comes next? Mil. Strateg. Aff. **5**, 45–58 (2013)

21. Skopik, F., Friedberg, I., Fiedler, R.: Dealing with advanced persistent threats in smart grid ICT networks. In: Innovative Smart Grid Technologies Conference (ISGT), 2014 IEEE PES, pp. 1–5. IEEE (2014)

22. Virvilis, N., Gritzalis, D.: The Big Four—What We Did Wrong in Advanced Persistent Threat Detection? Presented at the September (2013)

23. Ussath, M., Jaeger, D., Feng Cheng, Meinel, C.: Advanced persistent threats: Behind the scenes. In: 2016 Annual Conference on Information Science and Systems (CISS), pp. 181–186. IEEE, Princeton, NJ, USA (2016)

24. Paradise, A., Shabtai, A., Puzis, R., Elyashar, A., Elovici, Y., Roshandel, M., Peylo, C.: Creation and management of social network honeypots for detecting targeted cyber attacks. IEEE Trans. Comput. Soc. Syst. **4**, 65–79 (2017)

25. Bere, M., Bhunu-Shava, F., Gamundani, A., Nhamu, I.: How advanced persistent threats exploit humans. Int. J. Comput. Sci. **12**, 6 (2015)

26. Iwata, K., Nakamura, Y., Inamura, H., Takahashi, O.: An automatic training system against advanced persistent threat. In: 2017 Tenth International Conference on Mobile Computing and Ubiquitous Network (ICMU), pp. 1–2. IEEE, Toyama (2017)

27. Nicho, M., Khan, S.N.: A decision matrix model to identify and evaluate APT vulnerabilities at the user plane. In: 2018 41st International Convention on Information and Communication Technology, Electronics and Microelectronics (MIPRO), pp. 1155–1160. IEEE, Opatija (2018)

28. Baize, E.: Developing secure products in the age of advanced persistent threats. IEEE Secur. Priv. Mag. **10**, 88–92 (2012)

29. Giura, P., Wang, W.: A Context-Based Detection Framework for Advanced Persistent Threats. Presented at the December (2012)

30. Wang, X., Zheng, K., Niu, X., Wu, B., Wu, C.: Detection of command and control in advanced persistent threat based on independent access. In: 2016 IEEE International Conference on Communications (ICC), pp. 1–6. IEEE, Kuala Lumpur, Malaysia (2016)

31. Messaoud, B.I.D., Guennoun, K., Wahbi, M., Sadik, M.: Advanced persistent threat: new analysis driven by life cycle phases and their challenges. In: 2016 International Conference on Advanced Communication Systems and Information Security (ACOSIS), pp. 1–6. IEEE, Marrakesh, Morocco (2016)

32. Ghafir, I., Hammoudeh, M., Prenosil, V., Han, L., Hegarty, R., Rabie, K., Aparicio-Navarro, F.J.: Detection of advanced persistent threat using machine-learning correlation analysis. Future Gener. Comput. Syst. **89**, 349–359 (2018)

33. Marchetti, M., Pierazzi, F., Colajanni, M., Guido, A.: Analysis of high volumes of network traffic for advanced persistent threat detection. Comput. Netw. **109**, 127–141 (2016)

34. Lu, J., Chen, K., Zhuo, Z., Zhang, X.: A temporal correlation and traffic analysis approach for APT attacks detection. Cluster Comput. (2017)

35. Lerner, L.W., Farag, M.M., Patterson, C.D.: Run-time prediction and preemption of configuration attacks on embedded process controllers. In: Proceedings of the First International Conference on Security of Internet of Things, pp. 135–144. ACM (2012)

36. Lerner, L.W., Franklin, Z.R., Baumann, W.T., Patterson, C.D.: Application-Level Autonomic Hardware to Predict and Preempt Software Attacks on Industrial Control Systems. Presented at the June (2014)
37. Franklin, Z.R., Patterson, C.D., Lerner, L.W., Prado, R.J.: Isolating trust in an industrial control system-on-chip architecture. In: 2014 7th International Symposium on Resilient Control Systems (ISRCS), pp. 1–6. IEEE (2014)
38. QNCIS, Q.N.C. for I.S.: national ics security standard v.3 (2014), http://www.scadahacker. com/library/Documents/Standards/QCERT%20-%20National%20ICS%20Security% 20Standard%20v.3%20-%20March%202014.pdf
39. QCERT—National ICS Security Standard v.3—March 2014.pdf
40. Bequette, B.W.: Process Dynamics: Modeling, Analysis, and Simulation. Prentice Hall PTR, Upper Saddle River (2002)
41. screenshot.jpg (520 × 324), https://www.mathworks.com/matlabcentral/mlc-downloads/ downloads/submissions/13556/versions/1/screenshot.jpg
42. Blevins, T.: PID advances in industrial control. Presented at the Conference of Advances in PID Control, Brescia (2012)

Real Time, An IoT-Based Affordable Air Pollution Monitoring For Smart Home

Prashant Kumar, Gaurav Purohit, Pramod Tanwar, Chitra Gautam and Kota Solomon Raju

Abstract The emergence of the Internet of things (IoT) facilitates us to connect all the devices available at home. Also, we can control and access devices and appliances remotely in real time. This article, by taking the advantages of IoT, proposes an IoT-based affordable air quality monitoring with smart home system that brings controlling and monitoring of devices and appliances, and monitoring of home environment together in real time. The system uses the combination of wireless communication and cloud networking, to remote control various appliances (like fans, TV), also getting a glimpse of air quality. In order to monitor the home environment, we develop a practical information acquisition model, a network for indoor and outdoor air quality monitoring which can be carried anywhere as a portable device. The portable nodes can sense pollutants like CO, CO_2, PM 2.5, PM 10 with temperature and humidity and monitor the air quality. The collected data are sent to a cloud service platform like things speak and Firebase for real-time visualization and appropriate forecasting and prediction analytics. Also, a user-friendly Android app and web portal increase the usability of this system. Moreover, when critical conditions occur, an alert is reflected on the android app and on web portal in real time. Also, the system starts sending a warning ring on registered mobile.

P. Kumar · G. Purohit (✉) · P. Tanwar · C. Gautam · K. S. Raju
CSIR—Central Electronics Engineering Research Institute, Pilani, Rajasthan, India
e-mail: gp.bits@gmail.com

P. Kumar
e-mail: prashant.mnnit10@gmail.com

P. Tanwar
e-mail: pramod.tanwar@gmail.com

C. Gautam
e-mail: cgautam@ceeri.res.in

K. S. Raju
e-mail: kotasolomonraju@gmail.com

P. Kumar · G. Purohit · P. Tanwar · K. S. Raju
Academy of Scientific and Innovative Research (AcSIR), Ghaziabad, India

© Springer Nature Singapore Pte Ltd. 2020
A. K. Luhach et al. (eds.), *First International Conference on Sustainable Technologies for Computational Intelligence*, Advances in Intelligent Systems and Computing 1045, https://doi.org/10.1007/978-981-15-0029-9_65

Keywords Air quality monitoring (AQM) · Smart home · IoT · Raspberry Pi

1 Introduction

Home automation is a means that empowers individuals to control and monitor home appliances intelligently and automatically to make life effortless by providing amenity, security, ease and energy efficiency to its occupants. Recent advancement in technologies boost up the research and development of home automation. Home automation can assist the people who want to access electrical appliances while at a remote location. Also, it very much helps senior citizen and differently abled people. Moreover, over the past quarter century, the growth of industries has increased exponentially. These industries badly affect the environment and create serious problems. Stationary and mobile sources are also contributing to it and generate various chemical pollutants. World health organization (WHO) has established guidelines by seeing the significance of air quality in human lives, for reducing the health hazards of air pollution by setting the threshold values of various pollutants. Hence, the demand for air quality monitoring system is increasing with time and becomes the crucial part of the home automation system. Therefore, we need a system that could assist us in home appliances monitoring and controlling along with air quality monitoring. We develop such type of system by considering industrial class air quality monitoring sensors. It is an extended version of our previous work [1]; in addition, the proposed architecture is modified for adding a GSM module so that warning ring could be sent to registered mobile as critical conditions occur. To reduce the overall response time of the system, Honeywell HPMA-115S0 is used over SDS021 for PM 2.5 or PM 10 values. It takes less than 6 s to generate a new value. Moreover, we add our power supply module and display screen in extended version. Eventually, all components are accumulated and connected in a box and packed. Small box of same system is also obtained by replacing Arduino Mega with Node MCU for air quality monitoring.

2 Related Work

Home automation is the need of today's era. Small home to the cities having skyscrapers are being inclined towards automation. In this section, we review the various existing approaches, proposed by the authors.

In a paper [2], authors have proposed an IoT base home automation system, in which authors control home appliances by using Node MCU and Arduino UNO. Also, an alert message is transmitted for energy consumed regularly and for the gas level in cylinder when it goes down the certain limit [3]. The authors have proposed a home application using IoT, they have given one blynk app for controlling the home appliances. Also, they use Node MCU for interfacing the appliances and

Internet connectivity [4]. Authors have used Arduino Mega along with ESP8266 for interfacing electronic appliances and sensors. Moreover, they have developed a user interface for switching using virtuino [5]. Authors came with a new concept. They are controlling the home appliances using visual machine intelligence. In their project, they use Raspberry Pi along with Intel Galileo for image capturing and appliances interfacing [6]. Authors use Emon CMS platform for visualizing and collecting remote home data and remote controlling the devices [7]. An architecture is proposed, in which near field communication (NFC) is used as a central system to automate the home environment. Also, general packet radio service is used along with the mobile application [8]. Authors used Raspberry Pi along with Arduino Uno to interface home appliances and sensors. They also take the service of the real-time database to visualize the status of the devices in real time using android app [9]. A low-cost home automation system using Wi-fi is proposed, in which they are using ATmega for interfacing home appliances along with various sensors. Also, they gave the Android app for monitoring and controlling the appliances available at home [10]. An architecture is proposed, in which multiple Arduino Uno-based satellite nodes are used to interface home appliances and sensors. Also, they transmit their data to Arduino Mega-based base station for uploading the data on the server. They have also given one android app for monitoring and controlling the home environment [11]. Proposed TI-CC3200-based security and home automation system, in which CC3200 is used to interface various appliances and sensors. It also sends voice call at phone if the intruder is detected. In a paper [12], the authors explained home automation using Bluetooth in an indoor environment and home automation using ethernet in an outdoor environment. Authors have used Arduino Uno for interfacing home appliances.

In summary, most of the authors in their proposed systems are controlling and monitoring the various home appliances. Also, they are monitoring the home environment using various sensors like temperature, humidity, smoke, etc. Only a few of them have used MQ series sensors to keep an eye on air quality, even though it is a crucial component of home automation. Such type of gap encouraged us to develop a system that could give equal emphasis on air quality monitoring along with controlling and monitoring of home appliances. Moreover, the system must use industrial class air quality sensors so that accurate and stable values could be obtained.

3 Proposed Work

We propose a system that assists in monitoring and controlling the home appliances along with monitoring the home environment in real time. It is unique. Till now, almost all systems are giving high weightage to controlling and monitoring the electronic devices along with temperature, humidity sensors. These systems do not give equal emphasis to air quality monitoring even though it is an important aspect of home automation. However, separate articles are available on AQM in literature. In our work, we propose an IoT-based affordable air quality monitoring with smart

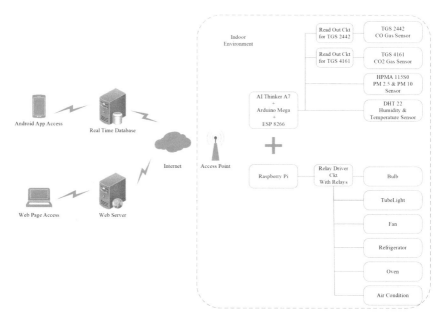

Fig. 1 Proposed architecture of intelligent home with AQM

home system. In order to monitor air quality, industrial class sensors are used. They give us precise values, long life, high stability, high sensitivity and ability to survive in the worst condition.

As shown in Fig. 1, we use Arduino Mega and Raspberry Pi for interfacing electronic devices and sensors. Here, Arduino Mega is used to monitoring the home environment, and Raspberry Pi takes the responsibility of controlling and monitoring the all electronic devices available at home. Although, it is possible to interface all devices and sensors with a single microcontroller, for reducing the complexity and easy troubleshooting, we use a combination of Arduino Mega and Raspberry Pi.

Initially, Arduino Mega reads the values of gas sensors like CO, CO_2, along with PM 2.5, PM 10, temperature and humidity, which senses pollutants and by using appropriate readout circuit the signals get extracted and conditioned appropriately using algorithm running on the controller. The collected data are sent to a cloud database like Google Firebase for real-time visualization. Similarly, Raspberry Pi controls all home appliances and uploads their status on the cloud database.

Hereafter, for controlling and visualizing the status of the home appliances, and monitoring of the home environment, we develop a user-friendly Android app and Web portal. Both types of interfaces fetch/upload the data in real time on the cloud database. Whenever change comes in the home environment and status of the home appliances, then Arduino Mega/Raspberry Pi updates the values in the real-time database. Consequently, the same is reflected in the Android app and Web portal. Also, we can change the status of home appliances by android app or Web portal.

4   System Implementation

The proposed architecture of the home automation that is used for monitoring and controlling the appliances along with air quality monitoring is implemented. To understand the implementation, we divide the system into three sections by the functionality provided. Initially, home environment monitoring is considered, in which data are collected from all sensors (CO, CO_2, PM 2.5, PM 10, humidity, temperature) continuously and then these values are compared with the predefined threshold values. If values go beyond the permissible limit, then alert is reflected in user interfaces as shown in Fig. 2 via red color. Also, system sends warning rings to registered number. To interface said sensors, Arduino Mega is used. The specifications of all the sensors are available in Table 1. Moreover, we also develop readout circuit for CO, CO_2 sensors. In the second section, home appliances monitoring and controlling is considered, and for interfacing various home appliances Raspberry Pi is used. Raspberry pi continuously monitors the status of the appliances and same is uploaded on the cloud database. Also, it controls the appliances as per user instructions. Eventually, interactive Web portal and Android app are developed for controlling and visualizing the status of the appliances along with air quality sensors as shown in Fig. 2.

In addition, all sensors of home environment monitoring are grouped and connected in a box and packed (see Fig. 4a) along with its power supply and display screen. In order to reduce the size of box (see Fig. 4b), all sensors of home environment monitoring are interfaced with Node MCU by replacing Arduino Mega.

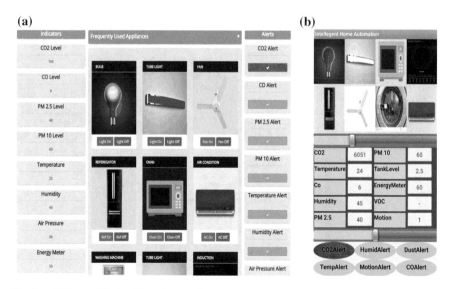

Fig. 2   a Web portal for intelligent home with AQM; **b** Android app for intelligent home with AQM

Table 1 Specifications of the sensors used

Sensor name	Sensor model	Nominal range (ppm)	Accuracy	Response time (s)	Sensor type
[a]CO	TGS 2442	30–1000	–	Approx. 1	MOS
[a]CO_2	TGS 4161	350–10,000	Approx. $\pm20\%$ at 1000 ppm	Approx. 90	Solid electrolyte
PM 2.5/PM 10	HPMA 115S0	0–1000 $\mu g/m^3$	0–100 $\mu g/m^3$ (±15 $\mu g/m^3$), 100–1000 $\mu g/m^3$ ($\pm$ 15%)	<6	Laser
Temperature	DHT 22	−40 to +80 °C	<±0.5 °C	Average 2	Capacitive
Humidity	DHT 22	0–100% RH	Max $\pm5\%$ RH	Average 2	Capacitive

[a]These sensors have very high selectivity to target gas

5 Result

This section displays real-time images produced by the system. Figure 3a, b displays CO and CO_2 concentration on sampled data, respectively; (c) demonstrates the humidity, temperature versus time graph. Android app, a Web portal for visualizing and controlling the appliances and home environment are shown in Fig. 2. In the case of unusual conditions, an alert is reflected in user interfaces as shown in Fig. 2 by red color.

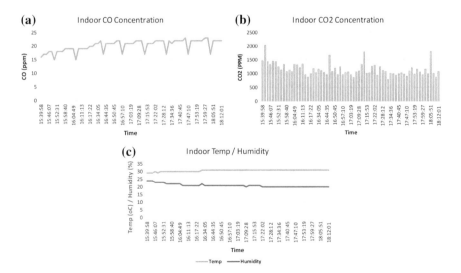

Fig. 3 **a** CO concentration on sampled data; **b** CO_2 concentration on sampled data; **c** temperature, humidity versus time

(a)   (b)

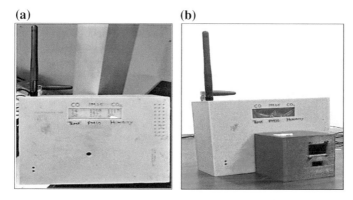

Fig. 4 **a** Portable edge node with display; **b** small edge node with display

We also packaged our edge node in two boxes of size $17 \times 6 \times 12$ cm and $6.5 \times 6.5 \times 6.5$ cm (see Fig. 4). Figure 4a is displaying an edge node in action and Fig. 4b displays small edge node of same system, obtained by replacing Arduino Mega with Node MCU, packaged in blue box.

6 Conclusion and Future Scope

A proposed system integrates various aspects of home automation using IoT, and it gives us a good understanding of environmental conditions in the home along with keeping an eye on appliances in real time. Moreover, the system also comes with good quality user interfaces, i.e., Web portal, an android app that gives a glimpse of air quality available at home and assists in controlling and visualizing the status of appliances. Eventually, the system can generate an alert, whenever unusual conditions occur, and the same is reflected on user interfaces. Also, it starts calling a registered number. In the future, we would like to deploy more standard sensors like VOC, NH3, Alcohol (TGS 2602), methane, and LPG (TGS 2612) at different locations of home and predict the air quality using machine learning.

References

1. Kumar, P., Agrawal, A., Ashish, A., Gaurav, P., Raju, K.S.: Intelligent home with air quality monitoring. In: 2019 Elsevier International Conference on Sustainable Computing in Science, Technology and Management, SUSCOM 2019 (2019)
2. Singh, H., Pallagani, V., Khandelwal, V., Venkanna, U.: IoT based smart home automation system using sensor node. In: 2018 Proceedings of the 4th IEEE International Conference on Recent Advances in Information Technology, RAIT 2018, 1–5 (2018). https://doi.org/10.1109/RAIT.2018.8389037

3. Durani, H., Sheth, M., Vaghasia, S.K.M.: Smart automated home application using IoT with Blynk app. In: 2018 IEEE Second International Conference on Inventive Communication and Computational Technologies (ICICCT), pp. 393–397 (2018). https://doi.org/10.1109/ICICCT.2018.8473224

4. Jabbar, W.A., Alsibai, M.H., Amran, N.S.S., Mahayadin, S.K.: Automation system for smart home. In: 2018 IEEE International Symposium on Networks, Computers and Communications (ISNCC), pp. 1–6 (2018). https://doi.org/10.1109/ISNCC.2018.8531006

5. Kool, I., Kumar, D., Barma, S.: Visual machine intelligence for home automation. In: 2018 IEEE 3rd International Conference on Internet of Things: Smart Innovation and Usages (IoT-SIU), pp. 1–6 (2018). https://doi.org/10.1109/IoT-SIU.2018.8519915

6. Al-Kuwari, M., Ramadan, A., Ismael, Y., Al-Sughair, L., Gastli, A., Benammar, M.: Smart-home automation using IoT-based sensing and monitoring platform. In: Proceedings—2018 IEEE 12th International Conference on Compatibility, Power Electronics and Power Engineering, CPE-POWERENG 2018, pp. 1–6 (2018). https://doi.org/10.1109/CPE.2018.8372548

7. Vagdevi, P., Nagaraj, D., Prasad, G.V.: Home: IOT based home automation using NFC. In: IEEE Proceedings of the International Conference on IoT in Social, Mobile, Analytics and Cloud, I-SMAC 2017, pp. 861–865 (2017). https://doi.org/10.1109/I-SMAC.2017.8058301

8. Sri Harsha, S.L.S., Chakrapani Reddy, S., Prince Mary, S.: Enhanced home automation system using internet of things. In: IEEE Proceedings of the International Conference on IoT in Social, Mobile, Analytics and Cloud, I-SMAC 2017, pp. 89–93 (2017). https://doi.org/10.1109/I-SMAC.2017.8058302

9. Vikram, N., Harish, K.S., Nihaal, M.S., Umesh, R., Shetty, A., Kumar, A.: A low-cost home automation system using wi-fi based wireless sensor network incorporating internet of things (IoT). In: Proceedings—7th IEEE International Advanced Computing Conference, IACC 2017, pp. 174–178 (2017). https://doi.org/10.1109/IACC.2017.0048

10. Govindraj, V., Sathiyanarayanan, M., Abubakar, B.: Customary homes to smart homes using Internet of Things (IoT) and mobile application. In: IEEE Proceedings of the 2017 International Conference on Smart Technology for Smart Nation, SmartTechCon 2017, pp. 1059–1063 (2017). https://doi.org/10.1109/SmartTechCon.2017.8358532

11. Kodali, R.K., Jain, V., Bose, S., Boppana, L.: IoT based smart security and home automation system. In: Proceeding—IEEE International Conference on Computing, Communication and Automation, ICCCA 2016, pp. 1286–1289 (2016). https://doi.org/10.1109/CCAA.2016.7813916

12. Mandula, K., Parupalli, R., Murty, C. H. A. S., Magesh, E., Lunagariya, R.: Mobile based home automation using Internet of Things (IoT). In: 2015 IEEE International Conference on Control Instrumentation Communication and Computational Technologies, ICCICCT 2015, pp. 340–343 (2015). https://doi.org/10.1109/ICCICCT.2015.7475301

Correction to: ConFuzz—A Concurrency Fuzzer

Nischai Vinesh and M. Sethumadhavan

Correction to:
Chapter "ConFuzz—A Concurrency Fuzzer" in:
A. K. Luhach et al. (eds.), *First International Conference on Sustainable Technologies for Computational Intelligence*, Advances in Intelligent Systems and Computing 1045, https://doi.org/10.1007/978-981-15-0029-9_53

In the original version of the book, the authors "Sanjay Rawat, Herbert Bos and Cristiano Giuffrida" were included erroneously as co-authors, which has now been removed. The chapter and book have been updated with the changes.

The updated version of this chapter can be found at
https://doi.org/10.1007/978-981-15-0029-9_53

© Springer Nature Singapore Pte Ltd. 2020
A. K. Luhach et al. (eds.), *First International Conference on Sustainable Technologies for Computational Intelligence*, Advances in Intelligent Systems and Computing 1045,
https://doi.org/10.1007/978-981-15-0029-9_66

Author Index

© Springer Nature Singapore Pte Ltd. 2020
A. K. Luhach et al. (eds.), *First International Conference on Sustainable Technologies for Computational Intelligence*, Advances in Intelligent Systems and Computing 1045,
https://doi.org/10.1007/978-981-15-0029-9